T0181909

Lecture Notes in Computer Science 9783

Commenced Publication in 1973
Founding and Former Series Editors:
Gerhard Goos, Juris Hartmanis, and Jan van Leeuwen

More information about this series at http://www.springer.com/series/7410

Thomas Peyrin (Ed.)

Fast
Software Encryption

23rd International Conference, FSE 2016
Bochum, Germany, March 20–23, 2016
Revised Selected Papers

 Springer

Editor
Thomas Peyrin
Nanyang Technological University
Singapore
Singapore

ISSN 0302-9743 ISSN 1611-3349 (electronic)
Lecture Notes in Computer Science
ISBN 978-3-662-52992-8 ISBN 978-3-662-52993-5 (eBook)
DOI 10.1007/978-3-662-52993-5

Library of Congress Control Number: 2016944480

LNCS Sublibrary: SL4 – Security and Cryptology

Printed on acid-free paper

This Springer imprint is published by Springer Nature
The registered company is Springer-Verlag GmbH Berlin Heidelberg

Preface

The 23rd International Conference on Fast Software Encryption (FSE 2016) was held at Bochum, Germany, during March 20–23, 2016. The conference was organized by Ruhr University Bochum with Gregor Leander serving as the general chair in collaboration with the International Association for Cryptologic Research (IACR). The conference had about 150 registered participants from 28 different countries. FSE 2016 received 91 submissions. The 25 members of the Program Committee were assisted by more than 80 external reviewers. In total, they delivered 304 reviews, with each submission being reviewed by at least three Program Committee members, five in the case of a submission co-authored by members of the Program Committee. The review process was double-blind, and conflicts of interest were handled carefully. It was managed through an online review system that supported discussions among Program Committee members. Eventually, the Program Committee selected 29 papers from 16 countries (a 31.9 % acceptance rate) for publication in the proceedings.

Besides the 29 selected talks, the program included one invited talk by Henri Gilbert from ANSSI, France, on white-box cryptography. The workshop also featured a rump session, chaired by Dan Bernstein and Tanja Lange, with several short informal presentations.

As in previous FSE events, the Program Committee identified the best submissions of the conference for their scientific quality, their originality, and their clarity. The FSE 2016 Best Paper Award went to José Bacelar Almeida, Manuel Barbosa, Gilles Barthe, and François Dupressoir, for their paper "Verifiable Side-Channel Security of Cryptographic Implementations: Constant-Time MEE-CBC." This paper, along with the article "Stream Ciphers: A Practical Solution for Efficient Homomorphic-Ciphertext Compression" by Anne Canteaut, Sergiu Carpov, Caroline Fontaine, Tancrède Lepoint, María Naya-Plasencia, Pascal Paillier, and Renaud Sirdey received a special invitation for submission to the *Journal of Cryptology*.

Many people contributed to FSE 2016. I would like to thank the authors for contributing their excellent research, but also the Program Committee members and their external reviewers, who spent a lot time and effort reading and analyzing the numerous submissions. I really enjoyed the discussions during the selection phase and I am particularly grateful to Alex Biryukov, Christina Boura, Svetla Nikova, Yu Sasaki, François-Xavier Standaert, and Marc Stevens for accepting to shepherd papers. Finally, I sincerely thank Gregor Leander, the general chair, and his organization team, who worked so hard for the conference to be pleasant for all attendees. Their smooth organization made the event a big success.

I was extremely honored to serve as Program Chair of FSE 2016. The program contained a wide spectrum of the latest research in symmetric cryptography, ranging from cryptanalysis to security proofs, practical implementation aspects to foundations, and considering various primitives such as block ciphers, stream ciphers, hash functions, authenticated encryption, MAC, etc. I hope the selected papers will consolidate our knowledge in symmetric cryptography, but also open new directions to continue making symmetric cryptography a vibrant research community.

May 2016   Thomas Peyrin

FSE 2016

23rd International Conference on Fast Software Encryption

Ruhr University Bochum, Germany
March 20–23, 2016

General Chair

Gregor Leander   Ruhr Universität Bochum, Germany

Program Chair

Thomas Peyrin   Nanyang Technological University, Singapore

Program Committee

Alex Biryukov	University of Luxembourg, Luxembourg
Christina Boura	University of Versailles, France
Itai Dinur	École Normale Supérieure, Paris, France
Orr Dunkelman	University of Haifa, Israel
Takanori Isobe	Sony Corporation, Japan
Tetsu Iwata	Nagoya University, Japan
Pascal Junod	HEIG-VD, Switzerland
Gaëtan Leurent	Inria, France
Florian Mendel	Graz University of Technology, Austria
Bart Mennink	KU Leuven, Belgium
Amir Moradi	Ruhr University Bochum, Germany
Mridul Nandi	Indian Statistical Institute, India
Ivica Nikolić	Nanyang Technological University, Singapore
Svetla Nikova	KU Leuven, Belgium
Kenny Paterson	Royal Holloway, University of London, UK
Thomas Peyrin (Chair)	Nanyang Technological University, Singapore
Christian Rechberger	DTU, Denmark
Yu Sasaki	NTT, Japan
Yannick Seurin	ANSSI, France
Thomas Shrimpton	University of Florida, USA
François-Xavier Standaert	Université Catholique de Louvain, Belgium
Marc Stevens	CWI Amsterdam, The Netherlands
Serge Vaudenay	Ecole Polytechnique Fédérale de Lausanne, Switzerland
Lei Wang	Shanghai Jiao Tong University, China
Meiqin Wang	Shandong University, China

Additional Reviewers

Divesh Aggarwal
Martin Albrecht
Elena Andreeva
Ralph Ankele
Tomer Ashur
Jean-Philippe Aumasson
Thomas Baignères
Subhadeep Banik
Achiya Bar-On
Georg T. Becker
Christof Beierle
Rishiraj Bhattacharaya
Ritam Bhaumik
Begül Bilgin
Sonia Bogos
Anne Canteaut
Carlos Cid
Joan Daemen
Nilanjan Datta
Jean Paul Degabriele
Daniel Dinu
Christoph Dobraunig
Alexandre Duc
Avijit Dutta
Maria Eichlseder
Sebastian Faust
Matthieu Finiasz
Thomas Fuhr
Peter Gazi

Lorenzo Grassi
Vincent Grosso
Jian Guo
Harunaga Hiwatari
Ashwin Jha
Anthony Journault
Pierre Karpman
Elif Bilge Kavun
Dmitry Khovratovich
Handan Kılınç
Miroslav Knezevic
Stefan Koelbl
Virginie Lallemand
Martin M. Lauridsen
Meicheng Liu
Yunwen Liu
Zhiqiang Liu
Atul Luykx
Marco Macchetti
Subhamoy Maitra
Santos Merino Del Pozo
Sean Murphy
Léo Paul Perrin
Peter Pessl
Jérôme Plût
Romain Poussier
Shahram Rasoolzadeh
Francesco Regazzoni
Jean-René Reinhard

Oscar Reparaz
Reza Reyhanitabar
Bastian Richter
Vincent Rijmen
Arnab Roy
Pascal Sasdrich
Falk Schellenberg
Tobias Schneider
Jacob Schuldt
Sourav Sengupta
Kyoji Shibutani
Siang Meng Sim
Valentin Suder
Tyge Tiessen
Elmar Tischhauser
Yosuke Todo
Aleksei Udovenko
Thomas Unterluggauer
Thyla van der Merwe
Kerem Varici
Vesselin Velichkov
Damian Vizár
Wei Wang
Alexander Wild
Hongjun Wu
Brecht Wyseur
Guoyan Zhang
Liting Zhang

Contents

Automated Tools for Cryptanalysis

Designs

Block-Cipher Cryptanalysis

Foundations and Theory

Authenticated-Encryption and Hash Function Cryptanalysis

Operating Modes

New Bounds for Keyed Sponges
with Extendable Output: Independence Between
Capacity and Message Length

Yusuke Naito[1]([⊠]) and Kan Yasuda[2]

[1] Mitsubishi Electric Corporation, Kanagawa, Japan
naito.yusuke@ce.mitsubishielectric.co.jp
[2] NTT Secure Platform Laboratories, Tokyo, Japan
yasuda.kan@lab.ntt.co.jp

Abstract. We provide new bounds for the pseudo-random function security of keyed sponge constructions. For the case $c \leq b/2$ (c the capacity and b the permutation size), our result improves over all previously-known bounds. A remarkable aspect of our bound is that dependence between capacity and message length is removed, partially solving the open problem posed by Gaži et al. at CRYPTO 2015. Our bound is essentially tight, matching the two types of attacks pointed out by Gaži et al. For the case $c > b/2$, Gaži et al.'s bound remains the best for the case of single-block output, but for keyed sponges with extendable outputs, our result partly (when query complexity is relatively large) provides better security than Mennink et al.'s bound presented at ASIACRYPT 2015.

Keywords: PRF · XOF · Game playing · Coefficient H technique · Lazy sampling · Multi-collision · Stirling's approximation

1 Introduction

The sponge construction today, though being originally introduced as a mode for keyless hash functions [7], is drawing more and more attention in the secret-key setting. The primary reason seems to lie in the flexibility: the *keyed* sponge construction has been modified in a variety of ways such as duplexing [6], parallelism [3] and full-state (i.e. the rate being equal to the permutation size) absorption [9,19]. However, one of the reasons why the sponge construction was so attractive in the first place was that it inherently possessed the capability of *extendable* output.

FIPS 202 [17] standardizes two sorts of extendable output functions (XOFs): SHAKE128 and SHAKE256, which have a permutation size of $b = 1600$ bits and capacity values of $c = 256, 512$ bits, respectively. FIPS 202 states:

XOFs are a powerful new kind of cryptographic primitive that offers the flexibility to produce outputs with any desired length. ... In practice, the use of an XOF as a key derivation function (KDF) could preclude the

© International Association for Cryptologic Research 2016
T. Peyrin (Ed.): FSE 2016, LNCS 9783, pp. 3–22, 2016.
DOI: 10.1007/978-3-662-52993-5_1

possibility of related outputs, by incorporating the length and/or type of the derived key into the message input to the KDF. In that case, a disagreement or misunderstanding between two users of the KDF about the type or length of the key they are deriving would almost certainly not lead to related outputs.

To confirm the above statement in a more formal way, we need to investigate the security of the KDF as a pseudo-random function (PRF).

Previous PRF Bounds. Several different types of PRF bounds are known for keyed sponges. Security parameters of keyed sponges include the permutation size b, the capacity c, the rate $r := b - c$, and the key length k. The main focus remains on the capacity value c, because usually it is this parameter that defines a dominant term in a bound. Nevertheless, none of the previous bounds has been shown to be strictly tight in relation to parameter c, as explained below.

The PRF security of keyed sponges can be derived from the indifferentiability of the sponge construction. The indifferentiability of the sponge construction [7] crucially depends on the capacity c, and hence so does the derived PRF bound. Roughly, the indifferentiability-based PRF bound has a dominant term of the form $(\ell q + Q)^2/2^c$, where parameter ℓ is the maximum length of an adversarial query, parameter q the maximum number of construction (online) queries to the keyed sponge $\mathcal{C}$, and parameter Q the maximum number of primitive (offline) queries to the underlying permutation P.

Note that we are working in the ideal model [1,13,16] where the underlying permutation P is regarded as a random permutation. In practice, P is a fixed permutation; hence Q corresponds to the time complexity of the adversary, measuring how many times the adversary could perform offline computation of P.

The above indifferentiability-based PRF bound is rather loose, and the actual PRF security of keyed sponges should be much higher, as first noticed by Bertoni et al. [8]. Later, Andreeva et al. [1] successfully removed the term $Q^2/2^c$ and obtained a bound which was basically $((\ell q)^2 + \mu Q)/2^c$. Here, μ is an adversarial parameter called "multiplicity" and lies somewhere between $2\ell q/2^r$ and $2\ell q$.

Concurrently, Gaži et al. [13] provided a "nearly tight" bound [16] which was roughly of the form $(q^2 + \ell q + qQ)/2^c$. Gaži et al. also pointed out two attacks matching $q^2/2^c$ and $qQ/2^c$, respectively. They observed that their bound "only mildly depends on the length" when ℓ is sufficiently small [13] but left it open whether their bound was tight for all cases, especially when ℓ is large. It should be noted that Gaži et al. [13] only treated the case of single-block output, and their method did not seem to be easily extendable to the case of multiple-block output [16].

For the case of extendable output, recently Mennink et al. [16] has provided another bound which is essentially $(\ell q^2 + \mu Q)/2^c$. While definitely improving Andreeva et al.'s $((\ell q)^2 + \mu Q)/2^c$, Mennink et al.'s bound does not come close to Gaži et al.'s $(q^2 + \ell q + qQ)/2^c$, at least for the case of single-block output.

Table 1. Comparison of target keyed sponge constructions

	Key		Extendable
	Inner	Outer	output
Bertoni et al. [8]	—	✓	✓
Chang et al. [11]	✓	✓	✓
Andreeva et al. [1]	✓	✓	✓
Gaži et al. [13][a]	—	✓	—
Mennink et al. [16][b]	✓	—	✓
This paper	✓	✓	✓

[a]Gaži et al. [13] treat the case where the rate values are different between absorbing and squeezing phases. Only the rate r for the squeezing phase appears in the bound; the rate for absorbing phase does not affect security in their analysis.

[b]Mennink et al. [16] study the case of full-state absorption, i.e. the rate for absorbing phase is equal to the permutation size except for the first call of the underlying permutation.

Consequently, it seems that there is still room for improvement. It might be possible to come up with a tighter PRF bound for keyed sponges, especially for the case of extendable output.

Inner- and Outer-Keying. There are two ways of keying the sponge construction. The difference between the two methods is analogous to the one between NMAC and HMAC [4]. The first method, which is like NMAC, is called the *inner-keyed* sponge [1]. This replaces (part of) the inner IV with a secret key $K \in \{0,1\}^k$, so that $k \leq c$. The inner-keyed sponge was proposed by Chang et al. [11] who showed that it has a certain advantage in the standard-model security.

The second method, which is like HMAC, is called the *outer-keyed* sponge [1]. This is nothing but the sponge construction itself that processes the input $K \| M$ (i.e. a message prefixed by a secret key K) and hence does not have a limitation on the key size k. A first analysis of the outer-keyed sponge was given by Bertoni et al. [8]. The obvious advantage of this method, besides key length, is that we can make use of existing sponge constructions that have been already implemented as hash functions.

Our Contributions. We provide new PRF bounds for keyed sponges with extendable output, under the condition that the rate and capacity remain the same for absorbing and squeezing phases. We treat both inner- and outer-keyed sponges (cf. Table 1). Previous PRF bounds and our results are summarized in Table 2.

- **Case $c \leq b/2$.** This case includes SHAKE128 and SHAKE256. In this case, our bound improves over all previously-known PRF bounds. For the inner-keyed sponge, our bound is qualitatively better than the previous two bounds by Andreeva et al. [1] and by Mennink et al. [16]. For example, if $k = c$ (which is the case that provides the highest security for the inner-keyed sponge), then the previous bounds contained $(\ell q^2 + \mu Q)/2^c$, whereas our bound only contains $(\ell q + q^2 + qQ)/2^c$. On the other hand, for the outer-keyed sponge, observe that the term related to capacity in our bound becomes roughly $(q^2 + qQ)/2^c$, which is dominant in many scenarios. Note the absence of ℓq here; we remove the dependence between capacity c and message length ℓ, partially answering the open question posed by Gaži et al. [13]. Together with the two attacks pointed out by Gaži et al. [13] whose complexities were roughly $q^2/2^c$ and $qQ/2^c$, we see that our bound is strictly tight in terms of parameters q and Q. Furthermore, for the outer-keyed sponge, the remaining parameter ℓ is restricted only by the term $\ell^2 q^2/2^b$, whereas previous bounds contained $\ell q/2^c$ or $\ell^2 q^2/2^c$. Hence, our bound has a qualitatively weaker restriction on ℓ, under the condition $c \leq b/2$.
- **Case $c > b/2$.** This is the case for lightweight hash functions, such as QUARK [2], SPONGENT [10] and PHOTON [14]. In this case, our contribution is more subtle. For single-block output, Gaži et al.'s bound [13] remains the best, beating our bound as well as Mennink et al.'s [16]. However, for multiple-block output, our result improves over Mennink et al.'s [16] which has been the best known bound for extendable output. The two bounds are incomparable due to the parameter μ, but roughly speaking, we see that our bound becomes better when query complexity is relatively large. For simplicity, assume $k = c$ and put $\mu = 2\ell q$. Then Mennink et al.'s bound becomes roughly $(\ell q^2 + \ell qQ)/2^c$, whereas our bound has a dominant term of $\left((\ell q^2 + \ell qQ)/2^b\right)^{1/2}$. By comparison, our bound becomes smaller when $\ell q^2 + \ell qQ > 2^{c-r}$.

For our proofs we take an approach different from previous work. We first make use of the game-playing technique, introducing just one intermediate game between the real and ideal worlds. Our transition between the games heavily relies on the coefficient H technique of Patarin [18]. To evaluate probabilities of "bad" events, we make extensive use of lazy sampling. As pointed out by Bellare and Rogaway [5], the lazy sampling of random functions with many constraints can be tricky. We show how to carefully lazy-sample input/output points for underlying permutations with certain restrictions. Lastly, we adopt techniques developed by Jovanovic et al. [15] for bounding the size of multi-collisions and for finally optimizing the bound (or "balancing" the terms).

2   Preliminaries

Notation. Let $\{0,1\}^*$ be the set of all bit strings, and for an integer $d \geq 0$, let $\{0,1\}^d$ be a set of d-bit strings. Let 0^d denotes the bit string of d-bit zeroes. For a bit string $x \in \{0,1\}^d$, let $x[i,j]$ be the substring of x from i-th bit to j-th bit, where $1 \leq i \leq j \leq d$. For a finite set X, $x \xleftarrow{\$} X$ means that an element is

Table 2. Comparison of PRF bounds for keyed sponges. In the bounds, parameter κ is key length in blocks, i.e. $\kappa := k/r$; parameter μ is the multiplicity, i.e. $2\ell q/2^r \leq \mu \leq 2\ell q$; parameter $t \geq 1$ can be arbitrary; the number e is Napier's constant $2.71828\cdots$; the function λ is defined as $\lambda(x) := x/2^k$ if $\kappa = 1$ and $\lambda(x) := \min\{\epsilon_1, \epsilon_2\}$ if $\kappa \geq 2$, where $\epsilon_1 := (x^2/2^{c+1}) + (x/2^k)$ and $\epsilon_2 := (1/2^b) + x(12b/2^r)^{\kappa/2}$.

Inner-keyed ($k \leq c$)	
Andreeva et al. [1]	$\dfrac{(\ell q)^2}{2^c} + \dfrac{\mu Q}{2^k}$
Mennink et al. [16]	$\dfrac{2\ell q^2}{2^c} + \dfrac{\mu Q}{2^k} + \dfrac{2(\ell q)^2}{2^b}$
This paper ($c \leq b/2$)	$\dfrac{3q^2 + qQ + 2r(q+Q)}{2^c} + \dfrac{\ell q + Q}{2^k} + \dfrac{(3 + 32e^2 r^{-2})\ell^2 q^2}{2^b}$
This paper ($c > b/2$)	$\left(\dfrac{18e\ell q(q+Q)}{2^b}\right)^{1/2} + \dfrac{3q^2 + qQ + 2r(q+Q)}{2^c}$
	$\qquad\qquad + \dfrac{\ell q + Q}{2^k} + \dfrac{3\ell^2 q^2}{2^b}$
Outer-keyed	
Indifferentiability [7]	$\dfrac{2(\kappa + \ell q + Q)^2}{2^c} + \dfrac{Q}{2^k}$
Andreeva et al. [1]	$\dfrac{(\ell q)^2 + 2\mu Q}{2^c} + \dfrac{2\kappa Q}{2^b} + \lambda(Q)$
Gaži et al. [13]	$\dfrac{6bq^2 + 8\ell q + qQ}{2^c} + \dfrac{(6t+17)\ell q^2 + 7\ell q Q + 2q}{2^b}$
	$\qquad\qquad + \dfrac{136\ell^4 q^2}{2^{2b}} + \dfrac{2(\ell q)^{t+1}}{2^{bt}} + \lambda(\ell q + Q)$
This paper ($c \leq b/2$)	$\dfrac{3q^2 + 2qQ + 2r(q+Q)}{2^c}$
	$\qquad + \dfrac{(3.5 + 32e^2 r^{-2})\ell^2 q^2 + 2qQ + 2\kappa Q}{2^b} + \lambda(Q)$
This paper ($c > b/2$)	$\left(\dfrac{18e\ell q(q+Q)}{2^b}\right)^{1/2} + \dfrac{3q^2 + 2qQ + 2r(q+Q)}{2^c}$
	$\qquad\qquad + \dfrac{3.5\ell^2 q^2 + 2qQ + 2\kappa Q}{2^b} + \lambda(Q)$

randomly drawn from X and is set to x. For a set X, $\mathsf{Perm}(X)$ is the set of all permutations on X. For sets X and Y, $\mathsf{Func}(X, Y)$ is the set of all functions: $X \to Y$. We denote by $\emptyset$ an empty set. For sets X and Y, $X \leftarrow Y$ means that set Y is assigned to set X, and $X \xleftarrow{\cup} Y$ means $X \leftarrow X \cup Y$.

PRF-Security. Through this paper, a distinguisher **D** is a computationally unbounded probabilistic algorithm. It is given query access to one or more oracles $\mathcal{O}$, denoted $\mathbf{D}^{\mathcal{O}}$. Its complexity is solely measured by the number of queries made to its oracles. For integers $k > 0$ and $\tau > 0$, let $\mathcal{F}_K : \{0,1\}^* \to \{0,1\}^\tau$ be a keyed hash function based on a permutation having keys $K \in \{0,1\}^k$. The security proof will be done in the ideal model, regarding the underlying permutation as a random permutation $\mathcal{P} \xleftarrow{\$} \mathsf{Perm}(\{0,1\}^b)$ for an integer $b > 0$. We denote by $\mathcal{P}^{-1}$ its inverse.

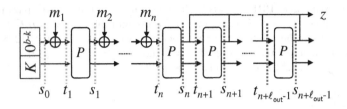

Fig. 1. IKSponge Construction

The PRF-security of $\mathcal{F}_K$ is defined in terms of indistinguishability between the real world and the ideal world. In the real world, **D** has query access to $\mathcal{F}_K$, $\mathcal{P}$, and $\mathcal{P}^{-1}$ for a key $K \xleftarrow{\$} \{0,1\}^k$ and $\mathcal{P} \xleftarrow{\$} \mathsf{Perm}(\{0,1\}^b)$. In the ideal world, it has query access to a random function $\mathcal{R}$, $\mathcal{P}$, and $\mathcal{P}^{-1}$, for $\mathcal{R} \xleftarrow{\$} \mathsf{Func}(\{0,1\}^*, \{0,1\}^\tau)$ and $\mathcal{P} \xleftarrow{\$} \mathsf{Perm}(\{0,1\}^b)$. After **D**'s interaction, it outputs $y \in \{0,1\}$. The event is denoted by $\mathbf{D} \Rightarrow y$. Then the advantage function is defined as

$$\mathbf{Adv}_{\mathcal{F}}^{\mathsf{prf}}(\mathbf{D}) = \Pr[\mathbf{D}^{\mathcal{F}_K, \mathcal{P}, \mathcal{P}^{-1}} \Rightarrow 1] - \Pr[\mathbf{D}^{\mathcal{R}, \mathcal{P}, \mathcal{P}^{-1}} \Rightarrow 1].$$

We call queries to $\mathcal{F}_K/\mathcal{R}$ "online queries" and queries to $(\mathcal{P}, \mathcal{P}^{-1})$ "offline queries." Though this paper, without loss of generality, assume that **D** is deterministic and makes no repeated query.

3   Inner Keyed Sponge and the PRF-Security

3.1   Inner Keyed Sponge Construction

The inner keyed sponge construction uses the sponge function as the underlying function. By IKSponge we denote the construction.

First we explain the sponge function. The sponge function is a permutation-based one. For an integer $b > 0$, let $P \in \mathsf{Perm}(\{0,1\}^b)$ be the underlying permutation. By Sponge^P, we denote the sponge function using P. For integers $r > 0$ and $c \geq 0$ with $r + c = b$, r is a bit length so-called rate and c is a bit length so-called capacity. For an input $m \in \{0,1\}^*$, the output $\mathsf{Sponge}^P(m) = z$ is calculated as follows. Firstly, a bit string $\mathsf{pad}(|m|)$ is appended to the suffix of m such that the bit length of $m\|\mathsf{pad}(|m|)$ becomes a multiple of r and the last r-bit block is not 0^r. The example of the padded string is $m\|\mathsf{pad}(|m|) = m\|1\|0^*$, which means that 1 and the minimum number of zeroes so that the bit length becomes a multiple of r. Secondly, the padded bit string is partitioned into r-bit blocks $m_1, \ldots, m_l$, where $m_l \neq 0^r$. Thirdly, b-bit internal state s is updated by the following procedure.

$$s \leftarrow 0^b; \text{ for } i = 1, \ldots l \text{ do } s \leftarrow P(m_i\|0^c \oplus s)$$

Finally, the $\ell_{\mathsf{out}} \times r$-bit string z is produced by the following procedure.

$$z \leftarrow s[1, r]; \text{ for } i = 1, \ldots \ell_{\mathsf{out}} - 1 \text{ do } s \leftarrow P(s); z \leftarrow z\|s[1, r]$$

Next we explain the IKSponge construction. For an integer k with $0 < k \leq c$, let $K \in \{0,1\}^k$ be a secret key. By IKSponge_K^P, we denote IKSponge with P having K. IKSponge equals Sponge with the initial value $0^{b-k} \| K$. Concretely, for a message m, the response $\text{IKSponge}_K^P(m) = z$ is denoted as follows, and the Fig. 1 shows the procedure.

1. Partition $m \| \text{pad}(|m|)$ into r-bit blocks $m_1, \ldots, m_n$
2. $s_0 \leftarrow 0^{b-k} \| K$
3. For $i = 1, \ldots, n$ do $t_i \leftarrow m_i \| 0^c \oplus s_{i-1}$; $s_i \leftarrow P(t_i)$
4. $z \leftarrow s_n[1, r]$
5. For $i = 1, \ldots, \ell_{\text{out}} - 1$ do $t_{n+i} \leftarrow s_{n+i-1}$; $s_{n+i} \leftarrow P(t_{n+i})$; $z \leftarrow z \| s_{n+i}[1, r]$
6. Return z

3.2 PRF-Security of the IKSponge Construction

We show the PRF-security of IKSponge in the ideal permutation model.

Theorem 1. *Let D be a distinguisher which makes q online queries of r-bit block length at most ℓ_{in} and Q offline queries. Then, for any parameter ρ, we have* $\mathbf{Adv}_{\text{IKSponge}}^{\text{prf}}(D) \leq \frac{\ell q + Q}{2^k} + \frac{3q^2 + qQ + 2\rho(q+Q)}{2^c} + \frac{3\ell^2 q^2}{2^b} + 2^{r+1} \times \left(\frac{2e\ell q}{\rho 2^r} \right)^\rho$, *where* $\ell = \ell_{\text{in}} + \ell_{\text{out}} - 1$ *and* $e = 2.71828\cdots$ *is Napier's constant.*

Corollary 1. *We assume $c \leq b/2$. Then, we put $\rho = r$, and without loss of generality, assume $r \geq 2$ (otherwise $r = c = 1$ and $b=2$). Since $r \geq b/2$, we have* $\mathbf{Adv}_{\text{IKSponge}}^{\text{prf}}(D) \leq \frac{3q^2 + qQ + 2r(q+Q)}{2^c} + \frac{(3+32e^2 r^{-2})\ell^2 q^2}{2^b} + \frac{\ell q + Q}{2^k}$.
We assume $c > b/2$, and put $\rho = \max\left\{ r, \left(\frac{2e \times \ell q}{2^{r-c}(q+Q)} \right)^{1/2} \right\}$. Then we have $\mathbf{Adv}_{\text{IKSponge}}^{\text{prf}}(D) \leq \left(\frac{32e\ell q(q+Q)}{2^b} \right)^{1/2} + \frac{3q^2 + qQ + 2r(q+Q)}{2^c} + \frac{3\ell^2 q^2}{2^b} + \frac{\ell q + Q}{2^k}$.

4 Proof of Theorem 1

We prove the PRF-security of IKSponge_K^P via three games. We denote these games by Game 1, Game 2, and Game 3. For $i \in \{1, 2, 3\}$, we let $G_i := (L_i, P, P^{-1})$ to which D has query access in Game i. Note that in each game, P is independently drawn as $P \xleftarrow{\$} \text{Perm}(\{0,1\}^b)$. We let $L_1 := \text{IKSponge}_K^P$ and $L_3 := \mathcal{R}$. Hence we have

$$\mathbf{Adv}_{\text{IKSponge}}^{\text{prf}}(D) = \sum_{i=1}^{2} \left(\Pr[D^{G_i} \Rightarrow 1] - \Pr[D^{G_{i+1}} \Rightarrow 1] \right). \tag{1}$$

Hereafter, we upper-bound $\Pr[D^{G_i} \Rightarrow 1] - \Pr[D^{G_{i+1}} \Rightarrow 1]$ for $i \in \{1, 2\}$. Note that we define L_2 before $\Pr[D^{G_1} \Rightarrow 1] - \Pr[D^{G_2} \Rightarrow 1]$ is evaluated.

In the following proof, for $\alpha \in \{1, \ldots, Q\}$, we denote an α-th offline query by x^α or y^α, and the response by y^α or x^α, where $y^\alpha = P(x^\alpha)$ or $x^\alpha = P^{-1}(y^\alpha)$. For $\alpha \in \{1, \ldots, q\}$, we denote an α-th online query by m^α and the response by z^α. We also use superscripts for other values defined by online queries, e.g., $n^1, t_1^1, s_1^1, n^2, t_1^2, s_1^2$, etc.

4.1   Upper-Bound of $\Pr[D^{G_1} \Rightarrow 1] - \Pr[D^{G_2} \Rightarrow 1]$

We start by defining L_2. Let $\mathcal{G}_1, \mathcal{G}_2, \ldots, \mathcal{G}_\ell \xleftarrow{\$} \mathsf{Func}(\{0,1\}^b, \{0,1\}^b)$ be random functions. Let $K \xleftarrow{\$} \{0,1\}^k$ be a secret key. For an online query $m \in \{0,1\}^*$, the response $L_2(m) = z$ is defined as follows.

1. Partition $m\|\mathsf{pad}(|m|)$ into r-bit blocks $m_1, \ldots, m_n$
2. $s_0 \leftarrow 0^{b-k}\|K$
3. For $i = 1, \ldots, n$ do $t_i \leftarrow m_i\|0^c \oplus s_{i-1}$; $s_i \leftarrow \mathcal{G}_i(t_i)$
4. $z \leftarrow s_n[1, r]$
5. For $i = 1, \ldots, \ell_{\mathrm{out}} - 1$ do $t_{n+i} \leftarrow s_{n+i-1}$; $s_{n+i} \leftarrow \mathcal{G}_{n+i}(t_{n+i})$; $z \leftarrow z\|s_{n+i}[1, r]$
6. Return z

Transcript. Let $\tau_L = \{(m^1, z^1), \ldots, (m^q, z^q)\}$ be the set of query-response pairs defined by online queries and $\tau_P = \{(x^1, y^1), \ldots, (x^Q, y^Q)\}$ be the set of query-response pairs defined by offline queries. Additionally, we define sets $\tau_1, \ldots, \tau_\ell$. For $i \in \{1, \ldots, \ell\}$, let $\tau_i = \bigcup_{\alpha=1}^q \{(t_i^\alpha, s_i^\alpha)\}$ be the set of all input-output pairs at the i-th block defined by online queries. Note that for $\alpha \in \{1, \ldots, q\}, i \in \{1, \ldots, \ell\}$ if (t_i^α, s_i^α) is not defined then $\{(t_i^\alpha, s_i^\alpha)\}$ is an empty set.

This proof permits D to obtain these sets and a secret key K after D's interaction but before it outputs a result. We let $\tau_{1..\ell} = \bigcup_{i=1}^\ell \tau_i$. Then D's transcript is summarized as $\tau = \{\tau_L, \tau_P, \tau_{1..\ell}, K\}$.

Let T_1 be the transcript in Game 1 obtained by sampling $K \xleftarrow{\$} \{0,1\}^k$ and $\mathcal{P} \xleftarrow{\$} \mathsf{Perm}(\{0,1\}^b)$. Let T_2 be the transcript in Game 2 obtained by sampling $K \xleftarrow{\$} \{0,1\}^k, \mathcal{P} \xleftarrow{\$} \mathsf{Perm}(\{0,1\}^b), \mathcal{G}_1, \mathcal{G}_2, \ldots, \mathcal{G}_\ell \xleftarrow{\$} \mathsf{Func}(\{0,1\}^b, \{0,1\}^b)$. We call τ *valid* if an interaction with their oracles could render this transcript, namely, $\Pr[T_i = \tau] > 0$ for $i \in \{1, 2\}$. Then $\Pr[D^{G_1} \Rightarrow 1] - \Pr[D^{G_2} \Rightarrow 1]$ is upper-bounded by the statistical distance of transcripts, i.e.,

$$\Pr[D^{G_1} \Rightarrow 1] - \Pr[D^{G_2} \Rightarrow 1] \leq \mathsf{SD}(T_1, T_2) = \frac{1}{2} \sum_\tau |\Pr[T_1 = \tau] - \Pr[T_2 = \tau]| ,$$

where the sum is over all valid transcripts.

Coefficient H Technique. We upper-bound the statistical distance by using the coefficient H technique [12,18]. In this technique, firstly, we need to partition valid transcripts into good transcripts $\mathcal{T}_{\mathsf{good}}$ and bad transcripts $\mathcal{T}_{\mathsf{bad}}$. Then we can upper-bound the statistical distance $\mathsf{SD}(T_1, T_2)$ by the following lemma.

Lemma 1 (Coefficient H Technique). *Let $0 \leq \varepsilon \leq 1$ be such that for all $\tau \in \mathcal{T}_{\mathsf{good}}, \frac{\Pr[T_1 = \tau]}{\Pr[T_2 = \tau]} \geq 1 - \varepsilon$. Then, $\mathsf{SD}(T_1, T_2) \leq \varepsilon + \Pr[T_2 \in \mathcal{T}_{\mathsf{bad}}]$.*

The proof of the lemma is given in [12]. Hence, we can upper-bound $\Pr[D^{G_1} \Rightarrow 1] - \Pr[D^{G_2} \Rightarrow 1]$ by defining good and bad transcripts and by evaluating ε and $\Pr[T_2 \in \mathcal{T}_{\mathsf{bad}}]$.

Good and Bad Transcripts. We define $\mathcal{T}_{\text{bad}}$ that satisfies one of the following conditions.

- $\text{hit}_{\text{tx,sy}} \Leftrightarrow \exists (t, s) \in \tau_{1..\ell}, (x, y) \in \tau_P$ s.t. $t = x \vee s = y$
- $\text{hit}_{\text{tt}} \Leftrightarrow \exists i, j \in \{1, \ldots, \ell\}$ with $i \neq j$ s.t. $\exists (t_i, s_i) \in \tau_i, (t_j, s_j) \in \tau_j$ s.t. $t_i = t_j$
- $\text{hit}_{\text{ss}} \Leftrightarrow \exists (t, s), (t', s') \in \tau_{1..\ell}$ s.t. $t \neq t' \wedge s = s'$

$\mathcal{T}_{\text{good}}$ is defined such that the above conditions are not satisfied.

Upper-Bound of $\mathbf{Pr[T_2 \in \mathcal{T}_{bad}]}$. We start by defining additional conditions mcoll_T, mcoll_S, and coll_{tt}. Firstly, we define mcoll_T and mcoll_S which are $(q + \rho)$- and ρ-multi-collision conditions for sets T and S, respectively. Here, T keeps all inputs to $\mathcal{G}_2, \ldots, \mathcal{G}_\ell$, and S keeps all outputs of $\mathcal{G}_1, \ldots, \mathcal{G}_\ell$, where $T := \bigcup_{\alpha=1}^{q} \bigcup_{i=2}^{n^\alpha + \ell_{\text{out}} - 1} \{t_i^\alpha\}$ and $S := \bigcup_{\alpha=1}^{q} \bigcup_{i=1}^{n^\alpha + \ell_{\text{out}} - 1} \{s_i^\alpha\}$. Note that sets T and S do not keep duplex elements, and T does not keep inputs to $\mathcal{G}_1$. Then the conditions are defined as

$$\text{mcoll}_T \Leftrightarrow \exists t^{(1)}, t^{(2)}, \ldots, t^{(q+\rho)} \in T \text{ s.t. } t^{(1)}[1, r] = t^{(2)}[1, r] = \cdots = t^{(q+\rho)}[1, r]$$
$$\text{mcoll}_S \Leftrightarrow \exists s^{(1)}, s^{(2)}, \ldots, s^{(\rho)} \in S \text{ s.t. } s^{(1)}[1, r] = s^{(2)}[1, r] = \cdots = s^{(\rho)}[1, r]$$

where ρ is a free parameter which was described in Theorem 1. We let $\text{mcoll} := \text{mcoll}_T \vee \text{mcoll}_S$. Secondly, we define coll_{tt} which is a collision condition for inputs to a random function in L_2. The condition is defined as follows.

$$\text{coll}_{\text{tt}} \Leftrightarrow \exists \alpha, \beta \in \{1, \ldots, q\} \text{ with } \alpha \neq \beta, i \in \{2, \ldots, \min\{n^\alpha, n^\beta\} + \ell_{\text{out}} - 1\}$$
$$\text{s.t. } t_{i-1}^\alpha \neq t_{i-1}^\beta \wedge t_i^\alpha = t_i^\beta.$$

Then we have

$$\Pr[T_2 \in \mathcal{T}_{\text{bad}}] \leq \Pr[\text{hit}_{\text{tx,sy}} \vee \text{hit}_{\text{tt}} \vee \text{hit}_{\text{ss}}]$$
$$\leq \Pr[\text{hit}_{\text{ss}}] + \Pr[\text{coll}_{\text{tt}}] + \Pr[\text{mcoll}_S] + \Pr[\text{mcoll}_T | \neg \text{coll}_{\text{tt}}]$$
$$+ \Pr[\text{hit}_{\text{tx,sy}} | \neg \text{mcoll}] + \Pr[\text{hit}_{\text{tt}} \wedge \neg(\text{coll}_{\text{tt}} \vee \text{mcoll})] . \quad (2)$$

▶ We upper-bound $\Pr[\text{hit}_{\text{ss}}]$. Note that $|\tau_{1..\ell}| \leq \ell q$ holds, and for all $(t, s) \in \tau_{1..\ell}$ s is randomly drawn from $\{0, 1\}^b$. Hence we have $\Pr[\text{hit}_{\text{ss}}] \leq \binom{\ell q}{2} \times \frac{1}{2^b} = \frac{0.5\ell^2 q^2}{2^b}$.
▶ We upper-bound $\Pr[\text{hit}_{\text{tx,sy}} | \neg \text{mcoll}]$. Note that $\text{hit}_{\text{tx,sy}}$ implies that

$$\exists \alpha \in \{1, \ldots, q\}, i \in \{1, \ldots, n^\alpha + \ell_{\text{out}} - 1\}, \beta \in \{1, \ldots, Q\} \text{ s.t. } t_i^\alpha = x^\beta \vee s_i^\alpha = y^\beta.$$

We then consider the following cases.

Case 1 $\Leftrightarrow \text{hit}_{\text{tx,sy}} \wedge t_i^\alpha = x^\beta \wedge i = 1$:
    Note that t_1^α has the form $t_1^\alpha = m_1^\alpha \| 0^c \oplus 0^{b-k} \| K$. Since K is randomly drawn from $\{0, 1\}^k$, the probability that Case 1 holds is at most $\frac{Q}{2^k}$.

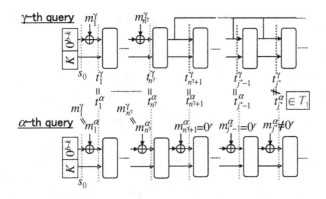

Fig. 2. Procedures for $\boxed{\text{set } T_1}$ and $\mathsf{prefix}^=_{m^\alpha}$

Case 2$\Leftrightarrow \mathsf{hit}_{\mathsf{tx,sy}} \wedge t_i^\alpha = x^\beta \wedge i \neq 1$:

By $\neg\mathsf{mcoll}_T$, the number of elements in T whose first r bits are equal to $x^\beta[1,r]$ is at most $q + \rho$. We note that for some r-bit block M^α, t_i^α has the form $t_i^\alpha = M^\alpha\|0^c \oplus s_{i-1}^\alpha$, where M^α is 0^r or a message block. Since $s_{i-1}^\alpha[r+1,b]$ is randomly drawn from $\{0,1\}^c$, the probability that Case 2 holds is at most $\frac{(q+\rho)Q}{2^c}$.

Case 3$\Leftrightarrow \mathsf{hit}_{\mathsf{tx,sy}} \wedge s_i^\alpha = y^\beta$:

By $\neg\mathsf{mcoll}_S$, the number of elements in S whose first r bits are equal to $y^\beta[1,r]$ is at most ρ. Since $s_i^\alpha[r+1,b]$ is randomly drawn from $\{0,1\}^c$, the probability that Case 3 holds is at most $\frac{\rho Q}{2^c}$.

Hence we have $\Pr[\mathsf{hit}_{\mathsf{tx,sy}}|\neg(\mathsf{hit}_{\mathsf{ux,wy}} \vee \mathsf{mcoll})] \leq \frac{Q}{2^k} + \frac{(q+2\rho)Q}{2^c}$.

▶ We upper-bound $\Pr[\mathsf{mcoll}_S]$. Fix $s \in \{0,1\}^r$ and $s^{(1)}, s^{(2)}, \ldots, s^{(\rho)} \in S$. Since they are randomly drawn from $\{0,1\}^b$, the probability that $s^{(1)}[1,r] = s^{(2)}[1,r] = \cdots = s^{(\rho)}[1,r] = s$ holds is at most $\left(\frac{1}{2^r}\right)^\rho$. By $s \in \{0,1\}^r$ and $|S| \leq \ell q$, we have $\Pr[\mathsf{mcoll}_S] \leq 2^r \times \binom{\ell q}{\rho} \times \left(\frac{1}{2^r}\right)^\rho \leq 2^r \times \left(\frac{e\ell q}{\rho} \times \frac{1}{2^r}\right)^\rho$, using Stirling's approximation $(x! \geq (x/e)^x$ for any $x)$.

▶ We upper-bound $\Pr[\mathsf{mcoll}_T|\neg\mathsf{coll}_{\mathsf{tt}}]$. First we partition set T into two sets T_1 and T_2. Roughly speaking, T_1 keeps all inputs to random functions whose first r bits can be controlled by message blocks. The Fig. 2 (with the boxed statement) depicts the procedure of L_2 corresponding with T_1, which considers γ-th and α-th online queries with $\gamma < \alpha$ and $n^\gamma < n^\alpha$ (n^γ and n^α are the query lengths in blocks at the γ-th and α-th online queries, respectively) such that these message blocks satisfy the condition: $\exists j^* \in \{n^\gamma + 1, \ldots, n^\gamma + \ell_{\mathsf{out}} - 1\}$ s.t. $m_1^\alpha = m_1^\gamma, m_2^\alpha = m_1^\gamma, \ldots, m_{n^\gamma}^\alpha = m_{n^\gamma}^\gamma, m_{n^\gamma}^\alpha = 0^r, \ldots, m_{j^*-1}^\alpha = 0^r, m_{j^*}^\alpha \neq 0^r$. We call the condition between the α-th and γ-th online queries "prefix condition."

In this case, $t_{j^*}^\alpha$ becomes an element of T_1. Since $s_{j^*-1}^\alpha = s_{j^*-1}^\gamma$ holds and before the α-th online query a distinguisher can find $s_{j^*-1}^\gamma[1,r]$ which is the part

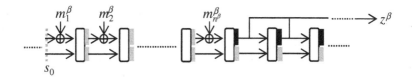

Fig. 3. Lazy sampling random functions in Case 2, where black boxes represent outputs defined at the β-th query and gray boxes represent outputs defined after **D**'s interaction.

of output blocks at the γ-th online query, he can assign any value to $t_{j*}^{\alpha}[1, r]$ by using the message block m_{j*}^{α}. We call the input t_{j*}^{α} "controllable input," and T_1 keeps all controllable inputs. The definitions of these sets are given as follows.

$$T_1 := \left\{ t_{j*}^{\alpha} \in T : (\alpha \in \{2, \ldots, q\}) \wedge \left(\exists \gamma \in \{1, \ldots, \alpha - 1\} \text{ s.t. } (n^{\gamma} < n^{\alpha}) \right. \right.$$
$$\wedge \left(\forall j \in \{1, \ldots, n^{\gamma}\} : m_j^{\alpha} = m_j^{\gamma} \right) \wedge \left(\exists j^* \in \{n^{\gamma} + 1, \ldots, n^{\gamma} + \ell_{\text{out}} - 1\} \text{ s.t. } \right.$$
$$\left. \left. (\forall j \in \{n^{\gamma} + 1, \ldots, j^* - 1\} : m_j^{\alpha} = 0^r) \wedge (m_{j*}^{\alpha} \neq 0^r) \right) \right) \right\} ,$$

and $T_2 := T \backslash T_1$. Note that for any $\alpha_1, \alpha_2, \ldots, \alpha_i \in \{1, \ldots, q\}$ with $\alpha_1 < \alpha_2 < \cdots < \alpha_i$ and with the prefix relations, the number of controllable inputs is at most $i - 1$, because set T_1 does not keep duplex elements. Hence, we have $|T_1| \leq q - 1$, and thereby $\Pr[\text{mcoll}_T | \neg \text{coll}_{\text{tt}}]$ is upper-bounded by the probability that a ρ-multi-collision occurs in T_2 under the condition $\neg \text{coll}_{\text{tt}}$, that is, $\exists t^{(1)}, t^{(2)}, \ldots, t^{(\rho)} \in T_2$ s.t. $t^{(1)}[1, r] = t^{(2)}[1, r] = \cdots = t^{(\rho)}[1, r]$. Hereafter, we upper-bound the ρ-multi-collision probability under the condition $\neg \text{coll}_{\text{tt}}$.

Fix $t \in \{0, 1\}^r$ and $t_i^{\alpha} \in T_2$ with $\alpha \in \{1, \ldots, q\}$ and $i \in \{2, \ldots, n^{\alpha} + \ell_{\text{out}} - 1\}$. We upper-bound the probability that $t_i^{\alpha}[1, r] = t$ holds under the condition $\neg \text{coll}_{\text{tt}}$. We consider the following cases.

Case 1 $\Leftrightarrow (t_i^{\alpha}[1, r] = t) \wedge (n^{\alpha} + 1 \leq i)$:
By $n^{\alpha} + 1 \leq i$, $t_i^{\alpha} = s_{i-1}^{\alpha}$ holds, where $s_{i-1}^{\alpha} = \mathcal{G}_{i-1}(t_{i-1}^{\alpha})$. By $\neg \text{coll}_{\text{tt}}$, s_{i-1}^{α} is randomly drawn from at least $2^b - q$ values. Thus, the probability that Case 1 holds is at most $\frac{2^c}{2^b - q}$.

Case 2 $\Leftrightarrow (t_i^{\alpha}[1, r] = t) \wedge (2 \leq i \leq n^{\alpha})$:
In the evaluation, we lazy sample random functions $\mathcal{G}_1, \ldots, \mathcal{G}_\ell$ that is consistent with the condition $\neg \text{coll}_{\text{tt}}$. The procedure is shown bellow.

– At the β-th online query with $\beta \in \{1, \ldots, q\}$, the following procedure is performed.
  • For $j \in \{n^{\beta}, \ldots, n^{\beta} + \ell_{\text{out}} - 1\}$, $s_j^{\beta}[1, r]$ is randomly drawn from $\{0, 1\}^r$.
– After **D**'s interaction, the following procedure is performed.
  • For all $\beta \in \{1, \ldots, q\}$ and $j \in \{1, \ldots, n^{\beta} - 1\}$, if t_j^{β} is a new input to $\mathcal{G}_j$ then s_j^{β} is randomly drawn from $\{0, 1\}^b$, keeping the condition $\neg \text{coll}_{\text{tt}}$.
  • For all $\beta \in \{1, \ldots, q\}$ and $j \in \{n^{\beta}, \ldots, n^{\beta} + \ell_{\text{out}} - 1\}$, $s_j^{\beta}[r + 1, b]$ is randomly drawn from $\{0, 1\}^c$, keeping the condition $\neg \text{coll}_{\text{tt}}$.

The Fig. 3 depicts the above procedure. Without loss of generality, assume that $q < 2^c$ (If $q \geq 2^c$ then the advantage of Theorem 1 becomes 1 or more). Note that for each random function, there are at most q inputs, and for $a \in \{0,1\}^r$, there are 2^c elements in $\{0,1\}^b$ whose first r bits are equal to a. Thus, for all $\beta \in \{1,\ldots,q\}$ and $j \in \{n^\beta,\ldots,n^\beta + \ell_{\text{out}} - 1\}$, $s_j^\beta[r+1,b]$ can be defined such that it is consistent with the condition $\neg\text{coll}_{\text{tt}}$. Thus, the above procedure realizes random functions $\mathcal{G}_1,\ldots,\mathcal{G}_\ell$ that are consistent with the condition $\neg\text{coll}_{\text{tt}}$.

For $2 \leq i \leq n^\alpha$, t_i^α has the form $t_i^\alpha = m_i^\alpha \| 0^c \oplus s_{i-1}^\alpha$. By the above procedure, s_{i-1}^α is randomly drawn from at least $2^b - q$ values after $\mathbf{D}$'s interaction (i.e., after m_i^α is determined). Hence, the probability that $t_i^\alpha[1,r] = t$ holds is at most $\frac{2^c}{2^b - q}$.

We next fix $t^{(1)}, t^{(2)},\ldots,t^{(\rho)} \in T_2$ and $t \in \{0,1\}^r$. By the above evaluations, the probability that $t^{(1)}[1,r] = t^{(2)}[1,r] = \cdots = t^{(\rho)}[1,r] = t$ holds is at most $\left(\frac{2^c}{2^b-q}\right)^\rho \leq \left(\frac{2}{2^r}\right)^\rho$, assuming $q \leq 2^{b-1}$. By $t \in \{0,1\}^r$ and $|T_2| \leq \ell q$, we have $\Pr[\text{mcoll}_T | \neg\text{coll}_{\text{tt}}] \leq 2^r \times \binom{\ell q}{\rho} \times \left(\frac{2}{2^r}\right)^\rho \leq 2^r \times \left(\frac{e\ell q}{\rho} \times \frac{2}{2^r}\right)^\rho$, using Stirling's approximation ($x! \geq (x/e)^x$ for any x).

▶ We upper-bound $\Pr[\text{coll}_{\text{tt}}]$. We denote by $\text{coll}_{\text{tt}}^\alpha$ the condition where at the α-th online query coll_{tt} holds. Then we have
$\Pr[\text{coll}_{\text{tt}}] \leq \sum_{\alpha=2}^q \Pr[\text{coll}_{\text{tt}}^\alpha \wedge \neg\text{coll}_{\text{tt}}^{\alpha-1}] \leq \sum_{\alpha=2}^q \Pr[\text{coll}_{\text{tt}}^\alpha | \neg\text{coll}_{\text{tt}}^{\alpha-1}]$.

Next we fix $\alpha \in \{2,\ldots,q\}$, and upper-bound $\Pr[\text{coll}_{\text{tt}}^\alpha | \neg\text{coll}_{\text{tt}}^{\alpha-1}]$, which is the probability that coll_{tt} holds at the α-th online query when it does not hold up to the $(\alpha-1)$-th online query. In order to upper-bound the probability, we consider two cases with respect to the following condition.

$$\text{prefix}_{m^\alpha}^{=} \Leftrightarrow \exists\gamma \in \{1,\ldots,\alpha-1\} \text{ s.t. } \left(n^\gamma < n^\alpha\right) \wedge \left(\forall j \in \{1,\ldots,n^\gamma\} : m_j^\gamma = m_j^\alpha\right)$$
$$\wedge \left(\exists j^* \in \{n^\gamma + 1,\ldots,n^\gamma + \ell_{\text{out}} - 1\} \text{ s.t.}\right.$$
$$\left. m_{n^\gamma+1}^\alpha = 0^r,\ldots,m_{j^*-1}^\alpha = 0^r, m_{j^*}^\alpha \neq 0^r\right).$$

We call such γ-th online query "prefix online query" of the α-th query, and such j^* "distinct point." The Fig. 2 (without the boxed statement) depicts the procedures of L_2 corresponding with the condition. In this evaluation, similar to Case 2 of $\Pr[\text{mcoll}_T | \neg\text{coll}_{\text{tt}}]$, we lazy sample random functions $\mathcal{G}_1,\ldots,\mathcal{G}_\ell$ that are consistent with the condition $\neg\text{coll}_{\text{tt}}^{\alpha-1}$. The procedure is shown bellow.

- At the β-th online query with $\beta \in \{1,\ldots,\alpha-1\}$, the following procedure is performed.
  - For all $j \in \{n^\beta,\ldots,n^\beta + \ell_{\text{out}} - 1\}$, $s_j^\beta[1,r]$ is randomly drawn from $\{0,1\}^r$.
- At the α-th online query, the following procedure is performed.
  - For all $\beta \in \{1,\ldots,\alpha-1\}$,
    * for all $j \in \{1,\ldots,n^\beta - 1\}$, if t_j^β is a new input to $\mathcal{G}_j$ then the response s_j^β is randomly drawn from $\{0,1\}^b$, keeping the condition $\neg\text{coll}_{\text{tt}}^{\alpha-1}$,
    * for all $j \in \{n^\beta,\ldots,n^\beta + \ell_{\text{out}} - 1\}$, $s_j^\beta[r+1,b]$ is randomly drawn from $\{0,1\}^c$, keeping the condition $\neg\text{coll}_{\text{tt}}^{\alpha-1}$.

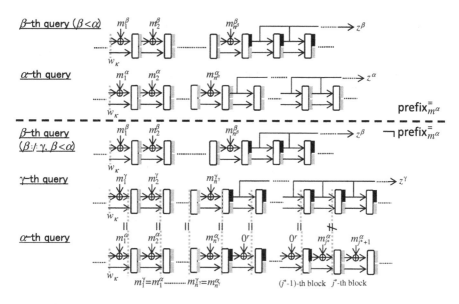

Fig. 4. Lazy sampling random functions in the evaluation of $\Pr[\mathsf{coll}_{\mathsf{tt}}^\alpha | \neg \mathsf{coll}_{\mathsf{tt}}^{\alpha-1}]$, where black boxes represent outputs defined up to the $(\alpha - 1)$-th query and gray boxes represent outputs defined at the α-th query.

- For $j \in \{1, \ldots, n^\alpha + \ell_{\mathrm{out}} - 1\}$, if t_j^α is a new input to $\mathcal{G}_j$ then the response s_j^α is randomly drawn from $\{0,1\}^b$.

The top (resp., the bottom) of the Fig. 4 depicts the above procedure under the condition $\mathsf{prefix}_{m^\alpha}^=$ (resp., $\neg\mathsf{prefix}_{m^\alpha}^=$). Then we evaluate the probability $\Pr[\mathsf{coll}_{\mathsf{tt}}^\alpha | \neg \mathsf{coll}_{\mathsf{tt}}^{\alpha-1}]$ as follows.

Case 1$\Leftrightarrow \mathsf{coll}_{\mathsf{tt}}^\alpha$ under the condition $\neg\mathsf{coll}_{\mathsf{tt}}^{\alpha-1} \wedge \neg\mathsf{prefix}_{m^\alpha}^=$:
For $i \in \{2, \ldots, n^\alpha + \ell_{\mathrm{out}} - 1\}$, let $\mathsf{coll}_{\mathsf{tt}}^{\alpha,i}$ be the condition where $\mathsf{coll}_{\mathsf{tt}}^\alpha$ holds at the i-th block of the α-th online query, and let $\mathsf{coll}_{\mathsf{tt}}^{\leq\alpha,i-1} := \mathsf{coll}_{\mathsf{tt}}^{\alpha,2} \vee \mathsf{coll}_{\mathsf{tt}}^{\alpha,3} \vee \cdots \vee \mathsf{coll}_{\mathsf{tt}}^{\alpha,i-1}$. Note that for $i \in \{2, \ldots, n^\alpha + \ell_{\mathrm{out}} - 1\}$, $\mathsf{coll}_{\mathsf{tt}}^{\alpha,i} \wedge \neg\mathsf{coll}_{\mathsf{tt}}^{\leq\alpha,i-1}$ is the condition where $\mathsf{coll}_{\mathsf{tt}}^\alpha$ holds at the i-th block of the α-th online query for the first time. (i.e., $\mathsf{coll}_{\mathsf{tt}}^\alpha$ does not hold up to the $(i-1)$-th block), and thus $\mathsf{coll}_{\mathsf{tt}}^\alpha \Leftrightarrow \bigvee_{i=2}^{n^\alpha+\ell_{\mathrm{out}}-1}(\mathsf{coll}_{\mathsf{tt}}^{\alpha,i} \wedge \neg\mathsf{coll}_{\mathsf{tt}}^{\leq\alpha,i-1})$, where $\mathsf{coll}_{\mathsf{tt}}^{\alpha,2} \wedge \neg\mathsf{coll}_{\mathsf{tt}}^{\leq\alpha,1} := \mathsf{coll}_{\mathsf{tt}}^{\alpha,2}$. In the following, for $i \in \{2, \ldots, n^\alpha + \ell_{\mathrm{out}} - 1\}$, we assume that $\mathsf{coll}_{\mathsf{tt}}^{\leq\alpha,i-1}$ does not hold, and thus upper-bound the probability that $\mathsf{coll}_{\mathsf{tt}}^{\alpha,i}$ holds under the condition $\neg\mathsf{coll}_{\mathsf{tt}}^{\alpha-1} \wedge \neg\mathsf{coll}_{\mathsf{tt}}^{\leq\alpha,i-1} \wedge \neg\mathsf{prefix}_{m^\alpha}^=$. By $p_{1,i}$, we denote the probability. Note that for some r-bit string M^α t_i^α has the form $t_i^\alpha = M^\alpha \| 0^c \oplus s_{i-1}^\alpha$, where M^α is a message block or 0^r. By the condition $\neg\mathsf{coll}_{\mathsf{tt}}^{\leq\alpha,i-1}$, t_{i-1}^α is a new input to $\mathcal{G}_{i-1}$, and thereby s_{i-1}^α is randomly drawn from $\{0,1\}^b$ after M^α is determined. Hence, we have $p_{1,i} \leq (\alpha - 1) \times \frac{1}{2^b}$, and thereby $\Pr[\textbf{Case 1}] \leq \ell \times (\alpha - 1) \times \frac{1}{2^b}$.

Case 2 $\Leftrightarrow$ $\mathsf{coll}^{\alpha}_{\mathsf{tt}}$ under the condition $\neg\mathsf{coll}^{\alpha-1}_{\mathsf{tt}} \wedge \mathsf{prefix}^{=}_{m^{\alpha}}$:

In this analysis, we use the conditions $\mathsf{coll}^{\alpha,i}_{\mathsf{tt}}$ and $\mathsf{coll}^{\leq\alpha,i-1}_{\mathsf{tt}}$ defined above. For $i \in \{2,\ldots,n^{\alpha}+\ell_{\mathsf{out}}-1\}$, we assume that $\mathsf{coll}^{\leq\alpha,i-1}_{\mathsf{tt}}$ does not hold, and thus upper-bound the probability that $\mathsf{coll}^{\alpha,i}_{\mathsf{tt}}$ holds under the condition $\neg\mathsf{coll}^{\alpha-1}_{\mathsf{tt}} \wedge \neg\mathsf{coll}^{\leq\alpha,i-1}_{\mathsf{tt}} \wedge \mathsf{prefix}^{=}_{m^{\alpha}}$. By $p_{2,i}$, we denote the probability. We assume that the γ-th online query ($\gamma \in \{1,\ldots,\alpha-1\}$) is the prefix online query of the α-th online query, and j^* is the distinct point. If there are two or more prefix online queries of the α-th online query then we consider the prefix online query such that the distinct point is maximum.

- Firstly, we consider the case of $i \in \{2,\ldots,j^*-1\}$. By $\mathsf{prefix}^{=}_{m^{\alpha}}$, $t^{\alpha}_i = t^{\gamma}_i$ holds. By the condition $\neg\mathsf{coll}^{\alpha-1}_{\mathsf{tt}} \wedge \neg\mathsf{coll}^{\leq\alpha,i-1}_{\mathsf{tt}}$, we have $p_{2,i} = 0$.
- Secondly, we consider the case of $i = j^*$. Note that $t^{\alpha}_{j^*}[r+1,b] = s^{\alpha}_{j^*-1}[r+1,b]$ holds, and by the lazy sampled random functions, $s^{\alpha}_{j^*-1}$ is randomly drawn from at least $2^b - q$ values. Thus we have $p_{2,i} \leq (\alpha-1) \times \frac{2^r}{2^b-q}$.
- Finally, we consider the case of $i \in \{j^*+1,\ldots,n^{\alpha}+\ell_{\mathsf{out}}-1\}$. In this case, for some r-bit string M^{α}, t^{α}_i has the form $t^{\alpha}_i = M^{\alpha}\|0^c \oplus s^{\alpha}_{i-1}$, where M^{α} is a message block or 0^r. Since j^* is maximum and by the condition $\neg\mathsf{coll}^{\leq\alpha,i-1}_{\mathsf{tt}}$ t^{α}_{i-1} is a new input to $\mathcal{G}_{i-1}$, s^{α}_{i-1} is randomly drawn from $\{0,1\}^b$ after M^{α} is determined. Hence, we have $p_{2,i} \leq (\alpha-1) \times \frac{1}{2^b}$.

Hence, we have $\Pr[\textbf{Case 2}] \leq (\alpha-1) \times \left(\frac{2^r}{2^b-q} + \frac{\ell_{\mathsf{out}}}{2^b}\right)$.

Finally, we assume that $q \leq 2^{b-1}$. We then have
$$\Pr[\mathsf{coll}_{\mathsf{tt}}] \leq \sum_{\alpha=2}^{q}(\alpha-1) \times \max\left\{\frac{\ell}{2^b}, \left(\frac{2^r}{2^b-q} + \frac{\ell_{\mathsf{out}}}{2^b}\right)\right\} \leq \frac{q^2}{2^c} + \frac{0.5\ell q^2}{2^b}.$$

▶ We upper-bound $\Pr[\mathsf{hit}_{\mathsf{tt}} \wedge \neg(\mathsf{coll}_{\mathsf{tt}} \vee \mathsf{mcoll})]$. We start by defining the following condition.

$$\mathsf{hit}_K \Leftrightarrow \exists\alpha \in \{1,\ldots,q\}, i \in \{2,\ldots,n^{\alpha}+\ell_{\mathsf{out}}-1\} \text{ s.t. } t^{\alpha}_i[r+1,b] = 0^{c-k}\|K$$

Then we have

$$\Pr[\mathsf{hit}_{\mathsf{tt}} \wedge \neg(\mathsf{coll}_{\mathsf{tt}} \vee \mathsf{mcoll})] \leq \Pr[\mathsf{hit}_K] + \Pr[\mathsf{hit}_{\mathsf{tt}} \wedge \neg(\mathsf{coll}_{\mathsf{tt}} \vee \mathsf{mcoll}) \wedge \neg\mathsf{hit}_K] .$$

Since K is randomly drawn from $\{0,1\}^k$, we have $\Pr[\mathsf{hit}_K] \leq \frac{\ell q}{2^k}$.

Next, we upper-bound $\Pr[\mathsf{hit}_{\mathsf{tt}} \wedge \neg(\mathsf{coll}_{\mathsf{tt}} \vee \mathsf{mcoll}) \wedge \neg\mathsf{hit}_K]$. Note that $\mathsf{hit}_{\mathsf{tt}}$ implies that

$$\exists\alpha,\beta \in \{1,\ldots,q\}, i \in \{1,\ldots,n^{\alpha}+\ell_{\mathsf{out}}-1\}, j \in \{1,\ldots,n^{\beta}+\ell_{\mathsf{out}}-1\}$$
$$\text{s.t. } i \neq j \wedge t^{\alpha}_i = t^{\beta}_j.$$

For $\alpha \in \{1,\ldots,q\}$, we define a condition where $\mathsf{hit}_{\mathsf{tt}}$ holds up to the α-th online query. The concrete definition is given bellow.

$$\mathsf{hit}^{\alpha}_{\mathsf{tt}} \Leftrightarrow \exists\beta,\gamma \in \{1,\ldots,\alpha\}, i \in \{1,\ldots,n^{\beta}+\ell_{\mathsf{out}}-1\}, j \in \{1,\ldots,n^{\gamma}+\ell_{\mathsf{out}}-1\}$$
$$\text{s.t. } i \neq j \wedge t^{\beta}_i = t^{\gamma}_j.$$

Then the following inequation holds.

$$\Pr[\mathsf{hit}_{\mathsf{tt}} \wedge \neg(\mathsf{coll}_{\mathsf{tt}} \vee \mathsf{mcoll}) \wedge \mathsf{hit}_K]$$

$$\leq \sum_{\alpha=1}^{q} \Pr[\mathsf{hit}_{\mathsf{tt}}^{\alpha} \wedge \neg\mathsf{hit}_{\mathsf{tt}}^{\alpha-1} \wedge \neg(\mathsf{mcoll} \vee \mathsf{coll}_{\mathsf{tt}}) \wedge \neg\mathsf{hit}_K]$$

$$\leq \sum_{\alpha=1}^{q} \Pr[\mathsf{hit}_{\mathsf{tt}}^{\alpha} \wedge \neg\mathsf{hit}_{\mathsf{tt}}^{\alpha-1} \wedge \neg\mathsf{mcoll} \wedge \neg\mathsf{hit}_K | \neg\mathsf{coll}_{\mathsf{tt}}] \ .$$

First fix $\alpha \in \{1, \dots, q\}$, and upper-bound the probability $\Pr[\mathsf{hit}_{\mathsf{tt}}^{\alpha} \wedge \neg\mathsf{hit}_{\mathsf{tt}}^{\alpha-1} \wedge \neg\mathsf{mcoll} \wedge \neg\mathsf{hit}_K | \neg\mathsf{coll}_{\mathsf{tt}}]$. In this evaluation, we lazy sample random functions $\mathcal{G}_1, \dots, \mathcal{G}_\ell$ by the similar way to the evaluation of $\Pr[\mathsf{coll}_{\mathsf{tt}}]$. The procedure is shown bellow, and the Fig. 4 depicts the procedure.

- At the β-th online query with $\beta \in \{1, \dots, \alpha - 1\}$, the following procedure is performed.
  - For all $j \in \{n^\beta, \dots, n^\beta + \ell_{\mathrm{out}} - 1\}$, $s_j^\beta[1, r]$ is randomly drawn from $\{0,1\}^r$.
- At the α-th online query, the following procedure is performed.
  - For all $\beta \in \{1, \dots, \alpha - 1\}$,
    * for all $j \in \{1, \dots, n^\beta - 1\}$, if t_j^β is a new input to $\mathcal{G}_j$ then the response s_j^β is randomly drawn from $\{0,1\}^b$, keeping the condition $\neg\mathsf{coll}_{\mathsf{tt}}$,
    * for all $j \in \{n^\beta, \dots, n^\beta + \ell_{\mathrm{out}} - 1\}$, $s_j^\beta[r+1, b]$ is randomly drawn from $\{0,1\}^c$, keeping the condition $\neg\mathsf{coll}_{\mathsf{tt}}$.
  - For $j \in \{1, \dots, n^\alpha + \ell_{\mathrm{out}} - 1\}$, if t_j^α is a new input to $\mathcal{G}_j$ then the response s_j^α is randomly drawn from $\{0,1\}^b$, keeping the condition $\neg\mathsf{coll}_{\mathsf{tt}}$.

In this evaluation, we consider two cases with respect to the condition $\mathsf{prefix}_{m^\alpha}^{=}$ which was defined in the analysis of $\Pr[\mathsf{coll}_{\mathsf{tt}}]$. In addition, the following analyses use the terms "prefix online query" and "distinct point."

Case 1 $\Leftrightarrow \mathsf{hit}_{\mathsf{tt}}^{\alpha} \wedge \neg\mathsf{hit}_{\mathsf{tt}}^{\alpha-1} \wedge \neg\mathsf{mcoll} \wedge \neg\mathsf{hit}_K$ under the condition $\neg\mathsf{coll}_{\mathsf{tt}} \wedge \neg\mathsf{prefix}_{m^\alpha}^{=}$: For $i \in \{1, \dots, n^\alpha + \ell_{\mathrm{out}} - 1\}$, let $\mathsf{hit}_{\mathsf{tt}}^{\alpha,i}$ be the condition where $\mathsf{hit}_{\mathsf{tt}}^{\alpha}$ holds at the i-th block of the α-th online query, that is,

$$\mathsf{hit}_{\mathsf{tt}}^{\alpha,i} \Leftrightarrow (\exists \beta \in \{1, \dots, \alpha - 1\}, j \in \{1, \dots, n^\beta + \ell_{\mathrm{out}} - 1\} \text{ s.t. } i \neq j \wedge t_i^\alpha = t_j^\beta)$$
$$\wedge (\exists j \in \{1, \dots, i - 1\} \text{ s.t. } t_i^\alpha = t_j^\alpha).$$

Then $\mathsf{hit}_{\mathsf{tt}}^{\alpha} \Rightarrow \bigvee_{i=1}^{n^\alpha + \ell_{\mathrm{out}} - 1} \mathsf{hit}_{\mathsf{tt}}^{\alpha,i}$. In the following, for $i \in \{1, \dots, n^\alpha + \ell_{\mathrm{out}} - 1\}$, we upper-bound the probability that $\mathsf{hit}_{\mathsf{tt}}^{\alpha,i} \wedge \neg\mathsf{hit}_{\mathsf{tt}}^{\alpha-1} \wedge \neg\mathsf{mcoll} \wedge \neg\mathsf{hit}_K$ holds under the condition $\neg\mathsf{coll}_{\mathsf{tt}} \wedge \neg\mathsf{prefix}_{m^\alpha}^{=}$. By $p_{1,i}$, we denote the probability.

- Firstly, we consider the case of $i = 1$. In addition to the condition $\neg\mathsf{coll}_{\mathsf{tt}} \wedge \neg\mathsf{prefix}_{m^\alpha}^{=}$, we assume that hit_K does not hold, and don't consider the condition $\neg\mathsf{hit}_{\mathsf{tt}}^{\alpha-1} \wedge \neg\mathsf{mcoll}$. Since t_1^α has the form $t_1^\alpha = (m_1^\alpha \| 0^c) \oplus (0^{b-k} \| K)$, the probability that $\mathsf{hit}_{\mathsf{tt}}^{\alpha,1}$ holds under the condition $\neg\mathsf{coll}_{\mathsf{tt}} \wedge \neg\mathsf{prefix}_{m^\alpha}^{=} \wedge \neg\mathsf{hit}_K$ is 0 and thus we have $p_{1,1} = 0$.

– Secondly, we consider the case of $i \geq 2$. In this case, we don't consider the condition $\neg \mathsf{hit}_{\mathsf{tt}}^{\alpha-1} \wedge \neg \mathsf{mcoll} \wedge \neg \mathsf{hit}_K$. Note that for an r-bit string M^α, t_i^α has the form $t_i^\alpha = M^\alpha \| 0^c \oplus s_{i-1}^\alpha$, where M^α is a message block or 0^r. Since s_{i-1}^α is randomly drawn from at least $2^b - q$ values after M^α is defined, the probability that $\mathsf{hit}_{\mathsf{tt}}^{\alpha,i}$ holds under the condition $\neg \mathsf{coll}_{\mathsf{tt}} \wedge \neg \mathsf{prefix}_{m^\alpha}^=$ is at most $\frac{(\ell-1)(\alpha-1)+(i-1)}{2^b-q} \leq \frac{(\ell-1)\alpha}{2^b-q}$, and thus we have $p_{1,i} \leq \frac{(\ell-1)\alpha}{2^b-q}$.

Hence, we have $\Pr[\textbf{Case 1}] \leq (\ell-1) \times \frac{(\ell-1)\alpha}{2^b-q}$.

Case 2 $\Leftrightarrow \mathsf{hit}_{\mathsf{tt}}^\alpha \wedge \neg \mathsf{hit}_{\mathsf{tt}}^{\alpha-1} \wedge \neg \mathsf{mcoll} \wedge \neg \mathsf{hit}_K$ under the condition $\neg \mathsf{coll}_{\mathsf{tt}} \wedge \mathsf{prefix}_{m^\alpha}^=$: In this analysis, we use the condition $\mathsf{hit}_{\mathsf{tt}}^{\alpha,i}$ for $i \in \{1,\dots,n^\alpha + \ell_{\mathsf{out}} - 1\}$, defined in Case 1. We let $\mathsf{hit}_{\mathsf{tt}}^{\leq \alpha,i-1} := \mathsf{hit}_{\mathsf{tt}}^{\alpha-1} \vee \mathsf{hit}_{\mathsf{tt}}^{\alpha,1} \vee \cdots \vee \mathsf{hit}_{\mathsf{tt}}^{\alpha,i-1}$, where $\mathsf{hit}_{\mathsf{tt}}^{\alpha,0} := \mathsf{hit}_{\mathsf{tt}}^{\alpha-1}$. Then the following holds: $\mathsf{hit}_{\mathsf{tt}}^\alpha \wedge \neg \mathsf{hit}_{\mathsf{tt}}^{\alpha-1} \Rightarrow \bigvee_{i=1}^{n^\alpha + \ell_{\mathsf{out}} - 1} (\mathsf{hit}_{\mathsf{tt}}^{\alpha,i} \wedge \neg \mathsf{hit}_{\mathsf{tt}}^{\leq \alpha,i-1})$. In this evaluation, we don't consider the condition $\neg \mathsf{hit}_K$, and thus for $i \in \{1,\dots,n^\alpha + \ell_{\mathsf{out}} - 1\}$, upper-bound the probability that $\mathsf{hit}_{\mathsf{tt}}^{\alpha,i} \wedge \neg \mathsf{hit}_{\mathsf{tt}}^{\leq \alpha,i-1} \wedge \neg \mathsf{mcoll}$ holds under the condition $\neg \mathsf{coll}_{\mathsf{tt}} \wedge \mathsf{prefix}_{m^\alpha}^=$. By $p_{2,i}$, we denote the probability. We assume that the γ-th online query ($\gamma \in \{1,\dots,\alpha-1\}$) is the prefix online query of the α-th online query, and j^* is the distinct point. If there are two or more prefix online queries of the α-th online query then we consider the prefix online query such that the distinct point is maximum.

– Firstly, we consider the case of $i < j^*$. In this case, we don't consider the condition $\neg \mathsf{mcoll}$, and assume that $\mathsf{hit}_{\mathsf{tt}}^{\leq \alpha,i-1}$ does not hold in addition to the condition $\neg \mathsf{coll}_{\mathsf{tt}} \wedge \mathsf{prefix}_{m^\alpha}^=$. By $\mathsf{prefix}_{m^\alpha}^=$, $t_i^\alpha = t_i^\gamma$ holds, and by $\neg \mathsf{hit}_{\mathsf{tt}}^{\leq \alpha,i-1}$, $\mathsf{hit}_{\mathsf{tt}}^\gamma$ does not hold. Hence, $\mathsf{hit}_{\mathsf{tt}}^{\alpha,i}$ does not hold under the condition $\neg \mathsf{coll}_{\mathsf{tt}} \wedge \mathsf{prefix}_{m^\alpha}^= \wedge \mathsf{hit}_{\mathsf{tt}}^{\leq \alpha,i-1}$, and thus we have $p_{2,i} = 0$.

– Secondly, we consider the case of $i = j^*$. In this analysis, we don't consider the condition $\neg \mathsf{hit}_{\mathsf{tt}}^{\leq \alpha,i-1}$, and assume that mcoll does not hold in addition to the condition $\neg \mathsf{coll}_{\mathsf{tt}} \wedge \mathsf{prefix}_{m^\alpha}^=$. Note that since j^* is the maximum distinct point, $t_{j^*}^\alpha$ is a new input to $\mathcal{G}_{j^*}$. By $\neg \mathsf{mcoll}_T$, the number of inputs to random functions whose first r bits are equal to $t_{j^*}^\alpha[1,r]$ is at most $(q + \rho)$. Note that $t_{j^*}^\alpha[r+1,b] = s_{j^*-1}^\alpha[r+1,b]$, and $s_{j^*-1}^\alpha[r+1,b]$ is randomly drawn from at least $2^c - q$ values. Hence, the probability that $\mathsf{hit}_{\mathsf{tt}}^{\alpha,i}$ holds under the condition $\neg \mathsf{coll}_{\mathsf{tt}} \wedge \mathsf{prefix}_{m^\alpha}^= \wedge \neg \mathsf{mcoll}$ is at most $\frac{q+\rho}{2^c-q}$, and thus we have $p_{2,i} \leq \frac{q+\rho}{2^c-q}$.

– Finally, we consider the case of $i > j^*$. In this analysis, we don't consider the conditions $\neg \mathsf{hit}_{\mathsf{tt}}^{\leq \alpha,i-1}$ and $\neg \mathsf{mcoll}_T$. Note that for an r-bit string M^α, t_i^α has the form $t_i^\alpha = M^\alpha \| 0^c \oplus s_{i-1}^\alpha$, where M^α is a message block or 0^r. By $\neg \mathsf{coll}_{\mathsf{tt}}$, s_{i-1}^α is randomly drawn from at least $2^b - q$ values after M^α is defined. We thus have $p_{2,i} \leq \frac{(\ell-2)\alpha}{2^b-q}$.

Hence, we have $\Pr[\textbf{Case 2}] \leq \frac{q+\rho}{2^c-q} + (\ell-2) \times \frac{(\ell-2)\alpha}{2^b-q}$.

Hence, we have

$$\Pr[\mathsf{hit_{tt}} \wedge \neg(\mathsf{coll_{tt}} \vee \mathsf{mcoll}) \wedge \neg\mathsf{hit}_K] \leq \sum_{\alpha=1}^{q} \max\left\{ \frac{(\ell-1)^2\alpha}{2^b - q}, \frac{q+\rho}{2^c - q} + \frac{(\ell-2)^2\alpha}{2^b - q} \right\}$$

$$\leq \frac{2(q+\rho)q}{2^c} + \frac{\ell^2 q^2}{2^b} \ , \ \text{assuming } q \leq 2^{c-1}.$$

Finally, we have $\Pr[\mathsf{hit_{tt}} \wedge \neg(\mathsf{coll_{tt}} \vee \mathsf{mcoll})] \leq \frac{\ell q}{2^k} + \frac{2(q+\rho)q}{2^c} + \frac{\ell^2 q^2}{2^b}$.

▶ We put the above bounds to the inequation (2). Then we have

$$\Pr[\mathsf{T}_2 \in \mathcal{T}_{\mathsf{bad}}] \leq \frac{\ell q + Q}{2^k} + \frac{2q^2 + qQ + 2\rho(q+Q)}{2^c} + \frac{2\ell^2 q^2}{2^b} + 2^{r+1} \times \left(\frac{2e\ell q}{\rho 2^r}\right)^{\rho}.$$

Upper-Bound of ε. Let $\tau \in \mathcal{T}_{\mathsf{good}}$. Let all$_i$ be the set of all oracles in Game i for $i = 1, 2$. Let $\mathrm{comp}_i(\tau)$ be the set of oracles compatible with τ in Game i for $i = 1, 2$. Then $\Pr[\mathsf{T}_1 = \tau] = \frac{|\mathrm{comp}_1(\tau)|}{|\mathrm{all}_1|}$ and $\Pr[\mathsf{T}_2 = \tau] = \frac{|\mathrm{comp}_2(\tau)|}{|\mathrm{all}_2|}$.

Firstly, we evaluate $|\mathrm{all}_1|$. Since $K \in \{0,1\}^k$ and $\mathcal{P} \in \mathsf{Perm}(\{0,1\}^b)$, we have $|\mathrm{all}_1| = 2^k \cdot 2^b!$.

Secondly, we evaluate $|\mathrm{all}_2|$. Since $K \in \{0,1\}^k$, $\mathcal{P} \in \mathsf{Perm}(\{0,1\}^b)$, and $\mathcal{G}_1, \mathcal{G}_2, \ldots, \mathcal{G}_\ell \in \mathsf{Func}(\{0,1\}^b, \{0,1\}^b)$, we have $|\mathrm{all}_2| = 2^k \cdot (2^b!) \cdot \left((2^b)^{2^b}\right)^\ell$.

Thirdly, we evaluate $|\mathrm{comp}_1(\tau)|$. For $i \in \{1, \ldots, \ell\}$, let γ_i be the number of pairs in τ_i. Let $\gamma_{\mathcal{P}}$ be the numbers of pairs in $\tau_{\mathcal{P}}$. Let $\gamma = \gamma_{\mathcal{P}} + \sum_{i=1}^{\ell} \gamma_i$. Since $\tau_1, \ldots, \tau_\ell$ and $\tau_{\mathcal{P}}$ are defined so that they do not overlap each other, we have $|\mathrm{comp}_1(\tau)| = (2^b - \gamma)!$.

Fourthly, we evaluate $|\mathrm{comp}_2(\tau)|$. Here, $\gamma_1, \ldots \gamma_\ell$, and $\gamma_{\mathcal{P}}$ are analogously defined. Then we have $|\mathrm{comp}_2(\tau)| = (2^b - \gamma_{\mathcal{P}})! \cdot \prod_{i=1}^{\ell} (2^b)^{2^b - \gamma_i} = (2^b - \gamma_{\mathcal{P}})! \cdot (2^b)^{\ell 2^b - \gamma + \gamma_{\mathcal{P}}}$.

Finally, we have

$$\frac{\Pr[\mathsf{T}_1 = \tau]}{\Pr[\mathsf{T}_2 = \tau]} = \frac{|\mathrm{comp}_1(\tau)|}{|\mathrm{all}_1|} \times \frac{|\mathrm{all}_2|}{|\mathrm{comp}_2(\tau)|} = \frac{(2^b - \gamma)!}{2^k \cdot (2^b!)} \times \frac{2^k \cdot (2^b!) \cdot (2^b)^{\ell 2^b}}{(2^b - \gamma_{\mathcal{P}})! \cdot (2^b)^{\ell 2^b - \gamma + \gamma_{\mathcal{P}}}}$$

$$= \frac{(2^b)^\gamma \cdot (2^b - \gamma)!}{(2^b)^{\gamma_{\mathcal{P}}} \cdot (2^b - \gamma_{\mathcal{P}})!} \geq 1 \ ,$$

and thus $\varepsilon = 0$.

Upper-Bound of $\Pr[\mathbf{D}^{\mathbf{G_1}} \Rightarrow 1] - \Pr[\mathbf{D}^{\mathbf{G_2}} \Rightarrow 1]$. Finally, by Lemma 1, the upper-bound of $\Pr[\mathsf{T}_2 \in \mathcal{T}_{\mathsf{bad}}]$ and ε yield the following bound.

$$\Pr[\mathbf{D}^{\mathbf{G_1}} \Rightarrow 1] - \Pr[\mathbf{D}^{\mathbf{G_2}} \Rightarrow 1]$$

$$\leq \frac{\ell q + Q}{2^k} + \frac{2q^2 + qQ + 2\rho(q+Q)}{2^c} + \frac{2\ell^2 q^2}{2^b} + 2^{r+1} \times \left(\frac{2e\ell q}{\rho 2^r}\right)^{\rho}. \tag{3}$$

4.2   Upper-Bound of $\Pr[\mathbf{D}^{G_2} \Rightarrow 1] - \Pr[\mathbf{D}^{G_3} \Rightarrow 1]$

Firstly, we prove the following lemma.

Lemma 2. G_2 and G_3 are indistinguishable unless the following condition holds in Game 2.[1]

$$\mathsf{coll} \Leftrightarrow \exists \alpha, \beta \in \{1, \dots, q\}, i \in \{\max\{n^\alpha, n^\beta\}, \dots, \min\{n^\alpha, n^\beta\} + \ell_{\text{out}} - 1\}$$
$$s.t.\ \alpha \neq \beta \wedge t_i^\alpha = t_i^\beta.$$

Proof. If coll does not hold then all blocks in outputs of L_2 are independently drawn by random functions. Hence the above lemma holds. $\square$

By the above lemma, $\Pr[\mathbf{D}^{G_2} \Rightarrow 1|\neg\mathsf{coll}] = \Pr[\mathbf{D}^{G_3} \Rightarrow 1]$ holds. Then we have

$$\Pr[\mathbf{D}^{G_2} \Rightarrow 1] - \Pr[\mathbf{D}^{G_3} \Rightarrow 1] \leq \Pr[\mathsf{coll}]\ .$$

Hereafter, we upper-bound $\Pr[\mathsf{coll}]$. In this evaluation, we use the condition $\mathsf{coll}_{\mathsf{tt}}$ given in Subsect. 4.1. Then we have $\Pr[\mathsf{coll}] \leq \Pr[\mathsf{coll}_{\mathsf{tt}}] + \Pr[\mathsf{coll}|\neg\mathsf{coll}_{\mathsf{tt}}]$ where the upper-bound of $\Pr[\mathsf{coll}_{\mathsf{tt}}]$ is given in Subsect. 4.1: $\Pr[\mathsf{coll}_{\mathsf{tt}}] \leq \frac{q^2}{2^c} + \frac{0.5\ell q^2}{2^b}$.

We thus upper-bound $\Pr[\mathsf{coll}|\neg\mathsf{coll}_{\mathsf{tt}}]$. First fix $\alpha, \beta \in \{1, \dots, q\}$ with $\alpha \neq \beta$, and upper-bound the probability that by the α-th and β-th online queries, coll holds. We consider the following cases.

Case 1 $\Leftrightarrow n^\alpha = n^\beta$: Since $m^\alpha \neq m^\beta$, there exists $j^* \in \{1, \dots, n^\alpha\}$ such that $t_{j^*}^\alpha \neq t_{j^*}^\beta$. By $\neg\mathsf{coll}_{\mathsf{tt}}$, for all $j \in \{j^* + 1, \dots, n^\alpha + \ell - 1\}$, $t_j^\alpha \neq t_j^\beta$ holds. Hence, in this case, coll does not hold.

Case 2 $\Leftrightarrow n^\alpha \neq n^\beta$: Without loss of generality, assume that $n^\alpha > n^\beta$. By $m_{n^\alpha}^\alpha \neq 0^r$ and $m^\alpha \neq m^\beta$, there exists $j^* \in \{1, \dots, n^\beta\}$ such that $t_{j^*}^\alpha \neq t_{j^*}^\beta$ holds. By $\neg\mathsf{coll}_{\mathsf{tt}}$, for all $j \in \{j^* + 1, \dots, n^\alpha + \ell - 1\}$, $t_j^\alpha \neq t_j^\beta$ holds. Hence, in this case, coll does not hold.

By the above evaluations, we have $\Pr[\mathsf{coll}|\neg\mathsf{coll}_{\mathsf{tt}}] = 0$.

Finally, we have

$$\Pr[\mathbf{D}^{G_2} \Rightarrow 1] - \Pr[\mathbf{D}^{G_3} \Rightarrow 1] \leq \Pr[\mathsf{coll}] \leq \frac{q^2}{2^c} + \frac{0.5\ell q^2}{2^b}\ . \tag{4}$$

4.3   Upper-Bound of the Advantage

We put the upper-bounds (3) and (4) into the inequation (1). Then we have

$$\mathbf{Adv}_{\text{IKSponge}}^{\text{prf}}(\mathbf{D}) \leq \frac{\ell q + Q}{2^k} + \frac{3q^2 + qQ + 2\rho(q + Q)}{2^c} + \frac{3\ell^2 q^2}{2^b} + 2^{r+1} \times \left(\frac{2e\ell q}{\rho 2^r}\right)^\rho.$$

---

[1] Note that in this condition we consider a collision at the same position for two online queries, where in the position the outputs of the queries are produced. Hence, the first point of i is $\max\{n^\alpha, n^\beta\}$ and the last point is $\min\{n^\alpha, n^\beta\} + \ell_{\text{out}} - 1$.

5   Outer Keyed Sponge and the PRF-Security

By OKSponge we denote the outer keyed sponge construction, and by OKSponge_K^P, denote OKSponge with P having K. For a message $m \in \{0,1\}^*$, the response is defined as $\text{OKSponge}_K^P(m) := \text{Sponge}^P(K^*\|m)$, where K^* is defined by appending some bit string to the suffix of K such that the bit length is a multiple of r, e.g., a zero string is appended. So the difference between OKSponge and IKSponge is the procedure to define the value s_0. In OKSponge_K^P, s_0 is defined as follows, where $\kappa := |K^*|/r$.

1. Partition K^* into r-bit blocks $K_1, \ldots, K_\kappa$;
   Partition $m\|\text{pad}(|K^*\|m|)$ into r-bit blocks $m_1, \ldots, m_n$
2. $w_0 \leftarrow 0^b$; For $i = 1, \ldots, \kappa$ do $u_i \leftarrow K_i\|0^c \oplus w_{i-1}$; $w_i \leftarrow P(u_i)$
3. $s_0 \leftarrow w_\kappa$

Basically, we can prove the PRF-security of OKSponge by the similar proof but need to consider the structural difference: $s_0 = 0^{b-k}\|K$ in IKSponge and $s_0 = w_\kappa$ in OKSponge. If $\mathbf{D}$ does not know w_κ, that is, $\mathbf{D}$ does not make an offline query $\mathcal{P}(u_\kappa)$ and $\mathcal{P}^{-1}(w_\kappa)$ then w_κ becomes a secret random value of b bits. Therefore, the upper-bound of the PRF-security of OKSponge can be obtained from that of IKSponge, where the probability for K, $\frac{\ell q + Q}{2^k}$, is replaced with the probability for the "bad" event where $\mathbf{D}$ knows w_κ. The probability for the bad event was considered in [1,13], and we use their bound. The concrete upper-bound is given as follows, where the probability for the bad event is $\lambda(Q) + \frac{2\kappa Q}{2^b}$.

Theorem 2. *Let $\mathbf{D}$ be a distinguisher which makes q online queries of r-bit block length at most ℓ_{in} and Q offline queries. Then for any ρ, we have*
$$\mathbf{Adv}_{\text{OKSponge}}^{\text{prf}}(\mathbf{D}) \le \lambda(Q) + \frac{2\kappa Q}{2^b} + \frac{2qQ + 3.5\ell^2 q^2}{2^b} + \frac{3q^2 + 2qQ + 2\rho(q+Q)}{2^c} + 2^{r+1} \times \left(\frac{2e\ell q}{\rho 2^r}\right)^\rho,$$
where $\ell = \ell_{\text{in}} + \ell_{\text{out}} - 1$, $e = 2.71828\cdots$ is Napier's constant, and $\lambda(Q) = \frac{Q}{2^k}$ if $k \le r$, and $\lambda(Q) = \min\left\{\frac{Q^2}{2^{c+1}} + \frac{Q}{2^k}, \frac{1}{2^b} + \frac{Q}{2^{\left(\frac{1}{2} - \frac{\log_2(3b)}{2r} - \frac{1}{r}\right)k}}\right\}$ otherwise.

Corollary 2. *We assume $c \le b/2$. Then, we put $\rho = r$, and without loss of generality, assume $r \ge 2$ (otherwise $r = c = 1$ and $b=2$). Since $r \ge b/2$, we have*
$$\mathbf{Adv}_{\text{OKSponge}}^{\text{prf}}(\mathbf{D}) \le \frac{3q^2 + 2qQ + 2r(q+Q)}{2^c} + \frac{(3.5 + 32e^2 r^{-2})\ell^2 q^2 + 2qQ + 2\kappa Q}{2^b} + \lambda(Q).$$
We assume $c > b/2$ and put $\rho = \max\left\{r, \left(\frac{2e \times \ell q}{2^{r-c}(q+Q)}\right)^{1/2}\right\}$. Then we have
$$\mathbf{Adv}_{\text{OKSponge}}^{\text{prf}}(\mathbf{D}) \le \left(\frac{18e\ell q(q+Q)}{2^b}\right)^{1/2} + \frac{3q^2 + 2qQ + 2r(q+Q)}{2^c} + \frac{3.5\ell^2 q^2 + 2qQ + 2\kappa Q}{2^b} + \lambda(Q).$$

References

1. Andreeva, E., Daemen, J., Mennink, B., Van Assche, G.: Security of keyed sponge constructions using a modular proof approach. In: Leander, G. (ed.) FSE 2015. LNCS, vol. 9054, pp. 364–384. Springer, Heidelberg (2015)

2. Aumasson, J.-P., Henzen, L., Meier, W., Naya-Plasencia, M.: QUARK: a lightweight hash. In: Mangard, S., Standaert, F.-X. (eds.) CHES 2010. LNCS, vol. 6225, pp. 1–15. Springer, Heidelberg (2010)
3. Aumasson, J.-P., Jovanovic, P., Neves, S.: NORX: parallel and scalable AEAD. In: Kutyłowski, M., Vaidya, J. (eds.) ICAIS 2014, Part II. LNCS, vol. 8713, pp. 19–36. Springer, Heidelberg (2014)
4. Bellare, M., Canetti, R., Krawczyk, H.: Keying hash functions for message authentication. In: Koblitz, N. (ed.) CRYPTO 1996. LNCS, vol. 1109, pp. 1–15. Springer, Heidelberg (1996)
5. Bellare, M., Rogaway, P.: The security of triple encryption and a framework for code-based game-playing proofs. In: Vaudenay, S. (ed.) EUROCRYPT 2006. LNCS, vol. 4004, pp. 409–426. Springer, Heidelberg (2006)
6. Bertoni, G., Daemen, J., Peeters, M., Van Assche, G.: Duplexing the sponge: single-pass authenticated encryption and other applications. In: Miri, A., Vaudenay, S. (eds.) SAC 2011. LNCS, vol. 7118, pp. 320–337. Springer, Heidelberg (2012)
7. Bertoni, G., Daemen, J., Peeters, M., Van Assche, G.: On the indifferentiability of the sponge construction. In: Smart, N.P. (ed.) EUROCRYPT 2008. LNCS, vol. 4965, pp. 181–197. Springer, Heidelberg (2008)
8. Bertoni, G., Daemen, J., Peeters, M., Van Assche, G.: On the security of the keyed sponge construction. In: SKEW 2011 (2011)
9. Bertoni, G., Daemen, J., Peeters, M., Van Assche, G.: Permutation-based encryption, authentication and authenticated encryption. In: DIAC 2012 (2012)
10. Bogdanov, A., Knežević, M., Leander, G., Toz, D., Varıcı, K., Verbauwhede, I.: SPONGENT: a lightweight hash function. In: Preneel, B., Takagi, T. (eds.) CHES 2011. LNCS, vol. 6917, pp. 312–325. Springer, Heidelberg (2011)
11. Chang, D., Dworkin, M., Hong, S., Kelsey, J., Nandi, M.: A keyed sponge construction with pseudorandomness in the standard model. In: Third SHA-3 Candidate Conference (2012)
12. Chen, S., Steinberger, J.: Tight security bounds for key-alternating ciphers. In: Nguyen, P.Q., Oswald, E. (eds.) EUROCRYPT 2014. LNCS, vol. 8441, pp. 327–350. Springer, Heidelberg (2014)
13. Gaži, P., Pietrzak, K., Tessaro, S.: The exact PRF security of truncation: tight bounds for keyed sponges and truncated CBC. In: Gennaro, R., Robshaw, M. (eds.) CRYPTO 2015. LNCS, vol. 9215, pp. 368–387. Springer, Heidelberg (2015)
14. Guo, J., Peyrin, T., Poschmann, A.: The PHOTON family of lightweight hash functions. In: Rogaway, P. (ed.) CRYPTO 2011. LNCS, vol. 6841, pp. 222–239. Springer, Heidelberg (2011)
15. Jovanovic, P., Luykx, A., Mennink, B.: Beyond $2^c/2$ security in sponge-based authenticated encryption modes. In: Sarkar, P., Iwata, T. (eds.) ASIACRYPT 2014. LNCS, vol. 8873, pp. 85–104. Springer, Heidelberg (2014)
16. Mennink, B., Reyhanitabar, R., Vizár, D.: Security of full-state keyed sponge and duplex: applications to authenticated encryption. In: Iwata, T., et al. (eds.) ASIACRYPT 2015. LNCS, vol. 9453, pp. 465–489. Springer, Heidelberg (2015)
17. NIST: SHA-3 standard: permutation-based hash and extendable-output functions. In: FIPS PUB 202 (2015)
18. Patarin, J.: The "Coefficients H" technique. In: Avanzi, R.M., Keliher, L., Sica, F. (eds.) SAC 2008. LNCS, vol. 5381, pp. 328–345. Springer, Heidelberg (2009)
19. Sasaki, Y., Yasuda, K.: How to incorporate associated data in sponge-based authenticated encryption. In: Nyberg, K. (ed.) CT-RSA 2015. LNCS, vol. 9048, pp. 353–370. Springer, Heidelberg (2015)

RIV for Robust Authenticated Encryption

Farzaneh Abed[1]([⊠]), Christian Forler[2], Eik List[1], Stefan Lucks[1],
and Jakob Wenzel[1]

[1] Bauhaus-Universität Weimar, Weimar, Germany
{farzaneh.abed,eik.list,stefan.lucks,jakob.wenzel}@uni-weimar.de
[2] Hochschule Schmalkalden, Schmalkalden, Germany
cforler@hs-schmalkalden.de

Abstract. Typical AE schemes are supposed to be secure when used as specified. However, they can – and often do – fail miserably when used improperly. As a partial remedy, Rogaway and Shrimpton proposed (nonce-)misuse-resistant AE (MRAE) and the first MRAE scheme SIV ("Synthetic Initialization Vector"). This paper proposes RIV ("Robust Initialization Vector"), which extends the generic SIV construction by an additional call to the internal PRF. RIV inherits the full security assurance from SIV, but unlike SIV and other MRAE schemes, RIV is also provably secure when releasing unverified plaintexts. This follows a recent line of research on *"Robust Authenticated Encryption"*, similar to the CAESAR candidate AEZ.

An AES-based instantiation of RIV runs at less than 1.5 cpb on current x64 processors. Unlike the proposed instantiation of AEZ, which gains speed by relying on reduced-round AES, our instantiation of RIV is provably secure under the single assumption of the AES being secure.

Keywords: Robustness · Subtle authenticated encryption · Provable security

1 Introduction

Authenticated Encryption. A secure authenticated encryption (AE) scheme generates ciphertexts that can not be efficiently distinguished from random bitstrings of the same length as the ciphertext and are infeasible to forge. Typical AE schemes are nonce-based [45], i.e., the user is responsible to supply an additional input that must be unique for every encryption. If a nonce ever repeats, the scheme's security may fully forfeit. While the concept of unique nonces is simple in theory, it is hard to ensure in practice [19], which led to severe security breaches in the past. Rogaway and Shrimpton [46] defined (nonce-)misuse-resistant AE (MRAE) as notion with the goal of providing full authenticity, and privacy up to the detection of repeated encryptions of the same associated data and message under the same nonce and key. Since then, the topic received significant attention by the community, resulting in a large corpus of MRAE schemes, e.g., [6, 10, 16, 20, 22, 27–30, 33, 43, 46].

© International Association for Cryptologic Research 2016
T. Peyrin (Ed.): FSE 2016, LNCS 9783, pp. 23–42, 2016.
DOI: 10.1007/978-3-662-52993-5_2

Robustness aspects of AE are not limited to nonce reuse. "One shortcoming of AE as commonly understood is its idealized, all-or-nothing decryption" [7]. Leaking any information about the message before its authentication has been verified breaks this assumption. At least five noteworthy recent works strengthened the existing security definitions of robustness.[1] Boldyreva et al. [15] (BDPS) studied the effects when multiple distinct error messages are distinguishable in probabilistic or stateful schemes. Andreeva et al. [4] formalized notions that capture the remaining security under *release of unverified plaintexts* (RUP). Hoang et al. [24] defined *robust AE* (RAE) as a notion for the *best achievable* security of an AE scheme with a user-chosen ciphertext expansion. Badertscher et al. [5] investigated RAE with the frameworks by Maurer and Renner [38,39]. Barwell et al. [7] defined *subtle AE* (SAE) as a reference framework for the BDPS, RUP, and RAE notions. The SAE definitions comprise leakage beyond information about the invalid plaintext, which allows to model leakage as a property of the decryption *implementation* rather than as a property of the scheme.

Previous Robust AE Schemes. In spite of so much progress regarding stricter security definitions, the portfolio of dedicated robust AE schemes remains still modest. Among the 57 CAESAR submissions, only four candidates consider robustness against leakage of invalid plaintexts: Julius [6] lacks a security proof; POET [1] and APE [3] concern on-line confidentiality, which cannot provide nonce-misuse resistance in the strong sense of Rogaway and Shrimpton, as has been criticized, e.g., by [25]. Only AEZ [24] provides robust AE. Though, AEZ follows a "proof-then-prune" approach: while the security proof assumes a strong block cipher, the performant instantiation employs four-round AES instead. Since AEZ also defines a key schedule, it appears more as a primitive of its own right than as a block-cipher-based AE scheme.

Beyond CAESAR, Bertoni et al. [12] proposed MR. MONSTER BURRITO, a four-round Feistel network with the round-reduced KECCAK-f permutation in duplex-wrap mode, and the sponge in counter mode for encryption. Shrimpton and Terashima [47] proposed Protected IV (PIV), a framework of strong tweakable ciphers (STPRPs), which generalized the Ψ_3 construction by Coron et al. [17]. PIV is fast (comparable with the construction proposed in this work); though, it requires the block-cipher inverse for decryption. Note that theoretically, more robust AE schemes could be constructed. Hoang et al. [24] showed that the well-known Encode-then-Encipher (EtE) [9] approach achieves RAE security when (a hash of) nonce and associated data are used as tweak. In theory, this implies that a secure STPRP can be transformed into a robust AE scheme, which allows to choose from the schemes that have been developed over the previous decade, e.g., in the domains of full-disk and format-preserving encryption.

Contribution. This work proposes a modular framework, called *Robust IV* (RIV), which provides provable SAE security. RIV is an extension of SIV [26,46]

---

[1] By robustness, we mean resistance against both nonce misuse and decryption leakage beyond the single error information.

that inherits both the simplicity and the naturally strong security properties of SIV and adds robustness against leakage of invalid plaintexts. We propose an instantiation which runs at less than 1.5 clock cycles per byte (cpb) on current x64 processors.

Outline. The remainder of this work is structured as follows: after Sect. 2 recalls the preliminaries, Sect. 3 describes the generic RIV framework. Section 4 recalls the relevant notions. Section 5 summarizes our formal security analysis. Section 6 details our instantiation, and Sect. 7 concludes this work.

2   Preliminaries

We use lowercase letters x, y for indices and integers, uppercase letters X, Y for binary strings and functions, and calligraphic uppercase letters $\mathcal{X}, \mathcal{Y}$ for sets. By ε we denote the empty string. We denote the concatenation of binary strings X and Y by $X \parallel Y$ and the result of their bitwise XOR by $X \oplus Y$. We indicate the length of X in bits by $|X|$, and write X_i for the i-th block, $X[i]$ for the i-th most significant bit of X, and $X[i..j]$ for the bit sequence $X[i], \ldots, X[j]$. $X \twoheadleftarrow \mathcal{X}$ denotes that X is chosen uniformly at random from the set $\mathcal{X}$. We define two sets of particular interest: $\mathsf{Perm}(\mathcal{X})$ be the set of all permutations on $\mathcal{X}$ and $\mathsf{Func}(\mathcal{X}, \mathcal{Y})$ the set of all functions $F : \mathcal{X} \to \mathcal{Y}$. A uniform random function $\rho : \mathcal{X} \to \mathcal{Y}$ with domain $\mathcal{X}$ and range $\mathcal{Y}$ is a random variable uniformly distributed over $\mathsf{Func}(\mathcal{X}, \mathcal{Y})$. We define by $X_1, \ldots, X_j \xleftarrow{x} X$ the injective splitting of the string X into x-bit blocks such that $X = X_1 \parallel \cdots \parallel X_j$, with $|X_i| = x$ for $1 \leq i \leq j - 1$, and $|X_j| \leq x$.

For an event E, we denote by $\Pr[E]$ the probability of E. We write $\langle x \rangle_m$ for the binary m-bit-string representation of an integer x and $\langle x \rangle$ for the binary n-bit-string representation of x for an integer n that is clear from the context. If not stated otherwise, we assume representations to be encoded in big-endian manner, i.e., the decimal $\langle 135 \rangle$ is encoded to the n-bit string 000..010000111.

Universal Hashing. Universal hash functions are well-known components for compressing a message while guaranteeing maximal probabilities about output relations. We briefly recall the definitions that are relevant in this work.

Definition 1 (ϵ-Almost-(XOR-)Universal Hash Functions). *Let $\mathcal{X}, \mathcal{Y} \subseteq \{0, 1\}^*$. Let $\mathcal{H} = \{H \mid H : \mathcal{X} \to \mathcal{Y}\}$ denote a family of hash functions. $\mathcal{H}$ is called ϵ-almost-universal (ϵ-AU) iff for all distinct elements $X, X' \in \mathcal{X}$, it holds that $\Pr_{H \leftarrow \mathcal{H}}[H(X) = H(X')] \leq \epsilon$. $\mathcal{H}$ is called ϵ-almost-XOR-universal (ϵ-AXU) iff for all distinct elements $X, X' \in \mathcal{X}$ and $Y \in \mathcal{Y}$, it holds that $\Pr_{H \leftarrow \mathcal{H}}[H(X) \oplus H(X') = Y] \leq \epsilon$.*

Theorem 1 (Theorem 3 from [14]). *Let $\mathcal{X}, \mathcal{Y} \subseteq \{0, 1\}^*$. Further, let $\mathcal{H} = \{H \mid H : \mathcal{X} \to \mathcal{Y}\}$ be a family of ϵ-AXU hash functions. Then, the family $\mathcal{H}' = \{H' \mid H' : \mathcal{X} \times \mathcal{Y} \to \mathcal{Y}\}$ with $H'(X, Y) := H(X) \oplus Y$, is ϵ-AU.*

Nonce-Based Encryption Schemes. A *nonce-based encryption scheme* [45] is a tuple $\Pi = (\mathcal{E}, \mathcal{D})$ of deterministic encryption and decryption algorithms $\mathcal{E} : \mathcal{K} \times \mathcal{N} \times \mathcal{M} \to \mathcal{C}$ and $\mathcal{D} : \mathcal{K} \times \mathcal{N} \times \mathcal{C} \to \mathcal{M}$, with associated non-empty key space $\mathcal{K}$, non-empty nonce space $\mathcal{N}$, and $\mathcal{M}, \mathcal{C} \subseteq \{0,1\}^*$ denoting message and ciphertext space, respectively. We often write $\mathcal{E}_K^N(M)$ and $\mathcal{D}_K^N(C)$ as short forms of $\mathcal{E}(K, N, M)$ and $\mathcal{D}(K, N, C)$. An adversary that never repeats a nonce over its encryption queries is called *nonce-respecting*, and *nonce-ignoring* otherwise. We assume for all $K \in \mathcal{K}$, $N \in \mathcal{N}$, $M \in \mathcal{M}$, and $C \in \mathcal{C}$ *length-preservation*, i.e., $|\mathcal{E}_K^N(M)| = |M|$, *correctness*, i.e., $\mathcal{D}_K^N(\mathcal{E}_K^N(M)) = M$, and *tidiness*, i.e., $\mathcal{E}_K^N(\mathcal{D}_K^N(C)) = C$. We call a nonce-based encryption scheme $\Pi = (\mathcal{E}, \mathcal{D})$ *nonce-keystream-based* iff its encryption algorithm derives a keystream $\kappa_N \subseteq \{0,1\}^*$, with $|\kappa_N| = |M|$, from the given nonce N and computes the ciphertext as $C \leftarrow \kappa_N \oplus M$. Naturally, the decryption algorithm of such an encryption scheme is identical to its encryption algorithm, i.e., $\mathcal{E}_K^N(M) := \mathcal{D}_K^N(M)$ for all $K \in \mathcal{K}$, $N \in \mathcal{N}$, and $M \in \mathcal{M}$.

Nonce-Based AE Schemes. A nonce-based authenticated encryption scheme (with associated data) [44] is a tuple $\widetilde{\Pi} = (\widetilde{\mathcal{E}}, \widetilde{\mathcal{D}})$ of a deterministic encryption algorithm $\widetilde{\mathcal{E}} : \mathcal{K} \times \mathcal{N} \times \mathcal{H} \times \mathcal{M} \to \mathcal{C} \times \mathcal{T}$, and a deterministic decryption algorithm $\widetilde{\mathcal{D}} : \mathcal{K} \times \mathcal{N} \times \mathcal{H} \times \mathcal{C} \times \mathcal{T} \to \mathcal{M} \cup \{\bot\}$, with associated non-empty key space $\mathcal{K}$, non-empty nonce space $\mathcal{N}$, and $\mathcal{H}, \mathcal{M}, \mathcal{C} \subseteq \{0,1\}^*$ denote the header, message, and ciphertext space, respectively. We define a tag space $\mathcal{T} = \{0,1\}^\tau$ for a fixed $\tau \geq 0$. We often write $\widetilde{\mathcal{E}}_K^{N,H}(M)$ and $\widetilde{\mathcal{D}}_K^{N,H}(C, T)$ as short forms of $\widetilde{\mathcal{E}}(K, N, H, M)$ and $\widetilde{\mathcal{D}}(K, N, H, C, T)$. If a given tuple (N, H, C, T) is valid, $\widetilde{\mathcal{D}}_K^{N,H}(C, T)$ returns the corresponding plaintext M, and $\bot$ otherwise. We assume that for all $K \in \mathcal{K}$, $N \in \mathcal{N}$, $H \in \mathcal{H}$, and $M \in \mathcal{M}$ holds *stretch-preservation*: if $\widetilde{\mathcal{E}}_K^{N,H}(M) = (C, T)$, then $|C| = |M|$ and $|T| = \tau$, *correctness*: if $\widetilde{\mathcal{E}}_K^{N,H}(M) = (C, T)$, then $\widetilde{\mathcal{D}}_K^{N,H}(C, T) = M$, and *tidiness*: if $\widetilde{\mathcal{D}}_K^{N,H}(C, T) = M \neq \bot$, then $\widetilde{\mathcal{E}}_K^{N,H}(M) = (C, T)$, for all $C \in \mathcal{C}$ and $T \in \mathcal{T}$. Note that some notions (e.g., [41]) regard an authenticated ciphertext C with $|C| = |M| + \tau$ instead of an explicitly separated tuple (C, T).

Subtle AE Schemes. Barwell et al. defined a subtle AE scheme $\widetilde{\Pi} = (\widetilde{\mathcal{E}}, \widetilde{\mathcal{D}}, \Lambda)$ as a tuple of deterministic encryption and decryption algorithms $\widetilde{\mathcal{E}}$ and $\widetilde{\mathcal{D}}$ as above[2], and an additional deterministic leakage algorithm $\Lambda : \mathcal{K} \times \mathcal{N} \times \mathcal{H} \times \mathcal{C} \times \mathcal{T} \to \{\top\} \cup \mathcal{L}$, with a non-empty leakage space $\mathcal{L}$ and a symbol $\top \notin \mathcal{L}$ to indicate a valid input. This means, for all $K \in \mathcal{K}$, $N \in \mathcal{N}$, $H \in \mathcal{H}$, $C \in \mathcal{C}$, and $T \in \mathcal{T}$ holds: if $\Lambda_K^{N,H}(C, T) = \top$, then $\widetilde{\mathcal{D}}_K^{N,H}(C, T) \neq \bot$; moreover, it holds that if $\Lambda_K^{N,H}(C, T) \neq \top$, then $\widetilde{\mathcal{D}}_K^{N,H}(C, T) = \bot$.

3   Definition of RIV

Definition 2 (RIV). *Let $d, n, \tau \geq 1$. Let $\mathcal{K}_1$, $\mathcal{K}_2$, and $\mathcal{K} = \mathcal{K}_1 \times \mathcal{K}_2$ be non-empty key sets, $\mathcal{N}$ a non-empty nonce space, $\{0,1\}^d$ the non-empty domain*

---

[2] Though, their definitions denote the authenticated ciphertext (C, T) as C.

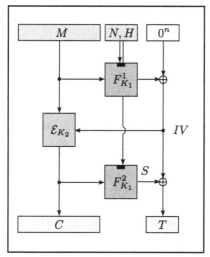

	1: **function** $\widetilde{\mathcal{E}}_{K_1,K_2}(N,H,M)$
	2: $IV \leftarrow F_{K_1}^1(N,H,M)$
	3: $C \leftarrow \mathcal{E}_{K_2}(IV,M)$
	4: $T \leftarrow F_{K_1}^2(N,H,C) \oplus IV$
	5: **return** (C,T)
	11: **function** $\widetilde{\mathcal{D}}_{K_1,K_2}(N,H,C,T)$
	12: $IV \leftarrow F_{K_1}^2(N,H,C) \oplus T$
	13: $M \leftarrow \mathcal{D}_{K_2}(IV,C)$
	14: $IV' \leftarrow F_{K_1}^1(N,H,M)$
	15: **if** $IV = IV'$ **then return** M
	16: **return** $\bot$
	21: **function** $\Lambda_{K_1,K_2}(N,H,C,T)$
	22: $IV \leftarrow F_{K_1}^2(N,H,C) \oplus T$
	23: $M \leftarrow \mathcal{D}_{K_2}(IV,C)$
	24: $IV' \leftarrow F_{K_1}^1(N,H,M)$
	25: **if** $IV = IV'$ **then return** $\top$
	26: **return** M

Fig. 1. Left: Schematic illustration of the encryption of $\mathrm{RIV}_{F,\Pi}$ with a PRF F and a nonce-based encryption scheme $\Pi = (\mathcal{E}, \mathcal{D})$. **Right:** Definition of encryption and decryption algorithms of $\mathrm{RIV}_{F,\Pi}$, and definition of a plaintext-leaking oracle Λ that will be used in our security analysis.

space, and $\mathcal{H}, \mathcal{M}, \mathcal{C} \subseteq \{0,1\}^*$ header, message, and ciphertext spaces, respectively, and $\mathcal{T} = \{0,1\}^\tau$ a tag space. Let further $F : \mathcal{K}_1 \times \{0,1\}^d \times \mathcal{N} \times \mathcal{H} \times \mathcal{M} \to \{0,1\}^n$ be a function and $\Pi = (\mathcal{E}, \mathcal{D})$ a nonce-based encryption scheme with associated key space $\mathcal{K}_2$ and nonce space $\{0,1\}^\tau$. Let $F_K^i(\cdot,\cdot,\cdot)$ denote $F_K(\langle i \rangle_d, \cdot, \cdot, \cdot)$. Then, we define the AE scheme $\mathrm{RIV}_{F,\Pi} = (\widetilde{\mathcal{E}}, \widetilde{\mathcal{D}})$ with encryption algorithm $\widetilde{\mathcal{E}} : \mathcal{K} \times \mathcal{N} \times \mathcal{H} \times \mathcal{M} \to \mathcal{C} \times \mathcal{T}$ and decryption algorithm $\widetilde{\mathcal{D}} : \mathcal{K} \times \mathcal{N} \times \mathcal{H} \times \mathcal{C} \times \mathcal{T} \to \mathcal{M} \cup \{\bot\}$, as given in Fig. 1.

Definition 3 ($\widehat{\mathrm{RIV}}$). We define the SAE scheme $\widehat{\mathrm{RIV}}_{F,\Pi} = (\widetilde{\mathcal{E}}, \widetilde{\mathcal{D}}, \Lambda)$ with an additional deterministic leakage algorithm $\Lambda : \mathcal{K} \times \mathcal{N} \times \mathcal{H} \times \mathcal{C} \times \mathcal{T} \to \mathcal{M} \times \{\top\}$, as given in Fig. 1.

Feistel Structure and Encode-then-Encipher (EtE). RIV can be seen as an application of the EtE [9] approach by Bellare et al. EtE can generically be used for constructing a robust AE scheme from a tweakable cipher, assuming its enciphering resists chosen-plaintext and chosen-ciphertext attacks [24]. The RIV cipher, however, is essentially an unbalanced three-round Feistel-network.[3] It is well-known that such ciphers are secure against chosen-plaintext, but vulnerable to chosen-ciphertext attacks [35] (see also [2,36,42]). RIV is robust *in spite* of its weak enciphering scheme, because its encoding operation has been chosen to specifically cover this weakness.

---

[3] If the used encryption scheme $\Pi = (\mathcal{E}, \mathcal{D})$ is nonce-*keystream*-based, the RIV cipher *is* a three-round Feistel network.

4   Security Notions

Adversaries and Advantages. An adversary $\mathbf{A}$ is an efficient Turing machine that interacts with a given set of oracles that appear as black boxes to $\mathbf{A}$. We use the notation $\mathbb{A}$ for the class of all computationally bounded adversaries and $\mathbf{A}^{\mathcal{O}}$ for the output of $\mathbf{A}$ after interacting with some oracle $\mathcal{O}$. We write $\Delta_{\mathbf{A}}(\mathcal{O}^L; \mathcal{O}^R) := \sup_{\mathbf{A} \in \mathbb{A}} |\Pr[\mathbf{A}^{\mathcal{O}^L} \Rightarrow 1] - \Pr[\mathbf{A}^{\mathcal{O}^R} \Rightarrow 1]|$ for the advantage of $\mathbf{A}$ to distinguish between oracles $\mathcal{O}^L$ and $\mathcal{O}^R$. All probabilities are defined over the random coins of the oracles and those of the adversary, if any. We write $\mathbf{Adv}_F^X(q, \ell, t) = \max_{\mathbf{A} \in \mathbb{A}} \{\mathbf{Adv}_F^X(\mathbf{A})\}$ to refer to the maximal advantage over all X-adversaries $\mathbf{A}$ on a given function F that run in time at most t and pose at most q queries consisting of at most ℓ blocks in total to the available oracles. If $\mathbf{A}$ shall distinguish between two sets of oracles $(\mathcal{O}_1^L, \ldots, \mathcal{O}_k^L)$ and $(\mathcal{O}_1^R, \ldots, \mathcal{O}_k^R)$, we refer to the i-th oracle that $\mathbf{A}$ interacts with by $\mathcal{O}_i \in \{\mathcal{O}_i^L, \mathcal{O}_i^R\}$. By $\mathcal{O}_i \hookrightarrow \mathcal{O}_j$, we denote that $\mathbf{A}$ first queries $\mathcal{O}_i$ and later $\mathcal{O}_j$ with the output of $\mathcal{O}_i$. Wlog., we assume that $\mathbf{A}$ never asks queries to which it already knows the answer. In the case when $\mathbf{A}$ has access to multiple oracles $\mathcal{O}_1, \ldots, \mathcal{O}_k$, we denote by q_i the number of queries and by ℓ_i the maximal number of blocks that $\mathbf{A}$ poses at most to oracle $\mathcal{O}_i$, $1 \leq i \leq k$.

If $\mathcal{O}_i$ and $\mathcal{O}_j$ represent a family of algorithms indexed by inputs, the indices must match, e.g., when $\widetilde{\mathcal{E}}_K^{N,H}(M)$ and $\widetilde{\mathcal{D}}_K^{N,H}(C)$ represent encryption and decryption algorithms with a fixed key K and indexed by N and H, then $\widetilde{\mathcal{E}}_K \hookrightarrow \widetilde{\mathcal{D}}_K$ says that $\mathbf{A}$ first queries $\widetilde{\mathcal{E}}_K^{N,H}(M)$ and later $\widetilde{\mathcal{D}}_K^{N,H}(C)$.

We define $\perp$, when in place of an oracle, to always return the invalid symbol $\perp$. We denote by $\$^{\mathcal{O}}$ an oracle that, given an input X, computes $Y \leftarrow \mathcal{O}(X)$, chooses uniformly at random a value Y' from the space of all possible outputs with $|Y'| = |Y|$, and returns Y'. We assume that $\$^{\mathcal{O}}$ performs lazy sampling, i.e., $\$^{\mathcal{O}}(X)$ returns the same value when queried with the same input X. We often omit the key for brevity, e.g., $\$^{\widetilde{\mathcal{E}}}(X)$ will be short for $\$^{\widetilde{\mathcal{E}}_K}(X)$.

4.1   Security Definitions for Encryption Schemes

Definition 4 (PRF Advantage). *Let $F : \mathcal{K} \times \mathcal{X} \to \mathcal{Y}$ be a function with non-empty key space $\mathcal{K}$, and $\mathbf{A}$ a computationally bounded adversary with access to an oracle, where $K \leftarrow \mathcal{K}$ and $\rho \leftarrow \mathsf{Func}(\mathcal{X}, \mathcal{Y})$. Then, the PRF advantage of $\mathbf{A}$ on F is defined as $\mathbf{Adv}_F^{\mathrm{PRF}}(\mathbf{A}) := \Delta_{\mathbf{A}}(F_K; \rho)$.*

Definition 5 (PRP Advantage). *Let $n, k \geq 1$ be fixed. Let $E : \{0,1\}^k \times \{0,1\}^n \to \{0,1\}^n$ be a block cipher and $\mathbf{A}$ a computationally bounded adversary with access to an oracle. Further, let $K \leftarrow \{0,1\}^k$ and $\pi \leftarrow \mathsf{Perm}(\{0,1\}^n)$. Then, the PRP advantage of $\mathbf{A}$ on E is defined as $\mathbf{Adv}_E^{\mathrm{PRP}}(\mathbf{A}) := \Delta_{\mathbf{A}}(E_K; \pi)$.*

Stinson [48] showed that one can construct an $(\epsilon_1 + \epsilon_2)$-AU family of hash functions from the consecutive application of an ϵ_1-AU and an ϵ_2-AU family of hash functions. From that we can derive the following theorem.

Theorem 2. *Let $\mathcal{X}, \mathcal{Y}, \mathcal{Z} \subseteq \{0,1\}^*$ and let $\mathcal{K}$ be a non-empty set. Further, let $\mathcal{H} = \{H : \mathcal{X} \to \mathcal{Y}\}$ be a family of ϵ-AU hash functions and let $G : \mathcal{K} \times \mathcal{Y} \to \mathcal{Z}$ be a function. Then, we can define $F_K(X) := G_K(H(X))$, with independent $K \twoheadleftarrow \mathcal{K}$ and $H \twoheadleftarrow \mathcal{H}$. Let $\mathbf{A}$ be a PRF adversary on F that asks at most q queries of at most ℓ blocks in total, and runs in time at most t. Then, there exists a PRF adversary $\mathbf{A}_1$ on G that asks at most q queries and runs in time $O(t)$ such that*

$$\mathbf{Adv}_F^{\mathrm{PRF}}(\mathbf{A}) \leq \mathbf{Adv}_G^{\mathrm{PRF}}(\mathbf{A}_1) + \epsilon \cdot q^2/2.$$

Theorem 2 follows from the fact, that the PRF advantage of F is upper bounded by the maximal PRF advantage on G plus the maximal probability of output collisions of the form $H(X) = H(X')$ over q queries.

Definition 6 (nE Advantage [41]). Let $\Pi = (\mathcal{E}, \mathcal{D})$ be a nonce-based encryption scheme and $K \twoheadleftarrow \mathcal{K}$. Let $\mathbf{A}$ be a nonce-respecting adversary with access to an oracle. Then, the NE advantage of $\mathbf{A}$ on Π is defined as $\mathbf{Adv}_\Pi^{\mathrm{NE}}(\mathbf{A}) := \Delta_{\mathbf{A}}(\mathcal{E}_K; \$^{\mathcal{E}})$.

We adapt the definition of indistinguishability from random bits from [23] for nonce-based encryption schemes. Note that we strengthen it to adversaries that do not repeat nonces over *all encryption and decryption queries*.

Definition 7 (SRND Advantage). *Let be $\Pi = (\mathcal{E}, \mathcal{D})$ a nonce-based encryption scheme and $K \twoheadleftarrow \mathcal{K}$. Let $\mathbf{A}$ be a nonce-respecting adversary with access to two oracles $\mathcal{O}_1$ and $\mathcal{O}_2$, s.t. $\mathbf{A}$ never asks for $\mathcal{O}_1 \hookrightarrow \mathcal{O}_2$ and never repeats a nonce over all its encryption and decryption queries. Then, we define the SRND advantage of $\mathbf{A}$ on Π as $\mathbf{Adv}_\Pi^{\mathrm{SRND}}(\mathbf{A}) := \Delta_{\mathbf{A}}(\mathcal{E}_K, \mathcal{D}_K; \$^{\mathcal{E}}, \$^{\mathcal{D}})$.*

4.2 Security Definitions for Nonce-Based AE Schemes

For this subsection, let $\widetilde{\Pi} = (\widetilde{\mathcal{E}}, \widetilde{\mathcal{D}})$ be a nonce-based AE scheme, $K \twoheadleftarrow \mathcal{K}$, and $\mathbf{A}$ be a computationally bounded adversary on $\widetilde{\Pi}$.

Definition 8 (IND-CPA Advantage). *Let $\mathbf{A}$ have access to an encryption oracle. Then, the IND-CPA advantage of $\mathbf{A}$ with respect to $\widetilde{\Pi}$ is defined as $\mathbf{Adv}_{\widetilde{\Pi}}^{\mathrm{IND\text{-}CPA}}(\mathbf{A}) := \Delta_{\mathbf{A}}(\widetilde{\mathcal{E}}_K; \$^{\widetilde{\mathcal{E}}})$.*

Definition 9 (INT-CTXT Advantage). *Let $\mathbf{A}$ have access to two oracles $\mathcal{O}_1$ and $\mathcal{O}_2$ such that $\mathbf{A}$ never queries $\mathcal{O}_1 \hookrightarrow \mathcal{O}_2$. Then, the INT-CTXT advantage of $\mathbf{A}$ on $\widetilde{\Pi}$ is defined as $\mathbf{Adv}_{\widetilde{\Pi}}^{\mathrm{INT\text{-}CTXT}}(\mathbf{A}) := \Pr[\mathbf{A}^{\widetilde{\mathcal{E}}_K, \widetilde{\mathcal{D}}_K} \text{ forges}]$, where "forges" means that $\widetilde{\mathcal{D}}_K$ returns anything other than $\perp$ for a query of $\mathbf{A}$.*

Definition 10 (nAE Advantage [41]). *Let $\mathbf{A}$ have access to two oracles $\mathcal{O}_1$ and $\mathcal{O}_2$ such that $\mathbf{A}$ never queries $\mathcal{O}_1 \hookrightarrow \mathcal{O}_2$. Then, the NAE advantage of $\mathbf{A}$ on $\widetilde{\Pi}$ is defined as $\mathbf{Adv}_{\widetilde{\Pi}}^{\mathrm{NAE}}(\mathbf{A}) := \Delta_{\mathbf{A}}(\widetilde{\mathcal{E}}_K, \widetilde{\mathcal{D}}_K; \$^{\widetilde{\mathcal{E}}}, \perp)$.*

Bellare and Namprempre showed for probabilistic AE that chosen-ciphertext security results from IND-CPA and INT-CTXT security [8]. Fleischmann et al. proved in [19] a generalized theorem for nonce-based AE.

Theorem 3 (Theorem 1 in [19]). Let $\mathbf{A}$ be a computationally bounded NAE adversary on $\widetilde{\Pi}$ with access to two oracles $\mathcal{O}_1$ and $\mathcal{O}_2$ such that $\mathbf{A}$ never queries $\mathcal{O}_1 \hookrightarrow \mathcal{O}_2$; $\mathbf{A}$ makes at most q queries of total length of at most ℓ blocks and runs in time at most t. Then, there exist an IND-CPA adversary $\mathbf{A}_1$ on $\widetilde{\Pi}$ and an INT-CTXT adversary $\mathbf{A}_2$ on $\widetilde{\Pi}$, both making at most q queries of at most ℓ blocks and running in time $O(t)$ each, such that

$$\mathbf{Adv}_{\widetilde{\Pi}}^{\mathrm{NAE}}(\mathbf{A}) \leq \mathbf{Adv}_{\widetilde{\Pi}}^{\mathrm{IND\text{-}CPA}}(\mathbf{A}_1) + \mathbf{Adv}_{\widetilde{\Pi}}^{\mathrm{INT\text{-}CTXT}}(\mathbf{A}_2).$$

4.3 Security Definitions for Subtle AE Schemes

Subtle AE (SAE) defines a compound security notion that provides guarantees for privacy and authenticity under the existence of a leakage oracle. It comprises the notions IND-CPA, INT-CTXT, and an additional notion ERR-CCA.

For this subsection, let $\widetilde{\Pi} = (\widetilde{\mathcal{E}}, \widetilde{\mathcal{D}}, \Lambda)$ be an SAE scheme, $K, K' \leftarrow \mathcal{K} \times \mathcal{K}$ independent keys, and $\mathbf{A}$ a deterministic adversary with access to three oracles $\mathcal{O}_1, \mathcal{O}_2,$ and $\mathcal{O}_3$ such that $\mathbf{A}$ neither queries $\mathcal{O}_1 \hookrightarrow \mathcal{O}_2$ nor $\mathcal{O}_1 \hookrightarrow \mathcal{O}_3$.

Definition 11 (ERR-CCA Advantage). *The* ERR-CCA *advantage of* $\mathbf{A}$ *on* $\widetilde{\Pi}$ *is defined as* $\mathbf{Adv}_{\widetilde{\Pi}}^{\mathrm{ERR\text{-}CCA}}(\mathbf{A}) := \Delta_{\mathbf{A}}(\widetilde{\mathcal{E}}_K, \widetilde{\mathcal{D}}_K, \Lambda_K; \widetilde{\mathcal{E}}_K, \widetilde{\mathcal{D}}_K, \Lambda_{K'}).$

Definition 12 (SAE Advantage). *The* SAE *advantage of* $\mathbf{A}$ *on* $\widetilde{\Pi}$ *is defined as* $\mathbf{Adv}_{\widetilde{\Pi}}^{\mathrm{SAE}}(\mathbf{A}) := \Delta_{\mathbf{A}}(\widetilde{\mathcal{E}}_K, \widetilde{\mathcal{D}}_K, \Lambda_K; \$^{\widetilde{\mathcal{E}}}, \perp, \Lambda_{K'}).$

In the full version of [7], Barwell et al. prove a statement equivalent to Theorem 4. We apply Theorem 3 to decompose their AE security advantage term into the separate advantages for IND-CPA and INT-CTXT.

Theorem 4. *Let* $\mathbf{A}$ *run in time at most* t *and ask at most* q *queries of at most* ℓ *blocks to its respective oracles. Then, there exist computationally bounded* IND-CPA, INT-CTXT, *and* ERR-CCA *adversaries* $\mathbf{A}_1, \mathbf{A}_2,$ *and* $\mathbf{A}_3,$ *respectively, on* $\widetilde{\Pi}$ *such that*

$$\mathbf{Adv}_{\widetilde{\Pi}}^{\mathrm{SAE}}(\mathbf{A}) \leq \mathbf{Adv}_{\widetilde{\Pi}}^{\mathrm{IND\text{-}CPA}}(\mathbf{A}_1) + \mathbf{Adv}_{\widetilde{\Pi}}^{\mathrm{INT\text{-}CTXT}}(\mathbf{A}_2) + \mathbf{Adv}_{\widetilde{\Pi}}^{\mathrm{ERR\text{-}CCA}}(\mathbf{A}_3),$$

where $\mathbf{A}_1, \mathbf{A}_2,$ *and* $\mathbf{A}_3$ *each make at most* q *queries of at most* ℓ *blocks and run in time* $O(t)$ *each.*

Since [4] omitted a compound notion for their security under release of unverified plaintexts, Barwell et al. defined RUPAE as $\Delta_{\mathbf{A}}(\widetilde{\mathcal{E}}_K, \widetilde{\mathcal{D}}_K, \mathcal{V}_K; \$^{\widetilde{\mathcal{E}}}, \widetilde{\mathcal{D}}_{K'}, \perp)$ [7, Theorem 3, Corollary 2]. They showed that the maximal SAE advantage on an AE scheme $\widetilde{\Pi}$ is, with a reduction term, also equivalent to the maximal RUPAE advantage. Moreover, they showed that – again with a reduction term – it is also equivalent to the maximal robust-AE advantage on $\widetilde{\Pi}$ with fixed stretch τ.

5   Security Results for Generic RIV

This section summarizes our security results. For the remainder of this section, let $d, n, \tau \geq 1$ be integers, $\mathcal{K}_1, \mathcal{K}_2$ be non-empty key spaces, and $K_1, K_2 \leftarrow \mathcal{K}_1 \times \mathcal{K}_2$ be independent keys, $F : \mathcal{K}_1 \times \{0,1\}^d \times \mathcal{N} \times \mathcal{H} \times \mathcal{M} \to \{0,1\}^n$, and $\Pi = (\mathcal{E}, \mathcal{D})$ be a nonce-based encryption scheme with associated key space $\mathcal{K}_2$.

Theorem 5. *Let* **A** *be a computationally bounded* SAE *adversary on* $\widehat{\mathrm{RIV}}_{F,\Pi}$ *which asks at most q queries of at most ℓ blocks in total and runs in time at most t. Then, there exists a computationally bounded* PRF *adversary* $\mathbf{A}_1$ *on F that asks at most $2q$ queries of at most $2(d + n\ell)$ bits and runs in time $O(t)$, and a computationally bounded* SRND *adversary* $\mathbf{A}_2$ *on Π that asks at most q queries of at most ℓ blocks in total and runs in time $O(t)$ such that*

$$\mathbf{Adv}^{\mathrm{SAE}}_{\widehat{\mathrm{RIV}}_{F,\Pi}}(\mathbf{A}) \leq \frac{8q^2 + 3q}{2^n} + 4 \cdot \left(\mathbf{Adv}^{\mathrm{PRF}}_F(\mathbf{A}_1) + \mathbf{Adv}^{\mathrm{SRND}}_\Pi(\mathbf{A}_2) \right).$$

Due to space limitations, the proof can be found in the full version of this paper[4]. We can derive the following corollary for the NAE advantage on $\mathrm{RIV}_{F,\Pi}$ in the absence of a plaintext-leaking oracle.

Corollary 1. *Let* **A** *be a computationally bounded* NAE *adversary on* $\mathrm{RIV}_{F,\Pi}$ *which asks at most q queries of at most ℓ blocks in total and runs in time at most t. Then, there exist a computationally bounded* PRF *adversary* $\mathbf{A}_1$ *on F that asks at most $2q$ queries of at most $2(d + n\ell)$ bits and runs in time $O(t)$, and a computationally bounded* SRND *adversary* $\mathbf{A}_2$ *on Π that asks at most q queries of at most ℓ blocks in total and runs in time $O(t)$, such that*

$$\mathbf{Adv}^{\mathrm{NAE}}_{\mathrm{RIV}_{F,\Pi}}(\mathbf{A}) \leq \frac{2q^2 + q}{2^n} + 2 \cdot \left(\mathbf{Adv}^{\mathrm{PRF}}_F(\mathbf{A}_1) + \mathbf{Adv}^{\mathrm{SRND}}_\Pi(\mathbf{A}_2) \right).$$

The proof can be found can be found in the full version of this paper.

Proof Ideas. The intuition of our proofs is the following: in encryption direction, for every fresh tuple of nonce, header, and message, F will produce a fresh $IV \leftarrow F^1_K(N, H, M)$ that has not occurred before with overwhelming probability. Since Π is SRND-secure, $\mathcal{E}$ will produce a randomly chosen ciphertext. The second invocation of F with a fresh ciphertext then produces a random tag. To determine the privacy advantage of the scheme, we have to bound only the PRF-advantage on F, the SRND-security of $\mathcal{E}$, and the probabilities of random collisions of IVs from the birthday paradox.

In decryption direction, whenever the nonce, header, or ciphertext changes, $IV \leftarrow F^2_K(N, H, C)$ will be a random value up to the birthday bound. Since Π is SRND-secure, a fresh IV (regarded over all encryption *and* decryption queries) will produce a fresh pseudorandom plaintext. Thus, even when the adversary learns the decrypted (invalid) message, M will provide it with no information

---

[4] The full version of this paper will soon appear on ePrint.

about other plaintexts as long as the IV does not repeat. When an adversary changes N, H, or C and manages to cancel the difference by a fresh tag, the second call to $F_K^1(N, H, M)$ will yield a random IV' that differs from IV with probability close to $1/2^n$. Thus, a similar argumentation as for the encryption also applies to the inverse direction. Finally, the domain separation from the first parameter to F protects against choices of $(N, H, M) = (N, H, C)$.

6   Instantiation

Pseudo-Dot-Product Hashing. Let $n, m \geq 1$ with even m and let $\mathcal{X} = \bigcup_{i=1}^{m/2} \{0,1\}^{2in}$. Given a set of m pair-wise independent key words $K = (K_1, \ldots, K_m)$ and an m-word input $M = (M_1, \ldots, M_m)$, with $M_i, K_i \in \{0,1\}^n$, $1 \leq i \leq m$, a *pseudo-dot-product* (PDP) family of hash functions $\mathcal{H} = \{H : \mathcal{X} \times \mathcal{X} \to \{0,1\}^{2n}\}$ is defined as

$$H_K(M) := \sum_{i=1}^{m/2} (M_{2i-1} + K_{2i-1}) \cdot (M_{2i} + K_{2i}).$$

Bernstein [11] credits it to Winograd [51] and classifies it as $(m, \lceil m/2 \rceil)$-design, i.e., it requires m independent key words and $\lceil m/2 \rceil$ multiplications to process m message words. If modular additions and multiplications are performed within the rings $\mathbb{Z}_{2^n}$ and $\mathbb{Z}_{2^{2n}}$, the construction is known as NH, to be $1/2^n$-AU, and is used in variants in UMAC [13], VMAC [18,32], and HS1 [33]. All these constructions employ a multi-stage hashing process: the input is first compressed with NH, before the results are used as inputs in a usual polynomial hash (and optionally further processed by an inner-product hash). To obtain a slightly higher security margin and efficiency, we consider a recently proposed variant, called CLHASH.

6.1   CLHASH

CLHASH [34] is a family of multi-stage hash functions that produces 64-bit hashes and employs a PDP family of hash functions CLNH, which resembles NH, but replaces modular additions and multiplications with XORs and carry-less multiplications in $\mathbb{GF}(2^{64})/p(\mathsf{x})$ with the irreducible polynomial $p(\mathsf{x}) = \mathsf{x}^{64} + \mathsf{x}^4 + \mathsf{x}^3 + \mathsf{x} + 1$. Therefore, CLNH can exploit the `vpclmulqdq` instruction for 64-bit carry-less multiplication which was originally introduced for boosting the performance of GCM [21].

CLHASH[m] splits a given message M into $(64m)$-bit blocks $(M_1, \ldots, M_s)$, and pads the final block with zeroes such that its length becomes a multiple of 128 bits. Each block M_j is compressed with CLNH to a 128-bit value A_j. If the message consists of only a single block, the message length $|M|$ is multiplied with an independent key $K_L \in \{0,1\}^{64}$ and XORed to the result; the result is reduced to a 64-bit value modulo $p(\mathsf{x}) = \mathsf{x}^{64} + \mathsf{x}^4 + \mathsf{x}^3 + \mathsf{x} + 1$ and returned.

Algorithm 1. Definition of $\text{CLHASH}^{\mathsf{T}}[m,t]$ with a hash length of $64t$ bits, a block length of $m/8$ bytes, and t Toeplitz iterations.

101: **function** $\text{CLHASH}^{\mathsf{T}}[m,t]_K(M)$	301: **function** $\text{KEYGEN}(K)$
102: $\quad (K_N, K_P, K_A, K_L) \leftarrow \text{KEYGEN}(K)$	302: $\quad \kappa \leftarrow 64(m + 2t - 2)$
103: $\quad s \leftarrow \max(\lceil 64 \cdot |M|/m \rceil, 1)$	303: $\quad K_N \leftarrow K[1..\kappa]$
104: $\quad (M_1, \dots, M_s) \xleftarrow{64m} M$	304: $\quad K_P \leftarrow K[(\kappa+1)..(\kappa+128t)]$
105: $\quad M_s \leftarrow \text{PAD}_{128}(M_s)$	305: $\quad \kappa \leftarrow \kappa + 128t$
106: $\quad$ **for** $i \leftarrow 1$ **to** t **do**	306: $\quad K_A \leftarrow K[(\kappa+1)..(\kappa+128t)]$
107: $\quad\quad$ **for** $j \leftarrow 1$ **to** s **do**	307: $\quad \kappa \leftarrow \kappa + 128t$
108: $\quad\quad\quad K_j \leftarrow K_{N(2i-1)..m+2(i-1)}$	308: $\quad K_L \leftarrow K[(\kappa+1)..(\kappa+64t)]$
109: $\quad\quad\quad A_j \leftarrow \text{CLNH}[m]_{K_j}(M_j)$	309: $\quad$ **return** (K_N, K_P, K_A, K_L)
110: $\quad\quad$ **if** $s = 1$ **then**	401: **function** $\text{POLY}_{K_P}(A_1, \dots, A_s)$
111: $\quad\quad\quad H_1 \leftarrow A_1$	402: $\quad$ **return** $\bigoplus_{i=1}^{s} A_i \cdot K_P^{s-i}$
112: $\quad\quad$ **else**	403: $\quad\quad \mod(2^{128} + 4 + 2)$
113: $\quad\quad\quad K_{Pi} \leftarrow K_{Pi} \mod 2^{126}$	501: **function** $\text{HASHLEN}_{K_L}(H_i, |M|)$
114: $\quad\quad\quad O_i \leftarrow \text{POLY}_{K_{Pi}}(A_1, \dots, A_s)$	502: $\quad$ **return** $(H_i \oplus (K_L \cdot |M|))$
115: $\quad\quad\quad H_i \leftarrow \text{CLNH}[2]_{K_{Ai}}(O_i)$	503: $\quad\quad \mod(2^{64} + 27)$
116: $\quad\quad H_i \leftarrow \text{HASHLEN}_{K_{Li}}(H_i, |M|)$	601: **function** $\text{PAD}_n(X)$
117: $\quad$ **return** $(H_1 \| \cdots \| H_t)$	602: $\quad$ **if** $(|X| \mod n = 0)$ **then**
201: **function** $\text{CLNH}[m]_{K_j}(M_j)$	603: $\quad\quad$ **return** X
202: $\quad$ **return** $\bigoplus_{i=1}^{m} (M_{j\,2i-1} \oplus K_{j\,2i-1})$	604: $\quad$ **return** $X \| 0^{n-|X| \mod n}$
203: $\quad\quad\quad\quad \cdot (M_{j\,2i} \oplus K_{j\,2i})$	

For longer messages, the values A_j are processed by a polynomial hash with an independent key $K_P \in \{0,1\}^{128}$ and reduced modulo $q(\mathsf{x}) = \mathsf{x}^{127} + \mathsf{x} + 1$. For efficiency, the two most significant bits of K_P are fixed to zero, and a lazy reduction modulo $\mathsf{x}^{128} + \mathsf{x}^2 + \mathsf{x}$ is used instead without affecting security.

The 128-bit result of the polynomial hash is then reduced to a 64-bit value by another application of CLNH with two further independent key words $K_{A_1}, K_{A_2} \in \{0,1\}^{64}$. The result H is finally XORed with the hashed length to account for inputs of variable lengths, and is reduced to a 64-bit value.

In [34], the authors show that CLHASH is XOR-universal for messages of up to $b = 8m$ bytes, and ϵ-AXU for messages of up to N bytes.

Theorem 6 (Lemma 9 in [34]). *Let $N \geq 1$ denote the maximal message length in bytes, $m \geq 2$ be even, and $b = 8m$ the key size of CLNH. Then, CLHASH as defined above is ϵ-AXU with*

$$\epsilon \leq \epsilon_{\text{CLNH}[m]} + \epsilon_{\text{POLY}} + \epsilon_{\text{CLNH}[2]} \leq \frac{1}{2^{64}} + \frac{N/b - 1}{2^{126}} + \frac{1}{2^{64}},$$

where the terms stem from the facts that $\text{CLNH}[m]$ is an $\epsilon_{\text{CLNH}[m]}$-AU, and the polynomial hash an ϵ_{POLY}-AXU family of hash functions.

The recommended values $N \leq 2^{64}$ and $b = 1024$ yield $\epsilon \leq 2.004/2^{-64}$. The construction requires $b + 40$ bytes of key material: b bytes for CLNH, a 16-byte value K_P for the polynomial hash, two eight-byte values $K_A[1], K_A[2]$ for the final call to CLNH, and an eight-byte value K_L for hashing the input length.

Toeplitz Extension. To obtain a hash function with 128-bit security, one can process the same message twice under independent keys and concatenate the results. Doubling the key lengths of K_P, K_A, and K_L increases their keys to 80 bytes. Since doubling the key length for CLNH would absurdly increase the key material, we use the Toeplitz extension [31,37] instead. Let $K_{i..j}$ be short for $K_i, \ldots, K_j$, $1 \leq i \leq j$. Given an ϵ-AU family of hash functions $H : \{0,1\}^{mn} \times \{0,1\}^{mn} \to \{0,1\}^n$ which compresses an m-word input with an m-word key, one can derive a hash function $H^t : \{0,1\}^{(m+2t-2)n} \times \{0,1\}^{mn} \to \{0,1\}^{tm}$ by

$$H^t_{K_{1..(m+2t-2)}}(M) := H_{K_{1..m}}(M) \parallel H_{K_{3..(m+2)}}(M) \parallel \cdots \parallel H_{K_{(2t-1)..(m+2t-2)}}(M).$$

So, the i-th call to H employs the key shifted by $2i-2$ words. In total, the key size increases slightly from m to $m + 2(t-1)$ words. We refer to the Toeplitz version of CLNH by $\text{CLNH}^T[m,t]$, and to that of $\text{CLHASH}[m]$ by $\text{CLHASH}^T[m,t]$. Algorithm 1 provides a specification. In total, $\text{CLHASH}^T[m,t]$ requires $(8m + 56t - 16)$ bytes of key material, which corresponds to $(8m + 96)$ bytes for $t = 2$.

Definition 13 (Toeplitz CLHASH). *Let $n = 64$, $t \geq 1$, $m \geq 2$ be even. Let $\mathcal{X} = \bigcup_{i=1}^{m/2} \{0,1\}^{2in}$. Let further $\mathcal{K}_N = \{0,1\}^{64m+128(t-1)}$, $\mathcal{K}_P = \{0,1\}^{128t}$, $\mathcal{K}_A = \{0,1\}^{128t}$, $\mathcal{K}_L = \{0,1\}^{64t}$, and $\mathcal{K} = \mathcal{K}_N \times \mathcal{K}_P \times \mathcal{K}_A \times \mathcal{K}_L$. The family of keyed hash functions $\text{CLHASH}^T[m,t] : \mathcal{K} \times \mathcal{X} \to \{0,1\}^{64t}$ is defined in Algorithm 1.*

Theorem 7. *For any fixed $n, t \geq 1$, and even $m \geq 2$, $\text{CLNH}^T[m,t]$ is 2^{-nt}-AU on equal-length strings.*

The proof of Theorem 7 can be found in the full version of this paper.

Theorem 8. *Let $N \leq 2^{64}$ be the maximal message length in bytes, $t \geq 1$, $m \geq 2$ be even, and $b = 8m$ the key length for CLNH in bytes. Then, $\text{CLHASH}^T[m,t]$ is an ϵ^t-AXU family of hash functions with*

$$\epsilon \leq \epsilon_{\text{CLNH}[m]} + \epsilon_{\text{POLY}} + \epsilon_{\text{CLNH}[2]} \leq \frac{1}{2^{64}} + \frac{N/b - 1}{2^{126}} + \frac{1}{2^{64}} \leq \frac{3}{2^{64}}.$$

The proof of Theorem 8 follows from Theorem 7 and the fact that the keys for the individual iterations of polynomial, inner-product, and length hashing steps are chosen uniformly from their respective spaces and pairwise independently for each iteration. We can derive that $\text{CLHASH}^T[m,2]$ is ϵ-AXU for $\epsilon \leq 9/2^{128}$ when $m \geq 2$.

6.2 Constructing a PRF

Let $n, d \geq 1$, and $\mathcal{N}$, $\mathcal{H}$, $\mathcal{M}$ be as in Sect. 3. For brevity, we define $\mathcal{Y} := \{0,1\}^d \times \mathcal{N} \times \mathcal{H} \times \mathcal{M}$. Let $\text{ENCODE} : \mathcal{Y} \to \{0,1\}^*$ define an injective encoding function. Then, we can construct a PRF from the composition of ENCODE,

---

Algorithm 2. Encryption of nonce-based XOR-CTR, instantiated with a block cipher $E : \{0,1\}^k \times \{0,1\}^n \to \{0,1\}^n$, with $n, k \geq 1$.

```
1: function XOR-CTR[E].E_K^N(M)
2:    IV ← E_K(N)
3:    m ← ⌈|M|/n⌉
4:    κ ← E_K(IV ⊕ ⟨0⟩) ‖ ··· ‖ E_K(IV ⊕ ⟨m − 1⟩)
5:    return C ← M ⊕ κ[first |M| bits]
```

---

a family of ϵ-AU hash functions $\mathcal{H}' = \{H'|H' : \{0,1\}^* \to \{0,1\}^n\}$, and a block cipher $E : \mathcal{K}_2 \times \{0,1\}^n \to \{0,1\}^n$, with independent keys $K_1 \in \mathcal{K}_1$ determining the hash function, and $K_2 \in \mathcal{K}_2$ for the cipher. We call the construction EHE[ENCODE, $\mathcal{H}', E$] : $\mathcal{Y} \to \{0,1\}^n$ (for *Encode-Hash-Encrypt*) and define it as

$$\text{EHE}[\text{ENCODE}, \mathcal{H}', E]_{K_1,K_2}(D, N, H, M) := E_{K_2}(\mathcal{H}'_{K_1}(\text{ENCODE}(D, N, H, M))).$$

We write EHE[$\mathcal{H}', E$] or even EHE as short forms of EHE[ENCODE, $\mathcal{H}', E$] when the components are clear from the context. The injective encoding excludes collisions between distinct inputs. From Theorem 2, and applying the PRF/PRP switching lemma, we can derive the following theorem.

Theorem 9. *Let* $\pi \twoheadleftarrow \text{Perm}(\{0,1\}^n)$. *Further, let* EHE[ENCODE, $\mathcal{H}', \pi$], $\mathcal{H}'$, *and* ENCODE *be defined as above. Let* **A** *be a computationally bounded adversary that asks at most q queries of at most ℓ blocks and runs in time at most t. Then*

$$\mathbf{Adv}_{\text{EHE}[\text{ENCODE}, \mathcal{H}', \pi]}^{\text{PRF}}(\mathbf{A}) \leq \binom{q}{2} \cdot \left(\frac{1}{2^n} + \epsilon\right).$$

6.3   Encryption

When starting counter-mode encryption from a random value and incrementing by modular addition, one has to either consider potential carry bits or to reduce the security by fixing a maximal message length. Wang et al. [50] proposed to replace modular addition by XOR, which avoids the need for concerning carry bits. Let $E : \{0,1\}^k \times \{0,1\}^n \to \{0,1\}^n$ be a block cipher. We define XOR-CTR[E] = $(\mathcal{E}, \mathcal{D})$ as the nonce-based encryption scheme with encryption algorithm XOR-CTR[E].$\mathcal{E}$: $\{0,1\}^k \times \mathcal{N} \times \{0,1\}^* \to \{0,1\}^*$ and associated non-empty nonce-space $\mathcal{N}$, as defined in Algorithm 2.

We denote by XOR-CTR[π, π'] a version of XOR-CTR with two independent n-bit permutations π and π', where π is used for encrypting the nonce and π' for producing the keystream. Then, XOR-CTR[π, π'] is almost identical to the CTR2[π, π'] construction in [45], with the difference that the former replaces the addition of IV and counter modulo 2^n by XOR. Since this change does not affect the probability of block-cipher inputs to repeat, the NE advantage of XOR-CTR is given by Theorem 10, which adapts Theorem 3 in [45].

Theorem 10. *Let* $\pi, \pi' \leftarrow \text{Perm}(\{0,1\}^n) \times \text{Perm}(\{0,1\}^n)$ *be independent permutations and* $\mathbf{A}$ *be a nonce-respecting* NE *adversary, which runs in time at most* t *and poses at most* q *queries to its oracles with at most* ℓ *blocks. Then*

$$\mathbf{Adv}^{\text{NE}}_{\text{XOR-CTR}[\pi,\pi']}(\mathbf{A}) \leq \frac{\ell^2}{2^n}.$$

From the fact that encryption and decryption of XOR-CTR$[\pi, \pi']$ are identical operations, we can derive the following theorem.

Theorem 11. *There exists a reduction of a nonce-respecting* SRND *adversary* $\mathbf{A}$ *with access to two oracles on* XOR-CTR$[\pi, \pi']$ *to a nonce-respecting* NE *adversary* $\mathbf{A}'$ *on* XOR-CTR$[\pi, \pi']$ *such that*

$$\mathbf{Adv}^{\text{SRND}}_{\text{XOR-CTR}[\pi,\pi']}(\mathbf{A}) \leq \mathbf{Adv}^{\text{NE}}_{\text{XOR-CTR}[\pi,\pi']}(\mathbf{A}'),$$

where both $\mathbf{A}$ *and* $\mathbf{A}'$ *ask at most* q *queries of at most* ℓ *blocks to their available oracle(s) and run in time* $O(t)$.

6.4   Instantiation of RIV

We instantiate RIV$_{F,\Pi}$ with EHE[ENCODE, $\mathcal{H}'$, E] for F, with CLHASH$^\top[m, 2]$ as family of universal hash functions $\mathcal{H}'$, and XOR-CTR$[E]$ for Π, with the AES-128 as E. Algorithm 3 provides a specification. Our instantiation RIV$_{F,\Pi}$ expects a 128-bit user-supplied secret key SK, from which the remaining key material is derived by calling $E_{SK}(\cdot)$ iteratively in counter mode. The secret key is not used further. RIV uses $n = \tau = 128$, i.e., n-bit tags, and n-bit IVs for the counter mode. Moreover, the nonce space is fixed to 128 bits: $\mathcal{N} = \{0,1\}^n$. For F, it employs a four-bit domain separation, i.e., $d = 4$, and an injective encoding function ENCODE $: \{0,1\}^d \times \mathcal{N} \times \mathcal{H} \times \mathcal{M} \rightarrow \{0,1\}^*$, as defined in Algorithm 3. Header and message lengths are restricted to multiple of eight bits. The maximal number of header and message bytes to be encrypted under the same key are 2^{60} bytes each. So, the maximal number of bytes for RIV is less than 2^{62} bytes. We recommend that at most 2^{50} bytes be encrypted under the same key.

Using a Single Key for the Block Cipher. There are four uses of the block cipher E in RIV: in the first invocation of EHE, for encrypting the IV, for generating the keystream in XOR-CTR$[E]$, and in the second invocation of EHE. If four more calls to the AES key schedule would be tolerable, one could use four independent keys. Alternatively, we use a single key for the uses of E, and have to consider the security impact in the following theorem. Its proof can be found in the full version of this paper.

Theorem 12. *Let* RIV$_{F,\Pi}$ *be defined as in Algorithm 3. Let* $K_1, K_2 \leftarrow \mathcal{K}$ *be independent keys. We replace the calls to* E *by independent random permutations* $\pi_1, \pi_2, \pi_3, \pi_4 \leftarrow \text{Perm}(\{0,1\}^n)^4$. *Let* $\mathbf{A}$ *be a computationally bounded adversary that has access to three oracles* $\mathcal{O}_1$, $\mathcal{O}_2$, *and* $\mathcal{O}_3$ *for encryption, decryption, and leakage, respectively.* $\mathbf{A}$ *shall distinguish between a real setting of* RIV$_{F,\Pi}$ *as*

Algorithm 3. Definition of our instantiation $\mathrm{RIV}_{F,\Pi}$. Message and header lengths are restricted to multiple of eight bits, and nonces/IVs/tags are 128 bits: $n = \tau = 128$, and $d = 4$. Here, we leave the key size of $\mathrm{CLHASH}^{\mathsf{T}}[m, 2]$, m, as a parameter to study its impact on performance later.

101: **function** $\widetilde{\mathcal{E}}_{SK}(N, H, M)$
102:     $(K_1, K_2) \leftarrow \mathrm{KeyGen}(SK)$
103:     $IV \leftarrow \mathrm{EHE}^1_{K_1, K_2}(N, H, M)$
104:     $C \leftarrow \mathrm{XOR\text{-}CTR}[E].\mathcal{E}_{K_2}(IV, M)$
105:     $T \leftarrow \mathrm{EHE}^2_{K_1, K_2}(N, H, C) \oplus IV$
106:     **return** (C, T)

201: **function** $\mathrm{KeyGen}(SK)$
202:     $K_2 \leftarrow E_{SK}(\langle 0 \rangle)$
203:     $\kappa \leftarrow (8m + 96)/16$
204:     $K_1 \leftarrow E_{SK}(\langle 1 \rangle) \| \cdots \| E_{SK}(\langle \kappa \rangle)$
205:     **return** (K_1, K_2)

301: **function** $\mathrm{EHE}^D_{K_1, K_2}(N, H, X)$
302:     $Y \leftarrow \mathrm{Encode}(D, N, H, X)$
303:     **return** $E_{K_2}(\mathcal{H}'_{K_1}(Y))$

401: **function** $\mathrm{PAD}_n(X)$
402:     **if** $(|X| \bmod n = 0)$ **then**
403:         **return** X
404:     **return** $X \| 0^{n - |X| \bmod n}$

501: **function** $\widetilde{\mathcal{D}}_{SK}(N, H, C, T)$
502:     $(K_1, K_2) \leftarrow \mathrm{KeyGen}(SK)$
503:     $IV \leftarrow \mathrm{EHE}^2_{K_1, K_2}(N, H, C) \oplus T$
504:     $M \leftarrow \mathrm{XOR\text{-}CTR}[E].\mathcal{D}_{K_2}(IV, C)$
505:     $IV' \leftarrow \mathrm{EHE}^1_{K_1, K_2}(N, H, M)$
506:     **if** $(IV = IV')$ **then**
507:         **return** M
508:     **return** $\perp$

601: **function** $\mathrm{Encode}(D, N, H, X)$
602:     $\overline{H} \leftarrow \mathrm{PAD}_{128}(H)$
603:     $\overline{X} \leftarrow \mathrm{PAD}_{128}(X)$
604:     $\overline{L} \leftarrow \langle D \rangle_d \| \langle |H|/8 \rangle_{60} \| \langle |X|/8 \rangle_{64}$
605:     **return** $(\overline{H} \| N \| \overline{X} \| \overline{L})$

701: **function** $\mathcal{H}'_{K_1}(X)$
702:     **return** $\mathrm{CLHASH}^{\mathsf{T}}[m, 2]_{K_1}(X)$

801: **function** $E_{K_2}(X)$
802:     **return** $\mathrm{AES\text{-}128}_{K_2}(X)$

---

above with a single-keyed block cipher E, and $\mathrm{RIV}_{F,\Pi}$ which uses four independent uniformly chosen permutations $\pi^1, \pi^2, \pi^3, \pi^4 \twoheadleftarrow \mathsf{Perm}(\{0,1\}^n)$ with π^1 used in EHE^1, π^2 used in EHE^2, and π^3, π^4 used for $\mathrm{XOR\text{-}CTR}[\pi^3, \pi^4]$. $\mathbf{A}$ asks at most q queries of at most ℓ blocks and runs in time at most t. Then, we can upper bound the distinguishing advantage of $\mathbf{A}$ by

$$16.5\ell^2 \cdot \max\{\epsilon, 1/2^n\} + \mathbf{Adv}^{\mathrm{PRP}}_E(\ell + 3q, O(t)).$$

Theorem 13. *Let $d = 4$, $n = \tau = 128$, and $m \geq 2$ be even. Let $\mathrm{RIV}_{F,\Pi}$ be as given in Algorithm 3 and let $\mathbf{A}_1, \mathbf{A}_2, \mathbf{A}_3$ be computationally bounded IND-CPA, INT-CTXT, and ERR-CCA adversaries on $\mathrm{RIV}_{F,\Pi}$, respectively, which run each in time at most t and ask at most q queries of at most ℓ blocks in total. Then, it holds that*

$$\mathbf{Adv}^{\mathrm{IND\text{-}CPA}}_{\mathrm{RIV}_{F,\Pi}}(\mathbf{A}) \leq \frac{2q^2 + \ell^2}{2^n} + q^2\epsilon + \delta_E,$$

$$\mathbf{Adv}^{\mathrm{INT\text{-}CTXT}}_{\mathrm{RIV}_{F,\Pi}}(\mathbf{A}) \leq \frac{2q^2 + q + \ell^2}{2^n} + q^2\epsilon + \delta_E,$$

$$\mathbf{Adv}^{\mathrm{ERR\text{-}CCA}}_{\mathrm{RIV}_{F,\Pi}}(\mathbf{A}) \leq \frac{8q^2 + 2q + 2\ell^2}{2^n} + 2q^2\epsilon + \delta_E,$$

where $\delta_E = 16.5\ell^2 \cdot \epsilon + \mathbf{Adv}^{\mathrm{PRP}}_E(\ell + 3q, O(t))$ and $\epsilon \leq 9/2^{128}$.

The proof follows from Theorems 5, 8, 9, and 11, and those of the lemmata from Sect. 5 that can be found in the full version of this paper.

6.5   Performance of RIV

We implemented reference and optimized versions of RIV in C.[5] Since the default key length for one iteration CLNH of $b = 1024$ bytes (which corresponds to $CLHASH^T[128, 2]$) appeared high, we tested also a variant with a smaller key size of $b = 256$ bytes for CLNH ($CLHASH^T[32, 2]$). Table 1 summarizes the results of our benchmarks. Our code was compiled using gcc v4.9.3 with options `-O3 -maes -mavx2 -mpclmul -march=native`, and run on (1) an Intel Core i5-4200M (Haswell) at 2.50 GHz, and (2) on an Intel i5-5200 (Broadwell) at 2.20 GHz, both with the TurboBoost, SpeedStep, and HyperThreading technologies *disabled*. For measuring, we used the median of 10000 encryptions, omitting the cost for key setup, using the `rdtsc` instruction.

Our results show that RIV can run at less than 1.5 cpb on Haswell. Interestingly, a *SIV-like* reduced version of RIV, which is an easily obtained byproduct that simply omits the second call to F, represents a performant MRAE scheme with ≤ 1.04 cpb. This is slightly faster than the $4867/4096 \approx 1.17$ cpb reported for the manually assembly-optimized AES-GCM-SIV [22] and 1.06 cpb for the version of MRO with four-round BLAKE2b in [20], concerning messages of at least four KiB length on Haswell. Clearly, the reported performance of AEZv4 of about 0.7 cpb is unrivaled. Though, our construction provides a slightly higher security margin. Moreover, the security of AEZv4 bases on heuristic assumptions on four-round AES.

Table 1. Performance results on Intel Haswell and Broadwell, respectively, in cycles per byte for the encryption with optimized implementations of RIV and a reduced version, which omits the second call to F. b denotes the key length for CLNH in bytes. Details regarding our setup are provided in the text.

Platform	Instance	b	Message length (bytes)							
			128	256	512	1024	2048	4096	8192	16384
Haswell	RIV	256	3.81	2.78	2.14	1.81	1.62	1.48	1.40	1.37
	RIV	1024	3.53	2.13	1.81	1.49	1.37	1.29	1.25	1.22
	RIV (2-pass)	256	1.71	1.40	1.26	1.14	1.08	1.04	1.01	0.99
	RIV (2-pass)	1024	2.20	1.60	1.17	1.08	1.01	0.97	0.94	0.92
Broadwell	RIV	256	3.16	2.41	1.84	1.49	1.38	1.26	1.20	1.15
	RIV	1024	3.13	2.11	1.56	1.34	1.16	1.09	1.04	1.02
	RIV (2-pass)	256	2.16	1.67	1.30	1.09	1.03	0.95	0.92	0.90
	RIV (2-pass)	1024	2.19	1.50	1.14	1.01	0.92	0.86	0.84	0.82

---

[5] Our code is open to the public domain: https://github.com/medsec/riv.

7   Conclusion

This work described a modular framework RIV for the construction of provably secure subtle AE schemes by extending the SIV framework from two to three passes. The obvious strength of RIV resides in the simplicity of its structure: it allows a straight-forward transformation of existing SIV-based constructions into subtle AE schemes. We proved the security in the standard model under notions that strive for ideal security goals; a further step could be to prove *achievable* security in the RAE setting with fixed stretch. Moreover, since the generic RIV construction bases only on PRF assumptions, this leaves open the possibility for proofs in the indifferentiability setting [40]. RIV is slightly less efficient than earlier STPRP constructions, i.e., it employs three additional calls to an n-bit PRP, compared to e.g., a single call in HCTR-based [50] constructions. Since the use of a nonce-based encryption scheme $(\mathcal{E}, \mathcal{D})$ poses only the requirement on the IV to be a nonce, it might look to be sufficient to have two calls to universal hash functions instead of to calls to a PRF F. Yet, at least the outputs from the first invocation of F, $F_{K_1}^1(\cdot, \cdot, \cdot)$ must be unpredictable in order to prevent leaking information about the message in the tag. A potential future work can be to further study reductions of the design to target even higher efficiency. Nevertheless, we proposed an instantiation that is highly efficient on current x64 platforms and avoids the weak-key issues that were reported for GHASH-based polynomials in HCTR instantiations [49].

Acknowledgments. We thank all reviewers of the FSE 2016 for their helpful comments, Daniel Lemire and Owen Kaser for valuable notes on CLHASH, and Guy Barwell, Daniel Page, and Martijn Stam for insights into their work on subtle authenticated encryption. Christian Forler received funding from the European Research Council under the European Union's Seventh Framework Programme (FP/2007–2013)/ERC Grant Agreement no. 307952 and from the Silicon Valley Community Foundation under the Cisco Systems project *Misuse-Resistant Authenticated Encryption for Complex and Low-End Systems* (MIRACLE).

References

1. Abed, F., Fluhrer, S., Forler, C., List, E., Lucks, S., McGrew, D., Wenzel, J.: Pipelineable on-line encryption. In: Cid, C., Rechberger, C. (eds.) FSE 2014. LNCS, vol. 8540, pp. 205–223. Springer, Heidelberg (2015)

2. Anderson, R.J., Biham, E.: Two practical and provably secure block ciphers: BEAR and LION. In: Gollmann, D. (ed.) FSE 1996. LNCS, vol. 1039, pp. 113–120. Springer, Heidelberg (1996)

3. Andreeva, E., Bilgin, B., Bogdanov, A., Luykx, A., Mendel, F., Mennink, B., Mouha, N., Wang, Q., Yasuda, K.: PRIMATES (2014). http://competitions.cr.yp.to/caesar-submissions.html

4. Andreeva, E., Bogdanov, A., Luykx, A., Mennink, B., Mouha, N., Yasuda, K.: How to securely release unverified plaintext in authenticated encryption. In: Sarkar, P., Iwata, T. (eds.) ASIACRYPT 2014. LNCS, vol. 8873, pp. 105–125. Springer, Heidelberg (2014)

5. Badertscher, C., Matt, C., Maurer, U., Rogaway, P., Tackmann, B.: Robust authenticated encryption and the limits of symmetric cryptography. In: Groth, J., et al. (eds.) IMACC 2015. LNCS, vol. 9496, pp. 112–129. Springer, Heidelberg (2015). doi:10.1007/978-3-319-27239-9_7

6. Bahack, L.: Julius (2014). http://competitions.cr.yp.to/caesar-submissions.html

7. Barwell, G., Page, D., Stam, M.: Rogue decryption failures: reconciling AE robustness notions. In: Groth, J., et al. (eds.) IMACC 2015. LNCS, vol. 9496, pp. 94–111. Springer, Heidelberg (2015). doi:10.1007/978-3-319-27239-9_6

8. Bellare, M., Namprempre, C.: Authenticated encryption: relations among notions and analysis of the generic composition paradigm. In: Okamoto, T. (ed.) ASIACRYPT 2000. LNCS, vol. 1976, p. 531. Springer, Heidelberg (2000)

9. Bellare, M., Rogaway, P.: Encode-then-encipher encryption: how to exploit nonces or redundancy in plaintexts for efficient cryptography. In: Okamoto, T. (ed.) ASIACRYPT 2000. LNCS, vol. 1976, pp. 317–330. Springer, Heidelberg (2000)

10. Bellare, M., Rogaway, P., Wagner, D.: The EAX mode of operation. In: Roy, B., Meier, W. (eds.) FSE 2004. LNCS, vol. 3017, pp. 389–407. Springer, Heidelberg (2004)

11. Bernstein, D.J.: Polynomial evaluation and message authentication (2007). http://cr.yp.to/papers, permanent ID: b1ef3f2d385a926123e1517392e20f8c, 2

12. Bertoni, G., Daemen, J., Peeters, M., Van Assche, G., Van Keer, R.: Using Keccak technology for AE: Ketje, Keyak and more. In: SHA-3 2014 Workshop, UC Santa Barbara, 22 August 2014

13. Black, J., Halevi, S., Krawczyk, H., Krovetz, T., Rogaway, P.: UMAC: fast and secure message authentication. In: Wiener, M. (ed.) CRYPTO 1999. LNCS, vol. 1666, pp. 216–233. Springer, Heidelberg (1999)

14. Boesgaard, M., Christensen, T., Zenner, E.: Badger – a fast and provably secure MAC. In: Ioannidis, J., Keromytis, A.D., Yung, M. (eds.) ACNS 2005. LNCS, vol. 3531, pp. 176–191. Springer, Heidelberg (2005)

15. Boldyreva, A., Degabriele, J.P., Paterson, K.G., Stam, M.: On symmetric encryption with distinguishable decryption failures. In: Moriai, S. (ed.) FSE 2013. LNCS, vol. 8424, pp. 367–390. Springer, Heidelberg (2014)

16. Chakraborty, D., Sarkar, P.: On modes of operations of a block cipher for authentication and authenticated encryption. In: IACR Cryptology ePrint Archive, 2014/627 (2014)

17. Coron, J.-S., Dodis, Y., Mandal, A., Seurin, Y.: A domain extender for the ideal cipher. In: Micciancio, D. (ed.) TCC 2010. LNCS, vol. 5978, pp. 273–289. Springer, Heidelberg (2010)

18. Dai, W., Krovetz, T.: VHASH security. In: IACR Cryptology ePrint Archive: 2007/338 (2007)

19. Fleischmann, E., Forler, C., Lucks, S.: McOE: a family of almost foolproof on-line authenticated encryption schemes. In: Canteaut, A. (ed.) FSE 2012. LNCS, vol. 7549, pp. 196–215. Springer, Heidelberg (2012)

20. Granger, R., Jovanovic, P., Mennink, B., Neves, S.: Improved masking for tweakable blockciphers with applications to authenticated encryption. In: Fischlin, M., Coron, J.-S. (eds.) EUROCRYPT 2016. LNCS, vol. 9665, pp. 263–293. Springer, Heidelberg (2016). doi:10.1007/978-3-662-49890-3_11

21. Kounavis, E.: Efficient implementation of the Galois Counter Mode using a carryless multiplier and a fast reduction algorithm. Inf. Process. Lett. 110(14–15), 549–553 (2010)

22. Gueron, S., Lindell, Y.: GCM-SIV: Full nonce misuse-resistant authenticated encryption at under one cycle per byte. In: Ray, I., Li, N., Kruegel, C. (eds.) ACM Conference on Computer and Communications Security, pp. 109–119. ACM (2015)

23. Halevi, S., Rogaway, P.: A parallelizable enciphering mode. In: Okamoto, T. (ed.) CT-RSA 2004. LNCS, vol. 2964, pp. 292–304. Springer, Heidelberg (2004)

24. Hoang, V.T., Krovetz, T., Rogaway, P.: Robust authenticated-encryption AEZ and the problem that it solves. In: Oswald, E., Fischlin, M. (eds.) EUROCRYPT 2015. LNCS, vol. 9056, pp. 15–44. Springer, Heidelberg (2015)

25. Hoang, V.T., Reyhanitabar, R., Rogaway, P., Vizár, D.: Online authenticated-encryption and its nonce-reuse misuse-resistance. In: Gennaro, R., Robshaw, M., et al. (eds.) CRYPTO (1). LNCS, vol. 9215, pp. 493–517. Springer, Heidelberg (2015)

26. ISO/IEC. 19772: 2009, Information technology - Security techniques - Authenticated Encryption (2009)

27. Iwata, T., Yasuda, K.: BTM: a single-key, inverse-cipher-free mode for deterministic authenticated encryption. In: Jacobson Jr., M.J., Rijmen, V., Reihaneh, S.-N. (eds.) Selected Areas in Cryptography, pp. 313–330 (2009)

28. Iwata, T., Yasuda, K.: HBS: a single-key mode of operation for deterministic authenticated encryption. In: Dunkelman, O. (ed.) FSE 2009. LNCS, vol. 5665, pp. 394–415. Springer, Heidelberg (2009)

29. Jean, J., Nikolić, I., Peyrin, T.: Deoxys (2014). http://competitions.cr.yp.to/caesar-submissions.html

30. Jean, J., Nikolić, I., Peyrin, T.: Joltik (2014). http://competitions.cr.yp.to/caesar-submissions.html

31. Krawczyk, H.: LFSR-based hashing and authentication. In: Desmedt, Y.G. (ed.) CRYPTO 1994. LNCS, vol. 839, pp. 129–139. Springer, Heidelberg (1994)

32. Krovetz, T.: Message authentication on 64-bit architectures. In: Biham, E., Youssef, A.M. (eds.) SAC 2006. LNCS, vol. 4356, pp. 327–341. Springer, Heidelberg (2007)

33. Krovetz, T.: HS1-SIV (2014). http://competitions.cr.yp.to/caesar-submissions.html

34. Lemire, D., Kaser, O.: Faster 64-bit universal hashing using carry-less multiplications. J. Crypt. Eng. 5, 1–15 (2015)

35. Luby, M., Rackoff, C.: How to construct pseudorandom permutations from pseudorandom functions. SIAM J. Comput. 17(2), 373–386 (1988)

36. Lucks, S.: Faster Luby-Rackoff ciphers. In: Gollmann, D. (ed.) FSE 1996. LNCS, vol. 1039, pp. 183–203. Springer, Heidelberg (1996)

37. Mansour, Y., Nisan, N., Tiwari, P.: The computational complexity of universal hashing. In: Ortiz, H. (ed.) STOC, pp. 235–243. ACM (1990)

38. Maurer, U.: Constructive cryptography – a new paradigm for security definitions and proofs. In: Mödersheim, S., Palamidessi, C. (eds.) TOSCA 2011. LNCS, vol. 6993, pp. 33–56. Springer, Heidelberg (2012)

39. Maurer, U., Renner, R.: Abstract cryptography. In: Chazelle, B. (ed.) ICS, pp. 1–21. Tsinghua University Press, Beijing (2011)

40. Maurer, U.M., Renner, R.S., Holenstein, C.: Indifferentiability, impossibility results on reductions, and applications to the random oracle methodology. In: Naor, M. (ed.) TCC 2004. LNCS, vol. 2951, pp. 21–39. Springer, Heidelberg (2004)

41. Namprempre, C., Rogaway, P., Shrimpton, T.: Reconsidering generic composition. In: Nguyen, P.Q., Oswald, E. (eds.) EUROCRYPT 2014. LNCS, vol. 8441, pp. 257–274. Springer, Heidelberg (2014)

42. Naor, M., Reingold, O.: On the construction of pseudorandom permutations: Luby-Rackoff revisited. J. Cryptol. **12**(1), 29–66 (1999)
43. Reyhanitabar, R., Vaudenay, S., Vizár, D.: Misuse-resistant variants of the OMD authenticated encryption mode. In: Chow, S.S.M., Liu, J.K., Hui, L.C.K., Yiu, S.M. (eds.) ProvSec 2014. LNCS, vol. 8782, pp. 55–70. Springer, Heidelberg (2014)
44. Rogaway, P.: Authenticated-encryption with associated-data. In: Atluri, V. (ed.) ACM Conference on Computer and Communications Security, pp. 98–107 (2002)
45. Rogaway, P.: Nonce-based symmetric encryption. In: Roy, B., Meier, W. (eds.) FSE 2004. LNCS, vol. 3017, pp. 348–359. Springer, Heidelberg (2004)
46. Rogaway, P., Shrimpton, T.: A provable-security treatment of the key-wrap problem. In: Vaudenay, S. (ed.) EUROCRYPT 2006. LNCS, vol. 4004, pp. 373–390. Springer, Heidelberg (2006)
47. Shrimpton, T., Terashima, R.S.: A modular framework for building variable-input-length tweakable ciphers. In: Sako, K., Sarkar, P. (eds.) ASIACRYPT 2013, Part I. LNCS, vol. 8269, pp. 405–423. Springer, Heidelberg (2013)
48. Stinson, D.R.: Universal hashing and authentication codes. Des. Codes Crypt. **4**(4), 369–380 (1994)
49. Sun, Z., Wang, P., Zhang, L.: Weak-key and related-key analysis of hash-counter-hash tweakable enciphering schemes. In: Foo, E., Stebila, D. (eds.) ACISP 2015. LNCS, vol. 9144, pp. 3–19. Springer, Heidelberg (2015)
50. Wang, P., Feng, D., Wu, W.: HCTR: a variable-input-length enciphering mode. In: Feng, D., Lin, D., Yung, M. (eds.) CISC 2005. LNCS, vol. 3822, pp. 175–188. Springer, Heidelberg (2005)
51. Winograd, S.: A new algorithm for inner product. IEEE Trans. Comput. **100**(7), 693–694 (1968)

A MAC Mode for Lightweight Block Ciphers

Atul Luykx[1,2]([⊠]), Bart Preneel[1,2], Elmar Tischhauser[3], and Kan Yasuda[4]

[1] Department of Electrical Engineering, ESAT/COSIC, KU Leuven, Leuven, Belgium
{atul.luykx,bart.preneel}@esat.kuleuven.be
[2] iMinds, Ghent, Belgium
[3] Department of Applied Mathematics and Computer Science,
Technical University of Denmark, Lyngby, Denmark
ewti@dtu.dk
[4] NTT Secure Platform Laboratories, NTT Corporation, Tokyo, Japan
yasuda.kan@lab.ntt.co.jp

Abstract. Lightweight cryptography strives to protect communication in constrained environments without sacrificing security. However, security often conflicts with efficiency, shown by the fact that many new lightweight block cipher designs have block sizes as low as 64 or 32 bits. Such low block sizes lead to impractical limits on how much data a mode of operation can process per key. MAC (message authentication code) modes of operation frequently have bounds which degrade with both the number of messages queried and the message length. We present a MAC mode of operation, LightMAC, where the message length has *no effect* on the security bound, allowing an order of magnitude more data to be processed per key. Furthermore, LightMAC is incredibly simple, has almost no overhead over the block cipher, and is parallelizable. As a result, LightMAC not only offers compact authentication for resource-constrained platforms, but also allows high-performance parallel implementations. We highlight this in a comprehensive implementation study, instantiating LightMAC with PRESENT and the AES. Moreover, LightMAC allows flexible trade-offs between rate and maximum message length. Unlike PMAC and its many derivatives, LightMAC is not covered by patents. Altogether, this makes it a promising authentication primitive for a wide range of platforms and use cases.

Keywords: Lightweight · MAC · LightMAC · Message length · Birthday bound · Integrity · Verification

1 Introduction

With the rise of the Internet of Things, connected devices are being placed everywhere, resulting in a wide variety of efficiency, robustness, and feature requirements for communication. Securing the communcation remains important, and as a result, many block ciphers have been created to work efficiently in constrained environments. These block ciphers offer a range of block and key sizes, from 128 to 32 bits; see Table 1 for a sample.

© International Association for Cryptologic Research 2016
T. Peyrin (Ed.): FSE 2016, LNCS 9783, pp. 43–59, 2016.
DOI: 10.1007/978-3-662-52993-5_3

The key size is often chosen carefully to ensure a sufficiently high security level, resulting in the block size becoming the dominant factor in determining security. As is well known, reducing block size can increase the chance of an inner state collision when block ciphers are used in so-called *modes of operation*: constructions which repeatedly apply a block cipher to achieve functionality beyond what a block cipher offers.

Consider MAC (Message Authentication Code) modes of operation, which aim to provide data authenticity for long messages. Common MAC modes, such as CBC-MAC [5], OMAC [24], and PMAC [10] have security bounds which degrade relative to both the number of messages tagged, q, and the length of the messages measured in blocks, ℓ; see Table 2 for a list of modes with their dependence on ℓ. For many modes, an adversary which is able to tag q messages of length ℓ blocks will have a success probability of roughly

$$\frac{q^2\ell}{2^n},\tag{1}$$

where n is the block size of the underlying block cipher. With a 32 bit block size and a guarantee that adversaries do not forge with probability more than one in a million, one gets a restriction of the form

$$\frac{q^2\ell}{2^{32}} \leq \frac{1}{2^{20}} \quad \text{or} \quad q^2\ell \leq 2^{12},\tag{2}$$

meaning 64 one-block messages can be tagged under the same key. But what if the messages are longer than one block? With conventional MACs only 32 four-block messages can be tagged, corresponding to $32 \cdot 2^2 \cdot 32 = 2^{12}$ bits, or 512 Bytes of data per key. If the messages are sixteen blocks long, only 16 messages can be tagged, which is $16 \cdot 2^4 \cdot 32 = 2^{13}$ bits, or 1 KiB of data per key. Figure 1 displays how much data the various modes from Table 2 can process per key, when the threshold success probability is set to $1/2^{20}$.

1.1   Contributions

We present a MAC mode, LightMAC, which enables one to tag much longer messages than typically possible. LightMAC is depicted in Fig. 2 and Algorithm 1.

The security upper bound for LightMAC is

$$(1+\epsilon) \cdot \frac{q^2}{2^n}, \quad \text{where } \epsilon \in O\left(\frac{1}{2^{n/2}-1}\right),\tag{3}$$

which is independent of the message length (see Sect. 4). In other words, with a 32 bit block size, and setting the message-length parameter s to 16, roughly 64 messages can be tagged with length up to 2^{15} blocks. Note that keys are used most efficiently when the messages are as long as possible: up to $64 \cdot 2^{15} \cdot 32 = 2^{26}$ bits, or *8 MiB* of data can be tagged per key. LightMAC uses two independent keys, but even after normalizing by the number of keys, the amount of data processed per key is still 4 MiB, a significant improvement over 1 KiB.

Table 1. Supported block sizes are often small, and can be as low as 32 bits.

Block size (bits)	32	48	64	80	96	128	256
AES [15]						×	
CLEFIA [38]						×	
DESLX [27]			×				
Fantomas [19]						×	
HIGHT [23]			×				
ITUbee [26]				×			
KLEIN [18]			×				
KATAN [13]	×	×	×				
LBlock [42]			×				
LED [21]			×				
LEA [22]						×	
mCrypton [28]			×				
Mysterion [25]						×	×
Noekeon [14]						×	
Piccolo [37]			×				
PRESENT [11]			×				
PRIDE [1]			×				
PRINCE [12]			×				
RC5 [36]	×		×			×	
Rectangle [48]			×				
RoadRunneR [2]			×				
Robin [19]						×	
SEA [39]					×		
SIMECK [43]	×	×	×				
Simon [3]	×	×	×		×	×	
Speck [3]	×	×	×		×	×	
TWINE [40]			×				
XTEA [33]			×				
Zorro [17]						×	

Figure 1 compares LightMAC to the other published modes from Table 2. The figure shows that LightMAC starts with a factor 2^4 improvement over many of the modes, which grows to roughly 2^{10} as the number of queries increases. Modes such as PMAC with Parity and PMACX were designed to handle long message lengths and offer competitive bounds, at the cost of increased design complexity. LightMAC's advantage over these modes is its simplicity and low overhead.

Table 2. The table below contains the coefficients of the powers of ℓ contained in the security bounds for adversaries making q queries of length ℓ, with block size n bits. References are to papers proving the bounds. In the bound for EMAC, the function $d'(\ell)$ has been replaced by ℓ.

Mode	1	ℓ	ℓ^2	ℓ^3	ℓ^4
3kf9 [47]	$\frac{4q}{2^n}+\frac{4q^3}{2^{2n}}$	$\frac{4q}{2^n}+\frac{4q^3}{2^{2n}}$	$\frac{2q^3}{2^{2n}}$	$\frac{4q^3}{2^{2n}}$	
CBC-MAC [6]		$\frac{12q^2}{2^n}$			$\frac{64q^2}{2^{2n}}$
EMAC [6]		$\frac{q^2}{2^n}$			$\frac{32q^2}{2^{2n}}$
OMAC [31]		$\frac{5q^2}{2^n}$			$\frac{8q^2}{2^{2n}}$
PMAC [32]	$\frac{-3.5q^2}{2^n}$	$\frac{5q^2}{2^n}$			
PMAC_Plus [45]		$\frac{3q}{2^n}$		$\frac{27q^3}{2^{2n}}$	
PMACX [49] $_{(m=14,l=12)}$	$\frac{72+1.5q^2}{2^n}+\frac{576q^2}{2^{2n}}$	$\frac{576q^2}{2^{2n}}$	$\frac{144q^2}{2^{2n}}$		
PMAC with Parity [46]	$\frac{q^2}{2^n}$		$\frac{q^2}{2^{2n}}$		
Sum of CBCs [44]					$\frac{12q^3}{2^{2n}}$

Like PMAC [10], LightMAC allows block cipher calls to be made in parallel, but unlike PMAC, LightMAC is based on Bernstein's *protected counter sum* [8], and hence should not suffer from patent issues.

A disadvantage of LightMAC is that its rate is low. In order to tag messages of length up to $2^{n/2-1}$ blocks, $n/2$ bits of the block must be sacrificed for a counter, hence two block cipher calls must be called per block of data. However, the rate can be improved: if the maximum message length that will be communicated is known to be less than $2^s(n-s)$ bits, then the rate can be set to $(n-s)/n$ blocks per block cipher call. For example, using a 32 bit block cipher, if the message lengths are less than 2^9 blocks, then the rate can be set to 2/3 blocks per call. Therefore, unlike other modes, LightMAC can be optimized according to the application: the shorter the messages, the more efficient LightMAC is, while allowing the same number of message to be queried. Section 5 presents implementation results for LightMAC instantiated with the AES [15] and PRESENT [11], and discusses LightMAC's efficiency in more detail.

1.2   Related Work

In 1995, Bellare et al. [4] described the *XOR MACs*, which XORed together finite-input-length pseudorandom functions (PRF) to create stateful and randomized MACs. In 1999, Bernstein [8] introduced the protected counter sum, which composes an XOR MAC with an independent PRF call to create a stateless, deterministic MAC. In 2012, Yasuda [46] explained the basic idea for LightMAC in his paper's introduction, which can be viewed as an adaptation of Bernstein's protected counter sum using block ciphers.

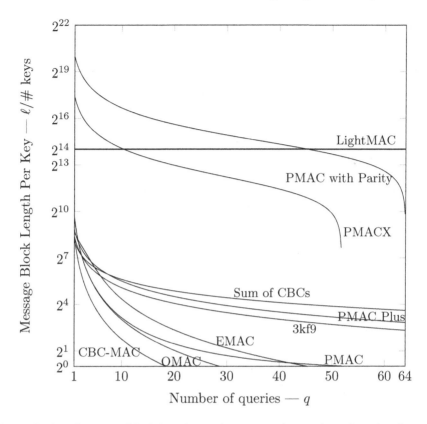

Fig. 1. A plot of message block lengths per key versus the number of queries that can be made in order to achieve the threshold success probability of 2^{-20}. In other words, if (x, y) is a point on the graph, then $x \cdot y$ represents the number of blocks that can be processed per key. The blocksize is set to 32 bits.

Another MAC algorithm designed for lightweight use is Chaskey [30]. The Chaskey paper includes a block cipher and a permutation mode, but both have bounds which deteriorate quadratically with respect to message length.

In certain cases the bounds in Table 2 can be improved. For example, for $\ell \leq 2^{n/8}$ and $q \geq \ell^2$, EMAC's bound becomes $\frac{16q^2}{2^n} + \frac{128q^2\ell^8}{2^{2n}}$ as shown by Pietrzak [34]. For the sum of CBCs, Yasuda [44] also showed that if $\ell \leq 2^{2n/5}$, the advantage becomes $\frac{40\ell^3 q^3}{2^{2n}}$.

2   Preliminaries

The set $\{0,1\}^n$ represents all bit-strings of length n; the set $\{0,1\}^{\leq n}$ is all bit-strings of length less than or equal to n. For two bit-strings A and B, we write $A\|B$ and AB interchangeably for the concatenation of A and B. Let r be an integer, then $M[1]M[2]\cdots M[\ell] \xleftarrow{r} M$ represents splitting M into r-bit blocks with the length of the last block, $M[\ell]$, being anywhere from zero to $r-1$ bits.

A block cipher is a function $E : \{0,1\}^k \times \{0,1\}^n \to \{0,1\}^n$ where $E(K,\cdot)$ defines a permutation for all $K \in \{0,1\}^k$. The integer n is the *block length* of E and we write $E_K(X)$ to mean $E(K,X)$. Given a block length n, concatenation of 10^* to a string means appending a one followed by the minimum number of zeros to make the total string length a multiple of n bits.

The symbol 0^n represents the n-bit string consisting of only zeros. Given a string A of length n, and an integer $t \le n$, then $\lfloor A \rfloor_t$ denotes the t least significant bits of A.

For an integer $1 \le i \le 2^s$, i_s represents some s-bit constant with the property that if $1 \le i < j \le 2^s$ then $i_s \ne j_s$. For example, i_s could be an s-bit representation of the integer i, or the ith s-bit Gray code.

3   LightMAC

Let $E : \{0,1\}^k \times \{0,1\}^n \to \{0,1\}^n$ be a block cipher. Let s and t be integers not greater than $n/2$ and n, respectively, and fix some representation for i_s (see Sect. 2). LightMAC accepts two independent and uniformly generated keys K_1 and K_2 from $\{0,1\}^k$, and a message M of length at most $2^s(n-s)$ bits. LightMAC produces an output of length t bits. Figure 2 and Algorithm 1 depict how the output is produced.

LightMAC can be used as either a pseudorandom function (PRF) or a MAC (see Sects. 4.2 and 4.3 for definitions). When used as a PRF, LightMAC is fully described by Algorithm 1. When used as a MAC, tags are generated using Algorithm 1, and verification of a message-tag pair (M,T) is done by comparing LightMAC (M) with T: if the two are equal, verification succeeds, otherwise not.

The parameters of LightMAC are the integers s and t, the representation of i_s, and the block cipher E, which implicity fixes k and n. The parameters must be agreed upon before a session starts, and remain constant during.

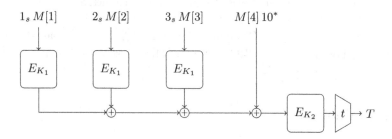

Fig. 2. LightMAC evaluated on a message $M[1]\,M[2]\,M[3]\,M[4] \xleftarrow{n-s} M$. The rounded squares represent block cipher calls and the trapezium is truncation to t bits.

Algorithm 1. LightMAC $_{K_1,K_2}(M)$

Input: $K_1, K_2 \in \{0,1\}^k$, $M \in \{0,1\}^{\leq 2^s(n-s)}$
Output: $T \in \{0,1\}^t$
1 $V \leftarrow 0^n \in \{0,1\}^n$
2 $M[1]M[2]\cdots M[\ell] \xleftarrow{n-s} M$
3 **for** $i = 1$ **to** $\ell - 1$ **do**
4 | $V \leftarrow V \oplus E_{K_1}(i_s\,M[i])$
5 **end**
6 $V \leftarrow V \oplus (M[\ell]\,10^*)$
7 $T \leftarrow \lfloor E_{K_2}(V) \rfloor_t$
8 **return** T

---

4 Security

Although Bellare, Guérin, and Rogaway [4] describe how to instantiate an XOR MAC using the Data Encryption Standard, they only provide proofs for pseudorandom functions, not pseudorandom permutations. Hence, even though the XOR MACs were proven to have bounds with no message length dependence, subsequent application of the PRP-PRF switching lemma would establish quadratic message length dependence. A similar explanation applies to the protected counter sum's security bound. Therefore a direct security proof is necessary for LightMAC.

The XOR MACs and protected counter sum did not exhibit any message length dependence because the XOR of independent, uniformly distributed random variables is still uniformly distributed. In this section we use the fact that roughly the same applies to the XOR of distinct block cipher outputs to achieve message length independence for LightMAC.

4.1 Block Cipher Security

The security of LightMAC is reduced to that of its underlying block cipher, that is, if an attack is found against LightMAC, then the attack can be reduced to an attack against the block cipher. The quality of the reduction is measured by the security bounds computed in Theorems 1 and 2.

The statements of the theorems include terms describing the quality of the underlying block cipher, which is measured as follows.

Definition 1. *Let $E : \mathsf{K} \times \mathsf{X} \to \mathsf{X}$ be a block cipher, and let π be a uniformly distributed random permutation over the set of permutations on X. Then the PRP-advantage against E of adversaries $\mathcal{A}$ making q queries and running in time τ is*

$$\mathsf{PRP}(q,\tau) := \sup_{A \in \mathcal{A}} \left| \mathbf{P}\left[A^{E_K} = 1\right] - \mathbf{P}\left[A^{\pi} = 1\right] \right|, \tag{4}$$

where $A^O = 1$ is the event that A outputs 1 when given access to oracle O, and K is uniformly distributed over K.

4.2   LightMAC as a PRF

A PRF $\Phi : \mathsf{K} \times \mathsf{M} \to \mathsf{T}$ is a construction which should be computationally indistinguishable from a uniformly distributed random function (URF), that is, a uniformly distributed random variable over the set of all functions from M to T. The quality of the PRF is measured via the PRF-advantage of adversaries.

Definition 2. *The PRF-advantage of an adversary A in distinguishing the PRF $\Phi : \mathsf{K} \times \mathsf{M} \to \mathsf{T}$ from the URF $\$: \mathsf{M} \to \mathsf{T}$ is*

$$\left| \mathbf{P}\left[A^{\Phi_K} = 1 \right] - \mathbf{P}\left[A^{\$} = 1 \right] \right|, \tag{5}$$

where $A^O = 1$ is the event that A outputs 1 when given access to oracle O, and K is uniformly distributed over K.

Theorem 1. *The PRF-advantage against LightMAC of any adversary running in time τ and making at most q queries of length at most $2^s (n-s)$ bits is bounded above by*

$$\left(1 + \frac{1}{2^{n/2} - 1} + \frac{1}{2(2^{n/2} - 1)^2}\right) \cdot \frac{q^2}{2^n} + \mathsf{PRP}(q \cdot (2^s - 1), \tau_1) + \mathsf{PRP}(q, \tau_2), \tag{6}$$

where n is the block size in bits, $\tau_1 \in \tau + O(q \cdot (2^s - 1))$, and $\tau_2 \in \tau + O(q)$.

Proof. Let A be a PRF-adversary against LightMAC running in time τ and making at most q queries of length at most $2^s (n - s)$ bits. Construct the PRP adversary B_1 against E_{K_1} as follows: B_1 simulates E_{K_2} by uniformly randomly choosing key K_2, runs A, and responds to A's queries using a combination of its own oracle and the simulated E_{K_2}; B_1 forwards A's response as its own. Construct the PRP adversary B_2 against E_{K_2} similarly. Then A's PRF-advantage against LightMAC is bounded above by

$$\alpha + \mathsf{PRP}(q \cdot (2^s - 1), \tau_1) + \mathsf{PRP}(q, \tau_2), \tag{7}$$

where α is A's PRF-advantage against LightMAC with its E_{K_1} and E_{K_2} calls replaced with π_1 and π_2 calls, respectively, where π_1 and π_2 are independent, uniformly distributed random permutations.

We replace π_2 with a uniformly distributed random function ϕ using the PRP-PRF switching lemma, at a cost of $q^2/2^{n+1}$ in advantage. The PRF we are left with is

$$\Phi(M) = \phi\left(M[\ell]10^* \oplus \bigoplus_{i=1}^{\ell-1} \pi_1(i_s M[i]) \right), \tag{8}$$

which is LightMAC instantiated with π_1 and ϕ, and

$$\alpha \leq \alpha' + \frac{q^2}{2^{n+1}}, \tag{9}$$

where α' is A's PRF-advantage against Φ.

Let F denote the function contained in the call to ϕ in Eq. 8. Then, as long as F's outputs are distinct, each input to ϕ is unique, meaning Φ will be indistinguishable from \$. In other words,

$$\alpha' \leq \sum_{i<j} \mathbf{P}\left[F(M_i) = F(M_j)\right] \leq \frac{q^2}{2} \max_{M_i \neq M_j} \mathbf{P}\left[F(M_i) = F(M_j)\right], \qquad (10)$$

where M_i for $i = 1, \ldots, q$ are the messages queried by A. The maximum on the right hand side is computed in Sect. 4.4, resulting in the bound

$$\alpha' \leq \frac{q^2}{2} \cdot \frac{1}{2^n - 2^{s+1} + 1}. \qquad (11)$$

Therefore, using the fact that $s \leq n/2$, we have

$$\alpha \leq \frac{q^2}{2^{n+1}} + \frac{q^2}{2} \cdot \frac{1}{2^n - 2^{s+1} + 1} \qquad (12)$$

$$\leq \frac{q^2}{2^n} \left(1 + \frac{1}{2^{n/2} - 1} + \frac{1}{2(2^{n/2} - 1)^2}\right), \qquad (13)$$

giving us our desired bound.   □

4.3 LightMAC as a MAC

A MAC consists of a tagging and a verification algorithm. The tagging algorithm accepts messages from some message set M and produces tags from a tag set T. The verification algorithm receives message-tag pairs (M, T) as input, and outputs 1 if the pair (M, T) is valid, and 0 otherwise. The insecurity of a MAC is measured as follows.

Definition 3. *Let A be an adversary with access to a MAC. The advantage of A in breaking the MAC is the probability that A is able to produce a message-tag pair (M, T) for which the verification algorithm outputs 1, where M has not been previously queried to the tagging algorithm.*

Theorem 2. *The MAC-advantage against LightMAC of any adversary running in time τ and making at most q tagging queries and v verification queries of length at most $2^s(n - s)$ bits, is bounded above by*

$$\left(1 + \frac{2}{2^{n/2} - 1} + \frac{1}{(2^{n/2} - 1)^2}\right) \cdot \left(\frac{q^2}{2^n} + \frac{v}{2^t}\right) +$$
$$\mathsf{PRP}(q \cdot (2^s - 1), \tau_1) + \mathsf{PRP}(q, \tau_2) + \mathsf{PRP}(v2^s, \tau_3), \qquad (14)$$

where n is the block size in bits, $\tau_1 \in \tau + O(q \cdot (2^s - 1))$, $\tau_2 \in \tau + O(q)$, and $\tau_3 \in \tau + O(v2^s)$.

Proof. We apply the same reduction as in the proof of Theorem 1 to replace LightMAC's E_{K_1} and E_{K_2} calls with π_1 and π_2 calls, respectively. As a MAC, LightMAC follows the hash-then-encrypt paradigm as described by Dodis and Pietrzak [16], with the function F from Sect. 4.4 as the "hash" part, hence applying Proposition 1 from their paper we get an upper bound of

$$\left(1 + \frac{2}{2^{n/2} - 1} + \frac{1}{(2^{n/2} - 1)^2}\right) \cdot \left(\frac{q^2}{2^n} + \frac{v}{2^t}\right). \tag{15}$$

$\square$

4.4   Collision Probability of F

Proposition 1. *Let* $m = 2^s(n - s)$. *Let* $M[1]M[2] \cdots M[\ell] \xleftarrow{n-s} M$ *for* $M \in \{0,1\}^{\leq m}$, *and define* F *to be*

$$F(M) = M[\ell]10^* \oplus \bigoplus_{i=1}^{\ell-1} \pi(i_s M[i]), \tag{16}$$

where π *is a uniformly distributed random permutation over* $\{0,1\}^n$, *then the probability that two distinct messages* $M_1, M_2 \in \{0,1\}^{\leq m}$ *collide is*

$$\mathbf{P}\left[F(M_1) = F(M_2)\right] \leq \frac{1}{2^n - \ell_1 - \ell_2 + 1}, \tag{17}$$

where ℓ_i *is the length of* M_i *in* $(n - s)$-*bit blocks rounded up.*

Proof. The equation $F(M_1) = F(M_2)$ can be rewritten as

$$\bigoplus_{i=1}^{\ell_1} \pi(i_s M_1[i]) \oplus \bigoplus_{i=1}^{\ell_2} \pi(i_s M_2[i]) = M_1[\ell_1]10^* \oplus M_2[\ell_2]10^*. \tag{18}$$

Since $M_1 \neq M_2$ there are two cases:

1. $\ell_1 = \ell_2$, $M_1[\ell_1]10^* \neq M_2[\ell_2]10^*$, and $M_1[i] = M_2[i]$ for all i, or
2. either $\ell_1 \neq \ell_2$ or there exists an i such that $M_1[i] \neq M_2[i]$.

In the first case there is no collision, hence we focus on the second case. Without loss of generality we can assume that $M_1[i] \neq M_2[i]$ for all i, and we can simplify the problem to calculating the probability that

$$\bigoplus_{i=1}^{\ell} \pi(x_i) = c, \tag{19}$$

where $\ell = \ell_1 + \ell_2$, $c = M_1[\ell_1]10^* \oplus M_2[\ell_2]10^*$, and $x_i \neq x_j$ for $i \neq j$.

Let $N = 2^n$, then $\mathbf{P}\left[\bigoplus_{i=1}^{\ell} \pi(x_i) = c\right]$ equals

$$\frac{1}{N!}\left|\left\{y_1,\ldots,y_N \left| \bigoplus_{i=1}^{\ell} y_i = c \text{ and } y_i \neq y_j \text{ for } i \neq j\right.\right\}\right|. \tag{20}$$

By Lemma 1 we have that the probability is bounded above by $N!/(N-\ell+1)$, giving us our desired result.   □

Lemma 1. *Let* $c \in \{0,1\}^n$ *and let* $N = 2^n$. *The number of sequences* $(y_1, y_2, \ldots, y_N) \in (\{0,1\}^n)^N$ *with* $y_i \neq y_j$ *for* $i \neq j$ *such that*

$$\bigoplus_{i=1}^{\ell} y_i = c, \tag{21}$$

is not greater than $N!/(N - \ell + 1)$.

Proof. We start by fixing y_1, for which there are N possibilities. Since y_2 cannot equal y_1, there are $N - 1$ possibilities for y_2. Continuing this way, we have that there are $N - i$ possibilities for y_{i+1}, with $i \leq \ell - 2$. For y_ℓ there is at most one possibility, namely $c \oplus y_1 \oplus y_2 \oplus \cdots y_{\ell-1}$. All y_j for $j > \ell$ must be distinct from all preceding y_i, hence in total there are at most

$$N \cdot (N-1) \cdot \cdots \cdot (N - \ell + 2) \cdot (N - \ell)! = \frac{N!}{N - \ell + 1} \tag{22}$$

possible sequences.   □

5   Implementation

In this section, we discuss the implementation characteristics of LightMAC and compare it to the serial two-key CBC-MAC with last block encryption, EMAC [6], and to PMAC with Parity (PMAC/P) [46], which provides a parallelizable rate 2/3 construction and can be considered its main competitor.

5.1   Implementation Characteristics of LightMAC

LightMAC is a mode with very low overhead: besides the block cipher calls, it only requires an s-bit counter generator and one additional n-bit state for summing the block cipher outputs.

This means that the code size (for embedded software or microcontrollers) and area requirements (for hardware implementations) of LightMAC can be estimated as roughly equivalent to CBC-MAC with encryption of the last block by a second key. Compared to PMAC with Parity, LightMAC uses only two keys instead of four. In comparison to all PMAC variants, the absence of finite field doubling further improves its implementation characteristics on embedded platforms or hardware.

In terms of throughput, a compact serial implementation of LightMAC will give a performance of about $n/(n-s)$ block cipher call equivalents per message block of $n-s$ bits, which means that the serial performance of LightMAC on a given platform can readily be evaluated based on the performance of the best available implementation of the chosen underlying block cipher. Except for very short messages, the overhead imposed by the final block cipher call is negligible.

Like PMAC and its derivatives, LightMAC has the advantage that the individual block cipher calls can be parallelized. While this is typically less important on lightweight platforms, where compactness and power/energy consumption are the prime concerns, this property enables high-performance implementations for the server side: since exactly the same lightweight algorithms used on small devices will also have to be used by the servers communicating with them, they should ideally also have good implementation characteristics in high-performance software environments. The importance of this was for instance pointed out in [29]. Many lightweight algorithms and modes of operation are inherently serial in nature and therefore inefficient in software. Our implementation study therefore focuses on this scenario.

5.2 The Setting

We explore the high-performance parallel software implementation possibilities for LightMAC, with the following choices regarding platform and instantiation parameters:

Underlying Block Ciphers. We use the block ciphers PRESENT [11] and AES [15] for our implementations. PRESENT is a lightweight 64-bit block cipher that was recently standardised by ISO, and AES serves as a baseline.

Choice of s and t. We always use full tag lengths $t = n$, meaning 64-bit tags for PRESENT and 128-bit tags for AES. We furthermore instantiate LightMAC with the following values of s:

1. $s = n/2$ for the maximum supported message length (and correspondingly lowest rate $1/2$);
2. $s = n/3$, rounded to the nearest multiple of 8, for a mode with rate $2/3$;
3. $s = 8$, for a short maximum message length with the highest rate $(1 - 8/n)$.

Altogether, these parameter choices illustrate a wide spectrum of use cases.

Platform. We implement LightMAC on Intel's recent Skylake microarchitecture, using the 256-bit AVX2 instruction set. PRESENT was implemented in a bitsliced fashion processing 8 blocks in parallel. Other implementation strategies are known to yield a significantly lower performance, see [7] for a comprehensive study. For the AES, the AES-NI instruction set [20] was used. The key scheduling was precomputed for both ciphers. Since byte-aligned s-bit addition is inexpensive on this platform, the counters i_s are implemented as the s-bit representation of the integer i.

Message Lengths. We provide performance data for all message lengths of $\ell = 2^b$ bytes, with $7 \leq b \leq 13$, wherever $8\ell \leq 2^s(n - s)$.

5.3   Performance Measurements

All measurements were taken on a single core of an Intel Core i7-6700 CPU at 3.4 GHz with Turbo Boost disabled, and averaged over 200000 repetitions. The performance of the block ciphers AES and PRESENT, both in serial and parallel implementations, is provided as a reference point in Table 3. Our findings on the performance of LightMAC and related MACs are summarised in Table 4. All performance numbers are given in cycles per byte (cpb).

Table 3. Baseline performance of ciphers PRESENT and AES on Skylake (AVX2, AES-NI).

Block cipher	Encryption [cycles/byte]	Key schedule [cycles]
PRESENT (table-based)	57.83	353
PRESENT (8 blocks bitsliced)	11.23	790
AES (AES-NI, serial)	2.57	116
AES (AES-NI, pipelined)	0.63	116

Table 4. Software performance of LightMAC, EMAC and PMAC with Parity (PMAC/P), instantiated with PRESENT and AES on the Intel Skylake platform (AVX2, AES-NI). All numbers are given in cycles per byte (cpb). Data is provided for message lengths smaller than $2^s(n - s)$ bits.

			Message length (bytes)						
Algorithm	s	Rate	128	256	512	1024	2048	4096	8192
EMAC-PRESENT	–	1	63.02	61.21	60.28	59.80	59.57	59.41	59.32
PMAC/P-PRESENT	–	2/3	39.62	32.44	28.82	27.07	26.48	26.14	26.00
LightMAC-PRESENT	32	1/2	25.50	23.67	22.75	22.32	22.08	21.97	21.92
LightMAC-PRESENT	24	2/3	25.70	21.21	20.17	19.03	18.09	17.80	17.80
LightMAC-PRESENT	8	7/8	20.31	18.34	14.65	13.48	–	–	–
EMAC-AES	–	1	3.42	3.19	3.03	2.91	2.74	2.68	2.67
PMAC/P-AES	–	2/3	1.53	1.48	1.33	1.24	1.17	1.15	1.14
LightMAC-AES	64	1/2	1.33	1.29	1.27	1.26	1.26	1.26	1.25
LightMAC-AES	40	2/3	1.37	1.31	1.12	1.04	0.95	0.95	0.92
LightMAC-AES	8	15/16	1.38	1.00	0.82	0.80	0.72	–	–

Discussion. One can observe that with both PRESENT and the AES as the underlying block ciphers, LightMAC provides a performance of about the inverse of its rate times the baseline block cipher speed. This confirms that LightMAC imposes very low overhead in addition to the block cipher invocations.

In contrast to the serial EMAC, LightMAC provides significantly greater performance despite featuring a smaller rate. This demonstrates the advantage of parallelisability over a sequential algorithm.

Comparing the LightMAC instantiations with rate 2/3 to PMAC with Parity (PMAC/P), we note that the use of the same key throughout the message processing (as opposed to three different keys in PMAC/P) significantly improves the performance for the PRESENT-based implementation: LightMAC is consistently around 50 % faster. This is largely due to the fact that the parts of each subkey of PMAC/P's three bitsliced keys have to be interleaved in an appropriate way. The effect is less pronounced for the AES where no conversion to bitsliced format is needed, and due to the AES-NI instructions which freely accept both registers and memory locations for the subkeys. Still, LightMAC is about 20 % faster, while additionally providing a flexible range of trade-offs between rate and maximum message length.

6   Conclusions

We proposed LightMAC, a new MAC mode of operation specifically suited to lightweight applications. Its security bound was shown in Sect. 4 to not depend on the message length, allowing an order of magnitude more data to be processed per key.

Featuring a simple design with very low overhead over the block cipher, it not only offers compact authentication for resource-constrained platforms, but also allows high-performance parallel implementations, as demonstrated by the implementation study of LightMAC instantiated with PRESENT and the AES in Sect. 5. Furthermore, the implementation results show how the s-parameter translates directly to a trade-off between rate and maximum message length.

Unlike PMAC and its many derivatives, LightMAC is not covered by patents. Altogether, this makes it a promising authentication solution for a wide range of platforms and use cases.

Acknowledgments. We would like to thank the various anonymous reviewers for providing useful comments. This work was supported in part by the Research Council KU Leuven: GOA TENSE (GOA/11/007). In addition, this work was supported by the Research Fund KU Leuven, OT/13/071. Atul Luykx is supported by a Ph.D. Fellowship from the Institute for the Promotion of Innovation through Science and Technology in Flanders (IWT-Vlaanderen).

References

1. Albrecht, M.R., Driessen, B., Kavun, E.B., Leander, G., Paar, C., Yalçın, T.: Block ciphers – focus on the linear layer (feat. PRIDE). In: Garay, J.A., Gennaro, R. (eds.) CRYPTO 2014, Part I. LNCS, vol. 8616, pp. 57–76. Springer, Heidelberg (2014). doi:10.1007/978-3-662-44371-2_4

2. Baysal, A., Sahin, S.: RoadRunneR: a small and fast bitslice block cipher for low cost 8-bit processors. In: Güneysu, T., Leander, G., Moradi, A. (eds.) Light-Sec 2015. LNCS, vol. 9542, pp. 58–76. Springer, Heidelberg (2016). doi:10.1007/978-3-319-29078-2_4

3. Beaulieu, R., Shors, D., Smith, J., Treatman-Clark, S., Weeks, B., Wingers, L.: The SIMON and SPECK families of lightweight block ciphers. Cryptology ePrint Archive, Report 2013/404 (2013). http://eprint.iacr.org/

4. Bellare, M., Guérin, R., Rogaway, P.: XOR MACs: new methods for message authentication using finite pseudorandom functions. In: Coppersmith, D. (ed.) CRYPTO 1995. LNCS, vol. 963, pp. 15–28. Springer, Heidelberg (1995). doi:10.1007/3-540-44750-4_2

5. Bellare, M., Kilian, J., Rogaway, P.: The security of cipher block chaining. In: Desmedt, Y.G. (ed.) CRYPTO 1994. LNCS, vol. 839, pp. 341–358. Springer, Heidelberg (1994). doi:10.1007/3-540-48658-5_32

6. Bellare, M., Pietrzak, K., Rogaway, P.: Improved security analyses for CBC MACs. In: Shoup, V. (ed.) CRYPTO 2005. LNCS, vol. 3621, pp. 527–545. Springer, Heidelberg (2005). doi:10.1007/11535218_32

7. Benadjila, R., Guo, J., Lomné, V., Peyrin, T.: Implementing lightweight block ciphers on x86 architectures. In: Lange, T., Lauter, K., Lisoněk, P. (eds.) SAC 2013. LNCS, vol. 8282, pp. 324–352. Springer, Heidelberg (2014). doi:10.1007/978-3-662-43414-7_17

8. Bernstein, D.J.: How to stretch random functions: the security of protected counter sums. J. Cryptology **12**(3), 185–192 (1999). doi:10.1007/s001459900051

9. Biryukov, A. (ed.): FSE 2007. LNCS, vol. 4593. Springer, Heidelberg (2007)

10. Black, J.A., Rogaway, P.: A block-cipher mode of operation for parallelizable message authentication. In: Knudsen, L.R. (ed.) EUROCRYPT 2002. LNCS, vol. 2332, pp. 384–397. Springer, Heidelberg (2002). doi:10.1007/3-540-46035-7_25

11. Bogdanov, A.A., Knudsen, L.R., Leander, G., Paar, C., Poschmann, A., Robshaw, M., Seurin, Y., Vikkelsoe, C.: PRESENT: an ultra-lightweight block cipher. In: Paillier, P., Verbauwhede, I. (eds.) CHES 2007. LNCS, vol. 4727, pp. 450–466. Springer, Heidelberg (2007). doi:10.1007/978-3-540-74735-2_31

12. Borghoff, J., Canteaut, A., Güneysu, T., Kavun, E.B., Knezevic, M., Knudsen, L.R., Leander, G., Nikov, V., Paar, C., Rechberger, C., Rombouts, P., Thomsen, S.S., Yalçin, T.: PRINCE - a low-latency block cipher for pervasive computing applications - extended abstract. In: Wang, X., Sako, K. (eds.) [41], pp. 208–225. http://dx.org/10.1007/978-3-642-34961-4_14

13. De Cannière, C., Dunkelman, O., Knežević, M.: KATAN and KTANTAN — a family of small and efficient hardware-oriented block ciphers. In: Clavier, C., Gaj, K. (eds.) CHES 2009. LNCS, vol. 5747, pp. 272–288. Springer, Heidelberg (2009). doi:10.1007/978-3-642-04138-9_20

14. Daemen, J., Peeters, M., Van Assche, G., Rijmen, V.: Nessie proposal: Noekeon. In: First Open Nessie Workshop (2000)

15. Daemen, J., Rijmen, V.: AES proposal: Rijndael. In: First Advanced Encryption Standard (AES) Conference (1998)

16. Dodis, Y., Pietrzak, K.: Improving the security of MACs via randomized message preprocessing. In: Biryukov, A. (ed.) [9], pp. 414–433. http://dx.org/10.1007/978-3-540-74619-5_26

17. Gérard, B., Grosso, V., Naya-Plasencia, M., Standaert, F.-X.: Block ciphers that are easier to mask: how far can we go? In: Bertoni, G., Coron, J.-S. (eds.) CHES 2013. LNCS, vol. 8086, pp. 383–399. Springer, Heidelberg (2013). doi:10.1007/978-3-642-40349-1_22

18. Gong, Z., Nikova, S., Law, Y.W.: KLEIN: a new family of lightweight block ciphers. In: Juels, A., Paar, C. (eds.) RFIDSec 2011. LNCS, vol. 7055, pp. 1–18. Springer, Heidelberg (2012). doi:10.1007/978-3-642-25286-0_1

19. Grosso, V., Leurent, G., Standaert, F.-X., Varıcı, K.: LS-designs: bitslice encryption for efficient masked software implementations. In: Cid, C., Rechberger, C. (eds.) FSE 2014. LNCS, vol. 8540, pp. 18–37. Springer, Heidelberg (2015). doi:10.1007/978-3-662-46706-0_2

20. Gueron, S.: Intel Advanced Encryption Standard (AES) Instructions Set. Intel White paper, September 2012

21. Guo, J., Peyrin, T., Poschmann, A., Robshaw, M.J.B.: The LED block cipher. In: Preneel, B., Takagi, T. (eds.) [35], pp. 326–341. http://dx.org/10.1007/978-3-642-23951-9_22

22. Hong, D., Lee, J.-K., Kim, D.-C., Kwon, D., Ryu, K.H., Lee, D.-G.: LEA: a 128-bit block cipher for fast encryption on common processors. In: Kim, Y., Lee, H., Perrig, A. (eds.) WISA 2013. LNCS, vol. 8267, pp. 1–24. Springer, Heidelberg (2014). doi:10.1007/978-3-319-05149-9_1

23. Hong, D., Sung, J., Hong, S.H., Lim, J.-I., Lee, S.-J., Koo, B.-S., Lee, C.-H., Chang, D., Lee, J., Jeong, K., Kim, H., Kim, J.-S., Chee, S.: HIGHT: a new block cipher suitable for low-resource device. In: Goubin, L., Matsui, M. (eds.) CHES 2006. LNCS, vol. 4249, pp. 46–59. Springer, Heidelberg (2006). doi:10.1007/11894063_4

24. Iwata, T., Kurosawa, K.: Stronger security bounds for OMAC, TMAC, and XCBC. In: Johansson, T., Maitra, S. (eds.) INDOCRYPT 2003. LNCS, vol. 2904, pp. 402–415. Springer, Heidelberg (2003). doi:10.1007/978-3-540-24582-7_30

25. Journault, A., Standaert, F.X., Varici, K.: Improving the security and efficiency of block ciphers based on LS-designs. In: Proceedings of the 9th International Workshop on Coding and Cryptography, WCC 2015 (2015)

26. Karakoç, F., Demirci, H., Harmancı, A.E.: ITUbee: a software oriented lightweight block cipher. In: Avoine, G., Kara, O. (eds.) LightSec 2013. LNCS, vol. 8162, pp. 16–27. Springer, Heidelberg (2013). doi:10.1007/978-3-642-40392-7_2

27. Leander, G., Paar, C., Poschmann, A., Schramm, K.: New lightweight DES variants. In: Biryukov, A. (ed.) [9], pp. 196–210. http://dx.org/10.1007/978-3-540-74619-5_13

28. Lim, C.H., Korkishko, T.: mCrypton – a lightweight block cipher for security of low-cost RFID tags and sensors. In: Song, J.-S., Kwon, T., Yung, M. (eds.) WISA 2005. LNCS, vol. 3786, pp. 243–258. Springer, Heidelberg (2006). doi:10.1007/11604938_19

29. Matsuda, S., Moriai, S.: Lightweight cryptography for the cloud: exploit the power of bitslice implementation. In: Prouff, E., Schaumont, P. (eds.) CHES 2012. LNCS, vol. 7428, pp. 408–425. Springer, Heidelberg (2012). doi:10.1007/978-3-642-33027-8_24

30. Mouha, N., Mennink, B., Van Herrewege, A., Watanabe, D., Preneel, B., Verbauwhede, I.: Chaskey: an efficient MAC algorithm for 32-bit microcontrollers. In: Joux, A., Youssef, A. (eds.) SAC 2014. LNCS, vol. 8781, pp. 306–323. Springer, Heidelberg (2014). doi:10.1007/978-3-319-13051-4_19

31. Nandi, M.: Improved security analysis for OMAC as a pseudorandom function. J. Math. Cryptology 3(2), 133–148 (2009)

32. Nandi, M., Mandal, A.: Improved security analysis of PMAC. J. Math. Cryptology 2(2), 149–162 (2008)

33. Needham, R.M., Wheeler, D.J.: Tea extensions. Computer Laboratory, University of Cambridge (Technical report), October 1997. http://www.cix.co.uk/~klockstone/xtea.pdf.

34. Pietrzak, K.: A tight bound for EMAC. In: Bugliesi, M., Preneel, B., Sassone, V., Wegener, I. (eds.) ICALP 2006. LNCS, vol. 4052, pp. 168–179. Springer, Heidelberg (2006). doi:10.1007/11787006_15

35. Preneel, B., Takagi, T. (eds.): CHES 2011. LNCS, vol. 6917. Springer, Heidelberg (2011). doi:10.1007/978-3-642-23951-9

36. Rivest, R.L.: The RC5 encryption algorithm. In: Preneel, B. (ed.) FSE 1994. LNCS, vol. 1008, pp. 86–96. Springer, Heidelberg (1995). doi:10.1007/3-540-60590-8_7

37. Shibutani, K., Isobe, T., Hiwatari, H., Mitsuda, A., Akishita, T., Shirai, T.: Piccolo: an ultra-lightweight blockcipher. In: Preneel, B., Takagi, T. (eds.) [35], pp. 342–357. http://dx.org/10.1007/978-3-642-23951-9_23

38. Shirai, T., Shibutani, K., Akishita, T., Moriai, S., Iwata, T.: The 128-bit blockcipher CLEFIA (extended abstract). In: Biryukov, A. (ed.) [9], pp. 181–195. http://dx.org/10.1007/978-3-540-74619-5_12

39. Standaert, F.-X., Piret, G., Gershenfeld, N., Quisquater, J.-J.: SEA: a scalable encryption algorithm for small embedded applications. In: Domingo-Ferrer, J., Posegga, J., Schreckling, D. (eds.) CARDIS 2006. LNCS, vol. 3928, pp. 222–236. Springer, Heidelberg (2006). doi:10.1007/11733447_16

40. Suzaki, T., Minematsu, K., Morioka, S., Kobayashi, E.: TWINE: a lightweight block cipher for multiple platforms. In: Knudsen, L.R., Wu, H. (eds.) SAC 2012. LNCS, vol. 7707, pp. 339–354. Springer, Heidelberg (2013). doi:10.1007/978-3-642-35999-6_22

41. Wang, X., Sako, K. (eds.): ASIACRYPT 2012. LNCS, vol. 7658. Springer, Heidelberg (2012). doi:10.1007/978-3-642-34961-4

42. Wu, W., Zhang, L.: LBlock: a lightweight block cipher. In: Lopez, J., Tsudik, G. (eds.) ACNS 2011. LNCS, vol. 6715, pp. 327–344. Springer, Heidelberg (2011). doi:10.1007/978-3-642-21554-4_19

43. Yang, G., Zhu, B., Suder, V., Aagaard, M.D., Gong, G.: The simeck family of lightweight block ciphers. In: Güneysu, T., Handschuh, H. (eds.) CHES 2015. LNCS, vol. 9293, pp. 307–329. Springer, Heidelberg (2015). doi:10.1007/978-3-662-48324-4_16

44. Yasuda, K.: The sum of CBC MACs is a secure PRF. In: Pieprzyk, J. (ed.) CT-RSA 2010. LNCS, vol. 5985, pp. 366–381. Springer, Heidelberg (2010). doi:10.1007/978-3-642-11925-5_25

45. Yasuda, K.: A new variant of PMAC: beyond the birthday bound. In: Rogaway, P. (ed.) CRYPTO 2011. LNCS, vol. 6841, pp. 596–609. Springer, Heidelberg (2011). doi:10.1007/978-3-642-22792-9_34

46. Yasuda, K.: PMAC with parity: minimizing the query-length influence. In: Dunkelman, O. (ed.) CT-RSA 2012. LNCS, vol. 7178, pp. 203–214. Springer, Heidelberg (2012). doi:10.1007/978-3-642-27954-6_13

47. Zhang, L., Wu, W., Sui, H., Wang, P.: 3kf9: enhancing 3GPP-MAC beyond the birthday bound. In: Wang, X., Sako, K. (eds.) [41], pp. 296–312. http://dx.org/10.1007/978-3-642-34961-4_19

48. Zhang, W., Bao, Z., Lin, D., Rijmen, V., Yang, B., Verbauwhede, I.: RECTANGLE: A Bit-slice Lightweight Block Cipher Suitable for Multiple Platforms. Cryptology ePrint Archive, Report 2014/084 (2014). http://eprint.iacr.org/

49. Zhang, Y.: Using an error-correction code for fast, beyond-birthday-bound authentication. In: Nyberg, K. (ed.) CT-RSA 2015. LNCS, vol. 9048, pp. 291–307. Springer, Heidelberg (2015). doi:10.1007/978-3-319-16715-2_16

Stream-Cipher Cryptanalysis

Cryptanalysis of the Full Spritz Stream Cipher

Subhadeep Banik[1,2]([✉]) and Takanori Isobe[3]

[1] DTU Compute, Technical University of Denmark, Lyngby, Denmark
subb@dtu.dk
[2] Temasek Labs, Nanyang Technological University, Singapore, Singapore
[3] Sony Corporation, Tokyo, Japan
takanori.isobe@jp.sony.com

Abstract. Spritz is a stream cipher proposed by Rivest and Schuldt at the rump session of CRYPTO 2014. It is intended to be a replacement of the popular RC4 stream cipher. In this paper we propose distinguishing attacks on the full Spritz, based on *a short-term bias* in the first two bytes of a keystream and *a long-term bias* in the first two bytes of every cycle of N keystream bytes, where N is the size of the internal permutation. Our attacks are able to distinguish a keystream of the *full* Spritz from a random sequence with samples of first two bytes produced by $2^{44.8}$ multiple key-IV pairs or $2^{60.8}$ keystream bytes produced by a single key-IV pair. These biases are also useful in the event of plaintext recovery in a broadcast attack. In the second part of the paper, we look at a state recovery attack on Spritz, in a special situation when the cipher enters a class of weak states. We determine the probability of encountering such a state, and demonstrate a state recovery algorithm that betters the 2^{1400} step algorithm of Ankele et al. at Latincrypt 2015.

Keywords: RC4 · Spritz · Stream cipher · Short-term bias · Long-term bias · Distinguishing attack · Plaintext recovery attack · State recovery attack

1 Introduction

RC4, designed by Rivest in 1987, is still one of most widely used stream ciphers in the world. It is adopted in many software applications and standard protocols such as SSL/TLS, WEP, Microsoft Lotus and Oracle secure SQL. After the disclosure of its algorithm in 1994, RC4 has attracted intensive cryptanalytic efforts over past 20 years. Finally, in 2013, practical plaintext recovery attacks on RC4 in SSL/TLS were proposed by AlFardan et al. [1] and Isobe et al. [9]. In the response to these results, usage of RC4 has drastically decreased, especially in TLS, and major companies such as Google, Microsoft, and Mozilla announced that they will officially remove the RC4 from web browsers by early 2016.

At the same time, there has been extensive research in recent years to come up with RC4-like stream ciphers that while marginally slower in software, would wipe out the known shortcomings of RC4. Many such ciphers like RC4A [17],

© International Association for Cryptologic Research 2016
T. Peyrin (Ed.): FSE 2016, LNCS 9783, pp. 63–77, 2016.
DOI: 10.1007/978-3-662-52993-5_4

NGG [14], GGHN [8], Quad-RC4 [16], RC4+ [10] and VMPC [24] have been proposed to fulfil this objective. However, all the aforementioned ciphers have had distinguishing attacks reported against them [3–5,12,18,20,21]. Spritz [19] is a stream cipher proposed by Rivest and Schuldt at the rump session of CRYPTO 2014. The authors intended Spritz to be a replacement for RC4, and hence the design for Spritz was chosen meticulously, with special attention given to the fact that known weaknesses of RC4 [11,13] do not carry over. The authors automatically examined many thousands of candidates to obtain cryptographically secure update functions and an estimated 5 "core-months" of CPU time were used in the statistical experiments performed by them. Their experiments suggested that 2^{81} samples were required to distinguish the output of Spritz from random.

1.1 Description of Spritz

Spritz consists of a permutation S over the set $\{0, 1, 2, \ldots, N-1\}$ (default value of N is 256) and six pointers i, j, k, w, a, z, where i, j, k are index pointers, w gives the step distance for i, a is a nibble counter, and z stores the output byte. The design specifies a number of modules that are executed for producing a keystream as defined in Fig. 1. The authors specify a number of modes of operation using the Spritz structure like a stream cipher, hash function, MAC etc. In the stream cipher mode of operation the keystream is produced in the following manner. First the permutation is initialized using the INITIALIZESTATE(N) routine. The secret key K is then absorbed into the state using the ABSORB(K) module. Additionally, if an IV is to be used, then the ABSORBSTOP() module is invoked and the IV is absorbed by calling the ABSORB(IV) function. Thereafter, the SQUEEZE module is invoked to produce keystream bytes.

1.2 Previous Work

The only published work on cryptanalysis of Spritz is presented in [2]. The authors tackle the problem of state recovery using three different approaches. The best algorithm they propose theoretically recovers the internal permutation used in Spritz in 2^{1400} steps. Additionally, in [23], the author proposed a distinguisher for a scaled down version of Spritz ($N = 8$). It was observed that the event $Z_i = Z_{i+2}$ was biased. However, the bias was not theoretically proven and no analogous result for the full Spritz ($N = 256$) was proposed.

1.3 Our Contribution and Organization

In this paper, we first show *a short-term bias* which is present in the first two bytes of a keystream and *a long-term bias* which appears in the first two bytes of every cycle of N keystream bytes. We theoretically prove that these biases exist in a keystream of Spritz regardless of the value of N. Based on these biases, we propose distinguishing attacks on the full Spritz ($N = 256$). Our attacks are able to distinguish a keystream of the full Spritz from a random sequence

INITIALIZESTATE(N)

1. $i = j = k = a = z = 0,\ w = 1$.
2. **for** $v \rightarrow 0$ to $N - 1$
   $S[v] = v$

ABSORB(I)

1. **for** $v \rightarrow 0$ to $I.length - 1$
   ABSORBBYTE($I[v]$)

ABSORBBYTE(b)

1. ABSORBNIBBLE($low(b)$)
2. ABSORBNIBBLE($high(b)$)

ABSORBNIBBLE(x)

1. **if** $a = \lfloor \frac{N}{2} \rfloor$
   SHUFFLE()
2. SWAP($S[a], S[\lfloor N/2 \rfloor + x]$)
3. $a = a + 1$

ABSORBSTOP()

1. **if** $a = \lfloor \frac{N}{2} \rfloor$
   SHUFFLE()
2. $a = a + 1$

SHUFFLE()

1. WHIP($2N$)
2. CRUSH()
3. WHIP($2N$)
4. CRUSH()
5. WHIP($2N$)
6. $a = 0$

WHIP(r)

1. **for** $v \rightarrow 0$ to $r - 1$
   UPDATE()
2. **do** $w = w + 1$
   until $gcd(w, N) = 1$

CRUSH()

1. **for** $v \rightarrow 0$ to $\lfloor N/2 \rfloor - 1$
   if $S[v] > S[N - 1 - v]$
   SWAP($S[v], S[N - 1 - v]$)

SQUEEZE(r)

1. **if** $a > 0$
   SHUFFLE()
2. $P = Array.New(r)$
3. **for** $v \rightarrow 0$ to $r - 1$
   $P[v] = $ DRIP()
4. **return** P

DRIP()

1. **if** $a > 0$
   SHUFFLE()
2. UPDATE()
3. **return** OUTPUT()

UPDATE()

1. $i = i + w$
2. $j = k + S[j + S[i]]$
3. $k = i + k + S[j]$
4. SWAP($S[i], S[j]$)

OUTPUT()

1. $z = S[j + S[i + S[z + k]]]$
2. **return** z

Fig. 1. Modules for Spritz. When N is a power of 2, the last two lines of WHIP are equivalent to $w = w + 2$.

Table 1. Summary of results on Spritz

	Type of attack	Complexity	Reference
1	Distinguishing attack on scaled down version ($N = 8$)	$2^{21.9}$ outputs	[23]
2	Distinguishing attack on full Spritz in multiple key-IV setting	$2^{44.8}$ outputs	Sect. 2
3	Distinguishing attack on full Spritz in single key-IV setting	$2^{60.8}$ outputs	Sect. 2
4	State recovery attack	2^{1400} steps	[2]
		2^{1247} steps	Sect. 3

with samples of first two bytes produced by $2^{44.8}$ multiple key-IV pairs or $2^{60.8}$ keystream bytes produced by a single key-IV pair. These biases are applicable to a plaintext recovery attack in a broadcast setting and multi-session setting in SSL/TLS.

Thereafter we show that under certain conditions, Spritz enters a weak class of states, during which, the odd and even elements of the permutation are never swapped with each other. In this case, the sequence constructed with the last bit of every keystream byte becomes periodic with period equal to 4. We show that in such an event, a state recovery attack on Spritz is more efficient and improves upon the 2^{1400} step algorithm proposed in [2]. Table 1 shows the summary of our results.

In Sect. 2, we will present the distinguisher on Spritz and study a few of its implications. In Sect. 3, we will present our state recovery attack on Spritz. Section 4 concludes the paper.

2   Distinguishing Attacks on Spritz

Before we proceed to outline the details of the distinguisher, let us present a few observations on how the various index pointers are used when Spritz is operated in the stream cipher mode. Note that when Spritz is used in the stream cipher mode: the sequence of execution of modules is

A. ABSORB(K)
B. ABSORBSTOP(), ABSORB(IV) **(optional, only if *IV* is used)**
C. SQUEEZE().

1. In the ABSORB(K) (and also ABSORB(IV)) phase, the internal permutation is swapped according to the nibble values of the key (IV). During this phase the index a is used only to keep track of the number of nibbles currently absorbed in the permutation. After the ABSORB phase, the index a plays no further role in the SQUEEZE phase when the cipher starts producing keystream bytes.
2. The index w, which is used to increment the index i, is constant during the SQUEEZE phase. The value of this index does not depend on the secret key, and hence is not secret. Its value can be deduced from the length of the secret

key and IV. If the length of key is limited to $\lfloor N/4 \rfloor$ bytes, and no IV is used, then the SHUFFLE procedure is executed only once. In that case, the value of w during the SQUEEZE phase is 7.

3. If the length of the Key is more than $\lfloor N/4 \rfloor$ bytes the value of w can be deduced by examining the number of times the SHUFFLE module has been called during the ABSORB phases. For example, if $N = 256$, and a Key of size 80 bytes, the SHUFFLE procedure gets called twice, at the end of the 64th byte and at the beginning of SQUEEZE. Each SHUFFLE call increases the value of w by 6 and so the value of w during the keystream generation is $1 + 6 + 6 = 13$.

4. The value of the index i at the beginning of the SQUEEZE phase is always 0, whatever be the the the size of the Key and IV used in the ABSORB phases. This is because whenever $\lfloor N/4 \rfloor$ bytes get absorbed, the value of the pointers i, j, k are altered by call to the SHUFFLE module. Each SHUFFLE module calls the WHIP($2N$) module thrice. Each WHIP module in turn updates i using the rule $i = i + w$ a total of $2N$ times. Whatever be the actual value of w, at the end of the any call to the WHIP module, the updated value of $i = 0 + 2wN \equiv 0 \bmod N$. And so the value of i remains 0 going in and out of the WHIP executions and hence also the SHUFFLE module.

5. The only indices that change during the SQUEEZE phase is i, j, k, z.

6. The sequence of updates during the SQUEEZE phase is therefore given as:
   (a) $i = i + w$
   (b) $j = k + S[j + S[i]]$
   (c) $k = k + i + S[j]$
   (d) SWAP $(S[i], S[j])$
   (e) **return** $z = S[j + S[i + S[z + k]]]$

2.1 Bias in First Two Output Bytes of a Keystream

We first prove that the first two output bytes produced by the Spritz stream cipher are biased towards the tuple $(-w, -w)$. For example, if $N = 256$, and if a 64 byte key is used, then $w = 7$, and then the first 2 bytes are biased towards the value $(249, 249)$.

Theorem 1. *The first two output bytes Z_1 and Z_2 produced by the Spritz stream cipher are biased towards $(-w, -w)$. The probability of this event is given by* $\Pr[Z_1 = Z_2 = -w] = \frac{1}{N^2} + \frac{3}{N^4}$.

Proof. We outline three mutually exclusive events **I, II** and **III**, each of which occurs with probability $\frac{1}{N^4}$, that guarantees that the first two output bytes produced by the cipher are both equal to $-w$. Each of the three events are denoted by the states of the permutation and the values of the index pointers before the beginning of the SQUEEZE phase.

I. $S[w] = -w$, $S[2w] = 0, k = 0, S[j - w] = 2w$
II. $k = 2w$, $S[j + S[w]] = -2w$, $S[2w] = w, S[0] = -w$
III. $k + S[j - w] = 2w$, $k + S[2w] = 0$, $S[w - k] = 0$, $S[w] = -w$

For example, when **I** occurs in the first round we have the following changes:
1. $i \leftarrow i + w = w$
2. $j \leftarrow 0 + S[j + S[w]] = S[j - w] = 2w$
3. $k \leftarrow k + i + S[j] = 0 + w + S[2w] = 0 + w + 0 = w$
4. $S[w] \leftarrow 0$, $S[2w] \leftarrow -w$ after SWAP
5. $z \leftarrow S[j + S[i + S[z + k]]] = S[2w + S[w + S[w]]] = S[2w + S[w]] = S[2w] = -w$

Similarly in the second round we have the following changes:
1. $i \leftarrow i + w = 2w$,
2. $j \leftarrow w + S[2w + S[2w]] = w + S[w] = w$
3. $k \leftarrow k + i + S[j] = w + 2w + S[w] = 3w + 0 = 3w$
4. $S[w] \leftarrow -w$, $S[2w] \leftarrow 0$ after SWAP
5. $z \leftarrow S[w + S[2w + S[3w - w]]] = S[w + S[2w + S[2w]]] = S[w] = -w$

We get similar results when we analyze **II** and **III**. Let us now denote by E the union of the events **I, II** and **III**. We have $\Pr[\mathsf{E}] = \frac{3}{N^4}$, and $\Pr[Z_1 = Z_2 = -w | \mathsf{E}] = 1$. We assume that when E does not occur $\Pr[Z_1 = Z_2 = -w | \mathsf{E}^c] = \frac{1}{N^2}$, and is more or less uniformly random. We were able to verify the assumption by running computer simulations. Therefore by Bayes theorem, we have:

$$\Pr[Z_1 = Z_2 = -w] = \Pr[Z_1 = Z_2 = -w | \mathsf{E}] \cdot \Pr[\mathsf{E}] + \Pr[Z_1 = Z_2 = -w | \mathsf{E}^c] \cdot \Pr[\mathsf{E}^c]$$

$$= 1 \cdot \frac{3}{N^4} + \frac{1}{N^2} \cdot \left[1 - \frac{3}{N^4}\right] \approx \frac{1}{N^2} + \frac{3}{N^4} \qquad \square$$

Experimental Results: By performing extensive computer simulations with **(a)** one billion random keys, and **(b)** a fixed key with one billion random IVs, the probability $\Pr[Z_1 = Z_2 = -w]$ was found to be around $\frac{1}{N^2} + \frac{2.9}{N^4}$ for $N = 16$ and $N = 32$. In Figs. 2 and 3, we plot $\left[\Pr[(Z_1, Z_2) = x] - \frac{1}{N^2}\right] \cdot N^4$ for all values of x when $N = 16$ and 32 respectively with $w = 7$. The x-axis is marked as $NZ_1 + Z_2$. We can see a sharp peak at the x-axis mark corresponding to $(-7, -7)$ (i.e. $9 * 16 + 9 = 153$ for $N = 16$ and $25 * 32 + 25 = 825$ for $N = 32$). The plot is not uniform and there seems to be some bias for other values of x too, but the most significant bias exists at the point corresponding to $(-w, -w)$.

2.2 Distinguishing Attack with Multiple Key-IV Pairs Based on a Short-Term Bias

We now state the following theorem from [11], which outlines the number of output samples required to distinguish two distributions X and Y.

Theorem 2. *(Mantin-Shamir [11]) Let X, Y be distributions, and suppose that the event e happens in X with probability p and in Y with probability $p(1 + q)$. Then for small p and q, $O\left(\frac{1}{pq^2}\right)$ samples suffice to distinguish X from Y with a constant probability of success.*

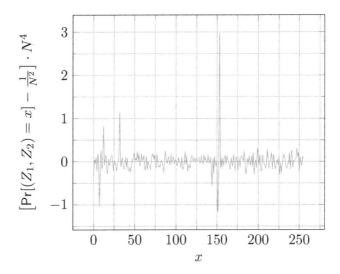

Fig. 2. $\left[\Pr[(Z_1, Z_2) = x] - \frac{1}{N^2}\right] \cdot N^4$ (for $N = 16$)

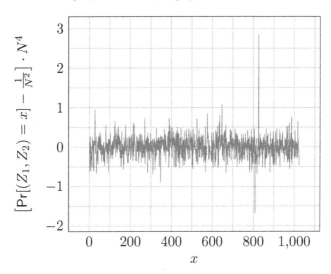

Fig. 3. $\left[\Pr[(Z_1, Z_2) = x] - \frac{1}{N^2}\right] \cdot N^4$ (for $N = 32$)

Let X be the probability distribution of Z_1 and Z_2 in an ideal random stream, and let Y be the probability distribution of Z_1 and Z_2 in streams produced by Spritz for randomly chosen keys. Let the event e denote $Z_1 = Z_2 = -w$, which occurs with probability of $\frac{1}{N^2}$ in X and $\frac{1}{N^2} + \frac{3}{N^4} = \frac{1}{N^2} \cdot \left(1 + \frac{3}{N^2}\right)$ in Y. By using the Theorem 2 with $p = \frac{1}{N^2}$ and $q = \frac{3}{N^2}$, we can conclude that we need about $\frac{1}{pq^2} = \frac{N^6}{9} \approx 2^{44.8}$ output samples to reliably distinguish the two distributions.

Therefore, we can mount a distinguishing attack with multiple key-IV pairs, if output samples of Z_1 and Z_2 produced by $2^{44.8}$ distinct key-IV pairs are available. In the single key setting, it requires samples of first two bytes Z_1 and Z_2 generated by $2^{44.8}$ different IVs.

2.3   Distinguishing Attack with a Single Key-IV Pair Based on a Long-Term Bias

The distinguishing attack on Spritz described in Theorem 1 requires that i and z are both zero at the beginning of the SQUEEZE phase. In general, during the production of a single stream of keystream bytes from any key or key/IV pair i and z are not both zero at the beginning of each round. This is why although the result in Theorem 1, holds for distinguishing the first 2 output bytes produced by multiple key/IV pairs, the same result can not be translated for a single keystream byte sequence using the event $Z_t = Z_{t+1} = -w$.

However i becomes 0 after every N rounds, and so in order to distinguish a single sequence of keystream bytes, one could look at the event $Z_{mN+1} = Z_{mN+2} = -w$ (for all integers $m \geq 0$) i.e. the first two of every cycle of N keystream bytes. However we still need $Z_{mN} = 0$ for the initial conditions of the distinguisher to be fulfilled and so we should really look at the event $\Pr[Z_{mN+1} = Z_{mN+2} = -w|Z_{mN} = 0]$. For the reasons outlined in Theorem 1, we also have

$$\Pr[Z_{mN+1} = Z_{mN+2} = -w|Z_{mN} = 0] = \frac{1}{N^2} + \frac{3}{N^4}$$

where the probability this time is calculated over several integral values of m. Note that we will need $T = \mathcal{O}(\frac{N^6}{9}) \approx 2^{44.8}$ samples to reliably distinguish the stream. However for this we need $T \cdot N$ cycle of keystream bytes (as $Z_{mN} = 0$ will on average occur once every N cycle) and hence $T \cdot N^2 = \mathcal{O}(\frac{N^8}{9}) \approx 2^{60.8}$ keystream bytes. The distinguishing attack was verified for 100 random keys for $N = 16, 32$.

2.4   Plaintext Recovery Attacks in the Broadcast Setting

These short- and long-term biases are also used for plaintext recovery attacks in the broadcast setting where the same plaintext is encrypted with different keys or/and IV in the same manner of previous attacks [1,9,11,15]. Note that the broadcast setting is converted into the multi-session setting where the target plaintext block are repeatedly sent in the same position in the plaintexts in multiple SSL/TLS sessions. According to Theorem 2, given $\frac{1}{pq^2}$ ciphertexts, we can distinguish the distribution of correct candidates of plaintext bytes (the biased distribution) from the distribution of wrong candidates of plaintext bytes (a random distribution) with a constant probability. It can be considered as the lower bound of the required number of ciphertexts for recovering biased bytes of a plaintext in this setting as mentioned in [11]. Recent statistical methods to detect a correct plaintext e.g. likelihood calculations of techniques [1,22] and Bayesian analysis [7] might help to reduce the required number of ciphertexts when mounting an actual attack.

3   State Recovery Attack on Spritz

We first look at a class of special states of the Spritz stream cipher that occurs just before the beginning of the SQUEEZE phase.

Definition 1. *Define a Spritz state as the 3-tuple* (S, j, k) *just at the beginning of the* SQUEEZE *phase. A Spritz state is called a* SPECIAL *state if all the following conditions hold simultaneously.*

1. $S[t] \equiv 0 \bmod 2,$ *if* $t \equiv 1 \bmod 2,$
2. $S[t] \equiv 1 \bmod 2,$ *if* $t \equiv 0 \bmod 2,$
3. $j \equiv 0 \bmod 2$ *and* $k \equiv 0 \bmod 2$

In other words a SPECIAL state occurs when all the even indexed positions of the S array hold odd values, all the odd indexed positions hold even values and additionally j and k are even. We will now show that if the state at the beginning of the SQUEEZE phase is a SPECIAL state, then the sequence $Z_t \bmod 2,\ t = 0, 1, 2, 3, \ldots$ is periodic with period equal to 4.

Lemma 1. *If the state at the beginning of the* SQUEEZE *phase is a* SPECIAL *state then the following hold (assuming N is even):*

(a) *The state after every four iterations is a* SPECIAL *state.*
(b) *In every iteration, the updated values of i and j are equal modulo 2. Hence no* SWAP *between odd and even values occur. And so, even and odd indexed positions of the S array will continue to hold odd and even values respectively.*
(c) $Z_t \equiv Z_{t+4} \bmod 2,$ *for all values of t.*

Proof. Note that i and z are 0 at the beginning of the SQUEEZE phase and so both are even to begin with. If N is even, the design of the WHIP module ensures that the value of w is odd, whatever be the length of key/IV. Thereafter, all the above claims can be verified by running four iterations of the UPDATE function. We summarize the modulo 2 values of the various indices over 4 iterations in Table 2. Note that the updated values of i, j in each round is either both odd or both even, which means that the odd and even values are never swapped during the SQUEEZE phase. At the end of round 4, i, j, k, z become even again and so the modulo values of the above indices will repeat every 4 cycles. And therefore, the sequence of the modulo 2 values of the keystream byte z becomes periodic with period 4: $0, 1, 1, 0,\ 0, 1, 1, 0,\ 0, 1, 1, 0 \ldots$ □

Probability of a SPECIAL state: Combinatorially, it is easy to see that the total number of SPECIAL states is $\left(\frac{N}{2}\right)^2 \cdot \left[\left(\frac{N}{2}\right)!\right]^2$. Therefore, if carry out the key/IV Setup operation with different keys/ single key and different IVs, then the probability that the state at the beginning of the SQUEEZE state is SPECIAL is given by

$$\rho = \frac{\left(\frac{N}{2}\right)^2 \cdot \left[\left(\frac{N}{2}\right)!\right]^2}{N^2 \cdot (N!)}$$

Table 2. The modulo 2 values of the various indices through 4 iterations. The ones marked with $*$ are used in the State recovery process in Algorithm 1

#	Index	$t = 1$	$t = 2$	$t = 3$	$t = 4$
1	$i = i + w^*$	1	0	1	0
2	$j + S[i]^*$	0	0	0	0
3	$j = k + S[j + S[i]]^*$	1	0	1	0
4	$k = k + i + S[j]$	1	0	1	0
5	$z + k^*$	1	0	0	1
6	$i + S[z + k]^*$	1	1	0	0
7	$j + S[i + S[z + k]]^*$	1	0	0	1
8	$z = S[j + S[i + S[z + k]]]$	0	1	1	0

For $N = 256$, $\rho \approx 2^{-253.7}$. So if one employs an IV of length more than 254 bits, it is likely that a SPECIAL state will be encountered in ρ^{-1} attempts. Using this, a state recovery attack can be mounted in a Multiple IV mode as follows:

1. For a fixed key, and Multiple IVs collect keystream of around $10 * N$ bytes and inspect the sequence $Z_t \bmod 2$.
2. If the sequence is $0, 1, 1, 0, \ 0, 1, 1, 0, \ 0, 1, 1, 0 \ldots$ i.e. periodic with period 4, then the attacker can conclude with high probability that he has encountered a SPECIAL state and he proceeds according to Algorithm 1.
3. The above technique is likely to succeed once in ρ^{-1} attempts.

3.1 State Recovery of SPECIAL states

Once the attacker is sure that he has encountered a SPECIAL state, he has the task of recovering a much simpler state and he proceeds in the same manner as in [2, Algorithm 1]. However, there a few differences as given in Algorithm 1.

The algorithm can be summarized in the following words: In each round, the attacker guesses the value of some of the elements of the internal permutation to determine the value of all the five indices required in the state update operation, each time making sure that odd indices get even values and vice versa. He then inspects the keystream byte produced in the round and tries to determine if the intermediate guessed permutation is consistent with the keystream byte observed. The attacker computes the index $d = j + S[i + S[z + k]]$ with the guessed values of the permutation and then performs the Verification step: Depending on the comparison between $S[d]$ and the current keystream byte Z_r he makes the following transitions:

If	$S[d] = $ NULL and $Z_r \notin S$	$\rightarrow$	Assign $S[d] = Z_r$, Go to next round $r + 1$
If	$S[d] = $ NULL and $Z_r \in S$	$\rightarrow$	Contradiction!! Try another assignment
If	$S[d] \neq $ NULL and $Z_r \neq S[d]$	$\rightarrow$	Contradiction!! Try another assignment
If	$S[d] \neq $ NULL and $Z_r = S[d]$	$\rightarrow$	Go to next round $r + 1$

Input: Keystream bytes Z_t for $t = 0$ to $10 * N$;
Output: Permutation S at the beginning of SQUEEZE stage;

$S[t] \leftarrow$ NULL for $t = 0$ to $N - 1$;
Run $StateRecovery(S, i, j, k, 0)$;

$StateRecovery(S, i, j, k, r)$;
$\overline{i_{next} \leftarrow i + w};$
if $S[i_{next}] =$ NULL $\wedge u_1$ is not in $S \wedge u_1 \not\equiv i_{next}$ mod 2 **then**
| Assign $S[i_{next}] \leftarrow u_1$ /* **for** $u_1 \leftarrow 0$ **to** $N - 1$ */
end

$a = j + S[i_{next}]$;
if $S[a] =$ NULL $\wedge u_2$ is not in $S \wedge u_2 \not\equiv a$ mod 2 **then**
| Assign $S[a] \leftarrow u_2$ /* **for** $u_2 \leftarrow 0$ **to** $N - 1$ */
end

$j_{next} \leftarrow j + S[a]$;
if $S[j_{next}] =$ NULL $\wedge u_3$ is not in $S \wedge u_3 \not\equiv j_{next}$ mod 2 **then**
| Assign $S[j_{next}] \leftarrow u_3$ /* **for** $u_3 \leftarrow 0$ **to** $N - 1$ */
end

$k_{next} \leftarrow k + i_{next} + S[j_{next}]$;
SWAP $(S[i_{next}], S[j_{next}])$;
$b \leftarrow Z_{r-1} + k_{next}$;
if $S[b] =$ NULL $\wedge u_4$ is not in $S \wedge u_4 \not\equiv b$ mod 2 **then**
| Assign $S[b] \leftarrow u_4$ /* **for** $u_4 \leftarrow 0$ **to** $N - 1$ */
end

$c \leftarrow i_{next} + S[b]$;
if $S[c] =$ NULL $\wedge u_5$ is not in $S \wedge u_5 \not\equiv c$ mod 2 **then**
| Assign $S[c] \leftarrow u_5$ /* **for** $u_5 \leftarrow 0$ **to** $N - 1$ */
end

$d \leftarrow j_{next} + S[c]$;
if $S[d]$ is NULL $\wedge Z_r$ is not in S **then**
| Assign $S[d] \leftarrow Z_r$;
| $StateRecovery(S, i_{next}, j_{next}, k_{next}, r + 1)$;
end
if $S[d]$ is NULL $\wedge Z_r$ is in S **then**
| Contradiction /*`Try another assignment` */;
end
if $S[d]$ is not NULL $\wedge S[d] \neq Z_r$ **then**
| Contradiction /*`Try another assignment` */;
end
if $S[d]$ is not NULL $\wedge S[d] = Z_r$ **then**
| $StateRecovery(S, i_{next}, j_{next}, k_{next}, r + 1)$;
end

Algorithm 1. State recovery algorithm for SPECIAL states

3.2   Complexity of the Algorithm

The complexity is given by the number of guesses or assignments made, until a solution is found. As in [2], we compute the complexity by splitting the algorithm in several cases $c_i(x)$ to which we assign probabilities according to the occurrence of each case. Note that we can view the above internal state recovery algorithm, as two modules each working to recover exactly one half of the elements of the permutation. This is true since, the odd and the even indices never swap among each other. Let us denote by T_1, T_2 as the average number of assignments that would made in recovering the odd/even indexed elements of the permutation, if they were operating independent of the other. Since for every assignment in T_1 we would need T_2 assignments to verify the correctness of the solution, the total complexity of our algorithm is $T = T_1 \cdot T_2$.

To estimate T_1, we have to note the parity of the the odd indices assigned in every cycle. We already know that the parity of all the indices will repeat after every 4 rounds, so observing the first 4 cycles is sufficient. As per Algorithm 1, the five indices that are used in the assignment process are $i_{next}, a, j_{next}, b, c$, and the index used in the verification process is d. It is easy to see that these correspond to $i, j + S[i], j, z + k, i + S[z + k]$ and $j + S[i + S[z + k]]$ respectively. A quick look at Table 2, tells us four of the assignment indices and the only verification index are odd in the first round. Thereafter the second and third rounds have one and two assignment indices odd. The fourth round has one assignment and one verification index odd. This means that there are four assignments followed by a verification, which is followed by another cycle of four assignments and a verification. Therefore in total we have 10 stages of assignment/verification. Let $c_i[x]$ $(1 \le i \le 10)$ denote the average complexity associated with each stage, assuming that x elements of the $N/2$ odd-indexed positions are already filled, then we have

$$
c_i[x] = \begin{cases}
\frac{x}{N/2} \cdot c_{i+1}[x] + (1 - \frac{x}{N/2}) \cdot (\frac{N}{2} - x) \cdot c_{i+1}[x+1], & \text{for } i \in [1, 10] \setminus \{5, 10\} \\
(\frac{x}{N/2})^2 \cdot c_{i+1}[x] + (1 - \frac{x}{N/2})^2 \cdot c_{i+1}[x+1], & \text{for } i = 5, 10 \\
& \text{/*} c_{11} \text{ denotes } c_1 \text{*/}.
\end{cases}
$$

In the above equation, when $i \in [1, 10] \setminus \{5, 10\}$, it denotes an assignment phase, when $i = 5, 10$, it denotes a verification phase. During an assignment, if x elements are already present in the permutation, then with probability $\frac{x}{N/2}$, the index to be assigned would be already filled, and in this case the algorithm would move on to stage $i + 1$ without assignment. Alternatively with probability $1 - \frac{x}{N/2}$, the index is empty and there are exactly $\frac{N}{2} - x$ ways to assign it, after which it moves to stage $i + 1$. During verification stage the analysis is as follows:

a. With probability $\frac{x}{N/2}$, the verification index d is already filled.

b. Therefore with probability $\frac{x}{N/2} \cdot (1 - \frac{x}{N/2})$, the index is already filled by a value other than Z_r. In this case the path is terminated.

c. With probability $(\frac{x}{N/2})^2$ the index is filled with Z_r and the algorithm moves to the next phase.

d. With probability $(1 - \frac{x}{N/2})$ the verification index d is empty.

e. Therefore with probability $(1 - \frac{x}{N/2}) \cdot (\frac{x}{N/2})$ it happens that Z_r exists in some other index of the permutation. In this case too the path is terminated.

f. With probability $(1 - \frac{x}{N/2})^2$, Z_r is not present in the permutation, and so after assigning $S[d] \leftarrow Z_r$ it moves to the next stage.

The complexity T_1 can be estimated as $c_1[0]$, with the boundary conditions $c_i[\frac{N}{2} - 1] = 1$. The above recurrence can be solved by a dynamic programming approach to find an estimate for $c_1[0]$. A similar recurrence relation can be deduced for estimating T_2 by keeping track of the even valued assignment/verification indices. We write the recurrence relation below for the benefit of the reader.

$$c_i[x] = \begin{cases} \frac{x}{N/2} \cdot c_{i+1}[x] + (1 - \frac{x}{N/2}) \cdot (\frac{N}{2} - x) \cdot c_{i+1}[x + 1], & \text{for } i \in [1, 14] \setminus \{6, 10\} \\ (\frac{x}{N/2})^2 \cdot c_{i+1}[x] + (1 - \frac{x}{N/2})^2 \cdot c_{i+1}[x + 1], & \text{for } i = 6, 10 \\ & \text{/*}c_{15} \text{ denotes } c_1\text{*/.} \end{cases}$$

Experimental Results: We performed the state recovery for $N = 14, 16, 18, 20$ for 100 random permutations. The algorithm was always able to recover the permutation. In Fig. 4, we plot the base 2 logarithm of the theoretical estimate T with the base 2 logarithm of the experimentally obtained average number of steps, for different even values of N. We can see that the theoretical value always overestimates the experimentally obtained complexity. For $N = 256$, the theoretical estimate for $T \approx 2^{1233}$. And so the estimated complexity of state recovery is given as $T \cdot (\frac{N}{2})^2 \approx 2^{1247}$ (taking into account the additional complexity of guessing the values of j, k at the beginning of the SQUEEZE phase). So the total complexity consists of ρ^{-1} encryptions plus $T \cdot (\frac{N}{2})^2$ assignments which again comes to approximately 2^{1247}.

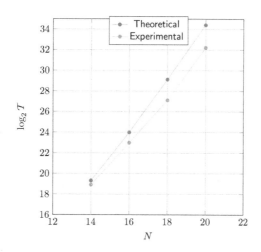

Fig. 4. Experimental and theoretical estimates of $\log_2 T$ (Color figure online)

4   Conclusion

In this paper, we analyzed the security of the stream cipher Spritz. We first proposed distinguishing attacks based on the short-term and the long-term biases in the keystream of Spritz. The distinguisher can be used both for distinguishing keystreams produced by multiple key-IVs and for distinguishing a keystream produced by a single key-IV pair. In the second half of the paper we looked at the state recovery attack on Spritz (in the multiple IV setting), in the situation when the cipher has entered a special class of SPECIAL states. We calculated the probability of such an event happening, and went on to outline an algorithm to recover the internal permutation. Our estimates suggest that in this case we need approximately 2^{1247} assignments to recover the internal state which is an improvement on the 2^{1400} step algorithm proposed in [2].

Acknowledgements. The authors would like to thank the anonymous reviewers who helped improve the quality of this paper.

References

1. AlFardan, N.J., Bernstein, D.J., Paterson, K.G., Poettering, B., Schuldt, J.C.N.: On the security of RC4 in TLS and WPA. In: Proceedings of 22nd USENIX Conference on Security, pp. 305–320 (2013)
2. Ankele, R., Kölbl, S., Rechberger, C.: State-recovery analysis of Spritz. In: Lauter, K., Rodríguez-Henríquez, F. (eds.) LatinCrypt 2015. LNCS, vol. 9230, pp. 204–221. Springer, Heidelberg (2015)
3. Banik, S., Sarkar, S., Kacker, R.: Security analysis of the RC4+ stream cipher. In: Paul, G., Vaudenay, S. (eds.) INDOCRYPT 2013. LNCS, vol. 8250, pp. 297–307. Springer, Heidelberg (2013)
4. Banik, S., Jha, S.: How not to combine RC4 states. In: Chakraborty, R.S., Schwabe, P., Solworth, J. (eds.) SPACE 2015. LNCS, vol. 9354, pp. 95–112. Springer, Heidelberg (2015)
5. Banik, S., Jha, S.: Some security results of the RC4+ stream cipher. Secur. Commun. Netw. **8**(18), 4061–4072 (2015). Wiley Online Publishing
6. Biham, E., Granboulan, L., Nguyen, P.Q.: Impossible fault analysis of RC4 and differential fault analysis of RC4. In: Gilbert, H., Handschuh, H. (eds.) FSE 2005. LNCS, vol. 3557, pp. 359–367. Springer, Heidelberg (2005)
7. Garman, C., Paterson, K.G., van der Merwe, T.: Attacks only get better: password recovery attacks against RC4 in TLS. In: Proceedings of 24th USENIX Conference on Security, pp. 113–128 (2015)
8. Gong, G., Gupta, K.C., Hell, M., Nawaz, Y.: Towards a general RC4-like keystream generator. In: Feng, D., Lin, D., Yung, M. (eds.) CISC 2005. LNCS, vol. 3822, pp. 162–174. Springer, Heidelberg (2005)
9. Isobe, T., Ohigashi, T., Watanabe, Y., Morii, M.: Full plaintext recovery attack on broadcast RC4. In: Moriai, S. (ed.) FSE 2013. LNCS, vol. 8424, pp. 179–202. Springer, Heidelberg (2014)
10. Maitra, S., Paul, G.: Analysis of RC4 and proposal of additional layers for better security margin. In: Chowdhury, D.R., Rijmen, V., Das, A. (eds.) INDOCRYPT 2008. LNCS, vol. 5365, pp. 27–39. Springer, Heidelberg (2008)

11. Mantin, I., Shamir, A.: A practical attack on broadcast RC4. In: Matsui, M. (ed.) FSE 2001. LNCS, vol. 2355, pp. 152–164. Springer, Heidelberg (2002)
12. Maximov, A.: Two linear distinguishing attacks on VMPC and RC4A and weakness of RC4 family of stream ciphers. In: Gilbert, H., Handschuh, H. (eds.) FSE 2005. LNCS, vol. 3557, pp. 342–358. Springer, Heidelberg (2005)
13. Maximov, A., Khovratovich, D.: New state recovery attack on RC4. In: Wagner, D. (ed.) CRYPTO 2008. LNCS, vol. 5157, pp. 297–316. Springer, Heidelberg (2008)
14. Nawaz, Y., Gupta, K.C., Gong, G.: A 32-bit RC4-like keystream generator. IACR Cryptology ePrint Archive 2005/175
15. Ohigashi, T., Isobe, T., Watanabe, Y., Morii, M.: How to recover any byte of plaintext on RC4. In: Lange, T., Lauter, K., Lisoněk, P. (eds.) SAC 2013. LNCS, vol. 8282, pp. 155–173. Springer, Heidelberg (2014)
16. Paul, G., Maitra, S., Chattopadhyay, A.: Quad-RC4: merging four RC4 states towards a 32-bit stream cipher. IACR Cryptology eprint Archive 2013/572 (2013)
17. Paul, S., Preneel, B.: A new weakness in the RC4 keystream generator and an approach to improve the security of the cipher. In: Roy, B., Meier, W. (eds.) FSE 2004. LNCS, vol. 3017, pp. 245–259. Springer, Heidelberg (2004)
18. Paul, S., Preneel, B.: On the (in)security of stream ciphers based on arrays and modular addition. In: Lai, X., Chen, K. (eds.) ASIACRYPT 2006. LNCS, vol. 4284, pp. 69–83. Springer, Heidelberg (2006)
19. Rivest, R., Schuldt, J.: Spritz - a spongy RC4-like stream cipher and hash function. https://people.csail.mit.edu/rivest/pubs/RS14.pdf
20. Tsunoo, Y., Saito, T., Kubo, H., Shigeri, M., Suzaki, T., Kawabata, T.: The most efficient distinguishing attack on VMPC and RC4A. In: SKEW 2005. http://www.ecrypt.eu.org/stream/papers.html
21. Tsunoo, Y., Saito, T., Kubo, H., Suzaki, T.: A distinguishing attack on a fast software-implemented RC4-like stream cipher. IEEE Trans. Inf. Theor. **53**(9), 3250–3255 (2007)
22. Vanhoef, M., Piessens, F.: All your biases belong to us: breaking RC4 in WPA-TKIP and TLS. In: Proceedings of 24th USENIX Conference on Security, pp. 97–112 (2015)
23. Zoltak, B.: Statistical weakness in Spritz against VMPC-R: in search for the RC4 replacement. http://eprint.iacr.org/2014/985.pdf
24. Zoltak, B.: VMPC one-way function and stream cipher. In: Roy, B., Meier, W. (eds.) FSE 2004. LNCS, vol. 3017, pp. 210–225. Springer, Heidelberg (2004)

Attacks Against Filter Generators Exploiting Monomial Mappings

Anne Canteaut$^{(\boxtimes)}$ and Yann Rotella

Inria, Paris, France
{Anne.Canteaut,Yann.Rotella}@inria.fr

Abstract. Filter generators are vulnerable to several attacks which have led to well-known design criteria on the Boolean filtering function. However, Rønjom and Cid have observed that a change of the primitive root defining the LFSR leads to several equivalent generators. They usually offer different security levels since they involve filtering functions of the form $F(x^k)$ where k is coprime to $(2^n - 1)$ and n denotes the LFSR length. It is proved here that this monomial equivalence does not affect the resistance of the generator against algebraic attacks, while it usually impacts the resistance to correlation attacks. Most importantly, a more efficient attack can often be mounted by considering non-bijective monomial mappings. In this setting, a divide-and-conquer strategy applies based on a search within a multiplicative subgroup of $\mathbb{F}_{2^n}^*$. Moreover, if the LFSR length n is not a prime, a fast correlation involving a shorter LFSR can be performed.

Keywords: Stream ciphers · Correlation attacks · LFSR · Filter generator · Nonlinear equivalence · Monomials

1 Introduction

The running-key used in a stream cipher is produced by a pseudo-random generator whose initialization is the secret key shared by the users. Linear feedback shift registers (LFSR) are building-blocks used in many keystream generators since they are appropriate to low-cost implementations, produce sequences with good statistical properties and have a simple mathematical description. While basic LFSR-based generators, like combination generators or filter generators, are not used directly as keystream generators in modern stream ciphers, they are still widely used either as a part of the generator or in modified form [13]. This situation then motivates an in-depth evaluation of the security of LFSR-based generators. Actually, several modern ciphers have been analyzed by enhanced variants of attacks, which were first dedicated to simple LFSR-based generators (e.g. [26,29,34]).

Partially supported by the French Agence Nationale de la Recherche through the BRUTUS project under Contract ANR-14-CE28-0015.

T. Peyrin (Ed.): FSE 2016, LNCS 9783, pp. 78–98, 2016.
DOI: 10.1007/978-3-662-52993-5_5

At this aim, our work investigates the security of the so-called filter generator, which consists of a single LFSR whose content is filtered by a nonlinear Boolean function. These generators have been extensively studied and are known to be vulnerable to several types of attacks, mainly algebraic attacks and their variants [9,10,17,38] and (fast) correlation attacks [32]. These attacks have led to the definition of design criteria, especially related to the choice of the filtering function, and they have initiated a whole line of research on the constructions of appropriate filtering functions. However, it has been observed more recently by Rønjom and Cid [36] that a simple change of the primitive characteristic polynomial of the LFSR (i.e., a change of the primitive root of the underlying finite field), may lead to an equivalent generator whose filtering function corresponds to the composition of a monomial permutation with the original filtering function, $x \mapsto F(x^k)$ for some k coprime to $(2^n - 1)$ where n is the LFSR length. This observation opens the door to new weaknesses since the main security criteria, like the nonlinearity, the degree or the algebraic immunity of the filtering function, are not invariant under this *nonlinear equivalence*. Hence, this raises many open questions about the relevance of the usual criteria, as noted by Rønjom and Cid. In this context, the objective of our paper is to answer most of these questions by evaluating the minimal security offered by all generators derived by monomial equivalence, and to further investigate the possibilities to transform the constituent LFSR by applying a monomial mapping, especially a *non-bijective* monomial mapping.

Our contributions. Our contributions are then two-fold: first, we show that, even if the degree and the algebraic-immunity of a Boolean function may highly vary within an equivalence class, the monomial equivalence defined by Rønjom and Cid has no impact on the resistance of a filter generator against algebraic attacks and their variants. The reason is that the degree and the algebraic immunity are not the relevant parameters for estimating the security of a filter generator as shown in [17,20,28]. Instead, the complexities of these attacks are determined by the linear complexity and the spectral immunity of the filtering function, which are derived from the univariate representation of the function and are therefore invariant under monomial equivalence. On the other hand, the second family of attacks, namely (fast) correlation attacks, are highly affected by monomial equivalence, implying that the associated criterion must be the generalized nonlinearity of the filtering function as defined in [41]. But we show that the non-bijective monomial mappings also play a very important role, usually much more important than monomial permutations, because the LFSR can then be transformed into an LFSR producing a sequence with smaller period τ. A divide-and-conquer attack can then be mounted exploiting this property, where the number of values to be examined decreases from $(2^n - 1)$ to τ. Moreover, if the LFSR length n is not a prime, the new LFSR involved in the attack may be shorter than the original one, leading to a much more efficient fast correlation attack.

Organization of the paper. We first introduce the monomial equivalence between filter generators as described by Rønjom and Cid [36] and show that the univariate representation of both the LFSR and the filtering function is well-suited for analyzing its impact. Section 3 then focuses on algebraic attacks and proves that all filter generators obtained by monomial equivalence have the same behaviour with respect to this family of attacks. Section 4 then investigates correlation attacks and their variants, and shows that the situation is very different. Also, we describe a new setting for (fast) correlation attacks where non-bijective monomials are used. Two types of attacks are then presented: fast correlation involving a shorter LFSR which can be mounted when the LFSR length is not a prime, and correlation attacks based on FFT which recover $\log_2 \tau$ bits of the initial state where τ is a divisor of $(2^n - 1)$.

2   Equivalence Between Filtered LFSR

2.1   Filtered LFSRs

In the following, we focus on binary filtered LFSRs. The binary LFSR of length n with *characteristic polynomial*, $P(X) = X^n + \sum_{i=0}^{n-1} c_i X^i \in \mathbb{F}_2[X]$, is the finite-state automaton which produces the binary sequences $\mathbf{s} = (s_t)_{t \geq 0}$, satisfying the linear recurrence relation

$$s_{t+n} = \sum_{i=0}^{n-1} c_i s_{t+i}, \quad \forall t \geq 0.$$

In this paper, we implicitly assume that the LFSRs we consider are non-singular, i.e., the constant term c_0 in the characteristic polynomial does not vanish. Otherwise the transition function of the LFSR is not bijective, leading to a possible loss of entropy of the internal state, which is clearly not suitable in cryptographic applications. Also, the characteristic polynomial is assumed to be irreducible, which guarantees that, for any nonzero initial state of the LFSR, the generated sequence cannot be produced by a shorter LFSR [42]. In other words, the *linear complexity* of any sequence generated by the LFSR from a nonzero initial state is equal to the LFSR length. A well-known property of LFSR sequences is that any sequence produced by an LFSR with an irreducible characteristic polynomial P (and a nonzero initial state) is periodic and its least period is equal to the order of P, i.e., to the smallest positive integer r for which $P(X)$ divides $X^r + 1$. Hence, the characteristic polynomials of LFSRs used in practical applications are chosen primitive. More details on the properties of LFSR sequences can be found e.g. in [19,25].

In this context, a *filter generator* (aka filtered LFSR), is a keystream generator composed of a single binary LFSR of length n whose content is filtered by a nonlinear Boolean function of n variables. More precisely, the output sequence $(s_t)_{t \geq 0}$ of the filter generator is given by

$$s_t = f(u_{t+n-1}, u_{t+n-2}, \ldots, u_t), \quad \forall t \geq 0 ,$$

where $(u_t)_{t \geq 0}$ denotes the sequence generated by the LFSR.

It is worth noticing that, in most practical proposals, the filtering function does not depend on all n bits of the internal state. For obvious implementation reasons, f is usually chosen in such a way that it depends on $m < n$ variables only. It can then be equivalently described by an m-variable Boolean function f' and a decreasing sequence $(\gamma_i)_{1 \leq i \leq m}$, with $1 \leq \gamma_i \leq n$, such that for any n-tuple $(x_1, \ldots, x_n)$,

$$f(x_1, \ldots, x_n) = f'(x_{\gamma_1}, \ldots, x_{\gamma_m}) .$$

Here, unless explicitly mentioned, the filtering function will be defined as a function of n variables, where n is the LFSR length, even if some (or most) of these variables are not involved in the evaluation of the function.

2.2 Univariate Representation of Filtered LFSRs

Filter generators have been extensively studied and are known to be vulnerable to several types of attacks which have led to the definition of some security criteria on the tapping sequence $(\gamma_i)_{1 \leq i \leq m}$ [14] and on the Boolean filtering function (see e.g. [4] for a survey). For instance, it is well-known that f must have a high algebraic degree in order to generate a keystream sequence with a high linear complexity [39], a high algebraic-immunity in order to resist algebraic attacks [10,31] and a high nonlinearity in order to resist fast correlation attacks [32]. These design criteria on the filtering function must be considered up to some equivalence in the sense that several filtered LFSR may generate the same set of sequences. This equivalence between filtered LFSR can be simply described by defining the LFSR next-state function over the finite field with 2^n elements instead of the vector space $\mathbb{F}_2^n$.

In this field-oriented description, we will use the following classical notation. The finite field with 2^n elements is denoted by $\mathbb{F}_{2^n}$. The multiplicative order of a nonzero element α in a finite field, $\mathsf{ord}(\alpha)$, is the smallest positive integer r such that $\alpha^r = 1$. The trace function from $\mathbb{F}_{2^n}$ into $\mathbb{F}_2$ is denoted by Tr^n, i.e.,

$$\mathsf{Tr}^n(x) = \sum_{i=0}^{n-1} x^{2^i} .$$

The index n will omitted if it is clear from the context.

Proposition 1 (Theorem 9.2 in [30]). *Let P be an irreducible polynomial in $\mathbb{F}_2[X]$ with degree n. Let $\alpha \in \mathbb{F}_{2^n}$ be a root of P and $\{\beta_0, \ldots, \beta_{n-1}\}$ denote the dual basis of $\{1, \alpha, \ldots, \alpha^{n-1}\}$, i.e.,*

$$\mathsf{Tr}^n(\alpha^i \beta_j) = \begin{cases} 0 & \text{if } i \neq j \\ 1 & \text{if } i = j \end{cases} .$$

Then, the content of the LFSR with characteristic polynomial P at time $(t+1)$ is equal to its content at time t multiplied by α, where these vectors are identified with elements in the field $\mathbb{F}_{2^n}$ decomposed on the basis $\{\beta_0, \ldots, \beta_{n-1}\}$.

With the notation of the previous proposition, we consider the isomorphism φ from $\mathbb{F}_2^n$ into $\mathbb{F}_{2^n}$ defined by the basis $\{\beta_0, \ldots, \beta_{n-1}\}$. Then, the internal state at time t of the LFSR initialized by $X_0 = \varphi(u_0, \ldots, u_{n-1})$ corresponds to

$$X_t = X_0 \alpha^t$$

and the keystream bit at time t is given by

$$s_t = f \circ \varphi^{-1}(X_0 \alpha^t) .$$

Therefore, any filter generator has an equivalent *univariate representation* defined by a root $\alpha \in \mathbb{F}_{2^n}$ of the LFSR characteristic polynomial, and a function F from $\mathbb{F}_{2^n}$ into $\mathbb{F}_2$. This generator produces from any initial state $X_0 \in \mathbb{F}_{2^n}$ the sequence $s_t = F(X_0 \alpha^t)$. For the sake of clarity, univariate functions defined over $\mathbb{F}_{2^n}$ will be denoted by capital letters, while small letters will be used for multivariate functions over $\mathbb{F}_2^n$. Clearly, the multivariate representation of a filter generator, (P, f), can be recovered from its univariate representation (α, F): since P is irreducible, it corresponds to the minimal polynomial of α and f is equal to $F \circ \varphi$ where φ is the isomorphism associated to the dual basis of $\{1, \alpha, \alpha^2, \ldots, \alpha^{n-1}\}$. Conversely, a given multivariate representation (P, f) corresponds to n univariate representations (α, F) since there are several possible values for α corresponding to the conjugate roots of P, i.e., $\alpha, \alpha^2, \alpha^{2^2}, \ldots, \alpha^{2^{n-1}}$. The univariate filtering functions F associated to the different choices for α are then linearly equivalent because they only differ from the composition with the Frobenius map. However, composing F with a linear permutation does not change its cryptographic properties (see the next section for details).

As a function from $\mathbb{F}_{2^n}$ into $\mathbb{F}_{2^n}$, F can be written as a univariate polynomial in $\mathbb{F}_{2^n}[X]$ and the coefficients of this polynomial are computed from the values of F by the discrete Fourier Transform (DFT) of F (aka Mattson-Solomon transform) (see e.g. [2,15,27]).

Proposition 2 (Discrete Fourier Transform of a Function). *Let F be a function from $\mathbb{F}_{2^n}$ into $\mathbb{F}_{2^n}$. Then, there exists a unique univariate polynomial in $\mathbb{F}_{2^n}[X]/(X^{2^n} + X)$ such that*

$$F(X) = \sum_{i=0}^{2^n - 1} A_i X^i .$$

Moreover, $A_0 = F(0)$, $A_{2^n-1} = \sum_{x \in \mathbb{F}_{2^n}} F(x)$ and the coefficients A_i, $1 \leq i \leq 2^n - 2$, are given by the discrete Fourier transform of the values of F at all nonzero inputs, namely

$$A_i = \sum_{k=0}^{2^n - 2} F(\gamma^k) \gamma^{-ki}, \ 1 \leq i \leq 2^n - 2$$

where γ is a primitive element in $\mathbb{F}_{2^n}$.

It is worth noticing that, in our context, the value of $F(0)$ does not affect the security of the filter generator: this value is only involved when the LFSR internal state vanishes, which is obviously always avoided since the sequence generated from the all-zero state is constant. Therefore, we will always consider in the following that the coefficient of degree $(2^n - 1)$ in the univariate form of F is equal to zero. In other words, the univariate form of F is identified with $(A_0, \ldots, A_{2^n-2})$ which is the DFT of the values of F. In our situation also, F takes its values in $\mathbb{F}_2$, implying that $A_{2i} = A_i^2$ for any $1 \leq i \leq 2^n - 2$. In this case, the coefficients A_i for all i in the same cyclotomic coset modulo $(2^n - 1)$, $\mathcal{C}(i) = \{i, 2i \bmod (2^n - 1), 2^2 i \bmod (2^n - 1), \ldots, 2^{n-1} i \bmod (2^n - 1)\}$ can be gathered, leading to the so-called trace representation:

$$F(X) = \sum_{k \in \Gamma} \mathsf{Tr}^{n_k}(A_k X^k),$$

where Γ is a set of representatives of all cyclotomic cosets modulo $(2^n - 1)$, n_k denotes the size of the cyclotomic coset of k and $A_k \in \mathbb{F}_{2^{n_k}}$.

2.3   Monomial Equivalence Between Filtered LFSR

Using the univariate representation, it is easy to observe that, for any nonzero $\lambda \in \mathbb{F}_{2^n}$, the sequence generated by the filtered LFSR with characteristic polynomial P and filtering function F from the initial state $X_0 \in \mathbb{F}_{2^n}$ is the same as the sequence obtained by filtering the same LFSR with $G(x) = F(\lambda x)$ from the initial state $Y_0 = \lambda^{-1} X_0$. It follows that not only F but also any function $G(x) = F(\lambda x)$ can be attacked when cryptanalyzing the generator. But, this equivalence does not affect the security of filter generators since all design criteria are known to be invariant under linear equivalence, i.e., under the composition of the filtering function by an $\mathbb{F}_2$-linear permutation of $\mathbb{F}_{2^n}$.

However, Rønjom and Cid [36] exhibited some nonlinear equivalence relations between filtered LFSR when the LFSR characteristic polynomial P is primitive. This equivalence relation boils down to changing the primitive root of $\mathbb{F}_{2^n}$ in the univariate representation of the generator. Let us consider two primitive elements in $\mathbb{F}_{2^n}$, namely α and β, implying that $\beta = \alpha^k$ for some integer k with $\gcd(k, 2^n - 1) = 1$. Let P_α and P_β denote their minimal polynomials. Then, we observe that, at any time $t \geq 0$, the internal state X_t of the LFSR with characteristic polynomial P_α and the internal state Y_t of the LFSR with characteristic polynomial P_β initialized with $Y_0 = X_0^k$ satisfy

$$Y_t = Y_0 \beta^t = (X_0 \alpha^t)^k = X_t^k.$$

This implies that the set of all sequences obtained by filtering by F the LFSR defined by α corresponds to the sequences generated by filtering by $G(x) = F(x^r)$ the LFSR defined by $\beta = \alpha^k$ where $rk \equiv 1 \bmod (2^n - 1)$. From now on, this equivalence between filter generators will be named *monomial equivalence*[1]. It

---

[1] Note that, among all monomials, only the *permutations* of $\mathbb{F}_{2^n}$, i.e., $X \mapsto X^k$ with $\gcd(k, 2^n - 1) = 1$ provide an equivalence relation.

follows that there exist $\frac{\Phi(2^n-1)}{n}$ monomial transformations which are not linearly equivalent and nevertheless provide equivalent filtering LFSR, where Φ is the Euler's totient function. Any attack against one among these $\frac{\Phi(2^n-1)}{n}$ generators then provides an attack against the whole class. Most notably, an initial-state recovery attack against the generator defined by β enables the attacker to recover the initial state X_0 of the LFSR defined by α by using that $X_0 = Y_0^r$. Therefore, the security level offered by a filter generator is clearly the minimal security among all generators in its equivalence class.

3   Monomial Equivalence and Algebraic Attacks

Determining the cryptographic properties of a Boolean function up to any change of the primitive element seems rather complicated, since the major properties of the function, like its degree or its nonlinearity, are not invariant under these nonlinear transformations (see e.g. [36, Appendix A]). However, the recent works by Gong et al. [17,20,37,38] point out that this difficulty mainly comes from the fact that the multivariate representation of the function is usually not relevant for evaluating its security level. Instead, the univariate representation provides a much more powerful tool which allows to directly determine the security offered by a generator against algebraic attacks (and its variants). Indeed, the action of the monomial equivalence can be described in a much simpler way when the univariate expression of the function is considered: the class of all filtering functions in the equivalence class of F consists of all functions $G = \sum_{i=0}^{2^n-2} B_i X^i$ whose univariate representation $(B_0, \ldots, B_{2^n-2})$ is obtained by decimating the univariate representation of F by some integer k coprime to $(2^n - 1)$, i.e., $B_i = A_{ik \bmod (2^n-1)}$. Using this simple transformation, it becomes possible to determine how the complexity of algebraic-type attacks varies within the equivalence class of a filtering function.

3.1   Linear Complexity

The simplest algebraic attack consists in writing the Boolean equations defining the successive keystream bits. We then obtain a multivariate system depending on n binary unknowns, which are the bits of the initial state. The degree of each equation is equal to the degree of the filtering function f, which tends to show that the complexity for solving this algebraic system highly depends on the degree of f. Instead of linearizing the system of degree $\deg(f)$ derived from f, another strategy consists in exploiting the fact that the keystream sequence produced by a filter generator can also be seen as the output of a single LFSR. The length of the shortest LFSR generating the sequence is its *linear complexity* Λ. It determines the complexity of solving the smallest linear system expressing each output bit of the generator as a linear function of its initial state. It is widely believed that, exactly as for the combination generator, the linear complexity of a filter generator increases with the degree of the filtering function (see e.g. [24, 39]). For instance, it has been shown by Rueppel that, when the LFSR length n

is a large prime, $\Lambda \geq \binom{n}{d}$ for most functions f of degree d [39, Chapter 5]. However, as explained in [28], the well-known Blahut's theorem [2] implies that Λ is entirely determined by the univariate form of the filtering function, $F(X) = \sum_{i=0}^{2^n-2} A_i X^i$:

$$\Lambda = \#\{0 \leq i \leq 2^n - 2 \ : \ A_i \neq 0\} \ .$$

Then, it clearly appears from this formula that the linear complexity of the filter generator is invariant under monomial equivalence since decimating the vector $(A_0, \ldots, A_{2^n-2})$ by some k coprime to $(2^n - 1)$ does not modify the number of its nonzero terms.

A major observation due to Rønjom and Helleseth [38] is that the linear complexity is always smaller than or equal to the number of unknowns we expect in a linearized version of the system of equations derived from the multivariate representation. Indeed, the resulting linear system considers as unknowns all monomials of degree at most $\deg(f)$ in the bits of the initial state, i.e. roughly

$$\Lambda = \Lambda(F) \triangleq \sum_{i=1}^{\deg f} \binom{n}{i} \quad \text{unknowns.}$$

Using that the multivariate degree of the univariate monomial X^k is the number of ones in the binary representation of k, which is identified with $\mathsf{w}_H(k)$, we get that all coefficients A_k with $\mathsf{w}_H(k) > \deg f$ vanish. Therefore, the linear complexity Λ of the generator, i.e., the number of nonzero A_k, is at most the number of k such that $\mathsf{w}_H(k) \leq \deg(f)$, which corresponds to the number of unknowns in the multivariate linear system. Therefore, for any filter generator obtained by monomial equivalence, the best basic algebraic attack has data complexity $\mathcal{O}(\Lambda)$. The on-line step of the attack has time complexity $\mathcal{O}(\Lambda)$ (since the knowledge of Λ keystream bits determines the initial state of the equivalent LFSR and the whole output sequence). The precomputation step consists in computing the linear complexity and the minimal polynomial of the keystream. This can be done by applying Berlekamp-Massey algorithm to the filter generator initialized by any chosen value, with time complexity $\mathcal{O}(\Lambda^2)$. This can also be done by inverting a $\Lambda \times \Lambda$ Vandermonde matrix, with time complexity $\mathcal{O}(\Lambda \log^2 \Lambda)$ as noticed in [17,35,38]. Another equivalent point of view, which yields the same complexity, is the so-called selective discrete Fourier spectra attack [16,17]. The complexities of all variants of this attack are then invariant under monomial equivalence.

3.2 Algebraic Attacks

The fact that algebraic attacks can be applied to any generator obtained by monomial equivalence has led Rønjom and Cid to define the *general algebraic immunity* of a filtering function F [36, Definition 6] as the smallest algebraic immunity for a function in the monomial equivalence class of F. But, exactly as algebraic attacks allow to decrease the degree of the equations below the degree of the filtering function by considering an annihilator g of f [10], the same idea can be used for improving the previously described attack based on the univariate approach [17]. Then, the complexity of the best attack is determined by the smallest linear complexity for an annihilator of F. This quantity has been named the *spectral immunity* of F [17, Definition 1]. As we discussed before, for any function G, including any annihilator of F,

$$\Lambda(G) \leq \sum_{i=0}^{\deg G} \binom{n}{i},$$

implying that this attack based on the univariate approach is always faster than the usual algebraic attack.

Suppose now that the previously described attack is applied to some equivalent filter generator involving the filtering function F' defined as $F'(x) = F(x^k)$, for some k with $\gcd(k, 2^n - 1) = 1$. The attack then exploits the linear complexity of an annihilator G' of F'. But, it can be observed that a function G' is an annihilator of F' if and only if $G(x) = G'(x^r)$ is an annihilator of F where $rk \equiv 1 \mod (2^n - 1)$. Then, the linear complexity of G' is then equal to the linear complexity of G, the corresponding annihilator of F. It follows that the attack applied to F' has the same complexity as the attack against the original filter generator. In other words, the spectral immunity of a filtering function F is invariant under monomial equivalence.

Therefore, it appears that the monomial equivalence does not affect the complexity of algebraic attacks since the optimal versions of these attacks are based on the univariate representation and involve the number of nonzero coefficients in this representation which is invariant under monomial equivalence.

4 Univariate Correlation Attacks

4.1 Correlation-Like Attacks on Filtered LFSR

Another type of attacks against LFSR-based stream ciphers is the correlation attack and its variants. For generators using many LFSR combined by a Boolean function, a divide-and-conquer technique can be used by exploiting an approximation of the combining function f by a function g with fewer variables [40]. The attack then consists in performing an exhaustive search for the internal state of the small generator (called the target generator) composed of fewer LFSR combined by g, and in deciding which one of the states gives an output sequence having the expected correlation with the keystream. A well-known improved variant, named *fast correlation attack* [32] applies when g is linear. It identifies the problem with a decoding problem. Then an exhaustive search for the initial state of the target generator is not required anymore. Instead, a decoding algorithm for a linear code is used, for instance an algorithm exploiting sparse parity-check relations [6,8,32]. In the case of filtered LFSR, the situation is different since the only relevant target generator producing sequences correlated to the keystream, consists of an LFSR of the same size as the original generator

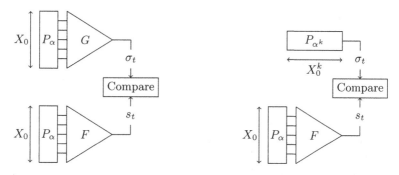

Fig. 1. Generalized correlation attack where $\gcd(k, 2^n - 1) = 1$ and $G(x) = \mathsf{Tr}^n(\lambda x^k)$.

filtered by a linear approximation of f. In this situation, the classical correlation attack cannot be faster than a brute-force attack, implying that only fast correlation attacks are relevant on filtered LFSR. To avoid these attacks, filtering functions must have a high nonlinearity.

Rønjom and Cid [36, Sect. 6.2] have then pointed out that the monomial equivalence requires extending the nonlinearity criterion. As the nonlinearity of a Boolean function f is the distance of f to all affine functions, the distance to all monomial functions with an exponent coprime to $(2^n - 1)$ must also be taken into account. Indeed, the fast correlation attack can be generalized as follows. Let us consider an LFSR of size n, of primitive root α and of initial state X_0, filtered by a Boolean function F. We suppose now that there exist $\lambda \in \mathbb{F}_{2^n} \backslash \{0\}$ and k coprime to $(2^n - 1)$ such that the function F is highly correlated to $G(x) = \mathsf{Tr}^n(\lambda x^k)$. Because k is coprime to $(2^n - 1)$, the monomial equivalence can be applied to the LFSR filtered by G, as depicted on Fig. 1. Then we can perform a fast correlation attack and recover the initial state of the LFSR defined by α^k, which corresponds to X_0^k. As k is coprime to $(2^n - 1)$, we then recover X_0. In other words, a fast correlation attack can be mounted even if the approximation G of F is nonlinear but has a trace representation with a single term, $\mathsf{Tr}^n(\lambda x^k)$ with $\gcd(k, 2^n - 1) = 1$. The corresponding design criterion is that the filtering function F must have a high generalized nonlinearity. This notion has been first introduced by Youssef and Gong in 2001 [41], but was not motivated by any attack.

Definition 1 (Extended Walsh-Transform [41]). *Let F a function from $\mathbb{F}_{2^n}$ into $\mathbb{F}_2$, then its extended Walsh transform is*

$$\widehat{F}(\lambda, k) = \sum_{x \in \mathbb{F}_{2^n}} (-1)^{F(x) + \mathsf{Tr}(\lambda x^k)}$$

where $\lambda \in \mathbb{F}_{2^n}$ and $\gcd(k, 2^n - 1) = 1$. Then, the generalized nonlinearity:

$$\mathsf{NLG}(F) = 2^{n-1} - \frac{1}{2} \max_{\substack{\lambda \in \mathbb{F}_{2^n} \\ k : \gcd(k, 2^n - 1) = 1}} |\widehat{F}(\lambda, k)|$$

is the distance of F to the components of all monomial permutations of $\mathbb{F}_{2^n}$.

4.2   A More Efficient Correlation Attack

The previously described attack applies when F is correlated with a monomial function whose exponent k is coprime to $(2^n - 1)$. However, the exponents k with $\gcd(k, 2^n - 1) > 1$ must also be taken into account even if they do not provide an equivalence relation. Let us now consider some k which is not coprime to $(2^n - 1)$ and some Boolean function H such that F is correlated to $G : x \mapsto H(x^k)$. We can then also apply some monomial transformation to the target generator which is composed of the LFSR defined by α filtered by G. Indeed, the LFSR internal state at time t is $X_0 \alpha^t$, implying that the sequence produced by the target generator is $\sigma_t = G(X_0 \alpha^t) = H(X_0^k \alpha^{kt})$ for all $t \geq 0$. On the other hand, the LFSR with characteristic polynomial P_{α^k} generates the successive internal states $(Y_0 \alpha^{kt})_{t \geq 0}$, implying that σ can also be generated by the LFSR defined by α^k filtered by H. In other words, the two generators produce exactly the same sequence if the initial state of the LFSR defined by α^k satisfies $Y_0 = X_0^k$, as depicted on Fig. 2. It is important to notice that the least period of the sequence generated by the LFSR defined by α^k is

$$\tau_k = \mathsf{ord}(\alpha^k) = \frac{2^n - 1}{\gcd(k, 2^n - 1)} \, .$$

We will see that this quantity plays a major role in the attack.

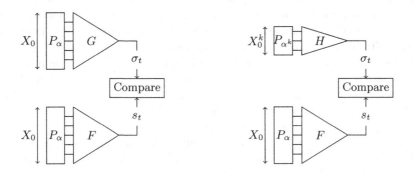

Fig. 2. Generalized correlation attack where $\gcd(k, 2^n - 1) > 1$.

Firstly, the number of possible values for an initial state of the target LFSR of the form $Y_0 = X_0^k$ is τ_k. As previously mentioned, the classical correlation attack described by Siegenthaler is not relevant against filter generators because it requires an exhaustive search over all possible initial states of the constituent LFSR, leading to a time complexity higher than or equal to the cost of a brute-force attack. But, in our new setting, the attacker needs to perform an exhaustive search over a set of size $\tau_k < 2^n$, implying that this exhaustive search may be faster than the brute-force attack. More precisely, the data complexity required for applying the optimal hypothesis test (i.e., defined by the Neyman-Pearson Lemma) and determining the correct initialization out of τ_k possibilities is

$$N = \frac{2\ln(\tau_k)}{\varepsilon^2}$$

where ε is the correlation between F and G (see e.g. [18, Sect. 4.1]). The time complexity of Siegenthaler's Algorithm is

$$\text{Time} = O\left(\frac{\tau_k \ln(\tau_k)}{\varepsilon^2}\right).$$

The counter-part of this attack compared to the case where k is coprime to $(2^n - 1)$ is that the knowledge of the quantity recovered in the attack, X_0^k, does not enable us to determine the whole initial state X_0 since k is no longer coprime to $(2^n - 1)$. However, we get some information on X_0.

Lemma 1. *The knowledge of X_0^k gives $\log_2(\tau_k)$ bits of information on X_0 where $\tau_k = (2^n - 1)/\gcd(k, 2^n - 1)$.*

Proof. Let X_0 be a non-zero element in the field $\mathbb{F}_{2^n}$ and α a primitive root. There is a unique $i \in [0, 2^n - 2]$ such that $X_0 = \alpha^i$. Then, $r = i \bmod \tau_k$ satisfies

$$X_0^k = \alpha^{qk\tau_k}\alpha^{rk} = \alpha^{rk}$$

by definition of τ_k. Moreover, r is the unique integer in $[0, \tau_k - 1]$ such that $X_0^k = \alpha^{rk}$. Indeed, if there exist r_1 and r_2, $r_1 > r_2$ such that $\alpha^{r_1 k} = \alpha^{r_2 k}$ then $\alpha^{(r_1 - r_2)k} = 1$. Then, $(r_1 - r_2)$ is a multiple of τ_k which is the order of α^k. This is impossible since $r_2 - r_1 \in [0, \tau_k - 1]$. Therefore, for $X_0 = \alpha^i$, the knowledge of X_0^k gives the value of the remainder of the Euclidean division of i by τ_k. It then provides $\log_2(\tau_k)$ bits of information on X_0. $\qquad\square$

4.3 Recovering the Remaining Bits of the Initial State

Once X_0^k has been recovered, the remaining $(n - \log(\tau_k))$ bits of X_0 can be found by an exhaustive search with time complexity proportional to

$$\frac{2^n - 1}{\tau_k} = \gcd(k, 2^n - 1).$$

Another method consists in combining several correlation attacks in a divide-and-conquer approach, exactly as against combination generators. Suppose that there exist two integers k_1 and k_2 such that the two distinct correlation attacks can be performed in order to successively recover $X_0^{k_1}$ and $X_0^{k_2}$. This means that we have found

$$r_1 = i \mod \tau_{k_1} \text{ and } r_2 = i \mod \tau_{k_2}.$$

By the Chinese remainder theorem, this leads to the value of the remainder of the Euclidean division of i by $\text{lcm}(\tau_{k_1}, \tau_{k_2})$. The best situation for the attacker is obviously the case where τ_{k_1} and τ_{k_2} are coprime, otherwise there is some redundancy between the information retrieved by the two distinct attacks.

4.4 Fast Correlation Attack When H is Linear

In the correlation attack, the target generator is composed of the LFSR defined by α^k filtered by a Boolean function H, and it generates sequences σ with period $\tau_k < (2^n - 1)$. Then, as noticed in the pioneer work by Meier and Staffelbach [32], any N-bit portion of σ can be seen as a codeword in a code of length N and size τ_k. Therefore, recovering the initial state of the target generator boils down to decoding the corresponding n-bit keystream with respect to this code since the keystream can be identified with the result of the transmission of σ through a binary symmetric channel with error-probability $\frac{1}{2}(1 - \varepsilon)$ where ε is the correlation between the two sequences.

In the specific case where the function H defining $G(x) = H(x^k)$ is linear, i.e., $H(x) = \mathsf{Tr}(\lambda x)$ for some $\lambda \in \mathbb{F}_{2^n}$, the involved code is a linear code. Some decoding algorithms dedicated to linear codes can then be used. These algorithms are faster than the exhaustive search (which corresponds to a maximum-likelihood decoding), at the price of a higher data complexity. The corresponding attack is then named *fast correlation attack* [32]. Obviously, a major parameter affecting the complexity of the decoding procedure is the dimension of the involved code. This dimension is the degree of the minimal polynomial of α^k, which may be smaller than n: it corresponds to the size n_k of the cyclotomic class of k. Equivalently, n_k is the smallest integer m such that $2^m \equiv 1 \bmod \tau_k$. In other words, if α^k belongs to a subfield $\mathbb{F}_{2^m}$ of $\mathbb{F}_{2^n}$, then the fast correlation attack consists in decoding a linear code of dimension m, instead of a code of dimension n. This may enable the attacker to recover $\log_2(\tau_k)$ bits of the initial state with a lower complexity than the fast correlation attack involving the original LFSR of length n. The optimal situation which maximizes the number of bits recovered by the attacker for a given complexity is then when $\tau_k = 2^m - 1$ for some divisor m of n, i.e., when k is such that $\gcd(k, 2^n - 1) = (2^n - 1)/(2^m - 1)$. Several decoding algorithms have been proposed in this context [6–8, 21, 22, 32, 33] which offer different trade-offs between the dimension of the code and the error probability (see [1] for a recent survey).

Example 1. Let us consider an LFSR of size 10 with primitive characteristic polynomial $P(X) = X^{10} + X^9 + X^7 + X^6 + X^5 + X^4 + X^3 + X^2 + 1$. We then use as a filtering function a balanced function of 10 variables with a high nonlinearity obtained by Dobbertin's construction [12]. As described by Dobbertin, we start from a bent function which is constant on a subspace of dimension $\frac{n}{2}$ and replace this constant restriction by a balanced function in order to make the whole function balanced. Here we start from $\mathsf{Tr}(\alpha x^{33})$ where α is a root of P since this function is bent, and modify it as in [12]. It is worth noticing that this modification makes the function much more complex. In particular, it increases its degree and its linear complexity, at the price of a very small degradation of its nonlinearity. We construct this way a balanced function F of 10 variables with nonlinearity 481 and algebraic immunity 3. By computing its univariate representation, we get that the linear complexity of the keystream is equal to 992. Therefore, this filtering function meets all design criteria related to algebraic-like attacks and to fast correlation attacks. However, by construction, our filtered function F is very close to the Boolean function $G(x) = \mathsf{Tr}(\alpha x^{33})$. This means that the keystream is highly correlated to the output of the LFSR defined by α^{33}. Indeed, the correlation between the two sequences equals $\varepsilon = 1 - 2^{-9} d_H(F, G) = 0.96$. We can mount a fast correlation attack on an LFSR of size 5, and we recover almost 5 bits of the internal state of the generator. This attack is obviously much faster than the usual fast correlation attack: in our new setting, the involved correlation is $\varepsilon = 0.96$ and the code dimension is $n_{33} = 5$, while the usual fast correlation attack corresponds to a correlation $\varepsilon' = 1 - 481 \times 2^{-9} = 0.06$ and code dimension $n = 10$. The remaining 5 bits of the initial state can be determined by an exhaustive search over 33 possible values.

The previous example was rather specific since the filtering function is designed from a component of a monomial mapping x^k with k of the form $k = \frac{(2^n - 1)}{(2^m - 1)}$. However, a similar situation may happen for many other filtering functions which do not have any such specific structure. In order to quantify the advantage of this new setting, we first need a closer look at the complexity of fast correlation attacks. The decoding algorithms used in this context include some methods exploiting the existence of low-weight parity-check relations for the LFSR sequence [6, 8, 21, 32]. These relations are derived from sparse multiples of the LFSR characteristic polynomial, implying that the data complexity which corresponds the degree of these multiples grows very fast with the LFSR length (unless the LFSR characteristic polynomial is very sparse). Once these relations have been found in a precomputation step, the attack consists in applying an iterative decoding algorithm. For instance, the complexity of the original attack based on parity-check relations with 3 terms is estimated by [6]:

$$\mathsf{Data} = \mathcal{O}\left(\frac{1}{\varepsilon} \times 2^{\frac{n}{2}}\right) \text{ and } \mathsf{Time} = \mathcal{O}\left(\left(\frac{1}{\varepsilon}\right)^3 \times 2^{\frac{n}{2}}\right).$$

Using parity-check relations with a higher weight w decreases the influence of the LFSR length by replacing $2^{n/2}$ by $2^{n/(w-1)}$, at the price of a higher influence of the correlation, i.e., in the data complexity ε is replaced by $\varepsilon^{2(w-2)/(w-1)}$. The time complexity can be improved by different techniques, but the data complexity of most of these algorithms has a similar behaviour.

Example 2. Let us consider the same LFSR of size 10 as in Example 1, but now filtered by a Boolean function which is not constructed from a monomial function. We choose as a filtering function the following function of 6 variables:
$f(x_0, x_1, x_2, x_3, x_4, x_5) = x_0 x_1 x_2 x_3 x_4 + x_0 x_1 x_2 x_3 x_5 + x_0 x_1 x_2 x_4 x_5 + x_0 x_1 x_2 x_4 + x_0 x_1 x_2 + x_0 x_1 x_3 x_4 + x_0 x_1 x_3 + x_0 x_1 x_4 + x_0 x_1 x_5 + x_0 x_1 + x_0 x_2 x_3 x_4 + x_0 x_2 x_3 x_5 + x_0 x_2 x_4 x_5 + x_0 x_2 x_4 + x_0 x_2 + x_0 x_3 x_4 + x_0 x_4 + x_0 + x_1 x_2 x_3 x_4 x_5 + x_1 x_2 x_3 x_4 + x_1 x_2 x_3 x_5 + x_1 x_2 x_3 + x_1 x_2 x_4 + x_1 x_2 + x_1 x_3 x_5 + x_1 x_3 + x_1 x_4 + x_1 x_5 + x_1 + x_2 x_3 x_4 x_5 + x_2 x_3 x_4 + x_2 x_3 x_5 + x_2 x_3 + x_2 + x_3 x_4 + x_4 x_5 + x_4$
and the inputs of f are given by the following tapping sequence $(\gamma_1, \ldots, \gamma_6) = (9, 8, 6, 3, 1, 0)$. The corresponding 10-variable function has nonlinearity 352, algebraic immunity 3 and the linear complexity of the generated keystream

is 637. But there exists a function G of the form $G(x) = \text{Tr}(\lambda x^{33})$ at distance 456 from F. The correlation between the keystream and the output of a non-filtered LFSR of size $n_{33} = 5$ is then equal to $\varepsilon = 0.11$. A fast correlation attack in this setting appears to be more efficient than the usual fast correlation attack, which has parameters $n = 10$ and $\varepsilon' = 0.31$. For instance, if the iterative algorithm with parity-check relations of weight 3 is used, the ratio between the data complexities of the two attacks is given by

$$\frac{\text{Data}}{\text{Data}'} = \left(\frac{\varepsilon'}{\varepsilon}\right) \times 2^{\frac{n_{33}-n}{2}} = 0.498 \ .$$

4.5   Correlation Attack Using a Fast Fourier Transform When H is Nonlinear

In the general case, i.e., when H is nonlinear, the correlation attack, as originally described in [40] corresponds to an exhaustive search over all initial states of the target generator of the form $Y_0 = X_0^k$. For each of these Y_0, the first N bits of the corresponding output sequence σ are generated and the correlation between σ and the keystream is computed, namely

$$\sum_{t=0}^{N-1} (-1)^{s_t + \sigma_t} \tag{1}$$

where N is the number of keystream bits we need to be able to detect the bias, i.e., $N = \frac{2\ln(\tau_k)}{\varepsilon^2}$ where ε is the expected correlation. The time complexity of this algorithm is therefore proportional to

$$\tau_k \times N = \frac{2\tau_k \ln(\tau_k)}{\varepsilon^2} \ .$$

We will now show that this time complexity can be improved by using a fast Fourier transform even when H is nonlinear[2]. A similar technique has been described in [5,34] but in an attack against combination generators. We now prove that it also applies in our context.

Let $\langle \alpha^k \rangle$ denote the multiplicative subgroup of $\mathbb{F}_{2^n}^*$ generated by α^k, i.e., the set with τ_k elements $\{1, \alpha^k, \alpha^{2k}, \cdots, \alpha^{(\tau_k-1)k}\}$. This set is composed of all possible internal states $Y_0 = X_0^k$ which must be examined in the attack. Then, the attacker aims at finding the initial state $Y_0 \in \langle \alpha^k \rangle$ which maximizes the correlation given by (1) where $\sigma_t = H(Y_0 \alpha^{kt})$. For any $Y_0 \in \langle \alpha^k \rangle$, we compute

$$\mathcal{Z}(Y_0) = \sum_{t=0}^{N-1} (s_t \oplus \sigma_t) = \sum_{r=0}^{\tau_k-1} \sum_{q=0}^{\lceil \frac{N-r}{\tau_k} \rceil - 1} (s_{q\tau_k+r} \oplus \sigma_r)$$

---

[2] The use of a fast Fourier transform for computing the correlation in the linear case has been pointed out by several authors including [8,26].

since for any t, $\sigma_t = \sigma_{t+\tau_k}$. We then deduce

$$\mathcal{Z}(Y_0) = \sum_{r=0}^{\tau_k-1} (\sigma_r \oplus 1) \left(\sum_{q=0}^{\lceil \frac{N-r}{\tau_k} \rceil - 1} s_{q\tau_k+r} \right) + \sum_{r=0}^{\tau_k-1} \sigma_r \left(\left\lceil \frac{N-r}{\tau_k} \right\rceil - \sum_{q=0}^{\lceil \frac{N-r}{\tau_k} \rceil - 1} s_{q\tau_k+r} \right) .$$

For any $0 \le r < \tau_k$, we set

$$\mathcal{S}(r) = \sum_{q=0}^{\lceil \frac{N-r}{\tau_k} \rceil - 1} s_{q\tau_k+r} .$$

Then, we have

$$\mathcal{Z}(Y_0) = \sum_{r=0}^{\tau_k-1} (\sigma_r \oplus 1)\mathcal{S}(r) + \sum_{r=0}^{\tau_k-1} \sigma_r \left(\left\lceil \frac{N-r}{\tau_k} \right\rceil - \mathcal{S}(r) \right)$$

$$= \sum_{r=0}^{\tau_k-1} (-1)^{\sigma_r} \left(\mathcal{S}(r) - \frac{1}{2}\left\lceil \frac{N-r}{\tau_k} \right\rceil \right) + \frac{N}{2} .$$

It follows that

$$\sum_{t=0}^{N-1} (-1)^{s_t + \sigma_t(Y_0)} = N - 2\mathcal{Z}(Y_0) = \sum_{r=0}^{\tau_k-1} (-1)^{\sigma_r(Y_0)} \left(\left\lceil \frac{N-r}{\tau_k} \right\rceil - 2\mathcal{S}(r) \right) .$$

We need to compute this value for $Y_0 = \alpha^{ik}$ for every $0 \le i < \tau_k$. But,

$$\sigma_t(\alpha^{ik}) = H(\alpha^{ik}\alpha^{tk}) = H(\alpha^{(t+i)k}) = \sigma_{t+i}(1) .$$

In other words, we search for the integer i, $0 \le i < \tau_k$ which maximizes the value

$$\sum_{r=0}^{\tau_k-1} (-1)^{\sigma_{r+i \bmod \tau_k}(1)} \left(\left\lceil \frac{N-r}{\tau_k} \right\rceil - 2\mathcal{S}(r) \right) ,$$

which corresponds to the convolution product of two vectors of length τ_k, namely $(\sigma_t(1))_{0 \le t < \tau_k}$ and $(\mathcal{S}(t))_{0 \le t < \tau_k}$. This can be done efficiently with a fast Fourier transform with time complexity $\mathcal{O}(\tau_k \log \tau_k)$ (see e.g. [3] or [23, p. 299]). The memory complexity of the attack is then $\mathcal{O}(\tau_k)$ and the overall time complexity (including the computation of all $\mathcal{S}(t)$) is then roughly

$$\mathsf{Time} = \tau_k \log \tau_k + \frac{2 \ln(\tau_k)}{\varepsilon^2}.$$

Example 3. Let us consider the LFSR of size 12 with characteristic polynomial $P(X) = X^{12} + X^{10} + X^9 + X^8 + X^7 + X^5 + X^4 + X^3 + X^2 + X + 1$ and filtered by the same 6-variable function as in Example 2, but where the inputs of F are now defined by the tapping sequence $(\gamma_1, \dots, \gamma_6) = (11, 10, 7, 5, 2, 0)$. Then, the correlation between F and any function of the form $G = \mathsf{Tr}(\lambda x^k)$

with $k = \ell\frac{2^n-1}{2^m-1}$ and $\gcd(\ell, 2^n - 1) = 1$ is too low for improving on the classical correlation attack. However, we can use $k = 45$ which satisfies $\mathrm{ord}(\alpha^k) = 91$. In this case, we are able to get a higher correlation since we allow all possible functions H, not only the linear ones. Here, the best approximation by a function of the form $G(x) = H(x^k)$ gives us a correlation equal to 0.125. With an FFT, the attack requires roughly $(592 + 574) = 1166$ operations, and 574 keystream bits. The whole initial state can then be recovered by an exhaustive search.

4.6   Approximation of the Filtering Function by $H(x^k)$

All previous correlation attacks exploit the existence of a function G of the form $G(x) = H(x^k)$ for some k with $\gcd(k, 2^n - 1) > 1$, which provides a good approximation of F. In particular, the fast correlation attacks involving a shorter LFSR point out that the notion of generalized nonlinearity as defined in [41] must be extended in order to capture these new attacks: it appears that the distance of the filtering function to all $\mathrm{Tr}(\lambda x^k)$ with $k = \ell \times \frac{2^n-1}{2^m-1}$ where m is a divisor of n and $\gcd(\ell, 2^n - 1) = 1$ is a much more relevant quantity than its distance to the components of monomial permutations.

Moreover, even if such a fast correlation attack is not feasible, for instance if n is a prime, an efficient correlation attack may be possible based on the approximation of F by $G(x) = H(x^k)$ for some k with $\gcd(k, 2^n - 1) > 1$. As observed in the previous example, the fact that H can be nonlinear usually yields a higher correlation. The best approximation of the form $G(x) = H(x^k)$ can be computed from F as follows. For the sake of simplicity, we now suppose that k is a divisor of $(2^n - 1)$, or equivalently that $\tau = (2^n - 1)/k$ (otherwise, we get similar results by replacing k by $\gcd(k, 2^n - 1)$). Let $\langle \alpha^\tau \rangle$ be the cyclic subgroup of $\mathbb{F}_{2^n}$ of order k. Then, by shifting this cyclic subgroup, we obtain the sets $E_i = \alpha^i \langle \alpha^\tau \rangle$, for $0 \leq i < \tau$ which provide the partition

$$\mathbb{F}_{2^n}^* = \bigcup_{i=0}^{\tau-1} E_i$$

where all sets E_i, for $0 \leq i < \tau$, are disjoint. It follows that G is constant on any set E_i since, for $x = \alpha^i \times \alpha^{j\tau}$, we have

$$G(x) = H((\alpha^i \alpha^{j\tau})^k) = H(\alpha^{ik}).$$

The correlation between F and G can therefore be expressed as follows:

$$\sum_{x \in \mathbb{F}_{2^n}} (-1)^{F(x)+H(x^k)} = 1 + \sum_{x \in \mathbb{F}_{2^n}^*} (-1)^{F(x)+H(x^k)}$$

$$= 1 + \sum_{i=0}^{\tau-1} (-1)^{H(\alpha^{ik})} \left(\sum_{y \in E_i} (-1)^{F(y)} \right). \qquad (2)$$

If $\gcd(k, \tau) = 1$, all values α^{ik}, for $0 \le i < \tau$ belong to different sets E_j. Hence, the function H which maximizes this correlation is the function defined by

$$H(\alpha^{ik}) = \begin{cases} 0 & \text{if } \sum_{y \in E_i} (-1)^{F(y)} > 0 \\ 1 & \text{if } \sum_{y \in E_i} (-1)^{F(y)} < 0 \end{cases}$$

In other words, $H(\alpha^{ik}) = 1$ if and only if the Hamming weight of the restriction of F to E_i is strictly greater than $k/2$. It can be observed that H is uniquely determined because the weight of the restriction of F cannot be equal to $k/2$ since k is odd. This also implies that, for the optimal choice of H, we obtain

$$\sum_{x \in \mathbb{F}_{2^n}} (-1)^{F(x)+H(x^k)} = 1 + \sum_{i=0}^{\tau-1} \left| \sum_{y \in E_i} (-1)^{F(y)} \right| \ge 1 + \tau$$

since each term in the sum is at least 1. Therefore, for any F, we can always find a function H such that the correlation between F and $G(x) = H(x^k)$ is at least $(1 + \tau)2^{-n} \simeq k^{-1}$. It is worth noticing that this lower bound on the correlation does not decrease when the LFSR length n increases.

In the case where $\gcd(k, \tau) = d > 1$, we have that α^{ik} and $\alpha^{(i+\frac{\tau}{d})k}$ belong to the same set E_j. Indeed, $\alpha^{\frac{k\tau}{d}} \in \langle \alpha^\tau \rangle$. Equation (2) can then be rewritten as

$$\sum_{x \in \mathbb{F}_{2^n}} (-1)^{F(x)+H(x^k)} = 1 + \sum_{i=0}^{\frac{\tau}{d}-1} (-1)^{H(\alpha^{ik})} \left(\sum_{j=0}^{d-1} \left(\sum_{y \in E_{i+j\frac{\tau}{d}}} (-1)^{F(y)} \right) \right).$$

In this case, the value of H at point α^{ik} is defined by the weight of the restriction of F to the set $\bigcup_{j=0}^{d-1} E_{i+j\frac{\tau}{d}}$. Using again that this set has an odd cardinality, we get that the correlation between F and $G(x) = H(x^k)$ is at least $(1 + \frac{\tau}{d})2^{-n}$.

While in usual (fast) correlation attacks, choosing a filtering function with a high nonlinearity guarantees that the attack will be infeasible, this is not the case here. For instance, some bent functions in the so-called class $\mathcal{PS}^-$ [11] are constant on all sets $\lambda \langle \alpha^\tau \rangle$ for $\tau = 2^{n/2} + 1$, while they have the best nonlinearity.

The previous results enable us to find the best approximation of F by a function of the form $H(x^k)$. However, improving the complexity of this search when n grows and F depends on a few inputs only remains an open issue. Indeed, it seems difficult to use this property of F to simplify the search for the optimal H. Another open problem is to be able to find in an efficient way the best approximation of the form $G(x) = \mathsf{Tr}(\lambda x^k)$.

5   Conclusions

While the monomial equivalence introduced by Rønjom and Cid does not affect the security of filter generators regarding algebraic attacks, it usually allows to decrease the complexity of correlation attacks and their variants. Most importantly, considering a non-bijective monomial mapping enables the attacker to mount a divide-and-conquer attack by decomposing the set of all nonzero initial states with respect to some multiplicative subgroup having a smaller order. If the LFSR length is not a prime, the involved subgroup may be a subfield and this divide-and-conquer attack can be further improved as in fast correlation attacks. A counter-measure to avoid these attacks then consists in choosing for the LFSR length a Mersenne prime, i.e. both n and $(2^n - 1)$ are prime.

References

1. Ågren, M., Löndahl, C., Hell, M., Johansson, T.: A survey on fast correlation attacks. Cryptogr. Commun. **4**(3–4), 173–202 (2012)
2. Blahut, R.E.: Theory and Practice of Error Control Codes. Addison-Wesley, Boston (1983)
3. Blahut, R.E.: Fast Algorithms for Digital Signal Processing. Addison-Wesley, Boston (1985)
4. Canteaut, A.: Filter generator. In: van Tilborg, H.C.A., Jajodia, S. (eds.) Encyclopedia of Cryptography and Security, pp. 726–729. Springer, Heidelberg (2011)
5. Canteaut, A., Naya-Plasencia, M.: Correlation attacks on combination generators. Cryptogr. Commun. **4**(3–4), 147–171 (2012)
6. Canteaut, A., Trabbia, M.: Improved fast correlation attacks using parity-check equations of weight 4 and 5. In: Preneel, B. (ed.) EUROCRYPT 2000. LNCS, vol. 1807, pp. 573–588. Springer, Heidelberg (2000)
7. Chepyzhov, V.V., Johansson, T., Smeets, B.: A simple algorithm for fast correlation attacks on stream ciphers. In: Schneier, B. (ed.) FSE 2000. LNCS, vol. 1978, pp. 181–195. Springer, Heidelberg (2001)
8. Chose, P., Joux, A., Mitton, M.: Fast correlation attacks: an algorithmic point of view. In: Knudsen, L.R. (ed.) EUROCRYPT 2002. LNCS, vol. 2332, pp. 209–221. Springer, Heidelberg (2002)
9. Courtois, N.T.: Fast algebraic attacks on stream ciphers with linear feedback. In: Boneh, D. (ed.) CRYPTO 2003. LNCS, vol. 2729, pp. 176–194. Springer, Heidelberg (2003)
10. Courtois, N.T., Meier, W.: Algebraic attacks on stream ciphers with linear feedback. In: Biham, E. (ed.) EUROCRYPT 2003. LNCS, vol. 2656, pp. 345–359. Springer, Heidelberg (2003)
11. Dillon, J.: Elementary Hadamard difference sets. Ph.D. thesis, University of Maryland (1974)
12. Dobbertin, H.: Construction of bent functions and balanced Boolean functions with high nonlinearity. In: Preneel, B. (ed.) FSE 1994. LNCS, vol. 1008, pp. 61–74. Springer, Heidelberg (1995)
13. ECRYPT - European Network of Excellence in Cryptology: The eSTREAM Stream Cipher Project (2005). http://www.ecrypt.eu.org/stream/
14. Golic, J.D.: On the security of nonlinear filter generators. In: Gollmann, D. (ed.) FSE 1996. LNCS, vol. 1039, pp. 173–188. Springer, Heidelberg (1996)
15. Golomb, S.W., Gong, G.: Signal Design for Good Correlation: For Wireless Communication, Cryptography, and Radar. Cambridge University Press, Cambridge (2004)
16. Gong, G.: A closer look at selective DFT attacks. CACR report 2011-35, University of Waterloo (2011)

17. Gong, G., Rønjom, S., Helleseth, T., Hu, H.: Fast discrete Fourier spectra attacks on stream ciphers. IEEE Trans. Inf. Theor. **57**(8), 5555–5565 (2011)
18. Hell, M., Johansson, T., Brynielsson, L.: An overview of distinguishing attacks on stream ciphers. Cryptogr. Commun. **1**(1), 71–94 (2009)
19. Helleseth, T.: Maximal-length sequences. In: van Tilborg, H.C.A., Jajodia, S. (eds.) Encyclopedia of Cryptography and Security, 2nd edn, pp. 763–766. Springer, Heidelberg (2011)
20. Helleseth, T., Rønjom, S.: Simplifying algebraic attacks with univariate analysis. In: Information Theory and Applications - ITA 2011, pp. 153–159. IEEE (2011)
21. Johansson, T., Jönsson, F.: Improved fast correlation attacks on stream ciphers via convolutional codes. In: Stern, J. (ed.) EUROCRYPT 1999. LNCS, vol. 1592, pp. 347–362. Springer, Heidelberg (1999)
22. Johansson, T., Jönsson, F.: Fast correlation attacks through reconstruction of linear polynomials. In: Bellare, M. (ed.) CRYPTO 2000. LNCS, vol. 1880, pp. 300–315. Springer, Heidelberg (2000)
23. Joux, A.: Algorithmic Cryptanalysis. Chapman & Hall/CRC, London (2009)
24. Key, E.L.: An analysis of the structure and complexity of nonlinear binary sequence generators. IEEE Trans. Inf. Theor. **22**, 732–736 (1976)
25. Lidl, R., Niederreiter, H.: Finite Fields. Cambridge University Press, Cambridge (1983)
26. Lu, Y., Vaudenay, S.: Faster correlation attack on bluetooth keystream generator E0. In: Franklin, M. (ed.) CRYPTO 2004. LNCS, vol. 3152, pp. 407–425. Springer, Heidelberg (2004)
27. MacWilliams, F.J., Sloane, N.J.: The Theory of Error-correcting Codes. North-Holland, Amsterdam (1977)
28. Massey, J.L., Serconek, S.: A Fourier transform approach to the linear complexity of nonlinearly filtered sequences. In: Desmedt, Y.G. (ed.) CRYPTO 1994. LNCS, vol. 839, pp. 332–340. Springer, Heidelberg (1994)
29. Maximov, A., Johansson, T., Babbage, S.: An improved correlation attack on A5/1. In: Handschuh, H., Hasan, M.A. (eds.) SAC 2004. LNCS, vol. 3357, pp. 1–18. Springer, Heidelberg (2004)
30. McEliece, R.J.: Finite Fields for Computer Scientists and Engineers. Kluwer, Dordrecht (1987)
31. Meier, W., Pasalic, E., Carlet, C.: Algebraic attacks and decomposition of Boolean functions. In: Cachin, C., Camenisch, J.L. (eds.) EUROCRYPT 2004. LNCS, vol. 3027, pp. 474–491. Springer, Heidelberg (2004)
32. Meier, W., Staffelbach, O.: Fast correlation attack on certain stream ciphers. J. Cryptol. **1**, 159–176 (1989)
33. Mihaljević, M.J., Fossorier, M.P.C., Imai, H.: A low-complexity and high-performance algorithm for the fast correlation attack. In: Schneier, B. (ed.) FSE 2000. LNCS, vol. 1978, pp. 196–212. Springer, Heidelberg (2001)
34. Naya-Plasencia, M.: Cryptanalysis of Achterbahn-128/80. In: Biryukov, A. (ed.) FSE 2007. LNCS, vol. 4593, pp. 73–86. Springer, Heidelberg (2007)
35. Rønjom, S.: Powers of subfield polynomials and algebraic attacks on word-based stream ciphers. IACR Cryptology ePrint Archive 2015/495 (2015)
36. Rønjom, S., Cid, C.: Nonlinear equivalence of stream ciphers. In: Hong, S., Iwata, T. (eds.) FSE 2010. LNCS, vol. 6147, pp. 40–54. Springer, Heidelberg (2010)
37. Rønjom, S., Gong, G., Helleseth, T.: On attacks on filtering generators using linear subspace structures. In: Golomb, S.W., Gong, G., Helleseth, T., Song, H.-Y. (eds.) SSC 2007. LNCS, vol. 4893, pp. 204–217. Springer, Heidelberg (2007)

38. Rønjom, S., Helleseth, T.: A new attack on the filter generator. IEEE Trans. Inf. Theor. **53**(5), 1752–1758 (2007)
39. Rueppel, R.A.: Analysis and Design of Stream Ciphers. Springer, Heidelberg (1986)
40. Siegenthaler, T.: Decrypting a class of stream ciphers using ciphertext only. IEEE Trans. Comput. **C–34**(1), 81–84 (1985)
41. Youssef, A.M., Gong, G.: Hyper-bent functions. In: Pfitzmann, B. (ed.) EURO-CRYPT 2001. LNCS, vol. 2045, pp. 406–419. Springer, Heidelberg (2001)
42. Zierler, N.: Linear recurring sequences. J. Soc. Indus. Appl. Math. **7**, 31–48 (1959)

Components

Lightweight MDS Generalized Circulant Matrices

Meicheng Liu[1,2(✉)] and Siang Meng Sim[1(✉)]

[1] Nanyang Technological University, Singapore, Singapore
ssim011@e.ntu.edu.sg
[2] State Key Laboratory of Information Security,
Institute of Information Engineering, Chinese Academy of Sciences,
Beijing 100093, People's Republic of China
meicheng.liu@gmail.com

Abstract. In this article, we analyze the circulant structure of generalized circulant matrices to reduce the search space for finding lightweight MDS matrices. We first show that the implementation of circulant matrices can be serialized and can achieve similar area requirement and clock cycle performance as a serial-based implementation. By proving many new properties and equivalence classes for circulant matrices, we greatly reduce the search space for finding lightweight maximum distance separable (MDS) circulant matrices. We also generalize the circulant structure and propose a new class of matrices, called *cyclic matrices*, which preserve the benefits of circulant matrices and, in addition, have the potential of being self-invertible. In this new class of matrices, we obtain not only the MDS matrices with the least XOR gates requirement for dimensions from 3×3 to 8×8 in $GF(2^4)$ and $GF(2^8)$, but also involutory MDS matrices which was proven to be non-existence in the class of circulant matrices. To the best of our knowledge, the latter matrices are the first of its kind, which have a similar matrix structure as circulant matrices and are involutory and MDS simultaneously. Compared to the existing best known lightweight matrices, our new candidates either outperform or match them in terms of XOR gates required for a hardware implementation. Notably, our work is generic and independent of the metric for lightweight. Hence, our work is applicable for improving the search for efficient circulant matrices under other metrics besides XOR gates.

Keywords: Lightweight cryptography · Diffusion layer · MDS · Circulant matrices

M. Liu—Supported by the National Natural Science Foundation of China (61303258, 61379139 and 11526215) and the Strategic Priority Research Program of the Chinese Academy of Sciences under Grant XDA06010701.
S.M. Sim—Supported by the Singapore National Research Foundation Fellowship 2012 (NRF-NRFF2012-06).

T. Peyrin (Ed.): FSE 2016, LNCS 9783, pp. 101–120, 2016.
DOI: 10.1007/978-3-662-52993-5_6

1   Introduction

In the designing of symmetric-key ciphers, there are two fundamental concepts required for the overall security of the cipher—the confusion and diffusion properties described by Shannon [21]. Informally, the latter is to spread the internal dependencies as much as possible [22]. The diffusion layer of a cipher is often achieved by a linear diffusion matrix that transforms an input vector to some output vector through linear operations. For the choice of the diffusion layer, there can be a trade-off between the security and computation efficiency. Several designs compromise the diffusion power for a faster diffusion layer, while another trend is to maximize the diffusion power with *maximum distance separable* (MDS) matrices. The diffusion power of a matrix is often quantified by the branch number of the matrix, and an MDS matrix achieves maximum branch number, also known as perfect diffusion property. MDS matrices are widely used in many ciphers like AES [9], LED [11], SQUARE [8]. However, very often the price for having strong diffusion property is the heavy implementation cost, in either software or hardware implementations. Therefore, there is a need to reduce the implementation cost when perfect diffusion property is desired.

Recently, the designing and improving of hardware efficiency become a major trend. Several lightweight block ciphers [5,7,11,24] and lightweight hash functions [2,6,10] are designed to minimize the implementation cost. Notably in the hash function PHOTON [10], a new type of MDS matrices that can be computed recursively were proposed, so-called *serial matrices*, where a serial matrix A of order k is raised to power k and the resultant matrix A^k is MDS. In comparison to round-based implementation, serial-based implementation trades more clock cycles for lesser hardware area requirement. Such matrices were later used in block ciphers like LED [11] and more recently in authentication encryption scheme like the PRIMATEs [1].

In a nutshell, a round-based implementation computes the entire diffusion matrix of order k and applies the diffusion layer in one clock cycle. Hence, it is necessary to have all, if not most, of the k^2 entries of the diffusion matrix to be lightweight. On the other hand, a serial-based implementation computes the non-trivial row of a serial matrix[1], and applies it for k times recursively. Therefore, the primary implementation cost is the k entries of the non-trivial row and the computation time takes k clock cycles. Although it is natural to perceive that these two implementations require very different matrices, there are a type of matrices that can achieve the best of both worlds—circulant matrices.

Circulant matrices are a common type of matrices for the diffusion layer, a typical example of which is the AES diffusion matrix. They have a simple structure that every row is a right-shift of the previous row. Hence, a circulant matrix can be defined by its first row of k entries. In addition, it is known that an MDS circulant matrix can contain repeated lightweight entries. For instance in the AES diffusion matrix, there are two 1's which practically has no implementation cost

---

[1] A serial matrix of order k consists of $k-1$ rows with a single 1 and $k-1$ many 0's and a row with non-trivial entries.

for multiplication. In comparison to Hadamard matrices, another common type of matrices for the diffusion layer [3,4], which must contain k distinct entries to be MDS, circulant matrices tend to achieve lower implementation cost in a round-based implementation. Although circulant matrices cannot be directly used in a serial-based implementation, their circulant structure can be implemented in a serialized manner and achieve similar performance as the serial-based implementation. In short, using a circulant matrix in the diffusion layer gives the flexibility to do a trade-off between the area requirement and the clock cycle, whereas most of the other matrix types are suitable for either one but not both implementations.

One approach to build lightweight MDS matrices from some matrix type is to focus on some subclass of such matrices that are MDS, based on some pre-defined metric for lightweight, then pick the lightest MDS matrices from this subclass. In [13,16], the authors chose to maximise the number of 1's for better efficiency and constructed circulant-like matrices that are MDS with as many 1's as possible, then searched for the lightest MDS circulant-like matrices. In another work [12], the authors quantified lightweight with low Hamming weight and focused on involutory (self-inverse) matrices, they proposed the construction of Hadamard-Cauchy matrices that are MDS and can be involutory, then minimized the Hamming weight of a few entries of the Hadamard-Cauchy matrices. Although this approach is efficient for finding lightweight MDS matrices, the matrices found are optimal among the subclasses rather than the whole population of the matrix type.

Another approach is to pick the lightest matrix from some matrix type and check for MDS, and extend the search to the next lightest matrix if it is not MDS. This approach, also often regarded as exhaustive search, can be seen in [17,22]. The clear advantage of the exhaustive search over the previous approach is that it guarantees optimal for the given matrix type. In addition, it has the freedom to change the metric for lightweight when necessary. Despite the advantages, this approach suffers from the large search space. In [22], the authors tackled this problem by introducing the concept of the equivalence classes of Hadamard matrices to significantly reduce the search space for finding lightweight involutory MDS (IMDS) matrices. However, the equivalence relation for circulant matrices has not yet been discovered in the literatures.

There are two main challenges in the second approach. Given a set of lightweight coefficients, the first challenge in finding MDS circulant matrices with these coefficients would be the large search space due to the necessity of checking the MDS property for all possible permutations. The second challenge is that MDS circulant matrices can have repeated entries which makes the search space larger than other types of matrices, for instance Hadamard matrices, of the same order. Perhaps due to these challenges, the existing work on circulant matrix used either the first approach to find lightweight MDS circulant matrix of order 8 from some subclass of circulant matrices [13,14], or the second approach but could not complete the search for lightweight MDS circulant matrix of order 8 [17]. Therefore, this paper is devoted to tackle these problems and reduce the search space for finding generic lightweight MDS circulant matrices through analyzing the circulant structure.

Contributions. In Sect. 2.3, we illustrate how circulant matrices can have a trade-off between the area requirement and clock cycle in hardware implementation. This shows that using circulant matrix in a diffusion layer gives the designer the flexibility to choose the implementation between lower area requirement and faster computation according to the needs. In Sect. 3, we tackle both challenges faced when using the second approach for finding lightweight MDS circulant matrices. In Sect. 3.1, we prove the existence of equivalence classes for circulant matrices in terms of the branch number. Since the circulant matrices within an equivalence class have the same branch number, it is sufficient to check one representative from each equivalence class and hence reduce the search space. In Sect. 3.2, we show that there are at most 5 types of MDS circulant matrices for order $k \leq 8$, namely circulant matrices whose first row has k distinct entries, 1, 2 or 3 pairs of repeated entries, or 3 repeated entries. This allows us to complete the search for lightweight MDS circulant matrix of order 8 which previously was not achievable by [17]. In Sect. 4, we generalize the circulant structure and propose a new type of matrices—cyclic matrices, which preserve the benefits and advantages of circulant matrices. Using group theory, we prove that, in terms of branch number, cyclic matrices are equivalent to circulant matrices. This greatly simplifies the understanding and analysis on the branch number of the cyclic matrices. In Sect. 5, we present the lightest MDS left-circulant matrices (where each row is a left rotation instead of right), for order $k \leq 8$, based on the same metric used in [17,22]. In addition, we overcome the constraint that circulant matrix cannot be involutory and MDS simultaneously, and also present the lightest involutory MDS left-circulant matrices. To the best of our knowledge, the latter matrices are the first of its kind. We would like to emphasize that all the techniques and most results presented in this paper are independent of the metric for lightweight. In other words, one can choose another metric and apply our techniques to reduce the search space for finding the desired matrices.

2 Preliminary

In this section, we first state some notations that will be frequently used for the rest of the paper. Next, we formally define what branch number of a matrix is, and provide two propositions that will be useful in the later proofs. Lastly, we give an introduction to circulant matrix, the advantages of using it and how the implementation of circulant matrix can be serialized. In this paper, we assume that the matrices are square matrices unless otherwise stated.

2.1   Notations

n : Dimension of the finite field

$GF(2^n)$: Finite field of order 2^n

0x : Prefix for hexadecimal, common notation for expressing binary polynomial coefficients or n-bit strings

k : Order of the square matrix

$M[i,j]$: (i,j)-entry of the matrix M, where $i, j \in \{0, 1, ..., k-1\}$

$wt(v)$: Number of nonzero components of the vector v

2.2   Branch Number of the Diffusion Layer

Recall that the diffusion power of the diffusion layer is often quantified by the branch number of the diffusion matrix.

Definition 1. *The branch number of a matrix M of order k over finite field $GF(2^n)$ is the minimum number of nonzero components in the input vector v and output vector $u = M \cdot v$ as we range over all nonzero $v \in [GF(2^n)]^k$. I.e., the branching number of matrix M is $\mathcal{B}_M = \min_{v \neq 0}\{wt(v) + wt(Mv)\}$.*

That is to say, for any nonzero input and output pair of a diffusion matrix, the number of nonzero components will be at least the branch number of the diffusion matrix. This is essential for protecting against the cryptanalysis like differential attack that exploits the differential patterns between the plaintext and the ciphertext. As the sum of nonzero components is lower bounded by the branch number, having a high branch number implies that a small input difference will inevitably lead to a large output difference, and to achieve a small output difference would require a large input difference.

Definition 2 [23]. *A maximum distance separable (MDS) matrix of order k is a matrix that attains the optimal branch number $k + 1$.*

When there is a single difference in the input vector, the best possible diffusion is to spread the difference to all k components of the output vector, hence the largest possible branch number is $k+1$. For instance, the AES diffusion matrix has order 4 and a branch number 5, hence it is MDS.

The following propositions are simple yet crucial building blocks for the results in this paper.

Proposition 1 [19, p. 321, Theorem 8]. *A matrix is MDS if and only if its square submatrices are all nonsingular.*

Proposition 2. *For any permutation matrices P and Q, the branch numbers of these two matrices M and PMQ are the same.*

Proof. Since P and Q are permutation matrices, there can be bijection mappings between the input vectors (resp. output vectors) of M and PMQ where the vectors differ by some permutation, hence the minimum number of nonzero components in the input and output pairs remains the same and they have the same branch number.   □

2.3 Circulant Matrices and Its Implementation

Circulant Matrices. Here, let us formally define circulant matrices and related notations.

Definition 3. *A circulant matrix C of order k is a matrix where each subsequent row is a right rotation of the previous row. We denote the matrix as $circ(c_0, c_1, ..., c_{k-1})$, where c_i's are the entries of the first row of the matrix. The (i, j)-entry of C can be expressed as $C[i, j] = c_{(j-i) \bmod k}$.*

There are several advantages of using circulant matrix in a diffusion layer:

1. It has a higher probability of finding an MDS matrix as compared to a randomized square matrix [8].
2. It has at most k distinct entries, and in addition it can be MDS and contain repeated lightweight entries, which tends to have lower implementation cost as compared to matrices like Hadamard and Cauchy matrices that must have at least k distinct entries in order to be MDS.
3. It has the flexibility to be implemented in both round-based and serialized implementations.

However, it was shown in [15] that involutory MDS (IMDS) circulant matrices of order 4 do not exist, and was further proved in [13] that IMDS circulant matrices of any order do not exist. To preserve the benefits of circulant matrices, we generalize the circulant structure in Sect. 4 and find lightweight IMDS matrices that are presented in Sect. 5.

Serialized Implementation of Circulant Matrices. First, let us illustrate the round-based implementation using an arbitrary circulant matrix $circ(a, b, c, d)$ of order 4, and an arbitrary input vector (w, x, y, z), we compute the output vector as follows,

$$\begin{pmatrix} a & b & c & d \\ d & a & b & c \\ c & d & a & b \\ b & c & d & a \end{pmatrix} \begin{pmatrix} w \\ x \\ y \\ z \end{pmatrix} = \begin{pmatrix} aw + bx + cy + dz \\ dw + ax + by + cz \\ cw + dx + ay + bz \\ bw + cx + dy + az \end{pmatrix}.$$

The entire diffusion matrix is implemented and the output components can be computed in parallel and in one clock cycle.

On the other hand, one clock cycle of a serial-based implementation is computed as follows,

$$\begin{pmatrix} 0 & 1 & 0 & 0 \\ 0 & 0 & 1 & 0 \\ 0 & 0 & 0 & 1 \\ a & b & c & d \end{pmatrix} \begin{pmatrix} w \\ x \\ y \\ z \end{pmatrix} = \begin{pmatrix} x \\ y \\ z \\ aw + bx + cy + dz \end{pmatrix},$$

where the output is fed back to the input and this process is repeated for another 3 times to get the final output. Excluding the control logics and memories required, serial-based implementation requires implementing one row of the matrix and takes k clock cycles to compute the output vector.

Clearly a circulant matrix can be implemented in the round-based manner. Although it is not in a form of a serial matrix that is required for serial-based implementation, implementation of a circulant matrix can still be serialized. The key observation is that the same permutation is applied to obtain each subsequent row. For a circulant matrix, the permutation is a right rotation. To serialize the implementation of circulant matrix, we implement the first row of the circulant matrix and compute the first output component.

$$\begin{pmatrix} a\ b\ c\ d \end{pmatrix} \begin{pmatrix} w \\ x \\ y \\ z \end{pmatrix} = \begin{pmatrix} aw + bx + cy + dz \end{pmatrix}.$$

Next, we update the input vector by applying the inverse permutation to obtain (x, y, z, w) and apply the first row of the matrix again,

$$\begin{pmatrix} a\ b\ c\ d \end{pmatrix} \begin{pmatrix} x \\ y \\ z \\ w \end{pmatrix} = \begin{pmatrix} ax + by + cz + dw \end{pmatrix},$$

and we obtain the second component of the output vector. We repeat the process to obtain the entire output vector in 4 clock cycles. Thus, similar to serial-based implementation, we only need to implement one row of the matrix and it takes k clock cycles to compute the output vector.

In fact, one can even achieve other area requirement and clock cycle trade-offs that are between the round-based and serial-based implementation performance. In the previous example, one can also implement 2 rows of the circulant matrix and compute 2 output components in parallel, this will take 2 clock cycles to complete the diffusion layer computation. More generally, we can have t-serialized implementation when we are using circulant matrices, where t divides k. The estimated implementation costs and clock cycles required for the implementations are summarized in Table 1. Note that this does not include the memory costs and control logics required for different implementations. From Table 1, it is clear that the round-based and serialized implementations are special case of t-serialized implementation where $t = 1$ and $t = k$ respectively.

Circulant matrices are not the only matrix type that can be serialized. In fact, if the same permutation, not necessarily being a right rotation, is applied to obtain each subsequent row, we can still serialize the implementation. This observation leads us to generalize the circulant matrices to cyclic matrices, see also Sect. 4, which can be serialized too.

Table 1. Estimated implementation costs and clock cycles for various implementations

Type of implementation	Matrix implementation (no. of entries)	Clock cycle
Round-based	k^2	1
Serial-based	k	k
Serialized	k	k
t-serialized	k^2/t	t

3   Properties of Circulant Matrices

There are mainly two challenges in the method of picking the lightest circulant matrix and checking the MDS property. Firstly, for a generic (not considering the values of the entries) circulant matrix of order k, $\mathrm{circ}(c_0, c_1, ..., c_{k-1})$, there are $k!$ ways to permute the entries, which can quickly be intractable. Secondly, the choice of the k lightweight nonzero entries need not be distinct, which potentially cause the search space to be much larger than just choosing k distinct entries and permuting them.

In Sect. 3.1, we first introduce an equivalence relation to partition the $k!$ circulant matrices into equivalence classes, where circulant matrices within an equivalence class share the same branch number. This allows us to reduce the search space by checking the MDS property for one representative from each equivalence class. Next in Sect. 3.2, we analyze the circulant structure and show that for order $k \leq 8$, there are at most 5 types of MDS circulant matrices, namely circulant matrices whose first row has k distinct entries, 1, 2 or 3 pairs of repeated entries, or 3 repeated entries. This shows that any MDS circulant matrix must belong to one of these 5 types.

3.1   Compact Equivalence Classes of Circulant Matrices

For the ease of our discussion on the permutation of the entries, we focus on the permutation of the index of the elements.

Definition 4. *An index permutation σ on an ordered set $\{c_0, c_1, ..., c_{k-1}\}$ is a permutation that permutes the index of the elements.*

Example 1. Let σ be an index permutation on an ordered set $\{c_0, c_1, c_2, c_3, c_4\}$ where $\sigma(i) = 4 - i$, the resultant ordered set will be $\{c_4, c_3, c_2, c_1, c_0\}$.

Definition 5. *Given a matrix M of order k that is defined by its first row under a rule, we denote by M^σ the matrix generated under the same rule by the first row of M modified by applying an index permutation σ.*

Definition 6. *Two matrices M and M' are called permutation-equivalent, denoted by $M \sim_\mathcal{B} M'$, if there exist two permutation matrices P and Q such that $M' = PMQ$.*

It is easy to verify that $\sim_{\mathcal{B}}$ is a well-defined equivalence relation. By Proposition 2, we know that the permutation-equivalent matrices have the same branch number. Using this equivalence relation, we partition the $k!$ possible circulant matrices into equivalence classes with respect to their branch number.

Definition 7. *An equivalence class of circulant matrices is a set of circulant matrices satisfying the equivalence relation $\sim_{\mathcal{B}}$.*

We first analyze what index permutation satisfies the relation, then we deduce the number of equivalence classes of circulant matrices.

Lemma 1. *Given two circulant matrices C and C^σ, $C \sim_{\mathcal{B}} C^\sigma$ if and only if σ is some index permutation satisfying $\sigma(i) = (bi + a) \bmod k$, $\forall i \in \{0, 1, ..., k-1\}$, where $a, b \in \mathbb{Z}_k$ and $\gcd(b, k) = 1$.*

Proof. The "if" direction is immediate once we have proven the "only if" direction. Assume that $C \sim_{\mathcal{B}} C^\sigma$. By Definition 6, there exists permutation matrices P and Q such that $C^\sigma = PCQ$, where P (resp. Q) is in fact a row (resp. column) permutation on C. Since C is circulant, one can observe that if $C^\sigma = PC$, then the first row of C^σ is some row of C and thus corresponds to some rotation of the first row of C, which shows that the index permutation σ can be expressed as $\pi_a(i) = (i + a) \bmod k$. That is, π_a corresponds to a row permutation P_a. Therefore, for any C^σ such that $C^\sigma = PCQ$, we can always apply some index permutation π_{-a} to fix the first element c_0 and accordingly pre-multiply C^σ by a corresponding row permutation P_{-a}, which gives $C^{\pi_{-a} \circ \sigma} = P_{-a}PCQ$, where $\pi_{-a}(\sigma(0)) = 0$.

Next, we consider index permutation that fixes 0. Note that this implies that the row and column permutations on C fix the first row and column. Suppose that $C^{\phi_b} = PCQ$, $\phi_b(0) = 0$ and $\phi_b(1) = b$, then the column permutation Q maps column b of C to column 1 of C^{ϕ_b}, and similarly the row permutation P maps row $k - b$ of C to row $k - 1$ of C^{ϕ_b}. By definition of circulant matrices, we know that $c_{\phi_b(2)}$, which is the third entry of C^{ϕ_b}, can be written as $C^{\phi_b}[0, 2] = C^{\phi_b}[(k-1), 1]$. Since the pre-image of row $k-1$ and column 1 of C^{ϕ_b} are row $k-b$ and column b of C, we can express that entry of C^{ϕ_b} as an entry of C, that is $C^{\phi_b}[(k-1), 1] = C[(k-b), b]$. And again by definition of circulant matrices, the entry $c_{b-(k-b) \bmod k} = c_{2b \bmod k}$. That is to say, by defining $\phi_b(1) = b$, we have restricted the permutation of the next index to be $\phi_b(2) = 2b \bmod k$. Following the same argument, we can conclude that $\phi_b(i) = bi \bmod k$. In addition, we must have $\gcd(b, k) = 1$ so that ϕ_b is a permutation on $\{0, 1, ..., k-1\}$.

Finally, we can see that if $C \sim_{\mathcal{B}} C^\sigma$ then $\sigma = \pi_a \circ \phi_b$, that is, $\sigma(i) = (bi + a) \bmod k$. $\qquad\square$

For simplicity, we call the permutations satisfying Lemma 1 the $\mathcal{C}$-permutations. That is to say, $C \sim_{\mathcal{B}} C^\sigma$ if and only if σ is a $\mathcal{C}$-permutation. We show in the full version of this paper [18] how to generate one representative for each equivalence class.

Theorem 1. *There are $\frac{(k-1)!}{\varphi(k)}$ equivalence classes of circulant matrices of order k, where $\varphi(k)$ is the Euler's totient function.*

Proof. It is clear that the cardinality of each equivalence class is the number of possible index permutation σ. By Lemma 1, we know that $\sigma(i) = (bi + a) \bmod k$, where $a, b \in \mathbb{Z}_k$ and $\gcd(b, k) = 1$. Since there are k possible values for a and b has to be coprime with k, there are $\varphi(k)$ possible values for b, and each equivalence class has cardinality of $k \cdot \varphi(k)$. Hence the number of equivalence classes is $\frac{k!}{k \cdot \varphi(k)} = \frac{(k-1)!}{\varphi(k)}$. $\square$

Note that the "only if" direction of the Lemma 1 implies that this is the most compact equivalence classes for generic circulant matrices in terms of branch number. In [22], the authors presented equivalence classes of Hadamard matrices to reduce the search space for checking the MDS property. But whether there exists larger (more compact) equivalence classes to further reduce the search space remains an open question. Observing its similarity with our work, we analyze the equivalence classes of Hadamard matrices in [22] and find that it is already the most compact equivalence class. The proof is included in the full version of this paper [18].

3.2   Types of MDS Circulant Matrices of Order $k \leq 8$

In short, this section proves the following theorem.

Theorem 2. *For order $k \leq 8$, there are at most 5 types of MDS circulant matrices, namely circulant matrices whose first row has:*

Type 0: *k distinct entries;*
Type 1: *1 pair of repeated entries;*
Type 2: *2 pairs of repeated entries;*
Type 3: *3 pairs of repeated entries;*
Type 4: *or 3 repeated entries.*

Given an ordered multi-set of entries $\{c_0, c_1, ..., c_{k-1}\}$, suppose that two entries of them are the same, denoted by $c_i = c_{(i+d) \bmod k}$ for some $i, d \in \{0, 1, ..., k-1\}$. From Sect. 3.1, we see that any rotation of the entries are permutation-equivalent. Hence, for any $d > \lfloor \frac{k}{2} \rfloor$, it is equivalent to considering $c_{(i-d) \bmod k} = c_{(i-d)+d}$ which is equal to $c_{i+(k-d) \bmod k} = c_i$, where $k - d \leq \lfloor \frac{k}{2} \rfloor$. Without loss of generality, we assume $i + d \leq k - 1$ and $d \leq \lfloor \frac{k}{2} \rfloor$.

First, we state two lemmas that will help us in proving Theorem 2.

Lemma 2. *An MDS circulant matrix of even order k does not have $c_i = c_{i+\frac{k}{2}}$.*

Proof. Suppose that there exists $c_i = c_{i+\frac{k}{2}}$. Considering the submatrix of order 2 by taking row 0 and $\frac{k}{2}$, and column i and $i + \frac{k}{2}$, we have

$$\begin{pmatrix} c_i & c_{i+\frac{k}{2}} \\ c_{(i-\frac{k}{2}) \bmod k} & c_i \end{pmatrix}.$$

Since $i - \frac{k}{2} \equiv i + \frac{k}{2}$ (mod k), we have a singular submatrix and by Proposition 1, there is a contradiction. □

Lemma 3. *An MDS circulant matrix does not have $c_i = c_{i+d}$ and $c_j = c_{j+d}$, where $i \neq j$.*

Proof. Suppose that there exist $c_i = c_{i+d}$ and $c_j = c_{j+d}$, where $i < j$. Consider the submatrix of order 2 by taking row 0 and $(i - j)$ mod k, and column i and $i + d$, we have

$$\begin{pmatrix} c_i & c_{i+d} \\ c_j & c_{j+d} \end{pmatrix},$$

Since these two columns are identical, we have a singular submatrix and by Proposition 1, there is a contradiction. □

From Lemmas 2 and 3, we can conclude that an MDS circulant matrix of order k allows at most $\lfloor \frac{k-1}{2} \rfloor$ possible distinct distances and thus has at least $\lceil \frac{k+1}{2} \rceil$ distinct elements. Specially for order $k = 8$, it allows 3 possible distinct distances and thus there are at most 3 pairs of repeated entries. If some entry has multiplicity 3, say $c_i = c_{i+d_1} = c_{i+d_2}$, then the three distances $d_1, d_2, d_2 - d_1$ are pairwise distinct. It also implies that any higher multiplicity is impossible for an MDS circulant matrix of order 8 as the number of pairwise equalities is more than 3 (a similar property that an MDS matrix of order 8 has at most 24 ones was proved in [16]). Similarly, for order $k < 8$, there are also at most 3 possible distances. Therefore, we obtain Theorem 2 that any MDS circulant matrix of order $k \leq 8$ is one of the 5 matrix types.

In Table 2, we list all the possible types of MDS circulant matrices for order $k \leq 8$. These results can also be extended to higher order circulant matrices. Note that this is a necessary condition for an MDS circulant matrix, it does not guarantee the existence of MDS circulant matrix for any of the circulant matrix type. For $k = 8$, we check that there are MDS matrices of each type, see also Sect. 5.

Table 2. Possible types of MDS circulant matrices of order $k \leq 8$

Order	Possible d	k distinct	1 pair	2 pairs	3 pairs	3 repeated
3	$\{1\}$	✓	✓			
4	$\{1\}$	✓	✓			
5	$\{1, 2\}$	✓	✓	✓		
6	$\{1, 2\}$	✓	✓	✓		
7	$\{1, 2, 3\}$	✓	✓	✓	✓	✓
8	$\{1, 2, 3\}$	✓	✓	✓	✓	✓

4   Cyclic Matrices

In this section, we generalize the circulant matrix structure and introduce a new type of matrices, we call them the *cyclic matrices*. Despite that cyclic matrices capture the essential requirement to have t-serialized implementation, analyzing all cyclic matrices is not feasible. Using results from elementary group theory, we can relate cyclic matrices to circulant matrices in terms of branch number. This allows us to apply the results on circulant matrices in Sect. 3 to the cyclic matrices as well.

Generalized Circulant Matrices. Recall from Sect. 2.3 that to serialize the implementation of a matrix, the same permutation is applied to obtain each subsequent row. Hence, we generalize the circulant structure by considering other permutations beside the right rotation.

Definition 8. *A cyclic matrix C_ρ of order k is a matrix where each subsequent row is some permutation ρ of the previous row, where ρ is a cycle of length k. We denote the matrix as $cyc_\rho(c_0, c_1, ..., c_{k-1})$, where c_i's are the entries of the first row of the matrix. The (i, j)-entry of C_ρ can be expressed as $C_\rho[i, j] = c_{\rho^i(j)}$.*

For example, the permutation of the circulant matrix structure can be expressed as a cycle $(0 \ 1 \ 2 \ ... \ k - 1)$, where $\rho = (i_0 \ i_1 \ i_2 \ ... \ i_{k-1})$ means $\rho(i_j) = i_{(j-1) \bmod k}$ for $0 \le j \le k - 1$. In the definition of cyclic matrix, we require the permutation to be a cycle of length k to avoid repeated rows and repeating elements in a column (which will not satisfy the property of MDS).

Since there are $(k - 1)!$ cycles of length k, it is infeasible to analyze every single the cyclic structures. However, using Proposition 2 and elementary group theory, we can elegantly reduce the problem to simply analyzing the circulant matrices. First, observe that the permutation ρ is an element of the symmetric group S_k, and the collection of the permutations of the k rows of the matrix forms a cyclic group, hence the name cyclic matrices.

Example 2. Considering the cycle permutation $\rho = (0 \ 2 \ 1 \ 3)$, we can express $cyc_\rho(a, b, c, d)$ as follows

$$\begin{pmatrix} (a, b, c, d) \\ \rho(a, b, c, d) \\ \rho^2(a, b, c, d) \\ \rho^3(a, b, c, d) \end{pmatrix} = \begin{pmatrix} a \ b \ c \ d \\ d \ c \ a \ b \\ b \ a \ d \ c \\ c \ d \ b \ a \end{pmatrix},$$

where the collection of the permutations of each row forms a cyclic group of order 4, $\langle (0 \ 2 \ 1 \ 3) \rangle = \{(), (0 \ 2 \ 1 \ 3), (0 \ 1)(2 \ 3), (0 \ 3 \ 1 \ 2)\}$.

Relation to Circulant Matrices. Next, we show that any cyclic matrix is permutation-equivalent to some circulant matrix. More preciously, there is a bijection between the cyclic and circulant matrices satisfying $\sim_B$. To prove this, we use the following proposition from elementary group theory.

Proposition 3 [20, Chap. 5.3]. *Any two permutations ρ, τ which have the same cycle type are conjugate in S_k.*

That is to say, there exists permutation $\sigma \in S_k$ such that $\sigma\rho = \tau\sigma$. In the nutshell, σ can be computed by placing one permutation above the other and view it as a Cauchy's 2-line notation for permutation.

Example 3. Let $\rho = (0\ 2\ 1\ 3)$ and $\tau = (0\ 1\ 2\ 3)$, viewing it as a Cauchy's 2-line notation, we have

$$\begin{pmatrix} 0\ 2\ 1\ 3 \\ 0\ 1\ 2\ 3 \end{pmatrix},$$

from which we see that 0 and 3 are fixed while 1 and 2 are swapped. Therefore, we obtain $\sigma = (1\ 2)$ and we can verify that $\sigma\rho = \tau\sigma$.

Theorem 3. *Given an ordered set S with k elements and some cyclic matrix structure, there exists a bijection between the cyclic matrices and the circulant matrices satisfying the relation $\sim_\mathcal{B}$, where both sets of matrices are generated by some index permutation on S.*

Proof. Let the permutation of some cyclic matrix be ρ and circulant matrix be $\tau = (0\ 1\ 2 \dots k-1)$. By Proposition 3, there exist some permutation σ such that $\sigma\rho = \tau\sigma$. Hence for any row $i \in \{0, 1, ..., k-1\}$, we have $\sigma\rho^i = \tau^i\sigma$. In the form of a matrix, the permutation for each row of the matrices can be expressed as

$$\begin{pmatrix} \sigma(S) \\ \sigma \circ \rho(S) \\ \sigma \circ \rho^2(S) \\ \vdots \\ \sigma \circ \rho^{k-1}(S) \end{pmatrix} = \begin{pmatrix} \sigma(S) \\ \tau \circ \sigma(S) \\ \tau^2 \circ \sigma(S) \\ \vdots \\ \tau^{k-1} \circ \sigma(S) \end{pmatrix},$$

where σ in the cyclic matrix can be viewed as a column permutation, while in the circulant matrix it is a index permutation on S. Therefore by Proposition 2, the cyclic matrix has the same branch number as a circulant matrix that undergoes index permutation σ.

Lastly, one can easily infer that for any index permutation π on the cyclic matrix, it corresponds to a circulant matrix that undergoes index permutation $\sigma \circ \pi$. □

Example 4. Consider a cyclic matrix of order 4 with the row permutation $\rho = (0\ 2\ 1\ 3)$, while the circulant matrix is $\tau = (0\ 1\ 2\ 3)$. From Example 3, we have $\sigma = (1\ 2)$ that satisfies $\sigma\rho = \tau\sigma$. Applying column permutation σ on the cyclic matrix and index permutation σ on circulant matrix, we obtain the same matrix as follows

$$\begin{pmatrix} a\ b\ c\ d \\ d\ c\ a\ b \\ b\ a\ d\ c \\ c\ d\ b\ a \end{pmatrix} \xrightarrow{\text{col perm } \sigma} \begin{pmatrix} a\ c\ b\ d \\ d\ a\ c\ b \\ b\ d\ a\ c \\ c\ b\ d\ a \end{pmatrix} \xleftarrow{\text{index perm } \sigma} \begin{pmatrix} a\ b\ c\ d \\ d\ a\ b\ c \\ c\ d\ a\ b \\ b\ c\ d\ a \end{pmatrix}.$$

This theorem shows that for any cyclic matrix, we have some column permutation σ that transforms it into a circulant matrix (or any other cyclic matrix) while preserving the branch number. However, the involution property of circulant matrix may not hold true for the cyclic matrices, which gives us an insight that there might exist IMDS cyclic matrices while it is not the case for the circulant matrices. And we indeed find IMDS cyclic matrices which are presented in Sect. 5.

Corollary 1. *Any cyclic matrix corresponds to some circulant matrix preserving the coefficients and the branch number.*

This is immediate from Theorem 3 and the fact that their entries are the same up to some permutation. In addition, we can draw the following corollary immediately from Theorems 2 and 3.

Corollary 2. *For order $k \leq 8$, there are at most 5 types of MDS cyclic matrices, namely cyclic matrices whose first row has:*

Type 0: *k distinct entries;*
Type 1: *1 pair of repeated entries;*
Type 2: *2 pairs of repeated entries;*
Type 3: *3 pairs of repeated entries;*
Type 4: *or 3 repeated entries.*

5   Results on Lightest (Involutory) MDS Matrices

There are different ways to define lightweight/efficient. For instance in AES, the diffusion matrix entries were chosen for its simplicity and low Hamming weight, while [14,16] defined efficiency by the number of 1's in the matrix. In hardware implementation, it is common to consider the area required and a simplified metric is to count the number of XOR gates needed for implementation. In [16, 17,22], the authors evaluate the number of XOR gates needed to implement the multiplication of the diffusion matrices. Detailed description of the XOR count can be found in [16,17,22]. In this paper, we quantify the weight of a diffusion matrix by the sum of XOR counts in its first row[2].

In this section, we mainly focus on a special case of cyclic matrices, called left-circulant matrices. First, we provide a strategy to search for MDS left-circulant matrices by exploiting the properties of the matrices, including the permutation-equivalence relationship. Then, we show that, though no circulant matrices are IMDS, there are IMDS left-circulant matrices. We also provide a strategy to search for such IMDS matrices. The experimental results show that all the lightest MDS matrices and IMDS matrices can be confirmed for $3 \leq k \leq 8$, by using our strategies.

---

[2] This is adapted from [17], in which the number of XOR counts of one row is given by $\sum_{i=1}^{k} \gamma_i + (\ell - 1) \cdot n$, where γ_i is the XOR count of the i-th entry and ℓ is the number of nonzero coefficients in the row. Since the latter term is fixed for any MDS matrix of order k over $\mathrm{GF}(2^n)$, we are only interested in the sum of the XOR counts of the coefficients in a row.

5.1   Lightweight MDS Left-Circulant Matrices

The definition of left-circulant matrices is given as follows.

Definition 9. *A left-circulant matrix L of order k is a matrix where each subsequent row is a left rotation of the previous row. We denote the matrix as $\ell\text{-}circ(c_0, c_1, ..., c_{k-1})$, where c_i's are the entries of the first row of the matrix. The (i, j)-entry of L can be expressed as $L[i, j] = c_{(i+j) \bmod k}$.*

It is infeasible to exhaust all the possible MDS left-circulant matrices over $GF(2^8)$ for $k = 8$. Notice that the permutation-equivalence relationship (Lemma 1) of circulant matrices also applies to left-circulant matrices. Combining Corollary 2 and permutation-equivalence relationship, we can exhaust all the possible MDS left-circulant matrices over $GF(2^n)$ with small XOR count for $n \leq 8$ and $k \leq 8$.

To efficiently determine whether a left-circulant matrix is MDS, we collect in advance the symbolic expressions of all determinants of its submatrices, and use them to compute the values of determinants. Once detecting that a determinant has value 0, the matrix is confirmed to be not MDS; otherwise, it is MDS. Using this method, the detection of MDS left-circulant matrices is speeded up (by dozens of times for $5 \leq k \leq 8$) since a lot of submatrices have the same determinants in terms of symbolic expressions.

We show in Table 3 our experimental results on MDS left-circulant $k \times k$ matrices over $GF(2^n)$ with smallest XOR count for $n = 4, 8$ and $3 \leq k \leq 8$. All the provided matrices are optimal among the MDS cyclic matrices in terms of the metric as used in [17, 22]. We also exhaust all the left-circulant matrices over $GF(2^4)$ for $k = 7, 8$, and the results show that no such matrices are MDS. It was also noted in [17] that there do not exist circulant $8 times 8$ matrices over $GF(2^4)$.

Table 3. Lightest MDS left-circulant matrices of order $3 \leq k \leq 8$

k	Polynomial	Left-circulant matrices	XOR count
$GF(2^8)$			
3	0x1c3	(0x1, 0x1, 0x2)	3
4	0x1c3	(0x1, 0x1, 0x2, 0x91)	8
5	0x1c3	(0x1, 0x1, 0x2, 0x91, 0x2)	11
6	0x1c3	(0x1, 0x2, 0xe1, 0x91, 0x1, 0x8)	18
7	0x1c3	(0x1, 0x1, 0x91, 0x2, 0x4, 0x2, 0x91)	21
8	0x1c3	(0x1, 0x1, 0x2, 0xe1, 0x8, 0xe0, 0x1, 0xa9)	30
$GF(2^4)$			
3	0x13	(0x1, 0x1, 0x2)	1
4	0x13	(0x1, 0x1, 0x9, 0x4)	3
5	0x13	(0x2, 0x2, 0x9, 0x1, 0x9)	4
6	0x13	(0x1, 0x1, 0x9, 0xc, 0x9, 0x3)	12

We list in Table 4 the lightest 8×8 MDS matrices for each type of left-circulant matrices as well as the lightest ones under the two commonly used irreducible polynomials, 0x11b and 0x11d, which are respectively adopted in AES and WHIRLPOOL, and we compare them with the WHIRLPOOL matrix and the MDS Hadamard matrix found in [22]. From this table, we can see that the lightest MDS left-circulant matrices of all types except Type 0 (in which all the coefficients are distinct) have XOR count smaller than the known best ones. For WHIRLPOOL, we also provide an MDS left-circulant matrix which has smaller XOR count using the same irreducible polynomial as in WHIRLPOOL.

We also compare in Table 5 our candidates with the previous lightweight MDS matrices for $n < 8$. It shows that all our candidates have the minimum XOR count, though some of them have the same XOR count as the known ones.

Table 4. Comparison of 8×8 MDS matrices

Type	Polynomial	Matrices	XOR count
4	0x1c3	(0x1, 0x1, 0x2, 0xe1, 0x8, 0xe0, 0x1, 0xa9)	30
3	0x1c3	(0x1, 0x1, 0x91, 0x2, 0x4, 0x2, 0x12, 0x91)	32
2	0x1c3	(0x1, 0x1, 0x4, 0x2, 0xa9, 0x91, 0x2, 0x3)	33
1	0x1c3	(0x1, 0x1, 0x2, 0xe0, 0x6, 0xe1, 0x91, 0x4)	35
0	0x1c3	(0x1, 0x2, 0x91, 0x8, 0x4, 0x6, 0xe1, 0x3)	42
4	0x11b	(0x1, 0x1, 0x2, 0x1, 0x74, 0x8d, 0x46, 0x4)	35
4	0x11d	(0x1, 0x1, 0x2, 0x8e, 0x47, 0x10, 0x1, 0x46)	34
4	0x11d	WHIRLPOOL	49
-	0x1c3	Hadamard [22]	40

5.2   Lightweight IMDS Left-Circulant Matrices

In this section, we first describe the involutory MDS left-circulant matrices and then show our experimental results.

Before showing our main results, we provide some useful properties for left-circulant matrices. It is known that the product of two circulant matrices is a circulant matrix. For left-circulant matrices, a similar property can be obtained. To simplify the presentation of the proofs, we omit "modulo k" from the indexes but it is expected that modulo k is applied when necessary.

Proposition 4. *The product of two left-circulant matrices is a circulant matrix.*

Proof. Let $A = \ell\text{-circ}(a_0, a_1, ..., a_{k-1})$ and $B = \ell\text{-circ}(b_0, b_1, ..., b_{k-1})$ be two left-circulant matrices. Then the (i, j)-entry of their product is $\sum_{t=0}^{k-1} A[i, t] \cdot B[t, j] = \sum_{t=0}^{k-1} a_{i+t} b_{t+j} = \sum_{t=0}^{k-1} a_t b_{t+(j-i)}$, which completes the proof.   □

It is shown in [14] that $C^{2^d} = (\sum_{i=0}^{2^d-1} c_i)^{2^d} I$ and $\det(C) = (\sum_{i=0}^{2^d-1} c_i)^{2^d}$ for any $2^d \times 2^d$ circulant matrix $C = \text{circ}(c_0, c_1, ..., c_{2^d-1})$ over $\text{GF}(2^n)$. Thus we have the following result for left-circulant matrices.

Table 5. Comparison of MDS matrices of order $k < 8$

k	Polynomial	Matrices	Matrix form	XOR count
		GF(2^8)		
4	0x1c3	[22]	Hadamard	13
4	0x11d	[17]	serial/circulant	9
4	0x1c3	this paper	left-circulant	8
6	0x11b	PHOTON P_{288}	serial	23
6	0x1c3	this paper	left-circulant	18
		GF(2^4)		
4	0x13	[22]	Hadamard	5
4	0x13	LED	serial	4
4	0x13	[17]	serial/circulant	3
4	0x13	this paper	left-circulant	3
5	0x13	PHOTON P_{100}	serial	4
5	0x13	this paper	left-circulant	4
6	0x13	PHOTON P_{144}	serial	14
6	0x13	this paper	left-circulant	12

Proposition 5. *For $2^d \times 2^d$ matrix $L = \ell\text{-}circ(c_0, c_1, ..., c_{2^d-1})$ over $GF(2^n)$,*
$L^{2^{d+1}} = (\sum_{i=0}^{2^d-1} c_i)^{2^{d+1}} I$ *and* $\det(L) = (\sum_{i=0}^{2^d-1} c_i)^{2^d}$.

Proof. By the proof of Propostion 4, we know L^2 is a circulant matrix with (i,j)-entry $\sum_{t=0}^{2^d-1} c_t c_{t+(j-i)}$, and thus $(L^2)^{2^d} = (\sum_{i=0}^{2^d-1} \sum_{t=0}^{2^d-1} c_t c_{t+i})^{2^d} I = ((\sum_{t=0}^{2^d-1} c_t)^2)^{2^d} I$, which also implies $\det(L) = (\sum_{i=0}^{2^d-1} c_i)^{2^d}$. $\square$

Proposition 6. *For matrix $L = \ell\text{-}circ(c_0, c_1, ..., c_{k-1})$ over $GF(2^n)$, L is involutory if and only if $\sum_{i=0}^{k-1} c_i = 1$ and $\sum_{i=0}^{k-1} c_i c_{i+j} = 0$ for all $1 \le j \le \lfloor \frac{k-1}{2} \rfloor$.*

Proof. Since the (i,j)-entry of L^2 is $\sum_{t=0}^{k-1} c_t c_{t+(j-i)}$, L is involutory if and only if $\sum_{t=0}^{k-1} c_t = 1$ and $\sum_{t=0}^{k-1} c_t c_{t+(j-i)} = 0$ for $j \ne i$. The proof is completed by the facts that $\sum_{t=0}^{k-1} c_t c_{t+(j-i)} = \sum_{t=0}^{k-1} c_t c_{t+(i-j)}$ and $\sum_{t=0}^{k-1} c_t c_{t+\frac{k}{2}} = 0$ for even k. $\square$

A left-circulant matrix is symmetric and thus an involutory left-circulant matrix is orthogonal. It was shown in [14] that a circulant matrix is not IMDS and an orthogonal circulant $2^d \times 2^d$ matrix is not MDS. Similarly, we can prove that an involutory (orthogonal) left-circulant $2^d \times 2^d$ matrix is not MDS.

Theorem 4. *If L is a $2^d \times 2^d$ left-circulant matrix over $GF(2^n)$, then L is not IMDS.*

Proof. It is sufficient to prove that if L is involutory then L is not MDS. Assume that $L = \ell\text{-}circ(c_0, c_1, ..., c_{2^d-1})$ is involutory. By Propostion 6, it holds that $\sum_{i=0}^{2^d-1} c_i c_{i+2t+1} = 0$ for $0 \le t \le 2^{d-2} - 1$, and thus $(\sum_{t=0}^{2^{d-1}-1} c_{2t})(\sum_{t=0}^{2^{d-1}-1} c_{2t+1}) = \sum_{t=0}^{2^{d-2}-1} \sum_{i=0}^{2^d-1} c_i c_{i+2t+1} = 0$. Note that

ℓ-circ$(c_0, c_2, ..., c_{2^d-2})$ and ℓ-circ$(c_1, c_3, ..., c_{2^d-1})$ are two submatrices of L. Therefore, according to Proposition 5, at least one of the determinants of these two submatrices equals 0, which shows L is not MDS.   □

Our computations also show that there are no IMDS cyclic matrices for $k = 4, 8$. Nevertheless, there are IMDS left-circulant matrices for $k = 3, 5, 6, 7$.

Next we explain how to search for IMDS left-circulant matrices. Notice that an IMDS left-circulant matrix must satisfy the $\lfloor \frac{k+1}{2} \rfloor$ equations mentioned in Proposition 6. Theoretically, we can solve the equations and then check whether the solutions satisfy the MDS property. However, it is unclear how to efficiently solve the equations in a straightforward way. Solving the equations over $GF(2^n)$ using Gröbner basis is very slow for $n = 8$ and is slow even for $n = 4$. To find the solutions faster, we first guess the values of about $\lfloor \frac{k-1}{2} \rfloor$ out of the k coefficients, then solve the equations. For $n = 4$, we guess all the possible values. For $n = 8$, we only guess some of the lightest elements. Our experiments show that it is sufficient to guess the lightest 9 elements to find the lightest IMDS left-circulant matrix.

We can check by Lemma 1 and Proposition 6 that if a left-circulant matrix is involutory then all its permutation-equivalent matrices are involutory. Thus we can use permutation-equivalence relationship to reduce the search space. In other words, once obtain an upper bound of the minimum XOR count, we can exhaust all the possible IMDS left-circulant matrices less than the threshold, and confirm the lightest one, as done for MDS left-circulant matrices.

We provide our results in Table 6. As shown in the table, there are no IMDS left-circulant matrices over $GF(2^4)$ for $k = 6$. All the listed matrices have been confirmed to achieve the smallest XOR count.

Table 6. Lightest IMDS left-circulant matrices of order $3 \le k \le 7$

k	Polynomial	Matrices	XOR count
$GF(2^8)$			
3	0x169	(0x5a, 0xa, 0x51)	30
4	-		
5	0x165	(0x1, 0x2, 0xb3, 0xbb, 0xa)	46
6	0x165	(0x1, 0x1, 0xb3, 0x2c, 0x4, 0x9a)	46
7	0x165	(0x1, 0x2, 0x5c, 0xb2, 0xa4, 0x10, 0x58)	68
7	0x139	(0x1, 0x1, 0x8, 0x96, 0x21, 0x98, 0x26)	68
$GF(2^4)$			
3	0x1f	(0x2, 0xf, 0xc)	12
4	-		
5	0x13	(0x1, 0x2, 0x5, 0x4, 0x3)	14
6	-		

6   Conclusion

In this paper, we have presented a series of theory on generalized circulant matrices, so-called cyclic matrices, and also exploited the technique to successfully find the lightest MDS and involutory MDS matrices among this class of matrices with small orders. On one hand, cyclic matrices maintain the characteristics of circulant matrices, such as compact and flexible implementations in hardware and branch number in diffusion layer. On the other hand, they possess some advantages that circulant matrices cannot provide, for instance, the existence of involutory MDS matrices. The discovery of properties and constructions of MDS cyclic matrices may provide practical significance as well as theory value. Before this work, searching for the lightest MDS circulant matrices of order 8 are widely believed to be infeasible. Our results demonstrate an opposite view on this—we make it feasible under a credible metric—despite no guarantee of general case. As such, we can find the lightest MDS circulant matrices of order 8 which have less XOR count than the previously known ones in the literatures. Specially for the hash function WHIRLPOOL, we also provide a better MDS matrix which has smaller XOR count under the same setting. Although it is proven that IMDS left-circulant matrix of order 2^d does not exist, we find IMDS matrices for the other orders which forms a complement to the work in [22], where there exist only IMDS Hadamard matrices of order 2^d. All in all, we have found new lightweight MDS matrices that are flexible in hardware implementation and also a complete set of lightweight IMDS matrices for order $k \leq 8$.

Acknowledgements. The authors would like to thank Jian Guo, Gregor Leander, Thomas Peyrin, Yu Sasaki and the anonymous reviewers for their valuable suggestions.

References

1. Andreeva, E., Bilgin, B., Bogdanov, A., Luykx, A., Mendel, F., Mennink, B., Mouha, N., Wang, Q., Yasuda, K.: PRIMATEs v1. Submission to the CAESAR Competition (2014). http://competitions.cr.yp.to/round1/primatesv1.pdf
2. Aumasson, J.-P., Henzen, L., Meier, W., Naya-Plasencia, M.: QUARK: a lightweight hash. In: Mangard, S., Standaert, F.-X. (eds.) CHES 2010. LNCS, vol. 6225, pp. 1–15. Springer, Heidelberg (2010)
3. Barreto, P., Rijmen, V.: The Anubis block cipher. Submission to the NESSIE Project (2000)
4. Barreto, P., Rijmen, V.: The Khazad legacy-level block cipher. In: First Open NESSIE Workshop (2000)
5. Beaulieu, R., Shors, D., Smith, J., Treatman-Clark, S., Weeks, B., Wingers, L.: The SIMON and SPECK families of lightweight block ciphers. Cryptology ePrint Archive, Report 2013/404 (2013)
6. Bogdanov, A., Knežević, M., Leander, G., Toz, D., Varıcı, K., Verbauwhede, I.: SPONGENT: a lightweight hash function. In: Preneel, B., Takagi, T. (eds.) CHES 2011. LNCS, vol. 6917, pp. 312–325. Springer, Heidelberg (2011)

7. Bogdanov, A.A., Knudsen, L.R., Leander, G., Paar, C., Poschmann, A., Robshaw, M., Seurin, Y., Vikkelsoe, C.: PRESENT: an ultra-lightweight block cipher. In: Paillier, P., Verbauwhede, I. (eds.) CHES 2007. LNCS, vol. 4727, pp. 450–466. Springer, Heidelberg (2007)
8. Daemen, J., Knudsen, L.R., Rijmen, V.: The block cipher SQUARE. In: Biham, E. (ed.) FSE 1997. LNCS, vol. 1267, pp. 149–165. Springer, Heidelberg (1997)
9. Daemen, J., Rijmen, V.: The Design of Rijndael: AES - The Advanced Encryption Standard. Springer, Berlin (2002)
10. Guo, J., Peyrin, T., Poschmann, A.: The PHOTON family of lightweight hash functions. In: Rogaway, P. (ed.) CRYPTO 2011. LNCS, vol. 6841, pp. 222–239. Springer, Heidelberg (2011)
11. Guo, J., Peyrin, T., Poschmann, A., Robshaw, M.: The LED block cipher. In: Preneel, B., Takagi, T. (eds.) CHES 2011. LNCS, vol. 6917, pp. 326–341. Springer, Heidelberg (2011)
12. Gupta, K.C., Ray, I.G.: On constructions of involutory MDS matrices. In: Youssef, A., Nitaj, A., Hassanien, A.E. (eds.) AFRICACRYPT 2013. LNCS, vol. 7918, pp. 43–60. Springer, Heidelberg (2013)
13. Gupta, K.C., Ray, I.G.: On constructions of circulant MDS matrices for lightweight cryptography. In: Huang, X., Zhou, J. (eds.) ISPEC 2014. LNCS, vol. 8434, pp. 564–576. Springer, Heidelberg (2014)
14. Gupta, K.C., Ray, I.G.: Cryptographically significant MDS matrices based on circulant and circulant-like matrices for lightweight applications. Cryptogr. Commun. 7(2), 257–287 (2015)
15. Jr, J.N., Abrahão, É.: A new involutory MDS matrix for the AES. Int. J. Netw. Secur. 9(2), 109–116 (2009)
16. Junod, P., Vaudenay, S.: Perfect diffusion primitives for block ciphers. In: Handschuh, H., Hasan, M.A. (eds.) SAC 2004. LNCS, vol. 3357, pp. 84–99. Springer, Heidelberg (2004)
17. Khoo, K., Peyrin, T., Poschmann, A.Y., Yap, H.: FOAM: searching for hardware-optimal SPN structures and components with a fair comparison. In: Batina, L., Robshaw, M. (eds.) CHES 2014. LNCS, vol. 8731, pp. 433–450. Springer, Heidelberg (2014)
18. Liu, M., Sim, S.M.: Lightweight MDS generalized circulant matrices. Cryptology ePrint Archive, Report 2016/186 (2016). http://eprint.iacr.org/
19. MacWilliams, F.J., Sloane, N.J.A.: The Theory of Error-Correcting Codes, 2nd edn. North-Holland Publishing Company, Amsterdam (1986)
20. Robinson, D.J.S.: An Introduction to Abstract Algebra. De Gruyter Textbook. Walter de Gruyter, Berlin (2003)
21. Shannon, C.E.: Communication theory of secrecy systems. Bell Syst. Tech. J. 28(4), 656–715 (1949)
22. Sim, S.M., Khoo, K., Oggier, F., Peyrin, T.: Lightweight MDS involution matrices. In: Leander, G. (ed.) FSE 2015. LNCS, vol. 9054, pp. 471–493. Springer, Heidelberg (2015)
23. Vaudenay, S.: On the need for multipermutations: cryptanalysis of MD4 and SAFER. In: Preneel, B. (ed.) FSE 1994. LNCS, vol. 1008, pp. 286–297. Springer, Heidelberg (1995)
24. Yang, G., Zhu, B., Suder, V., Aagaard, M.D., Gong, G.: The Simeck family of lightweight block ciphers. In: Güneysu, T., Handschuh, H. (eds.) CHES 2015. LNCS, vol. 9293, pp. 307–329. Springer, Heidelberg (2015)

On the Construction of Lightweight Circulant Involutory MDS Matrices

Yongqiang Li[1,2(✉)] and Mingsheng Wang[1]

[1] State Key Laboratory of Information Security,
Institute of Information Engineering, Chinese Academy of Sciences,
Beijing, China
`yongq.lee@gmail.com`, `wangmingsheng@iie.ac.cn`
[2] Science and Technology on Communication Security Laboratory,
Chengdu, China

Abstract. In the present paper, we investigate the problem of constructing MDS matrices with as few bit XOR operations as possible. The key contribution of the present paper is constructing MDS matrices with entries in the set of $m \times m$ non-singular matrices over $\mathbb{F}_2$ directly, and the linear transformations we used to construct MDS matrices are not assumed pairwise commutative. With this method, it is shown that circulant involutory MDS matrices, which have been proved do not exist over the finite field $\mathbb{F}_{2^m}$, can be constructed by using non-commutative entries. Some constructions of 4×4 and 5×5 circulant involutory MDS matrices are given when $m = 4, 8$. To the best of our knowledge, it is the first time that circulant involutory MDS matrices have been constructed. Furthermore, some lower bounds on XORs that required to evaluate one row of circulant and Hadamard MDS matrices of order 4 are given when $m = 4, 8$. Some constructions achieving the bound are also given, which have fewer XORs than previous constructions.

Keywords: MDS matrix · Circulant involutory matrix · Hadamard matrix · Lightweight

1 Introduction

Linear diffusion layer is an important component of symmetric cryptography which provides internal dependency for symmetric cryptography algorithms. The performance of a diffusion layer is measured by branch number. Using a diffusion layer with bigger branch number in cryptography provides better resistance to differential and linear attack. As for lightweight cryptography, which is aiming to provide security in a limited resource environment, the cost of implementing an linear diffusion layer is also of importance. With the rapid development of lightweight cryptography, it is of particular interest to investigate the problem of constructing lightweight linear diffusion with bigger branch number.

A linear diffusion layer is a linear transformation over $(\mathbb{F}_2^m)^n$, where m is the bit length of an S-box and n is the number of S-boxes that the linear diffusion layer acts on. Note that every linear transformation can be represented by a

© International Association for Cryptologic Research 2016
T. Peyrin (Ed.): FSE 2016, LNCS 9783, pp. 121–139, 2016.
DOI: 10.1007/978-3-662-52993-5_7

matrix, then a linear diffusion layer is often represented by a $n \times n$ matrix and the entries can be viewed as linear transformations over $\mathbb{F}_2^m$. The maximum branch number of a $n \times n$ matrix over $(\mathbb{F}_2^m)^n$ is $n+1$. A linear diffusion layer with maximum branch number is called a perfect diffusion layers or a Maximal Distance Separable (MDS) matrix. An MDS matrix is a linear multipermutation [22].

A common way to construct MDS matrices is using MDS codes over finite fields. Multiplication with elements in finite fields is a basic operation in the evaluation of a matrix over finite fields. Usually, this operation is heavy in implementation. To improve its implementation efficiency, it is often constructing a matrix with fewer different elements of finite fields and choosing elements of finite fields with lower Hamming weight. Therefore, some matrices can be defined by fewer elements are preferred, such as circulant matrix and Hadamard matrix. The diffusion layer of AES is an typical example of this construction method. It is a 4×4 circulant MDS matrix over $\mathbb{F}_{2^8}$.

Another main method to construct lightweight MDS matrices is recursive construction. The main idea is that firstly constructing a linear transformation which is sparse and compact in implementation, and then composing it several times to get an MDS matrix. This method is first used in the design of Photon lightweight hash family [10] and LED lightweight block cipher [9], and then attracted lots of attentions. The method is extended by using linear transformations instead of multiplications of elements in finite fields in [20]. Then the work is improved by using linear transformations with fewer XORs in [23], where some extreme lightweight MDS matrices are given. A method is given to get rid of expensive symbolic computations of the above method for constructing larger recursive MDS matrices in [1]. The method is also further investigated in [12]. The construction of recursive MDS matrices also has a relation with coding theory. It is shown that recursive MDS matrices can be constructed from Gabidulin codes [4], and also can be obtained directly from shortened MDS cyclic codes [2].

However, a recursive MDS matrix may leads to high latency since it has to run several rounds to get outputs. Then how to construct lightweight MDS matrices without using recursive construction is an interesting problem needs further study. Some works revisit the method of constructing MDS matrices over finite fields by choosing elements whose multiplication's implementation efficiency can be further improved. Recently, it is shown that the choice of the irreducible polynomial used to compute multiplication with elements over finite fields has a great influence of the efficiency [19]. This property is further investigated in [21], where algorithms are designed to search lightweight MDS matrices with few XORs that required to evaluate one row of the corresponding matrix. Several constructions and their comparisons with previous constructions are also given in [21].

Our Contributions. In the present paper, we investigate the problem of constructing MDS matrices with as few bit XOR operations as possible. Note that multiplication with elements of the finite field $\mathbb{F}_{2^m}$ is only a special type of linear transformations over $\mathbb{F}_2^m$. Moreover, there exist many other linear transformations over $\mathbb{F}_2^m$ which can not be represented by multiplication with elements over $\mathbb{F}_{2^m}$. Therefore, constructing matrices over the space of linear transformations over $\mathbb{F}_2^m$ may leads to new constructions of lightweight MDS matrices.

In previous constructions, the entries used to construct MDS matrices are pairwise commutative, such as MDS matrices over finite fields, or assumed pairwise commutative, such as recursive MDS matrices with elements being linear transformations [20, 23]. Note that a matrix over a commutative ring is nonsingular if and only if its determinant is a unity in the ring, then the assumption is convenient for charactering MDS matrices since the determinants of square sub-matrices can be computed.

However, the restriction of choosing commutative linear transformations may lose MDS matrices with fewer XORs. Then we do not assume the linear transformations over $\mathbb{F}_2^m$ that used to construct MDS matrices are pairwise commutative in the present paper.

The strategy we used to determine whether a construction is MDS is computing all its square sub-matrices' rank. Then it is too complex to construct MDS matrices with larger order. In symmetric cryptography algorithms, the most often used S-boxes are 4-bit and 8-bit S-boxes, and it is often use diffusion layers of order 4. Therefore, we focus on constructing 4×4 MDS matrices with entries in the space of linear transformations over $\mathbb{F}_2^4$ and $\mathbb{F}_2^8$ in the present paper.

The first result is that circulant involutory MDS matrices can be constructed with our method. Circulant involutory MDS matrices can be implemented efficiently and the same circuit can be used both in encryption and decryption. However, it has been proved in [13, 16] that there do not exist circulant involutory MDS matrices over the finite field $\mathbb{F}_{2^m}$. In fact, the proof is only valid when the entries of the matrix are pairwise commute. This property is satisfied by previous construction methods but not our method.

We show that there exist circulant involutory MDS matrices over the space of linear transformations over $\mathbb{F}_2^m$. Some constructions are also given. To the best of our knowledge, it is the first time that circulant involutory MDS matrices have been constructed. For 4×4 circulant involutory MDS matrices constructed in the present paper, the fewest sum of XORs of one row's entries is $m+1, m = 4, 8$. Moreover, we also construct 4×4 orthogonal circulant MDS matrix, which is also proved do not exist over finite fields [13].

Lower bounds on XORs that required to evaluate one row of circulant (non-involution) MDS matrices, involutory Hadamard MDS matrices and Hadamard (noninvolution) MDS matrices are also investigated. We show that for circulant MDS matrices with the first row's entries are $[I, I, A, B]$, the fewest sum of XORs of A and B is 3. For involutory Hadamard MDS matrices, the fewest sum (the fewest sum we get) of the XORs of entries in the first row is $m + 2$ for $m = 4$ ($m = 8$). For Hadamard MDS matrices, the fewest sum of XORs of one row's entries is 4 for $m = 4$ and the fewest sum we get of XORs of one row's entries is 5 for $m = 8$. Lower bounds on the entries of "optimal" 4×4 MDS matrices is also characterized.

Outline of This Paper. The present paper is organized as follows. In Sect. 2, we give some preliminaries. A general bound on XORs that required to evaluate one row of circulant and Hadamard MDS matrices is also given. In Sect. 3, we investigate the construction of lightweight involutory, non-involutory and orthogonal circulant MDS matrices. In Sect. 4, we investigate the construction of lightweight involutory and non-involutory Hadamard MDS matrices. Comparisons with previous constructions are given at the end of the section. In Sect. 5, we investigate the construction of lightweight "optimal" 4×4 MDS matrices. A short conclusion is given in Sect. 6.

2 Preliminaries and a General Bound

A map $A : \mathbb{F}_2^m \to \mathbb{F}_2^m$ is called linear if $A(x + y) = A(x) + A(y)$ for $x, y \in \mathbb{F}_2^m$. Fixed a basis of $\mathbb{F}_2^m$ over $\mathbb{F}_2$, a linear map over $\mathbb{F}_2^m$ can be represented by an $m \times m$ matrix over $\mathbb{F}_2$, which is also denoted by A. Then $A(x) = A \cdot x$, where $x = (x_1, \ldots, x_m) \in \mathbb{F}_2^m$ is viewed as a column vector throughout this paper. A linear map is a permutation over $\mathbb{F}_2^m$ if and only if its matrix representation is non-singular. The notation $GL(m, S)$ denotes the set of all $m \times m$ non-singular matrices with entries in S.

For $a, b \in \mathbb{F}_2$, $a + b$ is called the bit XOR operation. For $A \in GL(m, \mathbb{F}_2)$, $\#A$ denotes the number of XOR operations that required to evaluate $A \cdot x$ directly, where $x \in \mathbb{F}_2^m$, and we call A has $\#A$ XOR operations. It is easy to see that $\#A$ equals the number of XORs in $A(x)$ and hence

$$\#A = \sum_{i=1}^{m} (\omega(A[i]) - 1),$$

where $\omega(A[i])$ means the number of nonzero entries in the i-th row of A. For $A \in GL(m, \mathbb{F}_2)$, a simplified representation of A is given by extracting the nonzero positions in each row of A. For example, $[2, 3, 4, [1,4]]$ is the representation of the following matrix.

$$\begin{pmatrix} 0 & 1 & 0 & 0 \\ 0 & 0 & 1 & 0 \\ 0 & 0 & 0 & 1 \\ 1 & 0 & 0 & 1 \end{pmatrix},$$

and it is a matrix with 1 XOR operation.

Every linear diffusion can be represented by a matrix as follows

$$L = \begin{pmatrix} L_{1,1} & L_{1,2} & \cdots & L_{1,n} \\ L_{2,1} & L_{2,2} & \cdots & L_{2,n} \\ \vdots & \vdots & \cdots & \vdots \\ L_{n,1} & L_{n,2} & \cdots & L_{n,n} \end{pmatrix},$$

where $L_{i,j}$ is an $m \times m$ matrix over $\mathbb{F}_2$ for $1 \leq i, j \leq n$. For $X = (x_1, \ldots, x_n) \in (\mathbb{F}_2^m)^n$,

$$L(X) = \left(\sum_{i=1}^{n} L_{1,i}(x_i), \ldots, \sum_{i=1}^{n} L_{n,i}(x_i) \right),$$

where $L_{i,j}(x_k) = L_{i,j} \cdot x_k$, for $1 \le i,j \le n, 1 \le k \le m$. A linear diffusion L defined as above is called involutory if $L \circ L(X) = X$ for all $X \in (\mathbb{F}_2^m)^n$, which is equivalent to that L^2 is the identity matrix of order mn.

For $X = (x_1, \ldots, x_n) \in (\mathbb{F}_2^m)^n$, the bundle weight of X, which is denoted by $\omega_b(X)$, is defined as the number of nonzero entries of X. This means

$$\omega_b(X) = |\{x_i : x_i \ne 0, 1 \le i \le n\}|.$$

The branch number of L is defined as

$$\min\{\omega_b(X) + \omega_b(L(X)) \mid X \in (\mathbb{F}_2^m)^n, X \ne 0\}.$$

The upper bound on the branch number of L is $n + 1$, and a matrix achieved the bound is called an MDS matrix.

Square sub-matrices of L of order t means the following matrices

$$L(J, K) = (L_{j_l, k_p}, 1 \le l, p \le t)$$

where $J = [j_1, \ldots, j_t]$ and $K = [k_1, \ldots, k_t]$ are two sequence of length t, and $1 \le j_1 < \ldots < j_t \le n$, $1 \le k_1, \ldots, k_t \le n$. Note that $L(J, K) \cdot (x_1, \ldots, x_t) = 0$ does not have nonzero solutions if and only if $L(J, K)$ is of full rank. Then the following result holds, which is proved in [5].

Theorem 1. *Let* $L = (L_{i,j}), 1 \le i,j \le n$, *and the entries of* L *are* $m \times m$ *matrices over* $\mathbb{F}_2$. *Then* L *is an MDS matrix if and only if all square sub-matrices of* L *of order* t *are of full rank for* $1 \le t \le n$.

According to Theorem 1, the computation would be complicated when n is large. Then in the present paper we focus on 4×4 matrices, which are widely used in cryptography. More precisely, we construct lightweight MDS matrices using circulant matrix and Hadamard matrix. Both of them can be defined by the first row's entries and hence can be implemented efficiently.

2.1 A General Bound

In this subsection, we give a general bound of XORs on circulant and Hadamard MDS matrices.

A matrix is called circulant if each row is rotated to the right of the preceding row by one entry. Then for a 4×4 circulant matrix, we means

$$Circ(A, B, C, D) = \begin{pmatrix} A & B & C & D \\ D & A & B & C \\ C & D & A & B \\ B & C & D & A \end{pmatrix},$$

where $A, B, C, D \in GL(m, \mathbb{F}_2)$.

A $2^k \times 2^k$ matrix H is called a Hadamard matrix if it can be represented as

$$\begin{pmatrix} H_1, H_2 \\ H_2, H_1 \end{pmatrix},$$

where H_1, H_2 are two $2^{k-1} \times 2^{k-1}$ Hadamard matrices. Then for a 4×4 Hadamard matrix, we means

$$Had(A, B, C, D) = \begin{pmatrix} A, \ B, \ C, \ D \\ B, \ A, \ D, \ C \\ C, \ D, \ A, \ B \\ D, \ C, \ B, \ A \end{pmatrix},$$

where $A, B, C, D \in GL(m, \mathbb{F}_2)$.

Remember that our aim is constructing MDS matrices with as few XOR operations as possible. Then we prefer linear transformations with no XORs. However, the following results limits the amounts of such linear transformations used in our constructions.

Lemma 1. Let $L = \begin{pmatrix} L_1, L_2 \\ L_3, L_4 \end{pmatrix}$, $L_i \in GL(m, \mathbb{F}_2), 1 \leq i \leq 4$. If $\mathrm{rank}(L) = 2m$, then $\sum_{i=1}^{4} \#L_i \geq 1$.

Proof. Assume $\#L_i = 0$, $1 \leq i \leq 4$. Then for $1 \leq i \leq 4$, each row and each column of L_i has exactly one entry equals 1 since L_i are non-singular. This means every entry of $\sum_{j=1}^{m} L_i[j]$ equals to 1. Therefore, every entry of $\sum_{i=1}^{2m} L[i]$ equals to 0, which means $\mathrm{rank}(L) < 2m$ and we complete the proof.   □

Then we have the following result.

Theorem 2. *1. Let $L = Circ(A, B, C, D)$ be a circulant MDS matrix, where $A, B, C, D \in GL(m, \mathbb{F}_2)$. Then $\#A + \#B + \#C + \#D \geq 2$.*
2. Let $L = Had(A, B, C, D)$ be a Hadamard MDS matrix, where $A, B, C, D \in GL(m, \mathbb{F}_2)$. Then $\#A + \#B + \#C + \#D \geq 3$.

Proof. Let $L = Circ(A, B, C, D)$ be a circulant MDS matrix. Assume

$$\#A + \#B + \#C + \#D \leq 1.$$

Then there are at least 3 entries with 0 XORs in the first row. Without loss of generality, we suppose $\#A = \#B = \#C = 0$. Then according to Lemma 1, it holds

$$\mathrm{rank}(L([1, 2], [2, 3])) = \mathrm{rank}\left(\begin{pmatrix} B, C \\ A, B \end{pmatrix}\right) < 2m.$$

This is a contradiction since L is an MDS matrix. The other cases can be proved similarly.

Let $L = Had(A, B, C, D)$ be a Hadamard MDS matrix. Assume

$$\#A + \#B + \#C + \#D \leq 2.$$

Then there are at least 2 entries with 0 XORs in the first row. Without loss of generality, we suppose $\#A = \#C = 0$. Then according to Lemma 1, it holds

$$\text{rank}(L([1,3],[1,3])) = \text{rank}\left(\begin{pmatrix} A,C \\ C,A \end{pmatrix}\right) < 2m.$$

This is a contradiction since L is an MDS matrix. The other cases can be proved similarly. □

The above result means that there are at most two entries with no XORs in one row of a circulant MDS matrix, and there are at most one entry with no XORs in one row of a Hadamard MDS matrix. We suppose $L[1,1] = I$ in our constructions, where I denotes the identity matrix throughout this paper.

3 Lightweight Circulant MDS Matrices

In this section, we investigate the construction of lightweight circulant involutory, non-involutory and orthogonal MDS matrices respectively.

3.1 Constructing Circulant Involutory MDS Matrices

First, we have the following result.

Lemma 2. Let $L = Circ(I,A,B,C)$ be a circulant matrix, where $A, B, C \in GL(m, \mathbb{F}_2)$. Then L is an involution if and only if the following equalities hold:

$$AB = BA, BC = CB, A^2 = C^2, AC + CA = B^2.$$

Proof. By matrix multiplication, it can be checked that

$$L^2 = Circ(I,A,B,C) \cdot Circ(I,A,B,C)$$
$$= Circ(I + AC + CA + B^2, BC + CB, A^2 + C^2, AB + BA).$$

On the other hand, L is an involution if and only if $L^2 = Circ(I,0,0,0)$. Therefore, L is an involution if and only if

$$AB = BA, BC = CB, A^2 = C^2, AC + CA = B^2$$

hold simultaneously. □

We give a general construction of circulant involutory matrix in the following result. For $A \in GL(m, \mathbb{F}_2)$, the multiplication order of A is defined as the minimum positive integer d such that $A^d = I$.

Lemma 3. Suppose $A, C \in GL(m, \mathbb{F}_2)$ with $A^2 = C^2 = I$, and the multiplication order of $A+C$ equals $4k-2$ for some integer k with $k > 1$. Let $B = (A+C)^{2k}$. Then the matrix $Circ(I,A,B,C)$ is an involution.

Proof. Let $B = (A + C)^{2k}$. Note that

$$A^2 = C^2 = I,$$

then according to Lemma 2, we only need to prove that A, B, C satisfy the following equalities

$$AB = BA, BC = CB, AC + CA = B^2.$$

First, it is easy to see that

$$(A + C)^2 = A^2 + AC + CA + C^2 = AC + CA.$$

Then we have

$$B = (A + C)^{2k} = (AC + CA)^k.$$

Therefore,

$$\begin{aligned}
AB &= A(AC + CA)^k \\
&= A(AC + CA)(AC + CA)^{k-1} \\
&= (A^2C + ACA)(AC + CA)^{k-1} \\
&= (CA^2 + ACA)(AC + CA)^{k-1} \\
&= (CA + AC)A(AC + CA)^{k-1} \\
&= \cdots \\
&= (AC + CA)^k A \\
&= BA.
\end{aligned}$$

Similarly, it can be checked that

$$BC = CB.$$

Note that $(A + C)^{4k-2} = I$, then we have

$$B^2 = (A + C)^{4k} = (A + C)^2 = AC + CA.$$

According to Lemma 2, we have $Circ(I, A, (A + C)^{2k}, C)$ is an involution.   □

Remark 1. If $k = 1$, then the multiplication order of $A+C$ equals 2 and $B = (A + C)^2 = I$. In this case, $L = Circ(I, A, I, C)$ constructed as above is also a circulant involution. However, it is not an MDS matrix since $\operatorname{rank}(L([1, 3], [1, 3])) < 2m$. Then we always suppose $k > 1$ since we want to construct circulant involutory MDS matrices.

Using above results, our searching strategy is as follows. Firstly, we get the set S which contains all involutory matrix from the set which we want to search. Then for each pair of $(A, C) \in S \times S$, we compute the multiplication order d of $A + C$. If $d \bmod 4 = 2$, then let $B = (A + C)^{\frac{d}{2}+1}$, and test whether $Circ(I, A, B, C)$ is MDS by Theorem 1.

When $m = 4$, we search A, C over $GL(4, \mathbb{F}_2)$. There exist A, C such that $Circ(I, A, B, C)$ is MDS. The fewest sum of XORs of one rows' entries of an

MDS involutory $Circ(I, A, B, C)$ constructed as above is 5. There are 48 pairs of A, C with this property. These 48 matrices are of the type $Circ(I, A, B, C)$ and $Circ(I, C, B, A)$ for 24 different pairs of A, C.

When $m = 8$, we search A, C over all 8×8 non-singular matrices over $\mathbb{F}_2$ with less than or equal to 3 bit XOR operations. The fewest sum of XORs of one rows' entries of an MDS $Circ(I, A, B, C)$ constructed as above is 9. There are 40320 pairs of A, C satisfy this property. For all these pairs of A, C, $Circ(I, C, B, A)$ are also circulant involutory MDS matrices.

Theorem 3. *Their exist $A, B, C \in GL(m, \mathbb{F}_2)$, $m = 4, 8$, such that $Circ(I, A, B, C)$ is an involutory MDS matrix. Furthermore, the following statements hold.*

1. *When $m = 4$, circulant involutory MDS matrices constructed with the above method satisfy $\#A + \#B + \#C \geq 5$.*
2. *When $m = 8$, if $\#A \leq 3$ and $\#C \leq 3$, then circulant involutory MDS matrices constructed with the above method satisfy $\#A + \#B + \#C \geq 9$.*

Example 1. Examples of A, B, C such that $Circ(I, A, B, C)$ are circulant involutory MDS matrices with $\#A + \#B + \#C = m + 1$.[1]

(1) $m = 4$, $A = [1, 2, [1, 3], [1, 2, 4]]$, $C = [4, 3, 2, 1]$, $B = (A + C)^4 = [2, [1, 2], [3, 4], 3]$.
(2) $m = 8$, $A = [1, 2, [1, 3], [1, 2, 4], 6, 5, 8, 7]$, $C = [5, 8, [2, 6], 7, 1, [3, 8], 4, 2]$, and $B = (A + C)^{16} = [[7, 8], 1, 7, [3, 8], [2, 4], [1, 4], 6, 5]$.

We further investigate the construction of 5×5 circulant involutory MDS matrices. In order to simplify our characterization, we investigate 5×5 circulant matrices of the type $Circ(I, A, B, B, A)$, where $A, B \in GL(m, \mathbb{F}_2)$. Concerning the property of involutory of $Circ(I, A, B, B, C)$, it is easy to prove the following result.

Lemma 4. *Let $L = Circ(I, A, B, B, A)$ be a circulant matrix, where $A, B \in GL(m, \mathbb{F}_2)$. Then L is an involution if and only if $A^2 = AB + BA = B^2$.*

We give constructions by exhaustive searching for A, B with the following method. The method is often used hereafter in the paper, and we give a detailed general description here.

The following result is helpful. It can be proved via elementary linear algebra and we omit the proof here.

Lemma 5. *Suppose $A, B, C \in GL(m, \mathbb{F}_2)$ are $m \times m$ non-singular matrices over $\mathbb{F}_2$. Then the following statements hold.*

(1) $\begin{pmatrix} I, & A \\ B, & C \end{pmatrix}$ *is of full rank if and only if $\mathrm{rank}(BA + C) = m$.*

(2) $\begin{pmatrix} A, & I \\ B, & C \end{pmatrix}$ *is of full rank if and only if $\mathrm{rank}(CA + B) = m$.*

---

[1] More examples of circulant involutory MDS matrices with $\#A + \#B + \#C = m + 1$ are given in the appendix of the extended version of the paper [14].

(3) $\begin{pmatrix} A, B \\ I, C \end{pmatrix}$ is of full rank if and only if $\text{rank}(AC + B) = m$.

(4) $\begin{pmatrix} A, B \\ C, I \end{pmatrix}$ is of full rank if and only if $\text{rank}(BC + A) = m$.

Let $L = Circ(I, A, B, B, A)$. According to Theorem 1, if L is MDS, then all its square sub-matrices are of full rank. According to Lemma 5, we have the following fact by investigating all square sub-matrices of order 2. If L is MDS, then the following matrices are non-singular:

$$A + I, A^2 + I, B + I, B^2 + I, A^2 + B, A + B^2, A + B.$$

Note that $A^2 + I$ is non-singular if and only if $A + I$ is non-singular. Then the conditions can be simplified as the following matrices are non-singular:

$$A + I, B + I, A + B^2, A^2 + B, A + B.$$

Based on the above observations, we have the following searching strategy. First, note that both A and B should satisfy $\text{rank}(X + I) = m, X = A, B$. The equalities that both A and B satisfied are called general rules. Then we can select the candidate set of A and B from the set we want to search over by using general rules, which means

$$S_{A,B} := \{X : X \in S_{search} \mid \text{rank}(X + I) = m\}.$$

The for $A \in S_{A,B}$, we can get the candidate set of B by using the other conditions that should be satisfied, which means

$$S_B := \{B : B \in S_{A,B} \mid \text{rank}(A + B) = m \wedge \text{rank}(A^2 + B) = m \wedge \text{rank}(A + B^2) = m \\ \wedge A^2 = AB + BA \wedge A^2 = B^2\}.$$

At last, for $B \in S_B$, we test whether L is MDS by Theorem 1.

When $m = 4$, we search A, B over $GL(4, \mathbb{F}_2)$. The fewest XORs of one row's entries of an involutory MDS $Circ(I, A, B, B, A)$ is 4. There are 24 pairs of A, B such that $Circ(I, A, B, B, A)$ are involutory circulant MDS matrices with $\#A + \#B = 2$. These 24 MDS matrices are of the type $Circ(I, A, A^T, A^T, A)$ and $Circ(I, A^T, A, A, A^T)$ for 12 different A.

When $m = 8$, we search A, B over $GL(8, \mathbb{F}_2)$ with $\#A + \#B \le 3$. No involutory MDS matrix returns. Therefore, if $Circ(I, A, B, B, A)$ is an involutory MDS matrix, then $\#A + \#B \ge 4$.

Then we have the following result.

Theorem 4. *Their exist $A, B \in GL(m, \mathbb{F}_2)$, $m = 4, 8$, such that $Circ(I, A, B, B, A)$ is an 5×5 involutory MDS matrix. Furthermore, if $Circ(I, A, B, B, A)$ is an involutory MDS matrix, then $\#A + \#B \ge \frac{m}{2}$.*

Similar as the method "Subfield construction" that used in [6,19,21], it is easy to construct involutory MDS $Circ(I, A, B, B, A)$ over $\mathbb{F}_2^8$ with $\#A + \#B = 4$, since we have constructed involutory MDS $Circ(I, A, B, B, A)$ over $\mathbb{F}_2^4$ with

$\#A + \#B = 2$. Let $X \in GL(4, \mathbb{F}_2)$, $\#X = 1$ and $Circ(I, X, X^T, X^T, X)$ is an involutory MDS matrix. Then $Circ(I, A, A^T, A^T, A)$ is also an involutory MDS matrix, where $A \in GL(8, \mathbb{F}_2)$ of the following form

$$A = \begin{bmatrix} X, & 0 \\ 0, & X \end{bmatrix}.$$

Then we can construct 24 circulant involutory MDS by using the above method and the searching result when $m = 4$.

In order to get more circulant involutory MDS matrices, we searching A over $GL(8, \mathbb{F}_2)$ with $\#A = 2$. We get 20160 A such that $Circ(I, A, A^T, A^T, A)$ are involutory MDS matrices and $\#A + \#A^T = 4$.

Example 2. Examples of A, B such that $Circ(I, A, B, B, A)$ are circulant involutory MDS matrices with $\#A + \#B = \frac{m}{2}$.

(1) $m = 4$, $A = [2, 3, 4, [1, 3]]$, $B = A^T = [4, 1, [2, 4], 3]$.
(2) $m = 8$, $X = [2, 3, 4, [1, 3]]$, $A = \begin{bmatrix} X, & 0 \\ 0, & X \end{bmatrix} = [2, 3, 4, [1, 3], 6, 7, 8, [5, 7]]$, $B = A^T = [4, 1, [2, 4], 3, 8, 5, [6, 8], 7]$.
(3) $m = 8$, $A = [[3, 5], 8, 1, 3, 4, 2, 6, [2, 7]]$, $B = A^T = [3, [6, 8], [1, 4], 5, 1, 7, 8, 2]$.

It is interesting that 5×5 circulant involutory MDS matrices can be constructed with only 3 different entries. We have tried some other methods to construct circulant involutory MDS matrices with higher order. However, we do not get an circulant involutory MDS matrix with order large than or equal to 6 until present. We leave it as an open problem.

Problem 1. Construct $n \times n$ circulant involutory MDS matrices over $GL(m, \mathbb{F}_2)$ or prove that they do not exist, where $n \geq 6$, $m = 4, 8$.

3.2 Constructing Circulant Non-involutory MDS Matrices

In this subsection, we want to construct non-involutory MDS matrices with as few XORs as possible. We consider circulant matrices of the type

$$Circ(I, I, A, B),$$

since it has the most many entries with no XORs in one row.

The searching strategy is similar as previous subsection. If $Circ(I, I, A, B)$ is MDS, then the following matrices are non-singular:

$$A + I, B + I, A + B, AB + I, A^2 + B, A + B^2.$$

When $m = 4$, we search A, B over $GL(4, \mathbb{F}_2)$. The fewest XORs of one row's entries of an MDS $Circ(I, I, A, B)$ is 3. Their are 48 pair of (A, B) such that $Circ(I, I, A, B)$ are MDS matrices with $\#A + \#B = 3$. These 48 matrices are of the type $Circ(I, I, A, A^{-2})$ and $Circ(I, I, A^{-2}, A)$ for 24 different A.

When $m = 8$, we search A, B over all 8×8 non-singular matrices over $\mathbb{F}_2$ with 1 bit XOR. No MDS matrix returns. This means if $Circ(I, I, A, B)$ is an MDS matrix over $GL(8, \mathbb{F}_2)$, then either A or B has at least 2 XORs, and hence $\#A + \#B \geq 3$. Therefore, the following result hold.

Theorem 5. *Let* $L = Circ(I, I, A, B)$, *where* $A, B \in GL(m, \mathbb{F}_2)$, $m = 4, 8$. *If* L *is an MDS matrix, then* $\#A + \#B \geq 3$.

In order to get circulant MDS matrix with the above equality holds when $m = 8$, we let $B = A^{-2}$ and search A over all 8×8 non-singular matrices over $\mathbb{F}_2$ with 1 bit XOR. At last, we get 80640 A such that $Circ(I, I, A, A^{-2})$ are MDS matrices with $\#A + \#A^{-2} = 3$. Furthermore, $Circ(I, I, A^{-2}, A)$ are also MDS matrices for all these A.

Example 3. Examples of A, B such that $Circ(I, I, A, B)$ and $Circ(I, I, B, A)$ are MDS matrices with $\#A + \#B = 3$.

(1) $m = 4$, $A = [2, 3, 4, [1, 4]]$, $B = A^{-2} = [[2, 3], [3, 4], 1, 2]$.
(2) $m = 8$, $A = [2, 3, 4, 5, 6, 7, 8, [1, 3]]$, $B = A^{-2} = [[1, 7], [2, 8], 1, 2, 3, 4, 5, 6]$.

3.3   Constructing Circulant Orthogonal MDS Matrices

A square matrix L is called orthogonal if $L^{-1} = L^T$, where L^T is the transpose of L. It is proven in [13] there do not exist $2^d \times 2^d$ circulant orthogonal MDS matrix over finite fields. In this subsection, we show that 4×4 circulant orthogonal MDS matrices can also be constructed with non-commutative entries.

Firstly, note that for $L = Circ(I, A, B, C)$, where $A, B, C \in \mathbb{F}_{2^m}$, it holds $L^T = Circ(I, C^T, B^T, A^T)$. This means one have to implement new entries A^T, B^T, C^T in decryption circuit when L is orthogonal. In order to simplify implementation, we let $A, B, C \in GL(m, \mathbb{F}_2)$ are symmetric matrices, which means $A = A^T, B = B^T, C = C^T$. Then it holds

$$L^T = Circ(I, C^T, B^T, A^T) = Circ(I, C, B, A),$$

and it is easy to prove the following result.

Lemma 6. *Let* $L = Circ(I, A, B, C)$ *be a circulant matrix, where* $A, B, C \in GL(m, \mathbb{F}_2)$ *are symmetric matrices. Then* L *is orthogonal if and only if the following equalities hold:*

$$A^2 + B^2 = C^2, AC = CA, A + C = BA + CB, A + C = AB + BC.$$

If $L = Circ(I, A, B, C)$ is MDS, then the following matrices are non-singular:

$$B + I, B + A^2, B + C^2, AC + I, AB + C.$$

When $m = 4$, we search symmetric A, B, C over $GL(4, \mathbb{F}_2)$. The fewest XORs of one row's entries of an orthogonal MDS $Circ(I, A, B, C)$ is 8. Their are 24 triples of A, B, C such that $Circ(I, A, B, C)$ are orthogonal MDS matrices with $\#A + \#B + \#C = 8$. Then we have the following result.

Theorem 6. *There exist symmetric $A, B, C \in GL(4, \mathbb{F}_2)$ such that $Circ(I, A, B, C)$ is an orthogonal MDS matrix. Furthermore, if $Circ(I, A, B, C)$ is an orthogonal MDS matrix, then $\#A + \#B + \#C \geq 8$.*

Example 4. Example of A, B, C such that $Circ(I, A, B, C)$ is an orthogonal circulant MDS matrix $\#A + \#B + \#C = 2m$.

(1) $m = 4$, $A = [1, 2, 4, [3, 4]]$, $B = [[1, 4], [2, 3, 4], [2, 3], [1, 2, 4]]$, $C = [2, [1, 2], 3, 4]$.

(2) $m = 8$, $A = \begin{bmatrix} A_1, & 0 \\ 0, & A_1 \end{bmatrix}$, $B = \begin{bmatrix} B_1, & 0 \\ 0, & B_1 \end{bmatrix}$, $C = \begin{bmatrix} C_1, & 0 \\ 0, & C_1 \end{bmatrix}$, where A_1, B_1, C_1 are the A, B, C in the above item.

4   Lightweight Hadamard MDS Matrices

In this section, we investigate the construction of lightweight Hadamard involutory and non-involutory MDS matrices respectively.

4.1   Constructing Hadamard Involutory MDS Matrices

In the case of a, b, c are elements of finite fields, $Had(1, a, b, c)$ is an involution if and only if $a^2 + b^2 = c^2$. In the case of $A, B, C \in GL(m, \mathbb{F}_2)$, we have the following result.

Lemma 7. *Let $A, B, C \in GL(m, \mathbb{F}_2)$. Then $L = Had(I, A, B, C)$ is an involution if and only if A, B, C are pairwise commutative and $A^2 + B^2 = C^2$.*

Proof. By matrix multiplication, it can be checked that

$$L^2 = Had(I, A, B, C) \cdot Had(I, A, B, C)$$
$$= Had(I + A^2 + B^2 + C^2, BC + CB, AC + CA, AB + BA).$$

Therefore, L is an involution if and only if $L^2 = Had(I, 0, 0, 0)$, which is equivalent to

$$AB = BA, BC = CB, AC = CA, A^2 + B^2 = C^2$$

hold simultaneously.   □

When $m = 4$, we search A, B, C over $GL(4, \mathbb{F}_2)$ as previous. The fewest XORs of one row's entries of an involutory MDS $Had(I, A, B, C)$ is 6. There are 144 triples of A, B, C such that $Had(I, A, B, C)$ are involutory MDS matrices with $\#A + \#B + \#C = 6$. These 144 matrices are of the type $Had(I, A_1, A_2, A_3)$, where (A_1, A_2, A_3) is a permutation of $(A, A^{-1}, A + A^{-1})$ for 24 different A.

When $m = 8$, we also consider Hadamard matrix of the type

$$L = Had(I, A, A^{-1}, A + A^{-1}),$$

where $A \in GL(m, \mathbb{F}_2)$. According to the above lemma, L is an involution. We use the method in [20, 23] to characterize whether L is MDS. By computing the determinants of all the square sub-matrices of L and factorizing these polynomials, we get that L is an MDS matrix if and only if all the following matrices are non-singular:

$$A, A + I, A^2 + A + I, A^3 + A + I, A^3 + A^2 + I.$$

Then we search A over $GL(8, \mathbb{F}_2)$ with $\#A \leq 3$. The fewest XORs of one row's entries of an involutory MDS $Had(I, A, A^{-1}, A + A^{-1})$ is 10. We get 80640 A such that $Had(I, A, A^{-1}, A + A^{-1})$ are involutory MDS matrices with $\#A + \#A^{-1} + \#(A + A^{-1}) = 10$.

We also have searched some other types of Hadamard matrices. However, we do not get a Hadamard involutory matrix with one row's XORs less then 10 until present.

Theorem 7. *1. Let $A, B, C \in GL(4, \mathbb{F}_2)$. If $L = Had(I, A, B, C)$ is an MDS involution matrix, then $\#A + \#B + \#C \geq 6$.*
2. Let $A \in GL(8, \mathbb{F}_2)$ with $\#A \leq 3$. If $L = Had(I, A, A^{-1}, A + A^{-1})$ is an MDS involution matrix, then $\#A + \#A^{-1} + \#(A + A^{-1}) \geq 10$.

Example 5. Examples of A, B, C such that $Had(I, A, B, C)$ are involutory MDS matrices with $\#A + \#B + \#C = m + 2$.

(1) $m = 4$, $A = [2, [1, 3], 4, [2, 3]]$, $B = A^{-1} = [[1, 2, 4], 1, [1, 4], 3]$, $C = A + A^{-1} = [[1, 4], 3, 1, 2]$.
(2) $m = 8$, $A = [2, 3, 4, 5, 6, 7, 8, [1, 3]]$, $B = A^{-1} = [[2, 8], 1, 2, 3, 4, 5, 6, 7]$, $C = A + A^{-1} = [8, [1, 3], [2, 4], [3, 5], [4, 6], [5, 7], [6, 8], [1, 3, 7]]$.

4.2   Constructing Non-involutory Hadamard MDS Matrices

In this subsection, we want to construct non-involutory Hadamard MDS matrix with as few XORs as possible. The searching strategy is similar as previous. If $Had(I, A, B, C)$ is MDS, then the following matrices are non-singular:

$$A + I, B + I, C + I, AB + C, AC + B, BA + C, BC + A, CB + A, CA + B.$$

When $m = 4$, we search A, B, C over $GL(4, \mathbb{F}_2)$. The fewest XORs of one rows' entries of an MDS $Had(I, A, B, C)$ is 4. There are 72 triples of A, B, C such that $Had(I, A, B, C)$ are MDS matrices with $\#A + \#B + \#C = 4$. These 72 matrices are of the type $Had(I, A_1, A_2, A_3)$, where (A_1, A_2, A_3) is a permutation of $(A, A^T, A + A^T)$ for 12 different A.

When $m = 8$, we search A over $GL(8, \mathbb{F}_2)$ with $\#A \leq 2$. The fewest XORs of one rows' entries of an MDS $Had(I, A, A^T, A + A^T)$ is 8.

In order to get Hadamard MDS matrices with fewer XORs in one row, we investigate Hadamard matrices of the type $Had(I, A, A^T, B)$. According to our searching, if $\#A \leq 1$ and $\#B \leq 2$, then there are no MDS $Had(I, A, A^T, B)$. Then we have the following result.

Theorem 8. *1. Let $A, B, C \in GL(4, \mathbb{F}_2)$. If $L = Had(I, A, B, C)$ is an MDS matrix, then $\#A + \#B + \#C \geq 4$.*

2. Let $A, B \in GL(8, \mathbb{F}_2)$. If $L = Had(I, A, A^T, B)$ is an MDS matrix, then $\#A + \#A^T + \#B \geq 5$.

In order to get MDS $Had(I, A, A^T, B)$ with $\#A + \#A^T + \#B = 5$, we choose A with $\#A = 2$ and $\text{rank}(A + I) = 8$ randomly, and then test whether there exist B with $\#B = 1$ such that $Had(I, A, A^T, B)$ is MDS. We repeat the process several times and get 622 pairs of $A, B \in GL(8, \mathbb{F}_2)$, such that $Had(I, A, A^T, B)$ is MDS and $\#A + \#A^T + \#B = 5$.

Example 6. Examples of A, B, C such that $Had(I, A, B, C)$ are MDS matrices with the bounds in the above theorem hold.

(1) $m = 4$, $A = [2, 3, 4, [1, 3]]$, $B = A^T = [4, 1, [2, 4], 3]$, $C = A + A^T = [[2, 4], [1, 3], 2, 1]$.
(2) $m = 8$, $A = [2, 3, 4, [1, 5], 8, 7, 5, [3, 6]]$, $B = A^T = [4, 1, [2, 8], 3, [4, 7], 8, 6, 5]$, $C = [[4, 7], 6, 5, 8, 7, 1, 2, 3]$.

Table 1. Comparisons with previous constructions of non-involutory MDS matrices

Matrix type	Elements	The first row	XOR count	Ref.
Circulant	$GL(8, \mathbb{F}_2)$	$[I, I, A, B]$	$3 + 3 \times 8 = 27$	Subsect. 3.2
Circulant	$\mathbb{F}_{2^8}/0 \times 11b$	$(0 \times 02, 0 \times 03, 0 \times 01, 0 \times 01)$	$14 + 3 \times 8 = 38$	AES [8]
Hadamard	$GL(8, \mathbb{F}_2)$	$[I, A, A^T, B]$	$5 + 3 \times 8 = 29$	Subsect. 4.2
Hadamard	$\mathbb{F}_{2^8}/0 \times 1c3$	$(0 \times 01, 0 \times 02, 0 \times 04, 0 \times 91)$	$13 + 3 \times 8 = 37$	[21]
Subfield-Hadamard	$\mathbb{F}_{2^4}/0 \times 13$	$(0 \times 1, 0 \times 2, 0 \times 8, 0 \times 9)$	$2 \times (5 + 3 \times 4) = 34$	[21]

Table 2. Comparisons with previous constructions of involutory MDS matrices

Matrix type	Elements	The first row	XOR count	Ref.
Circulant	$GL(8, \mathbb{F}_2)$	$[I, A, B, C]$	$9 + 3 \times 8 = 33$	Subsect. 3.1
Hadamard	$GL(8, \mathbb{F}_2)$	$[I, A, A^{-1}, A + A^{-1}]$	$10 + 3 \times 8 = 34$	Subsect. 4.1
Subfield-Hadamard	$\mathbb{F}_{2^4}/0 \times 13$	$(0 \times 1, 0 \times 4, 0 \times 9, 0xd)$	$2 \times (6 + 3 \times 4) = 36$	[21]
Hadamard	$\mathbb{F}_{2^8}/0 \times 165$	$(0 \times 01, 0 \times 02, 0xb0, 0xb2)$	$16 + 3 \times 8 = 40$	[21]
Hadamard	$\mathbb{F}_{2^8}/0 \times 11d$	$(0 \times 01, 0 \times 02, 0 \times 04, 0 \times 06)$	$22 + 3 \times 8 = 46$	[3]
Compact Cauchy	$\mathbb{F}_{2^8}/0 \times 11b$	$(0 \times 01, 0 \times 12, 0 \times 04, 0 \times 16)$	$54 + 3 \times 8 = 78$	[7]
Hadamard-Cauchy	$\mathbb{F}_{2^8}/0 \times 11b$	$(0 \times 01, 0 \times 02, 0xfc, 0xfe)$	$74 + 3 \times 8 = 98$	[11]

Table 3. Comparisons of MDS matrices over $\mathbb{F}_2^4$ and $\mathbb{F}_{2^4}$

Matrix type	Elements	The first row	XOR count	Ref.
Circulant	$GL(4, \mathbb{F}_2)$	$[I, I, A, B]$	$3 + 3 \times 4 = 15$	Subsect. 3.2
Involutory circulant	$GL(4, \mathbb{F}_2)$	$[I, A, B, C]$	$5 + 3 \times 4 = 17$	Subsect. 3.1
Hadamard	$GL(4, \mathbb{F}_2)$	$[I, A, B, C]$	$4 + 3 \times 4 = 16$	Subsect. 4.2
Hadamard	$\mathbb{F}_{2^4}/0 \times 13$	$(0 \times 1, 0 \times 2, 0 \times 8, 0 \times 9)$	$5 + 3 \times 4 = 17$	[21]
Involutory Hadamard	$GL(4, \mathbb{F}_2)$	$[I, A, A^{-1}, A + A^{-1}]$	$6 + 3 \times 4 = 18$	Subsect. 4.1
Involutory Hadamard	$\mathbb{F}_{2^4}/0 \times 13$	$(0 \times 1, 0 \times 4, 0 \times 9, 0xd)$	$6 + 3 \times 4 = 18$	[15, 21]
Involutory Hadamard	$\mathbb{F}_{2^4}/0 \times 19$	$(0 \times 1, 0 \times 2, 0 \times 6, 0 \times 4)$	$6 + 3 \times 4 = 18$	[18]

We give comparisons of our constructions with previous constructions in Tables 1, 2 and 3 respectively.

The lower bounds on XORs of circulant and Hadamard MDS matrices given in Sects. 3 and 4 are under the supposition $L[1,1] = I$. Therefore, it is possible to improve the previous lower bounds when $L[1,1] \neq I$. However, we have the following result with searching, which shows that the lower bounds can not be improved when $m = 4$.

Theorem 9. *Let $A_i \in GL(4, \mathbb{F}_2)$, and $\mathcal{A} = \sum_{i=1}^{4} \#A_i$. Then the following statements hold.*

1. *If $Circ(A_1, A_2, A_3, A_4)$ is a circulant MDS matrix, then $\mathcal{A} \geq 3$.*
2. *If $Circ(A_1, A_2, A_3, A_4)$ is a circulant involutory MDS matrix, then $\mathcal{A} \geq 5$.*
3. *If $Had(A_1, A_2, A_3, A_4)$ is a Hadamard MDS matrix, then $\mathcal{A} \geq 4$.*
4. *If $Had(A_1, A_2, A_3, A_4)$ is a Hadamard involutory MDS matrix, then $\mathcal{A} \geq 6$.*

5   Lightweight "Optimal" 4×4 MDS Matrices

It is proven in [17] that the highest possible number of 1 and the lowest possible number of different entries for a 4×4 MDS matrix over finite fields are 9 and 3 respectively. The matrix with the two properties hold simultaneously are called "optimal" in their presentation slides. The following matrix

$$\begin{pmatrix} a & 1 & 1 & 1 \\ 1 & 1 & b & a \\ 1 & a & 1 & b \\ 1 & b & a & 1 \end{pmatrix}$$

is an example of "optimal" matrix which is given in [17]. Similarly as above, we investigate the following special matrix,

$$L = \begin{pmatrix} A & I & I & I \\ I & I & B & A \\ I & A & I & B \\ I & B & A & I \end{pmatrix},$$

where $A, B \in GL(m, \mathbb{F}_2)$ are $m \times m$ non-singular matrices over $\mathbb{F}_2$.

If L is MDS, then the following matrices are non-singular:

$$A + I, B + I, A + B, A + B^2, A^2 + B, AB + I.$$

When $m = 4$, we search A, B over $GL(4, \mathbb{F}_2)$, which is the set of all 4×4 non-singular matrices over $\mathbb{F}_2$. The fewest XORs of "optimal" MDS matrices is 13. There are 24 pairs of $A, B \in GL(m, \mathbb{F}_2)$ such that the corresponding constructions are MDS matrices with $4\#A + 3\#B = 13$. All these pairs satisfy $B = A^{-2}$.

When $m = 8$, we search A, B over the set of all 8×8 non-singular matrices over $\mathbb{F}_2$ with 1 bit XOR operation. No MDS matrix returns. This means if L is a "optimal" MDS matrix over $GL(8, \mathbb{F}_2)$, then either A or B has at least 2 XORs, and hence $\#L \geq 10$.

Then we have the following result.

Theorem 10. *Let L be a matrix constructed as above, where $A, B \in GL(m, \mathbb{F}_2)$, $m = 4, 8$. If L is an MDS matrix, then*

$$4\#A + 3\#B \geq \begin{cases} 13, & m = 4; \\ 10, & m = 8. \end{cases}$$

In order to get "optimal" matrices over $GL(8, \mathbb{F}_2)$ with 10 XORs, we let $B = A^{-2}$ and search A over all 8×8 non-singular matrices over $\mathbb{F}_2$ with 1 bit XOR operation. We get 40320 $A \in GL(8, \mathbb{F}_2)$ such that the corresponding constructions are "optimal" MDS matrices with 10 XORs.

It is interesting that "optimal" 4×4 MDS matrices over $GL(8, \mathbb{F}_2)$ has fewer XORs than "optimal" 4×4 MDS matrices over $GL(4, \mathbb{F}_2)$.

Example 7. Examples of A, B such that L are "optimal" MDS matrices with the bounds in the above result hold.

(1) Let $A = [[2,3], 4, 2, 1]$, $B = A^{-2} = [2, [1,3], [1,3,4], 3]$. Then L constructed as above is an MDS matrix with $4\#A + 3\#B = 13$.
(2) Let $A = [4, 5, 6, 8, 3, [4,7], 1, 2]$, $B = A^{-2} = [[1,6], 4, 2, 7, 8, 5, [3,7], 1]$. Then L constructed as above is an MDS matrix with $4\#A + 3\#B = 10$.

6   Conclusion

In the present paper, we mainly investigate the construction of 4×4 lightweight MDS matrices with entries in the set of $m \times m$ non-singular matrices over $\mathbb{F}_2$. With this method, circulant, Hadamard and involutory Hadamard MDS matrices with fewer XORs than previous constructions are given. Moreover, circulant involutory MDS matrices are also constructed with our method. Constructing lightweight MDS matrices of large order with the method of the present paper is an interesting problem need further study.

Acknowledgements. The authors are very grateful to the anonymous reviewers for their valuable comments. This work was supported by the 973 project under Grant (2013CB834203), by the National Science Foundation of China (No. 61303255, No. 61379142).

References

1. Augot, D., Finiasz, M.: Exhaustive search for small dimension recursive MDS diffusion layers for block ciphers and hash functions. In: Proceedings of 2013 IEEE International Symposium on Information Theory (ISIT), pp. 1551–1555. IEEE (2013)

2. Augot, D., Finiasz, M.: Direct construction of recursive MDS diffusion layers using shortened BCH codes. In: Cid, C., Rechberger, C. (eds.) FSE 2014. LNCS, vol. 8540, pp. 3–17. Springer, Heidelberg (2015)
3. Barreto, P., Rijmen, V.: The anubis block cipher. Submission to the NESSIE Project (2000)
4. Berger, T.P.: Construction of recursive MDS diffusion layers from Gabidulin codes. In: Paul, G., Vaudenay, S. (eds.) INDOCRYPT 2013. LNCS, vol. 8250, pp. 274–285. Springer, Heidelberg (2013)
5. Blaum, M., Roth, R.M.: On lowest density MDS codes. IEEE Trans. Inf. Theory **45**(1), 46–59 (1999)
6. Choy, J., Yap, H., Khoo, K., Guo, J., Peyrin, T., Poschmann, A., Tan, C.H.: SPN-Hash: improving the provable resistance against differential collision attacks. In: Mitrokotsa, A., Vaudenay, S. (eds.) AFRICACRYPT 2012. LNCS, vol. 7374, pp. 270–286. Springer, Heidelberg (2012)
7. Cui, T., Jin, C.I., Kong, Z.: On compact Cauchy matrices for substitution permutation networks. IEEE Trans. Comput. **99**, 1 (2014). Preprint
8. Daemen, J., Rijmen, V.: The Design of Rijndael: AES - The Advanced Encryption Standard. Springer, Heidelberg (2002)
9. Guo, J., Peyrin, T., Poschmann, A., Robshaw, M.: The LED block cipher. In: Preneel, B., Takagi, T. (eds.) CHES 2011. LNCS, vol. 6917, pp. 326–341. Springer, Heidelberg (2011)
10. Guo, J., Peyrin, T., Poschmann, A.: The PHOTON family of lightweight hash functions. In: Rogaway, P. (ed.) CRYPTO 2011. LNCS, vol. 6841, pp. 222–239. Springer, Heidelberg (2011)
11. Chand Gupta, K., Ghosh Ray, I.: On constructions of involutory MDS matrices. In: Youssef, A., Nitaj, A., Hassanien, A.E. (eds.) AFRICACRYPT 2013. LNCS, vol. 7918, pp. 43–60. Springer, Heidelberg (2013)
12. Gupta, K.C., Ray, I.G.: On constructions of MDS matrices from companion matrices for lightweight cryptography. In: Cuzzocrea, A., Kittl, C., Simos, D.E., Weippl, E., Xu, L. (eds.) CD-ARES Workshops 2013. LNCS, vol. 8128, pp. 29–43. Springer, Heidelberg (2013)
13. Gupta, K.C., Ray, I.G.: Cryptographically significant MDS matrices based on circulant and circulant-like matrices for lightweight applications. Crypt. Commun. **7**, 257–287 (2015)
14. Li, Y., Wang, M.: On the construction of lightweight circulant involutory MDS matrices (Extend version). In: FSE 2016. http://eprint.iacr.org/2016/406
15. Jean, J., Nikolić, I., Peyrin, T.: Joltik v1.1. Submission to the CAESAR competition (2014). http://www1.spms.ntu.edu.sg/~syllab/Joltik
16. Nakahara Jr., J., Abraho, I.: A new involutory MDS matrix for the AES. Int. J. Netw. Secur. **9**(2), 109–116 (2009)
17. Junod, P., Vaudenay, S.: Perfect diffusion primitives for block ciphers. In: Handschuh, H., Hasan, M.A. (eds.) SAC 2004. LNCS, vol. 3357, pp. 84–99. Springer, Heidelberg (2004)
18. Kavun, E.B., Lauridsen, M.M., Leander, G., Rechberger, C., Schwabe, P., Yalc, T.: Prøstv1.1. Submission to the CAESAR competition (2014). http://competitions.cr.yp.to/round1/proestv11.pdf
19. Khoo, K., Peyrin, T., Poschmann, A.Y., Yap, H.: FOAM: searching for hardware-optimal SPN structures and components with a fair comparison. In: Batina, L., Robshaw, M. (eds.) CHES 2014. LNCS, vol. 8731, pp. 433–450. Springer, Heidelberg (2014)

20. Sajadieh, M., Dakhilalian, M., Mala, H., Sepehrdad, P.: Recursive diffusion layers for block ciphers and hash functions. In: Canteaut, A. (ed.) FSE 2012. LNCS, vol. 7549, pp. 385–401. Springer, Heidelberg (2012)
21. Sim, S.M., Khoo, K., Oggier, F., Peyrin, T.: Lightweight MDS involution matrices. In: Leander, G. (ed.) FSE 2015. LNCS, vol. 9054, pp. 471–493. Springer, Heidelberg (2015)
22. Vaudenay, S.: On the need for multipermutations: cryptanalysis of MD4 and SAFER. In: Preneel, Bart (ed.) FSE 1994. LNCS, vol. 1008, pp. 286–297. Springer, Heidelberg (1994)
23. Wu, S., Wang, M., Wu, W.: Recursive diffusion layers for (lightweight) block ciphers and hash functions. In: Knudsen, L.R., Wu, H. (eds.) SAC 2012. LNCS, vol. 7707, pp. 355–371. Springer, Heidelberg (2013)

Optimizing S-Box Implementations for Several Criteria Using SAT Solvers

Ko Stoffelen$^{(\boxtimes)}$

Digital Security, Radboud University, Nijmegen, The Netherlands
k.stoffelen@cs.ru.nl

Abstract. We explore the feasibility of applying SAT solvers to optimizing implementations of small functions such as S-boxes for multiple optimization criteria, e.g., the number of nonlinear gates and the number of gates. We provide optimized implementations for the S-boxes used in Ascon, ICEPOLE, Joltik/Piccolo, Keccak/Ketje/Keyak, LAC, Minalpher, PRIMATEs, Prøst, and RECTANGLE, most of which are candidates in the second round of the CAESAR competition. We then suggest a new method to optimize for circuit depth and we make tooling publicly available to find efficient implementations for several criteria. Furthermore, we illustrate with the 5-bit S-box of PRIMATEs how multiple optimization criteria can be combined.

Keywords: S-box · SAT solvers · Implementation optimization · Multiplicative complexity · Circuit depth complexity · Shortest linear straight-line program

1 Introduction

Implementations of cryptographic algorithms are typically optimized for one or multiple criteria, such as latency, throughput, power consumption, memory consumption, etc., but also criteria such as the cost of adding masking countermeasures to protect against side-channel attacks. It is worthwhile to spend time on this optimization, as the implementations are typically used many times. It is usually a hard problem to find an implementation that is actually theoretically minimal with respect to the criteria, e.g., general circuit minimization is $\sum_2^p$-complete [10]. However, for small functions this is still possible, using, for instance, SAT solvers. Especially for building blocks that can be used in multiple cryptographic algorithms, such as S-boxes, it is useful to look at methods for finding minimal implementations with respect to some given criteria.

In Sect. 2, we first discuss the simpler problem of finding minimal implementations of linear functions. We give a brief overview of methods for finding the shortest linear straight-line program.

This work was supported by the European Commission through the Horizon 2020 program under project number ICT-645622 (PQCRYPTO).

T. Peyrin (Ed.): FSE 2016, LNCS 9783, pp. 140–160, 2016.
DOI: 10.1007/978-3-662-52993-5_8

We then move towards S-boxes and in Sect. 3 we consider known methods [13,20] that manage to find minimal implementations for the relevant optimization criteria of multiplicative complexity [9], bitslice gate complexity [12], and gate complexity. The definitions of these criteria are given in Sect. 3. We study how feasible the methods actually are by applying them to S-boxes that are used in recent cryptographic algorithms, such as several candidates in the CAESAR competition and lightweight block ciphers. Additionally, we provide tools that allow anyone to conveniently do the same to other small S-boxes.

Then we look at another optimization criterion: the circuit depth complexity. This is relevant in hardware implementations to decrease the delay and to be able to increase the clock frequency. We suggest a new method for encoding the circuit depth complexity decision problem in SAT and we show how feasible this method is in practice by providing efficient low-depth S-box implementations for Joltik [17], Piccolo [22], LAC [23], Prøst [18], and RECTANGLE [24] in Sect. 3.5.

Finally, in Sect. 4 it is discussed how several optimization criteria can be combined, by first optimizing the S-box used by the PRIMATEs [2] for multiplicative complexity and then for gate complexity. This is done by taking the intermediate result after optimizing for multiplicative complexity, identifying the linear parts of this, and by treating these as instances of the shortest linear straight-line program problem.

Contributions of This Paper. To summarize, the contributions of this paper are

- implementations of the S-boxes in Ascon, ICEPOLE, Joltik/Piccolo, Keccak/Ketje/Keyak, LAC, Minalpher, Prøst, and RECTANGLE with a provably minimal number of nonlinear gates;
- a new method for encoding the circuit depth complexity decision problem as an instance of SAT;
- optimized and sometimes even provably minimal implementations of the S-boxes in Joltik/Piccolo, LAC, Prøst, and RECTANGLE with respect to bitslice gate complexity, gate complexity, and circuit depth complexity;
- a method to combine multiple optimization criteria;
- an implementation of the S-box used by the PRIMATEs that is first optimized for multiplicative complexity and then for (bitslice) gate complexity;
- tools and documentation to optimize implementations of small nonlinear functions such as S-boxes using SAT solvers, with respect to multiplicative complexity, bitslice gate complexity, gate complexity, or circuit depth complexity, are put into the public domain. These tools are available online.

2   The Shortest Linear Straight-Line Program Problem

Before tackling the optimization of S-boxes, let us restrict ourselves to linear functions and let us consider the Shortest Linear Program (SLP) problem over $GF(2)$. Let $\boldsymbol{A}$ be an $m \times n$ matrix of constants over $GF(2)$ and let $\boldsymbol{x}$ be a vector of n variables over $GF(2)$. The SLP problem is to find the program with the smallest number of lines that computes $\boldsymbol{Ax}$, where every program line is of a certain form.

Let Z be a set of variables over $GF(2)$, that initially contains the input variables $\{x_0, \ldots, x_{n-1}\}$. Let $z_i, z_j \in Z$. Then every program line is of the form

$$z' := z_i + z_j.$$

After executing this program line, the new variable z' is added to the set, $Z := Z \cup \{z'\}$. The new variable z' can therefore be used in the next program line. The program is said to compute $\boldsymbol{Ax}$ when $\exists(z_1, \ldots, z_m) \in Z^m \{\boldsymbol{Ax} = (z_1, \ldots, z_m)^\mathsf{T}\}$ holds.

Being able to find the shortest straight-line linear program has obvious applications to cryptology. Solving the SLP over $GF(2)$ is equivalent to finding the shortest circuit to compute a function using only XOR gates. Optimizing implementations of linear operations, such as MixColumns in AES and the linear transformation in certain implementations of SubBytes, can therefore be seen as instances of the SLP problem over $GF(2)$. However, this method does not apply to nonlinear operations such as S-boxes. We show in Sect. 3 what kind of methods can be used in such cases.

Solving the SLP Problem. Boyar, Matthews, and Peralta showed in [7] that the SLP problem over $GF(2)$ is NP-hard. Off-the-shelf SAT solvers can be used to find solutions for small instances of this problem. Fuhs and Schneider-Kamp presented a method [16] to encode the SLP problem as an instance of SAT and they show how this can be used to optimize the affine transformation of AES's SubBytes [15,16].

For larger instances, exact methods will quickly become infeasible. Alternatively, Boyar and Peralta published an approach to solve the SLP problem over $GF(2)$ based on a heuristic [8]. In short, the heuristic method uses a base vector set S, initialized with unit vectors for all variables in $\boldsymbol{x}$, and a distance vector $Dist[]$ that keeps track of the minimal Hamming distance to S for each row in $\boldsymbol{A}$. Repeatedly, the sum of the pair of base vectors in S that minimizes the sum of $Dist[]$ is added to S and $Dist[]$ is updated, until $Dist[]$ is the all-zero vector. If there is a tie between two pairs of base vectors, the pair that maximizes the Euclidean length of the new $Dist[]$ vector is chosen. This algorithm makes it possible to find solutions to larger instances of the SLP problem.

3 Optimizing S-Box Implementations using SAT-Solvers

For nonlinear functions such as S-boxes, known approaches based on heuristics [8] all exploit additional algebraic structure that may be available, e.g., as for the S-box of AES. However, in general this additional structure may not exist and one may need to fall back to generic methods such as SAT solvers.

S-box implementations in both software and hardware can be optimized with SAT solvers according to several criteria. In this paper we consider the following optimization goals:

Multiplicative complexity. The multiplicative complexity of a function [9] is defined as the smallest number of nonlinear gates with fan-in 2 required to compute this function. If we restrict our S-box implementations to the {AND, OR, XOR, NOT} operations, we only need to consider the number of ANDs and ORs. Optimizing for this goal is useful in the case of protecting against side-channel attacks using random masks, where nonlinear gates are typically more expensive to mask. There are also applications in multi-party computation and fully homomorphic encryption, where the cost of nonlinear operations is even more significant [1].

Bitslice gate complexity. The bitslice gate complexity of a function [12] is defined as the smallest number of operations in {AND, OR, XOR, NOT} required to compute this function. This translates directly to efficient bitsliced software implementations, as on most common CPU architectures, there are no instructions for computing NAND, NOR, or XNOR immediately.

Gate complexity. The gate complexity of a function is defined as the smallest number of logic gates required to compute this function. Unlike for bitslice gate complexity, NAND, NOR, and XNOR gates are now also allowed. This translates to efficient hardware implementations, although the different amounts of area required by these types of gates and the different delays still need to be taken into account. Note that we only consider gates with a fan-in of at most 2.

Circuit depth complexity. The depth of a circuit is defined as the length of the longest paths from an input gate to an output gate. Every function can be computed by a circuit with depth 2, e.g., by expressing the function in conjunctive or disjunctive normal form. However, this can lead to very wide circuits with a lot of gates, which is typically not desirable. There is somewhat of a trade-off between circuit depth and number of gates. Still, optimizing for this goal is useful in the case of hardware implementations, to be able to decrease the total delay and therefore to be able to increase the clock frequency. Again, only gates with a fan-in of at most 2 are considered.

These criteria come with corresponding decision problems. For example, given a function f and some positive integer k, the **multiplicative complexity decision problem** is defined as:

"Is there a circuit that implements f and that uses at most k nonlinear operations?"

The decision problems for the other three optimization goals can be defined analogously. Off-the-shelf SAT solvers can be used to solve these decision problems. When a SAT solver successfully finds a circuit for some value k but outputs UNSAT for $k-1$, it is proven that k is the minimum value. Note that when a SAT solver outputs SAT for some value k, it also provides a satisfying valuation that can be used to reconstruct an implementation of f.

In order to use SAT solvers to solve these decision problems, the problems first have to be encoded in logical formulas in conjunctive normal form (CNF), because that is the input format that the SAT solver requires.

3.1   Notation

For the encoding, we use the notation of [20]. We consider systems of multivariate equations over $GF(2)$. In these equations, let:

- x_i be variables representing S-box inputs;
- y_i be variables representing S-box outputs;
- q_i be variables representing gate inputs;
- t_i be variables representing gate outputs;
- a_i be variables representing wiring between gates;
- b_i be variables representing wiring 'inside' gates. This will become more clear when they are first used in Sect. 3.3.

In the implementations the *logical connectives* are used to denote the types of operations, i.e., let $\wedge$, $\vee$, $\oplus$, $\neg$ denote AND, OR, XOR, NOT, respectively, and let $\uparrow$, $\downarrow$, $\leftrightarrow$ denote NAND, NOR, XNOR, respectively.

3.2   Optimizing for Multiplicative Complexity

Courtois, Mourouzis and Hulme [13,20] suggested a method to encode the multiplicative complexity decision problem. Let $f : \mathbb{F}_2^n \rightarrow \mathbb{F}_2^m$ be an S-box and let k be the multiplicative complexity that we want to test for. Then first create a set of equations C in ANF consisting of:

- $\forall i \in \{0, \ldots, k-1\}$: $t_i = q_{2i} \cdot q_{2i+1}$, to encode the k AND gates.
- $\forall i \in \{0, \ldots, 2k-1\}$: $q_i = a_l + \left(\sum_{j=0}^{n-1} a_{l+j+1} \cdot x_j \right) + \left(\sum_{j=0}^{\lfloor \frac{i}{2} \rfloor - 1} a_{l+n+j+1} \cdot t_j \right)$,

  where $l = i(n+1) + \left\lfloor \frac{i^2 - 2i + 1}{4} \right\rfloor$, to encode that the inputs of the AND gates can be any linear combination of S-box inputs and previous AND gate outputs. The single a represents an optional NOT gate.
- $\forall i \in \{0, \ldots, m-1\}$: $y_i = \left(\sum_{j=0}^{n-1} a_{s+j} \cdot x_j \right) + \left(\sum_{j=0}^{k-1} a_{s+n+j} \cdot t_j \right)$, where $s = 2k(n+1) + k(k-1) + i(n+k)$, to encode that the S-box outputs can be any linear combination of S-box inputs and AND gate outputs.

For example, when $n = m = 4$ and $k = 3$, this leads to the following set of equations C:

$$q_0 = a_0 + a_1 \cdot x_0 + a_2 \cdot x_1 + a_3 \cdot x_2 + a_4 \cdot x_3$$
$$q_1 = a_5 + a_6 \cdot x_0 + a_7 \cdot x_1 + a_8 \cdot x_2 + a_9 \cdot x_3$$
$$t_0 = q_0 \cdot q_1$$
$$q_2 = a_{10} + a_{11} \cdot x_0 + a_{12} \cdot x_1 + a_{13} \cdot x_2 + a_{14} \cdot x_3 + a_{15} \cdot t_0$$
$$q_3 = a_{16} + a_{17} \cdot x_0 + a_{18} \cdot x_1 + a_{19} \cdot x_2 + a_{20} \cdot x_3 + a_{21} \cdot t_0$$
$$t_1 = q_2 \cdot q_3$$

$$q_4 = a_{22} + a_{23} \cdot x_0 + a_{24} \cdot x_1 + a_{25} \cdot x_2 + a_{26} \cdot x_3 + a_{27} \cdot t_0 + a_{28} \cdot t_1$$

$$q_5 = a_{29} + a_{30} \cdot x_0 + a_{31} \cdot x_1 + a_{32} \cdot x_2 + a_{33} \cdot x_3 + a_{34} \cdot t_0 + a_{35} \cdot t_1$$

$$t_2 = q_4 \cdot q_5$$

$$y_0 = a_{36} \cdot x_0 + a_{37} \cdot x_1 + a_{38} \cdot x_2 + a_{39} \cdot x_3 + a_{40} \cdot t_0 + a_{41} \cdot t_1 + a_{42} \cdot t_2$$

$$y_1 = a_{43} \cdot x_0 + a_{44} \cdot x_1 + a_{45} \cdot x_2 + a_{46} \cdot x_3 + a_{47} \cdot t_0 + a_{48} \cdot t_1 + a_{49} \cdot t_2$$

$$y_2 = a_{50} \cdot x_0 + a_{51} \cdot x_1 + a_{52} \cdot x_2 + a_{53} \cdot x_3 + a_{54} \cdot t_0 + a_{55} \cdot t_1 + a_{56} \cdot t_2$$

$$y_3 = a_{57} \cdot x_0 + a_{58} \cdot x_1 + a_{59} \cdot x_2 + a_{60} \cdot x_3 + a_{61} \cdot t_0 + a_{62} \cdot t_1 + a_{63} \cdot t_2$$

This set of equations does not depend on f yet, but only on the values of n and m. The equations in C have to be satisfied for all possible S-box inputs. An equation set C' is created that contains 2^n copies of the equations in C, in which all x_i, y_i, q_i, t_i are renumbered, but in which all a_i, b_i remain the same. f is 'bound' to the problem description by adding its truth table as $2^n(n + m)$ constant equations, i.e., one for every bit in both the S-box input and the S-box output, to C'.

C' is in ANF. The method by Bard, Courtois, and Jefferson [3] for converting sparse systems of low-degree multivariate polynomials over $GF(2)$ is used to convert C' to CNF, such that it is understood by the SAT solver.

Results. This method makes it feasible to find the multiplicative complexity of several 4-bit and 5-bit S-boxes. Finding the multiplicative complexity comes with an actual implementation that uses this minimal number of nonlinear gates. After Courtois, Hulme, and Mourouzis applied this method to the S-boxes of PRESENT and GOST [12], we show that we can also find results for more recently introduced 4-bit and 5-bit S-boxes.

We consider the S-boxes, and if applicable, their inverses (denoted by $^{-1}$), in Ascon [14], ICEPOLE [19], Keccak [4]/Ketje [5]/Keyak [6], all PRIMATEs [2], Joltik [17]/Piccolo [22], LAC [23], Minalpher [21], Prøst [18], and RECTAN-GLE [24]. Minalpher's and Prøst's S-boxes are involutory, which is why their inverses are not listed separately. The inverse S-boxes in Ascon, ICEPOLE, Keccak, Ketje, and Keyak are not actually used in decryption and are there-fore not considered.

For all S-boxes except the one used by the PRIMATEs we are able to prove the multiplicative complexity. The results are summarized in Table 1. The actual implementations can be found in Appendix A, but note that these should not be used by themselves as we are being very generous with XOR gates. The linear parts should be optimized separately, as we will demonstrate in Sect. 4.

These and subsequent results are obtained using MINISAT 2.2.0[1] and CRYP-TOMINISAT 2.9.10[2] using default parameters on a single core of an Intel Xeon E7-4870 v2 running at 2.30 GHz.

---

[1] http://www.minisat.se/MiniSat.html.

[2] http://www.msoos.org/cryptominisat2/.

Table 1. Multiplicative complexity of S-boxes

S-box	Size $n \times m$	Multiplicative complexity
Ascon	5×5	5
ICEPOLE	5×5	6
Keccak/Ketje/Keyak	5×5	5
PRIMATEs	5×5	$\in \{6, 7\}$
PRIMATEs^{-1}	5×5	$\in \{6, 7, 8, 9, 10\}$
Joltik/Piccolo	4×4	4
Joltik^{-1}/Piccolo^{-1}	4×4	4
LAC	4×4	4
Minalpher	4×4	5
Prøst	4×4	4
RECTANGLE	4×4	4
RECTANGLE^{-1}	4×4	4

For the PRIMATEs S-box and inverse S-box, we find solutions for $k = 7$ and $k = 10$, respectively. Furthermore, we find for both S-boxes that the case for $k = 5$ yields UNSAT. We have started several attempts to find a decisive answer for $k = 6$, including

- reducing the CNF, e.g., using NICESAT [11];
- fine-tuning SAT solver parameters;
- trying other SAT solvers;
- trying other SAT solvers that can run in parallel on many cores, such as PLINGELING and TREENGELING[3]; and
- letting all of this run for several months on a machine with 120 cores and 3 TB of RAM.

Unfortunately, none of these attempts resulted in an answer as no solver instance has terminated yet. As these SAT solvers typically have much more difficulty with proving the UNSAT case than proving the SAT case, and as the SAT proof for $k = 7$ was found in less than 40 hours, we expect the $k = 6$ case to yield UNSAT and we therefore conjecture the multiplicative complexity of the PRIMATEs S-box to be 7. In Sect. 4 we go into more detail on optimizing the PRIMATEs S-box. For the inverse S-box, we did not manage to find solutions for $k \in \{6, 7, 8, 9\}$.

3.3   Optimizing for Bitslice Gate Complexity

In [13, 20], a method is also given to optimize for bitslice gate complexity. However, it is only applied on the small CTC2 toy cipher and therefore it remains unclear how practical this method is for real-world ciphers. We investigate this by applying the method to the same S-boxes as in the previous section.

---

[3] http://fmv.jku.at/lingeling/.

The encoding scheme for the bitslice gate complexity decision problem is slightly different compared to the multiplicative complexity decision problem. Let $f : \mathbb{F}_2^n \to \mathbb{F}_2^m$ again be an S-box and let k now be the bitslice gate complexity that we want to test for. Then our first set of equations C in ANF consists of:

- $\forall i \in \{0, \ldots, k-1\}$: $t_i = b_{3i} \cdot q_{2i} \cdot q_{2i+1} + b_{3i+1} \cdot q_{2i} + b_{3i+1} \cdot q_{2i+1} + b_{3i+2} + b_{3i+2} \cdot q_{2i}$, to encode the k AND, OR, XOR or NOT gates. The b_i determine what kind of gate this will represent, as can be seen in Table 2.
- $\forall i \in \{0, \ldots, k-1\}$: $0 = b_{3i} \cdot b_{3i+2}$ and $0 = b_{3i+1} \cdot b_{3i+2}$, to make sure that the gate is either a unary NOT or a binary AND/OR/XOR, but not the XOR of them. This excludes NAND/NOR/XNOR gates.
- $\forall i \in \{0, \ldots, 2k-1\}$: $q_i = \left(\sum_{j=0}^{n-1} a_{l+j} \cdot x_j \right) + \left(\sum_{j=0}^{\lfloor \frac{i}{2} \rfloor - 1} a_{l+n+j} \cdot t_j \right)$, where $l = in + \left\lfloor \frac{i^2 - 2i + 1}{4} \right\rfloor$, to encode that the inputs of the gates can be any S-box input bit or any previously computed bit.
- $\forall i \in \{0, \ldots, 2k-1\}, \forall j \in \{l, \ldots, l+n+\lfloor \frac{i}{2} \rfloor - 2\}, \forall u \in \{j+1, \ldots, l+n+\lfloor \frac{i}{2} \rfloor - 1\}$: $0 = a_j \cdot a_u$, to encode an 'at most one' constraint on the gate inputs.
- $\forall i \in \{0, \ldots, m-1\}$: $y_i = \left(\sum_{j=0}^{n-1} a_{s+j} \cdot x_j \right) + \left(\sum_{j=0}^{k-1} a_{s+n+j} \cdot t_j \right)$, where $s = 2kn + k(k-1) + i(n+k)$, to encode that the S-box output bit can be any S-box input bit or any gate output.
- $\forall i \in \{0, \ldots, m-1\}, \forall j \in \{s, \ldots, s+n+k-2\}, \forall u \in \{j+1, \ldots, s+n+k-1\}$: $0 = a_j \cdot a_u$, to encode an 'at most one' constraint on the S-box outputs.

Table 2. Encoding of different types of gates (bitslice gate complexity)

$b_{3i} b_{3i+1} b_{3i+2}$	Gate t_i function
000	0
001	$\neg q_{2i}$
010	$q_{2i} \oplus q_{2i+1}$
011	Prevented by constraint on b_{3i+2}
100	$q_{2i} \wedge q_{2i+1}$
101	Prevented by constraint on b_{3i+2}
110	$q_{2i} \vee q_{2i+1}$
111	Prevented by constraint on b_{3i+2}

Converting C to C' and then to CNF is the same process as with the multiplicative complexity decision problem. Note that the 'constraint equations' on a_i and b_j do not have to be duplicated 2^n times for C', as they are not renumbered. This saves a lot of redundant clauses.

Results. As the amount of CNF clauses that is necessary to describe the bitslice gate complexity decision problem becomes much larger compared to the multiplicative complexity decision problem, it can take much more time for a

SAT solver to actually solve a problem instance. Still, for some 4-bit and 5-bit S-boxes results can be obtained within minutes or within a few hours. Table 3 contains some examples. If a bitslice gate complexity is listed as $\leq k$, a solution was found for k, but we were unable to prove that this is the minimum because the SAT solver did not terminate within a reasonable amount of time for $k - 1$. The actual implementations with the given number of operations can be found in Appendix A.

Table 3. Bitslice gate complexity of S-boxes

S-box	Size $n \times m$	Bitslice gate complexity	Implementation
Keccak/Ketje/Keyak	5×5	≤ 13	3 AND, 2 OR, 5 XOR, 3 NOT
Joltik/Piccolo	4×4	10	1 AND, 3 OR, 4 XOR, 2 NOT
Joltik^{-1}/Piccolo^{-1}	4×4	10	1 AND, 3 OR, 4 XOR, 2 NOT
LAC	4×4	11	2 AND, 2 OR, 6 XOR, 1 NOT
Minalpher	4×4	≥ 11	
Prøst	4×4	8	4 AND, 4 XOR
RECTANGLE	4×4	$\in \{11, 12\}$	1 AND, 3 OR, 7 XOR, 1 NOT
RECTANGLE^{-1}	4×4	$\in \{10, 11, 12\}$	4 OR, 7 XOR, 1 NOT

For Prøst and the (forward) S-box of RECTANGLE, it is interesting to note that the SAT solvers are able to find the same implementations as the corresponding authors already suggested. We have proven that their bitsliced implementations are indeed minimal.

3.4 Optimizing for Gate Complexity

A method to encode the gate complexity decision problem was also provided in [13,20], but again, actual results were only given for the CTC2 toy cipher. We show that it is feasible to compute the gate complexity for real-world 4-bit S-boxes as well.

The encoding is very similar to the bitslice gate complexity decision problem. The first set of equations C in ANF only differs in two places:

- Instead of the previous rule for t_i, the gates are encoded differently:
  $\forall i \in \{0, \ldots, k - 1\}$: $t_i = b_{3i} \cdot q_{2i} \cdot q_{2i+1} + b_{3i+1} \cdot q_{2i} + b_{3i+1} \cdot q_{2i+1} + b_{3i+2}$, to encode the k gates. The b_i determine what kind of gate this will represent, as can be seen in Table 4.
- The additional constraints on the b_i are completely omitted.

Converting C to C' and then to CNF is similar to the previous optimization goals.

Table 4. Encoding of different types of gates (gate complexity)

$b_{3i}b_{3i+1}b_{3i+2}$	Gate t_i function
000	0
001	1
010	$q_{2i} \oplus q_{2i+1}$
011	$q_{2i} \leftrightarrow q_{2i+1}$
100	$q_{2i} \wedge q_{2i+1}$
101	$q_{2i} \uparrow q_{2i+1}$
110	$q_{2i} \vee q_{2i+1}$
111	$q_{2i} \downarrow q_{2i+1}$

Results. Our results on real-world 4-bit S-boxes are summarized in Table 5. The full implementations can be found in Appendix A. For our 5-bit S-boxes we did not manage to retrieve results. Note that all types of logic gates are considered equally expensive. There is no type of gate that is preferred over the other, because information such as differences in area consumption or time delay are not taken into account. The implementations found by the SAT solver should therefore not be used directly for hardware implementations. However, they serve as an optimal starting point from where to swap 'expensive' gates for cheaper ones, depending on the specific technology that is to be used. For example, the designers of Piccolo suggested a hardware implementation [22] of their S-box that may or may not be more efficient than the implementation given here, depending on the specific technology.

Table 5. Gate complexity of S-boxes

S-box	Gate complexity	Implementation
Joltik/Piccolo	8	2 OR, 1 XOR, 2 NOR, 3 XNOR
Joltik^{-1}/Piccolo^{-1}	8	2 OR, 1 XOR, 2 NOR, 3 XNOR
LAC	10	1 AND, 3 OR, 2 XOR, 4 XNOR
Prøst	8	4 AND, 4 XOR
RECTANGLE	$\in \{10, 11\}$	1 AND, 1 OR, 2 XOR, 1 NAND, 1 NOR, 5 XNOR
RECTANGLE^{-1}	$\in \{10, 11\}$	1 AND, 1 OR, 6 XOR, 1 NAND, 1 NOR, 1 XNOR

3.5   Optimizing for Depth Complexity

There are many situations in high-speed hardware implementations where the implementer wants to keep the depth of the circuit as low as possible, in order to be able to increase the clock frequency, without having to use significantly more gates. We provide a novel method to find low-depth implementations of small functions such as S-boxes using SAT solvers. This method is inspired by the encoding of the gate complexity decision problem, but modified in some important ways.

In the encoding of the gate complexity decision problem, we expressed that every gate can use the S-box input and the outputs of previous gates as its input. The key idea here is to divide the circuit into *depth layers* and to encode the notion that a gate can only use the S-box input and the output of gates in the previous layers as its input. This is made more precise later.

First we note that it is necessary to limit the potential increase of the number of gates when reducing the depth of a circuit. We introduce a fixed maximum layer width w to address this, so we allow at most w gates to be executed in parallel. For some function f, we want to be able to answer questions such as: "is there a circuit implementing f with depth k and with at most w gates on each depth layer?".

Using this fixed maximum layer width, we make our encoding method more precise by once more creating a set C of multivariate equations over $GF(2)$ in ANF that consists of:

- $\forall i \in \{0, \ldots, kw - 1\}$: $t_i = b_{3i} \cdot q_{2i} \cdot q_{2i+1} + b_{3i+1} \cdot q_{2i} + b_{3i+1} \cdot q_{2i+1} + b_{3i+2}$, to encode the kw gates. The b_i determine what kind of gate this will represent, as can be seen in Table 4.
- $\forall i \in \{0, \ldots, 2kw - 1\}$: $q_i = \left(\sum_{j=0}^{n-1} a_{l+j} \cdot x_j\right) + \left(\sum_{j=0}^{v-1} a_{l+n+j} \cdot t_j\right)$, where $v = \lfloor \frac{i}{2w} \rfloor w$ and $l = in + v(i - v - w)$, to encode that the inputs of the gates can be any S-box input bit or any previously computed bit.
- $\forall i \in \{0, \ldots, 2kw-1\}, \forall j \in \{l, \ldots, l+n+v-2\}, \forall u \in \{j+1, \ldots, l+n+v-1\}$: $0 = a_j \cdot a_u$, to encode an 'at most one' constraint on the gate inputs.
- $\forall i \in \{0, \ldots, m-1\}$: $y_i = \left(\sum_{j=0}^{n-1} a_{s+j} \cdot x_j\right) + \left(\sum_{j=0}^{kw-1} a_{s+n+j} \cdot t_j\right)$, where $s = kw(2n + kw - w) + i(n + kw)$, to encode that the S-box output bit can be any S-box input bit or any gate output.
- $\forall i \in \{0, \ldots, m-1\}, \forall j \in \{s, \ldots, s+n+kw-2\}, \forall u \in \{j+1, \ldots, s+n+kw-1\}$: $0 = a_j \cdot a_u$, to encode an 'at most one' constraint on the S-box outputs.

Converting C to C' and subsequently expressing this in CNF is again the same process as before.

Results. Using our method, we are able to find low-depth implementations for our 4-bit S-boxes. The results are summarized in Table 6 and the corresponding implementations can be found in Appendix A. The last column in Table 6 lists scenarios that yield UNSAT, to show boundaries on what is possible. The trade-off between circuit depth and the number of gates is made here in such a way that reducing the depth by 1 would imply the implementation to have at least twice as many gates as is required by the gate complexity.

4   Combining Criteria: Optimizing the PRIMATEs S-Box

So far, we have seen how to optimize for one specific goal. However, a result that is optimized for multiplicative complexity may contain more XOR gates than is

Table 6. Depth complexity of S-boxes

S-box	Depth complexity	w	Implementation	UNSAT boundaries
Joltik/Piccolo	4	2	2 OR, 1 XOR,	$k = 4, w = 1$
			2 NOR, 3 XNOR	$k = 3, w = 10$
Joltik^{-1}/Piccolo^{-1}	4	3	3 OR, 5 XOR,	$k = 4, w = 2$
			1 NOR, 3 XNOR	$k = 3, w = 10$
LAC	3	6	3 OR, 4 XOR,	$k = 3, w = 4$
			4 NAND, 4 XNOR	$k = 2, w = 10$
Prøst	4	3	4 AND, 1 OR, 4 XOR,	$k = 4, w = 2$
			1 NAND, 1 XNOR	$k = 3, w = 10$
RECTANGLE	3	6	2 AND, 3 OR, 5 XOR,	$k = 3, w = 4$
			1 NAND, 1 NOR, 3 XNOR	$k = 2, w = 10$
RECTANGLE^{-1}	3	6	1 OR, 8 XOR,	$k = 3, w = 4$
			3 NAND, 2 NOR, 2 XNOR	$k = 2, w = 10$

desired, and a result that is optimized for gate complexity may contain more nonlinear gates than is desired for a masked implementation. Here we show how multiple optimization goals can be combined by looking at the 5-bit PRIMATEs S-box. We first optimize for multiplicative complexity to have a minimal number of nonlinear gates, and subsequently we minimize the number of linear gates. The result is an implementation that has 4 AND, 3 OR, 31 XOR, and 5 NOT gates.

The PRIMATEs S-box is an almost bent permutation with a maximum linear and differential probability of 2^{-4}. It is chosen because of its low area consumption in hardware implementations.

When the optimization method for multiplicative complexity is applied, we find a solution with multiplicative complexity 7 as follows:

$$q_0 = x_0 \oplus x_3$$
$$q_1 = x_1$$
$$t_0 = q_0 \vee q_1$$
$$q_2 = \neg(x_1 \oplus x_3)$$
$$q_3 = x_0 \oplus x_2$$
$$t_1 = q_2 \wedge q_3$$
$$q_4 = x_0 \oplus x_1 \oplus x_4$$
$$q_5 = x_0 \oplus x_2 \oplus x_3$$
$$t_2 = q_4 \wedge q_5$$
$$q_6 = \neg(x_0 \oplus x_2 \oplus x_3 \oplus x_4)$$
$$q_7 = x_1 \oplus x_2 \oplus x_4$$
$$t_3 = q_6 \vee q_7$$
$$q_8 = x_0 \oplus x_1 \oplus x_2 \oplus x_3 \oplus x_4$$

$$q_9 = x_2 \oplus t_0 \oplus t_3$$
$$t_4 = q_8 \wedge q_9$$
$$q_{10} = x_0 \oplus x_3 \oplus x_4$$
$$q_{11} = \neg(x_0 \oplus x_4)$$
$$t_5 = q_{10} \vee q_{11}$$
$$q_{12} = \neg(x_1 \oplus x_2 \oplus t_0 \oplus t_2 \oplus t_3 \oplus t_4)$$
$$q_{13} = x_2 \oplus x_3$$
$$t_6 = q_{12} \wedge q_{13}$$
$$y_0 = x_1 \oplus x_3 \oplus t_2 \oplus t_3 \oplus t_5 \oplus t_6$$
$$y_1 = x_0 \oplus x_4 \oplus t_1 \oplus t_2 \oplus t_3 \oplus t_4 \oplus t_5 \oplus t_6$$
$$y_2 = x_1 \oplus x_2 \oplus x_4 \oplus t_1 \oplus t_3 \oplus t_4 \oplus t_5$$
$$y_3 = x_0 \oplus x_2 \oplus x_3 \oplus x_4 \oplus t_3 \oplus t_4 \oplus t_5 \oplus t_6$$
$$y_4 = \neg(x_2 \oplus t_0 \oplus t_2 \oplus t_3 \oplus t_4 \oplus t_5 \oplus t_6)$$

It is not hard to see that there are a lot of redundant XOR operations in this implementation. We distinguish between XOR operations before the nonlinear gates (on x_i) and XOR operations after the nonlinear gates (on t_i). It is possible to see them as two straight-line linear programs, where the first describes the linear part of the S-box approached from the input and the second describes the linear part approached from the S-box output.

The shortest linear straight-line program problem $A_1 x_1$ can be given by

$$
A_1 = \begin{array}{c} q_0 \\ q_1 \\ q_2 \\ q_3 \\ q_4 \\ q_5 \\ q_6 \\ q_7 \\ q_8 \\ q_9 \\ q_{10} \\ q_{11} \\ q_{12} \\ q_{13} \\ y_0 \\ y_1 \\ y_2 \\ y_3 \\ y_4 \end{array}
\begin{pmatrix}
1 & 0 & 0 & 1 & 0 \\
0 & 1 & 0 & 0 & 0 \\
0 & 1 & 0 & 1 & 0 \\
1 & 0 & 1 & 0 & 0 \\
1 & 1 & 0 & 0 & 1 \\
1 & 0 & 1 & 1 & 0 \\
1 & 0 & 1 & 1 & 1 \\
0 & 1 & 1 & 0 & 1 \\
1 & 1 & 1 & 1 & 1 \\
0 & 0 & 1 & 0 & 0 \\
1 & 0 & 0 & 1 & 1 \\
1 & 0 & 0 & 0 & 1 \\
0 & 1 & 1 & 0 & 0 \\
0 & 0 & 1 & 1 & 0 \\
0 & 1 & 0 & 1 & 0 \\
1 & 0 & 0 & 0 & 1 \\
0 & 1 & 1 & 0 & 1 \\
1 & 0 & 1 & 1 & 1 \\
0 & 0 & 1 & 0 & 0
\end{pmatrix}
\qquad
x_1 = \begin{pmatrix} x_0 \\ x_1 \\ x_2 \\ x_3 \\ x_4 \end{pmatrix}.
$$

The shortest linear straight-line program problem $A_2 x_2$ can be given by

$$
A_2 = \begin{array}{c} q_9 \\ q_{12} \\ y_0 \\ y_1 \\ y_2 \\ y_3 \\ y_4 \end{array}
\begin{pmatrix}
1 & 0 & 0 & 1 & 0 & 0 & 0 \\
1 & 0 & 1 & 1 & 1 & 0 & 0 \\
0 & 0 & 1 & 1 & 0 & 1 & 1 \\
0 & 1 & 1 & 1 & 1 & 1 & 1 \\
0 & 1 & 0 & 1 & 1 & 1 & 0 \\
0 & 0 & 0 & 1 & 1 & 1 & 1 \\
1 & 0 & 1 & 1 & 1 & 1 & 1
\end{pmatrix}
\qquad
x_2 = \begin{pmatrix} t_0 \\ t_1 \\ t_2 \\ t_3 \\ t_4 \\ t_5 \\ t_6 \end{pmatrix}.
$$

We are able to find a minimal straight-line program computing $A_2 x_2$ using SAT solvers. We use the method suggested by Fuhs and Schneider-Kamp [16] to encode the SLP problem as a SAT instance in CNF. This yields a result that is incorporated in our implementation of the PRIMATEs S-box. Finding a minimal straight-line program computing $A_1 x_1$ turned out to be infeasible using SAT solvers within a reasonable amount of time. Therefore, we apply the heuristic approach as suggested by Boyar and Peralta [8]. This does provide us with a short straight-line program. We combine both results and amend the original PRIMATEs S-box implementation to get the more efficient implementation below, where z_i represent helper variables.

$$z_0 = x_0 \oplus x_4 \qquad q_7 = x_4 \oplus z_1 \qquad z_5 = t_2 \oplus z_4$$
$$z_1 = x_1 \oplus x_2 \qquad t_3 = q_6 \vee q_7 \qquad z_6 = t_1 \oplus t_6$$
$$z_2 = x_2 \oplus x_3 \qquad q_8 = q_4 \oplus z_2 \qquad z_7 = t_4 \oplus z_5$$
$$q_0 = x_0 \oplus x_3 \qquad z_9 = t_0 \oplus t_3 \qquad z_8 = t_1 \oplus z_7$$
$$t_0 = q_0 \vee x_1 \qquad q_9 = x_2 \oplus z_9 \qquad z_{10} = t_0 \oplus z_7$$
$$q_2 = x_1 \oplus x_3 \qquad t_4 = q_8 \wedge q_9 \qquad z_{11} = t_4 \oplus z_4$$
$$q_3 = \neg(x_0 \oplus x_2) \qquad q_{10} = \neg(x_3 \oplus z_0) \qquad z_{12} = z_6 \oplus z_{11}$$
$$t_1 = q_2 \vee q_3 \qquad t_5 = q_{10} \wedge z_0 \qquad y_0 = \neg(q_2 \oplus z_5)$$
$$q_4 = x_1 \oplus z_0 \qquad q_{12} = \neg(z_1 \oplus z_9 \oplus t_2 \oplus t_4) \qquad y_1 = z_0 \oplus z_8$$
$$q_5 = x_0 \oplus z_2 \qquad t_6 = q_{12} \wedge z_2 \qquad y_2 = q_7 \oplus z_{12}$$
$$t_2 = q_4 \wedge q_5 \qquad z_3 = t_5 \oplus t_6 \qquad y_3 = q_6 \oplus z_{11}$$
$$q_6 = \neg(x_4 \oplus q_5) \qquad z_4 = t_3 \oplus z_3 \qquad y_4 = x_2 \oplus z_{10}$$

We are able to decrease the previous result of 58 XOR gates to only 31 XOR gates.

Tools. We provide tools to generate C' in ANF for all discussed optimization goals and to convert a SAT solver solution back to an S-box implementation. We place those tools into the public domain. They and additional documentation are available online at https://github.com/Ko-/sboxoptimization.

5 Conclusion

SAT solvers can be used to find minimal implementations for small functions such as S-boxes with respect to criteria as the multiplicative complexity, bitslice gate complexity, gate complexity, and circuit depth complexity. We have shown how this can be done and how multiple criteria can be combined. However, for 8-bit S-boxes and larger functions these methods quickly become infeasible. One will then have to resort to approaches based on heuristics.

A Optimized S-Box Implementations

For all given implementations, x_0 and y_0 denote the most significant bit of the S-box input x and the S-box output y, respectively.

A.1   Optimized for Multiplicative Complexity

Only implementations that do not reach the minimal number of nonlinear operations when optimizing for other criteria are listed here. The implementations below serve as a demonstration of what kind of output can be expected from SAT solvers when optimizing for multiplicative complexity. To increase the amount of solutions and therefore the likelihood that we will find one fast, we do not put restrictions on the number of linear gates, which is why the implementations below are not very efficient. The number of linear gates can be reduced further as shown in Sect. 4.

Ascon

k = 5

$q_0 = \neg(x_3 \oplus x_4)$

$q_1 = \neg x_4$

$t_0 = q_0 \wedge q_1$

$q_2 = x_0 \oplus x_2 \oplus x_4$

$q_3 = x_1$

$t_1 = q_2 \wedge q_3$

$q_4 = x_0 \oplus x_1 \oplus x_4$

$q_5 = x_1$

$t_2 = q_4 \wedge q_5$

$q_6 = x_3 \oplus x_4$

$q_7 = x_0$

$t_3 = q_6 \wedge q_7$

$q_8 = x_3 \oplus t_1 \oplus t_2$

$q_9 = x_1 \oplus x_2$

$t_4 = q_8 \wedge q_9$

$y_0 = x_0 \oplus x_1 \oplus x_2 \oplus x_3 \oplus t_1$

$y_1 = x_0 \oplus x_2 \oplus x_3 \oplus x_4 \oplus t_4$

$y_2 = x_1 \oplus x_2 \oplus x_3 \oplus t_0$

$y_3 = x_0 \oplus x_1 \oplus x_2 \oplus x_3 \oplus x_4 \oplus t_3$

$y_4 = x_3 \oplus x_4 \oplus t_2$

ICEPOLE

k = 6

$q_0 = x_0 \oplus x_3 \oplus x_4$

$q_1 = x_0 \oplus x_3$

$t_0 = q_0 \wedge q_1$

$q_2 = \neg(x_2 \oplus x_4)$

$q_3 = x_2 \oplus x_3 \oplus x_4$

$t_1 = q_2 \wedge q_3$

$q_4 = x_2 \oplus t_0 \oplus t_1$

$q_5 = x_0 \oplus x_2 \oplus x_3 \oplus x_4 \oplus t_1$

$t_2 = q_4 \wedge q_5$

$q_6 = x_0 \oplus x_1 \oplus x_4$

$q_7 = x_1 \oplus x_4$

$t_3 = q_6 \wedge q_7$

$q_8 = x_1 \oplus x_2 \oplus t_0 \oplus t_1 \oplus t_2$

$q_9 = x_0 \oplus x_1 \oplus t_0 \oplus t_1 \oplus t_2$

$t_4 = q_8 \wedge q_9$

$q_{10} = \neg(x_2 \oplus t_1 \oplus t_3 \oplus t_4)$

$q_{11} = \neg(x_0 \oplus t_4)$

$t_5 = q_{10} \wedge q_{11}$

$y_0 = x_0 \oplus t_0 \oplus t_1 \oplus t_2 \oplus t_5$

$y_1 = x_0 \oplus x_1 \oplus x_2 \oplus x_3 \oplus x_4 \oplus t_2 \oplus \cdots$
$\cdots \oplus t_3 \oplus t_4 \oplus t_5$

$y_2 = x_0 \oplus x_3 \oplus t_1 \oplus t_2 \oplus t_3 \oplus t_4 \oplus t_5$

$y_3 = x_0 \oplus x_2 \oplus x_3 \oplus x_4 \oplus t_0 \oplus t_1 \oplus \cdots$
$\cdots \oplus t_2 \oplus t_3 \oplus t_4 \oplus t_5$

$y_4 = x_2 \oplus x_4 \oplus t_0 \oplus t_1 \oplus t_2 \oplus t_4 \oplus t_5$

PRIMATEs

k = 7

See Sect. 4.

PRIMATEs^{-1}
k = 10

$q_0 = x_0 \oplus x_2 \oplus x_3$

$q_1 = \neg(x_2 \oplus x_4)$

$t_0 = q_0 \wedge q_1$

$q_2 = x_0$

$q_3 = x_1$

$t_1 = q_2 \wedge q_3$

$q_4 = x_2 \oplus x_3 \oplus t_0$

$q_5 = \neg x_1$

$t_2 = q_4 \wedge q_5$

$q_6 = x_1 \oplus t_1 \oplus t_2$

$q_7 = x_2 \oplus x_4$

$t_3 = q_6 \wedge q_7$

$q_8 = x_2 \oplus t_0 \oplus t_2 \oplus t_3$

$q_9 = x_0 \oplus x_3 \oplus x_4 \oplus t_1 \oplus t_2 \oplus t_3$

$t_4 = q_8 \wedge q_9$

$q_{10} = x_0 \oplus x_2 \oplus x_3 \oplus t_1 \oplus t_2 \oplus t_3$

$q_{11} = x_1 \oplus x_3 \oplus t_0 \oplus t_2$

$t_5 = q_{10} \wedge q_{11}$

$q_{12} = x_0 \oplus x_4$

$q_{13} = t_0 \oplus t_3 \oplus t_4 \oplus t_5$

$t_6 = q_{12} \wedge q_{13}$

$q_{14} = \neg(x_0 \oplus x_1 \oplus x_2 \oplus x_4 \oplus t_0 \oplus \cdots$
$\cdots \oplus t_1 \oplus t_3 \oplus t_4 \oplus t_5 \oplus t_6)$

$q_{15} = x_0 \oplus x_3 \oplus t_0 \oplus t_1 \oplus t_2 \oplus t_4 \oplus t_6$

$t_7 = q_{14} \wedge q_{15}$

$q_{16} = \neg(x_2 \oplus x_3 \oplus t_2 \oplus t_5)$

$q_{17} = \neg(x_0 \oplus x_1 \oplus x_4 \oplus t_0 \oplus t_1 \oplus \cdots$
$\cdots \oplus t_2 \oplus t_3 \oplus t_6 \oplus t_7)$

$t_8 = q_{16} \wedge q_{17}$

$q_{18} = x_4 \oplus t_2 \oplus t_5 \oplus t_6 \oplus t_8$

$q_{19} = \neg(x_0 \oplus x_1 \oplus x_4 \oplus t_4 \oplus t_7 \oplus t_8)$

$t_9 = q_{18} \wedge q_{19}$

$y_0 = x_0 \oplus x_1 \oplus t_0 \oplus t_6 \oplus t_7 \oplus t_9$

$y_1 = t_0 \oplus t_3 \oplus t_6$

$y_2 = t_3 \oplus t_5 \oplus t_6 \oplus t_7$

$y_3 = t_1 \oplus t_2 \oplus t_4$

$y_4 = x_1 \oplus t_0 \oplus t_4 \oplus t_8$

Minalpher
k = 5

$q_0 = x_1 \oplus x_2 \oplus x_3$

$q_1 = x_1$

$t_0 = q_0 \wedge q_1$

$q_2 = x_0 \oplus x_1 \oplus x_3$

$q_3 = x_1 \oplus x_2 \oplus t_0$

$t_1 = q_2 \wedge q_3$

$q_4 = x_0 \oplus t_0$

$q_5 = x_0 \oplus x_1 \oplus x_2 \oplus t_0$

$t_2 = q_4 \wedge q_5$

$q_6 = \neg(x_0 \oplus x_1 \oplus x_2 \oplus t_0 \oplus t_2)$

$q_7 = \neg(x_0 \oplus x_1 \oplus t_1)$

$t_3 = q_6 \wedge q_7$

$q_8 = x_0 \oplus x_2 \oplus x_3 \oplus t_0 \oplus t_1 \oplus t_2 \oplus t_3$

$q_9 = x_1 \oplus x_2 \oplus t_0 \oplus t_2 \oplus t_3$

$t_4 = q_8 \wedge q_9$

$y_0 = x_2 \oplus t_4$

$y_1 = x_0 \oplus x_2 \oplus t_1$

$y_2 = t_0 \oplus t_3$

$y_3 = t_1 \oplus t_2 \oplus t_3$

A.2 Optimized for Bitslice Gate Complexity

Keccak/Ketje/Keyak
k = 13

$t_0 = \neg x_2$

$t_1 = t_0 \wedge x_3$

$y_1 = t_1 \oplus x_1$

$t_3 = \neg x_4$

$t_4 = t_3 \wedge x_0$

$y_3 = x_3 \oplus t_4$

$t_6 = x_3 \vee t_3$

$y_2 = t_0 \oplus t_6$

$t_8 = \neg x_0$

$t_9 = y_1 \vee t_0$

$t_{10} = t_8 \wedge x_1$

$y_0 = t_9 \oplus t_8$

$y_4 = x_4 \oplus t_{10}$

Joltik/Piccolo
k = 10

$t_0 = x_0 \vee x_1$

$t_1 = t_0 \oplus x_3$

$y_0 = \neg t_1$

$t_3 = x_2 \vee y_0$

$y_2 = t_3 \oplus x_1$

$t_5 = x_1 \vee x_2$

$t_6 = t_5 \oplus x_0$

$t_7 = t_1 \wedge t_6$

$y_3 = x_2 \oplus t_7$

$y_1 = \neg(t_6)$

Joltik^{-1}/Piccolo^{-1}
k = 10

$t_0 = \neg x_1$

$t_1 = \neg x_0$

$t_2 = t_1 \wedge t_0$

$y_2 = t_2 \oplus x_3$

$t_4 = x_0 \vee y_2$

$y_1 = x_2 \oplus t_4$

$t_6 = y_2 \vee y_1$

$y_0 = t_6 \oplus t_0$

$t_8 = y_0 \vee y_1$

$y_3 = t_8 \oplus t_1$

LAC
k = 11

$t_0 = x_3 \oplus x_2$

$t_1 = x_1 \vee x_0$

$y_3 = t_1 \oplus t_0$

$t_3 = x_1 \wedge y_3$

$t_4 = \neg x_3$

$t_5 = t_4 \oplus t_3$

$y_2 = t_5 \oplus x_0$

$t_7 = t_5 \wedge y_2$

$t_8 = y_3 \vee y_3$

$y_1 = t_7 \oplus x_1$

$y_0 = t_8 \oplus x_0$

Prøst
k = 8

$t_0 = x_2 \wedge x_1$

$y_1 = t_0 \oplus x_3$

$t_2 = x_0 \wedge x_1$

$y_0 = x_2 \oplus t_2$

$t_4 = y_1 \wedge y_0$

$y_2 = x_0 \oplus t_4$

$t_6 = y_1 \wedge y_2$

$y_3 = x_1 \oplus t_6$

RECTANGLE
k = 12

$t_0 = x_3 \vee x_0$

$t_1 = x_1 \oplus t_0$

$y_1 = x_2 \oplus t_1$

$t_3 = x_3 \wedge t_1$

$t_4 = x_0 \oplus t_3$

$y_2 = y_1 \oplus t_4$

$t_6 = x_3 \oplus x_2$

$t_7 = \neg y_2$

$t_8 = t_7 \vee t_1$

$y_0 = t_8 \oplus t_6$

$t_{10} = t_7 \vee y_0$

$y_3 = t_{10} \oplus t_1$

RECTANGLE^{-1}
k = 12

$t_0 = \neg x_2$

$t_1 = x_0 \vee t_0$

$t_2 = x_3 \oplus t_1$

$y_2 = t_2 \oplus x_1$

$t_4 = t_0 \vee t_2$

$t_5 = x_0 \oplus t_4$

$y_3 = t_5 \oplus y_2$

$t_7 = t_2 \vee y_3$

$t_8 = t_7 \oplus t_5$

$y_0 = t_8 \oplus x_2$

$t_{10} = y_0 \vee y_3$

$y_1 = t_{10} \oplus t_2$

A.3   Optimized for Gate Complexity

Joltik/Piccolo
k = 8

$t_0 = x_1 \vee x_0$

$t_1 = x_1 \downarrow x_2$

$y_0 = x_3 \leftrightarrow t_0$

$y_1 = t_1 \oplus x_0$

$t_4 = y_1 \vee y_0$

$t_5 = y_0 \downarrow x_2$

$y_2 = t_5 \leftrightarrow x_1$

$y_3 = t_4 \leftrightarrow x_2$

Joltik^{-1}/Piccolo^{-1}
k = 8

$t_0 = x_1 \downarrow x_0$

$y_2 = t_0 \oplus x_3$

$t_2 = y_2 \downarrow x_0$

$y_1 = x_2 \leftrightarrow t_2$

$t_4 = y_1 \vee y_2$

$y_0 = t_4 \leftrightarrow x_1$

$t_6 = y_0 \vee y_1$

$y_3 = t_6 \leftrightarrow x_0$

LAC
k = 10

$t_0 = x_2 \leftrightarrow x_3$

$t_1 = x_1 \wedge t_0$

$t_2 = t_1 \oplus x_3$

$y_2 = x_0 \leftrightarrow t_2$

$t_4 = x_0 \vee x_1$

$y_3 = t_4 \leftrightarrow t_0$

$t_6 = t_3 \vee y_3$

$t_7 = x_0 \vee t_2$

$y_0 = t_6 \oplus x_0$

$y_1 = x_1 \leftrightarrow t_7$

Prøst
k = 8

$t_0 = x_2 \wedge x_1$

$y_1 = t_0 \oplus x_3$

$t_2 = x_0 \wedge x_1$

$y_0 = x_2 \oplus t_2$

$t_4 = y_1 \wedge y_0$

$y_2 = x_0 \oplus t_4$

$t_6 = y_1 \wedge y_2$

$y_3 = x_1 \oplus t_6$

RECTANGLE
k = 11

$t_0 = x_3 \downarrow x_0$

$t_1 = x_1 \oplus t_0$

$t_2 = x_2 \leftrightarrow x_0$

$y_1 = t_1 \leftrightarrow x_2$

$t_4 = t_1 \wedge t_2$

$t_5 = y_1 \leftrightarrow x_3$

$t_6 = t_1 \vee x_3$

$y_2 = t_2 \oplus t_6$

$t_8 = y_2 \uparrow t_5$

$y_3 = t_1 \leftrightarrow t_8$

$y_0 = t_5 \leftrightarrow t_4$

RECTANGLE^{-1}
k = 11

$t_0 = x_3 \vee x_2$

$t_1 = x_0 \oplus t_0$

$t_2 = t_1 \downarrow x_1$

$t_3 = t_2 \oplus x_3$

$y_1 = x_2 \oplus t_3$

$t_5 = t_1 \oplus x_2$

$y_3 = t_5 \oplus x_1$

$t_7 = y_1 \uparrow y_3$

$t_8 = y_3 \wedge t_1$

$y_0 = t_7 \leftrightarrow t_1$

$y_2 = t_8 \oplus t_3$

A.4   Optimized for Depth Complexity

The extra whitespace separates the different depth layers.

Joltik/Piccolo
k = 4, w = 2

$t_0 = x_1 \vee x_0$

$t_1 = x_1 \downarrow x_2$

$y_0 = x_3 \leftrightarrow t_0$

$y_1 = t_1 \oplus x_0$

$t_4 = y_1 \vee y_0$

$t_5 = y_0 \downarrow x_2$

$y_2 = t_5 \leftrightarrow x_1$

$y_3 = t_4 \leftrightarrow x_2$

Joltik^{-1}/Piccolo^{-1}
k = 4, w = 3

$t_0 = x_1 \vee x_0$

$t_1 = x_2 \leftrightarrow x_0$

$t_2 = x_3 \oplus x_1$

$t_3 = t_2 \oplus t_0$

$y_2 = x_3 \leftrightarrow t_0$

$t_5 = t_0 \oplus t_1$

$t_6 = t_3 \downarrow t_5$

$t_7 = y_2 \vee t_1$

$t_8 = y_2 \vee x_0$

$y_1 = x_2 \oplus t_8$

$y_3 = x_0 \oplus t_6$

$y_0 = t_3 \leftrightarrow t_7$

LAC
k = 3, w = 6

$t_0 = x_0 \uparrow x_1$

$t_1 = x_3 \oplus x_0$

$t_2 = x_3 \leftrightarrow x_2$

$t_3 = x_2 \oplus x_0$

$t_4 = x_2 \vee x_0$

$t_5 = x_1 \vee x_0$

$y_3 = t_5 \leftrightarrow t_2$

$t_7 = t_0 \leftrightarrow t_4$

$t_8 = t_5 \uparrow x_3$

$t_9 = t_3 \uparrow t_2$

$t_{10} = t_5 \vee t_2$

$t_{11} = x_1 \uparrow t_2$

$y_1 = t_{10} \oplus t_7$

$y_2 = t_{11} \oplus t_1$

$y_0 = t_9 \leftrightarrow t_8$

Prøst
k = 4, w = 3

$t_0 = x_1 \wedge x_2$

$t_1 = x_1 \wedge x_0$

$t_2 = x_3 \wedge x_0$

$y_1 = t_0 \oplus x_3$

$t_4 = t_2 \oplus x_1$

$y_0 = x_2 \oplus t_1$

$t_6 = y_0 \uparrow y_1$

$t_7 = y_1 \wedge x_2$

$t_8 = t_4 \vee t_2$

$y_2 = x_0 \leftrightarrow t_6$

$y_3 = t_7 \oplus t_8$

RECTANGLE
k = 3, w = 6

$t_0 = x_0 \downarrow x_3$

$t_1 = x_1 \oplus x_2$

$t_2 = x_3 \leftrightarrow x_2$

$t_3 = x_0 \wedge x_1$

$t_4 = x_1 \wedge x_2$

$t_5 = x_1 \oplus x_0$

$t_6 = t_4 \vee t_2$

$t_7 = x_3 \uparrow t_5$

$t_8 = t_4 \oplus t_3$

$t_9 = t_1 \vee t_5$

$y_1 = t_0 \leftrightarrow t_1$

$t_{11} = t_0 \vee t_2$

$y_3 = t_9 \oplus t_6$

$y_2 = t_1 \leftrightarrow t_7$

$y_0 = t_8 \oplus t_{11}$

RECTANGLE^{-1}
k = 3, w = 6

$t_0 = x_0 \oplus x_1$

$t_1 = x_0 \uparrow x_2$

$t_2 = x_3 \leftrightarrow x_2$

$t_3 = x_2 \downarrow x_3$

$t_4 = x_2 \oplus x_1$

$t_5 = x_1 \vee x_0$

$t_6 = t_3 \oplus x_2$

$t_7 = t_3 \downarrow x_1$

$t_8 = t_0 \uparrow t_2$

$t_9 = t_4 \oplus t_1$

$t_{10} = t_4 \uparrow t_1$

$t_{11} = t_2 \oplus t_5$

$y_1 = t_7 \oplus t_{11}$

$y_2 = t_9 \oplus x_3$

$y_3 = t_0 \leftrightarrow t_6$

$y_0 = t_{10} \oplus t_8$

References

1. Albrecht, M.R., Rechberger, C., Schneider, T., Tiessen, T., Zohner, M.: Ciphers for MPC and FHE. In: Oswald, E., Fischlin, M. (eds.) EUROCRYPT 2015. LNCS, vol. 9056, pp. 430–454. Springer, Heidelberg (2015)
2. Andreeva, E., Bilgin, B., Bogdanov, A., Luykx, A., Mendel, F., Mennink, B., Mouha, N., Wang, Q., Yasuda, K.: PRIMATEs v1.02. CAESAR submission (2015). http://competitions.cr.yp.to/round2/primatesv102.pdf, http://primates.ae/
3. Bard, G.V., Courtois, N.T., Jefferson, C.: Efficient methods for conversion and solution of sparse systems of low-degree multivariate polynomials over GF(2) via SAT-solvers. IACR Cryptology ePrint Archive, Report 2007/024 (2007). http://eprint.iacr.org/

4. Bertoni, G., Daemen, J., Peeters, M., Van Assche, G.: The Keccak reference, January 2011. http://keccak.noekeon.org/
5. Bertoni, G., Daemen, J., Peeters, M., Van Assche, G., Van Keer, R.: Ketje v1. CAESAR submission (2014). http://competitions.cr.yp.to/round1/ketjev11.pdf, http://ketje.noekeon.org/
6. Bertoni, G., Daemen, J., Peeters, M., Van Assche, G., Van Keer, R.: Keyak v2. CAESAR submission (2015). http://competitions.cr.yp.to/round2/keyakv2.pdf, http://keyak.noekeon.org/
7. Boyar, J., Matthews, P., Peralta, R.: On the shortest linear straight-line program for computing linear forms. In: Ochmański, E., Tyszkiewicz, J. (eds.) MFCS 2008. LNCS, vol. 5162, pp. 168–179. Springer, Heidelberg (2008)
8. Boyar, J., Peralta, R.: A new combinational logic minimization technique with applications to cryptology. In: Festa, P. (ed.) SEA 2010. LNCS, vol. 6049, pp. 178–189. Springer, Heidelberg (2010)
9. Boyar, J., Peralta, R., Pochuev, D.: On the multiplicative complexity of Boolean functions over the basis $(\wedge, \oplus, 1)$. Theoret. Comput. Sci. **235**(1), 43–57 (2000)
10. Buchfuhrer, D., Umans, C.: The complexity of Boolean formula minimization. In: Aceto, L., Damgård, I., Goldberg, L.A., Halldórsson, M.M., Ingólfsdóttir, A., Walukiewicz, I. (eds.) ICALP 2008, Part I. LNCS, vol. 5125, pp. 24–35. Springer, Heidelberg (2008)
11. Chambers, B., Manolios, P., Vroon, D.: Faster SAT solving with better CNF generation. In: Proceedings of the Conference on Design, Automation and Test in Europe, DATE 2009, 3001 Leuven, Belgium, Belgium, pp. 1590–1595. European Design and Automation Association (2009)
12. Courtois, N., Hulme, D., Mourouzis, T.: Solving circuit optimisation problems in cryptography and cryptanalysis. Cryptology ePrint Archive, Report 2011/475 (2011). http://eprint.iacr.org/
13. Courtois, N., Mourouzis, T., Hulme, D.: Exact logic minimization and multiplicative complexity of concrete algebraic and cryptographic circuits. Int. J. Adv. Intell. Syst. **6**(3 and 4), 165–176 (2013)
14. Dobraunig, C., Eichlseder, M., Mendel, F., Schläffer, M.: Ascon v1.1. CAESAR submission (2015). http://competitions.cr.yp.to/round2/asconv11.pdf, http://ascon. iaik.tugraz.at
15. Fuhs, C., Schneider-Kamp, P.: Optimizing the AES S-box using SAT. In: IWIL@ LPAR, pp. 64–70. Citeseer (2010)
16. Fuhs, C., Schneider-Kamp, P.: Synthesizing shortest linear straight-line programs over GF(2) using SAT. In: Strichman, O., Szeider, S. (eds.) SAT 2010. LNCS, vol. 6175, pp. 71–84. Springer, Heidelberg (2010)
17. Jean, J., Nikolic, I., Peyrin, T.: Joltik v1.3. CAESAR submission (2015). http://competitions.cr.yp.to/round2/joltikv13.pdf, http://www1.spms.ntu.edu. sg/~syllab/m/index.php/Joltik
18. Kavun, E.B., Lauridsen, M.M., Leander, G., Rechberger, C., Schwabe, P., Yalçın, T.: Prøst v1.1. CAESAR submission (2014). http://competitions.cr.yp.to/round1/ proestv11.pdf
19. Morawiecki, P., Gaj, K., Homsirikamol, E., Matusiewicz, K., Pieprzyk, J., Rogawski, M., Srebrny, M., Wójcik, M.: ICEPOLE v2. CAESAR submission (2015). http://competitions.cr.yp.to/round2/icepolev2.pdf
20. Mourouzis, T.: Optimizations in Algebraic and Differential Cryptanalysis. PhD thesis, UCL (University College London) (2015)

21. Sasaki, Y., Todo, Y., Aoki, K., Naito, Y., Sugawara, T., Murakami, Y., Matsui, M., Hirose, S.: Minalpher v1.1. CAESAR submission (2015). http://competitions. cr.yp.to/round2/minalpherv11.pdf

22. Shibutani, K., Isobe, T., Hiwatari, H., Mitsuda, A., Akishita, T., Shirai, T.: *Piccolo*: an ultra-lightweight blockcipher. In: Preneel, B., Takagi, T. (eds.) CHES 2011. LNCS, vol. 6917, pp. 342–357. Springer, Heidelberg (2011)

23. Zhang, L., Wenling, W., Wang, Y., Shengbao, W., Zhang, J.: LAC: A lightweight authenticated encryption cipher. CAESAR submission (2014). http:// competitions.cr.yp.to/round1/lacv1.pdf

24. Zhang, W., Bao, Z., Lin, D., Rijmen, V., Yang, B., Verbauwhede, I.: RECTANGLE: a bit-slice ultra-lightweight block cipher suitable for multiple platforms. Cryptology ePrint Archive, Report 2014/084 (2014). http://eprint.iacr.org/

Side-Channels and Implementations

Verifiable Side-Channel Security
of Cryptographic Implementations:
Constant-Time MEE-CBC

José Bacelar Almeida[1,2], Manuel Barbosa[1,3(✉)], Gilles Barthe[4],
and François Dupressoir[4(✉)]

[1] HASLab – INESC TEC, Porto, Portugal
[2] University of Minho, Braga, Portugal
jba@di.uminho.pt
[3] DCC-FC, University of Porto, Porto, Portugal
mbb@dcc.fc.up.pt
[4] IMDEA Software Institute, Madrid, Spain
fdupress@gmail.com

Abstract. We provide further evidence that implementing software
countermeasures against timing attacks is a non-trivial task and requires
domain-specific software development processes: we report an implementation bug in the s2n library, recently released by AWS Labs.
This bug (now fixed) allowed bypassing the balancing countermeasures
against timing attacks deployed in the implementation of the MAC-then-
Encode-then-CBC-Encrypt (MEE-CBC) component, creating a timing
side-channel similar to that exploited by Lucky 13.

Although such an attack could only be launched when the MEE-CBC
component is used in isolation – Albrecht and Paterson recently confirmed in independent work that s2n's second line of defence, once rein-
forced, provides adequate mitigation against current adversary capabilities – its existence serves as further evidence to the fact that conventional
software validation processes are not effective in the study and validation of security properties. To solve this problem, we define a methodology for proving security of implementations in the presence of timing
attackers: first, prove *black-box security* of an algorithmic description of
a cryptographic construction; then, establish *functional correctness* of an
implementation with respect to the algorithmic description; and finally,
prove that the implementation is *leakage secure*.

We present a proof-of-concept application of our methodology to
MEE-CBC, bringing together three different formal verification tools to
produce an assembly implementation of this construction that is ver-
ifiably secure against adversaries with access to some timing leakage.
Our methodology subsumes previous work connecting provable security
and side-channel analysis at the implementation level, and supports the
verification of a much larger case study. Our case study itself provides
the first provable security validation of complex timing countermeasures
deployed, for example, in OpenSSL.

© International Association for Cryptologic Research 2016
T. Peyrin (Ed.): FSE 2016, LNCS 9783, pp. 163–184, 2016.
DOI: 10.1007/978-3-662-52993-5_9

1   Introduction

There is an uncomfortable gap between provable security and practical implementations. Provable security gives strong guarantees that a cryptographic construction is secure against efficient *black-box* adversaries. Yet, implementations of provably secure constructions may be vulnerable to practical attacks, due to implementation errors or side-channels. The tension between provable security and cryptographic engineering is illustrated by examples such as the MAC-then-Encode-then-CBC-Encrypt construction (MEE-CBC), which is well-understood from the perspective of provable security [22,26], but whose implementation has been the source of several practical attacks in SSL or TLS implementations. These security breaks are, in the case of MEE-CBC, due to vulnerable implementations providing the adversary with padding oracles, either through error messages [29], or through observable non-functional behaviours such as execution time [2,16]. These examples illustrate two shortcomings of provable security when it comes to dealing with implementations. First, the algorithmic descriptions used in proofs elide many potentially critical details; these details must be filled by implementors, who may not have the specialist knowledge required to make the right decision. Second, attackers targeting real-world platforms may break a system by exploiting side-channel leakage, which is absent in the black-box abstractions in which proofs are obtained.

These shortcomings are addressed independently by *real-world cryptography* and *secure coding methodologies*, both of which have their own limitations. Real-world cryptography [18] is a branch of provable security that incorporates lower-level system features in security notions and proofs (for example, precise error messages or message fragmentation). Real-world cryptography is a valuable tool for analyzing the security of real-world protocols such as TLS or SSH, but is only now starting to address side-channels [8,15] and, until now, has stayed short of considering actual implementations. Secure coding methodologies effectively mitigate side-channel leakage; for instance, the constant-time methodology [13,21] is consensual among practitioners as a means to ensure a *good* level of protection against timing and cache-timing attacks. However, a rigorous justification of such techniques and their application is lacking and they are disconnected from provable security, leaving room for subtle undetected vulnerabilities even in carefully tailored implementations.

In this paper we show how the real-world cryptography approach can be extended – with computer-aided support – to formally capture the guarantees that implementors empirically pursue using secure coding techniques.

1.1   Our Contributions

Recent high-visibility attacks such as Lucky 13 [2] have shown that timing leakage can be exploited in practice to break the security of pervasively used protocols such as TLS, and have led practitioners to pay renewed attention to software countermeasures against timing attacks. Two prominent examples of this are the recent reimplementation of MEE-CBC decryption in OpenSSL [23], which enforces a constant-time coding policy as mitigation for the Lucky 13 attack, and the *defense in depth* mitigation strategy adopted by Amazon Web Services Labs (AWS Labs) in a new implementation of TLS called s2n, where various fuzzing- and balancing-based timing countermeasures are combined to reduce the amount of information leaked through timing. However, the secure-coding efforts of cryptography practitioners are validated using standard software engineering techniques such as testing and code reviews, which are *not* well-suited to reasoning about non-functional behaviours or cryptography.

As a first contribution and motivation for our work, we provide new evidence of this latent problem by recounting the story of Amazon's recently released s2n library, to which we add a new chapter.

NEW EVIDENCE IN S2N. In June 2015, AWS-Labs made public a new open-source implementation of the TLS protocol, called s2n [28] and designed to be "small, fast, with simplicity as a priority". By excluding rarely used options and extensions, the implementation can remain small, with only around 6 K lines of code. Its authors also report extensive validation, including three external security evaluations and penetration tests. The library's source code and documentation are publicly available.[1]

Recently, Albrecht and Paterson [1] presented a detailed analysis of the countermeasures against timing attacks in the original release of s2n, in light of the lessons learned in the aftermath of Lucky 13 [2]. In their study, they found that the implementation of the MEE-CBC component was not properly balanced, and exposed a timing attack vector that was exploitable using Lucky 13-like techniques. Furthermore, they found that the second layer of countermeasures that randomizes error reporting delays was insufficient to remove the attack vector. Intuitively, the granularity of the randomized delays was large enough in comparison to the data-dependent timing variations generated by the MEE-CBC component that they could be 'filtered out' leaving an exploitable side-channel. As a response to these findings, the s2n implementation was patched,[2] and both layers of countermeasures were improved to remove the attack vector.[3]

Unfortunately, this is not the end of the story. In this paper we report an implementation bug in this "fixed" version of the library, as well as a timing attack akin to Lucky 13 that bypasses once more the branch-balancing timing countermeasures deployed in the s2n implementation of MEE-CBC. This implementation bug was subtly hidden in the implementation of the timing countermeasures themselves, which were added as mitigation for the attack reported

---

[1] https://github.com/awslabs/s2n.

[2] See the details of the applied fixes in https://github.com/awslabs/s2n/commit/ 4d3729.

[3] We note that the delay randomization countermeasure was further improved since the attacks we describe to sampling the delay between 10 s and 30 s (https://github. com/awslabs/s2n/commit/731e7d). Further, measures were added to prevent careless or rogue application code from forcing s2n to signal decryption failures to the adversary before that delay had passed (https://github.com/awslabs/s2n/commit/ f8a155).

by Albrecht and Paterson [1]. We show that the bug rendered the countermeasure code in the MEE-CBC component totally ineffective by presenting a timing attack that breaks the MEE-CBC implementation when no additional timing countermeasures were present. Due to space constraints, details of the attack are given in the full version of the paper.[4]

Disclosure Timeline and Recommendations. The implementation bug and timing attack were reported to AWS Labs on September 4, 2015. The problem was promptly acknowledged and the current head revision of the official s2n repository no longer exhibits the bug and potential attack vector from the MEE-CBC implementation. Subsequent discussions with Albrecht and Paterson and AWS Labs lead us to believe that s2n's second line of defence (the finer grained error reporting delay randomization mechanism validated by Albrecht and Paterson [1]) is currently sufficient to thwart potential exploits of the timing side-channel created by the bug. Therefore, systems relying on unpatched but *complete* versions of the library are safe. On the other hand, any system relying directly on the unpatched MEE-CBC implementation, without the global randomized delay layer, will be vulnerable and should upgrade to the latest version.

THE NEED FOR FORMAL VALIDATION. The sequence of events reported above[5] shows that timing countermeasures are extremely hard to get right and very hard to validate. Our view is that implementors currently designing and deploying countermeasures against side-channel attacks face similar problems to those that were faced by the designers of cryptographic primitives and protocols before the emergence of provable security. On the one hand, we lack a methodology to rigorously characterize and prove the soundness of existing designs such as the ones deployed, e.g., in OpenSSL; on the other hand, we have no way of assessing the soundness of new designs, such as those adopted in s2n, except via empirical validation and trial-and-error. This leads us to the following question: *can we bring the mathematical guarantees of provable security to cryptographic implementations?* We take two steps towards answering this question.

A CASE STUDY: CONSTANT-TIME MEE-CBC. Our second and main contribution is the first formal and machine-checked proof of security for an x86 implementation of MEE-CBC in an attack model that includes control-flow and cache-timing channels. In particular, our case study validates the style of countermeasures against timing attacks currently deployed in the OpenSSL implementation of MEE-CBC. We achieve this result by combining three state-of-the-art formal verification tools: i. we rely on EasyCrypt [6,7] to formalize a specification of MEE-CBC and some of the known provable security results for

---

[4] https://eprint.iacr.org/2015/1241.

[5] The very interesting blog post in http://blogs.aws.amazon.com/security/post/ TxLZP6HNAYWBQ6/s2n-and-Lucky-13 analyses these events from the perspective of the AWS Labs development team.

this construction;[6] ii. we use Frama-C to establish a functional equivalence result between EasyCrypt specifications and C implementations; and iii. we apply the CompCert certified compiler [24] and the certified information-flow type-system from [4] to guarantee that the compiled implementation does not leak secret information through the channels considered, and that the compiled x86 code is correct with respect to the EasyCrypt specification proved secure initially.

A FRAMEWORK FOR IMPLEMENTATION SECURITY. To tie these verification results together, we introduce — as our third contribution — a framework of definitions and theorems that abstracts the details of the case study. This framework yields a general methodology for proving security properties of low-level implementations in the presence of adversaries that may observe leakage. This methodology relies on separating three different concerns: i. *black-box specification security*, which establishes the computational security of a functional specification (here one can adopt the real-world cryptography approach); ii. *implementation correctness*, which establishes that the considered implementation behaves, as a black-box, exactly like its functional specification; and iii. *leakage security*, which establishes that the leakage due to the execution of the implementation code *in some given leakage model* is independent from its secret inputs. Our main theorem, which is proven using the previous methodology, establishes that our x86 implementation retains the black-box security properties of the MEE-CBC specification, i.e., it is a secure authenticated encryption scheme, even in the presence of a strong timing attacker, and based on standard black-box cryptographic assumptions.

We insist that we do *not* claim to formally or empirically justify the validity of any particular leakage model: for this we rely on the wisdom of practitioners. What we *do* provide is a means to take a well-accepted leakage model, and separately and formally verify, through leakage security, that a concrete deployment of a particular countermeasure in a given implementation does in fact guarantee the absence of any leakage that would weaken a particular security property in the chosen leakage model.

Outline. In Sect. 2, we describe the MEE-CBC construction and informally discuss its security at specification- and implementation-level. We then present the definitions for implementation-level security notions and the statement of our main theorem (Sect. 3). In Sect. 4, we introduce our methodology, before detailing its application to MEE-CBC in Sect. 5. We then present and discuss some benchmarking results in Sect. 6. Finally, we discuss potential extensions to our framework not illustrated by our case study (Sect. 7). We conclude the paper and discuss directions for future work in Sect. 8. A long version of this paper,

---

[6] Formalizing all known results for MEE-CBC would be beyond the scope of this paper, and we assume that our EasyCrypt specification of the construction inherits all the security properties that have been proved in the literature. In other words, in addition to the properties we formalize, we assume that our MEE-CBC specification satisfies the standard notions of security for authenticated encryption as proved, e.g., by Paterson, Ristenpart and Shrimpton [26].

with appendices including code snippets, formal definitions of standard black-box specification-level security notions, and a discussion of further related work appears on the IACR eprint server.[7]

2   Case Study: MEE-CBC

MAC-then-Encode-then-CBC-Encrypt (MEE-CBC) is an instance of the MAC-then-Encrypt generic construction that combines a block cipher used in CBC mode with some padding and a MAC scheme in order to obtain an authenticated encryption scheme. We consider the specific instantiation of the construction that is currently most widely used within TLS: i. A MAC tag of length tlen is computed over the TLS record header hdr, a sequence number seq and the payload pld. The length of the authenticated string is therefore the length of the payload plus a small and fixed number of bytes. Several MAC schemes can be used to authenticate this message, but we only consider HMAC-SHA256. ii. The CBC-encrypted message m comprises the payload pld concatenated with the MAC tag (the sequence number is not transmitted and the header is transmitted in the clear). iii. The padding added to m comprises plen bytes of value plen − 1, where plen may be any value in the range [1..256], such that plen + |m| is a multiple of the cipher's block size. iv. We use AES-128 as block cipher, which fixes a 16-byte block size.

At the high level, the HMAC construction computes

$$H((\text{key}_{\text{MAC}} \oplus \text{opad}) \mathbin{\|} H((\text{key}_{\text{MAC}} \oplus \text{ipad}) \mathbin{\|} \text{hdr} \mathbin{\|} \text{seq} \mathbin{\|} \text{pld})) \,.$$

We consider a hash function such as SHA-256, which follows the Merkle-Damgård paradigm: a compression function is iterated to gradually combine the already computed hash value with a new 64-byte message block (hash values are tlen bytes long).

INFORMAL SECURITY DISCUSSION. The theoretical security of MEE-CBC has received a lot of attention in the past, due to its high-profile usage in the SSL/TLS protocol. Although it is well-known that the MAC-then-Encrypt construction does *not* generically yield a secure authenticated encryption scheme [9], the particular instantiation used in TLS has been proven secure [22,25,26]. The most relevant result for this paper is that by Paterson, Ristenpart and Shrimpton [26]. Crucially, their high-level proof explicitly clarifies the need for the implementation to not reveal, in any way, which of the padding or MAC check failed on decryption failures. This is exactly the kind of *padding oracles* exploited in practical attacks against MEE-CBC such as Lucky 13 [2].

After the disclosure of the Lucky 13 attack [2], significant effort was invested into identifying all potential sources of timing leakage in the MEE-CBC decryption algorithm. The implementation subsequently incorporated into OpenSSL, for example, deploys constant-time countermeasures that guarantee the following

---

[7] https://eprint.iacr.org/2015/1241.

behaviours [23]: i. removing the padding and checking its well-formedness occurs in constant-time; ii. the MAC of the unpadded message is always computed, even for bad padding; iii. the MAC computation involves the same number of calls to the underlying compression function regardless of the number of hash input blocks in the decoded message, and regardless of the length of the final hash block (which may cause an additional block to be computed due to the internal Merkle-Damgård length padding); and iv. the transmitted MAC is compared to the computed MAC in constant-time (the transmitted MAC's location in memory, which may be leaked through the timing of memory accesses, depends on the plaintext length). *Constant-time*, here and in the rest of this paper, is used to mean that the trace of program points and memory addresses accessed during the execution is independent from the initial value of secret inputs. In particular, we note that the OpenSSL MEE-CBC implementation is *not* constant time following this definition: the underlying AES implementation uses look-up table optimizations that make secret-dependent data memory accesses and may open the way to cache-timing attacks.

OUR IMPLEMENTATION. The main result of this paper is a security theorem for an x86 assembly implementation of MEE-CBC ($MEE\text{-}CBC_{x86}$). The implementation is compiled using CompCert from standard C code that replicates the countermeasures against timing attacks currently implemented in the OpenSSL library [23]. We do not use the OpenSSL code directly because the code style of the library (and in particular its lack of modularity) makes it a difficult target for verification. Furthermore, we wish to fully prove constant-time security, which we have noted is not achieved by OpenSSL. However, a large part of the code we verify is existing code, taken from the NaCl library [14] without change (for AES, SHA256 and CBC mode), or modified to include the necessary countermeasures (HMAC, padding and MEE composition). Our C code is composed of the following modules, explicitly named for later reference: i. $AES128_{NaCl}$ contains the NaCl implementation of AES128; ii. $HMACSHA256_{NaCl}$ contains a version of the NaCl implementation of HMAC-SHA256 extended with timing countermeasures mimicking those described in [23]; and iii. $MEE\text{-}CBC_C$ contains an implementation of MEE-CBC using $AES128_{NaCl}$ and $HMACSHA256_{NaCl}$. We do not include the code in the paper due to space constraints.

As we prove later in the paper, a strict adherence to the coding style adopted in OpenSSL is indeed sufficient to guarantee security against attackers that, in addition to input/output interaction with the MEE-CBC implementation, also obtain full traces of program counter and memory accesses performed by the implementation. However, not all TLS implementations have adopted a strict adherence to constant-time coding policies in the aftermath of the Lucky 13 attack. We now briefly present the case of Amazon's s2n library, discussing their choice of countermeasures, and describing a bug in their implementation that leads to an attack. A more detailed discussion can be found in the long version of this paper.

BREAKING THE MEE-CBC IMPLEMENTATION IN S2N. Although parts of the s2n code for MEE-CBC are written in the constant-time style, there are many (intentional) deviations from a strict constant-time coding policy. For example, no attempt is made to de-correlate memory accesses from the padding length value that is recovered from the decrypted (but not yet validated) plaintext. As an alternative, the code includes countermeasures that intend to balance the execution time of secret-dependent conditional branches that might lead to significant variability in the execution time. Roughly, the goal of these countermeasures is to ensure that the total number of calls to the hash compression function is always the same, independently of the actual padding length or validity.

The bug we found resides in a special routine that aims to guarantee that a dummy compression function computation is performed whenever particular padding patterns might lead to shorter execution times. An off-by-one error in the checking of a boundary condition implied that the dummy compression function would be invoked unnecessarily for some padding values (more precisely, there are exactly 4 such padding values, which are easily deduced from the (public) length of the encrypted record).

The leakage the bug produces is similar in size to that exploited by AlFardan and Paterson [2] to recover plaintexts. We have implemented a padding-oracle-style attack on the MEE-CBC decryption routine to recover single plaintext bytes from a ciphertext: one simply measures the decryption time to check if the recovered padding length causes the bug to activate and proceeds by trial and error.[8] The attack can be extended to full plaintext recovery using the same techniques reported in [2].

We already discussed the real-world impact of our attack and our disclosure interaction with AWS Labs in the introduction of this paper. However, we insist that for the purpose of this paper it is *not* the real-world impact of our attack that matters, but the software bug that gave rise to it in the first place. Indeed the existence of such a programming bug and the fact that it remained undetected through AWS Labs' code validation process (and in particular despite unit testing specifically designed to detect timing side-channels) reveal that there is a need for a formal framework in which to rigorously prove that an implementation is secure against timing attacks. This is what we set out to do in the rest of the paper.

3   Security Definitions and Main Theorem

After a brief reminder of the syntax and security notions for secret key encryption relevant to our case study, we introduce and discuss the corresponding implementation-level security notions for the constant-time leakage model and state our main theorem. Cryptographic implementations are often hardwired at a particular security level, which means that asymptotic security notions are not adequate to capture the security guarantees provided by software. We therefore

---

[8] Plaintext recovery is easier than in Lucky 13, since leakage occurs whether or not the padding string is correct.

omit the security parameter in all our definitions. For simplicity we also keep the running time of algorithms implicit in our notations, although we take care to account for it in our security proofs and to show that there is no hidden slackness in our reductions.

3.1 Secret Key Encryption

We recall that a secret-key encryption scheme $\Pi = (\mathsf{Gen}, \mathsf{Enc}, \mathsf{Dec})$ is specified as three algorithms: i. a probabilistic key generation algorithm $\mathsf{Gen}(; r)$ that returns a secret key SK on input some random coins r; ii. a probabilistic encryption algorithm $\mathsf{Enc}(m, \mathsf{SK}; r)$ that returns a ciphertext c on input a message m, the secret key SK, and some random coins r; and iii. a deterministic decryption algorithm $\mathsf{Dec}(c, \mathsf{SK})$ that returns either a message m or a failure symbol $\perp$ on input a ciphertext c and secret key SK. We denote the set of valid messages with MsgSp and adopt standard notions of correctness, confidentiality (IND\$-CPA) and integrity (INT-PTXT and INT-CTXT) for authenticated symmetric encryption schemes.

Our goal in the rest of this section is to adapt these standard notions to formally capture implementation-level security. In particular, we wish to give the adversary the ability to observe the leakage produced by the computation of its oracle queries. We first give generic definitions for some core concepts.

3.2 Implementation: Languages, Leakage and Generation

For the sake of generality, our definitions abstract the concrete implementation languages and leakage models adopted in our case study. We later instantiate these definitions with a black-box security model for C implementations and a timing leakage model for x86 assembly implementations.

LANGUAGE, LEAKAGE AND MACHINE. Given an implementation language $\mathcal{L}$, we consider a machine $\mathbb{M}$ that animates its semantics. Such a machine takes as input a program P written in $\mathcal{L}$, an input i for P, and some randomness r and outputs both the result o of evaluating P with i and r, and the leakage ℓ produced by the evaluation. We use the following notation for this operation $o \leftarrow \mathbb{M}(P, i; r)_{\leadsto \ell}$. We make the assumption that the machine is deterministic, so that all randomness required to execute programs is given by the input r. However, our security experiments are probabilistic, and we write $o \leftarrow_\$ \mathbb{M}(P, i)_{\leadsto \ell}$ to denote the probabilistic computation that first samples the random coins r that must be passed as randomness input of P, and then runs $\mathbb{M}(P, i; r)$. This approach agrees with the view that the problem of randomness generation is orthogonal to the one of secure implementation [14]. We discuss this further in Sect. 7.

We note that the definition of $\mathbb{M}$ makes three implicit assumptions. First, the semantics of a program must always be defined, since $\mathbb{M}$ always returns a result; termination issues can be resolved easily by aborting computations after a fixed number of steps. Second, our view of $\mathbb{M}$ does not allow an adversary to influence a program's execution other than through its queries. Finally, our model implies that the semantics of $\mathcal{L}$ can be equipped with meaningful notions of leakage. In the context of our use case, we adopt the common view of practical cryptography that timing leakage can be captured via the code-memory and data-memory accesses performed while executing a program. These can be sensibly formalized over assembly implementations, but not over higher-level implementations (e.g., over C implementations), not least because there is no guarantee that optimizing compilers do not introduce leakage. For this reason, in our case study, we consider the following two implementation models:

- a C-level model using a machine $\mathbb{M}_C^\emptyset$ (or simply $\mathbb{M}_C$) that animates the C language semantics with no leakage;
- an assembly-level model using a machine $\mathbb{M}_{x86}^{CT}$ that animates (a subset of) the x86 assembly language, and produces leakage traces in the constant-time leakage model as detailed below.

In both languages, we adopt the semantic definitions as formalized in the CompCert certified compiler.

CONSTANT-TIME LEAKAGE TRACES. Formally, we capture the constant-time leakage model by assuming that each semantic step extends the (initially empty) leakage trace with a pair containing: i. the program point corresponding to the statement being executed; and ii. the (ordered) sequence of memory accesses performed during the execution step. We specify when this particular leakage model is used by annotating the corresponding notion with the symbol CT.

3.3 Authenticated Encryption in the Implementation Model

Given a language $\mathcal{L}$ and a (potentially leaking) machine $\mathbb{M}$ animating its semantics, we now define $\mathbb{M}$-correctness, $\mathbb{M}$-IND\$-CPA and $\mathbb{M}$-INT-PTXT security for $\mathcal{L}$-implementations of SKE schemes in the leakage model defined by $\mathbb{M}$. In what follows, we let $\Pi^* = (\text{Gen}^*, \text{Enc}^*, \text{Dec}^*)$ be an SKE implementation in language $\mathcal{L}$.

SKE IMPLEMENTATION CORRECTNESS. We say that Π^* is $\mathbb{M}$-correct if, for all $m \in \text{MsgSp}$, random coins r_{gen}, r_{enc}, and $SK = \mathbb{M}(\text{Gen}^*; r_{gen})$, we have that

$$\mathbb{M}(\text{Dec}^*, \mathbb{M}(\text{Enc}^*, m, SK; r_{enc}), SK) = m.$$

SKE IMPLEMENTATION SECURITY. The $\mathbb{M}$-IND\$-CPA *advantage of an adversary* $\mathcal{A}$ *against* Π^* *and public length function* ϕ is defined as the following (concrete) difference

$$\mathbf{Adv}_{\Pi^*,\phi,\mathcal{A}}^{\mathbb{M}\text{-ind\$-cpa}} := \Big| \Pr\Big[\mathbb{M}\text{-IND\$-CPA}_{\Pi^*,\phi}^{\mathcal{A}}(\text{Real}) \Rightarrow \text{true} \Big]$$
$$- \Pr\Big[\mathbb{M}\text{-IND\$-CPA}_{\Pi^*,\phi}^{\mathcal{A}}(\text{Ideal}) \Rightarrow \text{true} \Big] \Big|,$$

where implementation-level game $\mathbb{M}$-IND\$-CPA is shown in Fig. 1. Here, public length function ϕ is used to capture the fact that SKEs may partially hide the

Game M-IND\$-CPA$_{\Pi^*,\phi}^{\mathcal{A}}(b)$:	proc. RoR(m):	proc. Dec(c):
SK $\leftarrow\!\!\$ \, M(Gen^*)_{\leadsto\ell_g}$	c $\leftarrow\!\!\$ \, M(Enc^*, m, SK)_{\leadsto\ell_e}$	m $\leftarrow M(Dec^*, c, SK)_{\leadsto\ell_d}$
$b' \leftarrow\!\!\$ \, \mathcal{A}^{\mathbf{RoR},\mathbf{Dec}}(\ell_g)$	If $(b = \mathsf{Ideal})$ Then c $\leftarrow\!\!\$ \, \{0,1\}^{\phi(|m|)}$	Return $(\bot, \ell_d)$
Return $(b' = b)$	Return (c, ℓ_e)	

Fig. 1. M-IND\$-CPA experiment.

Game M-INT-PTXT$_{\Pi^*}^{\mathcal{A}}$:	proc. Enc(m):	proc. Ver(c):
List $\leftarrow []$; win $\leftarrow \bot$	c $\leftarrow\!\!\$ \, M(Enc^*, m, SK)_{\leadsto\ell_e}$	m $\leftarrow M(Dec^*, c, SK)_{\leadsto\ell_d}$
SK $\leftarrow\!\!\$ \, M(Gen^*)_{\leadsto\ell_g}$	List $\leftarrow$ m : List	win $\leftarrow$ win $\vee$ (m $\neq \bot \wedge$ m $\notin$ List)
$\mathcal{A}^{\mathbf{Enc},\mathbf{Ver}}(\ell_g)$	Return (c, ℓ_e)	Return $(m \neq \bot, \ell_d)$
Return win		

Fig. 2. M-INT-PTXT experiment.

length of a message. If ϕ is the identity function or is efficiently invertible, then the message length is trivially leaked by the ciphertext. In the case of our MEE-CBC specification, for example, the message length is revealed only up to AES block alignment.

We observe that in this refinement of the IND\$-CPA security notion for implementations, the adversary may learn information about the secrets via the leakage produced by the decryption oracle Dec*, even if its functional input-output behaviour reveals nothing. In particular, in a black-box adversary model where leakage traces are always empty, the **Dec** oracle can be perfectly implemented by the procedure that ignores its argument and returns $(\bot, \epsilon)$, and the **RoR** oracle can be simulated without any dependency on m in the Ideal world; this allows us to recover the standard computational security experiment for IND\$-CPA. On the other hand, in models where leakage traces are not always empty, the adversary is given the ability to use the decryption oracle with invalid ciphertexts and recover information through its leakage output.

We extend standard INT-PTXT security in a similar way and define the M-INT-PTXT *advantage of an adversary $\mathcal{A}$ against Π^** as the following (concrete) probability:

$$\mathbf{Adv}_{\Pi^*,\mathcal{A}}^{\mathsf{M\text{-}int\text{-}ptxt}} := \Pr\left[\mathsf{M\text{-}INT\text{-}PTXT}_{\Pi^*}^{\mathcal{A}}() \Rightarrow \mathsf{true}\right],$$

where implementation-level game M-INT-PTXT is shown in Fig. 2.

We similarly "lift" INT-CTXT, PRP (pseudorandomness of a permutation) and UF-CMA (existential MAC unforgeability) security experiments and advantages to implementations. This allows us to state our main theorem.

3.4   Main Theorem

The proof of Theorem 1 is fully machine-checked. However, foregoing machine-checking of the specification's security theorems allows us to strengthen the results we obtain on the final implementations. We discuss this further after we present our proof strategy.

Theorem 1 (CT security of MEE-CBC$_{\text{x86}}$). *MEE-CBC$_{\text{x86}}$ is $\mathbb{M}_{\text{x86}}^{\text{CT}}$-correct and provides $\mathbb{M}_{\text{x86}}^{\text{CT}}$-IND\$-CPA and $\mathbb{M}_{\text{x86}}^{\text{CT}}$-INT-PTXT security if the underlying components AES128$_{\text{NaCl}}$ and HMACSHA256$_{\text{NaCl}}$ are black-box secure as a PRP and a MAC, respectively. More precisely, let $\phi(i) = \lceil (i+1)/16 \rceil + 3$, then*

- *For any $\mathbb{M}_{\text{x86}}^{\text{CT}}$-IND\$-CPA adversary $\mathcal{A}^{\text{cpa}}$ that makes at most q queries to its **RoR** oracle, each of length at most n octets, there exists an (explicitly constructed) $\mathbb{M}_{\text{C}}^{\emptyset}$-IND\$-CPA adversary $\mathcal{B}^{\text{prp}}$ that makes at most $q \cdot \lceil (n+1)/16 \rceil + 2$ queries to its forward oracle and such that*

$$\mathbf{Adv}_{\text{MEE-CBC}_{\text{x86}}, \phi, \mathcal{A}^{\text{cpa}}}^{\mathbb{M}_{\text{x86}}^{\text{CT}}\text{-ind\$-cpa}} \leq \mathbf{Adv}_{\text{AES128}_{\text{NaCl}}, \mathcal{B}^{\text{prp}}}^{\mathbb{M}_{\text{C}}^{\emptyset}\text{-prp}} + 2 \cdot \frac{(q \cdot (\lceil \frac{n+1}{16} \rceil + 2))^2}{2^{128}}.$$

- *For any $\mathbb{M}_{\text{x86}}^{\text{CT}}$-INT-PTXT adversary $\mathcal{A}^{\text{ptxt}}$ that makes at most q_E queries to its **Enc** oracle and q_V queries to its **Ver** oracle, there exists an (explicitly constructed) $\mathbb{M}_{\text{C}}^{\emptyset}$-UF-CMA adversary $\mathcal{B}^{\text{cma}}$ that makes at most q_E queries to its **Tag** oracle and q_V queries to its **Ver** oracle and such that*

$$\mathbf{Adv}_{\text{MEE-CBC}_{\text{x86}}, \mathcal{A}^{\text{ptxt}}}^{\mathbb{M}_{\text{x86}}^{\text{CT}}\text{-int-ptxt}} \leq \mathbf{Adv}_{\text{HMACSHA256}_{\text{NaCl}}, \mathcal{B}^{\text{cma}}}^{\mathbb{M}_{\text{C}}^{\emptyset}\text{-uf-cma}}.$$

In addition, the running time of our constructed adversaries is essentially that of running the original adversary plus the time it takes to emulate the leakage of the x86 implementations using dummy executions in machine $\mathbb{M}_{\text{x86}}$. Under reasonable assumptions on the efficiency of $\mathbb{M}_{\text{x86}}$, this will correspond to an overhead that is linear in the combined inputs provided by an adversary to its oracles (the implementations are proven to run in constant time under the semantics of $\mathcal{L}$ when these inputs are fixed).

Note that the security assumptions we make are on C implementations of AES (AES128$_{\text{NaCl}}$) and HMAC-SHA256 (HMACSHA256$_{\text{NaCl}}$). More importantly, they are made in a *black-box* model of security where the adversary gets empty leakage traces.

The proof of Theorem 1 is detailed in Sect. 5 and relies on the general framework we now introduce. Rather than reasoning directly on the semantics of the executable x86 program (and placing our assumptions on objects that may not be amenable to inspection), we choose to prove complex security properties on a clear and simple functional specification, and show that each of the refinement steps on the way to an x86 assembly executable preserves this property, or even augments it in some way.

4   Formal Framework and Connection to PL Techniques

Our formal proof of implementation security follows from a set of conditions on the software development process. We therefore introduce the notion of an implementation generation procedure.

IMPLEMENTATION GENERATION. An implementation generation procedure $\mathcal{C}^{\mathcal{L}_1 \to \mathcal{L}_2}$ is a mapping from specifications in language $\mathcal{L}_1$ to implementations in

Game $\mathrm{Corr}^{\mathcal{A}}_{\mathbb{M},\Pi,\mathcal{C}}()$:	**proc. Eval**(k, i, r):
bad $\leftarrow$ false	$o \leftarrow \Pi[k](i; \mathsf{r})$
$\Pi^* \leftarrow \mathcal{C}(\Pi)$	$o' \leftarrow \mathbb{M}(\Pi^*[k], i; \mathsf{r})_{\leadsto\ell}$
$\mathcal{A}^{\mathbf{Eval}}(\Pi^*)$	If $o \neq o'$ then bad $=$ true
Return $\neg$ bad	

Fig. 3. Game defining correct implementation generation. For compactness, we use notation $\Pi[k]$ (resp. $\Pi^*[k]$) for $k \in \{1, 2, 3\}$ to denote the k-th algorithm in scheme Π (resp. implementation Π^*), corresponding to key generation (1), encryption (2) and decryption (3).

language $\mathcal{L}_2$. For example, in our use case, the top-level specification language is the expression language $\mathcal{L}_{EC}$ of EasyCrypt (a polymorphic and higher-order λ-calculus) and the overall implementation generation procedure $\mathcal{C}^{\mathcal{L}_{EC} \rightarrow \mathcal{L}_{x86}}$ is performed by a verified manual refinement of the specification into C followed by compilation to x86 assembly using CompCert (here, $\mathcal{L}_{x86}$ is the subset of x86 assembly supported by CompCert).

We now introduce two key notions for proving our main result: *correct implementation generation* and *leakage security*, which we relate to standard notions in the domain of programming language theory. This enables us to rely on existing formal verification methods and tools to derive intermediate results that are sufficient to prove our main theorem. In our definitions we consider two arbitrary languages $\mathcal{L}_1$ and $\mathcal{L}_2$, a (potentially leaking) machine $\mathbb{M}$ animating the semantics of the latter, and an implementation generation procedure $\mathcal{C}^{\mathcal{L}_1 \rightarrow \mathcal{L}_2}$. In this section, $\mathcal{L}_1$ and $\mathcal{L}_2$ are omitted when denoting the implementation generation procedure (simply writing $\mathcal{C}$ instead). In the rest of the paper, we also omit them when clear from context.

CORRECT IMPLEMENTATION GENERATION. Intuitively, the minimum requirement for an implementation generation procedure is that it preserves the input-output functionality of the specification. We capture this in the following definition.

Definition 1 (Correct implementation generation). *The implementation generation procedure $\mathcal{C}$ is correct if, for every adversary $\mathcal{A}$ and primitive specification Π, the game in Fig. 3 always returns* true.

For the programming languages we are considering (deterministic, I/O-free languages) this notion of implementation generation correctness is equivalent to the standard language-based notion of simulation, and its specialization as semantic preservation when associated with general-purpose compilers. A notable case of this is CompCert [24] for which this property is formally proven in Coq. Similarly, as we discuss in Sect. 5, a manual refinement process can be turned into a correct implementation generation procedure by requiring a total functional correctness proof. This is sufficient to guarantee *black-box* implementation security. However, it is not sufficient in general to guarantee implementation security in the presence of leakage.

LEAKAGE SECURITY. In order to relate the security of implementations to that of black-box specifications, we establish that leakage does not depend on secret inputs. We capture this intuition via the notion of *leakage security*, which imposes that all the leakage produced by the machine $\mathbb{M}$ for an implementation is benign. Interestingly from the point of view of formal verification, leakage security is naturally related to the standard notion of non-interference [19]. In its simplest form, non-interference is formulated by partitioning the memory of a program into *high-security* (or secret) and *low-security* (or public) parts and stating that two executions that start in states that agree on their low-security partitions end in states that agree on their low-security partitions.

We define what the public part of the input means by specifying a function τ that parametrizes our definition of leakage security. For the case of symmetric encryption, for example, τ is defined to tag as public the inputs to the algorithms an attacker has control over through its various oracle interfaces (in IND\$-CPA, INT-PTXT and INT-CTXT). More precisely, we define a specific projection function τ_{SKE} as follows:

$$\tau_{\mathsf{SKE}}(\mathsf{Gen}) = \epsilon \qquad \tau_{\mathsf{SKE}}(\mathsf{Enc}, \mathsf{key}, m) = (|\mathsf{key}|, |m|) \qquad \tau_{\mathsf{SKE}}(\mathsf{Dec}, \mathsf{key}, c) = (|\mathsf{key}|, c)$$

Our definition of leakage security then consists in constraining the information-flow into the leakage due to each algorithm, via the following non-interference notion.[9]

Definition 2 (($\mathbb{M}, \tau$)-non-interference). *Let P be a program in $\mathcal{L}_2$ and τ be a projection function on P's inputs. Then, P is ($\mathbb{M}, \tau$)-non-interferent if, for any two executions $o \leftarrow \mathbb{M}(P, i; r) \leadsto \ell$ and $o' \leftarrow \mathbb{M}(P, i'; r') \leadsto \ell'$, we have $\tau(P, i) = \tau(P, i') \Rightarrow \ell = \ell'$.*

Intuitively, ($\mathbb{M}, \tau$)-non-interference labels the leakage ℓ as a public output (which must be proved independent of secret information), whereas τ is used to specify which inputs of P are considered public. By extension, those inputs that are *not* revealed by τ are considered secret, and are not constrained in any way during either executions. Note that the leakage produced by a ($\mathbb{M}, \tau$)-non-interferent program for some input i can be predicted given only the public information revealed by $\tau(P, i)$: one can simply choose the remaining part of the input arbitrarily, constructing some input i' such that $\tau(P, i) = \tau(P, i')$. In this case, ($\mathbb{M}, \tau$)-non-interference guarantees that the leakage traces produced by $\mathbb{M}$ when executing P on i and i' are equal.

We can now specialize this notion of leakage security to symmetric encryption.

Definition 3 (Leakage-secure implementation generation for SKE). *An implementation generation procedure $\mathcal{C}$ produces $\mathbb{M}$-leakage-secure implementations for SKE if, for all SKE specifications Π written in $\mathcal{L}_1$, we have that the generated $\mathcal{L}_2$ implementation $(\mathsf{Gen}^*, \mathsf{Enc}^*, \mathsf{Dec}^*) = \mathcal{C}(\Pi)$ is $(\mathbb{M}, \tau_{\mathsf{SKE}})$-non-interferent.*

---

[9] For simplicity, the length of random inputs is assumed to be fixed by the algorithm itself.

PUTTING THE PIECES TOGETHER. The following lemma, shows that applying a correct and leakage secure implementation generation procedure to a black-box secure SKE specification is sufficient to guarantee implementation security.

Theorem 2. *Let $\mathcal{C}$ be correct and produce $\mathbb{M}$-leakage-secure implementations. Then, for all SKE scheme Π that is correct, IND\$-CPA-, INT-PTXT- and INT-CTXT-secure, the implementation $\Pi^* = \mathcal{C}(\Pi)$ is $\mathbb{M}$-correct, $\mathbb{M}$-IND\$-CPA-, $\mathbb{M}$-INT-PTXT- and $\mathbb{M}$-INT-CTXT-secure with the same advantages.*

Proof. Correctness of Π^* follows directly from that of $\mathcal{C}$ and Π. The security proofs are direct reductions. We only detail the proof of $\mathbb{M}$-IND\$-CPA, but note that a similar proof can be constructed for $\mathbb{M}$-INT-PTXT and $\mathbb{M}$-INT-CTXT. Given an implementation adversary $\mathcal{A}$, we construct an adversary $\mathcal{B}$ against Π as follows. Adversary $\mathcal{B}$ runs Gen^* on an arbitrary randomness of appropriate size to obtain the leakage ℓ_{Gen} associated with key generation and runs adversary $\mathcal{A}$ on ℓ_{Gen}. Oracle queries made by $\mathcal{A}$ are simulated by using $\mathcal{B}$'s specification oracles to obtain outputs, and the same leakage simulation strategy to present a perfect view of the implementation leakage to $\mathcal{A}$. When $\mathcal{A}$ outputs its guess, $\mathcal{B}$ forwards it as its own guess. We now argue that $\mathcal{B}$'s simulation is perfect. The first part of the argument relies on the correctness of the implementation generation procedure, which guarantees that the values obtained by $\mathcal{B}$ from its oracles in the CPA-game are identically distributed to those that $\mathcal{A}$ would have received in the implementation game. The second part of the argument relies on the fact that leakage-secure implementation generation guarantees that $\mathcal{B}$ knows enough about the (unknown) inputs to the black-box algorithms (the information specified by τ_{SKE}) to predict the exact leakage that such inputs would produce in the implementation model. Observe for example that, in the case of decryption leakage, the adversary $\mathcal{B}$ only needs the input ciphertext c to be able to exactly reproduce the leakage ℓ_{Dec}. Finally, note that the running time of the constructed adversary $\mathcal{B}$ is that of adversary $\mathcal{A}$ where each oracle query $\mathcal{A}$ introduces an overhead of one execution of the implementation in machine $\mathbb{M}$ (which can reasonably be assumed to be close to that of the specification).   □

5   Implementation Security of MEE-CBC

We now return to our case study, and explain how to use the methodology from Sect. 4, instantiated with existing verification and compilation tools, to derive assembly-level correctness and security properties for MEE-CBC$_{\mathrm{x86}}$.

PROOF STRATEGY. We first go briefly over each of the steps in our proof strategy, and then detail each of them in turn in the remainder of this section. In the first step, we specify and verify the correctness and black-box computational security of the MEE-CBC construction using EasyCrypt. In a second step, we use Frama-C to prove the functional correctness of program MEE-CBC$_{\mathrm{C}}$ with respect to the EasyCrypt specification. Finally, we focus on the x86 assembly code generated by CompCert (MEE-CBC$_{\mathrm{x86}}$), and prove: i. its functional correctness with respect to the C code (and thus the top-level EasyCrypt specification); and ii. its leakage security. An instantiation of Theorem 2 allows us to conclude the proof of Theorem 1.

BLACK-BOX SPECIFICATION SECURITY. We use EasyCrypt to prove that the MEE-CBC construction provides IND$-CPA security (when used with freshly and uniformly sampled IVs for each query) and INT-PTXT security.

Lemma 1 (Machine-checked MEE-CBC security). *The following two results hold:*

- *For all legitimate* IND$-CPA *adversary* $\mathcal{A}^{cpa}$ *that makes at most q queries, each of length at most n octets, to its* **RoR** *oracle, there exists an explicitly constructed PRP adversary* $\mathcal{B}^{prp}$ *that makes $q \cdot \lceil (n+1) / \lambda \rceil + 2$ queries to its forward oracle and such that:*

$$\mathbf{Adv}_{\Pi,\phi,\mathcal{A}}^{ind\$-cpa} \leq \mathbf{Adv}_{Perm,\mathcal{B}^{prp}}^{prp} + 2 \cdot \frac{(q \cdot \lceil \frac{n+1}{\lambda} \rceil + 2)^2}{2^{8 \cdot \lambda}},$$

  where $\phi(i) = \lceil (i+1) / \lambda \rceil + 3$ reveals only the number of blocks in the plaintext (and adds to it the fixed number of blocks due to IV and MAC tag).
- *For all* PTXT *adversary* $\mathcal{A}$ *that makes q_V queries to its* **Dec** *oracle, there exists an explicitly constructed* SUF-CMA *adversary* $\mathcal{B}^{cma}$ *that makes exactly q_V queries to its* **Ver** *oracle and such that:*

$$\mathbf{Adv}_{\Pi,\mathcal{A}}^{int-ptxt} \leq \mathbf{Adv}_{Mac,\mathcal{B}^{cma}}^{uf-cma}.$$

Our EasyCrypt specification relies on abstract algorithms for the primitives. More precisely, it is parameterized by an abstract, stateless and deterministic block cipher Perm with block size λ octets, and by an abstract, stateless and deterministic MAC scheme Mac producing tags of length $2 \cdot \lambda$.[10] The proofs, formalized in EasyCrypt, are fairly standard and account for all details of padding and message formatting in order to obtain the weak length-hiding property shown in this lemma. Running times for $\mathcal{B}^{prp}$ and $\mathcal{B}^{cma}$ are as usual.

We note that, although we have not formalized in EasyCrypt the proof of INT-CTXT security (this would imply a significant increase in interactive theorem proving effort) the known security results for MEE-CBC also apply to this specification and, in particular, it follows from [26] that it also achieves this stronger level of security when the underlying MAC and cipher satisfy slightly stronger security requirements.

IMPLEMENTATION GENERATION. Using Frama-C, a verification platform for C programs,[11] we prove functional equivalence between the EasyCrypt specification and our C implementation. Specifically, we use the deductive verification (WP) plugin to check that our C code fully and faithfully implements a functionality described in the ANSI/ISO C Specification Language (ACSL). To make sure

---

[10] This is only for convenience in these definitions.

[11] http://frama-c.com/.

that the ACSL specification precisely corresponds to the EasyCrypt specification on which black-box security is formally proved, we rely on Frama-C's ability to link ACSL logical constructs at the C annotation level to specific operators in underlying Why3 theories, which we formally relate to those used in the Easy-Crypt proof. This closes the gap between the tools by allowing us to refer to a common specification. Note that, since the abstract block cipher Perm and MAC scheme Mac are concretely instantiated in the C implementation, we instantiate $\lambda = 16$ (the AES block length in bytes) in this common specification and lift the assumptions on Perm and Mac to the C implementation of their chosen instantiation. We then use the CompCert certified compiler [24] to produce our final x86 assembly implementation.

To prove leakage security, we use the certifying information-flow type system for x86 built on top of CompCert [4], marking as public those inputs that correspond to values revealed by τ_{SKE}. Obtaining this proof does not put any additional burden on the user—except for marking program inputs as secret or public. However, the original C code must satisfy a number of restrictions in order to be analyzed using the dataflow analysis from [4]. Our C implementations were constructed to meet these restrictions, and lifting them to permit a wider applicability of our techniques is an important challenge for further work.[12]

PROOF OF THEOREM 1. Let us denote by $\mathcal{C}^{\mathcal{L}_{\mathsf{EC}} \to \mathsf{x86}}$ the implementation generation procedure that consists of hand-crafting a C implementation (annotated with τ_{SKE} consistent security types), equivalence-checking it with an EasyCrypt specification using Frama-C, and then compiling it to assembly using CompCert (accepting only assembly implementations that type-check under the embedded secure information-flow type system), as we have done for our use case. We formalize the guarantees provided by this procedure in the following lemma.

Lemma 2 (Implementation generation). $\mathcal{C}^{\mathcal{L}_{\mathsf{EC}} \to \mathsf{x86}}$ *is a* $\mathbb{M}^{\mathsf{CT}}_{\mathsf{x86}}$-*correct implementation generation procedure that produces* $\mathbb{M}^{\mathsf{CT}}_{\mathsf{x86}}$-*leakage secure SKE implementations.*

Proof. Correctness follows from the combination of the Frama-C functional correctness proof and the semantic preservation guarantees provided by CompCert. CompCert's semantics preservation theorem implies that the I/O behaviour of the assembly program exactly matches that of the C program. Functional equivalence checking using Frama-C yields that the C implementation has an I/O behaviour that is consistent with that of the EasyCrypt specification (under the C semantics adopted by Frama-C). Finally, under the reasonable assumption that

---

[12] In a recent development in this direction, Almeida et al. [3] describe a method, based on limited product programs, for verifying constant-time properties of LLVM code. Their method and the implementation they describe can deal with many examples that the type system from [4] cannot handle, including a less ad hoc version of our code and some of the OpenSSL code for MEE-CBC, whilst preserving a high degree of automation. In addition, their construction easily extends to situations where public outputs are needed to simulate the leakage trace.

the CompCert semantics of C are a sound refinement of those used in Frama-C, we obtain functional correctness of the assembly implementation with respect to the EasyCrypt specification. For leakage security, we rely on the fact that the information-flow type system of [4] enforces τ_{SKE}-non-interference and hence only accepts $(\mathbb{M}^{\mathsf{CT}}_{\mathsf{x86}}, \tau_{\mathsf{SKE}})$-leakage secure implementations. $\square$

Theorem 1 follows immediately from the application of Theorem 2 instantiated with Lemmas 1 and 2. Furthermore, foregoing machine-checking of the black-box specification security proof and simply accepting known results on MEE-TLS-CBC [26], we can also show that MEE-CBC$_{\mathsf{x86}}$ is $\mathbb{M}^{\mathsf{CT}}_{\mathsf{x86}}$-INT-CTXT-secure under slightly stronger black-box assumptions on AES128$_{\mathsf{NaCl}}$ and HMACSHA256$_{\mathsf{NaCl}}$.

6   Performance Comparison

We now characterize the different assurance/performance trade-offs of the timing mitigation strategies discussed in this paper. Figure 4 shows the time taken by 5 different implementations of MEE-CBC (one of them compiled in different ways) when decrypting a 1.5KB TLS1.2 record using the AES128-SHA256 ciphersuite.[13] More specifically, we consider code from s2n (#1) and OpenSSL (#2), and five different compilations of our formally verified MEE-CBC implementation (#3-7), focusing on raw invocations of MEE-CBC. It is clear that the s2n code (#1) benefits from its less strict timing countermeasures, gaining roughly 1.8× performance over OpenSSL's (semi-)constant-time implementation approach (#2). The figures for our verified implementation of MEE-CBC show both the cost of formal verification and the cost of full constant-time guarantees. Indeed, the least efficient results are obtained when imposing full code and data memory access independence from secret data (#4-6).

#	Implementation	Compiler	Clock Cycles	Time
1	s2n	GCC x86-64 -O2	14K	5μs
2	OpenSSL	GCC x86-64 -O2	23K	9μs
3	MEE-CBC$_\mathsf{C}$ (AES-NI)	CompCert x86-32	51K	21μs
4	MEE-CBC$_\mathsf{C}$	GCC x86-64 -O2	59M	25ms
5	MEE-CBC$_\mathsf{C}$	GCC x86-64 -O1	62M	26ms
6	MEE-CBC$_{\mathsf{x86}}$	CompCert x86-32	101M	42ms
7	MEE-CBC$_\mathsf{C}$	GCC x86-64 -O0	237M	99ms

Fig. 4. Performance comparison of various MEE-CBC implementations. (Median over 500 runs.)

---

[13] The numbers were obtained in a virtualized Intel x86-64 Linux machine with 4 GB RAM.

The assembly implementation produced using the constant-time version of CompCert (#6), is roughly 8400× slower than s2n, but still over twice as fast as unoptimized GCC. However, the fact that the same C code compiled with GCC -O2 (#4) is only 1.7× faster[14] than the fully verified CompCert-generated code shows that the bottleneck does not reside in verification, but in the constant-time countermeasures. Indeed, profiling reveals that NaCl's constant-time AES accounts for 97 % of the execution time. These results confirm the observations made in [12] as to the difficulties of reconciling resistance against cache attacks and efficiency in AES implementations. To further illustrate this point, we also include measurements corresponding to a modification of our MEE-CBC implementation that uses hardware-backed AES (#3). This cannot, in fairness, be compared to the other implementations, but it does demonstrate that, with current verification technology, the performance cost of a fully verified constant-time MEE-CBC implementation is not prohibitive.

7   Discussions

ON RANDOMNESS. Restricting our study to deterministic programs with an argument containing random coins does not exclude the analysis of real-world systems. There, randomness is typically scarce and pseudorandom generators are used to expand short raw high-entropy bitstrings into larger random-looking strings that are fed to deterministic algorithms, and it is common to assume that the small original seed comes from an ideal randomness source, as is done in this paper. Our approach could therefore be used to analyze the entire pseudo-random generation implementation, including potential leakage-related vulnerabilities therein.

ON LENGTH-HIDING SECURITY. Existing implementations of MEE-TLS-CBC (and indeed our own implementation of MEE-CBC) are not length-hiding as defined in [26] in the presence of leakage. Indeed, the constant-time countermeasures are only applied in the decryption oracle and precise information about plaintext lengths may be leaked during the execution of the encryption oracle. Carrying length-hiding properties down to the level of those implementations may therefore require, either the implementation to be modified (and the Frama-C equivalence proof adapted accordingly), or the specification of implementation security to more closely reflect particular scenarios–such as the TLS record layer–where it may be difficult for the adversary to make chosen-plaintext queries, but easy to make padding and verification oracle queries. In any case, Lemma 1 does capture the length-hiding property given by our choice of minimal padding, and could be adapted to capture the more general length-hiding property of Paterson, Ristenpart and Shrimpton [26] by making padding length a public choice.

---

[14] This is in line with general CompCert benchmarks.

LEAKAGE SIMULATION AND WEAKER NON-INTERFERENCE NOTIONS. Our use of leakage security in proving that leakage is not useful to an adversary naturally generalizes to a notion of *leakage simulation*, whereby an implementation is secure as long as its leakage can be efficiently and perfectly simulated from its public I/O behaviour, including its public *outputs*. For example, an implementation of Encrypt-then-MAC that aborts as soon as MAC verification fails, but is otherwise fully constant-time should naturally be considered secure,[15] since the information gained through the leakage traces is less than that gained by observing the output of the **Ver** oracle. The more general notion of leakage simulation informally described here would capture this and can be related to weaker notions of non-interference, where equality on low outputs is only required on traces that agree on the value of public outputs. Theorem 2 can be modified to replace leakage security with the (potentially weaker) leakage simulation hypothesis.

8 Conclusions and Directions for Future Work

Our proposed methodology allows the derivation of strong security guarantees on assembly implementations from more focused and tractable verification tasks. Each of these more specialized tasks additionally carries its own challenges.

Proving security in lower-level leakage models for assembly involves considering architectural details such as memory management, scheduling and data-dependent and stateful leakage sources. Automatically relating source and *existing* assembly implementations requires developing innovative methods for checking (possibly conditional or approximate) equivalences between low-level probabilistic programs. Finally, obtaining formal proofs of computational security and functional correctness in general remain important bottlenecks in the proof process, requiring high expertise and effort. However, combining formal and generic composition principles (such as those used in our case study) with techniques that automate these two tasks for restricted application domains [5,11,20] should enable the formal verification of extensive cryptographic libraries, in the presence of leakage. We believe that this goal is now within reach.

On the cryptographic side, the study of computational security notions that allow the adversary to tamper with the oracle implementation [10] may lead to relaxed functional correctness requirements that may be easier to check, for example by testing. Extensions of our framework to settings where the adversary has the ability to tamper with the execution of the oracle are possible, and would allow it to capture recent formal treatments of countermeasures against fault injection attacks [27].

---

[15] Some anonymity properties, such as untraceability, may require the cause of decryption failure to remain secret in the black-box model, in which case leakage must not reveal it either [17].

Acknowledgements. The first two authors were funded by Project "TEC4Growth - Pervasive Intelligence, Enhancers and Proofs of Concept with Industrial Impact/ NORTE-01-0145-FEDER-000020", which is financed by the North Portugal Regional Operational Programme (NORTE 2020), under the PORTUGAL 2020 Partnership Agreement, and through the European Regional Development Fund (ERDF). The third and fourth authors were supported by projects S2013/ICE-2731 N-GREENS Software-CM and ONR Grants N000141210914 (AutoCrypt) and N00014151 2750 (SynCrypt). The fourth author was supported by FP7 Marie Cure Actions-COFUND 291803 (Amarout II). The machine-checked proof for CBC improves on a script by Benjamin Grégoire and Benedikt Schmidt. Pierre-Yves Strub provided support for extracting Why3 definitions from **EasyCrypt** specifications. We thank Mathias Pedersen and Bas Spitters for useful comments.

References

1. Albrecht, M.R., Paterson, K.G.: Lucky microseconds: a timing attack on Amazon's s2n implementation of TLS. Cryptology ePrint Archive, report 2015/1129 (2015). http://eprint.iacr.org/
2. AlFardan, N.J., Paterson, K.G.: Lucky thirteen: breaking the TLS and DTLS record protocols. In: IEEE Symposium on Security and Privacy SP 2013, pp. 526–540. IEEE Computer Society (2013)
3. Almeida, J., Barbosa, M., Barthe, G., Dupressoir, F., Emmi, M.: Verifying constant-time implementations. Manuscript (2015). https://fdupress.net/files/ctverif.pdf
4. Barthe, G., Betarte, G., Campo, J.D., Luna, C.D., Pichardie, D.: System-level non-interference for constant-time cryptography. In: Ahn, G.-J., Yung, M., Li, N. (eds.) ACM CCS 2014, pp. 1267–1279. ACM Press, November 2014
5. Barthe, G., Crespo, J.M., Grégoire, B., Kunz, C., Lakhnech, Y., Schmidt, B., Béguelin, S.Z.: Fully automated analysis of padding-based encryption in the computational model. In: Sadeghi, A.-R., Gligor, V.D., Yung, M. (eds), ACM CCS 2013, pp. 1247–1260. ACM Press, November 2013
6. Barthe, G., Dupressoir, F., Grégoire, B., Kunz, C., Schmidt, B., Strub, P.-Y.: EasyCrypt: a tutorial. In: Aldini, A., Lopez, J., Martinelli, J. (eds.) FOSAD 2013, pp. 146–166. Springer, Heidelberg (2014)
7. Barthe, G., Grégoire, B., Heraud, S., Béguelin, S.Z.: Computer-aided security proofs for the working cryptographer. In: Rogaway, P. (ed.) CRYPTO 2011. LNCS, vol. 6841, pp. 71–90. Springer, Heidelberg (2011)
8. Barwell, G., Page, D., Stam, M.: Rogue decryption failures: reconciling AE robustness notions. In: Groth, J., et al. (eds.) IMACC 2015. LNCS, vol. 9496, pp. 94–111. Springer, Heidelberg (2015). doi:10.1007/978-3-319-27239-9_6
9. Bellare, M., Namprempre, C.: Authenticated encryption: relations among notions and analysis of the generic composition paradigm. In: Okamoto, T. (ed.) ASIACRYPT 2000. LNCS, vol. 1976, pp. 531–545. Springer, Heidelberg (2000)
10. Bellare, M., Paterson, K.G., Rogaway, P.: Security of symmetric encryption against mass surveillance. In: Garay, J.A., Gennaro, R. (eds.) CRYPTO 2014, Part I. LNCS, vol. 8616, pp. 1–19. Springer, Heidelberg (2014)
11. Bernstein, D., Schwabe, P.: Cryptographic software, side channels, and verification. In: COST CryptoAction WG3 Meeting, April 2015
12. Bernstein, D.J.: AES timing variability at a glance (2015). http://cr.yp.to/mac/variability1.html. Accessed 25 Oct 2015

13. Bernstein, D.J.: Cache-timing attacks on AES (2005). Author's webpage
14. Bernstein, D.J., Lange, T., Schwabe, P.: The security impact of a new cryptographic library. In: Hevia, A., Neven, G. (eds.) LatinCrypt 2012. LNCS, vol. 7533, pp. 159–176. Springer, Heidelberg (2012)
15. Boldyreva, A., Degabriele, J.P., Paterson, K.G., Stam, M.: On symmetric encryption with distinguishable decryption failures. In: Moriai, S. (ed.) FSE 2013. LNCS, vol. 8424, pp. 367–390. Springer, Heidelberg (2014)
16. Canvel, B., Hiltgen, A.P., Vaudenay, S., Vuagnoux, M.: Password interception in a SSL/TLS channel. In: Boneh, D. (ed.) CRYPTO 2003. LNCS, vol. 2729, pp. 583–599. Springer, Heidelberg (2003)
17. Chothia, T., Smirnov, V.: A traceability attack against e-Passports. In: Sion, R. (ed.) FC 2010. LNCS, vol. 6052, pp. 20–34. Springer, Heidelberg (2010)
18. Degabriele, J.-P., Paterson, K.G., Watson, G.J.: Provable security in the real world. IEEE Secur. Priv. **9**(3), 33–41 (2011)
19. Goguen, J.A., Meseguer, J.: Security policies and security models. In: 1982 IEEE Symposium on Security and Privacy, Oakland, CA, USA, April 26–28, pp. 11–20. IEEE Computer Society (1982)
20. Hoang, V.T., Katz, J., Malozemoff, A.J.: Automated analysis and synthesis of authenticated encryption schemes. Cryptology ePrint Archive, report 2015/624 (2015). http://eprint.iacr.org/2015/624
21. Käsper, E., Schwabe, P.: Faster and timing-attack resistant AES-GCM. In: Clavier, C., Gaj, K. (eds.) CHES 2009. LNCS, vol. 5747, pp. 1–17. Springer, Heidelberg (2009)
22. Krawczyk, H.: The order of encryption and authentication for protecting communications (or: how secure is SSL?). In: Kilian, J. (ed.) CRYPTO 2001. LNCS, vol. 2139, pp. 310–331. Springer, Heidelberg (2001)
23. Langley, A.: Lucky thirteen attack on TLS CBC. Imperial violet, February 2013. https://www.imperialviolet.org/2013/02/04/luckythirteen.html. Accessed 25 Oct 2015
24. Leroy, X.: Formal certification of a compiler back-end, or: programming a compiler with a proof assistant. In: ACM Symposium on Principles of Programming Languages POPL 2006 (2006)
25. Maurer, U., Tackmann, B.: On the soundness of Authenticate-then-Encrypt: formalizing the malleability of symmetric encryption. In: Al-Shaer, E., Keromytis, A.D., Shmatikov, V. (eds.) ACM CCS 2010, pp. 505–515. ACM Press, October 2010
26. Paterson, K.G., Ristenpart, T., Shrimpton, T.: Tag size *Does* matter: attacks and proofs for the TLS record protocol. In: Lee, D.H., Wang, X. (eds.) ASIACRYPT 2011. LNCS, vol. 7073, pp. 372–389. Springer, Heidelberg (2011)
27. Rauzy, P., Guilley, S.: A formal proof of countermeasures against fault injection attacks on CRT-RSA. J. Crypt. Eng. **4**(3), 173–185 (2014)
28. Schmidt, S.: Introducing s2n, a new open source TLS implementation, June 2015. https://blogs.aws.amazon.com/security/post/TxCKZM94ST1S6Y/Introducing-s2n-a-New-Open-Source-TLS-Implementation. Accessed 25 Oct 2015
29. Vaudenay, S.: Security flaws induced by CBC padding - applications to SSL, IPSEC, WTLS. In: Knudsen, L.R. (ed.) EUROCRYPT 2002. LNCS, vol. 2332, pp. 534–546. Springer, Heidelberg (2002)

White-Box Cryptography in the Gray Box
– A Hardware Implementation and its Side Channels –

Pascal Sasdrich[1(✉)], Amir Moradi[1], and Tim Güneysu[2]

[1] Horst Görtz Institute for IT Security, Ruhr-Universität Bochum, Bochum, Germany
{Pascal.Sasdrich,Amir.Moradi}@rub.de
[2] University of Bremen and DFKI, Bremen, Germany
tim.gueneysu@uni-bremen.de

Abstract. Implementations of white-box cryptography aim to protect a secret key in a white-box environment in which an adversary has full control over the execution process and the entire environment. Its fundamental principle is the map of the cryptographic architecture, including the secret key, to a number of encoded tables that shall resist the inspection and decomposition of an attacker. In a gray-box scenario, however, the property of hiding required implementation details from the attacker could be used as a promising mitigation strategy against side-channel attacks (SCA). In this work, we present a first white-box implementation of AES on reconfigurable hardware for which we evaluate this approach assuming a gray-box attacker. We show that – unfortunately – such an implementation does not provide sufficient protection against an SCA attacker. We continue our evaluations by a thorough analysis of the source of the observed leakage, and present additional results which can be used to build stronger white-box designs.

1 Introduction

Initially the field of white-box cryptography was mainly motivated by applications of the field of Digital Rights Management (DRM) that aims to protect a secret key in a white-box environment, where an adversary has full control over the execution process and the environment of a cryptographic implementation. However, with the widespread emerging of embedded and pervasive computing devices implementing cryptographic functions and primitives, the threat of white-box adversaries is no longer limited to cryptographic software implementations. Although, an adversary might be limited by the gray-box model in practice (i.e., he cannot control the execution process and the environment entirely), Side-Channel Analysis (SCA) attacks are well-known to be used to exploit information leakage related to the device internals e.g., by analyzing power consumption or electromagnetic radiations (EM). Still, for successfully mounting such physical attacks, the attacker requires at least some knowledge about the internals in order to build adequate hypotheses that can be used, for example, for key extraction. In this context the nature of white-box cryptography that effectively disguising all internals and the secret key from the attacker

© International Association for Cryptologic Research 2016
T. Peyrin (Ed.): FSE 2016, LNCS 9783, pp. 185–203, 2016.
DOI: 10.1007/978-3-662-52993-5_10

by encoding them into tables, seems to yield some inherent resistance against such physical attacks.

Previous Works: In 2002, first white-box implementations for DES [9] and AES [10] were proposed by Chow et al. in order to protect a secret key within a cryptographic implementation in presence of a white-box adversary. However, these seminal proposals and their implementations were soon shown to be vulnerable to differential cryptanalysis [13,22] as well as algebraic cryptanalytic attacks [3,16,17]. This led to some new proposals for white-box implementations of AES. In 2009, Xiao et al. in [23] proposed a variant of the design of Chow et al. using larger linear encodings, for which again a vulnerability against algebraic cryptanalytic attacks was identified in [20]. Other approaches suggest to build white-box AES implementations using pertubations [7] (which was broken in [21]) or based on the concept of dual-ciphers [14].

Recent work in [2] aims to generalize and formalize notions for white-box cryptography and related attacks for any SLT cipher presenting general attack strategies and upper bounds for their complexity. Besides the vulnerabilities against differential and algebraic cryptanalysis, Bos et al. in [4] showed that secret keys of existing white-box implementations can be extracted by observing the addresses which are accessed during the execution if the external encodings are known to the adversary. The underlying so-called Differential Computational Analysis (DCA) applies the concept of Differential Power Analysis (DPA) [15] on eavesdropped address bits.

A first white-box implementation in hardware has been proposed for the NOEKEON cipher in [6,8] using 1-bit linear nibble encodings (i.e. masking with deterministic masks).

Our Contribution: In this work we propose a white-box implementation of AES dedicated to reconfigurable hardware. Although the white-box implementation of Chow et al. initially was proposed for software implementations, we show that the implementation can be mapped to existing reconfigurable hardware architectures. Note that only recent generations of reconfigurable hardware devices provide adequate amounts of resources to cope with the large memory requirements of white-box implementations.

For this hardware implementation we next examine the vulnerability to SCA attacks assuming a gray-box adversary model. These results, obtained from an FPGA platform extend the observation by Bos et al. (in [4]). We show that SCA attacks such as classical DPA can reveal the secrets in hardware implementations applying white-box cryptography even in gray-box settings.

Finally, we perform a thorough mathematical investigation and analysis of the construction of look-up tables used in white-box cryptography. We explain and verify the reason behind the success of such (DCA and DPA) attacks what has not been addressed in the seminal work of Bos et al.. Our results give a better understanding of the mathematical foundations of these attacks which can pave the way for improved future white-box designs and implementations that are resistant against such analyses and threats.

Outline: The remainder of this article is organized as follows: Sect. 2 introduces the basic concept of white-box cryptography and gives a detailed explanation of the white-box implementation of Chow et al. including design and construction approach and known attacks and vulnerabilities. The process of transforming this white-box implementation into a hardware architecture (realized on an FPGA) is described in Sect. 3. In Sect. 4 we deal with gray-box adversary model and SCA attacks. We recap the concept of DCA and pinpoint the source of leakage of the given AES white-box implementation before we conclude in Sect. 5.

2 Background

This section introduces the basic concept of white-box cryptography and gives a detailed description of the seminal AES white-box implementation of Chow et al.

2.1 White-Box Cryptography

Cryptographic algorithms are designed to enable a secure communication even in the presence of an attacker. Nowadays, cryptographers differentiates between three common attacker models which try to estimate and model the capabilities of an adversary. Usually, modern cryptographic algorithms and their implementations are analyzed within such attacker models in order to deduce and estimate their security.

The traditional security and attacker model is the so-called black-box model which assumes a trusted execution environment and secure communication endpoints. In this model, cryptographic implementations are considered as black-box where an adversary can only observe the input and output behavior.

Since the development and deployment of embedded systems for security purposes the black-box model has been superseded by the gray-box model. This model includes the black-box settings but in addition assumes some expanded capabilities of a possible attacker. Cryptographic implementations are no longer considered as black-box but instead an adversary has limited access to the implementation internals which can be used to break the implementation. Note that gray-box attacks (e.g., SCA attacks) usually focus and target cryptographic implementations rather than cryptographic algorithms which still should be secure under the assumption of the black-box model.

However, another attacker model called white-box model has been introduced in particular for software implementations of cryptographic algorithms. For this model, the capabilities of an adversary are virtually unlimited since the attacker is assumed to have full control over the implementation and its execution environment. Aim of any implementation considered to be secure under the white-box model is to behave as a virtual black-box to any kind of attacker such that even a white-box attacker should not have any additional advantage over black-box attackers. The ideal white-box implementation would consists of a single look-up table mapping a plaintext to its specific ciphertext already including a (hidden) secret key. Obviously this is impractical for modern ciphers with block and key sizes of 128 bits or more. An alternative approach is to transform the cryptographic primitive into a functionally-equivalent implementation using a series of smaller look-up tables. In a further step, secret and invertible encodings are applied to each look-up table individually in order to protect and hide secret key materials.

In general, the strategy for the design of white-box implementations of a round-based symmetric block cipher can be depicted as:

$$\underbrace{(f^{(r+1)})^{-1} \circ E^r \circ f^r}_{table} \circ \cdots \circ \underbrace{(f^{(3)})^{-1} \circ E^2 \circ f^2}_{table} \circ \underbrace{(f^{(2)})^{-1} \circ E^1 \circ f^1}_{table}$$

$$= (f^{(r+1)})^{-1} \circ E^r \circ \cdots \circ E^2 \circ E^1 \circ f^1 = (f^{(r+1)})^{-1} \circ E_K \circ f^1,$$

where $E^{i \in \{1 \dots r\}}$ is a single round instance of the block cipher and f^1 respectively $(f^{r+1})^{-1}$ are considered as external input and output encoding of the white-box implementation (in order to prevent Code Lifting attacks [11]).

The white-box model has initially been proposed by Chow et al. [9] in 2002 when focusing on a fixed key implementation of the DES algorithm, and shortly afterwards a white-box implementation of the AES algorithm was presented [10]. In the following, we first introduce this seminal AES white-box implementation and discuss the design principles and known attacks and vulnerabilities under the white-box model before we show how to implement this design in hardware.

2.2   White-Box Implementation of AES

The architecture presented in [10] is a fixed key implementation with a fully unrolled design merging the atomic operations into a series of look-up tables. Basic design goals of this construction are to hide the key and algorithm structure through implementing the algorithm as a network of randomized look-up tables. Each look-up table is encoded and protected individually using random linear and non-linear bijections. Since a detailed discussion of the design would exceed the scope of this work we refer the interested reader to [19] and restrict the discussion of the white-box implementation to its basic design principle and construction.

Design and Construction: The transformation of an unprotected AES implementation (independently of the used key size) into a white-box protected fixed key implementation according to the scheme of Chow et al. can be achieved in two phases: first, the AES algorithm has to be rewritten and translated as a series of look-up tables and second, secret but invertible encodings have to be applied to all look-up tables in order to build a white-box implementation.

The following section will describe this process exemplary for the case of AES-128 as presented in [10], but again subdividing each phase into two steps.

In the following, we use the lower-case letter x for single bytes of the intermediate round state, $\hat{k}$ for a single byte of a round key, a raising index r for the current round and lowering indices (i,j) for the current byte position in the state matrix, where i denotes the row index and j the column index. Functions are represented with sans serif fonts. The AES S-box is denoted with S(.) and the matrix of the *MixColumns* operation is denoted by MC.

Step 1: Partial Evaluation. In the first step, the S-box computation is combined with the preceding addition of the round key. Merging both operations yields into a single look-up table defined as T-box:

$$\mathsf{T}_{i,j}^r(x) = \mathsf{S}(x \oplus \hat{k}_{i,j}^r) \qquad\qquad \text{for } 0 \leq i,j \leq 3 \text{ and } 1 \leq r \leq 9$$

$$\mathsf{T}_{i,j}^{10}(x) = \mathsf{S}(x \oplus \hat{k}_{i,j-i}^{10}) \oplus \hat{k}_{i,j}^{11} \qquad\qquad \text{for } 0 \leq i,j \leq 3$$

This step results in 160 different key-dependent T-boxes. It should be noted, that the T-boxes of the last round incorporate two bytes of two different round keys. This is due to the missing *MixColumns* operation and the final post-whitening key addition.

Step 2: Matrix Partitioning. A well-known implementation technique for the *MixColumns* operation is to decompose it into four different 8×32-bit look-up tables using the matrix partitioning strategy. Eventually, four 32-bit table outputs are added, resulting in the original *MixColumns* transformation. Applying this approach to our previously constructed T-boxes gives us a new set of different TMC tables, where MC_i denotes the i-th column of the MC matrix:

$$\mathsf{TMC}_{i,j}^r(x) = \mathsf{MC}_i \circ \mathsf{T}_{i,j}^r(x) \qquad\qquad \text{for } 0 \leq i,j \leq 3 \text{ and } 1 \leq r \leq 9$$

Finally, this results in 144 different 8×32-bit TMC look-up tables and additionally 16 different 8×8-bit T-boxes for the last round. Since all look-up tables comprise a small portion of the secret key, they have to be protected against attackers aiming at extracting the secret. For a better illustration, the key-dependent tables can be seen as miniature block ciphers that have to be enhanced by well-known techniques such as *diffusion* and *confusion* for protection purposes. Before applying randomly chosen invertible non-linear white-box encodings to the key-dependent tables in order to achieve *confusion*, *diffusion* is achieved through the application of linear transformations[1] called mixing bijections.

Step 3: Mixing Bijections. To add diffusion to each key-dependent table, two different linear transformations are necessary: an 8×8-bit linear transformations

---

[1] Note that originally affine and non-affine transformations were considered. However, since the constant of any affine transformation can be combined with the non-affine mapping, this eventually behaves as linear transformations.

$L_{i,j}^r$ is inserted before $TMC_{i,j}^r$, and a 32×32-bit transformation R_i^r is applied afterwards. In order to cancel out the effect of the transformation R_i^r after the addition of the TMC output values, another untwist table is introduced after each TMC table. This untwist table takes care of canceling the effect of the transformation R_i^r and applying new 8×8-bit transformations $(L_{i,j}^{r+1})^{-1}$ to keep the encryption process consistent during all rounds. These transformations can be found by randomly creating linear matrices and checking for invertibility.

Step 4: Nibble Encodings. Eventually, non-linear white-box encodings are applied to all table inputs and outputs. For the sake of efficiency, concatenation of 4-bit nibble encodings were chosen rather than 8-bit byte encodings. Since these non-linear encodings avoid linear operation over the TMC table outputs, dedicated tables for the XOR operations have to be introduced. These nibble encodings can be found by constructing random 4-bit permutations. All in all, this design strategy results in five different look-up tables that are defined as follows:

$\mathcal{L}$-Ia: $N_{out} \circ R_i^1 \circ TMC_{i,j}^1 \circ (F_{i,j})^{-1}$ (8×32-bit)

$\mathcal{L}$-Ib: $G_{i,j} \circ T_{i,j}^{10} \circ L_{i,j}^{10} \circ (N_{in})^{-1}$ (8×8-bit)

$\mathcal{L}$-II: $N_{out} \circ R_i^r \circ TMC_{i,j}^r \circ L_{i,j}^r \circ (N_{in})^{-1}$ (8×32-bit)

$\mathcal{L}$-III: $N_{out} \circ (L_{i,j}^{r+1})^{-1} \circ (R_i^r)^{-1} \circ (N_{in})^{-1}$ (8×32-bit)

$\mathcal{L}$-IV: $N_{in} \circ \mathcal{L}_{\oplus} \circ (N_{out})^{-1}$ (8×4-bit)

Combining these tables in their designated way (a single round is depicted in Fig. 1) results in an encoded fixed-key white-box AES instantiation

$$AES_K' = G \circ AES_K \circ F^{-1},$$

where F^{-1} and G are responsible for external input and output encodings respectively.

Known Attacks and Vulnerabilities: Below we briefly outline the known attacks and vulnerabilities of the above presented white-box AES implementation. Some of the threats were already considered during its design. For those, we additionally explain how the attacks were targeted and how the countermeasures were integrated.

Code Lifting Attacks. Since the secret key is hidden and integrated into the white-box implementation, the goal of an attacker is obviously to extract the secret key. However, such fixed-key white-box implementations suffer from another kind of threat where an attacker is not interested in extracting the secret key but instead cloning the entire white-box implementation in order to use it at another place. This threat is known as "'Code Lifting"' where the entire white-box application is seen as a single key that is cloned and misused by an attacker to encrypt and decrypt data without being in possession of the

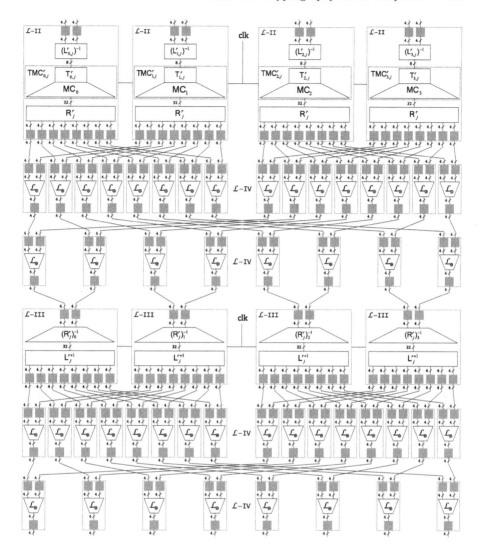

Fig. 1. White-box implementation of a quarter AES round

secret key. To avoid such kind of attacks, external encodings (F and G) are introduced, turning an white-box implementation E_K into an obfuscated encryption function $E'_K = G \circ E_K \circ F^{-1}$ with hidden external encodings. By pushing the white-box implementation boundaries, the attacker is no longer able to misuse the white-box implementation as long as the external encodings are unknown.

White-Box Inversion. Besides cloning the white-box implementation through Code Lifting, inverting the encryption (or decryption) function is another practical issue in particular for white-box implementations of AES. Since the entire algorithm is implemented through look-up tables, any white-box attacker would be able to extract the tables and compute the inverses of all rounds. This allows to turn any implemented encryption (respectively decryption) function into an decryption (respectively encryption) without knowing the secret key. In fact, this issue cannot be prevented but mitigated by external encodings since it prevents the attacker to use the inverted function in a meaningful way. In particular the application of non-invertible external encodings can prevent the inversion of white-box implementations ensuring the property of *one-wayness*.

Stripping of Non-Linear Encodings. A first algebraic analysis of the above-explained white-box AES implementation has been presented by Billet et al. [3] which revealed serious vulnerabilities of this design approach by stripping of the non-linear encodings of the look-up tables and allowing a white-box attacker to efficiently extract the embedded secret key. Later, Michiels et al. [17] generalized this attack for any cipher following the substitution-linear transformation (SLT) approach. In general, Billet et al.'s approach considers a quarter of the AES round function (depicted in Fig. 1) as a single 32×32-bit function rather than a decomposition into a series of look-up tables. Following this strategy, the influence of the mixing bijection R_i^r and any other internal (non-linear) encoding are canceled out.

It was observed, that with moderate computational effort, the non-linear encodings at the beginning and end of each quarter AES round can be removed, so that only some (unknown) affine transformation will remain. Applying this technique to three subsequent rounds, thus removing the non-linear encodings up to an affine part, the secret key eventually can be retrieved with a complexity of at maximum 2^{30} (cf. [3]). Note, however, that this attack is only possible in the setting of white-box adversaries, since an attacker needs to have full access to the tables and control over their inputs and outputs.

3   FPGA Implementation

This section briefly introduces modern reconfigurable hardware architectures exemplary considering Xilinx FPGAs and describes necessary hardware resources to implement white-box cryptography efficiently in reconfigurable hardware. Afterwards, the approach of transforming the white-box AES implementation of Chow et al. into an efficient hardware architecture for recent Xilinx Kintex-7 FPGAs is outlined. Finally, we give performance and implementation results on the area and throughput efficiency of the proposed architecture.

3.1   Hardware Resources

Modern FPGAs consist of a sea of general-purpose logic resources that can implement arbitrary circuits of Boolean functions using small look-up tables. The logic resources are arranged in an extremely regular array-like structure and enhanced by special purpose units e.g., Digital Signal Processors (DSPs) or

Block Memories (BRAMs). The reconfigurable devices are programmed using a configuration file called *bitstream* that contains all configuration information for implemented hardware resources, i.e., the programmable interconnections, the general purpose logic and the special purpose resources.

General Purpose Logic Resources: Xilinx decided to cluster several general purpose logic resources as Configurable Logic Blocks (CLBs) and arrange them in a grid-like structure of rows and columns. Starting with the Virtex-5 family of Xilinx devices, each CLB constitutes two slices each equipped with four 6-input Look-Up Tables (LUTs) and eight adjacent Flip-Flops (FFs) to implement any circuit of Boolean functions. Starting with the newer 7-series devices, only two different types of slices (Slice-L and Slice-M) were implemented which only differ in capabilities of using LUTs as distributed memory instead of function generators. Both, Slice-L and Slice-M instances, provide some wide multiplexers that allow to connect the outputs of the LUTs in order to implement any 8 × 1-bit Boolean function efficiently into a single slice.

Dedicated Block Memory Resources: Besides general purpose logic that can also serve as (distributed) memory, modern FPGAs provide larger amounts of data storage in terms of dedicated BRAMs. These flexible, low-power memory units can be configured by the user and provide between 16-Kbits to 32-Kbits accessible in single or dual port mode (additionally, 2-Kbits respectively 4-Kbits memory for parity check purposes are available). In dual port mode, two fully independent ports providing read and write access (even with different clocks) can be used to access or manipulate data that is stored in memory. In addition, each BRAM can be configured individually and used in different configurations considering port width and memory depth, ranging from 32K × 1-bit to 1K × 32-bit entries.

3.2 White-Box Architecture in Hardware

White-box cryptography was initially proposed to protect software implementations. In this context we like to remark that bitstream configuration files of FPGA designs are digital binary files that are stored in external memory (that are accessible for an attacker) and thus exposed to very similar threats. Further, the basic idea of white-box implementations is to transform a cryptographic implementation into a series of look-up tables. This perfectly fits the regular structure of FPGAs implementing arrays of look-up tables with programmable interconnections. Hence we can conclude that FPGAs seem to be a very good fit for cryptographic white-box implementations in hardware.

However, since every individual look-up table of the white-box implementation is different (due to different round keys and randomly chosen encodings), we cannot implement any area-efficient round-based or serialized architecture of the AES algorithm nor reuse any of the look-up tables. Instead, we have to implement an entirely unrolled implementation with every round instantiated separately. Due to the application of BRAM primitives, which have a latency of a single clock cycle, this causes an initial latency of 19 clock cycles (due to 19 stages of 16 parallel look-up tables in the proposed white-box implementation) but in order to increase the throughput it is possible to operate the encryption architecture in a pipeline fashion providing ciphertexts at each clock cycle (after the initial latency).

Mapping Tables into CLBs: Besides the implementation of the key depending TMC-Tables and T-boxes, the encoded look-up tables to perform the XOR operations consume a large part of the required storage. Although modern FPGAs provide large amounts of general purpose data storage in terms of BRAM, implementing all look-up tables using these dedicated memory primitives is still not feasible. Therefore, some tables have to be transferred to the general purpose logic in order to fit the design into an FPGA. Since any 8×1-bit Boolean function can be implemented efficiently into a single slice and each XOR operation and its corresponding look-up table can be decomposed into four different 8×1-bit functions, it is a natural choice to implement these tables in general purpose logic. In total, each XOR-table can be implemented using four slices equipped with 4 LUTs each, thus in total 16 LUT instances are required (this equals 1024-bit memory). Fortunately, the last round can do without XOR operations, so we only have to implement these tables for 9 rounds. As depicted in Fig. 1, a quarter round of the AES white-box implementation implements 48 XOR-tables which results in 192 tables per full round and 1728 tables in total.

Mapping Tables into Block Memory: The remaining look-up tables can be implemented in BRAM primitives. Most of the tables, except for the T-boxes of the last round, 8×32-bit functions are implemented which requires 8192-bit of memory. Since we can use the BRAM in dual port mode, two tables can be implemented in a single BRAM which allows us to entirely use the 16-Kbit BRAMs resulting in a very dense and efficient implementation. In total, as depicted in Fig. 1, 8 different look-up tables with 32-bit output values are implemented in a quarter AES round, thus 32 tables are necessary to build a full round function (except for the last round). In total, 176 different such tables have to be instantiated along with 16 different 8×8-bit T-boxes for the last round. Note, that all BRAM tables have a similar shape except for the first and last round.

3.3   Performance Evaluation

Table 1 provides the memory consumption of our white-box implementation of AES-128 broken down to different look-up table types and their implementation size (resources and memory). In total, 536KB of memory are required to implement this white-box implementation on an FPGA, whereby 41 % of the memory is required for tables of type $\mathcal{L}$-IV implemented in logic and the remaining 59 % of memory is necessary to store tables of type $\mathcal{L}$-I to $\mathcal{L}$-III in BRAMs.

Table 1. Area and memory consumption of different table types

Look-up tables			Resources		Memory
No.	Type	Size	LUT	BRAM	Byte
16	$\mathcal{L}$-Ia	(8 × 32-bit)	-	8	16 384
16	$\mathcal{L}$-Ib	(8 × 8-bit)	-	8	4 096
144	$\mathcal{L}$-II	(8 × 32-bit)	-	72	147 456
144	$\mathcal{L}$-III	(8 × 32-bit)	-	72	147 456
1728	$\mathcal{L}$-IV	(8 × 4-bit)	27 648	-	221 184
Total			27 648	160	536 576
Utilization (for XC7K160T)			28 %	46 %	40 %

As mentioned before, the design has an initial latency of 19 clock cycles introduced by the BRAM stages. If operated in pipelined mode, this architecture can return one ciphertext per clock cycle after the initial 19 clock cycles. Due to the pipelined architecture and small critical paths, the entire design can operate at a maximum frequency of 100 MHz, resulting in a final throughput of 12.8 Gbit/s. Implementing this on a recent Xilinx Kintex-7 XC7K160T, this design occupies roughly 28 % of the available slices and 46 % of provided BRAM ressources.

4   Side-Channel Analysis

4.1   Differential Computational Analysis Attack

Recently, Bos et al. introduced a new analysis methodology for cryptographic white-box implementations in [4] which requires neither knowledge or possession of the implemented and used look-up tables nor reverse-engineering the tables during the attack process. The following section briefly introduces the methodology of the DCA attack in order to extract secret keys from unknown white-box implementations.

Methodology: DCA primarily targets software-based white-box implementations. In order to successfully perform a key-recovery attack the following two conditions have to be fulfilled:

1. The attacker is able to execute the white-box implementation several times, with different (randomly chosen) plaintexts.
2. Either input- or output external encodings are known to the attacker.

In particular the second requirement is of major importance since it already implies that this attack can be prevented if external encodings are applied and kept secret. However, in practice, at least one encoding (either the initial encoding or the final decoding) usually is known by the user in order to allow a meaningful application of the encryption (or decryption) function. If both afore-mentioned conditions are fulfilled, assuming that the underlying cryptographic algorithm is known to the attacker, the following three steps can be followed to perform a DCA attack.

Step 1: Record Multiple Measurements. It is assumed, that the adversary can execute the white-box implementation in a fully controlled environment. During multiple execution of the encryption algorithm with randomly chosen plaintexts, all accessed memory addresses and any data written to or read from memory are recorded.

Step 2: Conversion to Ideal Traces. A certain type of information is extracted from the recorded data. Common examples of promising information are data read from memory (corresponding to the look-up table outputs), data written to stack (intermediate values of the encryption process) or parts of memory addresses (corresponding to inputs of the look-up tables). The extracted data is converted to a format that can be used by common DPA tools. The authors proposed to serialize the recorded data into a binary string and append the results according to their temporal occurrence. This final binary string is handled as a kind of side-channel trace that we denote to as Ideal Trace since it refers to the result of a fully noise-free probing process.

Step 3: Perform DPA Attack. Following the concept of classical DPA, by guess-ing a key byte k^* and knowing the corresponding plaintext bytes p, the output bits of the S-box, i.e., $S(p \oplus k^*)$, are predicted. Using these models (8 for each key byte) DPA attacks are performed on the Ideal Traces to distinguish the correct key guess amongst the others.

Although the authors of [4] reported successful key recoveries, the reason behind such a success has not be clearly stated. Below we first address our observations from an SCA adversary point of view, and later deal with the leakage source.

4.2   Differential Power Analysis Attack

In this scenario we supposed a gray-box adversary model, where the underly-ing cryptographic algorithm (e.g., AES) is known, but no information about the type of the implementation and its structure (e.g., white-box or ordinary design) is known to the attacker. Further, we suppose that there is no external encoding in the design, e.g., the gray-box seen by the attacker performs stan-dard AES encryption (or decryption). However, the adversary is able to observe side-channel information (e.g., power consumption) of the implementation while it is operated.

Measurement Setup. We made use of a SAKURA-X FPGA board [1] equipped with a Kintex-7 XC7K160T FPGA to practically examine the vulnerability of our white-box design with respect to such an SCA adversary. By means of a digital oscilloscope, the side-channel traces have been collected by measuring the voltage drop over a $1\,\Omega$ resistor in the V_{dd} path of the FPGA during the operation of the design. The sampling was performed at a rate of $500\,\mathrm{MS/s}$ and a bandwidth limit of $20\,\mathrm{MHz}$ while the design was running at a stable, jitter-free, but low clock frequency of $3\,\mathrm{MHz}$ to mitigate the noise. During the measurement phase, our hardware implementation of white-box AES was provided by fully random plaintexts. A sample power trace, where the rounds (19 clock cycles) are clearly distinguishable, is shown in Fig. 2.

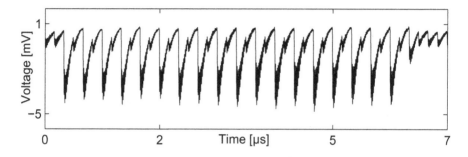

Fig. 2. A sample power trace.

Evaluation. We have collected 10 million power traces of encryptions while the plaintexts were selected randomly. In fact, we have applied several different variants of power analysis attacks including CPA [5], DPA [15] and collision ones [18] with different hypothetical models. The best result has been achieved by means of the classical DPA, which is the same as CPA with single-bit power model. Similar to the case of DCA, for each key byte candidate k^* the output bits of the S-box at the first round, i.e., $\mathsf{S}(p \oplus k^*)$, have been predicted and correlated to the power traces. The results of such 8 different CPA attacks on each bit of one of the S-box outputs are shown in Fig. 3. As shown by the graphics, only one of the attacks (bit 2) is able to recover the secret. We have performed the same attacks on all 16 S-boxes of the first round. Although the attacks on different S-boxes did not show identical results, at least one of the output bits of each S-box led to a successful key recovery, hence full 128-bit key could be recovered.

We would like to note that DCA [4] is indeed a CPA with single-bit power model, assuming the identity function as the actual leakage model of the device and noise-free measurements. Hence, we have shown that the attack is still feasible in case of imperfect (i.e., noisy) measurements and a more complex side-channel leakage function.

4.3   Mathematical Foundations

In order to discuss about the reason behind such a leakage, we first need to give the following definitions.

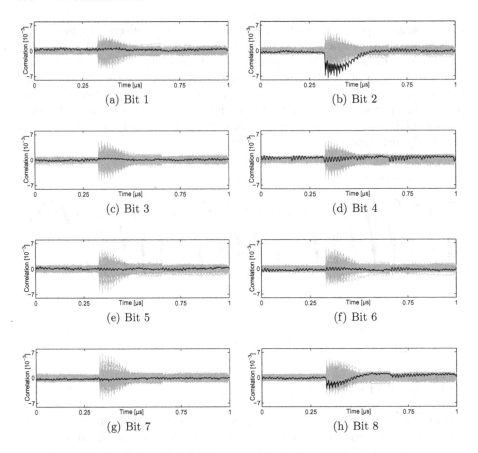

Fig. 3. CPA results, S-box output bit model, 10 million traces

Definition 1. *Let* $x = <x_1, ..., x_n>$, $\omega = <\omega_1, ..., \omega_n>$ *be elements of* $\{0,1\}^n$ *and* $x \cdot \omega = x_1\omega_1 \oplus ... \oplus x_n\omega_n$. *Let* $f(x)$ *be a Boolean function of* n *variables. Then the Walsh transform of the function* $f(x)$ *is a real valued function over* $\{0,1\}^n$ *that can be defined as* $W_f(\omega) = \sum_{x \in \{0,1\}^n} (-1)^{f(x) \oplus x \cdot \omega}$.

Definition 2. *Iff the Walsh transform* W_f *of a Boolean function* $f(x_1, ..., x_n)$ *satisfies* $W_f(\omega) = 0$, *for* $0 \leq HW(\omega) \leq m$, *it is called a balanced* m-*th order correlation immune (CI) function or an* m-*resilient function, where HW stands for Hamming weight.*

For the sake of simplicity, we consider Fig. 4 as one of the 8-to-32 bit $\mathcal{L}$-Ia look-up tables used at the first round of our white-box implementation. As stated before, it is supposed that no external encodings exist in the design (or they are known to the adversary), hence we did not draw them in the figure. Let us denote the output of the S-box by x and the combination of MC and linear encoding R and non-linear 4-to-4 bit encodings by 32 Boolean functions $f_{i \in \{1,...,32\}}(x)$: $\{0,1\}^8 \rightarrow \{0,1\}$. The results of CPA and DCA indicate that at least one of these

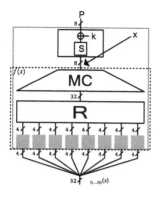

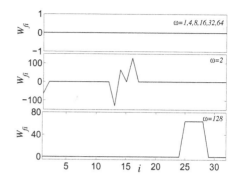

Fig. 4. Detailed representation of an 8×32 look-up table at the first round of our white-box design

Fig. 5. Walsh transforms for all 32 functions $f_{i \in \{1,\ldots,32\}}(\cdot)$ with $HW(\omega) = 1$.

functions $f_i(\cdot)$ is not first-order correlation immune. In order to investigate this, we calculated the Walsh transform of all these functions for all $\omega \in \{0,1\}^8$. The results for 8 cases, where $HW(\omega) = 1$, are shown in Fig. 5.

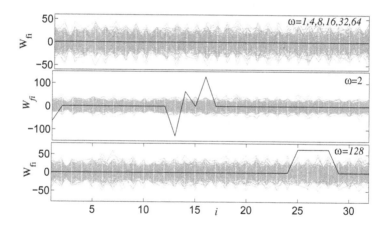

Fig. 6. Walsh transforms for all 32 functions $f_{i \in \{1,\ldots,32\}}(\cdot)$ with $HW(\omega) = 1$ for all key candidates $k^* \in \{0,1\}^8$.

As shown by the graphics, Walsh transform of a couple of functions for two particular ω show an extreme imbalance. However, this fact does not guarantee that a CPA or DPA leads to a successful key recovery. To clarify this fact, we suppose that the linear encoding R and non-linear 4-to-4 bit encodings are unknown, and for each key candidate k^* we derive $f_{i \in \{1,\ldots,32\}}(x)$ by 32-bit output of $\mathcal{L}\text{-}\mathrm{Ia}\left(p = \mathsf{S}^{-1}(x) \oplus k^*\right)$. For each key candidate $k^* \in \{0,1\}^8$ we again calculated the Walsh transforms for all $\omega \in \{0,1\}^8$. Figure 6 represents the results of each ω; $HW(\omega) = 1$ over all key candidates. As shown by the figures, for $\omega = 2$ the extreme imbalance of some functions $f_{i \in \{1,...,32\}}(\cdot)$ for the correct key can be detected amongst that for other key candidates. This indeed justifies why DCA and CPA led to successful key recoveries as this observation perfectly fits to the result of CPA on the same key byte (as shown in Fig. 3), where similarly only second bit of the S-box output (compatible with $\omega = 2$) led to successful key recovery. It is noteworthy that we have similarly examined all other look-up tables of the first cipher round, and for each of them the Walsh transform of at least one ω; $HW(\omega) = 1$ for the correct key showed extremely high imbalance (compared to that for other key candidates). We should stress that all linear and non-linear encodings used in our design have been randomly generated as stated in Step 3 and Step 4 of Sect. 2.2.

4.4   How to Avoid Such Attacks

At the first glance, it can be concluded that if any imbalances is avoided in functions $f_{i \in \{1,...,32\}}(\cdot)$, i.e., all f_i to be first-order correlation immune, DPA and DCA can be avoided. However, it should be noted that such a correlation immunity is valid only in case of classical DPA. In other words, if any of the functions f_i has an extremely high imbalance for any $\omega \in \{0,1\}^8$, that makes it recognizable compared to other key candidates, there exists an attack which can recover the correct key. Such an attack can make use of a power model (or distinguisher) corresponding to that ω. Alternatively, those power analysis attacks which consider the distribution of the leakages, e.g., Mutual Information Analysis [12] which relaxes the power model, can be applied.

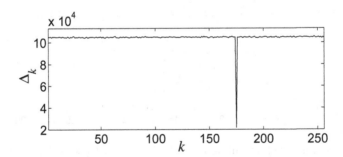

Fig. 7. Sum of all imbalances Δ_k for all key candidates.

In contrary, if many of the functions f_i are m-correlation immune (for any arbitrary m), this opens another door to recover the key. Suppose that for all key candidates k^* and for all ω we calculated the Walsh transforms W_{f_i}. If we sum up all the imbalances for each key candidate as

$$\Delta_{k \in \{0,1\}^8} = \sum_{\forall \omega \in \{0,1\}^8} \sum_{i=1,...,32} \left| W_{f_i}(\omega) \right| \; ; \; k^* = k,$$

the Δ_k for the correct key candidate might be distinguishable (though minimum). In case of our design (the same look-up table which have been considered above), Fig. 7 shows Δ_k for all key candidates, where the correct key is obviously distinguishable. In fact, these results indicate that the linear and non-linear encodings cannot be arbitrary (randomly) selected. Otherwise, the key can be easily revealed by the above explained procedure. This raises a question as what should be the characteristics of such random encodings in such a way that these attacks are not applicable. At least, it can be said that $\forall \omega$ the distribution of Walsh transforms of all f_i should be not distinguishable from that of other key candidates. But how to define the corresponding characteristics to fulfill such a property is considered as future works.

5   Conclusion

In this paper, we presented the first white-box implementation of AES realized in reconfigurable hardware. Assuming a gray-box adversary model, we have practically examined the resistance of our architecture against side-channel attacks. Unfortunately, we were able to successfully perform attacks using classical DPA. However, our observations approve previous results on software-based white-box implementations and extend these results to hardware implementations and physical side-channel attacks. Finally, we provide a to-date missing thorough mathematical analysis of the underlying reasons that enable attacks on such white-box implementations even assuming a gray-box model in case of a lack of unknown external encodings.

Directions for future works include (i) specifying the requirements of linear and non-linear encodings in such a way that the tables cannot be analyzed through their imbalances and (ii) developing designs of new white-box implementations to provide resistance against side-channel attacks. In practice, a conceivable approach to avoid vulnerabilities of white-box implementations in a gray-box adversary model might be a dynamic update of intermediate encodings. In particular for reconfigurable devices, which offer partial reconfiguration abilities, this might be an interesting approach to make side-channel attacks practically infeasible.

Acknowledgment. The authors would like to thank Gregor Leander from Ruhr University Bochum (Germany) for helpful discussions and his comments on the application of Walsh transform.

References

1. Side-channel AttacK User Reference Architecture. http://satoh.cs.uec.ac.jp/SAKURA/index.html
2. Baek, C.H., Cheon, J.H., Hong, H.: Analytic toolbox for white-box implementations: limitation and perspectives. IACR Cryptol. ePrint Arch. **2014**, 688 (2014)

3. Billet, O., Gilbert, H., Ech-Chatbi, C.: Cryptanalysis of a white box AES implementation. In: Handschuh, H., Hasan, M.A. (eds.) SAC 2004. LNCS, vol. 3357, pp. 227–240. Springer, Heidelberg (2004)

4. Bos, J.W., Hubain, C., Michiels, W., Teuwen, P.: Differential computation analysis: hiding your white-box designs is not enough. IACR Cryptol. ePrint Arch. **2015**, 753 (2015)

5. Brier, E., Clavier, C., Olivier, F.: Correlation power analysis with a leakage model. In: Joye, M., Quisquater, J.-J. (eds.) CHES 2004. LNCS, vol. 3156, pp. 16–29. Springer, Heidelberg (2004)

6. Bringer, J., Chabanne, H., Danger, J.: Protecting the NOEKEON cipher against SCARE attacks in fpgas by using dynamic implementations. In: Prasanna, V.K., Torres, L., Cumplido, R. (eds.) 2009 Proceedings of the International Conference on Reconfigurable Computing and FPGAs, ReConFig 2009, Cancun, Quintana Roo, Mexico, pp. 9–11, pp. 183–188. IEEE Computer Society, December 2009

7. Bringer, J., Chabanne, H., Dottax, E.: White box cryptography: another attempt. IACR Cryptol. ePrint Arch. **2006**, 468 (2006)

8. Cherif, Z., Flament, F., Danger, J., Bhasin, S., Guilley, S., Chabanne, H.: Evaluation of white-box and grey-box noekeon implementations in FPGA. In: Prasanna, V.K., Becker, J., Cumplido, R. (eds.) 2010 Proceedings of the International Conference on Reconfigurable Computing and FPGAs, ReConFig 2010, Cancun, Quintana Roo, Mexico, pp. 13–15, pp. 310–315. IEEE Computer Society, December 2010

9. Chow, S., Eisen, P.A., Johnson, H., van Oorschot, P.C.: A white-box DES implementation for DRM applications. In: Security and Privacy in Digital Rights Management, ACM CCS-9 Workshop, DRM 2002, Washington, DC, USA, November 18, 2002, Revised Papers, pp. 1–15 (2002)

10. Chow, S., Eisen, P.A., Johnson, H., van Oorschot, P.C.: White-box cryptography and an AES implementation. In: Selected Areas in Cryptography, 9th Annual International Workshop, SAC 2002, St. John's, Newfoundland, Canada, August 15–16, 2002, Revised Papers, pp. 250–270 (2002)

11. Delerablée, C., Lepoint, T., Paillier, P., Rivain, M.: White-box security notions for symmetric encryption schemes. In: Lange, T., Lauter, K., Lisoněk, P. (eds.) SAC 2013. LNCS, vol. 8282, pp. 247–264. Springer, Heidelberg (2014)

12. Gierlichs, B., Batina, L., Tuyls, P., Preneel, B.: Mutual information analysis. In: Oswald, E., Rohatgi, P. (eds.) CHES 2008. LNCS, vol. 5154, pp. 426–442. Springer, Heidelberg (2008)

13. Goubin, L., Masereel, J.-M., Quisquater, M.: Cryptanalysis of white box DES implementations. In: Adams, C., Miri, A., Wiener, M. (eds.) SAC 2007. LNCS, vol. 4876, pp. 278–295. Springer, Heidelberg (2007)

14. Karroumi, M.: Protecting white-box AES with dual ciphers. In: Information Security and Cryptology - ICISC 2010–13th International Conference, Seoul, Korea, December 1–3, 2010, Revised Selected Papers, pp. 278–291 (2010)

15. Kocher, P.C., Jaffe, J., Jun, B.: Differential power analysis. In: Wiener, M. (ed.) CRYPTO 1999. LNCS, vol. 1666, pp. 388–397. Springer, Heidelberg (1999)

16. Lepoint, T., Rivain, M., Mulder, Y.D., Roelse, P., Preneel, B.: Two attacks on a white-box AES implementation. In: Selected Areas in Cryptography - SAC 2013–20th International Conference, Burnaby, BC, Canada, August 14–16, 2013, Revised Selected Papers, pp. 265–285 (2013)

17. Michiels, W., Gorissen, P., Hollmann, H.D.L.: Cryptanalysis of a generic class of white-box implementations. In: Avanzi, R.M., Keliher, L., Sica, F. (eds.) SAC 2008. LNCS, vol. 5381, pp. 414–428. Springer, Heidelberg (2009)

18. Moradi, A., Mischke, O., Eisenbarth, T.: Correlation-enhanced power analysis collision attack. In: Mangard, S., Standaert, F.-X. (eds.) CHES 2010. LNCS, vol. 6225, pp. 125–139. Springer, Heidelberg (2010)
19. Muir, J.A.: A tutorial on white-box AES. IACR Cryptol. ePrint Arch. **2013**, 104 (2013)
20. Mulder, Y.D., Roelse, P., Preneel, B.: Cryptanalysis of the Xiao - Lai white-box AES implementation. In: Selected Areas in Cryptography, 19th International Conference, SAC 2012, Windsor, ON, Canada, August 15–16, 2012, Revised Selected Papers, pp. 34–49 (2012)
21. De Mulder, Y., Wyseur, B., Preneel, B.: Cryptanalysis of a perturbated white-box AES implementation. In: Gong, G., Gupta, K.C. (eds.) INDOCRYPT 2010. LNCS, vol. 6498, pp. 292–310. Springer, Heidelberg (2010)
22. Wyseur, B., Michiels, W., Gorissen, P., Preneel, B.: Cryptanalysis of white-box DES implementations with arbitrary external encodings. In: Adams, C., Miri, A., Wiener, M. (eds.) SAC 2007. LNCS, vol. 4876, pp. 264–277. Springer, Heidelberg (2007)
23. Xiao, Y., Lai, X.: A secure implementation of white-box AES. In: 2009 2nd International Conference on Computer Science and its Applications (2009)

Detecting Flawed Masking Schemes
with Leakage Detection Tests

Oscar Reparaz[✉]

KU Leuven Department of Electrical Engineering-ESAT/COSIC and iMinds,
Kasteelpark Arenberg 10, 3001 Leuven-Heverlee, Belgium
oscar.reparaz@esat.kuleuven.be

Abstract. Masking is a popular countermeasure to thwart side-channel attacks on embedded systems. Many proposed masking schemes, even carrying "security proofs", are eventually broken because they are flawed by design. The security validation process is nowadays a lengthy, tedious and manual process. In this paper, we report on a method to verify the soundness of a masking scheme before implementing it on a device. We show that by instrumenting a high-level implementation of the masking scheme and by applying leakage detection techniques, a system designer can quickly assess at design time whether the masking scheme is flawed or not, and to what extent. Our method requires not more than working high-level source code and is based on simulation. Thus, our method can be used already in the very early stages of design. We validate our approach by spotting in an automated fashion first-, second- and third-order flaws in recently published state-of-the-art schemes in a matter of seconds with limited computational resources. We also present a new second-order flaw on a table recomputation scheme, and show that the approach is useful when designing a hardware masked implementation.

1 Introduction

Since Kocher published the seminal paper on side-channel attacks [Koc96], cryptographic embedded systems have been broken using some auxiliary timing information [Koc96], the instantaneous power consumption of the device [KJJ99] or the EM radiation [AARR02], among others. An attack technique of particular interest, due to its inherent simplicity, robustness and efficiency to recover secrets (such as cryptographic keys or passwords) on embedded devices is Differential Power Analysis (DPA), introduced in [KJJ99]. DPA relies on the fact that the instantaneous power consumption of a device running a cryptographic implementation is somehow dependent on the intermediate values occurring during the execution of the implementation. An especially popular countermeasure to thwart power analysis attacks, including DPA, is masking [CJRR99, GP99]. Masking works by splitting every sensitive variable appearing during the computation of a cryptographic primitive into several shares, so that any proper subset of shares is independent of any sensitive variable. This, in turn, implies that the instantaneous power consumption of the device is independent of any sensitive

© International Association for Cryptologic Research 2016
T. Peyrin (Ed.): FSE 2016, LNCS 9783, pp. 204–222, 2016.
DOI: 10.1007/978-3-662-52993-5_11

variable, and thus vanilla DPA cannot be mounted. In theory, a $(d + 1)$-order DPA attack can still be mounted against a d-th order masked implementation; however, in practice higher order DPA attacks are exponentially more difficult to carry out [CJRR99].

Crucially, in many cases the attacker is not required to perform a higher order attack because the masking is imperfect and thus does not provide the claimed security guarantees. The causes of the imperfections can be manifold: from implementation mistakes to more fundamental flaws stemming from the masking scheme itself. There are many examples in the literature of such flawed schemes: a "provably secure" scheme published in 2006 [PGA06] based on FFT and broken two years later [CGPR08], a scheme published in 2006 [SP06] and broken one year later [CPR07], another "provably secure" scheme published in 2010 [RP10] and (academically) broken three years later [CPRR13]; a scheme published in 2012 [BFGV12] and broken in 2014 [PRR14].

The verification process of a masking scheme is nowadays a very lengthy manual task, and the findings are published in solid papers involving convoluted probability arguments at leading venues, some years later after the scheme is published. Some even won a best paper award as [CPR07]. From the stand point of a system designer, it is often not acceptable to wait for a public scrutiny of the scheme or invest resources in a lengthy, expensive, evaluation.

Our Contribution. In this paper we provide an automated method to test whether the masking scheme is sound or not, and to what extent. The method is conceptually very simple, yet powerful and practically relevant. We give experimental evidence that the technique works by reproducing state-of-the-art first-, second- and third-order flaws of masking schemes with very limited computational resources. Our approach is fundamentally different from previously proposed methodologies and is based on sampling and leakage detection techniques.

2   Leakage Detection for Masked Schemes in Simulation

Core Idea. In a nutshell, our approach to detect flawed masking schemes is to simulate power consumption traces from a high-level implementation of the masking scheme and then perform leakage detection tests on the simulated traces to verify the first- and higher-order security claims of the masking scheme.

Input and Output of the Tool. The practitioner only ought to provide working source code of the masked implementation. The tool instruments the code, performs leakage detection tests and outputs whether the scheme meets its security goals or not. In addition, should a problem be detected, the tool pinpoints the variables causing the flaw and quantifies the magnitude of the statistical bias.

Security Claims of Masking Schemes. We use in this paper the conventional notions for expressing the security claim of a masking scheme. Namely, a masking scheme provides first-order security if the expected value of each single intermediate does not depend on the key. More generally, a masking scheme provides

k-order security if the k-th order statistical moment of any combination of k intermediates does not depend on the key. This formulation is convenient since leakage detection tests are designed specifically to test these claims.

Three Steps. Our tool has three main ingredients: trace generation, trace pre-processing and leakage detection. We describe each one in detail in the sequel.

2.1   Trace Generation

The first step of our approach is to generate simulated power traces in a noise-free environment.

Implementation. To accomplish this, the masking scheme is typically implemented in a high-level software language. The implementation is meant to generically reproduce the intermediate values present in the masking scheme, and can be typically written from the pseudo-code description of the masking scheme. (Alternatively, the description of the masking scheme can be tailored to a specific software or hardware implementation and incorporate details from those.)

Execution. This implementation is executed many times, and during each execution, the instrumentation environment observes each variable V that the implementation handles at time n. At the end of each execution, the environment emits a leakage trace $c[n]$. Each time sample n within this trace consists of leakage $L(V)$ of the variable V handled at time n. The leakage function L is predefined; typical instantiations are the Hamming weight, the least significant bit, the so-called zero-value model or the identity function.

Randomness. The high-level implementation may consume random numbers (for example, for remasking.) This randomness is provided by a PRNG.

2.2   Trace Pre-processing

This step is only executed if the masking scheme claims higher-order security. The approach is similar to higher-order DPA attacks [CJRR99] and higher-order leakage detection [SM15]. Suppose the scheme claims security at order k. We pre-processes each simulated trace $c[n]$ to yield $c'[n_1, \ldots, n_k]$ through a combination function as

$$c'[n_1, \ldots, n_k] = \prod_{i=1}^{i=k} (c[n_i] - \bar{c}[n_i]). \tag{1}$$

The result is a preprocessed trace c'. The length of the trace is expanded from N to $\binom{N}{k}$ unique time samples. (It is normally convenient to treat c' as a uni-dimensional trace.)

2.3   Leakage Detection

The next step of our approach is to perform a leakage detection test on the (potentially pre-processed) simulated traces. In its simplest form, a leakage detection test [CKN00,CNK04,GJJR11,CDG+13,SM15] tries to locate and potentially quantify information leakage within power traces, by detecting statistical dependencies between sensitive data and power consumption. In our context, if the test detects information leakage on the simulated traces, this means that the masking scheme fails to provide the promised security guarantees.

Procedure. The instrumentation environment performs a fixed-vs-fixed leakage detection test using the T-test distinguisher [CDG+13].

The process begins by simulating a set of power traces with fixed unmasked intermediate $z = z_0$ and another set of traces with different unmasked intermediate value $z = z_1$. Typical choices for the intermediate z are the full unmasked state or parts of it. Then, a statistical hypothesis test (in this case, T-test) is performed per time sample for the equality of means. The T-test [Stu08,Wel47] first computes the following statistic

$$t[n] = \frac{m_0[n] - m_1[n]}{\sqrt{\frac{s_0^2[n]}{N_0} + \frac{s_1^2[n]}{N_1}}} \tag{2}$$

where $m_i[n]$, $s_i^2[n]$, N_i are respectively the sample mean, variance and number of traces of population $i \in \{0,1\}$ and n is the time index. This statistic $t[n]$ is compared against a predefined threshold C. A common choice is $C = \pm 4.5$, corresponding to a very high statistical significance level of $\alpha = 0.001$. If the statistic $t[n]$ surpasses the threshold C, the test determines that the means of the two distributions are significantly different, and thus the mean power consumption of (potentially pre-processed) simulated power traces carry information on the intermediate z. In this case, we say that the masking scheme exhibits leakage at time sample n and flunks the test. Otherwise, if no leakage is detected, another test run is executed with different specific values for z_0 and z_1. The test is passed only if no leakage is detected for any value of z_0 and z_1. (Typically, there are only a couple dozen of (z_0, z_1) pairs if the optimizations described in the next section are applied.) Note that a time sample n may correspond to a single variable (first-order leakage) or a combination of variables (higher-order leakage), if a pre-processing step is executed.

On Fixed-vs-Fixed. Using fixed-vs-fixed instead of fixed-vs-random has the advantage of faster convergence of the statistic (at the expense of leakage behavior assumptions that are benign in our context). (This has been previously observed by Durvaux and Standaert [DS15] in a slightly different context.) One could also use a fix-vs-random test. This usually results in a more generic evaluation.

2.4 Optimizations

We note that the following "engineering" optimizations allow to lower the computational complexity so that it is becomes very fast to test relevant masking schemes.

Online Algorithms. There is certainly no need to keep in memory the complete set of simulated power traces. For the computation of the T-test as Eq. 2, one can use online formulas to compute means and variances present in the formula. These algorithms traverse only once through each trace, so that a simulated power consumption trace can be generated, processed and thrown away. This makes the memory consumption of the approach independent of the number of traces used. More number of traces would require just more computational time, but not more memory. We note that the same is possible in higher-order scenarios. Lengthy but straightforward calculations show that a T-test on pre-processed traces can be computed online using well-known online formulae for (mixed) higher-order moments [P08]. (This was previously reported by Schneider and Moradi [SM15].)

Scale Down the Masking Scheme. It is usually possible to extrapolate the masking scheme to analogous, trimmed down, cryptographic operations that work with smaller bit-widths or smaller finite fields. For example, when masking the AES sbox, many masking schemes [RP10, CPRR13] rely on masked arithmetic (masked multiplication and addition blocks) in $GF(2^8)$ to carry out the inversion in $GF(2^8)$. It is often convenient to scale down the circuit to work on, say, $GF(2^4)$ for testing purposes –since the masking approach normally does not rely on the specific choice of field size, any flaw exhibited in the smaller $GF(2^4)$ version is likely to be exhibited in the $GF(2^8)$ version of the algorithm (and vice versa). By experience we have observed that statistical biases tend to be more pronounced in smaller field sizes, and thus are more easily detectable. (See for instance [PRR14].) We suggest the use of this heuristic whenever possible for an early alert of potential problems.

Reduce the Number of Rounds. There is little sense to check for a first-order leak in more than a single round of an iterative cryptographic primitive, such as AES. If the implementation is iterative, any first-order flaw is likely to show up in all rounds. When testing for higher order security, however, one should take into account that the flaw may appear from the combination of variables belonging to different rounds.

Judiciously Select the Components to Check. For first-order security it is sufficient to check each component of the masking scheme one by one in isolation. The situation is slightly different in the multivariate scenario, where multiple components can interfere in a way that degrades security. Still, the practitioner can apply some heuristics to accelerate the search, such as testing for second-order leakage first only in contiguous components. For example, second-order leakage is likely to appear earlier between two variables within the same round or belonging to two consecutive rounds.

Deactivate Portions of the Plaintext. To accelerate the leakage search, a substantial portion of the plaintext can be deactivated, that is, fixed to a constant value or even directly taken out from the algorithm. For example, in the case of AES-128 one could deactivate 3 columns of the state, test only 4 active plaintext bytes and still test for the security of all the components within one round.

Carefully Fix the Secret Intermediate Values. As we described, the framework fixes two values z_0, z_1 for the unmasked sensitive intermediate, and then compares the simulated traces distributions conditioned on z_0 and z_1. Depending on the algorithm, concrete choices for z_i (such as fixed points of the function being masked) can produce "special" leakage. For example, in AES if we choose z_1 such that the input to the inversion is 0×00, we can hit faster zero-value type flaws.

3   Results

In this section we provide experimental results. We first begin by testing the first-order security claim of two schemes, one that fails the claim (Sect. 3.1) and another that fulfills it (Sect. 3.2). Then we will focus on second- and third- order claims (Sects. 3.3 and 3.4 respectively). We point out a new second-order flaw in Sect. 3.5, we elaborate on how previously published flaws were discovered in Sect. 3.6. Finally in Sect. 3.7 we report on the use of the tool when designing masked circuits.

3.1   Smoke Test: Reproducing a First-Order Flaw

As a first test, we test the first-order security of the scheme published in [BFGV12]. We will refer to this scheme as IP in the sequel. We focus on reproducing the results from [PRR14],

Test Fixture. We study first the IPRefresh procedure. This procedure performs a refreshing operation on the input IP shares. We scale down the scheme to work in $GF(2^2)$ following Sect. 2.4. The instrumentation framework finds 141 intermediate variables within a single execution of IPRefresh. The chosen leakage function is Hamming weight, and there is no pre-processing involved.

Leakage Detection. We ran the $\binom{4}{2} = 6$ possible fixed-vs-fixed tests covering all possible combinations of pairs of different unshared input values (z_1, z_0). (Here z_i is the input to IPRefresh.) For each test, the maximum absolute t-value, across all time samples, is plotted in the y-axis of Fig. 1 as a function of the number of simulated traces (x-axis). A threshold for the T-test at 4.5 is also plotted as a dotted line. This threshold divides the graph into two regions: a

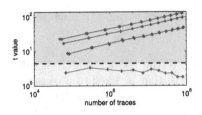

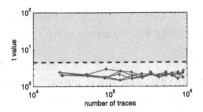

Fig. 1. T-statistic (absolute values) of the IP masking scheme, under a HW leakage model. Deemed insecure (clearly exceeds the threshold at $t =$ 4.5.) (Color figure online)

Fig. 2. T-statistic (absolute values) applied to the Coron table recomputation masking scheme, under an Identity leakage model. First order test. Deemed secure (no value beyond the threshold at $t = 4.5$.)

t-statistic greater than $|C| = 4.5$ (in red) means that the implementation fails the test, while a t-statistic below 4.5 (area in green) does not provide enough evidence to reject the hypothesis that the scheme is secure. We can see that 5 out of 6 tests clearly fail in Fig. 1, since they attain t-values around 100 greater than C. Thus, the IPRefresh block is deemed insecure. (Similar observations apply to the IPAdd procedure.)

It is also possible to appreciate the nature of the T-test statistic: the t-statistic grows with the number of traces N as of $\sqrt{N}$ in the cases that the implementation fails the test (note that the y-axis is in logarithmic scale.) This can be interpreted as follows: as we have more measurements, we build more confidence to reject the null hypothesis (in our context being that the masking is effective.) If the number of simulated traces is large enough and no significant t-value has been observed, the practitioner can gain confidence on the scheme not being flawed. We will find this situation in the next subsection and elaborate on this point.

3.2   A First-Order Secure Implementation

We tested the table recomputation scheme of Coron [Cor14]. This scheme passes all fixed-vs-fixed tests with the identity leakage model. The results are plotted in Fig. 2. We can observe that the t-statistic never crosses the threshold of 4.5 for any test, and thus we cannot reject the null hypothesis that the implementation is secure (i.e., the implementation is deemed secure, *"on the strength of the evidence presented"* [CKN00]). Although it is theoretically possible that the masking scheme exhibits a small bias that would only be detectable when using more than 10^6 traces, that flaw would be negligible from a practical point of view when using $\leq 10^6$ traces, and definitely impossible to exploit in a noisy environment if it is not even detectable at a given trace count, in a noiseless scenario.

```
70 void MaskRefresh(u8 *s) {
71   u8 r;
72   for (int i = 1; i < number_shares; i++) {
73     r = rnd ();
74     s[0] ^= r;
75     s[i] ^= r;
76   }
77 }
...
110 void SecMult (u8 *out, u8 *a, u8 *b) {
111   u8 aibj,ajbi;
...
114   for (int i = 0; i < number_shares; i++) {
115     for (int j = i + 1; j < number_shares; j++) {
...
119       aibj = mult(a[i], b[j]);
120       ajbi = mult(a[j], b[i]);
------------------------------------------------------
$ ./run
entering fixed_vs_fixed(00,01)
> leakage detected with 1.20k traces
  higher order leakage between
    line 74 and
    line 120
  with tvalue of -7.03
```

Fig. 3. Excerpts of the code and output of the leakage detection for the RP scheme.

3.3   Reproducing a Second-Order Flaw

To show that our proposed tool can also detect higher-order flaws, we implemented the scheme of Rivain and Prouff (RP) from [RP10]. For the allegedly second-order secure version of this scheme, there is a second-order flaw as spotted by Coron et al. in [CPRR13] between two building blocks: MaskRefresh and SecMult. We will see that we can easily spot this flaw with the methodology proposed in this paper.

Text Fixture. We implemented the second-order masked inversion $x \mapsto x^{-1}$ in $GF(2^n)$ as per [RP10] with $n = 3$. This inversion uses the MaskRefresh and SecMult procedures. In this case, we enable second-order pre-processing (on the fly), expanding 135 time samples to $\binom{135}{2} = 9045$ time samples. Some excerpts of the implementation are shown in Fig. 3.

Results. The instrumentation frameworks takes less than 5 s to determine that there is a second order leakage between the variable handled at line 74 (inside MaskRefresh) and 120 (inside SecMult), as Fig. 3, bottom, shows. Note that it is trivial to backtrack to which variables corresponds a leaking time sample, and thus determine the exact lines that leak jointly.

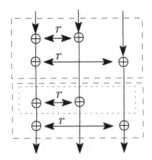

Fig. 4. Two `MaskRefresh` concatenated. As explained in the text, the second refresh can be optimized to reduce the randomness requirements yet still achieving second order security. (Color figure online)

Fixing the Second-Order Flaw. The folklore solution to fix the previous second-order flaw is to substitute each `MaskRefresh` module by two consecutive `MaskRefresh` invocations, as shown in Fig. 4. Applying the leakage detection tests to this new construction shows that the leak is effectively gone. However, it is quite reasonable to suspect that this solution is not optimal in terms of randomness requirements. We can progressively strip down this design by eliminating some of the randomness of the second refreshing and check if the design is still secure. We verified in this very simple test fixture that if we omit the last randomness call (that is, we only keep the dotted red box instead of the second dashed box in Fig. 4), the higher-order leaks are no longer present.

3.4   Reproducing a Third Order Flaw

Schramm and Paar published at CT-RSA 2006 [SP06] a masked table lookup method for Boolean masking claiming higher-order security. This countermeasure was found to be flawed by Coron et al. at CHES 2007. Coron et al. found a third-order flaw irrespective of the security parameter of the original scheme. We reproduced their results by setting $k = 3$ when preprocessing the traces as in Eq. 1. The flaw of [CPR07] was detected in less than one second, which demonstrates that the tool is also useful to test the higher-order security claims of masking schemes.

3.5   Schramm–Paar Second-Order Leak

Here we report on a new second-order flaw that we found with the presented tool in the masked table recomputation method of Schramm and Paar when used with unbalanced sboxes.

Schramm–Paar Method. The goal of the masked table recomputation is to determine the sbox output shares $N_0, N_1, \ldots, N_d$ from the sbox input shares $M_0, M_1, \ldots, M_d$. Schramm–Paar proceed as follows (we borrow the notation from [CPR07]):

1. Draw d output shares $N_1, \ldots, N_d$ at random
2. Compute from $N_1, \ldots, N_d$ a table S^* such that

$$S^*(x) = S\left(x \oplus \bigoplus_{i=1}^{d} M_i\right) \oplus \bigoplus_{i=1}^{d} N_i \tag{3}$$

3. Set $N_0 := S^*(M_0)$

We set here $d = 2$, and aim at second-order security. An important part of the procedure is to build the table S^* in a way that the higher-order security claims are fulfilled. [SP06] proposes several methods. However, for the purposes of this paper the details of the recomputation method are not important.

Test Fixture. Following the guidelines of Sect. 2.4, we implement a very stripped down version of the table recomputation method. We fix the simplest unbalanced sbox $S = (0,0,0,1)$ (an AND gate), and work with 2-bit inputs and outputs leaking Hamming weights. In a couple of seconds the tool outputs 4 different bivariate second-order leakages, corresponding to the pairs $(S^*(i), N_0)$ for each i in the domain of S^*. Here $S^*(i)$ is the i-th entry on the S^* table, and N_0 is one output mask.

Once these leaks are located, proving them becomes an easy task. And also it becomes easy to generalize and see that the flaw appears whenever S is unbalanced. (We verified that second-order attacks using the leakage of $S^*(0)$ and N work as expected.)

3.6   Higher-Order Threshold Implementations

Here we report on how the observations from [RBN+15] regarding the security of higher-order threshold implementations [BGN+14] were found. The results of this section are obviously not new; the focus here is on the methodology carried out to find those.

Intuition. The first suspicion stems from the fact that higher-order threshold implementations originally claimed that the composition of sharings provides higher-order security, if the sharings satisfy some property, namely uniformity. This is a very surprising result, since it would imply that there is no need to inject fresh randomness during the computation, minimizing overheads. In contrast, all other previously published higher-order masking schemes need to inject randomness from time to time as the computation progresses. For example, the security proof of private circuits (one of the earliest masking schemes) [ISW03] critically relies on the fresh randomness to provide security.

Test Fixture. The hypothesis is that the previous security claim does not hold, that is, the concatenation of uniform sharings do not provide higher-order security. To test this, we design a minimal working test fixture consisting of a 32-round Feistel cipher with a blocksize of 4 bits. For more details see [RBN+15].

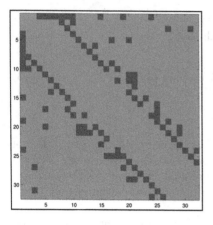

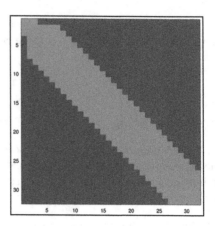

Fig. 5. Pairs of rounds with $|t| > 80$ (Color figure online)

Fig. 6. Pairs of rounds with $|t| > 5$ (Color figure online)

The shared version aims at providing second-order security, and shares each native bit into 5 shares. The traces consist of 225 "timesamples" (each one comprising one leaked bit, including initialization.) This spans to 25650 timesamples after second-order pre-processing.

Cranking it up. We run the simulation for a night (about 8 h), having simulated 200 million traces. We performed a fixed-vs-fixed test with unshared initial state 0000 and 1111. (There is no key in this cipher, the initial state is considered to be the secret.) (This is grossly unoptimized code.) The results of the leakage detection test is drawn in Fig. 5. We plot on the x- and y-axes the round index, and each pixel in red if the t statistic surpasses the value 80, green otherwise. We can see that many pairs of rounds leak jointly, in contradiction with the security claims of the scheme. In Fig. 6 the same information is plotted but changing the threshold to $|t| > 5$. We can see, surprisingly, that *almost all* pairs of rounds lead to second-order leakage. A bit of manual mechanical effort is required to prove this, but not more than taking a covariance.

3.7   Refreshing in Higher-Order Threshold AES Sbox

The designers from [CBR+15] had access to the tool presented in this paper. They performed several design iterations, and verified the design on each iteration. The final evaluation was performed on an FPGA.

Text Fixture. We implemented the whole sbox, with no downscaling of the components to work in smaller fields. We leak register bits and the input value (identity leakage function) to combinatorial logic blocks. (This is to account for glitches as will be explained below.)

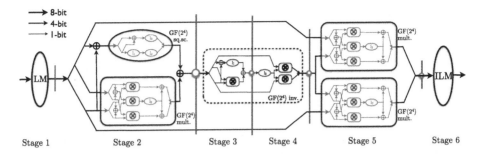

Fig. 7. Higher-order masked AES sbox from de Cnudde et al.

First-Order Leaks. Within one day, a first-order leak was identified due to a design mistake. This design error considered the concatenation of two values $a\|b$ as input to the next stage; each value a and b considered independently is a uniform sharing but its concatenation $a\|b$ is not, and hence the first order leak. This first-order leak disappears if a refresh is applied to the inputs of one $GF(2^2)$ multiplier using 4 units of randomness (here 1 unit = 1 random field element = 2 bits). This refresh block is similar to the 2010 Rivain–Prouff refresh block [RP10], we remind it uses $n-1$ units of randomness to refresh n shares (in our particular case here $n = 5$). We will see later that this refresh is problematic in the higher-order setting.

Second-Order Leaks. Subsequently, two second-order bivariate leaks were identified between register values. This was solved by placing a refresh block between stage 2 and 3 from Fig. 7 (taken from [CBR+15]).

In addition, many second-order bivariate leaks were identified between input values to combinatorial logic blocks. In theory, hardware glitches could express these leakages. Those disappear if one uses a "full refresh" using 5 units of randomness. This effect was previously observed [BBD+15, RBN+15] and is a reminiscent of [CPRR13].

Other Uses. We also used a preliminary version of this tool in [RRVV15].

4   Discussion

4.1   Implementing the Framework

We implemented the instrumentation framework on top of `clang`-LLVM. The whole implementation (including leakage detection code) takes around 700 lines of C++ code, which shows the inherent simplicity of our approach. It is easy to audit and maintain.

4.2   Time to Discover Flaw, Computational Complexity and Scaling

The computational requirements of the proposed approach are very low. In Fig. 8 we write the elapsed execution times required to spot the flaws from Sects. 3.1, 3.3 and 3.4. We can see that the flaws were identified in a matter of seconds on a standard computer. All the experiments on this paper were carried out on a modest 1.6 GHz laptop with 2 GB of RAM.

Bottlenecks. The main bottleneck on the running time of the methodology is the first step: trace generation. The RP scheme is the one that took longer to detect the flaw (5 s), presumably because of two reasons: (a) the scheme is more inherently complex and thus it takes more time to simulate each trace and (b) the bias exhibited in the scheme is smaller than the bias of other schemes, and thus more traces are required to detect such a bias. We note that no special effort on optimizing the implementations was made, yet, an average throughput of 5 k trace per second (including instrumentation) was achieved. The overhead of instrumentation in the running time was estimated to make the implementation on average $\approx \times 1.6$ slower.

Time to Pass. The time it takes to discover a flaw is normally less than the time it takes to deem a masking scheme secure. For example, to assess that the patch of Sect. 3.3 is indeed correct, it took about 6 min to perform a fix-vs-fix test with up to 1 million traces (no leakage was detected). All possible 6 tests take around 37 min. (The threshold of 1 million traces was chosen arbitrarily in this example.)

Parallelization. We remark that this methodology is embarrassing parallel. Thus, it is much easier to parallelize to several cores or machines than other approaches based on SAT.

Memory. The memory requirements for this method are also negligible, taking less than 4.5 MB of RAM on average. More interestingly, memory requirements are constant and do not increase with the number of simulated traces, thanks to online algorithms.

Scheme	Flaw order	Field size	Time	Traces needed
IP	1	4	0.04s	1k
RP	2	4	5s	14k
SP	3	4	0.2s	2k

Fig. 8. Running time to discover flaw in the studied schemes, and number of traces needed to detect the bias.

Scaling. The execution time of our approach scales linearly with the number of intermediates when testing for first-order leakage, quadratically when testing for second-order leakage and so on. This scaling property is exactly the same as for DPA attacks. We could benefit from performance improvements that are typically used to mitigate scaling issues in DPA attacks such as trace compression, but did not implemented those yet.

4.3 Limitations

Risk of False Negatives. Our tool should not be taken as the only test when assessing a masked implementation, and is not meant to substitute practical evaluation with actual measurements. Our tool provides an early warning that the masking scheme may be structurally flawed, "by design". However, even when the masking scheme is theoretically secure, it is still possible to implement it in an insecure way. This will not be detected with the proposed tool. For example, in the case of a first-order masked software implementation, an unfortunate choice of register allocation may cause distance leakage between shares, leading to first-order leakage. Without register allocation information, our tool will not detect this issue. One could provide this kind of extra information to our tool. We left this as future work.

4.4 Related Works

There are already some publications that address the problem of automatic verification of power analysis countermeasure.

SAT-based. Sleuth [BRNI13] is a SAT-based methodology that outputs a hard yes/no answer to the question of whether the countermeasures are effective or not. A limitation of [BRNI13] is that it does not attempt to quantify the degree of (in)security. A first approximation to the problem was tackled in [EWTS14, ABMP13].

MiniCrypt-based. Barthe et al. [BBD+15] use program verification techniques to build a method prints a proof of security for a given masking scheme. It is very hard to compare our tool with theirs since they are fundamentally different. The goal is also different: while our results are probabilistic, the goal of Barth et al. is to categorically prove the security of the scheme. Depending on the context, one might be preferrable over the other. The two approaches are also very different. Barthe et al. base their approach on EasyCrypt, a sophisticated "toolset for reasoning about relational properties of probabilistic computations with adversarial code."

Considerations Related to Other Approaches. While our approach does certainly not carry the beauty of proofs and formal methods, it offers a very practice-oriented methodology to test the soundness of masking schemes. Our approach is in nature statistical, and is a necessary condition for a masked scheme to be sound. It can be thought of a worst-case scenario, where the adversary has access to noiseless and synchronized traces. A more formal study can then be performed with the methods of Barthe et al. to gain higher confidence, since the output of the tool of Barthe et al. is a hard proof.

4.5 Which Leakage Function to Select?

In previous Sect. 2 we mentioned that the practitioner has to choose a leakage function to generate the simulated traces. It turns out that the specific choice of leakage function seems not to be crucial —any reasonable choice will work. Figure 9 compares different leakage functions: Hamming weight, identity, least-significant bit and zero-value. The test fixture is the same one as in Sect. 3.1. For each leakage function, we performed all possible fixed-vs-fixed tests. Figure 9 is composed of 4 plots, one per leakage function. We can see that for any leakage function, there is at least one fixed-vs-fixed test that fails. For the identity leakage function, *all* tests fail. Thus, it is often convenient to use it to detect flaws faster (more fixed-vs-fixed tests fail.) We speculate that this behavior may depend on the concrete masking method used, and leave a detailed study as future work.

Glitches and Identity Leakage. We note that we can include the effect of hardware glitches in our tool. Note that the information leaked by a combinatorial logic block F on input x due to glitches is contained already in the input x. Thus, we can simulate the information leaked by hardware glitches, even if we do not have a precise timing model of the logic function, by leaking the whole input value x (that is, using the identity leakage model.)

This would correspond to an extremely glitchy implementation of F where glitches would allow to observe the complete input. This is certainly a worst-case scenario. Crucially, glitches would not reveal *more* information than x. This trick

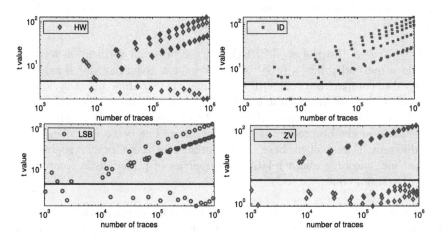

Fig. 9. Influence of leakage function.

of using the identity leakage model on inputs of combinatorial blocks is helpful when evaluating, for example, masked threshold implementations.

Another alternate approach is to add a detailed gate-level timing model to simulate glitches. If such timing model is available, the detection quality can be substantially enhanced.

5   Conclusion

We described a methodology to test in an automated way the soundness of a masking scheme. Our methodology enjoys several attractive properties: simplicity, speed and scalability. Our methodology is based on established and well-understood tools (leakage detection). We demonstrated the usefulness of the tool by detecting state-of-the-art flaws of modern masking designs in a matter of seconds with modest computational resources. In addition, we showed how the tool can assist the design process of masked implementations.

Acknowledgements. We thank an anonymous reviewer that found a mistake in Sect. 3.5, François-Xavier Standaert for extensive comments and Ingrid Verbauwhede. The author is funded by a PhD fellowship of the Fund for Scientific Research - Flanders (FWO). This work was funded also by Flemish Government, FWO G.0550.12N, G.00130.13N, Hercules Foundation AKUL/11/19, and through the Horizon 2020 research and innovation programme under grant agreement 644052 HECTOR.

Auxiliary Supporting Material

A   MATLAB Code

This code prints the distribution of $Z = S(M \oplus M_0) \oplus S(M)$ for a fixed M and varying M_0.

```
% the sbox
S=[0 0 0 1];
% number of samples
N=10000;

% the sbox input
for M=0:3
    M0=floor(4.*rand(1,N));
    Z =bitxor(S(bitxor(M,M0)+1),S(M+1));

    for i=0:1
        fprintf(' p(Z=%d|M=%d) = %1.2f\n', i, M, sum(Z==i)./length(Z))
    end
    fprintf('\n')
end
```

B   Examplary Output

This is the distribution of Z when the secret M takes different values. We can see that the expected value of Z is different when conditioned on $M = 0$ than when $M = 3$. This means that there is a second-order information leak between $(S^*(0), N_0)$ and the secret M.

```
p(Z=0|M=0) = 0.75
p(Z=1|M=0) = 0.25


p(Z=0|M=1) = 0.75
p(Z=1|M=1) = 0.25


p(Z=0|M=2) = 0.75
p(Z=1|M=2) = 0.25


p(Z=0|M=3) = 0.25
p(Z=1|M=3) = 0.75
```

References

[AARR02] Agrawal, D., Archambeault, B., Rao, J.R., Rohatgi, P.: The EM side-channel(s). In: Kaliski Jr., B.S., Koç, Ç.K., Paar, C. (eds.) CHES 2002. LNCS, vol. 2523. Springer, Heidelberg (2003)

[ABMP13] Agosta, G., Barenghi, A., Maggi, M., Pelosi, G.: Compiler-based side channel vulnerability analysis and optimized countermeasures application. In: 2013 50th ACM/EDAC/IEEE Design Automation Conference (DAC), pp. 1–6, May 2013

[BBD+15] Barthe, G., Belaïd, S., Dupressoir, F., Fouque, P.-A., Grégoire, B., Strub, P.-Y.: Verified proofs of higher-order masking. In: Oswald, E., Fischlin, M. (eds.) EUROCRYPT 2015. LNCS, vol. 9056, pp. 457–485. Springer, Heidelberg (2015)

[BFGV12] Balasch, J., Faust, S., Gierlichs, B., Verbauwhede, I.: Theory and practice of a leakage resilient masking scheme. In: Wang, X., Sako, K. (eds.) ASIACRYPT 2012. LNCS, vol. 7658, pp. 758–775. Springer, Heidelberg (2012)

[BGN+14] Bilgin, B., Gierlichs, B., Nikova, S., Nikov, V., Rijmen, V.: Higher-order threshold implementations. In: Sarkar, P., Iwata, T. (eds.) ASIACRYPT 2014, Part II. LNCS, vol. 8874, pp. 326–343. Springer, Heidelberg (2014)

[BRNI13] Bayrak, A.G., Regazzoni, F., Novo, D., Ienne, P.: Sleuth: automated verification of software power analysis countermeasures. In: Bertoni, G., Coron, J.-S. (eds.) CHES 2013. LNCS, vol. 8086, pp. 293–310. Springer, Heidelberg (2013)

[CBR+15] De Cnudde, T., Bilgin, B., Reparaz, O., Nikov, V., Nikova, S.: Higher-order threshold implementation of the AES S-box. In: Homma, N., Medwed, M. (eds.) CARDIS 2015. LNCS, vol. 9514, pp. 259–272. Springer, Bochum (2015)

[CDG+13] Cooper, J., DeMulder, E., Goodwill, G., Jaffe, J., Kenworthy, G., Rohatgi, P.: Test Vector Leakage Assessment (TVLA) methodology in practice. In: International Cryptographic Module Conference (2013)

[CGPR08] Coron, J.-S., Giraud, C., Prouff, E., Rivain, M.: Attack and improvement of a secure S-Box calculation based on the Fourier Transform. In: Oswald, E., Rohatgi, P. (eds.) CHES 2008. LNCS, vol. 5154, pp. 1–14. Springer, Heidelberg (2008)

[CJRR99] Chari, S., Jutla, C.S., Rao, J.R., Rohatgi, P.: Towards sound approaches to counteract power-analysis attacks. In: Wiener, M. (ed.) CRYPTO 1999. LNCS, vol. 1666, pp. 398–412. Springer, Heidelberg (1999)

[CKN00] Coron, J.-S., Kocher, P.C., Naccache, D.: Statistics and secret leakage. In: Frankel, Y. (ed.) FC 2000. LNCS, vol. 1962, pp. 157–173. Springer, Heidelberg (2001)

[CNK04] Coron, J.-S., Naccache, D., Kocher, P.C.: Statistics and secret leakage. ACM Trans. Embed. Comput. Syst. 3(3), 492–508 (2004)

[Cor14] Coron, J.-S.: Higher order masking of look-up tables. In: Nguyen, P.Q., Oswald, E. (eds.) EUROCRYPT 2014. LNCS, vol. 8441, pp. 441–458. Springer, Heidelberg (2014)

[CPR07] Coron, J.-S., Prouff, E., Rivain, M.: Side channel cryptanalysis of a higher order masking scheme. In: Paillier, P., Verbauwhede, I. (eds.) CHES 2007. LNCS, vol. 4727, pp. 28–44. Springer, Heidelberg (2007)

[CPRR13] Coron, J.-S., Prouff, E., Rivain, M., Roche, T.: Higher-order side channel security and mask refreshing. In: Moriai, S. (ed.) FSE 2013. LNCS, vol. 8424, pp. 410–424. Springer, Heidelberg (2014)

[DS15] Durvaux, F., Standaert, F.-X.: From improved leakage detection to the detection of points of interests in leakage traces. IACR Cryptology ePrint Archive, 2015:536 (2015)

[EWTS14] Eldib, H., Wang, C., Taha, M.M.I., Schaumont, P.: QMS: evaluating the side-channel resistance of masked software from source code. In: The 51st Annual Design Automation Conference 2014, DAC 2014, San Francisco, 1–5 June 2014, pp. 1–6. ACM (2014)

[GH15] Güneysu, T., Handschuh, H. (eds.): CHES 2015. LNCS, vol. 9293, pp. 517–534. Springer, Heidelberg (2015)

[GJJR11] Goodwill, G., Jun, B., Jaffe, J., Rohatgi, P.: A testing methodology for side channel resistance validation. In: NIST Non-invasive Attack Testing Workshop (2011)

[GP99] Goubin, L., Patarin, J.: DES and differential power analysis. In: Koç, Ç.K., Paar, C. (eds.) CHES 1999. LNCS, vol. 1717, pp. 158–172. Springer, Heidelberg (1999)

[ISW03] Ishai, Y., Sahai, A., Wagner, D.: Private circuits: securing hardware against probing attacks. In: Boneh, D. (ed.) CRYPTO 2003. LNCS, vol. 2729, pp. 463–481. Springer, Heidelberg (2003)

[KJJ99] Kocher, P.C., Jaffe, J., Jun, B.: Differential power analysis. In: Wiener, M. (ed.) CRYPTO 1999. LNCS, vol. 1666, pp. 388–397. Springer, Heidelberg (1999)

[Koc96] Kocher, P.C.: Timing attacks on implementations of Diffie-Hellman, RSA, DSS, and other systems. In: Koblitz, N. (ed.) CRYPTO 1996. LNCS, vol. 1109, pp. 104–113. Springer, Heidelberg (1996)

[P08] Pébay, P.: Formulas for robust, one-pass parallel computation of co-variances and arbitrary-order statistical moments. Technical report SAND2008-6212, Sandia National Laboratory (2008)

[PGA06] Prouff, E., Giraud, C., Aumônier, S.: Provably secure S-Box implementation based on Fourier Transform. In: Goubin, L., Matsui, M. (eds.) CHES 2006. LNCS, vol. 4249, pp. 216–230. Springer, Heidelberg (2006)

[PRR14] Prouff, E., Rivain, M., Roche, T.: On the practical security of a leakage resilient masking scheme. In: Benaloh, J. (ed.) CT-RSA 2014. LNCS, vol. 8366, pp. 169–182. Springer, Heidelberg (2014)

[RBN+15] Reparaz, O., Bilgin, B., Nikova, S., Gierlichs, B., Verbauwhede, I.: Consolidating masking schemes. In: Gennaro, R., Robshaw, M. (eds.) CRYPTO 2015. LNCS, vol. 9216. Springer, Heidelberg (2015)

[RP10] Rivain, M., Prouff, E.: Provably secure higher-order masking of AES. In: Mangard, S., Standaert, F.-X. (eds.) CHES 2010. LNCS, vol. 6225, pp. 413–427. Springer, Heidelberg (2010)

[RRVV15] Reparaz, O., Roy, S.S., Vercauteren, F., Verbauwhede, I.: A masked ring-LWE implementation. In: Güneysu, T., Handschuh, H. [GH15], pp. 683–702

[SM15] Schneider, T., Moradi, A.: Leakage assessment methodology - a clear roadmap for side-channel evaluations. In: Güneysu, T., Handschuh, H. [GH15], pp. 495–513

[SP06] Schramm, K., Paar, C.: Higher order masking of the AES. In: Pointcheval, D. (ed.) CT-RSA 2006. LNCS, vol. 3860, pp. 208–225. Springer, Heidelberg (2006)

[Stu08] Student. The probable error of a mean. Biometrika, pp. 1–25 (1908)

[Wel47] Welch, B.L.: The generalization ofstudent's' problem when several different population variances are involved. Biometrika **34**, 28–35 (1947)

There Is Wisdom in Harnessing the Strengths of Your Enemy: Customized Encoding to Thwart Side-Channel Attacks

Houssem Maghrebi[(⊠)], Victor Servant, and Julien Bringer

SAFRAN Morpho, 18, Chaussée Jules César, 95520 Osny, France
{houssem.maghrebi,victor.servant,julien.bringer}@morpho.com

Abstract. Side-channel attacks are an important concern for the security of cryptographic algorithms. To counteract it, a recent line of research has investigated the use of software encoding functions such as dual-rail rather than the well known masking countermeasure. The core idea consists in encoding the sensitive data with a fixed Hamming weight value and perform all operations following this fashion. This new set of countermeasures applies to all devices that leak a function of the Hamming weight of the internal variables. However when the leakage model deviates from this idealized model, the claimed security guarantee vanishes. In this work, we introduce a framework that aims at building customized encoding functions according to the precise leakage model based on stochastic profiling. We specifically investigate how to take advantage of adversary's knowledge of the physical leakage to select the corresponding optimal encoding. Our solution has been evaluated within several security metrics, proving its efficiency against side-channel attacks in realistic scenarios. A concrete experimentation of our proposal to protect the PRESENT Sbox confirms its practicability. In a realistic scenario, our new custom encoding achieves a hundredfold reduction in leakage compared to the dual-rail, although using the same code length.

Keywords: Constant weight countermeasures · Stochastic characterization · Customized encoding function · Security metrics · Information theoretic analysis · Side-channel analysis

1 Introduction

Side-Channel Attacks. Side-Channel attacks (SCA) are nowadays well known and most designers of secure embedded systems are aware of them. Since the first public reporting of these threats [15], a lot of effort has been devoted towards the research about side-channel attacks and the development of corresponding countermeasures. Side-channel attacks exploit information leaked from the physical implementations of cryptographic algorithms. Since, this leakage (*e.g.* the power consumption or the electromagnetic emanations) depends on the internally used secret key, the adversary may perform an efficient key-recovery attack to reveal

© International Association for Cryptologic Research 2016
T. Peyrin (Ed.): FSE 2016, LNCS 9783, pp. 223–243, 2016.
DOI: 10.1007/978-3-662-52993-5_12

this sensitive data. As this property can be exploited with relatively cheap equipment, these attacks pose a serious practical threat to cryptographic embedded systems. Amongst the side-channel attacks, two classes may be distinguished:

- The set of so-called *profiling SCA*: is the most powerful kind of SCA attacks and consists of two steps. First, the adversary procures a copy of the *target device* and uses it to characterize the physical leakage. Second, he performs a key-recovery attack on the target device. This set of profiled attacks includes Template attacks [5] and Stochastic models (*a.k.a.* Linear Regression Analysis) [10,22,23].
- The set of so-called *non-profiling SCA*: corresponds to a much weaker adversary who has only access to the physical leakage captured on the target device. To recover the secret key used, he performs some statistical analyses to detect the dependency between the leakage measurements and this sensitive variable. This set of non-profiled attacks includes Differential Power Analysis (DPA) [15], Correlation Power Analysis (CPA) [3] and Mutual Information Analysis (MIA) [13].

Side-Channel Countermeasures. A deep look at the state-of-the-art shows that several countermeasures have been published to deal with side-channel attacks. Amongst SCA countermeasures, two classes may be distinguished [18]:

- The set of so-called *masking countermeasures*: the core principle of masking is to ensure that every sensitive variable is randomly split into at least two shares so that the knowledge of a strict sub-part of the shares does not give information on the shared variable itself. Masking can be characterized by the number of random masks used per sensitive variable. So, it is possible to give a general definition for a dth-order masking scheme: every sensitive variable Z is randomly split into $d + 1$ shares $M_0, \cdots, M_d$ in such a way that the relation $M_0 \perp \cdots \perp M_d = Z$ is satisfied for a group operation $\perp$ (*e.g.* the XOR operation used in the *Boolean masking*, denoted as $\oplus$) and no tuple of strictly less than $d+1$ shares depends on Z. In the literature, several provably secure higher-order masking schemes have been proposed, see for instance [9,12,21].
- The set of so-called *hiding countermeasures*: the core idea consists in making the activity of the physical implementation constant by either adding complementary logic to the existing logic [7] (in a hardware setting) or by using a *specific encoding* of the sensitive data [6,14,24] (in a software setting). Therefore, making this activity constant would theoretically remove the correlation between the leakage measurements and the secret key.

Constant Weight Countermeasures. A recent line of works has investigated possibilities to protect block ciphers in software implementations using *constant weight coding* rather than using masking techniques. It is a specific encoding that has the particularity that all its codewords have a constant Hamming weight. More precisely, Hoogvorst *et al.* in [14] have presented a *dual-rail* implementation of PRESENT [2]. The idea was straightforwardly taken from existing dual-rail hardware, and consists in encoding one bit $s.t.$ the logical value 0 is represented as 01 and the logical value 1 is represented as 10 (or the inverse). Another idea derived from dual-rail can be found in a work by Chen *et al.* [6]. Several encodings are used, by reordering the bits and their complements in a word, in order to ensure constant Hamming weight and distance leakage for all operations of the block cipher PRINCE.

Recently, at CARDIS 2014, Servant *et al.* in [24] have proposed a new constant weight implementation of the AES extending the idea of the software dual-rail countermeasure proposed by Hoogvorst *et al.* in [14]. The core idea consists in encoding efficiently the sensitive data as a whole (*i.e.* not bit per bit) with a fixed Hamming weight value and then performing the AES internal operations following this fashion. When assuming a Hamming weight leakage model, the authors proved that their proposal is a leak-free countermeasure. However real world devices do not fit this model, as explained hereafter.

Stochastic Characterization of the Leakage. It is often assumed that a device leaks information based on the Hamming weight of the processed data. This assumption is quite realistic and many security analyses in the literature have been conducted following this model [3,19]. However, this assumption is not complete in real hardware [28], due to small load imbalances, process variations, routing, *etc.* For instance, authors in [16] have characterized, using a *stochastic approach*, the leakage of four AES Sbox outputs when implemented in three different devices. The obtained results prove that the leakage is very unbalanced for each Sbox output and hence, the Hamming weight assumption is unsound in practice. This imbalance always leaks some information that can be exploited by a SCA adversary. Hence, the security guarantee claimed by constant weight countermeasures does not necessarily hold in real world.

Our Contribution. In this paper, we refine the notion of data encoding as a countermeasure to thwart side-channel attacks. In fact, we try to bridge the gap between the physical leakage characteristics and the optimal encoding which balances at best the data leakage. This work exposes a method based on a first precise stochastic characterization of the target device, followed by the generation of a specific encoding according to this model. To do so, we propose an algorithm to select the best encoding function according to the physical leakage characterized on the target device. Our experiments show that the proposed encoding framework is more efficient than the existing constant weight countermeasures. We theoretically prove that our proposal reduces the Signal-to-Noise Ratio and hence, an adversary requires more traces to disclose the secret key than on the existing constant weight countermeasures. Furthermore, the security evaluation conducted illustrates that the leaked information is minimal and the efficiency of stochastic attack in exploiting this leakage is reduced drastically. Finally, the practical assessment of our proposal confirms its practicability to protect cryptographic operations. When device registers leak roughly the same function, our proposal could be applied to fully protect a block cipher. This assumption is not fully realistic, meanwhile, our work is a first step towards protecting block ciphers by involving customized encoding. If registers happen to leak vastly differently, then we need different encodings for each register and code conversions between memory accesses to protect a whole block cipher.

Paper Outline. The paper is organized as follows. In Sect. 2, we first detail two published constant weight implementations to protect a block cipher in a software setting. Then, we describe our new encoding framework in Sect. 3 and provide a theoretical analysis of it in Sect. 4. Furthermore, an information theoretic analysis and a security evaluation are conducted in Sect. 5. This is followed by some practical experiments applied on real devices in Sect. 6. Finally, Sect. 7 draws general conclusions and opens some perspectives for future work.

2 Existing Works on Leakage Balancing by Involving Encoding Functions

This principle of data internal encodings has already been proposed by Chow *et al.* in [8] in the context of *white-box cryptography*. Since then, several countermeasures have been proposed aiming at balancing the leakage by using some constant Hamming weight encodings in a *grey-box context*[1]. For instance, Hoogvorst *et al.* [14] have adapted the hardware dual-rail countermeasure to protect a software implementation of PRESENT. To do so, the authors suggest to duplicate the bit values representation, *i.e.* to use two bits to represent the logical value of one bit. For instance, one can encode the logical value 0 as 01 and the logical value 1 as 10 (or the inverse). When applying such an encoding to protect a n-bit variable, all codewords generated have a constant Hamming weight of n. Hence, assuming a Hamming weight leakage model, the power consumption provides no sensitive information. In the sequel, it will be referred as the dual-rail code.

The dual-rail representation is a specific case of this class of constant weight codes, but it is not the only option one should consider in a software setting. As a first example, authors of [6] propose a variation of the dual-rail applied to the block cipher PRINCE. Another example is [24], in which the authors propose a new balancing strategy based on the use of a code with the smallest cardinal available to encode the sensitive data. To protect a 4-bit variable, one can use 16 codewords of 6-bit length, each with a constant Hamming weight of 3. This code will be referred as the $(3,6)$-code in the rest of this paper. The security analysis conducted in [24] proves that this constant weight implementation is a leak-free countermeasure under a Hamming weight leakage model assumption. However, when the leakage function deviates from this idealized model, the security guarantee provided by this countermeasure vanish as discussed in [24].

To sum up, all these investigations on how to balance the physical leakage were conducted under the Hamming weight leakage model and with no prior characterization of the target device to incorporate the precise leakage model. Moreover, the choice of the code is made independently of the real bit leakage

---

[1] The adversary has access to the inputs and outputs of the cryptographic algorithm plus extra side-channel information.

(for example, in dual-rail representation, the logical value 1 is usually encoded as 10). Therefore, the claimed security level of these countermeasures could not be obtained in practice, where the bits may leak differently [16].

In the following section, we propose a framework for protecting sensitive data by using specific encoding. It is aimed to bridge the physical leakage characteristics to the choice of an optimal encoding function.

3   Towards a New Encoding Procedure for Leakage Balancing

Unlike previous works in which the Hamming weight model is often assumed, our solution is essentially based on harnessing the leakage characteristics and building a customized encoding accordingly to obtain the best balanced leakage. So, our framework is composed of two steps detailed in the following subsections.

3.1   First Step: Stochastic Characterization of the Leakage Function

A primordial step in our proposed framework is to take advantage of the adversary's knowledge of the target device during a *stochastic characterization phase*, a.k.a. *leakage profiling*.

Let Z be a sensitive variable defined over $\mathbb{F}_2^n$, then a stochastic characterization assumes that the leakage function $\mathcal{L}(Z)$ can be expressed as the sum of two *mutually independent* parts:

- a deterministic part $\mathcal{D}(Z)$: a function representing the power consumption during the processing of the sensitive variable Z and,
- a random part $\mathcal{R}$: a Gaussian noise with null mean and standard deviation σ.

Hence, the leakage function can be rewritten: $\mathcal{L}(Z) = \mathcal{D}(Z) + \mathcal{R} = \sum_{i=1}^{u} \alpha_i \delta_i(Z) + \mathcal{N}(0, \sigma)$, where α_i are some weighting coefficients and δ_i are some well chosen *basis functions*. Besides, we stress the fact that the basis choice is essential since it directly impacts the profiling efficiency.

For the sake of simplicity, in this work, we assume a linear basis. This choice is also motivated by the fact that higher-order basis functions are playing a minor role despite their better representation of the reality [10]. Moreover, the deterministic part of the leakage in practice is very close to the value of the linear part as discussed in [10]. So, our goal here is to characterize the leakage function when its deterministic part deviates from the Hamming weight model, but keeps the same degree. The study of higher-order basis functions (*e.g.* quadratic, cubic,...) is out of the scope of this paper.

This implies that every bit of the sensitive variable leaks independently. This assumption is often used in SCA context to characterize the perceived device leakage and sometimes referred as *Independent Bit Leakage* (IBL) assumption [10]. We recall hereafter this assumption.

Assumption 1 (IBL Assumption). *Let Z be a sensitive variable defined over $\mathbb{F}_2^n$, then the deterministic part of the leakage function can be rewritten: $\mathcal{D}(Z) = \sum_{i=1}^{n} \alpha_i Z[i]$, where $Z[i]$ denotes the ith bit of the sensitive variable Z.*

Under Assumption 1, the leakage function can be rewritten:

$$\mathcal{L}(Z) = \sum_{i=1}^{n} \alpha_i Z[i] + \mathcal{N}(0, \sigma) \ . \tag{1}$$

So to recover the leakage function, one can apply a *linear regression* [10,16] to obtain a precise estimation of the α_i coefficients under the IBL assumption.

3.2 Second Step: Encoding Function Selection

Once the leakage function is characterized, the second step of our framework consists in applying Algorithm 1 to obtain the optimal encoding function *w.r.t.* the profiled leakage.

Algorithm 1. Selection of the optimal encoding function

Input: m: the codeword bit-length, n: the sensitive variable bit-length and α_i: the leakage bit weights, where i in $[\![1, m]\!]$
Output: 2^n codewords of m-bit length
1: **for** X in $[\![0, 2^m - 1]\!]$ **do**
2:     Compute the power consumption for each codeword X and store the result in table D: $D[X] = \sum_{i=1}^{m} \alpha_i X[i]$
3:     Store the corresponding value of the codeword in the index table I: $I[X] = X$
4: **end for**
5: Sort the power consumption stored in table D and the index table I accordingly
6: **for** j in $[\![0, 2^m - 2^n]\!]$ **do**
7:     Find the *argmin* of $|D[j] - D[j + 2^n]|$
8: **end for**
9: **return** 2^n codewords corresponding to $[\![I[argmin], I[argmin + 2^n]]\!]$

Our Algorithm 1 takes as inputs: the length in bits of respectively the codewords and the sensitive data and, for each bit, the corresponding leakage weight obtained during a stochastic profiling as explained in the previous subsection. Then, it outputs 2^n codewords such that the delta consumption is the lowest among all subsets of 2^n codewords. Since the bit weights are unbalanced in practice, we argue that finding a code that guarantees a perfectly constant leakage remains an unreachable goal in most of cases.

Given the output codewords length, we compute the expected power consumption for each codeword and we store the result and the codeword value in table D and table I respectively (*c.f.* the first loop from Line 1 to Line 4 in Algorithm 1). Then, we sort table D (in ascending or descending order) and the index table I accordingly (*c.f.* Line 5 in Algorithm 1). Finally, since our goal is to choose a subset of 2^n codewords such that the delta consumption is the lowest one, we compute the delta of consumption for each subset of 2^n elements (*c.f.* the last loop from Line 6 to Line 8 in Algorithm 1) and later we select 2^n indexes from table I that minimize this delta. Thus, we obtain a code that ensures the best balancing of the leakage *w.r.t.* the stochastic profiling result.

A clustering Algorithm [1] would also give good results for this problem, but we explain hereafter why we chose this algorithm which is somewhat simpler to analyze. Let d be the maximum distance between two elements of a set S of n elements. One can show that $Var(S) < n.d^2$, so that intuitively, minimizing this distance d gives a subset with one of the lowest variances (and hence, one of the lowest SNR). There might be a set S' with lower variance but higher distance d', but in that case it would be easier to distinguish the two extreme values of this set. Some attacks might use this fact to improve the success rate.

Our framework consequently helps building properly encoding function customized for the physically observable leakage. It acts as an interface between the adversary's knowledge of the physical leakage and the optimal encoding to be used accordingly. We stress the fact that our Algorithm 1 is still applicable if the IBL assumption is not respected. To do so, one should inject the obtained leakage function in Line 2 and execute the algorithm to carry out the code.

4   Theoretical Analysis of the New Customized Encoding

In what follows, we provide a theoretical analysis of our solution. Namely, we will show that to succeed a *first-order univariate correlation attack* on our proposal, an adversary requires much more traces than on the existing constant weight countermeasures. This is due to the fact that the selected subset of codewords has a close-to-lowest power consumption variance among all possible subsets.

Let us start our analysis by exhibiting the explicit relationship between two *security metrics*: the Minimum number of Traces to Disclose the key with a given percentage of success rate (MTD), and the Signal-to-Noise Ratio (SNR). This link has already been demonstrated by Mangard in [17] for unprotected implementation. Our purpose is to provide the link between these two security metrics for encoding-based countermeasures.

To do so, we first recall how the number of traces to disclose the key is connected to the Correlation Power Analysis (CPA).

4.1   Analytical Derivation of the Security Level for Correlation Attacks

The CPA attack [3] is based on the computation of the *Pearson correlation coefficient* between the leakage function $\mathcal{L}(Z)$ and a *prediction function* $f(Z)$ chosen according to some assumptions on the device leakage model (*e.g.* the Hamming weight function). Hence, the Pearson correlation coefficient can be rewritten:

$$\rho[\mathcal{L}(Z); f(Z)] = \frac{\mathsf{Cov}[\mathcal{L}(Z); f(Z)]}{\sigma_{\mathcal{L}(Z)}\sigma_{f(Z)}} \ ,$$

where $\mathsf{Cov}[.;.]$ is the covariance and $\sigma_{\mathcal{L}(Z)}$ and $\sigma_{f(Z)}$ are respectively the standard deviation of the physical leakage and the prediction leakage function. Besides, in [17] the author demonstrated that the number of curves required to break a cryptographic implementation by CPA is equal to:

$$N_{1-\beta} = 3 + 8 \left(\frac{Z_{1-\beta}}{\ln\left(\frac{1+\rho}{1-\rho}\right)} \right)^2 , \tag{2}$$

where $Z_{1-\beta}$ is a quantile of a normal distribution for the 2-sided confidence interval with error $1 - \beta$.

We introduce hereafter the *optimal correlation function* and exhibit its relationship to the SNR security metric. Then, we deduce the explicit link between the number of traces to disclose the key and the SNR.

4.2   From Optimal Correlation Function to the SNR

The optimal correlation function is defined as the function that maximizes the correlation $\rho[\mathcal{L}(Z); f(Z)]$ and can be obtained from Corollary 8 in [20]:

$$\rho_{\mathrm{opt}} = \sqrt{\frac{\mathsf{Var}[\mathsf{E}[\mathcal{L}(Z) \mid Z = z]]}{\mathsf{Var}[\mathcal{L}(Z)]}} , \tag{3}$$

where $\mathsf{E}[.]$ and $\mathsf{Var}[.]$ denote the mean and the variance function respectively. Based on this definition, we introduce the following proposition.

Proposition 1. *Let $\mathcal{L}(Z)$ satisfies Eq. (1). Then, the optimal correlation function satisfies:*

$$\rho_{opt} = \sqrt{\frac{1}{1 + \frac{1}{SNR}}} , \tag{4}$$

where the SNR can be rewritten:

$$SNR = \left(\mathsf{Var}[\sum_{i=1}^{n} \alpha_i Z[i]] \right) / \sigma^2 . \tag{5}$$

As a direct consequence of Proposition 1, one can inject Eq. (4) into Eq. (2) to find the number of traces required by a CPA attack to succeed according to the SNR. Thus, assuming ρ is small[2], it yields the number of traces to achieve a success rate of 90 %, denoted $N_{90\%}$:

$$N_{90\%} \approx 8 \left(\frac{Z_{90\%}}{2\rho} \right)^2 \approx 8 \left(\frac{Z_{90\%}}{2\sqrt{\frac{1}{1+\frac{1}{SNR}}}} \right)^2 \approx \frac{2Z_{90\%}^2}{SNR} \tag{6}$$

---

[2] In fact, we can approximate $\ln\left(\frac{1+\rho}{1-\rho}\right) = \ln(1 + \rho) - \ln(1 - \rho) \approx \rho - (-\rho) \approx 2\rho$.

From Eq. (6) one can conclude that the smaller the SNR is, the more traces are required to achieve a success rate of 90 % for a CPA attack. As a direct consequence, if we decrease the SNR by a factor X, then the required number of traces to succeed the CPA attack will be multiplied by X.

In the next subsection, we evaluate our proposal by computing the SNR and then deducing the $N_{90\%}$.

4.3   Evaluation of Our Proposal Within the SNR and the $N_{90\%}$ Security Metrics

We recall that the deterministic part of the leakage function, defined under the IBL assumption, satisfies $\mathcal{D}(Z) = \sum_{i=1}^{n} \alpha_i Z[i]$. In the sequel, we make an additional assumption on the statistical distribution of the bit leakage weights α_i. In fact, for the sake of simplicity, the distribution of the α_i coefficients can fairly be approximated by a Gaussian law. This assumption that we shall call Gaussian Bit Leakage Weight (GBLW) assumption is formalized hereafter.

Assumption 2 (GBLW Assumption). *The bit leakage weights α_i are mutually independent random variables drawn from a Gaussian distribution with unity mean and standard deviation σ_α.*

Under Assumption 2, the leakage function can be rewritten $\mathcal{L}(Z) = \boldsymbol{\alpha} \cdot Z + \mathcal{N}(0, \sigma)$, where $(\cdot)$ denotes the scalar product operation and $\boldsymbol{\alpha} = [\alpha_1, \alpha_2, \dots, \alpha_n]$ denotes the bit leakage weight vector such that for every i in $[\![1, n]\!]$ we have $\alpha_i \sim \mathcal{N}(1, \sigma_\alpha)$. Let $\mathcal{C}$ be a (n, m)-function, *i.e.* $\mathcal{C} : \mathbb{F}_2^n \mapsto \mathbb{F}_2^m$ *s.t.* $n \leq m$, denoting the encoding operation used to protect a sensitive variable Z in $\mathbb{F}_2^n$. Then, the leakage function can be expressed as:

$$\mathcal{L}(Z) = \boldsymbol{\alpha} \cdot \mathcal{C}(Z) + \mathcal{N}(0, \sigma) = \sum_{i=1}^{m} \alpha_i \mathcal{C}(Z)[i] + \mathcal{N}(0, \sigma). \tag{7}$$

In the next proposition, we give an explicit formula of the SNR when an encoding function is involved to thwart SCA attacks.

Proposition 2. *Let $\mathcal{L}(Z)$ satisfy Eq. (7). Then, for every Z in $\mathbb{F}_2^n$, the Signal-to-Noise Ratio satisfies:*

$$SNR = \left(\sum_{\substack{i,j=1 \\ i \neq j}}^{m} \mathsf{E}[\mathcal{C}(Z)[i]\mathcal{C}(Z)[j]] + (\sigma_\alpha^2 + 1) \sum_{i=1}^{m} \mathsf{E}[\mathcal{C}(Z)[i]] - \left(\sum_{i=1}^{m} \mathsf{E}[\mathcal{C}(Z)[i]] \right)^2 \right) / \sigma^2. \tag{8}$$

Using the result of Proposition 2 and Eq. (6), one can evaluate the amount of traces required to reach a 90 % of success rate when an encoding is applied to protect a sensitive data. For the sake of comparison, we will also evaluate this metric for some well known countermeasures. We list hereafter the leakage functions we consider:

– Unprotected: $\mathcal{L}_{\mathrm{unpro}}(Z) = \boldsymbol{\alpha} \cdot \mathcal{C}_{\mathrm{unpro}}(Z) + \mathcal{N}(0, \sigma)$, where $\mathcal{C}_{\mathrm{unpro}}$ is the identity function.
– Software dual-rail [14]: $\mathcal{L}_{\mathrm{dual}}(Z) = \boldsymbol{\alpha} \cdot \mathcal{C}_{\mathrm{dual}}(Z) + \mathcal{N}(0, \sigma)$, where $\mathcal{C}_{\mathrm{dual}}$ is the dual-rail code.
– Software constant weight [24]: $\mathcal{L}_{\mathrm{cstHW}}(Z) = \boldsymbol{\alpha} \cdot \mathcal{C}_{\mathrm{cstHW}}(Z) + \mathcal{N}(0, \sigma)$, where $\mathcal{C}_{\mathrm{cstHW}}$ is the $(3,6)$-code.
– Our proposed customized encoding: $\mathcal{L}(Z)_{\mathrm{cust}} = \boldsymbol{\alpha} \cdot \mathcal{C}_{\mathrm{cust}}(Z) + \mathcal{N}(0, \sigma)$, where $\mathcal{C}_{\mathrm{cust}}$ is the code generated using Algorithm 1 for different codeword lengths.

In the sequel, we consider that the sensitive variable Z is a 4-bit variable, (*i.e.* $n = 4$). Then, for each of the above described leakage functions, we have computed the SNR over a set of 5.000 independent experiments using the result of Proposition 2. The standard deviation of the bit leakage weights σ_{α} was fixed at 0.25 and 0.5. Finally, we have deduced the $N_{90\%}$ using Eq. (6).

In Fig. 1, we plot the number of traces to achieve a success rate of 90 % according to the noise standard deviation σ. For our customized encoding functions, we show the results for different codewords lengths, *i.e.* $\mathcal{C}_{\mathrm{cust}} : \mathbb{F}_2^4 \mapsto \mathbb{F}_2^m$ with m in $[\![5, 10]\!]$.

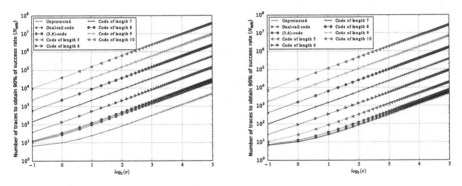

Fig. 1. Evolution of the number of traces to achieve a success rate of 90 % (y-axis) according to an increasing noise standard deviation σ (x-axis in log scale base 2). Left: for $\sigma_{\alpha} = 0.25$. Right: for $\sigma_{\alpha} = 0.5$.

From Fig. 1, the following observations could be emphasized:

– As expected the constant weight encoding countermeasures are less efficient than our customized encoding functions. For instance, when the noise standard deviation equals 16, about 10.000 and 2.000 are sufficient to reach a success rate of 90 % when σ_{α} equals 0.25 and 0.5 respectively. This is due to the fact that these codes are generated independently of the physical leakage by simply assuming a Hamming weight leakage model.

- The longer the code is, the more resistant the countermeasure is. In fact, our Algorithm 1 generates an encoding function such that the delta consumption of the selected codewords, the corresponding physical leakage variance, and the SNR are the lowest among all the subsets of codewords. So, the longer the code is, the more efficient our proposed algorithm is in selecting an encoding function that minimizes further the SNR. For instance, when $\sigma_\alpha = 0.25$ and $\sigma = 2$, the SNR decreases from about 0.032 for the $(3,6)$-code to 7.8×10^{-5} for the customized code of length 10.
- For a fixed noise standard deviation, one can notice that if σ_α increases, the adversary will need less traces to achieve a success rate of 90 %. For instance, when $\sigma = 8$ and the customized encoding of length 6 is used, the $N_{90\%}$ equals approximately 10.000 and 8.000 traces when σ_α varies from 0.25 to 0.5 as shown in Fig. 1. This observation is in-line with Eq. (8). In fact, when σ_α increases, the SNR increases accordingly and hence the $N_{90\%}$ decreases. To sum up, the degree of randomness of the leakage function has a noticeable impact on the amount of traces required by an adversary to achieve a success rate of 90 %. So, the higher σ_α is, the longer encoding function a designer should use.
- It is noteworthy that the code of length 5 is less efficient than the state-of-the-art countermeasures when $\sigma_\alpha = 0.25$. However, when $\sigma_\alpha = 0.5$, this customized code achieves a better result than the dual-rail and the $(3,6)$-code.

To conclude, our proposed encodings bring an overwhelming gain in terms of number of traces to succeed the CPA attack. For instance, to break the code of length 7, an adversary requires about 12 and 50 times more traces to achieve a CPA success rate of 90 % compared to the dual-rail countermeasure when σ_α equals 0.25 and 0.5 respectively.

5   Security Evaluation of the New Customized Encoding

As argued on the evaluation framework introduced in [25], the robustness of a countermeasure encompasses two dimensions: its amount of leakage irrespective of any attack strategy and its resistance to specific attacks. So, the evaluation of protected implementations should hold in two steps. First, an information theoretic analysis determines the actual information leakage. Second, a security analysis determines the efficiency of various attacks in exploiting this leakage.

Following this evaluation framework, we start with an information theoretic analysis in the following subsection.

5.1   Information Theoretic Analysis

To evaluate the information revealed by our proposed encoding functions, we compute the *Mutual Information Metric* (MIM) between the sensitive variable Z and the leakage function: $\mathsf{I}[\mathcal{L}_{\text{cust}}(Z); Z] = \mathsf{H}[\mathcal{L}_{\text{cust}}(Z)] - \mathsf{H}[\mathcal{L}_{\text{cust}}(Z) \mid Z]$, where $\mathsf{H}[.]$ denotes the entropy function. For the sake of comparison, we evaluate the

MIM for the leakage functions listed in Sect. 4.3 as well. Besides, we compute this metric also for a first-order masking leakage function:

$$\mathcal{L}_{\mathrm{mask}}(Z) = (\boldsymbol{\alpha_1} \cdot (Z \oplus M)) \times (\boldsymbol{\alpha_2} \cdot M) + \mathcal{N}(0, \sigma) , \qquad (9)$$

where M denotes a random mask defined over $\mathbb{F}_2^4$ and $(\boldsymbol{\alpha_1}, \boldsymbol{\alpha_2})$ are respectively the bit leakage weight vector of the masked data $(Z \oplus M)$ and the mask (M) such that $\boldsymbol{\alpha_1} \neq \boldsymbol{\alpha_2}$. Put differently, we assume that the masked data bits and the mask bits leak independently[3]. From Eq. (9), one can conclude that for masking we consider a *bivariate* leakage, *i.e.* a product combination of the two leakages (the masked data and the mask) is exploited by the adversary.

For each leakage function, the MIM was computed for several standard deviations of the bit leakage weights (σ_α in $\{0.05, 0.25, 0.75, 1\}$) and over a set of 200 independent experiments. The MIM is computed via numerical integration (Sect. 4.1.b of [10]). This method is accurate when the leakage is mathematically generated to perform simulations. The obtained results are shown in Table 1.

Table 1. Evolution of the MIM (y-axis in log scale base 2) according to an increasing noise standard deviation σ (x-axis in log scale base 2).

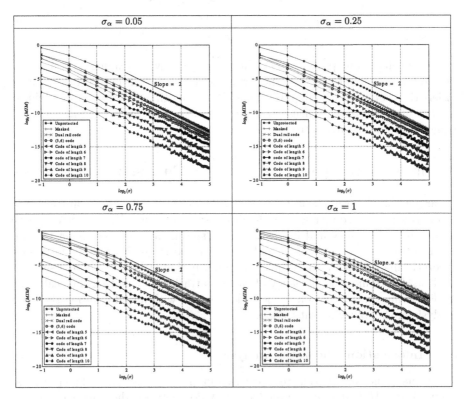

---

[3] Our goal here is to analyze the masking countermeasure in the worst case scenario (*i.e.* the mask register and the masked data register have different leakage functions).

From Table 1, the following observations can be emphasized:

- Interestingly, all MIM curves are parallel and have the same slope (-2). In fact, it has been demonstrated that the mutual information is proportional to $\sigma^{(-2d)}$ for large enough noises, where d denotes the order of the smallest statistical moment in the leakage distribution depending on the secret key and corresponds also to the number of shares used to represent the sensitive data [4,11,26]. Since for all the considered leakage functions the sensitive data is represented with a single share (*i.e.* $d = 1$)[4], then the corresponding mutual information decrease exponentially following a curve with slope (-2) when the noise standard deviation σ increases. As expected, this confirms that the unprotected implementation and the encoding-based countermeasures lead to first-order univariate weaknesses and that the masking countermeasure leads to first-order bivariate weakness.
- Despite having the same gradient, the amount of information leaked differs from a leakage function to another. For instance, one can see that whatever the σ_α value is, our proposed encoding functions of length superior to 6 leak less than the other encoding countermeasures and the first-order masking. This result is in-line with that of Sect. 4.3. In fact, the longer the code is, the less information is leaked, the lowest the SNR is, and the more traces are needed to break the implementation.
- For $\sigma_\alpha \leq 0.25$, our customized code of length 5 performs worse since it leaks more than the dual-rail and the $(3, 6)$-code. This result is also in-line with that shown in Fig. 1. In fact, an adversary requires less traces to break the optimal code of length 5 than the $(3, 6)$-code. This could be simply explained by the fact that for small σ_α the best code is a constant weight one and no such a code exists for length 5 to generate 16 codewords.
- It is noteworthy that the first-order masking performs worse when $\sigma_\alpha \geq 0.75$. It leaks slightly more information than an unprotected implementation. This result can be explained by the fact that when the bits of the two shares (the masked data and the mask) leak "very" differently, the countermeasure is doubly impacted (*i.e.* unbalance of the masked data leakage and unbalance of the mask leakage). This implies that the security guarantee by masking vanish in such a scenario. This result is in-line with that obtained in [11], where the MIM has been evaluated when the masking and the unprotected leakage functions radically deviate from the idealized Hamming weight model.
- It appears also that the degree of the deviation from the Hamming weight model (*i.e.* σ_α) has a noticeable influence on the amount of information leaked. In fact, for a fixed noise standard deviation σ, the higher σ_α, the larger the leakage. The same observation has been pinpointed in [11], *i.e.* the quantity of information leaked is strongly affected by the degree of randomness of the leakage function. Moreover, this result is in-line with that

---

[4] For the masking leakage function, we stress the fact that we have used one share which corresponds to the product combination of the masked data share and the mask share (*i.e.* a second-order analysis of the first-order masking).

discussed in Sect. 4.3, *i.e.* the higher σ_α, the less number of traces are needed to achieve a 90 % of success rate.

5.2   Side-Channel Security Analysis

To complete the security evaluation of our proposal, we conduct in this subsection a security analysis to evaluate its resistance to thwart SCA attacks. Namely, we perform a security evaluation of the stochastic attacks, for which a strong consistency with the previous security metrics analyses (*i.e.* the information theoretic analysis, the SNR and the MTD) should hold. To do so, we detail hereafter the attack simulation setup.

Simulation Setup. The leakage measurements have been simulated as samples of all the leakage functions listed in Sect. 4.3 and that detailed in Eq. (9) for the first-order masking countermeasure. Moreover, the sensitive variable Z was chosen to be a PRESENT Sbox output of the form $S(X \oplus k)$, where X represents a varying 4-bit plaintext and k represents the key nibble to recover.

Attack Scenarios. For our simulation attacks, we focus on two scenarios:

- *The best-case scenario:* we consider a powerful adversary who has access to the bit leakage weights and the characteristics of the optimal used code (*i.e.* the code length and the subset of the codewords). Then, he performs a stochastic attack by targeting the protected variable $\mathcal{C}(S(X \oplus k))$.
- *The worst-case scenario:* we consider a more realistic (and much weaker) adversary who has only the control on the target device to characterize the physical leakage. However, the characteristics of the used code are unknown. So, the adversary performs a linear regression over a 4-bit variable, *i.e.* the PRESENT Sbox output $S(X \oplus k)$.

For each scenario and for each leakage function, we compute the success rate of the stochastic attack [16] over 200 independent experiments. Moreover, this security metric was computed for several standard deviations of the bit leakage weights (σ_α in $\{0.05, 0.25, 0.75, 1\}$). The noise standard deviation was fixed at $\sigma = 0.25$. The simulation results in the best-case scenario are plotted in Table 2.

Simulation Results. For the best-case scenario, the results shown in Table 2 are in-line with those obtained during the information theoretic evaluation. In fact, when the $\sigma_\alpha \leq 0.25$, the optimal code of length 5 performs worse since an adversary requires less traces to achieve a 100 % of success rate than the constant weight countermeasures. Besides, we conclude again that the longer the code is, the more resistant the implementation is. Moreover, the higher the standard deviation of the bit leakage weights is, the less efficient the encoding function is.

For the worst-case scenario, as expected, the stochastic attack performs worse since the adversary does not have the control on the code length and the subset of codewords used for the protection. So, the profiling phase outputs an imprecise leakage model which impacts the attack efficiency. These simulation results also

Table 2. Stochastic attack results in the best-case scenario for a noise standard deviation $\sigma = 0.25$.

highlight the inefficiency of fixed constant weight codes such as the dual-rail in all the presented models. Customized encodings of the same length of 8 bits exhibit a much higher resistance.

6 Practical Evaluation of the New Customized Encoding

In the previous sections, we have confronted our theoretical analyses based on the SNR and the MTD security metrics with simulations based on the security evaluation framework proposed in [25]. In the following, we aim to confront these results against real measurements.

6.1 Implementation Considerations and Memory-Security Trade-Off

Encoding of sensitive data with codewords of longer length, *e.g.* representing the PRESENT Sbox output nibble as a byte, seems unreasonable for embedded software products at first, as the computation tables grow quadratically in size with the length of the code. In order to avoid large memory penalties when implementing our solution, a trick detailed in [14] and [24] has to be used. It consists in encoding a n-bit variable as two separate halves. This way, the linear and non-linear operations of a block-cipher can be performed at a much lower memory cost than with a double-length encoding. We begin with a quick reminder of this trick. Listing 1.1 shows how to perform an encoded memory access with an 8-bit input, encoded in two words of 7 bits each. This kind of operation could be an AES Sbox or a XOR operation between two nibbles for example.

```
1  // R3 = @table_msb, R4 = @table_lsb,  R5 = @shift1_table, R6 =
        @shift7_table
2  // R0 = operand MSB = 0xxxxxxx, R1 = operand LSB = 0yyyyyyy
3  LDRB    R2, [R5,R0]        // R2 = 00000000xxxxxxx0
4  EOR     R0,R0,R0           // Clear R0
5  LDRH    R0, [R6,R2]        // R0 = 00xxxxxxx0000000
6  EOR     R0,R0,R1           // R0 = 00xxxxxxxyyyyyyy
7  EOR     R1,R1,R1           // Clear R1
8  EOR     R2,R2,R2           // Clear R2
9  LDRB    R1, [R3,R0]        // R1 =table_msb[operand] (7 bits)
10 LDRB    R2, [R4,R0]        // R2 =table_lsb[operand] (7 bits)
```

Listing 1.1. Double-length encoded access for a code of length 7 (ARM assembly)

This procedure works if we assume all registers leak more or less the same function of their inputs. If registers happen to leak vastly differently, then we need different encodings for each register and code conversions between memory accesses. This study is out of the scope of this paper. As the code for the most significant bits of a register could be different than the one for the least significant bits, we require to tabulate the shift operation so that it outputs the correct code for the given destination within the register (R5 and R6 in Listing 1.1).

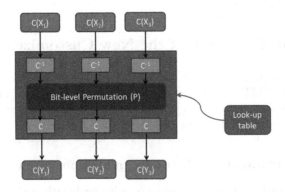

Fig. 2. Protecting bit-level permutation with encoding functions.

In the end this shows all classical operations of a block cipher (e.g. XOR, Sbox) can be covered. Regarding bit-level operations (e.g. permutation in DES, PRESENT), a solution may consist in tabulating these operations too, if there is enough memory available. As shown in Fig. 2, inside the look-up table we decode the inputs $\mathcal{C}(x_i)$ by computing $(\mathcal{C}^{-1})$, apply the bit-level permutation (P) and encode the result $(Y_i = P(x_i))$. In the former case, the overhead compared to an unprotected implementation would be the same as the one obtained in [24]. This means that an encoded AES would execute only roughly 4 times slower than its unprotected version.

Regarding the choice of the code length, it is up to the designer to choose the suitable length that guarantees the best performance-security trade-off according to the perceived physical leakage. Perhaps a recommendation could be to estimate the minimum number of traces to disclose the key (MTD) for different code lengths (as investigated in Sect. 4.3) then select the encoding function according to the available memory and the required level of security.

6.2   Experimental Setup

We have performed several practical experiments using a Micro-Controller integrated circuit from STMicroelectronics. Namely, we choose the $STM32F3$ circuit [27] based on the 32-bit RISC ARM Cortex-M4F processor core with 90nm CMOS process. In order to assess the practicability of our new framework in realistic case, we use 4 different copies of the $STM32F3$ circuit (referred as copy #1, #2, #3 and #4). Our goal is to provide an answer to the following question: *Does a customized encoding for one circuit ensure the same security level when implemented on a different circuit of the same family?*

So, the idea behind using four copies is to enable us to apply our framework (*i.e.* stochastic profiling of the leakage and customized encoding generation) on one copy and to use the same encoding functions to protect the other copies without a prior profiling. The target operation is a PRESENT Sbox computation protected by a customized encoding function with different codeword lengths. The side-channel traces were obtained by measuring the electromagnetic radiations (EM) emitted by the target device.

6.3   Attack Experiments and Results

To perform our profiling phase, we have first acquired 25.000 EM traces recording an AES Sbox computation when implemented on copy #1. The use of the AES Sbox (8 bits output) rather than the PRESENT Sbox (4 bits output) was necessary to extract the weights of all the 8 bits of a register by a stochastic approach. To do so, we have performed a linear regression attack and we have captured the averaged bit weights returned when the attack succeeds to find key[5]. For this circuit, we have observed that it leaks closely to the Hamming

---

[5] As detailed in [16], the stochastic attack does not only return the best key candidate but also a linear regression of the leakage.

weight model which implies an exploitable penalty in the security of constant weight countermeasures.

Second, we have executed Algorithm 1 to obtain the optimal encoding functions of length varying in $[\![5,7]\!]$ to protect the 4-bit PRESENT Sbox output. Third, for each code length, we have implemented the protected PRESENT Sbox on each copy. We stress the fact that we have used the obtained encoding functions (for copy #1) to protect the three other copies without a prior leakage characterization. For fair comparison, 50.000 EM traces were acquired with a fixed experimental setup: *i.e.* the same electromagnetic probe, the same probe's position, the same oscilloscope configuration to sample the measurements and the same temporal acquisition window. The code setup is a simple Sbox access in RAM which overwrites a register containing zero. The Sbox was aligned in memory for every encoding and the same registers were used for each copies.

Finally, we conducted 10 independent *enhanced CPA attacks*[6] against each implementation of the four copies (*i.e.* we used 10 independent set of 5.000 EM traces). The evolution of the averaged rank of the correct key among 16 (4-bit keys) is plotted in Fig. 3 for each circuit and code length.

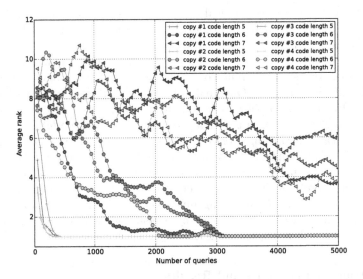

Fig. 3. Evolution of the correct key rank according to the number of observations.

The various encodings perform as expected on all circuits, although they were constructed from the profiling of only one of them. Longer codes provide higher resistance, but only very slightly for a code of length 5. These practical results are in-line with the simulation ones shown on Sect. 5.2. Overall, these

---

[6] We assume a powerful adversary who has access to the used encoding function $\mathcal{C}_{\mathrm{cust}}$ and the bit leakage weight vector $\boldsymbol{\alpha}$. Hence, he is able to compute $\rho[\mathcal{L}(Z); \boldsymbol{\alpha} \cdot \mathcal{C}_{\mathrm{cust}}(Z)]$. We emphasize the fact each attack was processed on the whole trace.

results confirm that one can profile a single device, devise the corresponding encodings, then use them to protect another device of the same family without loosing much in resistance. More studies should be performed in order to fully assess the generic side of the countermeasure nonetheless.

7   Conclusion

In this paper, we have proposed a new framework for building customized encoding function according to the physical leakage characteristics of the target device. It gives assurance that even under good profiling conditions for an attacker, the Signal-to-Noise Ratio is close to minimal. We also showed how much leakage reduction is to be expected for previous constant weight countermeasures in the case of an imbalanced leakage. The security evaluation conducted has shown the overwhelming advantages of our proposal compared to the existing constant weight countermeasures in more realistic scenarios. It is also more difficult to attack than a first-order masking when the latter's shares can be easily combined by an attacker. It is also possible to obtain the same performance impact as constant weight implementations, making customized encodings faster than known second-order masking schemes. Besides, the obtained results within the four considered security metrics (*i.e.* the SNR, the MTD, the information theoretic, the success rate of stochastic attack) are in-line, proving the tightness of our security evaluation process. Finally, the practical assessment of our solution have enabled us to confirm its practicability to protect cryptographic operations when applied on four different copies of the same device.

Our work opens avenues for further research of new encoding functions when assuming a higher-order leakage model (*e.g.* quadratic, cubic,...) and also the study of new designs combining both masking and encodings. Another future work will consist in studying the inter-conversion of encoding functions when the registers of a circuit have different leakage model and then, several customized codes have to be used to protect a block cipher.

Acknowledgments. This work has been partially funded by the ANR project SER-TIF. We thank anonymous reviewers for the various constructive comments and suggestions.

References

1. Batina, L., Gierlichs, B., Lemke-Rust, K.: Differential cluster analysis. In: Clavier, C., Gaj, K. (eds.) CHES 2009. LNCS, vol. 5747, pp. 112–127. Springer, Heidelberg (2009)
2. Bogdanov, A.A., Knudsen, L.R., Leander, G., Paar, C., Poschmann, A., Robshaw, M., Seurin, Y., Vikkelsoe, C.: PRESENT: an ultra-lightweight block cipher. In: Paillier, P., Verbauwhede, I. (eds.) CHES 2007. LNCS, vol. 4727, pp. 450–466. Springer, Heidelberg (2007)

3. Brier, E., Clavier, C., Olivier, F.: Correlation power analysis with a leakage model. In: Joye, M., Quisquater, J.-J. (eds.) CHES 2004. LNCS, vol. 3156, pp. 16–29. Springer, Heidelberg (2004)

4. Carlet, C., Danger, J.-L., Guilley, S., Maghrebi, H., Prouff, E.: Achieving side-channel high-order correlation immunity with leakage squeezing. J. Cryptographic Eng. 4(2), 107–121 (2014)

5. Chari, S., Rao, J.R., Rohatgi, P.: Template attacks. In: Kaliski, B.S., Koç, Ç.K., Paar, C. (eds.) CHES 2002. LNCS, vol. 2523, pp. 13–28. Springer, Heidelberg (2003)

6. Chen, C., Eisenbarth, T., Shahverdi, A., Ye, X.: Balanced encoding to mitigate power analysis: a case study. In: Joye, M., Moradi, A. (eds.) CARDIS 2014. LNCS, vol. 8968, pp. 49–63. Springer, Heidelberg (2015)

7. Chen, Z., Zhou, Y.: Dual-rail random switching logic: a countermeasure to reduce side channel leakage. In: Goubin, L., Matsui, M. (eds.) CHES 2006. LNCS, vol. 4249, pp. 242–254. Springer, Heidelberg (2006)

8. Chow, S., Eisen, P., Johnson, H., van Oorschot, P.C.: A white-box DES implementation for DRM applications. In: Feigenbaum, J. (ed.) DRM 2002. LNCS, vol. 2696, pp. 1–15. Springer, Heidelberg (2003)

9. Coron, J.-S.: Higher order masking of look-up tables. In: Nguyen, P.Q., Oswald, E. (eds.) EUROCRYPT 2014. LNCS, vol. 8441, pp. 441–458. Springer, Heidelberg (2014)

10. Doget, J., Prouff, E., Rivain, M., Standaert, F.-X.: Univariate side channel attacks and leakage modeling. J. Cryptographic Eng. 1(2), 123–144 (2011)

11. Duc, A., Faust, S., Standaert, F.-X.: Making masking security proofs concrete. In: Oswald, E., Fischlin, M. (eds.) EUROCRYPT 2015. LNCS, vol. 9056, pp. 401–429. Springer, Heidelberg (2015)

12. Genelle, L., Prouff, E., Quisquater, M.: Thwarting higher-order side channel analysis with additive and multiplicative maskings. In: Preneel, B., Takagi, T. (eds.) CHES 2011. LNCS, vol. 6917, pp. 240–255. Springer, Heidelberg (2011)

13. Gierlichs, B., Batina, L., Tuyls, P., Preneel, B.: Mutual information analysis. In: Oswald, E., Rohatgi, P. (eds.) CHES 2008. LNCS, vol. 5154, pp. 426–442. Springer, Heidelberg (2008)

14. Hoogvorst, P., Danger, J.-L., Duc, G.: Software implementation of dual-rail representation. In: COSADE, Darmstadt, Germany, 24–25 February 2011

15. Kocher, P.C., Jaffe, J., Jun, B.: Differential power analysis. In: Wiener, M. (ed.) CRYPTO 1999. LNCS, vol. 1666, pp. 388–397. Springer, Heidelberg (1999)

16. Lomné, V., Prouff, E., Roche, T.: Behind the scene of side channel attacks. In: Sako, K., Sarkar, P. (eds.) ASIACRYPT 2013, Part I. LNCS, vol. 8269, pp. 506–525. Springer, Heidelberg (2013)

17. Mangard, S.: Hardware countermeasures against DPA – a statistical analysis of their effectiveness. In: Okamoto, T. (ed.) CT-RSA 2004. LNCS, vol. 2964, pp. 222–235. Springer, Heidelberg (2004)

18. Mangard, S., Oswald, E., Popp, T.: Power Analysis Attacks: Revealing the Secrets of Smart Cards. Springer, Berlin (2006)

19. Peeters, E., Standaert, F.-X., Donckers, N., Quisquater, J.-J.: Improved higher-order side-channel attacks with FPGA experiments. In: Rao, J.R., Sunar, B. (eds.) CHES 2005. LNCS, vol. 3659, pp. 309–323. Springer, Heidelberg (2005)

20. Prouff, E., Rivain, M., Bevan, R.: Statistical analysis of second order differential power analysis. IEEE Trans. Comput. 58(6), 799–811 (2009)

21. Rivain, M., Prouff, E.: Provably secure higher-order masking of AES. In: CHES, pp. 413–427 (2010)

22. Schindler, W.: Advanced stochastic methods in side channel analysis on block ciphers in the presence of masking. J. Math. Cryptology **2**(3), 291–310 (2008)
23. Schindler, W., Lemke, K., Paar, C.: A stochastic model for differential side channel cryptanalysis. In: Rao, J.R., Sunar, B. (eds.) CHES 2005. LNCS, vol. 3659, pp. 30–46. Springer, Heidelberg (2005)
24. Servant, V., Debande, N., Maghrebi, H., Bringer, J.: Study of a novel software constant weight implementation. In: Joye, M., Moradi, A. (eds.) CARDIS 2014. LNCS, vol. 8968, pp. 35–48. Springer, Heidelberg (2015)
25. Standaert, F.-X., Malkin, T.G., Yung, M.: A unified framework for the analysis of side-channel key recovery attacks. In: Joux, A. (ed.) EUROCRYPT 2009. LNCS, vol. 5479, pp. 443–461. Springer, Heidelberg (2009)
26. Standaert, F.-X., Veyrat-Charvillon, N., Oswald, E., Gierlichs, B., Medwed, M., Kasper, M., Mangard, S.: The world is not enough: another look on second-order DPA. In: Abe, M. (ed.) ASIACRYPT 2010. LNCS, vol. 6477, pp. 112–129. Springer, Heidelberg (2010)
27. STMicroelectronics designer. http://www.st.com/web/en/catalog/mmc/FM141/SC1169/SS1576?icmp=ss1576_pron_pr_jan2015&sc=stm32f3-pr3
28. Veyrat-Charvillon, N., Standaert, F.-X.: Mutual information analysis: how, when and why? In: Clavier, C., Gaj, K. (eds.) CHES 2009. LNCS, vol. 5747, pp. 429–443. Springer, Heidelberg (2009)

Automated Tools for Cryptanalysis

Automatic Search for Key-Bridging Technique: Applications to LBlock and TWINE

Li Lin[1,2,3], Wenling Wu[1,2,3]([✉]), and Yafei Zheng[1,2,3]

[1] TCA Laboratory, SKLCS, Institute of Software, Chinese Academy of Sciences, Beijing, China
{linli,wwl,zhengyafei}@tca.iscas.ac.cn
[2] State Key Laboratory of Cryptology, P.O. Box 5159, Beijing 100878, China
[3] Graduate University of Chinese Academy of Sciences, Beijing 100190, China

Abstract. Key schedules in block ciphers are often highly simplified, which causes weakness that can be exploited in many attacks. At ASIACRYPT 2011, Dunkelman et al. proposed a technique using the weakness in the key schedule of AES, called key-bridging technique, to improve the overall complexity. The advantage of key-bridging technique is that it allows the adversary to deduce some sub-key bits from some other sub-key bits, even though they are separated by many key mixing steps. Although the relations of successive rounds may be easy to see, the relations of two rounds separated by some mixing steps are very hard to find. In this paper, we describe a versatile and powerful algorithm for searching key-bridging technique on word-oriented and bit-oriented block ciphers. To demonstrate the usefulness of our approach, we apply our tool to the impossible differential and multidimensional zero correlation linear attacks on 23-round LBlock, 23-round TWINE-80 and 25-round TWINE-128. To the best of our knowledge, these results are the currently best results on LBlock and TWINE in the single-key setting.

Keywords: Block ciphers · Key-bridging technique · Automatic search · Impossible differential cryptanalysis · Zero-correlation linear cryptanalysis · LBlock · TWINE

1 Introduction

A key schedule is an algorithm that expands a relatively short master key to a relatively large expanded key for later use in encryption and decryption algorithms. The key schedules in block ciphers are often highly simplified, which causes weakness that can be exploited in many attacks, especially for lightweight block ciphers. In these lightweight block ciphers, the security margin that conventional block ciphers are equipped with is reduced as much as possible in order to optimize the software and hardware efficiency. One obvious sacrifice is that the key schedules are highly simplified for saving memory. Some key schedules have round-by-round iterations with low diffusion [4, 21, 25]. Some key schedules do simple permutations or linear operations with low diffusion [1, 13]. Some have no

© International Association for Cryptologic Research 2016
T. Peyrin (Ed.): FSE 2016, LNCS 9783, pp. 247–267, 2016.
DOI: 10.1007/978-3-662-52993-5_13

key schedules, and just use master keys directly in each round [12,15]. These key schedules are succinct but responsible for many attacks, especially related-key attacks [3,16] and meet-in-the-middle attacks [2,5].

AES [8] is the most significant standard for block ciphers, so its security is of paramount importance. However, the key schedule of AES has clear weakness that directly assists the execution of some effective attacks. Especially in recent years, meet-in-the-middle cryptanalysis with differential enumeration technique [11] has shown to be a very powerful form of cryptanalysis against 7-round AES-128 [9], 9-round AES-192 [17] and 10-round AES-256 [18], which are the best single-key attacks on all versions of AES so far. A technique using the weakness of the key schedule on AES, called key-bridging technique, is used in these attacks to improve the overall complexity. Key-bridging technique is proposed by Dunkelman et al. at ASIACRYPT 2011 [11]. The advantage of key-bridging technique is that it allows the adversary to deduce some sub-key bytes from some other sub-key bytes, even though they are separated by many key mixing steps. Although the relations of successive rounds may be easy to see, the relations of two rounds separated by some mixing steps are very hard to find. The main novelty in this observation is that it exploits the weak key schedule of AES-192 in order to provide a surprisingly long "bridge" for two sub-keys which are separated by 8 key mixing steps. The key-bridging technique considerations reduce the time complexity in the online phase of the attack on 8-round AES-192 by a factor of 2^{32} and 8-round AES-256 by a factor of 2^8. At FSE 2014, Li et al. introduce a new application of key-bridging technique called key-dependent sieve technique, which filters the wrong states based on the key relations, to further reduce the complexity in the precomputation phase [17]. Besides, they introduce another application of key-bridging technique to split the whole attack into some weak-key attacks according to the relations between the sub-keys in the online phase and the precomputation phase.

Besides AES, the key-bridging technique helps improve the attack complexities of other block ciphers. For example, at FSE 2015, Biryukov et al. apply the key-bridging technique to 25-round TWINE-128, and get a meet-in-the-middle attack and an impossible differential attack [2]. At ACISP 2014, Wang et al. give multidimensional zero-correlation linear attacks on LBlock and TWINE. In the online phase of their attacks, the key-bridging technique is used to improve the attack complexity [22].

Our Contribution. In this paper, we describe versatile and powerful algorithms for searching key-bridging technique on word-oriented and bit-oriented block ciphers. Our tool tries to find key-bridges automatically by dealing with a system of equations. It takes as input a system of equations that describes the key schedule and a set $\mathbb{K}_0$ of some key variables that we want to build key-bridges among. It is made up of two phases: knowledge-propagation phase and relation-derivation phase. In the knowledge-propagation phase, we can derive a set $\mathbb{K}$ that $\mathbb{K}_0$ can propagate to. In the relation-derivation phase, the relations of the variables in $\mathbb{K}_0$ can be known from $\mathbb{K}$.

To demonstrate the usefulness of our approach, we apply our tool to LBlock and TWINE. We automatize the search for the best impossible differential attacks by combining our key-bridging tool with the tool of Wu et al. [24]. Using Wu's tool, we can get all the impossible differential distinguishers with certain rounds [20]. Using our key-bridging tool, we can get all the key-bridges to reduce the complexity in the key-sieving phase. With these two tools, we get a 23-round impossible differential attack on LBlock with time complexity of $2^{74.5}$ 23-round LBlock encryptions, memory complexity of $2^{74.3}$ bytes and data complexity of $2^{59.5}$ chosen plaintexts. For TWINE-128, we get in total twelve 25-round impossible differential attacks with the same complexity as Biryukov et al.'s attack in [2].

For multidimensional zero-correlation linear cryptanalysis, we use the same attack model Wang et al. proposed in [22] and get more key-bridges to improve the overall complexity with our key-bridging tool. For the 23-round attack on TWINE-80, we find that the key-bridges Wang et al. used in their attack do not exist. This will make the time complexity of their attack greater than exhaustively search. We use another zero-correlation linear distinguisher to fix this error and get an attack on 23-round TWINE-80 with time complexity of 2^{73} 23-round TWINE-80 encryptions, memory complexity of 2^{60} bytes and data complexity of $2^{62.1}$ known plaintexts. For the 25-round attack on TWINE-128, we first get some more key-bridges to improve the time complexity of Wang's work. Then, we use another distinguisher with more key-bridges in the first two steps of the attack, and get an attack with time complexity of 2^{119} 25-round TWINE-128 encryptions, memory complexity of 2^{60} bytes and data complexity of $2^{62.1}$ known plaintexts. For the 23-round multidimensional zero-correlation linear attack on LBlock, we find a distinguisher with more key-bridges than Wang et al. in [24], and get an attack with time complexity of 2^{72} 23-round LBlock encryptions, memory complexity of 2^{60} bytes and data complexity of $2^{62.1}$ known plaintexts. For 25-round TWINE-128, we also find some meet-in-the-middle attacks with the same complexity as Biryukov et al.'s attack in [2], even with one attack which starts with two inactive nibbles at the beginning of distinguisher. This distinguisher is useful when we want to get less false positive. To the best of our knowledge, these results are the currently best results on LBlock and TWINE.

We present here a summary of our attack results on LBlock and TWINE, and compare them to the best attacks known for them. This summary is given in Table 1. The source code of some of these attacks is available at https://onedrive.live.com/redir?resid=20C3554F0C8B0806!108& authkey=!AJPOWJTJ4mSLrbI&ithint=folder%2c7z.

Organization of This Paper. The rest of this paper is organized as follows. Section 2 presents the input of our tool and the previous works on key-bridging technique. Section 3 gives our automatic search tool for key-bridging technique. Section 4 (resp. Sect. 5) applies our tool to the impossible differential and multidimensional zero-correlation linear cryptanalysis on LBlock (resp. TWINE). Finally, Sect. 6 concludes this paper.

Table 1. Summary of the best attacks on LBlock and TWINE-80/128.

Cipher	Attack type	Rounds	Data	Memory (Bytes)	Time (Enc)	Source
LBlock	Impossible Diff.	23	2^{59} CPs	2^{74}	$2^{75.36}$	[6]
	Impossible Diff	23	$2^{59.5}$ CPs	$2^{74.3}$	$2^{74.5}$	Sect. 4.2
	Multidim. ZC	23	$2^{62.1}$ KPs	2^{60}	2^{76}	[22]
	Multidim. ZC	23	$2^{62.1}$ KPs	2^{60}	2^{72}	Sect. 4.3
TWINE-80	Impossible Diff	23	$2^{57.85}$ CPs	$2^{84.06}$	$2^{79.09}$	[26]
	Multidim. ZC	23	$2^{62.1}$ KPs	2^{60}	2^{73}	Sect. 5.2
TWINE-128	Impossible Diff.★	25	$2^{59.1}$ CPs	$2^{78.1}$	$2^{124.5}$	[2]
	MITM ★	25	2^{48} CPs	2^{109}	$2^{124.7}$	[2]
	Multidim. ZC	25	$2^{62.1}$ KPs	2^{60}	$2^{122.12}$	[22]
	Multidim. ZC	25	$2^{62.1}$ KPs	2^{60}	2^{119}	Sect. 5.2

KPs: Known-Plaintexts. CPs: Chosen-Plaintexts.
★: Find the other attacks with the same complexity in Sect. 5.3.

2   Preliminaries

In this section, we introduce the definitions and related works of key-bridging technique. First of all, let's give some notations used throughout this paper.

2.1   Notations

In this paper, WK_i denotes the ith round key register, WK_i^j denotes the jth word of WK_i, $WK_i^{j_0-j_1}$ denotes the j_0^{th} word to j_1^{th} word of WK_i, $WK_i[k]$ denotes the kth bit of WK_i, $WK_i^j[k]$ denotes the kth bit of WK_i^j and $WK_i \lll b$ denotes b-bit left cyclic shift of WK_i.

2.2   The Key Schedule Functions

The input of our tool is a system of equations that describes the key schedule and the key variables which we want to find relations among. Since our tool is useful not only for the word-oriented key schedules (e.g., AES), but also for the bit-oriented key schedules (e.g., PRESENT), we describe the systems of equations for these two kinds of key schedules here. We take the key schedules of AES-192 and PRESENT-80 as examples.

The key schedule of AES-192 takes the 192-bit master key WK_0 and extends it into 9 key registers $WK_0, WK_1, \cdots, WK_8$ of 192-bit each using a key schedule algorithm given by the following equations [8]:

$$KS_i : \begin{cases} WK_i^j + WK_i^{j-4} + WK_{i-1}^j = 0, j = 4, \cdots, 23, \\ WK_i^0 + WK_{i-1}^0 + S(WK_{i-1}^{21}) + RCON_i = 0, \\ WK_i^1 + WK_{i-1}^1 + S(WK_{i-1}^{22}) = 0, \\ WK_i^2 + WK_{i-1}^2 + S(WK_{i-1}^{23}) = 0, \\ WK_i^3 + WK_{i-1}^3 + S(WK_{i-1}^{20}) = 0, \end{cases}$$

where S represents the S-box of the SubBytes transformation and "+" represents xor. These 9 key registers are used to get 13 sub-keys $RK_{-1}, RK_0, \cdots, RK_{11}$ of 128-bit each (only the first 128-bit of WK_8 is used to get RK_{11}).

In some cases, we are interested in interchanging the order of the MixColumns and AddRoundKey operations. As these operations are linear, they can be interchanged by first xoring the data with an equivalent key u_i and then applying the MixColumns operation.

The key schedule of PRESENT-80 takes the 80-bit master key WK_0 and extends it into 32 key registers $WK_0, WK_1, \cdots, WK_{31}$ of 80 bits each using a key schedule algorithm given by the following equations [4]:

$$KS_i : \begin{cases} WK_i[0-3] + S(WK_{i-1}[61-64]) = 0, \\ WK_i[60-64] + WK_{i-1}[41-45] + [i-1] = 0, \\ WK_i[j] + WK_{i-1}[(j+19)mod\ 80] = 0, \qquad j = 4, \cdots, 40, 46, \cdots, 79 \end{cases}$$

At round i, the 64-bit round key $RK_i = RK_i[0]RK_i[1] \cdots RK_i[63]$ consists of the 64 leftmost bits of the current content of register WK_i.

The key schedules of other bit-oriented and word-oriented block ciphers can be treated as before. To simplify the statement, we ignore the round constants in this paper since they are known to us.

2.3   Key-Bridging Technique on AES

In [11], Dunkelman et al. proposed the key bridging technique on AES-192. The advantage of key-bridging technique is that it allows the adversary to deduce some sub-key bytes from some other sub-key bytes, even though they are separated by many key mixing steps. Although the relations of successive rounds may be easy to see, the relations of two rounds separated by some mixing steps are very hard to find. The main novelty in this observation is that it exploits the weak key schedule of AES-192 in order to provide a surprisingly long "bridge" for two sub-keys which are separated by 8 key mixing steps (applied in reverse direction). This observation is shown in Observation 1.

Observation 1 (Key-Bridging Technique on AES, [11]). *By the key schedule of AES-192, knowledge of columns 0, 1, 3 of the sub-key RK_7 allows to deduce column 3 of the whitening key RK_{-1} (which is actually column 3 of the master key).*

Given RK_7^{0-3} and RK_7^{4-7}, it is possible to compute RK_5^{12-15}; given RK_7^{4-7} and RK_7^{12-15}, it is possible to compute RK_4^{12-15}. From these two values, it is possible to compute RK_{-1}^{12-15}. We refer to [10] the detailed proof and reasoning.

The key-bridging technique considerations reduce the time complexity of the online phase of the attacks on 8-round AES-192 by a factor of 2^{32} and 8-round AES-256 by a factor of 2^8 [10], and also improve the SQUARE attack and related-key impossible differential attack on AES-192.

At EUROCRYPT 2013, Derbez et al. gave improved attacks on 7-round AES-128, 8-round AES-192 and 9-round AES-256 [9]. In the online phase of their attack on 8-round AES-192, the use of the key-bridging technique saves a large amount of time.

At FSE 2014, Li et al. introduced a new application of key-bridging technique called **key-dependent sieve technique**, which filters the wrong states based on the key relations, to further reduce the complexity in the precomputation phase [17]. Besides, they found that the whole attack can be split up into some weak-key attacks according to the relations between the sub-keys in the online phase and the precomputation phase. These can be seen as other applications of key-bridging technique.

In [18], Li and Jen gave an attack on 10-round AES-256. In their works, they use key-bridging technique both in the precomputation phase and the online phase.

2.4   Key-Bridging Technique on Other Block Ciphers

At FSE 2015, Biryukov et al. applied the key-bridging technique to 25-round TWINE-128, and got a meet-in-the-middle attack and an impossible differential attack [2].

In the meet-in-the-middle attack, 58 state nibbles are needed to perform the online phase. Hopefully, the key schedule equations reduce the amount of possible values from $2^{4 \times 58} = 2^{232}$ to 2^{124}. Indeed, knowing 23 out of 24 nibbles of one sub-key leads to the knowledge of enough key material to partially encrypt and decrypt the plaintext and the ciphertext in order to obtain the value of the required state variables. This can be seen as 37 key-bridges among the 68 relevant sub-key nibbles. The same technique is applied to the impossible differential attack.

At ACISP 2014, Wang et al. gave multidimensional zero-correlation linear attacks on LBlock and TWINE. In the online phase of their attacks, the key-bridging technique is used to reduce the overall complexity [22].

Most attacks on block ciphers can be split into three consecutive parts of r_1, r_2 and r_3 rounds, $r = r_1 + r_2 + r_3$, such that a particular set of messages may verify a certain property in the middle r_2 rounds by guessing some key-bits in the first r_1 and last r_3 rounds. These key-bits may have some relations by the key schedule. If we can get these relations automatically, it can not only give better attacked-rounds and complexity, but also a better understanding of the design of block ciphers. Therefore, we give our automatic search tool for key-bridging technique in the next section.

3   An Automatic Search Tool for Key-Bridging Technique

In this section, we introduce our automatic search tool for key-bridging technique on word-oriented and bit-oriented block ciphers.

3.1   Outline of the Tool

Let us denote by $\mathcal{V}(X)$ the vector space spanned by $1, x, S(x)$ for all $x \in X$, for any set of variables X. If we denote by $\mathbb{X}$ the set of all internal key variables, then the key schedule equations can be seen as a subspace of $\mathcal{V}(\mathbb{X})$. We introduce the notation $\mathbb{K}_0$ to denote the set of original variables that we want to build key-bridges among. We also introduce $\mathbb{K}$ to denote the set of variables that $\mathbb{K}_0$ can propagate to. And $|X|$ means the number of variables in a set X.

Our goal is to find relations among a set of variables. The difficulty in finding such relations is how to get more information from $\mathbb{K}_0$ and how to use this information to retrieve the relations. In this section, we present a tool that finds such attacks automatically. It takes as input a system of equations $\mathbb{E} \subseteq \mathcal{V}(\mathbb{X})$ that describes the key schedule and a set of variables $\mathbb{K}_0$ that we want to find relations among. This tool consists of two phases: knowledge-propagation phase and relation-derivation phase. In the knowledge-propagation phase, we can derive a set $\mathbb{K}$ that $\mathbb{K}_0$ can propagate to. In the relation-derivation phase, the relations of the variables in $\mathbb{K}_0$ can be known from $\mathbb{K}$.

In the knowledge-propagation phase, if we substitute the values of $\mathbb{K}$ into the original equations $\mathbb{E}$, we would indeed get a system of equations with less variables. In fact, this reduced system is the subspace $(\mathbb{E} + \mathcal{V}(\mathbb{K}))/\mathcal{V}(\mathbb{K})$ of the quotient space $\mathcal{V}(\mathbb{X})/\mathcal{V}(\mathbb{K})$: starting from an equation $f \in \mathbb{E}$, its equivalence class $[f]$ in the quotient contains a representative where all the variables in $\mathbb{K}$ have disappeared. Let's denote by $\mathscr{L}$ a linear combination of some variables in $\mathcal{V}(\mathbb{K})$. The variable x can be deduced from $\mathbb{K}$ if there exists an $\mathscr{L}$ such that $x + \mathscr{L} \in \mathbb{E}$, $S(x) + \mathscr{L} \in \mathbb{E}$ or the linear combination of x, $S(x)$ and $\mathscr{L}$ belongs to $\mathbb{E}$, and we will write $x \in \text{PROPAGATE}(\mathbb{K})$ when it is the case. It follows that in any solution of the equations $\mathbb{E}$, the value of x (resp. $S(x)$) is the value of $\mathscr{L}$. Therefore, it just has to evaluate $\mathscr{L}$ to uniquely determine the value of x.

In the relation-derivation phase, the subspace $\mathbb{E} \cap \mathcal{V}(\mathbb{K})$ of $\mathcal{V}(\mathbb{K})$ should be derived. Then the linear relations among $\mathbb{K}_0$ can be known by dealing with the quotient space $(\mathbb{E} \cap \mathcal{V}(\mathbb{K}))/\mathcal{V}(\mathbb{K}_0)$.

3.2   A Tool for Word-Oriented Ciphers

Knowledge-Propagation Phase.[1] Let's denote by $\mathbb{M}$ the coefficient matrix made by the key schedule equations $\mathbb{E}$. Each row is a function, and each column is a variable. The order of variables is $(\mathbb{X} - \mathbb{K}, \mathbb{K}, c)$, where $\mathbb{X} - \mathbb{K}$ means the supplementary set of $\mathbb{K}$ in $\mathbb{X}$. We ignore the constant column in the matrix to better describe and express our tool in the rest of this paper. We can also view the constant column as a special column vector which always exists in the last of the matrix.

Given $\mathbb{K}_0$, we may propagate knowledge and derive the values of new variables, and this yields a new set $\mathbb{K}$. But it may turn out that new variables may again be derived from $\mathbb{K}$. The problem boils down to getting new variables and using these variables to get more information.

---

[1] A similar idea of this phase was proposed in [14] by Khovratovich et al..

Gauss-Jordan Elimination (GJE) is introduced to propagate knowledge [23,27]. GJE is an algorithm for solving systems of linear equations. It is usually understood as a sequence of elementary row operations performed on the associated matrix of coefficients. This method can also be used to find the rank of a matrix, and to convert a matrix into reduced row echelon form. $GJE(\mathbb{M})$ means that we convert a matrix $\mathbb{M}$ into reduced row echelon form by Gauss-Jordan Elimination. $GJE_n(\mathbb{M})$ means that we only convert the first n columns into reduced row echelon form by the row operations of the whole $\mathbb{M}$.

Since the equations $\mathbb{E}$ can be completely linear (e.g., key schedule of Simon) or partial-nonlinear (e.g., key schedule of AES), some variables appear both linearly and under the S-box. The following three situations can be used to propagate knowledge:

1. If either x or $S(x)$ belongs to $\mathbb{K}$, then the other one can be deduced.
2. If there exists a linear combination $\mathscr{L}$ of $\mathcal{V}(\mathbb{K})$ such that for one variable $x \notin \mathbb{K}$, $x + \mathscr{L} \in \mathbb{E}$, then x can be deduced from $\mathbb{K}$.
3. If there is a linear combination $\mathscr{L}$ of $\mathcal{V}(\mathbb{K})$ such that there is a linear combination of x ($x \notin \mathbb{K}$), $S(x)$ ($S(x) \notin \mathbb{K}$) and $\mathscr{L}$ belonging to $\mathbb{E}$, then x can be deduced from $\mathbb{K}$.

Gauss-Jordan Elimination is used to deal with situation 2 and situation 3 as follows. The proof of the following two lemmas will be presented in the full version of this paper.

Lemma 1. *Situation 2 holds if and only if there is only one non-zero variable in the first $|\mathbb{X} - \mathbb{K}|$ columns of one row in $GJE_{|\mathbb{X}-\mathbb{K}|}(\mathbb{M})$.*

Lemma 2. *Situation 3 holds if and only if one of the following two cases holds in $GJE_{|\mathbb{X}-\mathbb{K}|}(\mathbb{M})$ (for x and $S(x)$):*

(i) The coefficients of x and $S(x)$ are both pivot elements, and the corresponding rows are $(\underbrace{0, \cdots, 0}_{t_1}, e_x, e_{t_1+1}, \cdots, e_{n-1})$ and $(\underbrace{0, \cdots, 0}_{t_2}, e'_{S(x)}, e'_{t_2+1}, \cdots, e'_{n-1})$ $(t_1 < t_2)$, where $e_{t_1+1} = e'_{t_1+1} = \cdots = e_{t_2-1} = e'_{t_2-1} = 0$, $e_i = c \cdot e'_i$ for $i = t_2 + 1, \cdots, n - |\mathbb{K}| - 1$.

(ii) One of the coefficients of x and $S(x)$ is pivot element (e.g., x) and the corresponding row is $(\underbrace{0, \cdots, 0}_{t_1}, e_x, \underbrace{0, \cdots, 0}_{t_2}, e_{S(x)}, \underbrace{0, \cdots, 0}_{n-2-t_1-t_2-|\mathbb{K}|}, e_{n-|\mathbb{K}|}, \cdots, e_{n-1})$.

When a new variable x (resp. $S(x)$) becomes a member of $\mathbb{K}$, we have to move the column that x (resp. $S(x)$) represents in $\mathbb{M}$ to the last few columns to make the order of variables $(\mathbb{X} - \mathbb{K}, \mathbb{K}, c)$ unchanged. If one of these coefficients is pivot element, then moving it may leave the matrix not in reduced row echelon form. This can be fixed through simple column permutations in some cases. In some other cases, a new column has to be recomputed. The following lemma will make sure that the column operations don't change the property of linear relations.

Lemma 3. *Column operations keep the linear relations we get from situation 2 and 3 unchanged, i.e., these relations can be recovered from the last $|\mathbb{K}| + 1$ columns of $GJE_{|\mathbb{X}-\mathbb{K}|}(\mathbb{M})$.*

Proof. If situation 2 holds, then there is one new $x \in \textsc{Propagate}(\mathbb{K})$. After moving the corresponding column and getting a new $\mathbb{K}$, there exists a vector α such that $\mathbb{K} \cdot \alpha^T = c$ and the component of α for x is non-zero (here we treat $\mathbb{K}$ as a vector and α is not unique).

After $GJE_{|\mathbb{X}-\mathbb{K}|}$, the matrix can be represented as block matrix $\begin{pmatrix} \overbrace{A_0}^{|\mathbb{X}-\mathbb{K}|} & \overbrace{A_1}^{|\mathbb{K}|} \\ 0 & A_2 \end{pmatrix}$.

Since $\mathbb{K} \cdot \alpha^T$ has no variables in $\mathbb{X} - \mathbb{K}$, the row represents this equation must exist in A_2. If not, there must be a pivot element in $\mathbb{X} - \mathbb{K}$.

Situation 3 can be got for the same reason.

$\square$

The pseudo-code of the knowledge-propagation phase is shown in Algorithm 1. The inputs are a set of all internal key variables, a set of original variables that we want to build key-bridges among and a coefficient matrix made by the key schedule equations. The algorithm returns a set of variables that $\mathbb{K}_0$ can propagate to and a block matrix $\begin{pmatrix} \overbrace{A_0}^{|\mathbb{X}-\mathbb{K}|} & \overbrace{A_1}^{|\mathbb{K}|} \\ 0 & A_2 \end{pmatrix}$ in order to recover all the relations in $\mathbb{K}_0$.

From Lemmas 1 and 2, we can conclude that $\mathbb{K}$ is the maximum set $\mathbb{K}_0$ can propagate to. From Lemma 3, all the relations can be recovered from $\mathbb{M}$.

Relation-Derivation Phase. The input of this phase is the output of Algorithm 1. First of all, we should derive the linear relations among $\mathbb{K}$. Since the output matrix $\mathbb{M}$ of Algorithm 1 has the form $\begin{pmatrix} \overbrace{A_0}^{|\mathbb{X}-\mathbb{K}|} & \overbrace{A_1}^{|\mathbb{K}|} \\ 0 & A_2 \end{pmatrix}$, due to the proof of Lemma 3, all the linear relations among $|\mathbb{K}|$ exist in A_2. Meanwhile, $Rank(\mathbb{M}) = Rank(A_0) + Rank(A_2)$. We should test the existence of linear relations among $\mathbb{K}$ by testing whether $Rank(\mathbb{M})$ equals $Rank(A_0)$. If $Rank(A_2) \neq 0$, change the order of columns in A_2 to make sure that the order of variables is $(\mathbb{K} - \mathbb{K}_0, \mathbb{K}_0, c)$.

Algorithm 1. Pseudo-Code for Knowledge-Propagation Phase

```
 1: function PROPAGATE(X, K, M)
 2:     Flag ← true
 3:     while Flag do
 4:         Flag ← false
 5:         M ← GJE_{|X−K|}(M)
 6:         for all rows r in M do
 7:             if only one non-zero variable in the first |X − K| columns then
 8:                 Flag ← true                                          ▷ situation 2
 9:                 if S(x) ∈ X then
10:                     Change columns for x and S(x) in M
11:                     K ← K ∪ {x, S(x)}
12:                     go to line 3
13:                 else
14:                     Change columns for x in M
15:                     K ← K ∪ {x}
16:                     go to line 3
17:                 end if
18:             end if
19:             if case (ii) of Lemma 2 happens in r then
20:                 Flag ← true                                          ▷ situation 3
21:                 Change columns for x and S(x) in M
22:                 K ← K ∪ {x, S(x)}
23:                 go to line 3
24:             end if
25:         end for
26:         for all pairs (x, S(x)) in X do
27:             if case (i) of Lemma 2 happens in M then
28:                 Flag ← true                                          ▷ situation 3
29:                 Change columns for x and S(x) in M
30:                 K ← K ∪ {x, S(x)}
31:                 go to line 3
32:             end if
33:         end for
34:     end while
35:     return (K, M)
36: end function
```

Denote by $\mathbb{E}_{\mathbb{K}}$ the subspace of $\mathbb{K}$ spanned by the row vectors of A_2. Indeed, $\mathbb{E}_{\mathbb{K}}$ is the subspace $\mathbb{E} \cap \mathcal{V}(\mathbb{K})$ of $\mathcal{V}(\mathbb{K})$. By Gauss-Jordan Elimination, the linear relations among $\mathbb{K}_0$ can be known by block matrix. However, more information can be known by S-box operations.

If there exist an $x \in \mathbb{K} - \mathbb{K}_0$ and a linear combination $\mathscr{L}' \in \mathcal{V}(\mathbb{K}_0)$ such that $x + \mathscr{L}' \in \mathbb{E}_{\mathbb{K}}$, then one can get $S(x) = S(\mathscr{L}')$. Since $\mathscr{L}' \in \mathcal{V}(\mathbb{K}_0)$, $S(\mathscr{L}')$ can be also deduced by variables in $\mathbb{K}_0$. Add $S(\mathscr{L}')$ to $\mathbb{K}$ and $\mathbb{K}_0$ (also add a new column corresponding to $S(\mathscr{L}')$), and add a new row corresponding to $S(x) + S(\mathscr{L}')$ to A_2 at the same time (make sure the order of variables is still $(\mathbb{K} - \mathbb{K}_0, \mathbb{K}_0, c)$). The reason to do this is that if there is a linear combination of $\mathscr{L}'' \in \mathcal{V}(\mathbb{K}_0)$ such that $S(x) + \mathscr{L}'' \in \mathbb{E}_{\mathbb{K}}$, then $S(\mathscr{L}')$ and $\mathscr{L}''$ can form a linear relation we want. Besides, if there is a linear combination of $\mathscr{L}'' \in \mathcal{V}(\mathbb{K}_0)$ such that $S(x) + e_y \cdot y + \mathscr{L}'' \in \mathbb{E}_{\mathbb{K}}$ ($y \notin \mathbb{K}_0$), then $e_y \cdot y + S(\mathscr{L}') + \mathscr{L}'' \in \mathbb{E}_{\mathbb{K}}$. So y can be used to gain more information just as x. This can be also applied to $S(x)$. This step is called **new-variable-adding**.

Algorithm 2. Pseudo-Code for Relation-Derivation Phase

```
 1: function DERIVATION(𝕂₀, 𝕂, A₂)
 2:     Flag ← true
 3:     Change the order of columns in A₂
 4:     while Flag do
 5:         Flag ← false
 6:         A₂ ← GJE_{|𝕂−𝕂₀|}(A₂)
 7:         for all rows r in A₂ do                              ▷ new-variable-adding
 8:             if only one non-zero variable x in the first |𝕂 − 𝕂₀| columns then
 9:                 Flag ← true
10:                 if x is input of S-box then
11:                     Let S(ℒ') be a new variables
12:                     𝕂 ← 𝕂 ∪ {S(ℒ')} and 𝕂₀ ← 𝕂₀ ∪ {S(ℒ)}
13:                     Add a new column for S(ℒ') and a new row for S(x) + S(ℒ,)
14:                     go to line 4
15:                 else
16:                     Let S⁻¹(ℒ) be a new variables
17:                     𝕂 ← 𝕂 ∪ {S⁻¹(ℒ')} and 𝕂₀ ← 𝕂₀ ∪ {S⁻¹(ℒ')}
18:                     Add a new column for S⁻¹(ℒ') and a new row for S⁻¹(x) + S⁻¹(ℒ')
19:                     go to line 4
20:                 end if
21:             end if
22:         end for
23:     end while
24:     RelationSet ← ∅
25:     A₂ ← GJE(A₂)
26:     for all row r in B₂ do
27:         Derive relation from r
28:         Add this relation to RelationSet
29:     end for
30:     return RelationSet
31: end function
```

After the step above, a matrix as $\begin{pmatrix} \overbrace{B_0}^{|\mathbb{K}-\mathbb{K}_0|} & \overbrace{B_1}^{|\mathbb{K}_0|} \\ 0 & B_2 \end{pmatrix}$ can be known, $Rank(B_2)$

linear independent relations among $\mathbb{K}_0$ can be recovered from B_2.

The pseudo-code of the relation-derivation phase is shown in Algorithm 2. The inputs are the outputs of Algorithm 1. The function returns a set of relations among the variables of $\mathbb{K}_0$.

3.3 A Tool for Bit-Oriented Ciphers

The key schedules of some block ciphers have operations on word-level (e.g., S-box) and bit-level (e.g., cyclic shift), such as PRESENT, LBlock and so on. This tool is slightly different from the tool for word-oriented ciphers since it has operations both on words and bits. It also consists of two phases: knowledge-propagation phase and relation-derivation phase.

In the knowledge-propagation phase, since S-box permutation treats b bits as a union, situations 1 and 3 of Sect. 3.2 are no longer suitable for bit-oriented ciphers. The following lemma is used to deal with this situation.

Lemma 4. *Let $S[w_0^I \cdots w_{b-1}^I] = [w_0^O \cdots w_{b-1}^O]$, where w_i^I and w_i^O are 1-bit variables, respectively. If the values in any b out of $2b$ input/output bits of one S-box are known, then the values in the other b bits are uniquely determined, and can be computed efficiently.*

This situation can be dealt with by Gauss-Jordan Elimination as follows.

Lemma 5. *Let $\mathbb{S}$ be a set of input and output bit-variables of one S-box. If the order of variables in $\mathbb{M}$ is $(\mathbb{X} - \mathbb{K} - \mathbb{S}, \mathbb{S}, \mathbb{K})$ and $GJE(\mathbb{M})$ can be represented as:*

$$\begin{pmatrix} \overbrace{D_0}^{|\mathbb{X}-\mathbb{K}-\mathbb{S}|} & \overbrace{D_1}^{|\mathbb{S}|} & \overbrace{D_2}^{|\mathbb{K}|} \\ 0 & D_3 & D_4 \end{pmatrix}$$

then the bit-variables of $\mathbb{S}$ can be uniformly determined if and only if $Rank(D_3) \geq b - n_k$, where n_k is the number of bits in $\mathbb{S}$ which are already in $\mathbb{K}$.

It is easy to see that since $\mathbb{S}$ is the set of input and output bit-variables of one S-box, the entropy of these bits is b. $Rank(D_3) = b - n_k$ means $b - n_k$ linearly independent relations can be built among variables in $\mathbb{S}$ and $\mathbb{K}$, and these relations are enough to reduce the entropy to 0.

This property is used in Algorithm 1 to get more information from S-box instead of situation 1 and 3 of Sect. 3.2

In the relation-derivation phase, suppose the order of variables in $\mathbb{K}$ is $(\mathbb{K} - \mathbb{K}_0 - \mathbb{S}, \mathbb{S}, \mathbb{K}_0)$ and $GJE(\mathbb{A}_2)$ can be represented as:

$$\begin{pmatrix} \overbrace{E_0}^{|\mathbb{K}-\mathbb{K}_0-\mathbb{S}|} & \overbrace{E_1}^{|\mathbb{S}|} & \overbrace{E_2}^{|\mathbb{K}_0|} \\ 0 & E_3 & E_4 \end{pmatrix}$$

If $Rank(E_3) \geq b - n_k$, since $b - n_k$ linearly independent relations are enough to reduce the entropy to 0, $2b - n_k - Rank(E_3)$ new functions with the form $x_i + \mathcal{L}'$ can be added to A_2, where $x_i \in \mathbb{S}$ is not a pivot element of E_3 and $\mathcal{L}'$ is a variable denoting how x_i can be known from $\mathbb{S}$ and $\mathbb{K}_0$. Add $\mathcal{L}'$ to $\mathbb{K}_0$ and $\mathbb{K}$ at the same time. This step is used to replace the new-variable-adding step of Algorithm 2.

If $Rank(E_3) > b - n_k$, since $b - n_k$ linearly independent relations are enough to reduce the entropy to 0, the other $Rank(E_3) - (b - n_k)$ relations can be used to filter the variables in $\mathbb{K}_0$. Use the variables in $\mathbb{K}_0$ to deduce these relations, only $2^{b-Rank(E_3)}$ of them can satisfy the S-box table. For example, if we can get $b + 1$ S-box input/output bit-variables from $\mathbb{K}_0$, we can get 1 bit relation among the variables in $\mathbb{K}_0$, and it is obliviously one key-bridge we want. This property is used to get key-bridges in Algorithm 2.

We apply our automatic search tool to the attacks on LBlock and TWINE in the following sections.

4   Applications to LBlock

4.1   Description of LBlock

LBlock is a lightweight 64-bit block cipher designed by Wu et al. in 2011 [25] and is based on a variant of Feistel Network. It supports key size of 80 bits and the total number of iterations is 32. The Feistel function of LBlock is made up of a key addition AK, an S-box layer S made up of 8 4-bit S-boxes and a nibble permutation P. LBlock's function design can be visualized in Fig. 1. The key schedule of LBlock is rather simple. The 80-bit master key WK_0 is stored in a key register and represented as $WK_0 = WK_0[0] \cdots WK_0[79]$. At round i, the leftmost 32-bit of current content of register is output as round key. The key schedule of round i can be shown as follows ($i = 1, \cdots 31$):

$$WK_i \leftarrow WK_{i-1} \lll 29,$$
$$WK_i[0 - 3] \leftarrow S_9(WK_i[0 - 3]),$$
$$WK_i[4 - 7] \leftarrow S_8(WK_i[4 - 7]),$$
$$WK_i[29 - 33] \leftarrow WK_i[29 - 33] \oplus [i]_2.$$

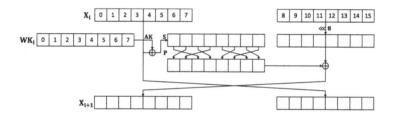

Fig. 1. Round function of LBlock block cipher

4.2   Impossible Differential Cryptanalysis on 23-Round LBlock

In INDOCRYPT 2012, Wu et al. presented an automatic search tool to search for the best impossible differential distinguishers [24]. In CRYPTO 2015, Sun et al. proved that this tool could find all impossible differentials of a cipher that are independent of the choices of the S-boxes [20]. In this paper, we automatize the search of the best impossible differential attacks by combining Wu's tool with our tool. Using Wu's tool, we can get all distinguishers with certain rounds. Using our key-bridging tool, we can get all the key-bridges to reduce the complexity in key-sieving phase.

Recently, Boura et al. [7] proposed a generic version of impossible differential attacks with the aim of simplifying and helping the construction and verification of this type of cryptanalysis. In particular, they provided a formula to compute the complexity of such an attack according to its parameters. To understand the formula, we first briefly review how an impossible differential attack is constructed. It starts by splitting the cipher into three parts: $E = E_3 \circ E_2 \circ E_1$ and finding an impossible differential $(\Delta_X \nrightarrow \Delta_Y)$ through E_2. Then Δ_X (resp. Δ_Y) is propagated through E_1^{-1} (resp. E_3) with probability 1 to obtain Δ_{in} (resp. Δ_{out}). We denote by c_{in} and c_{out} the log_2 of the probability of the transitions $\Delta_{in} \rightarrow \Delta_X$ and $\Delta_{out} \rightarrow \Delta_Y$, respectively. Finally we denote by k_{in} and k_{out} the key materials involved in those transitions. All in all, the attack consists in discarding the keys k for which at least one pair follows the characteristic through E_1 and E_3 and in exhausting the remaining ones. The complexity of doing so is as follows:

- **data:** C_{N_α}
- **memory:** N_α
- **time:** $C_{N_\alpha} + (1 + 2^{|k_{in} \cup k_{out}| - c_{in} - c_{out}}) N_\alpha C_{E'} + 2^{|k| - \alpha}$

where N_α is such that $(1 - 2^{-c_{in} - c_{out}})^{N_\alpha} < (1 - 2^{-\alpha})$, C_{N_α} is the number of chosen plaintexts required to generate N_α pairs satisfying $(\Delta_{in}, \Delta_{out})$, $|k|$ is the key size and $C_{E'}$ is the ratio of the cost of partial encryption to the full encryption.

We use this framework to mount an impossible differential attack on 23-round LBlock. First we find an impossible differential distinguisher through 14 rounds of LBlock. The input (resp. output) inactive nibble of this distinguisher is at position 12 (resp. 5). It is extended by 4 rounds at the beginning and by 5 rounds at the end in order to attack 23 rounds of the cipher. It can be seen in Fig. 2 that the difference in the plaintexts has to be zero in 8 nibbles such that $c_{in} + c_{out} = 28 + 44 = 72$. The key material $k_{in} \cup k_{out}$ is composed of 36 round-key nibbles which can assume 2^{73} values thanks to our key-bridging tool. Specifically, we can find 71 linear independent key-bridges among these 36 round-key nibbles. We show parts of the key-bridges we found in the full version of this paper.

As a consequence, and according to the above formula, the memory complexity of our attack is $\alpha \cdot 2^{71.5}$, the time complexity is $\alpha \cdot 2^{73} \cdot C_{E'} + 2^{80 - \alpha}$. As we estimate the ratio $C_{E'}$ to $36/184 \approx 2^{-2.4}$, the value of α minimizing the overall complexity is 6.8. So the memory complexity of our attack is $2^{74.3}$ bytes, the time complexity is $2^{74.5}$ 23-round LBlock encryptions and the data complexity is $2^{59.5}$ chosen plaintexts.

Besides this attack, we can get another 2 impossible differential attacks on LBlock with the same complexity, i.e., $((12, 0)5, 14, 4)$, $((12, 5)5, 14, 4)$, where $((l_a, l_b)R_b, R_d, R_e)$ means that the position of input (resp. output) inactive nibble of R_d-round distinguisher is l_a (resp. l_b), and the number of rounds before (resp. after) the distinguisher is R_b (resp. R_e).

4.3   Zero-Correlation Cryptanalysis on 23-Round LBlock

At ACISP 2014, Wang et al. gave a multidimensional zero-correlation linear attack on 23-round LBlock [22]. The main technique they used to improve the

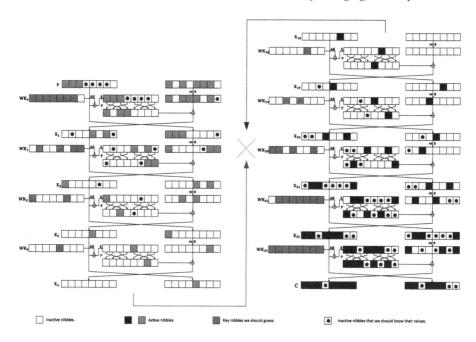

Fig. 2. Impossible differential attack on 23 rounds LBlock (Color figure online)

overall complexity is the partial compression technique, i.e., they reduce the complexity of online phase by guessing each sub-key nibble one after another. Since the time complexity of this attack is still greater than exhaustive search, they use 13 key-bridges to make this attack available.

According to their paper, for 14-round LBlock, if the input mask a of the first round locates at the left branch and the output mask b of the last round locates in the right branch, then the correlation of the linear approximation is zero, where $a, b \in F_2^4, a \neq 0$ and $b \neq 0$. Indeed, we find in total 21 key-bridges to reduce the overall complexity.

Combining this observation with our key-bridging tool, we find that $((1, 12)4, 14, 5)$ can get a better overall complexity. This is for the reason that we find 21 key-bridges thanks to our key-bridging tool. Since the major complexity of this attack comes from Step 4.1 and 4.2 of their paper, we explain the key-bridges of these 2 steps in detail.

The nibble X_4^1 corresponding to the input non-zero linear mask is affected by 32 bits of plaintext X_0 and 28 bits of round keys and the expression can be shown as:

$$X_4^1 = X_0^5 \oplus S(X_0^{12} \oplus S(X_0^0 \oplus WK_0^0) \oplus WK_1^2) \oplus S(X_0^{15} \oplus S(X_0^7 \oplus WK_0^7) \oplus$$
$$S(X_0^4 \oplus S(X_0^{10} \oplus S(X_0^1 \oplus WK_0^1) \oplus WK_1^0) \oplus WK_2^2) \oplus WK_3^3)$$

Similarly, the nibble X_{18}^{12} corresponding to the output non-zero linear mask is affected by 48 bits of plaintext X_0 and 48 bits of round keys:

$$X_{18}^{12} = X_{23}^6 \oplus S(X_{23}^{12} \oplus WK_{22}^4) \oplus S(X_{23}^{15} \oplus S(X_{23}^4 \oplus S(X_{23}^{13} \oplus WK_{22}^5) \oplus WK_{21}^6) \oplus WK_{20}^1)$$
$$\oplus S(X_{23}^{12} \oplus S(X_{23}^3 \oplus S(X_{23}^{10} \oplus WK_{22}^2) \oplus WK_{21}^5) \oplus S(X_{23}^0 \oplus S(X_{23}^9 \oplus WK_{22}^1) \oplus$$
$$S(X_{23}^{14} \oplus S(X_{23}^2 \oplus S(X_{23}^8 \oplus WK_{22}^0) \oplus WK_{21}^4) \oplus WK_{20}^0) \oplus WK_{19}^0) \oplus WK_{18}^0)$$

Step 4.1. The guessed-keys of Step 4.1 are WK_0^1, WK_0^7, $WK_1^0[3]$, WK_0^0 and WK_{22}^1. Since $WK_0^7 \Rightarrow WK_1^0[0-2]$, WK_1^0 can be known. Meanwhile, since WK_{22}^1, $WK_{21}[33]$, $WK_{21}[34]$, $WK_{21}[35]$ and $WK_{21}[36]$ are inputs/outpus bits of one S-box, $WK_{21}[33]$, $WK_{21}[34]$, $WK_{21}[35]$ and $WK_{21}[36]$ can be known. Since $WK_{21}[34] \Rightarrow WK_{11}[4]$, $WK_{21}[35] \Rightarrow WK_{11}[5]$, $WK_{21}[36] \Rightarrow WK_{11}[6]$, $WK_0^1[0] \Rightarrow WK_{10}[34]$, $WK_0^1[1] \Rightarrow WK_{10}[35]$, $WK_0^1[2] \Rightarrow WK_{10}[36]$, $WK_0^1[3] \Rightarrow WK_{10}[33]$, and $WK_{11}[4]$, $WK_{11}[5]$, $WK_{11}[6]$, $WK_{10}[34]$, $WK_{10}[35]$, $WK_{10}[36]$, $WK_{10}[33]$ are input/output bits of one S-box, we can get 3-bit information to restrain the values of $WK_0^0[3]$, $WK_0^1[0-2]$ and WK_{22}^1. This is easily done by making a small lookup table.

As the following four equations:

$$X_1^5 = X_0^{15} \oplus S(X_0^7 \oplus WK_0^7),$$
$$X_2^2 = X_0^4 \oplus S(X_0^{10} \oplus S(X_0^1 \oplus WK_0^1) \oplus WK_1^0),$$
$$X_1^2 = X_0^{12} \oplus S(X_0^0 \oplus WK_0^0),$$
$$X_{22}^{10} = X_{23}^0 \oplus S(X_{23}^9 \oplus WK_{22}^1),$$

are true for LBlock, the 80-bit plaintext and ciphertext state value which affects the value of $X_4^1 || X_{18}^{14}$ can be reduced to 60-bit after guessing the 14-bit equivalent key. The time complexity of this step is $N \cdot 2^{14} \cdot 5$ S-box accesses.

Step 4.2. The guessed-key of Step 4.2 is WK_{22}^0. Since $WK_0[2] \Rightarrow WK_{10}[32]$, $WK_0[1] \Rightarrow WK_{10}[31]$, $WK_0[0] \Rightarrow WK_{10}[30]$, $WK_{19}[11] \Rightarrow WK_{11}[3]$, $WK_{19}[10] \Rightarrow WK_{11}[2]$, $WK_{19}[9] \Rightarrow WK_{11}[1]$, $WK_{19}[8] \Rightarrow WK_{11}[0]$ and $WK_{10}[32]$, $WK_{10}[31]$, $WK_{10}[30]$, $WK_{11}[3]$, $WK_{11}[2]$, $WK_{11}[1]$, $WK_{11}[0]$ are input/output bits of one S-box, 3-bit information of $WK_{19}[11]$, $WK_{19}[10]$, $WK_{19}[9]$ and $WK_{19}[8]$ can be known. Meanwhile, $WK_{22}^1 \Rightarrow WK_{19}^2[3]$, $WK_{19}[11]$, $WK_{19}[10]$, $WK_{19}[9]$ and $WK_{19}[8]$ can be known.[2] Since $WK_{19}[9] \Rightarrow WK_{21}[32]$, $WK_{19}[8] \Rightarrow WK_{21}[31]$, $WK_{19}[7] \Rightarrow WK_{21}[30]$, and $WK_{22}[0]$, $WK_{22}[1]$, $WK_{22}[2]$, $WK_{22}[3]$, $WK_{21}[32]$, $WK_{21}[31]$, $WK_{21}[30]$ are input/output bits of one S-box, 3-bit information of WK_{22}^0 can be known. Since $X_{22}^{12} = X_{23}^2 \oplus S(X_{23}^8 \oplus WK_{22}^0)$, we can obtain a new state with 56-bit length. The time complexity of this step is $2^{60} \cdot 2^{14+1}$ S-box accesses.

The time complexity of the following sub-steps will become less and less thanks to our key-bridges. Let $N = 2^{62.1}$ as [22] shows, the time complexity of

---

[2] WK_0^0 and WK_{22}^1 are known from Step 4.1.

this attack is manipulated by Step 4.1, which is about $2^{62.1+14} \cdot 5 \cdot 1/8 \cdot 1/23 \approx 2^{71}$. The total time complexity is $2^{71} + 2^{71} = 2^{72}$ 23-round LBlock encryptions. The data complexity and memory complexity are the same as [22], i.e., the data complexity is $N = 2^{62.1}$ known plaintexts, the memory complexity is about 2^{60} bytes.

5   Applications to TWINE

5.1   Description of TWINE

TWINE is a lightweight 64-bit block cipher designed by Suzaki et al. in 2013 [21] and is based on a variant of Type-2 generalized Feistel structure. One version of TWINE uses an 80-bit key, another uses an 128-bit key and we denote these versions TWINE-80 and TWINE-128. The Feistel function of TWINE consists of an xor of a sub-key and a call to a unique S-box. TWINE's function design can be visualized in Fig. 3. The key schedule of TWINE-80 is quite simple. The 80-bit master key WK_0 is stored in a key register and represented as $WK_0 = WK_0^0 \cdots WK_0^{20}$.

The key schedule of round i can be shown as follows ($i = 1, \cdots 35$):

$$WK_{i-1}^1 \leftarrow WK_{i-1}^1 \oplus S(WK_{i-1}^0), WK_{i-1}^4 \leftarrow WK_{i-1}^4 \oplus S(WK_{i-1}^{16}),$$
$$WK_{i-1}^7 \leftarrow WK_{i-1}^7 \oplus 0||CONST_i^H, WK_{i-1}^{19} \leftarrow WK_{i-1}^{19} \oplus 0||CONST_i^L),$$
$$WK_{i-1}^{0-3} \leftarrow WK_{i-1}^{0-3} \lll 4,$$
$$WK_i \leftarrow WK_{i-1} \lll 16.$$

Then $WK_i^1||WK_i^3||WK_i^4||WK_i^6||WK_i^{13}||WK_i^{14}||WK_i^{15}||WK_i^{16}$ is used as the 8-nibble round key of round i. We use RK_i to denote the round key of round i. We refer to [21] for the 128-bit version of key schedule.

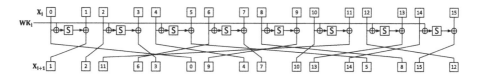

Fig. 3. Round function of TWINE block cipher

5.2   Zero-Correlation Cryptanalysis on TWINE

In [22], Wang et al. also gave multidimensional zero-correlation linear attacks on 23-round TWINE-80 and 25-round TWINE-128 using the partial compression technique.

However, using our automatic search tool, we find that the key-bridges they used in the attack on 23-round TWINE-80 do not exist. In their paper, they say that $RK_3^5 \Rightarrow RK_0^3$, $RK_2^6 \Rightarrow RK_0^1$ and $RK_{20}^1 \Rightarrow RK_{22}^6$. By the key schedule of TWINE-80,

$$RK_0^3 \Rightarrow WK_0^6 \Rightarrow WK_1^2 \Rightarrow WK_2^{17} \Rightarrow WK_3^{13} \Rightarrow RK_3^4,$$
$$RK_0^1 \Rightarrow WK_0^3 \Rightarrow WK_1^{18} \Rightarrow WK_2^{14} \Rightarrow RK_2^5,$$
$$RK_{20}^1 \Rightarrow WK_{20}^3 \Rightarrow WK_{21}^{18} \Rightarrow WK_{22}^{14} \Rightarrow RK_{22}^5.$$

So the key-bridges they used are not true.

Since RK_3^4 and RK_2^5 do not exist in the set of related round keys, this will make the time complexity of Step 4.1 in their paper greater than exhaustive search. So their attack on 23-round TWINE-80 is not available.

According to their paper, if the input mask a of the first round locates at the even nibble and the output mask b of the last round locates in the odd nibble for 14-round TWINE, then the correlation of the linear approximation is zero, where $a, b \in F_2^4, a \neq 0, b \neq 0$. Among these distinguishers, we find 4 of them with the minimal number of guessed-keys (with some key-bridges) by our automatic search tool. We use $((6, 9)4, 14, 5)$ to get multidimensional zero-correlation linear attack on 23-round TWINE-80.

Three key-bridges we found are $RK_{22}^5 \Rightarrow RK_{20}^1$, $RK_{21}^0 \Rightarrow RK_{18}^4$ and $RK_{22}^6 \oplus S(RK_{22}^2) \oplus S(RK_{22}^5) \Rightarrow RK_2^6$.

Since the major complexity of this attack comes from Step 4.1 and 4.2 , we explain the key-bridges of these 2 steps. Assuming N known plaintexts are used.

Step 4.1. The distinguisher input nibble X_4^6 is affected by 32 bits of plaintext X_0 and 28 bits of round keys, and the distinguisher output nibble X_{18}^9 is affected by 48 bits of ciphertext X_{23} and 48 bits of round keys. Since $X_{22}^{11} = X_{23}^2 \oplus S(X_{23}^9 \oplus RK_{22}^9)$, X_{23}^9 and X_{23}^2 can be compressed to X_{22}^{11} by guessing RK_{22}^9. Since $X_{22}^{13} = X_{23}^{10} \oplus S(X_{23}^{15} \oplus RK_{22}^6)$, X_{23}^{15} and X_{23}^{10} can be compressed to X_{22}^{13} by guessing RK_{22}^6. Since $RK_{22}^5 \Rightarrow RK_{20}^1$, let $A = X_{23}^8 \oplus S(X_{23}^5 \oplus S(X_{23}^{12} \oplus S(X_{23}^7 \oplus RK_{22}^2) \oplus RK_{20}^0) \oplus RK_{20}^1)$, X_{23}^8, X_{23}^5, X_{23}^{12} and X_{23}^7 can be compressed to A by guessing RK_{22}^2 and RK_{21}^0. The time complexity of this step is $N \cdot 2^{16} \cdot 5$ S-box accesses.

Step 4.2. By guessing RK_{22}^2, one more nibble can be compressed. The time complexity of this step is 2^{64+16} S-box accesses.

Let $N = 2^{62.1}$, the time complexity of this attack is 2^{73} 23-round TWINE-80 encryptions, the data complexity is $2^{62.1}$, and the memory complexity is 2^{60} bytes.

For their attack on 25-round TWINE-128, we find another key-bridge besides their four key-bridges, i.e., $S(RK_{21}^1 \oplus S^{-1}(RK_0^3 \oplus RK_{24}^2)) \Rightarrow RK_{24}^3$. So after Step 4.3 of their attack, one more key nibble RK_0^3 can be known, and one more state nibble can be compressed in this step. The time complexity of the following steps is much smaller than the above 3 steps, so the time complexity of this attack is $(2^{62+60} \cdot 17 + 2 \cdot 2^{124})/(25 \times 8) + 2^{119} \approx 2^{120}$.

Besides, if $((12,3)5,14,6)$ is used to mount this attack, a better result can be got. Using our tool, four key-bridges can be found, i.e., $RK_0^3 \Rightarrow RK_3^1$, $RK_{21}^3 \Rightarrow RK_{24}^1$, $S(RK_{24}^0 \oplus S^{-1}(RK_2^5 \oplus RK_{23}^7 \oplus S(RK_{22}^2))) \Rightarrow RK_{22}^1$, $S(RK_{22}^3 \oplus S(RK_{19}^1)) \oplus S^{-1}(RK_0^7 \oplus RK_{24}^6 \oplus S(RK_{24}^7)) \Rightarrow RK_{22}^2$. The overall time complexity is determined by searching the remaining key candidates. So the time complexity becomes 2^{119}.

5.3 Impossible Differential and Meet-in-the-Middle Cryptanalysis on TWINE

At FSE 2015, Biryukov et al. gave impossible differential cryptanalysis and meet-in-the-middle cryptanalysis on 25-round TWINE-128 [2].

Combining Wu's tool for impossible differential distinguisher and our automatic tool, we find in total 12 attacks with the same time complexity as [2].

Combining Lin's propagate-then-prune tool for meet-in-the-middle distinguisher [19] and our automatic tool, we find some attacks with the same complexity as [2]. One of these attacks is $(6, 10) \rightarrow 5$, i.e., the distinguisher starts with two inactive nibbles at position $(6,10)$ and ends with one nibble at position 5. This attack is useful when we want to get less false positive.

6 Conclusions

In this paper, we studied the key-bridging technique Dunkelman et al. proposed to deduce some sub-key bits from some other sub-key bits. We presented a versatile and powerful algorithm for searching key-bridging technique on word-oriented and bit-oriented block ciphers. This tool can not only give better attacked-rounds and complexity, but also a better understanding of the design of block ciphers. To demonstrate the usefulness of our approach, we used our tool to the impossible differential and multidimensional zero-correlation linear attacks on 23-round LBlock, 23-round TWINE-80 and 25-round TWINE-128. To the best of our knowledge, these results are the currently best results on LBlock and TWINE in the single-key setting.

Acknowledgements. We would like to thank the anonymous reviewers for providing valuable comments. The research presented in this paper is supported by the National Basic Research Program of China (No. 2013CB338002) and National Natural Science Foundation of China (No. 61272476, No. 61232009 and No. 61202420).

References

1. Beaulieu, R., Shors, D., Smith, J., Treatman-Clark, S., Weeks, B., Wingers, L.: The SIMON and SPECK Families of Lightweight Block Ciphers. Cryptology ePrint Archive, Report 2013/404 (2013). http://eprint.iacr.org/
2. Biryukov, A., Derbez, P., Perrin, L.: Differential analysis and meet-in-the-middle attack against round-reduced TWINE. In: Leander, G. (ed.) FSE 2015. LNCS, vol. 9054, pp. 3–27. Springer, Heidelberg (2015)

3. Biryukov, A., Khovratovich, D., Nikolić, I.: Distinguisher and related-key attack on the full AES-256. In: Halevi, S. (ed.) CRYPTO 2009. LNCS, vol. 5677, pp. 231–249. Springer, Heidelberg (2009)

4. Bogdanov, A.A., Knudsen, L.R., Leander, G., Paar, C., Poschmann, A., Robshaw, M., Seurin, Y., Vikkelsoe, C.: PRESENT: an ultra-lightweight block cipher. In: Paillier, P., Verbauwhede, I. (eds.) CHES 2007. LNCS, vol. 4727, pp. 450–466. Springer, Heidelberg (2007)

5. Bogdanov, A., Rechberger, C.: A 3-subset meet-in-the-middle attack: cryptanalysis of the lightweight block cipher KTANTAN. In: Biryukov, A., Gong, G., Stinson, D.R. (eds.) SAC 2010. LNCS, vol. 6544, pp. 229–240. Springer, Heidelberg (2011)

6. Boura, C., Minier, M., Naya-Plasencia, M., Suder, V.: Improved Impossible Differential Attacks against Round-Reduced LBlock. Cryptology ePrint Archive, Report 2014/279 (2014). http://eprint.iacr.org/

7. Boura, C., Naya-Plasencia, M., Suder, V.: Scrutinizing and improving impossible differential attacks: applications to CLEFIA, Camellia, LBlock and SIMON. In: Sarkar, P., Iwata, T. (eds.) ASIACRYPT 2014. LNCS, vol. 8873, pp. 179–199. Springer, Heidelberg (2014)

8. Daemen, J., Rijmen, V.: The Design of Rijndael: AES-the Advanced Encryption Standard. Springer, Berlin (2002)

9. Derbez, P., Fouque, P.-A., Jean, J.: Improved key recovery attacks on reduced-round AES in the single-key setting. In: Johansson, T., Nguyen, P.Q. (eds.) EURO-CRYPT 2013. LNCS, vol. 7881, pp. 371–387. Springer, Heidelberg (2013)

10. Dunkelman, O., Keller, N., Shamir, A.: Improved Single-Key Attacks on 8-Round AES. Cryptology ePrint Archive, Report 2010/322 (2010). http://eprint.iacr.org/

11. Dunkelman, O., Keller, N., Shamir, A.: Improved single-key attacks on 8-round AES-192 and AES-256. In: Advances in Cryptology-ASIACRYPT 2010, pp. 158–176. Springer, Berlin (2010)

12. Guo, J., Peyrin, T., Poschmann, A., Robshaw, M.: The LED block cipher. In: Preneel, B., Takagi, T. (eds.) CHES 2011. LNCS, vol. 6917, pp. 326–341. Springer, Heidelberg (2011)

13. Hong, D., et al.: HIGHT: a new block cipher suitable for low-resource device. In: Goubin, L., Matsui, M. (eds.) CHES 2006. LNCS, vol. 4249, pp. 46–59. Springer, Heidelberg (2006)

14. Khovratovich, D., Biryukov, A., Nikolic, I.: Speeding up collision search for byte-oriented hash functions. In: Fischlin, M. (ed.) CT-RSA 2009. LNCS, vol. 5473, pp. 164–181. Springer, Heidelberg (2009)

15. Knudsen, L., Leander, G., Poschmann, A., Robshaw, M.J.B.: PRINTCIPHER: a block cipher for IC-printing. In: Mangard, S., Standaert, F.-X. (eds.) CHES 2010. LNCS, vol. 6225, pp. 16–32. Springer, Heidelberg (2010)

16. Ko, Y., Hong, S.H., Lee, W.I., Lee, S.-J., Kang, J.: Related key differential attacks on 27 rounds of XTEA and full-round GOST. In: Roy, B., Meier, W. (eds.) FSE 2004. LNCS, vol. 3017, pp. 299–316. Springer, Heidelberg (2004)

17. Li, L., Jia, K., Wang, X.: Improved single-key attacks on 9-round AES-192/256. In: Cid, C., Rechberger, C. (eds.) FSE 2014. LNCS, vol. 8540, pp. 127–146. Springer, Heidelberg (2015)

18. Li, R., Jin, C.: Meet-in-the-middle attacks on 10-round AES-256. In: Designs, Codes and Cryptography, pp. 1–13 (2015)

19. Lin, L., Wu, W., Wang, Y., Zhang, L.: General model of the single-key meet-in-the-middle distinguisher on the word-oriented block cipher. In: Lee, H.-S., Han, D.-G. (eds.) ICISC 2013. LNCS, vol. 8565, pp. 203–223. Springer, Heidelberg (2014)

20. Sun, B., Liu, Z., Rijmen, V., Li, R., Cheng, L., Wang, Q., Alkhzaimi, H., Li, C.: Links among impossible differential, integral and zero correlation linear cryptanalysis. In: Gennaro, R., Robshaw, M. (eds.) CRYPTO 2015. LNCS, vol. 9215, pp. 95–115. Springer, Heidelberg (2015)
21. Suzaki, T., Minematsu, K., Morioka, S., Kobayashi, E.: TWINE: a lightweight block cipher for multiple platforms. In: Wu, H., Knudsen, L.R. (eds.) SAC 2012. LNCS, vol. 7707, pp. 339–354. Springer, Heidelberg (2013)
22. Wang, Y., Wu, W.: Improved multidimensional zero-correlation linear cryptanalysis and applications to LBlock and TWINE. In: Susilo, W., Mu, Y. (eds.) ACISP 2014. LNCS, vol. 8544, pp. 1–16. Springer, Heidelberg (2014)
23. Wikipedia. Invariant Subspace –Wikipedia, The Free Encyclopedia (2015). https://en.wikipedia.org/wiki/Invariant_subspace
24. Wu, S., Wang, M.: Automatic search of truncated impossible differentials for word-oriented block ciphers. In: Galbraith, S., Nandi, M. (eds.) INDOCRYPT 2012. LNCS, vol. 7668, pp. 283–302. Springer, Heidelberg (2012)
25. Wu, W., Zhang, L.: LBlock: a lightweight block cipher. In: Lopez, J., Tsudik, G. (eds.) ACNS 2011. LNCS, vol. 6715, pp. 327–344. Springer, Heidelberg (2011)
26. Zheng, X., Jia, K.: Impossible differential attack on reduced-round TWINE. In: Lee, H.-S., Han, D.-G. (eds.) ICISC 2013. LNCS, vol. 8565, pp. 123–143. Springer, Heidelberg (2014)
27. Zhijie, C.: Higher Algebra and Analytic Geometry. Springer, Berlin (2001). (in Chinese)

MILP-Based Automatic Search Algorithms for Differential and Linear Trails for Speck

Kai Fu[1], Meiqin Wang[1,2(✉)], Yinghua Guo[1], Siwei Sun[3,4], and Lei Hu[3,4]

[1] Key Laboratory of Cryptologic Technology and Information Security,
Ministry of Education, Shandong University, Jinan 250100, China
mqwang@sdu.edu.cn
[2] State Key Laboratory of Cryptology, P.O. Box 5159, Beijing 100878, China
[3] State Key Laboratory of Information Security, Institute of Information
Engineering, Chinese Academy of Sciences, Beijing 100093, China
[4] Data Assurance and Communication Security Research Center,
Chinese Academy of Sciences, Beijing 100093, China

Abstract. In recent years, Mixed Integer Linear Programming (MILP) has been successfully applied in searching for differential characteristics and linear approximations in block ciphers and has produced the significant results for some ciphers such as SIMON (a family of lightweight and hardware-optimized block ciphers designed by NSA) etc. However, in the literature, the MILP-based automatic search algorithm for differential characteristics and linear approximations is still infeasible for block ciphers such as ARX constructions. In this paper, we propose an MILP-based method for automatic search for differential characteristics and linear approximations in ARX ciphers. By researching the properties of differential characteristic and linear approximation of modular addition in ARX ciphers, we present a method to describe the differential characteristic and linear approximation with linear inequalities under the assumptions of independent inputs to the modular addition and independent rounds. We use this representation as an input to the publicly available MILP optimizer Gurobi to search for differential characteristics and linear approximations for ARX ciphers. As an illustration, we apply our method to Speck, a family of lightweight and software-optimized block ciphers designed by NSA, which results in the improved differential characteristics and linear approximations compared with the existing ones. Moreover, we provide the improved differential attacks on Speck48, Speck64, Speck96 and Speck128, which are the best attacks on them in terms of the number of rounds.

Keywords: Automatic search · Differential characteristic · Linear approximation · ARX · Speck

1 Introduction

Differential attacks [3] and linear attacks [15] are the most fundamental cryptanalytic methods. They have been used in the cryptanalysis of numerous symmetric ciphers. Since the first and most important thing for the two methods is

© International Association for Cryptologic Research 2016
T. Peyrin (Ed.): FSE 2016, LNCS 9783, pp. 268–288, 2016.
DOI: 10.1007/978-3-662-52993-5_14

to identify differential characteristics and linear approximations, the automatic search algorithms for differential characteristics and linear approximations have been a focus of cryptographer's concern. At EUROCRYPT'94, Matsui [16] presented the branch-and-bound search algorithm and found the differential characteristics and linear approximations for DES block cipher. The branch-and-bound search algorithm is one of the most powerful and classic search tools and is still widely used now. Another research line for the application of automatic search algorithm is to provide the provable security against differential cryptanalysis and linear cryptanalysis, which is usually achieved by automatic searching for the minimal number of active S-boxes.

Mixed-Integer Linear Programming (MILP) has been explicitly applied in constructing automatic search algorithm in differential and linear cryptanalysis. The problem of MILP is a class of optimization problems derived from Linear Programming which aims to optimize an objective function under certain constrains. Mouha *et al.* [18] and Wu *et al.* [28] translated the problem of counting the minimal number of differentially active S-boxes to an MILP problem which can be solved automatically with open source or commercially available MILP solvers. Their method has been applied in searching for the differential and linear characteristics with specific patterns [14, 29] and counting the minimal number of active S-boxes of bit-oriented block ciphers by introducing bit-level representations [21, 27].

Recently, the MILP-based method has been developed to be a general method to automatically search for the real differential characteristics. Sun *et al.* [22] constructed the MILP-based model to search for (related-key) differential characteristics by generating linear inequalities from the differential distribution table of S-box, where only partial linear inequalities are used in MILP model to make it solvable in practical time. Their search algorithm, however, is heuristic, since the identified differential characteristics may not be consistent. By computing a small number of inequalities which can exactly describe the differential distribution table of an S-box with the greedy algorithm, Sun *et al.* [24] transformed the heuristic searching method to the exact and practical searching method. Moreover, they constructed the MILP-based model for automatically searching for linear approximations and extended these models to search for differential and linear hull. Sun *et al.*'s method [22, 24] is applicable to block ciphers involving bitwise XOR, S-box operation and the linear layer with bit permutation[1]. Although the general linear layer can be transformed into bit XOR operations, it makes the MILP problem much more difficult to be solved in practical time since more XOR operations result in more variables and constraints.

Due to the excellent performance of ARX-based ciphers in software, many symmetric-key ciphers are designed based on ARX operations[2]. It is worth noting that the cryptanalytic techniques for ARX ciphers are very different from those for ciphers with S-boxes such as AES and DES. In particular, the search

---

[1] Although SIMON has no S-box, the And and XOR operations for SIMON could be regarded as one S-box. So they also applied the method to SIMON.

[2] ARX operation: modular addition, bit rotation and XOR.

algorithms for differential characteristics and linear approximations for ARX cipher utilize the different principle compared with those for ciphers with S-boxes. In [10,12,17], the methods of automatic search for differential characteristics in ARX designs are provided, but the methods are only compatible with ARX-based Hash functions where the key is known and can be freely chosen. By using the partial differential distribution table and Matsui's branch-and-bound algorithm, Biryukov and Velichkov [4] presented the first automatic search algorithm for differential characteristics in ARX block ciphers, such as (X)TEA and Speck. In a very recent paper [5] appearing in this volume of FSE'16, Biryukov *et al.* proposed the first adaptation of Matsui's algorithm for finding the best differential and linear trails in ARX ciphers.

Although MILP-based search algorithm has got extremely remarkable application for some block ciphers, the current method cannot be applied to ARX block ciphers. A straightforward method to apply MILP model for ARX constructions is to regard the modular addition in $\mathbb{F}_2^n$ as a $2n \times n$ S-box and compute a small number of linear inequalities to exactly represent the differential or linear pattern of the modular addition. However, in this way the number of linear inequalities is too large to be solved in practical time for real ARX ciphers where n is typically at least 16. This motivates us to study MILP-based search method for ARX block ciphers.

1.1 Our Contributions

In this paper, we revisit the differential property and linear property for modular addition and provide a new framework of constructing the MILP model. Concretely, we transform the differential property of modular addition shown in [13] into linear inequalities to describe all possible differential patterns and the corresponding differential probabilities. Moreover, we use linear inequalities to capture all possible linear patterns and the corresponding correlations based on the automaton algorithm for correlation of modular addition in [19,25]. The number of the resulting linear inequalities is significantly less than that of linear inequalities produced by regarding modular addition as one S-box. With the linear inequalities, we can construct the MILP model to automatically search for differential characteristics and linear approximations using the the commercial optimizer Gurobi, where the object function is the probability of differential characteristic or the correlation of linear approximation.

During constructing MILP models, we assume that the two inputs to modular addition and the consecutive rounds are independent. However, as demonstrated in [26], for some ARX constructions, the inputs to modular addition and the consecutive rounds are not independent, which will result that the practical probability (resp. correlation) of our identified differential (resp. linear) tails for some fixed key may vary significantly from that derived from our model. This deviation will have effect on the success rate of the attacks from practitioner's perspective.

As an illustration, we apply our method to the block cipher Speck, which is a family of lightweight block ciphers publicly released by the National Security Agency (NSA) and has been optimized for performance in software implementations [2]. A variety of block sizes and key sizes for different implementations are provided for it. Since its publication, Speck has received much attention and many cryptanalytic results have been given. Abed et al. presented differential and rectangle attacks for almost all variants of Speck [1]. At FSE'14 [6], Biryukov *et al.* searched for the differential characteristics, which cover 9, 11 and 14 rounds for Speck32, Speck48 and Speck64, respectively, and are better than the differential characteristics in [1]. In [9], Dinur proposed the sub-cipher attack and improved the key recovery attacks on all variants of Speck using the differential characteristics in [6]. In [5], Biryukov *et al.* presented the probabilities of the best differential trails for up to 10, 9, 8, 7, and 6 rounds of Speck32, Speck48, Speck64, Speck96 and Speck128 respectively and evaluate the security bounds of Speck against single-trail differential cryptanalysis under the Markov assumption. As regards to linear cryptanalysis, Yao *et al.* identified 9, 9, 12, 6 and 6 rounds linear approximations for Speck32, Speck48, Speck64, Speck96 and Speck128, respectively [30], and gave the key recovery attacks.

We use our models to search for the differential and linear trails for Speck. In order for the MILP tool to run in reasonable time for larger block sizes (>48 bits), we split the block cipher into two or three parts – upper (middle) and lower. We then search for trails independently in each part, by ensuring that the output difference (mask) for one part is the same as the input difference (mask) for its following part. For Speck48, Speck64, Speck96 and Speck128, we find better differential characteristics and linear approximations than those of previous works under the assumptions of independent inputs to the modular addition and independent rounds. With the new differential characteristics, we improve the differential attacks on the four variants of Speck. Comparing with the previous best attacks for them [9], we can attack one, one, three and five more rounds for Speck48, Speck64, Speck96 and Speck128 with any key size, respectively. We summarize known attacks on Speck in Table 1. We compare our identified differential characteristics and linear approximations with those of previous works in Table 2.

Outline. The remainder of this paper is organized as follows. Section 2 gives a brief description of the existing MILP-based search methods for block cipher. Sections 3 and 4 introduce MILP-based algorithm for automatic searching for differential characteristics and linear approximations for ARX ciphers. We apply the new search tools in Speck and give the improved differential attacks on all variants of Speck except Speck32 in Sect. 5. Finally, we conclude the paper in Sect. 6.

Table 1. Summary of attacks on speck

Variant $2n/mn$	Rounds attacked/ Total rounds	Time	Data	Memory	Method	Ref.
48/72	11/22	$2^{67.93}$	$2^{43.727}$	-	Linear	[30]
	12/22	$2^{58.8}$	$2^{43.2}$	$2^{45.8}$	Rectangle	[1]
	12/22	$2^{45.3}$	2^{45}	2^{24}	Differential	[1]
	14/22	2^{65}	2^{41}	2^{22}	Differential	[9]
	15/22	$\mathbf{2^{70}}$	$\mathbf{2^{46}}$	$\mathbf{2^{22}}$	**Differential**	**This Paper**
48/96	12/23	$2^{91.93}$	$2^{43.727}$	-	Linear	[30]
	12/23	$2^{58.8}$	$2^{43.2}$	$2^{45.8}$	Rectangle	[1]
	12/23	$2^{45.3}$	2^{45}	2^{24}	Differential	[1]
	15/23	2^{89}	2^{41}	2^{22}	Differential	[9]
	16/23	$\mathbf{2^{94}}$	$\mathbf{2^{46}}$	$\mathbf{2^{22}}$	**Differential**	**This Paper**
64/96	14/26	$2^{94.9}$	$2^{62.7}$	-	Linear	[30]
	14/26	$2^{89.4}$	$2^{63.6}$	$2^{65.6}$	Rectangle	[1]
	15/26	$2^{61.1}$	2^{61}	2^{32}	Differential	[1]
	18/26	2^{93}	2^{61}	2^{22}	Differential	[9]
	19/26	$\mathbf{2^{95}}$	$\mathbf{2^{63}}$	$\mathbf{2^{22}}$	**Differential**	**This Paper**
64/128	15/27	$2^{126.9}$	$2^{62.7}$	-	Linear	[30]
	14/27	$2^{89.4}$	$2^{63.6}$	$2^{65.6}$	Rectangle	[1]
	15/27	$2^{61.1}$	2^{61}	2^{32}	Differential	[1]
	19/27	2^{125}	2^{61}	2^{22}	Differential	[9]
	20/27	$\mathbf{2^{127}}$	$\mathbf{2^{63}}$	$\mathbf{2^{22}}$	**Differential**	**This Paper**
96/96	8/28	$2^{74.7}$	$2^{27.6}$	-	Linear	[30]
	15/28	$2^{89.1}$	2^{89}	2^{48}	Differential	[1]
	16/28	2^{85}	2^{85}	2^{22}	Differential	[9]
	19/28	$\mathbf{2^{88}}$	$\mathbf{2^{88}}$	$\mathbf{2^{22}}$	**Differential**	**This Paper**
96/144	9/29	$2^{122.7}$	$2^{27.6}$	-	Linear	[30]
	16/29	$2^{135.9}$	$2^{90.9}$	$2^{94.5}$	Rectangle	[1]
	15/29	$2^{89.1}$	2^{89}	2^{48}	Differential	[1]
	17/29	2^{133}	2^{85}	2^{22}	Differential	[9]
	20/29	$\mathbf{2^{136}}$	$\mathbf{2^{88}}$	$\mathbf{2^{22}}$	**Differential**	**This Paper**
128/128	8/32	$2^{92.7}$	$2^{28.3}$	-	Linear	[30]
	16/32	$2^{111.1}$	2^{116}	2^{64}	Differential	[1]
	17/32	2^{113}	2^{113}	2^{22}	Differential	[9]
	22/32	$\mathbf{2^{120}}$	$\mathbf{2^{120}}$	$\mathbf{2^{22}}$	**Differential**	**This Paper**
128/192	9/33	$2^{156.7}$	$2^{28.3}$	-	Linear	[30]
	16/33	$2^{111.1}$	2^{116}	2^{64}	Differential	[1]
	18/33	$2^{182.7}$	$2^{125.9}$	$2^{121.9}$	Rectangle	[1]
	18/33	2^{177}	2^{113}	2^{22}	Differential	[9]
	23/33	$\mathbf{2^{184}}$	$\mathbf{2^{120}}$	$\mathbf{2^{22}}$	**Differential**	**This Paper**
128/256	7/34	$2^{220.7}$	$2^{28.3}$	-	Linear	[30]
	16/34	$2^{111.1}$	2^{116}	2^{64}	Differential	[1]
	18/34	$2^{182.7}$	$2^{125.9}$	$2^{121.9}$	Rectangle	[1]
	19/34	2^{241}	2^{113}	2^{22}	Differential	[9]
	24/34	$\mathbf{2^{248}}$	$\mathbf{2^{120}}$	$\mathbf{2^{22}}$	**Differential**	**This Paper**

Table 2. Summary of differential characteristics and linear approximations for speck

Cipher	Differential characteristic			Linear approximation		
	# Rounds	$\log_2 p$	Ref.	# Rounds	$\log_2 c$	Ref.
Speck32	9	−31	[1]	9	−14	[30]
	9	−30	[6]	9	−14	This paper
	9	−30	This paper			
Speck48	10	−41	[1]	9	−20	[30]
	11	−47	[6]	10	−22	This paper
	11	−45	This paper			
Speck64	13	−59	[1]	11	−25	[30]
	13	−51	This paper	12	−31	[30]
	14	−60	[6]	13	−30	This paper
	14	−56	This paper			
	15	−62	This paper			
Speck96	13	−84	[1]	6	−11	[30]
	13	−67	This paper	15	−45	This paper
	16	−87	This paper			
Speck128	14	−112	[1]	6	−11	[30]
	14	−90	This paper	16	−58	This paper
	19	−119	This paper			

2 Sun *et al.*'s MILP-Based Automatic Search for (Related-Key) Differential and Linear Trails (Hull)

In this section, we briefly recall Sun *et al.*'s algorithm. For more details of their algorithm, we refer to [22, 24].

Objective Function of Differential Model. Let x_i denote the difference variable for the bit i. That is, $x_i = 0$ if there is no difference at bit i; Otherwise, $x_i = 1$. Another bit variable A_j is used to denote the activity of an S-box, *i.e.*, $A_j = 0$ if the S-box is non-active; Otherwise, $A_j = 1$. The objective function is to minimize the sum of all variables $\sum_j A_j$, which indicates the activities of the S-boxes appearing in the schematic description of the encryption and key schedule algorithm.

Constraints of Differential Model. For every XOR operation with bit-level input differences a, b and bit-level output difference c, the constraints include

$$\begin{cases} d_\oplus \geq a, d_\oplus \geq b, d_\oplus \geq c \\ a + b + c \geq 2d_\oplus \\ a + b + c \leq 2 \end{cases} \tag{1}$$

where $d_\oplus$ is a dummy bit variable.

Next, we describe the constraints of the differential properties of an S-box in a more accurate way. For an $\omega \times \nu$ S-box denoted by A_t, the input and output differences are $(x_0, \ldots, x_{\omega-1})$ and $(y_0, \ldots, y_{\nu-1})$, respectively. Then

$$\begin{cases} A_t - x_k \geq 0, k \in \{0, \ldots, \omega - 1\} \\ -A_t + \sum_{j=0}^{\omega-1} x_j \geq 0 \end{cases} \qquad (2)$$

which ensures that nonzero input difference must activate the S-box.

Let $(x_0, \ldots, x_{\omega-1}, y_0, \ldots, y_{\nu-1}) \in \{0, 1\}^{\omega+\nu} \subseteq \mathbb{R}^{\omega+\nu}$ denote an $(\omega + \nu)$-dimensional vector, where $\mathbb{R}$ is the real number field. By computing the H-Representation of the convex hull of all possible input-output differential patterns of an S-box, many linear inequalities which can be used to remove some impossible differential patterns of the S-box are obtained. The greedy algorithm in [24] is applied to select a subset of the H-Representation of the convex hull with less inequalities. As a result, they generate only a small number of linear inequalities, which can be used to exactly describe the differential pattern of S-box and construct the MILP problem. Using any MILP optimizer such as Gurobi [11], good differential characteristics can be found. If we set the value of the object function as N, finish the solving process and output the current solution till the value of object function is reduced to N. The corresponding solution is the identified differential characteristic with N active S-boxes.

Note that this exact searching method is also applicable to searching for the linear approximations.

Objective Function of Linear Model. Some notations for differential model are also used in linear model, e.g., A_j denotes the activity of an S-box and the objective function is to minimize $\sum_j A_j$.

Constraints of Linear Model. For every XOR operation with input masks a, b and output mask c, the constraints should be

$$a = b = c.$$

For every three-forked branch with input mask a and output masks b and c, the constraints should be

$$\begin{cases} d_\curvearrowright \geq a, d_\curvearrowright \geq b, d_\curvearrowright \geq c \\ a + b + c \geq 2d_\curvearrowright \\ a + b + c \leq 2 \end{cases} \qquad (3)$$

where $d_\curvearrowright$ is a dummy bit variable.

For an $\omega \times \nu$ S-box denoted by A_t, the input and output masks are $(x_0, \ldots, x_{\omega-1})$ and $(y_0, \ldots, y_{\nu-1})$, respectively. If the output mask is nonzero, $A_t = 1$; Otherwise, $A_t = 0$. Then, we have

$$\begin{cases} A_t - y_k \geq 0, k \in \{0, \ldots, \nu - 1\} \\ -A_t + \sum_{j=0}^{\nu-1} y_j \geq 0 \end{cases}$$

which ensures that nonzero output mask must activate the S-box.

For an $(\omega + \nu)$-dimensional vector $(x_0, \ldots, x_{\omega-1}, y_0, \ldots, y_{\nu-1}) \in \{0,1\}^{\omega+\nu} \subseteq \mathbb{R}^{\omega+\nu}$, compute a small number of linear inequalities to exactly represent the linear pattern of S-box. The other processes are similar to those in the model of searching for differential characteristics.

In addition, the technique has been extended to find differential or linear hull [24]. By adding the constraints imposed by the given properties (such as fixed difference or linear mask), they updated the MILP model and obtained all trails which consist of the given differential or linear hull.

3   MILP-Based Algorithm for Automatic Search for Differential Characteristics in ARX Ciphers

In this section, we analyze the differential characteristics of modular addition and identify important properties, which are crucial to the construction of our MILP-based models for ARX ciphers. Using our method, we can give the linear inequalities which can exactly describe all differential patterns for modular addition.

3.1   XOR-Differential Characteristics of Modular Addition

Definition 1. *Let α, β and γ be fixed n-bit **XOR** differences. The **XOR**-differential probability (DP) of addition modulo 2^n (xdp^+) is the probability with which α and β propagate to γ through the **ADD** operation, computed over all pairs of n-bit inputs (x,y):*

$$xdp^+(\alpha, \beta \to \gamma) = 2^{-2n} \cdot \#\{(x,y) : ((x \oplus \alpha) + (y \oplus \beta)) \oplus (x + y) = \gamma\}.$$

In [13], Lipmaa *et al.* showed Algorithm 2 to compute $xdp^+(\alpha, \beta \to \gamma)$ which consists of two steps: the first step is to verify if the differential characteristic is possible and the second step is to compute the differential probability xdp^+. More precisely, the above two steps are shown in Theorems 1 and 2, respectively.

Theorem 1 (see [13]). *The differential $(\alpha, \beta \to \gamma)$ is possible iff $(\alpha[0] \oplus \beta[0] \oplus \gamma[0]) = 0$ and $\alpha[i-1] = \beta[i-1] = \gamma[i-1] = \alpha[i] \oplus \beta[i] \oplus \gamma[i]$ for $\alpha[i-1] = \beta[i-1] = \gamma[i-1]$, $i \in [1, n-1]$.*

Theorem 2 (see [13]). *Assume that $(\alpha, \beta \to \gamma)$ is a possible differential characteristic, then the differential probability $xdp^+ = 2^{-\sum_{i=0}^{n-2} \neg eq(\alpha[i], \beta[i], \gamma[i])}$, where*

$$eq(\alpha[i], \beta[i], \gamma[i]) = \begin{cases} 1 & \alpha[i] = \beta[i] = \gamma[i] \\ 0 & others \end{cases}.$$

Theorem 1 can be used to decide if the differential characteristic $(\alpha, \beta \to \gamma)$ for modular addition is possible. For instance, the differential $(\alpha, \beta \to \gamma) = (11100, 11100 \to 11110)$ is impossible as $\alpha[0] = \beta[0] = \gamma[0] \neq \alpha[1] \oplus \beta[1] \oplus \gamma[1]$. Using Theorem 2, the probability of the differential characteristic can be computed efficiently. For example, for the differential $(\alpha, \beta \to \gamma) = (11100, 00110 \to 10110)$, the probability $xdp^+(\alpha, \beta \to \gamma) = 2^{-(\neg eq(0,0,0) + \neg eq(0,1,1) + \neg Eq(1,1,1) + \neg Eq(1,0,0))} = 2^{-2}$.

From Theorem 2, if the n-bit differential characteristic is possible, the probability is only related with $(\alpha[i], \beta[i], \gamma[i])$ for $i \in [0, n-2]$. Taking advantage of this property, we can construct the MILP model to compute the differential probability xdp^+. More details are shown in the following.

3.2 MILP Model for Differential Characteristics of Modular Addition

In order to append the first condition $\alpha[0] \oplus \beta[0] \oplus \gamma[0] = 0$ in Theorem 1 to the set of the linear inequalities, we derive five linear inequalities satisfying the first condition. The five linear inequalities are listed as follows,

$$\begin{cases} d_\oplus \geq \alpha[0], d_\oplus \geq \beta[0], d_\oplus \geq \gamma[0] \\ \alpha[0] + \beta[0] + \gamma[0] - 2d_\oplus \geq 0 \\ \alpha[0] + \beta[0] + \gamma[0] \leq 2 \end{cases} \tag{4}$$

where $d_\oplus$ is a dummy bit variable.

Let the vector $(\alpha[i], \beta[i], \gamma[i], \alpha[i+1], \beta[i+1], \gamma[i+1])$ denote the relation of the differential values for the i-th and the $(i+1)$-th bits. We have that there are totally 56 possible patterns for the vector in accordance with Theorem 1. For example, the differential pattern $(0, 0, 0, 1, 1, 1)$ is impossible as $\alpha_i = \beta_i = \gamma_i \neq \alpha_{i+1} \oplus \beta_{i+1} \oplus \gamma_{i+1}$. Moreover, in order to compute the differential probability efficiently, we append $\neg eq(\alpha[i], \beta[i], \gamma[i])$ to the vector. As described in [23], using the **inequality_generator**() function in the **sage. geometry. polyhedron** class of the SAGE Computer Algebra System [20], we get 65 linear inequalities satisfying all 56 possible patterns. Based on the greedy algorithm in [24], the number of linear inequalities can be reduced from 65 to 13. Furthermore, the 13 linear inequalities can be used to compute the probability of $(\alpha[i]\|\beta[i]\|\gamma[i] \to \alpha[i+1]\|\beta[i+1]\|\gamma[i+1])$ as the variable $\neg eq(\alpha[i], \beta[i], \gamma[i])$ is involved.

Actually, the 13 linear inequalities only represent the second condition $\alpha[i] = \beta[i] = \gamma[i] = \alpha[i+1] \oplus \beta[i+1] \oplus \gamma[i+1]$, $i \in [0, n-2]$ in Theorem 1. In total, there are $(13 \times (n-1) + 5)$ linear inequalities to represent the differential property of addition modulo 2^n with two inputs of n-bit length. As described in Theorem 2, the differential probability $xdp^+ = 2^{-\sum_{i=0}^{n-2} \neg eq(\alpha[i], \beta[i], \gamma[i])}$.

3.3 MILP Model for Differential Characteristics of ARX Ciphers

Besides modular addition, the XOR operations, three-forked branch and the circular shift operations are also involved in ARX ciphers. For each XOR operation, we can also use Inequalities (4). For each three-forked branch operation with input differences a, b and output difference c, the constraints should be

$$a = b = c.$$

For the case of circular shift, we can also list some equations for the related bits. So far, we have finished the construction of all linear inequalities or equations for each operations in ARX ciphers which compose the **constraints** of MILP model for differential characteristics of ARX ciphers.

As the differential probability $xdp^+ = 2^{-\sum_{i=0}^{n-2} \neg eq(\alpha[i], \beta[i], \gamma[i])}$ can be computed using the method described in Sect. 3.2, we set the **objective function** for the r-round differential characteristic as the $\sum_{j=1}^{r} \sum_{i=0}^{n-2} \neg eq(\alpha_j[i], \beta_j[i], \gamma_j[i])$ where α_j, β_j and γ_j are the input differences and output difference of modular addition for the j-th round. We aim to find the minimal value of $\sum_{j=1}^{r} \sum_{i=0}^{n-2} \neg eq(\alpha_j[i], \beta_j[i], \gamma_j[i])$ which represents the differential probability of the best identified differential characteristic. We can use the Gurobi optimizer to solve the system of inequalities to search for differential characteristics for ARX ciphers. However, just being able to obtain one differential characteristic may be not enough. We can apply Sun's method [24] to our new MILP model and find the differential of ARX ciphers.

Note that in the above model, we assume that the two inputs to modular addition and the consecutive rounds are independent. However, for some ARX constructions, they are not independent, which will result that the practical probability of our identified differential characteristics for some fixed key may vary significantly from that derived from our model.

4 MILP Models for Automatic Search for Linear Approximations in ARX Ciphers

In this section, we revisit the property of linear approximations for modular addition operation and develop a new MILP-based tool to search for the linear approximations for ARX ciphers.

4.1 Linear Approximations for Modular Addition

Let n be a non-negative integer. Given two vectors $x = (a_{n-1}, \ldots, a_0)$ and $y = (b_{n-1}, \ldots, b_0) \in \mathbb{F}_2^n$, let $x \cdot y$ denote the standard inner product $x \cdot y = a_{n-1}b_{n-1} \oplus \cdots \oplus a_0 b_0$.

Definition 2 *Let Λ_α, Λ_β and Γ be fixed n-bit linear masks. The correlation of addition modulo 2^n ($cor_\boxplus$) with input masks $\Lambda_\alpha, \Lambda_\beta$ and output mask Γ can be computed over all pairs of n-bit inputs (x, y):*

$$cor_\boxplus(\Gamma, \Lambda_\alpha, \Lambda_\beta) = cor(\Gamma \cdot (x+y) \oplus \Lambda_\alpha \cdot x \oplus \Lambda_\beta \cdot y)$$
$$= 2^{-2n}(\#\{x, y \in \mathbb{F}_2^n : \Gamma \cdot (x+y) \oplus \Lambda_\alpha \cdot x \oplus \Lambda_\beta \cdot y = 0\}$$
$$- \#\{x, y \in \mathbb{F}_2^n : \Gamma \cdot (x+y) \oplus \Lambda_\alpha \cdot x \oplus \Lambda_\beta \cdot y = 1\}).$$

Based on a fairly simple classification of the linear approximations of the carry function, Nyberg and Wallén derive an efficient algorithm for computing the correlation of linear approximation of addition modulo 2^n with k inputs in [19,25]. Since we only consider modular addition with two inputs, we describe this method only for two inputs in [19,25] as follows.

Theorem 3 (see [19,25]). *For the linear approximation of addition modulo 2^n of two inputs with input masks $\Lambda_\alpha, \Lambda_\beta$ and output mask Γ, $\Lambda_\alpha, \Lambda_\beta, \Gamma \in \mathbb{F}_2^n$ and $\Lambda_\alpha = (\Lambda_\alpha[n-1], \ldots, \Lambda_\alpha[0])$, $\Lambda_\beta = (\Lambda_\beta[n-1], \ldots, \Lambda_\beta[0])$, $\Gamma = (\Gamma[n-1], \ldots, \Gamma[0])$, we define the vector $u = (u[n-1], \ldots, u[0])$ where $u[i] = 4\Gamma[i] + 2\Lambda_\alpha[i] + \Lambda_\beta[i], 0 \leq u[i] < 8, 0 \leq i < n$. The correlation can be computed with the following linear representation,*

$$cor_\boxplus(\Gamma, \Lambda_\alpha, \Lambda_\beta) = LA_{u[n-1]}A_{u[n-2]} \cdots A_{u[1]}A_{u[0]}C, \tag{5}$$

where $A_r, r = 0, \ldots, 7$, is 2×2 matrice,

$$A_0 = \frac{1}{2}\begin{bmatrix} 2 & 0 \\ 0 & 1 \end{bmatrix}, A_1 = A_2 = -A_4 = \frac{1}{2}\begin{bmatrix} 0 & 0 \\ 1 & 0 \end{bmatrix},$$

$$A_7 = \frac{1}{2}\begin{bmatrix} 0 & 2 \\ 1 & 0 \end{bmatrix}, -A_3 = A_5 = A_6 = \frac{1}{2}\begin{bmatrix} 0 & 0 \\ 0 & 1 \end{bmatrix},$$

L is a row vector $L = (1\ 0)$, and C is a column vector $C = (1\ 1)^T$.

For example, for the linear approximation with binary vector masks ($\Gamma = 10100, \Lambda_\alpha = 11110, \Lambda_\beta = 11000$), $u = 73620_8$ and $cor_\boxplus = LA_7A_3A_6A_2A_0C = -2^{-3}$.

In order to provide a fast implementation for Theorem 3, Nyberg and Wallén utilized the automaton to calculate $LA_{u[n-1]}A_{u[n-2]} \cdots A_{u[1]}A_{u[0]}C$ by multiplying from left to right. Let $e_0 = L = (1\ 0)$ and $e_1 = (0\ 1)$, then we can show the state transitions in Fig. 1. When reading u from left to right, if the automaton ends up in state 0, then $LA_{u[n-1]}A_{u[n-2]} \cdots A_{u[1]}A_{u[0]}C = 0$. If the automaton ends up in state e_0 or e_1, then $LA_{u[n-1]}A_{u[n-2]} \cdots A_{u[1]}A_{u[0]}C = \pm 2^{-t}$, where t is the number of transitions marked by a solid arrow, and the sign is determined by the number of occurrences of $\{3, 4\}$:

$$LA_{u[n-1]}A_{u[n-2]} \cdots A_{u[1]}A_{u[0]}C > 0$$

if and only if the number of occurrences is even. For example, as $u = 73620_8$, $LA_7A_3A_6A_2A_0C = -2^{-3}$.

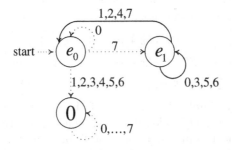

Fig. 1. State transitions for $u = 73620_8$

Fig. 2. State transitions for addition modulo 2^n

Based on the Fig. 1, we will give Proposition 1 to calculate the absolute value of the correlation $cor_{\boxplus}(\Gamma, \Lambda_\alpha, \Lambda_\beta)$.

Proposition 1. *For the linear approximation of addition modulo 2^n of two inputs with input masks $\Lambda_\alpha, \Lambda_\beta$ and output mask Γ, the state transitions of the automaton are shown in Fig. 2, where $u[i] = 4\Gamma[i] + 2\Lambda_\alpha[i] + \Lambda_\beta[i], 0 \leq u[i] < 8, 0 \leq i < n$ and $\epsilon_j \in \{e_0, e_1\}, 0 \leq j < n$. If the correlation for the linear approximation is non-zero, the absolute value of the correlation can be computed as follows,*

$$|cor_{\boxplus}(\Gamma, \Lambda_\alpha, \Lambda_\beta)x| = 2^{-\#\{0 < i < n | \epsilon_i = e_1\}}. \tag{6}$$

Based on Proposition 1, we construct the MILP model to compute the absolute value of correlation of modular addition with two inputs in the following.

4.2   MILP Model for Linear Approximations of Modular Addition

In this part, we will introduce a method to describe linear property of modular addition in Theorem 3 and Proposition 1 as linear inequalities.

For the state transition from ϵ_{i+1} to ϵ_i under $u[i]$, $0 \leq i < n$, the bit variable s_i is defined as follows, $s_i = 0$ if $\epsilon_i = e_0$, and $s_i = 1$ if $\epsilon_i = e_1$. We utilize the vector $(s_{i+1}, \Gamma[i], \Lambda_\alpha[i], \Lambda_\beta[i], s_i)$ to denote the state transition, so $e_{s_{i+1}} A_{u[i]} = e_{s_i}$. For the vector $(s_{i+1}, \Gamma[i], \Lambda_\alpha[i], \Lambda_\beta[i], s_i)$, there are only 10 possible transitions for the vector. As described in Sect. 3.2, we also use the **inequality_generator()** function in the **sage. geometry. polyhedron** class of SAGE and the greedy algorithm in [24] to get eight linear inequalities satisfying all 10 possible transitions. Note that there is an additional constraint $\epsilon_n = e_0$ according to Fig. 2. Hence, for linear approximation of addition modulo 2^n with two inputs, the constraints contain $8 \times n + 1$ linear inequalities and the absolute value of correlation is $|cor_{\boxplus}| = 2^{-\sum_{i=1}^{n-1} s_i}$.

4.3   MILP Model for Linear Approximations for ARX Ciphers

For each XOR, three-forked branch and circular shift operations, we can also use the method in Sect. 2 to produce the linear inequalities or equations. All linear inequalities or equations for each operations in ARX ciphers compose the **constraints** of MILP model for linear approximations of ARX ciphers. We can set the **objective function** for r-round linear approximation as the $\sum_{j=1}^{r} \sum_{i=1}^{n-1} s_i$ to find the minimal value of it. It means to find the optimized linear approximation. We use the Gurobi optimizer to solve the system of inequalities to search for the linear approximations.

As describe in the model of searching for differential characteristics, we assume that the two inputs to modular addition and the consecutive rounds are independent. However, for some ARX constructions, they are not independent. So the practical correlation of our identified linear approximations for some fixed key may vary significantly from that derived from our model.

5   Application to Speck

5.1   Description of Speck

Speck is a family of ARX-based block ciphers proposed by the National Security Agency of the USA in [2], which contains 10 variants. The variants are characterized by the block size of $2n$ bits (where n is the word size) and the key size of mn bits. The Speck block cipher variant with block size $2n$ and key size mn is denoted as Speck$2n/mn$. The rotation constants α, β used in round functions and the number of rounds r are listed in Table 3 for all variants of the Speck.

The round function of Speck consists of XOR, modular addition in $\mathbb{F}_2^n$ and rotation operations. If we denote the subkey in the i-th round as k_i, the input and output of the i-th round as (x_{i-1}, y_{i-1}) and (x_i, y_i), the round function is operated as follows,

$$x_i = ((x_{i-1} \ggg \alpha) \boxplus y_{i-1}) \oplus k_i, \qquad y_i = (y_{i-1} \lll \beta) \oplus x_i,$$

where α and β are rotation constants listed in Table 3.

Table 3. Parameters for speck family of block ciphers

Variant $2n/mn$	Block size $2n$	Word size n	Key size mn	Key words m	Rounds r	α	β
32/64	32	16	64	4	22	7	2
48/72	48	24	72	3	22	8	3
48/96	48	24	96	4	23	8	3
64/96	64	32	96	3	26	8	3
64/128	64	32	128	4	27	8	3
96/96	96	48	96	2	28	8	3
96/144	96	48	144	3	29	8	3
128/128	128	64	128	2	32	8	3
128/192	128	64	192	3	33	8	3
128/256	128	64	256	4	34	8	3

5.2   Differential Characteristics and Linear Approximations of Speck

In this subsection, we will give the details how to use the models in Sects. 3 and 4 to search for the differential characteristics and linear approximations for Speck.

Firstly, we produce the system of inequalities to construct the model for the differential or linear trails for Speck based on the methods in Sects. 3 and 4. Then we use Gurobi optimizer to solve our MILP model as [21–24]. Indeed, other MILP optimizers, such as CPLEX [8], can also be used. The procedure of our method is outlined as follows.

Step 1: Convert the system of inequalities describing r rounds of Speck into a format that is readable by Gurobi.

Step 2: Use Gurobi to search for the trails with the input from *Step* 1.

Without loss of generality, we describe how to construct the model to search for the differential characteristic of r-round Speck32. We denote the input difference and the output difference for the i-th round as Δz_{i-1} and Δz_i, respectively, $\Delta z_i = (\Delta z_i^{31}, \ldots, \Delta z_i^0), 0 < i \leq r$.

If we denote the two input differences of modular addition in the i-th round as α_i and β_i and the output difference as γ_i, then we have $\alpha_i = (\Delta z_{i-1}^{22}, \ldots, \Delta z_{i-1}^{16}, \Delta z_{i-1}^{31}, \ldots, \Delta z_{i-1}^{23}), \beta_i = (\Delta z_{i-1}^{15}, \ldots, \Delta z_{i-1}^0), \gamma_i = (\Delta z_i^{31}, \ldots, \Delta z_i^{16})$. According to Sect. 3, we can produce $13 \times (16 - 1) + 5 = 200$ linear inequalities to represent the differential property of $(\alpha_i, \beta_i \to \gamma_i)$ for modular addition in the i-th round.

For the XOR operation of two branches in the i-th round, the two input differences are $(\Delta z_{i-1}^{13}, \ldots, \Delta z_{i-1}^0, \Delta z_{i-1}^{15}, \Delta z_{i-1}^{14})$ and $(\Delta z_i^{31}, \ldots, \Delta z_i^{16})$, and the output difference is $(\Delta z_i^{15}, \ldots, \Delta z_i^0)$. Thus we have

$$\Delta z_{i-1}^{13} \oplus \Delta z_i^{31} = \Delta z_i^{15},$$

$$\vdots$$

$$\Delta z_{i-1}^{14} \oplus \Delta z_i^{16} = \Delta z_i^0.$$

Then we use Inequalities (4) to describe the differential property of XOR operation of two branches in the i-th round, so $5 \times 16 = 80$ linear inequalities are produced. Therefore, we use the above produced $200 + 80 = 280$ linear inequalities to describe the differential property of $(\Delta z_{i-1} \to \Delta z_i)$.

In the similar way, $280 \cdot r$ linear inequalities are derived to describe the differential property of r rounds $(\Delta z_0 \to \Delta z_1 \to \ldots \to \Delta z_r)$. Moreover, one additional condition $\sum_{j=0}^{31} \Delta z_0^j > 0$ is required to ensure the non-zero plaintext difference.

We set the objective function as the minimal value of $\sum_{i=1}^r \sum_{j=0}^{32-2} \neg eq(\alpha_i[j], \beta_i[j], \gamma_i[j])$ and convert the above $280 \cdot r + 1$ linear inequalities as constraints into the LP format that is readable by Gurobi. Finally, we use the Gurobi to find the the differential characteristic of r rounds of Speck32.

Similarly, the process of constructing the model to search for the linear approximations for Speck can be implemented using the model in Sect. 4. Here we will not provide the details about it due to the limited space. The source code is published in https://github.com/fukai6/milp_speck.git.

As we search for the differential and linear trails for Speck with block size greater than 48, we use the splicing heuristic in order to speed up the search process. The splicing heuristic is to search two short trails and concatenate them to produce a long trail. For example, we can search r_1-round differential characteristic with output difference δ and r_2-round differential characteristic with input difference δ to construct an $(r_1 + r_2)$-round differential characteristic. Based on the observation from our identified differential characteristic for small number of rounds and differential characteristics presented in [4], we find that the differential probability probably is better when the left of δ is $0x80$ and the right of δ is 0. In this way, we manually choose different values of r_1 and r_2, and set $\delta = 0x80\|0$ as the output difference or input difference to search two differential trails, then concatenate them to produce an $(r_1 + r_2)$-round differential characteristic. For the linear approximation, we set the input mask or output mask as $0x1\|0$.

Finally, the best differential characteristics and linear approximations we found are listed in Tables 4, 5, 6 and 7, where $\sum_{i=1}^{r} \log_2 p_i$ and $\sum_{i=1}^{r} \log_2 c_i$ are the probability of differential characteristic and the correlation of linear approximation, respectively.

Note that the differential characteristics in Table 5 have been produced with $r_1 = 11, r_2 = 4$ for Speck64 and $r_1 = 11, r_2 = 5$ for Speck 96, respectively. For Speck128, with the splicing heuristic, we can only get an 18-round differential characteristic with the probability 2^{-126} by setting $r_1 = r_2 = 9$ with reason-

Table 4. Differential characteristics for Speck32, Speck48 and Speck64

i	Speck32			Speck48			Speck64		
	Δ_L	Δ_R	$\log_2 p_i$	Δ_L	Δ_R	$\log_2 p_i$	Δ_L	Δ_R	$\log_2 p_i$
0	0211	0A04		001202	020002		04092400	20040104	
1	2800	0010	-4	000010	100000	-3	20000820	20200001	-6
2	0040	0000	-2	000000	800000	-1	00000009	01000000	-4
3	8000	8000	0	800000	800004	-0	08000000	00000000	-2
4	8100	8102	-1	808004	808020	-2	00080000	00080000	-1
5	8004	840E	-3	8400A0	8001A4	-4	00080800	00480800	-2
6	8532	9508	-8	608DA4	608080	-9	00480008	02084008	-4
7	5002	0420	-7	042003	002400	-11	06080808	164A0848	-7
8	0080	1000	-3	012020	000020	-5	F2400040	40104200	-13
9	1001	5001	-2	200100	200000	-3	00820200	00001202	-8
10				202001	202000	-3	00009000	00000010	-4
11				210020	200021	-4	00000080	00000000	-2
12							80000000	80000000	0
13							80800000	80800004	-1
14							80008004	84008020	-3
15							808080A0	A08481A4	-5
$\sum_{i=1}^{r} \log_2 p_i$			-30			-45			-62

Table 5. Differential characteristic for Speck96 and Speck128

i	Δ_L	Δ_R	$\log_2 p_i$	Δ_L	Δ_R	$\log_2 p_i$
0	240004000009	010420040000	.	0124000400000000	0801042004000000	
1	082020000000	000120200000	-6	0800202000000000	4808012020000000	-6
2	000900000000	000001000000	-4	4800010000000000	0840080100000002	-6
3	000008000000	000000000000	-2	0808080000000006	4A08480800000016	-7
4	000000080000	000000080000	-1	4000400000000032	1042004000000080	-12
5	000000080800	000000480800	-2	0202000000000080	8012020000000480	-7
6	000000480008	000002084008	-4	0010000000000480	0080100000002084	-5
7	0800FE080808	0800EE4A0848	-12	8080000000002080	84808000000124A0	-5
8	000772400040	400000104200	-21	0400000000012440	2004000000080144	-9
9	000000820200	000000001202	-11	2000000000080220	2020000000480801	-9
10	000000009000	000000000010	-4	0000000000480001	0100000002084008	-7
11	000000000080	000000000000	-2	000000000E080808	080000001E4A0848	-8
12	800000000000	800000000000	0	00000000F2400040	4000000000104200	-15
13	808000000000	808000000004	-1	0000000000820200	0000000000001202	-8
14	800080000004	840080000020	-3	0000000000009000	0000000000000010	-4
15	808080800020	A08480800124	-5	0000000000000080	0000000000000000	-2
16	800400008124	842004008801	-9	8000000000000000	8000000000000000	0
17				8080000000000000	8080000000000004	-1
18				8000800000000004	8400800000000020	-3
19				8080808000000020	A084808000000124	-5
$\sum_{i=1}^{r} \log_2 p_i$			-87			-119

Table 6. Linear approximations of Speck32, Speck48 and Speck64

i	Speck32			Speck48			Speck64		
	Γ_L	Γ_R	$\log_2 c_i$	Γ_L	Γ_R	$\log_2 c_i$	Γ_L	Γ_R	$\log_2 c_i$
0	0380	5224		000131	050021		18600010	10724800	
1	4880	4885	-1	018100	200101	-2	1B000000	03104000	-3
2	20A0	2071	-2	000100	000001	-1	18000000	18120000	-2
3	40A0	00C1	-2	000001	000000	0	C0000000	C0100000	-1
4	0080	4001	-3	0D0000	0C0000	-1	04000006	04800006	-1
5	0000	0001	0	606100	606C00	-2	00260030	04200030	-2
6	0004	0004	0	00024A	00620B	-2	01010501	21013181	-5
7	3810	3010	-1	181040	731042	-4	01800126	00018021	-4
8	2180	01C0	-3	D812C0	9802D0	-3	00018100	20000101	-5
9	066A	0608	-2	040600	C4961A	-5	00000100	00000001	-1
10				2484F2	2480F6	-2	00000001	00000000	0
11							09800000	08000000	-1
12							40610000	40680000	-2
13							00024982	00420802	-3
$\sum_{i=1}^{r} \log_2 c_i$	2^{-14}			2^{-22}			2^{-30}		

Table 7. Linear approximation of Speck96 and Speck128

i	Γ_L	Γ_R	$\log_2 c_i$	Γ_L	Γ_R	$\log_2 c_i$
0	000001000130	040000010021		0001010000018798	6A800101300601C1	
1	000000018100	200000000101	-2	0000018000300720	9400000180300625	-7
2	000000000100	000000000001	-1	0000000181818100	200000000181B105	-4
3	000000000001	000000000000	0	0000000001800120	0000000000018021	-3
4	0D0000000000	0C0000000000	-1	0000000000018100	2000000000000101	-4
5	604500000000	604C00000000	-2	0000000000000100	0000000000000001	-1
6	00224D000003	006228000003	-4	0000000000000001	0000000000000000	0
7	181070001018	1B105A680018	-12	0980000000000000	0800000000000000	-1
8	001200000010	180210400000	-6	4045000000000000	4048000000000000	-2
9	101000000000	00101A000000	-3	00224D0000000002	0042280000000002	-4
10	001800000000	000010000000	-2	1810600000000010	1A10536C00000010	-9
11	000010000000	000000000000	-1	0012000000000080	1002186000000080	-4
12	000000D00000	000000C00000	-1	0010000000000406	8010130000000406	-3
13	000006041800	000006048000	-2	3680000000002000	3080180000002004	-3
14	000030003490	000030043080	-3	8500000000010181	8524C000000101A1	-4
15	180181910500	800181A10526	-5	8002000000080001	2106000000080100	-6
16				01301A0000404401	0030180000404801	-3
$\sum_{i=1}^{r} \log_2 c_i$	2^{-45}			2^{-58}		

able time. Thus, in order to find a better trail, we firstly search for a 13-round differential trail with the splicing heuristic by setting $r_1 = 9$ and $r_2 = 4$, then extend six rounds before the 13-round differential trial to get the 19-round differential characteristic in Table 5, where the 6-round trail has been also found with Gurobi.

The linear approximations in Tables 6 and 7 have been identified with the parameters: $r_1 = 10, r_2 = 3$ for Speck64, $r_1 = 3, r_2 = 12$ for Speck96 and $r_1 = 6, r_2 = 10$ for Speck128.

For the runtime of the searching algorithm, we spent about several hours on personal computer (4 Core, Intel(R) Core(TM) i5 CPU 650 @3.20 GHz) for Speck32 and about one day on IBM server (64 Core, Intel(R)Xeon(R) CPU E7330, 2.40 GHz) for other variants of Speck. Note that we have searched for all the differential characteristics and linear approximations for Speck32, however, for other variants we aim to only find better trails than the previous ones but we cannot guarantee they are the best trails.

A summary of the differential characteristics and the linear approximations for Speck is provided in Table 2, which shows that we got better differential characteristics and linear approximations for Speck48, Speck64, Speck96 and Speck128.

In order to check the effect of the assumptions of independent inputs to the modular addition and independent rounds for Speck, we experimentally calculate the probability for our identified differential characteristics in Tables 4 and 5. As it is not feasible to do this for many rounds, we break down the differential characteristics to small overlapping segments according to the differential probability of the segments. The calculated probability of each one of these segments has been verified experimentally by encrypting sufficiently many random plaintext pairs for some arbitrary keys. The test results are shown in Table 8. In Table 8, the first column is the tested cipher, the second column shows rounds covered by the segment of differential characteristic, the third column is the theoretic differential probability of the corresponding segment of differential characteristic, the fourth column is the total number of random chosen plaintext pairs used in the test, the fifth column is the total number of tested key values, and the last column is the number of keys which can get the calculated differential probability no less than the theoretic differential probability. Note that we only test the last segment from round 12 to round 19 for Speck-128 because the theoretic differential probabilities for the previous segments are too low to be tested. From Table 8, we can see that the number of good keys significantly deviates from the mean for some cases, which is due to the independent assumptions for Speck cipher. Such deviation will have effect on the success rate of the attacks in the practitioner's perspective.

5.3   Key Recovery Attack on Speck

In [9] Dinur presented a generic key recovery framework for Speck which can extend the differential attack for more rounds. The idea of the framework uses the guess-and-determine technique instead of counting technique for standard key recovery attack. Furthermore, the cryptanalytic technique of ARX cipher is utilized to speed up the attack on Speck.

For Speck$2n/mn$, if the differential characteristic with probability p for $r-1$ rounds has been found, the attacker can add one round at the top of the differential characteristic and m rounds at the bottom of the differential characteristic

Table 8. Test for differential characteristics in Tables 4 and 5

Cipher	Rounds of segment	Differential probability	Number of plaintext pairs	Total number of keys	Number of good keys
Speck32/64	0–9	2^{-30}	2^{32}	7040	3456
Speck48/96	0–6	2^{-19}	2^{22}	10000	4093
Speck48/96	6–11	2^{-26}	2^{30}	6400	2989
Speck64/128	0–6	2^{-19}	2^{22}	10000	5513
Speck64/128	6–8	2^{-20}	2^{24}	10000	4887
Speck64/128	8–15	2^{-23}	2^{26}	10000	4918
Spec96/144	0–5	2^{-15}	2^{18}	10000	5123
Spec96/144	5–7	2^{-16}	2^{20}	10000	5039
Spec96/144	7–8	2^{-21}	2^{24}	10000	5454
Spec96/144	8–11	2^{-17}	2^{20}	10000	5020
Spec96/144	11–16	2^{-18}	2^{22}	10000	4645
Spec128/256	12–19	2^{-23}	2^{26}	10000	4876

to cover $r + m$ rounds in total. It is not necessary to guess the subkey in the first round as it has no effect on the input difference. At last, the attacker can recover the mn-bit secret key of a variant with $r + m$ rounds of Speck$2n/mn$ using $2 \cdot p^{-1}$ chosen plaintexts with time complexity of $2 \cdot p^{-1} \cdot 2^{(m-2)n}$ and memory complexity of 2^{22} bytes.

With our identified differential characteristics for Speck, we can give the improved key recovery attack. Since the attack is same as that in [9], details are omitted. For each variant of Speck, we summarize our attacks and the previous differential attacks in Table 1, which shows that our attacks for the variants of Speck with block size 48, 64, 96 and 128 are best attacks in terms of the number of rounds.

6   Conclusion

In this paper, we construct the MILP model to automatically search for differential and linear approximations for ARX ciphers by researching the differential and linear property of modular addition under the assumptions of independent inputs to the modular addition and independent rounds. Then we use the new MILP model to search for the differential characteristics and linear approximations for Speck cipher. Compared with the previous best differential characteristics for them, our identified differential characteristics for Speck64, Speck96 and Speck128 are extended for one, three and five rounds, respectively, and our differential characteristic for Speck48 has higher probability. We use those new differential characteristics to improve the currently best public attacks on the four variants of Speck. In addition, we searched for the linear approximations for Speck cipher and improved the previous linear approximations for Speck variants with block size greater than 32.

Acknowledgments. This work has been supported by 973 Program (No. 2013C B834205), NSFC Projects (No. 61133013, No. 61572293), Program for New Century Excellent Talents in University of China (NCET- 13-0350).

References

1. Abed, F., List, E., Lucks, S., Wenzel, J.: Differential cryptanalysis of round-reduced simon and speck. In: Cid, C., Rechberger, C. (eds.) FSE 2014. LNCS, vol. 8540, pp. 525–545. Springer, Heidelberg (2015)
2. Beaulieu, R., Shors, D., Smith, J., Treatman-Clark, S., Weeks, B., Wingers, L.: The SIMON and SPECK famillies of lightweight block ciphers. Cryptology ePrint Archive, report 2013/543 (2013). http://eprint.iacr.org/
3. Biham, E., Shamir, A.: Differential cryptanalysis of DES-like cryptosystems. J. Cryptol. 4(1), 3–72 (1991)
4. Biryukov, A., Velichkov, V.: Automatic search for differential trails in ARX ciphers. In: Benaloh, J. (ed.) CT-RSA 2014. LNCS, vol. 8366, pp. 227–250. Springer, Heidelberg (2014)

5. Biryukov, A., Velichkov, V., Corre, Y.: Automatic search for the best trails in ARX: application to block cipher speck. In: FSE 2016, Bochum, Germany, 20–23 March (2016, to appear)
6. Biryukov, A., Roy, A., Velichkov, V.: Differential analysis of block ciphers SIMON and SPECK. In: Cid, C., Rechberger, C. (eds.) FSE 2014. LNCS, vol. 8540, pp. 546–570. Springer, Heidelberg (2015)
7. Biryukov, A., Velichkov, V., Corre, Y.L.: Automatic search for the best trails in ARX: application to block cipher speck. In: FSE 2016. LNCS. Springer (to appear)
8. Cplex, I.I.: IBM software group. User-Manual CPLEX (2011)
9. Dinur, I.: Improved differential cryptanalysis of round-reduced speck. In: Joux, A., Youssef, A. (eds.) SAC 2014. LNCS, vol. 8781, pp. 147–164. Springer, Heidelberg (2014)
10. De Cannière, C., Rechberger, C.: Finding SHA-1 characteristics: general results and applications. In: Lai, X., Chen, K. (eds.) ASIACRYPT 2006. LNCS, vol. 4284, pp. 1–20. Springer, Heidelberg (2006)
11. Gurobi Optimization: Gurobi Optimizer Reference Manual (2013). http://www.gurobi.com
12. Leurent, G.: Construction of differential characteristics in ARX designs-application to skein. Cryptology ePrint Archive, 2012/668 (2012). https://eprint.iacr.org/
13. Lipmaa, H., Moriai, S.: Efficient algorithms for computing differential properties of addition. In: Matsui, M. (ed.) FSE 2001. LNCS, vol. 2355, pp. 336–350. Springer, Heidelberg (2002)
14. Liu, M., Chen, J.: Improved linear attacks on the chinese block cipher standard. JCST **29**(6), 1123–1133 (2014). Springer
15. Matsui, M.: Linear cryptanalysis method for DES cipher. In: Helleseth, T. (ed.) EUROCRYPT 1993. LNCS, vol. 765, pp. 386–397. Springer, Heidelberg (1994)
16. Matsui, M.: On Correlation between the order of s-boxes and the strength of DES. In: De Santis, A. (ed.) EUROCRYPT 1994. LNCS, vol. 950, pp. 366–375. Springer, Heidelberg (1995)
17. Mendel, F., Nad, T., Schläffer, M.: Finding SHA-2 characteristics: searching through a minefield of contradictions. In: Lee, D.H., Wang, X. (eds.) ASIACRYPT 2011. LNCS, vol. 7073, pp. 288–307. Springer, Heidelberg (2011)
18. Mouha, N., Wang, Q., Gu, D., Preneel, B.: Differential and linear cryptanalysis using mixed-integer linear programming. In: Wu, C.-K., Yung, M., Lin, D. (eds.) Inscrypt 2011. LNCS, vol. 7537, pp. 57–76. Springer, Heidelberg (2012)
19. Nyberg, K., Wallén, J.: Improved linear distinguishers for SNOW 2.0. In: Robshaw, M. (ed.) FSE 2006. LNCS, vol. 4047, pp. 144–162. Springer, Heidelberg (2006)
20. Stein, W., et al.: Sage: Open Source Mathematical Software (2008)
21. Sun, S., Hu, L., Song, L., Xie, Y., Wang, P.: Automatic security evaluation of block ciphers with S-bP structures against related-key differential attacks. In: Lin, D., Xu, S., Yung, M. (eds.) Inscrypt 2013. LNCS, vol. 8567, pp. 39–51. Springer, Heidelberg (2014)
22. Sun, S., Hu, L., Wang, P., Qiao, K., Ma, X., Song, L.: Automatic security evaluation and (related-key) differential characteristic search: application to SIMON, PRESENT, LBlock, DES(L) and other bit-oriented block ciphers. In: Sarkar, P., Iwata, T. (eds.) ASIACRYPT 2014. LNCS, vol. 8873, pp. 158–178. Springer, Heidelberg (2014)
23. Sun, S., Hu, L., Qiao, K., Ma, X., Song, L.: Automatic security evaluation and (related-key) differential characteristic search: application to SIMON, PRESENT, LBlock, DES(L) and other bit-oriented block ciphers. Cryptology ePrint Archive, report 2013/676 (2013). https://eprint.iacr.org/

24. Sun, S., Hu, L., Wang, M., Wang, P., Qiao, K., Ma, X., Shi, D., Song, L., Fu, K.: Towards finding the best characteristics of some bit-oriented block ciphers and automatic enumeration of (related-key) differential and linear characteristics with predefined properties. Cryptology ePrint Archive, report 2014/747 (2014). https://eprint.iacr.org/

25. Wallén, J.: Linear approximations of addition modulo 2^n. In: Johansson, T. (ed.) FSE 2003. LNCS, vol. 2887, pp. 261–273. Springer, Heidelberg (2003)

26. Wang, G., Keller, N., Dunkelman, O.: The delicate issues of addition with respect to XOR differences. In: Adams, C., Miri, A., Wiener, M. (eds.) SAC 2007. LNCS, vol. 4876, pp. 212–231. Springer, Heidelberg (2007)

27. Winnen, L.: Sage S-box MILP toolkit. http://www.ecrypt.eu.org/tools/sage-s-box-milp-toolkit

28. Wu, S., Wang, M.: Security Evaluation against differential cryptanalysis for block cipher structures. Cryptology ePrint Archive, report 2011/551 (2011). https://eprint.iacr.org/

29. Wu, S., Wu, H., Huang, T., Wang, M., Wu, W.: Leaked-state-forgery attack against the authenticated encryption algorithm ALE. In: Sako, K., Sarkar, P. (eds.) ASIACRYPT 2013, Part I. LNCS, vol. 8269, pp. 377–404. Springer, Heidelberg (2013)

30. Yao, Y., Zhang, B., Wu, W.: Automatic search for linear trails of the SPECK family. In: López, J., Mitchell, C.J. (eds.) ISC 2015. LNCS, vol. 9290, pp. 158–176. Springer, Heidelberg (2015)

Automatic Search for the Best Trails in ARX: Application to Block Cipher SPECK

Alex Biryukov$^{(\boxtimes)}$, Vesselin Velichkov$^{(\boxtimes)}$, and Yann Le Corre$^{(\boxtimes)}$

Laboratory of Algorithmics, Cryptology and Security (LACS),
University of Luxembourg, Luxembourg, Luxembourg
{Alex.Biryukov,Vesselin.Velichkov,Yann.LeCorre}@uni.lu

Abstract. We propose the first adaptation of Matsui's algorithm for finding the best differential and linear trails to the class of ARX ciphers. It is based on a branch-and-bound search strategy, does not use any heuristics and returns optimal results. The practical application of the new algorithm is demonstrated on reduced round variants of block ciphers from the SPECK family. More specifically, we report the probabilities of the best differential trails for up to 10, 9, 8, 7, and 7 rounds of SPECK32, SPECK48, SPECK64, SPECK96 and SPECK128 respectively, together with the exact number of differential trails that have the best probability. The new results are used to compute bounds, under the Markov assumption, on the security of SPECK against single-trail differential cryptanalysis. Finally, we propose two new ARX primitives with provable bounds against single-trail differential and linear cryptanalysis – a long standing open problem in the area of ARX design.

Keywords: Symmetric-key · Cryptanalysis · ARX · SPECK

1 Introduction

ARX stands for Addition/Rotation/XOR and denotes a class of cryptographic algorithms based on the simple arithmetic operations: modular addition, bitwise rotation (and bitwise shift) and exclusive-OR. Although the acronym has gained popularity only recently, algorithms using these operations have been designed ever since the 80s.

Some notable historical examples of ARX designs are the block ciphers FEAL (1987), RC5 (1994), and TEA (1994) (with its modified versions XTEA (1997) and XXTEA (1998)). More recent proposals include the stream cipher Salsa20 (2008) and its variant ChaCha (2008); the hash functions BLAKE (2008) (using a modified version of ChaCha) and Skein [12] (2008) (with its underlying block cipher Threefish); the hash function for short messages SipHash (2012) and the block cipher SPECK [2] (2013) (both using a variant of Threefish's MIX operation); the lightweight block cipher LEA (2013) and the MAC algorithm for 32-bit microcontrollers Chaskey (2014) (based on a reduced word-size variant of SipHash's round function).

© International Association for Cryptologic Research 2016
T. Peyrin (Ed.): FSE 2016, LNCS 9783, pp. 289–310, 2016.
DOI: 10.1007/978-3-662-52993-5_15

All mentioned ARX designs are also called *pure*, since they are exclusively composed of the three basic ARX operations. In addition, there is also the subclass of *augmented* ARX designs that consists of a combination of the ARX operations with other bitwise operations such as Boolean operators, Boolean functions, etc. The most eminent representatives of this group are the hash functions from the MD and SHA families.

As evidenced by the long list of proposals, there is a steady interest in the ARX design philosophy. The reason is the simplicity and efficiency in both software and hardware of these designs. In recent years ARX algorithms have become especially attractive in the area of lightweight cryptography for environments with highly constrained resources. According to new results from the Framework for Fair Evaluation of Lightweight Cryptographic Systems (FELICS) [6], presented at the NIST Lightweight Cryptography Workshop 2015 [25], the most efficient lightweight designs have ARX structure.

The ARX class of primitives is often seen as an alternative to the well-established class of S-box based algorithms, among whose most notable representatives are the block cipher AES [8] and the historically significant block cipher DES [24]. While primitives from this class make use of substitution tables (S-boxes) as a source of non-linearity, the only non-linear component in ARX is the modular addition operation. Due to the latter, these primitives are also less vulnerable to cache-timing and side-channel attacks.

While ARX algorithms provide level of security comparable to S-box based ones, they suffer from a major drawback – the methods for their analysis and design are far less rigorous and mature. For S-box based ciphers it is possible to compute provable bounds on the security against the two most powerful cryptanalytic attacks – differential cryptanalysis [3] and linear cryptanalysis [19] (see e.g. [7]). In contrast, the state of the art in the design of ARX can be summarized in the following heuristic common-sense rule: mix the basic arithmetic operations in a *reasonable* way and iterate them over *sufficient* number of rounds. While this strategy seems to be largely successful in practice, it is based more on experience and intuition, rather than on sound scientific arguments.

In this paper we address the mentioned problem by proposing for the first time an algorithm that finds the best differential and linear trails of an ARX cipher for a given number of rounds. It is based on a branch-and-bound search strategy similar to Matsui's search algorithm that was applied to DES [18] and is inspired by the threshold search technique proposed in [5]. While the latter uses heuristics in order to find high-probability trails that are not necessarily optimal, our algorithm does not use any heuristics and finds optimal results.

The trails found with the described method are optimal under the *Markov assumption* [14, Sect. 3, Theorem 2] (see also [8, Sect. 6.2, pp. 84]). The Markov assumption ensures that (a) the analyzed primitive is a Markov cipher in the sense of the definition in [14, Sect. 3] and (b) it can be assumed that its round keys are chosen at random independently (i.e. the Hypothesis of independent round keys [8, Sect. 8.7.2] holds). The Markov assumption allows to treat the rounds of an iterated cipher independently and thus to compute the

Table 1. Probabilities of the best (under the Markov assumption) differential trails for SPECK found with Best Search (BS) (Sect. 5) versus best probabilities found with Threshold Search (TS) [4]. The column # lists the number of trails having the best probability. The column R contains the number of rounds.

R	SPECK32			SPECK48			SPECK64			SPECK96		SPECK128	
	TS	BS	#	TS	BS	#	TS	BS	#	BS	#	BS	#
1	-0	-0	3	-0	-0	3	-0	-0	3	-0	3	-0	3
2	-1	-1	3	-1	-1	3	-1	-1	3	-1	3	-1	3
3	-3	-3	3	-4	-3	2	-3	-3	2	-3	2	-3	2
4	-5	-5	1	-7	-6	2	-7	-6	2	-6	2	-6	2
5	-9	-9	2	-10	-10	4	-13	-10	2	-10	2	-10	2
6	-15	-13	1	-14	-14	2	-21	-15	2	-15	2	-15	2
7	-22	-18	2	-20	-19	2	-27	-21	3	-21	2	-21	≥ 1
8	-26	-24	7	-27	-26	12	-32	-29	≥ 1	< -27		< -26	
9	-30	-30	15	-33	-33	≥ 1	-36	< -31					
10		-34	1	-40	< -34		-40						
11				-47			-44						
12							-47						
13							-52						
14							-60						

differential probability (resp. absolute linear correlation) of an r-round trail as the product of the probabilities (resp. absolute correlations) of its corresponding 1-round trails. **For ciphers that do not satisfy the Markov assumption, fixed keys may exist for which the probability (resp. correlation) of the best differential (resp. linear) trail may significantly deviate from the optimal one as computed with our algorithm.**

As a demonstration of the effectiveness of the technique we apply it to block cipher SPECK and we report for the first time all provably best (under the Markov assumption) differential trails for reduced number of rounds. We also demonstrate that in some cases the threshold search algorithm returns sub-optimal results. These new results are summarized in Table 1.

As noted, the results shown in Table 1 are to be interpreted under the Markov assumption. In Appendix 8 we show for the first time that SPECK is not, in fact, a Markov cipher. We stress, however, that making the Markov assumption even for non-Markov ciphers is the best that a cryptanalyst can do in order to be able to analyze such constructions. Furthermore, we have experimentally checked that the reported differentials hold for most of the keys and therefore the results shown in Table 4 are meaningful from a practitioner's perspective.

The new technique can also be used to design new ARX primitives with provable security bounds against linear and differential cryptanalysis – a long standing problem in the area of ARX design. Our main contributions can be summarized as follow:

1. An algorithm for finding the best differential and linear trails in ARX ciphers that satisfy the Markov Assumption.
2. The probabilities of the best differential trails for up to 10, 9, 8, 7, and 7 rounds of SPECK32, SPECK48, SPECK64, SPECK96 and SPECK128 respectively, together with the exact number of differential trails that have the best probability.
3. A better choice of rotation constants for SPECK w.r.t. single-trail differential cryptanalysis.
4. Bounds on the security of SPECK, under the Markov assumption, against differential cryptanalysis, based on the reported best trails.
5. Two atomic ARX constructions with provable bounds against single-trail differential and linear cryptanalysis.

The paper is organized as follows. We begin in Sect. 2 with a review of previous work on techniques for searching for differential and linear trails in ARX. Section 3 provides basic definitions and propositions, necessary to follow the exposition in subsequent sections. A general strategy for searching for the best trails in ARX is described in Sect. 4 and the results from its application to SPECK are given in Sect. 5. Two new primitives – MARX and SPECKEY – with provable bounds against single trail differential and linear cryptanalysis are proposed in Sects. 6 and 7 concludes the paper. The notation used in the paper is summarized in Table 2.

Table 2. Notation.

Symbol	Meaning
w	Word size in bits
n	Total number of rounds
r	Iterator over the number of rounds: $1 \leq r \leq n$
N	Cipher block size (in bits)
LSB, MSB	Least Significant Bit, Most Significant Bit
$x[i]$	The i-th bit of w-bit word x: $0 \leq i < w$: $x[0] = \text{LSB}$, $x[w-1] = \text{MSB}$
$x[i:j]$	The sequence of bits $x[i], x[i+1], \ldots, x[j]$ (if $i < j$) or $x[i], x[i-1], \ldots, x[j]$ (if $i > j$)
$\boxplus$	Addition modulo 2^w
α_r, β_r	Input XOR differences (resp. linear masks) to $\boxplus$ at round r
γ_r	Output XOR difference (resp. linear mask) of $\boxplus$ at round r
$(\alpha_r, \beta_r \rightarrow \gamma_r)$	A differential or a linear approximation of $\boxplus$
$|c|$	Absolute value of c

2   Previous Work

Finding high probability (resp. high absolute correlation) trails for ARX has traditionally been a difficult task. The lack of S-boxes in this class of primitives does not allow to efficiently compute the probabilities (resp. correlations) of all possible differential transitions (resp. linear approximations) by the means of the difference distribution table – DDT (resp. linear approximation table – LAT) of the non-linear elements. This makes the construction of trails in ARX a tedious and especially error-prone process as shown in [15]. Furthermore, while most S-box designs are word-based with relatively small word sizes of 4 and 8 bits, all ARX designs are bit-based with typical size of the words 32 and 64 bits. As a consequence it is not possible to apply elegant design strategies such as the wide-trail [7] to design primitives with provable bounds against differential and linear cryptanalysis. Indeed the design of such an ARX construction is still an open problem.

The described difficulties in the analysis and design of ARX have been addressed by several researchers in the past. Depending on the angle from which they approach the problem, their work can broadly be divided into three categories: bottom-up, top-down and approximation-based techniques. We briefly describe these categories below.

Bottom-up Techniques. This category is by far the largest and encompasses methods for the (automatic) construction of differential and linear trails in ARX. Arguably the first such techniques date back to the collisions on the MD and SHA families of hash functions by Wang et al. [34–36]. While these results were reportedly developed by hand, subsequent methods were proposed for the fully automatic construction of differential paths in ARX all of which were applied to *augmented* ARX designs such as SHA1, SHA2, MD4 and MD5. In [16] was proposed a method for the automatic construction of differential trails in *pure* ARX designs and applied to the hash function Skein. While many of the mentioned techniques are general and potentially applicable to any ARX primitive, all of them were applied exclusively to hash functions. To fill the gap, in [5] was proposed the threshold search method for searching for differential trails in ARX ciphers such as TEA, XTEA and SPECK. This method was subsequently extended to the case of differentials in [4]. Most recently, in 2015, two new techniques for automatic search for linear trails have been proposed. One has been applied to SPECK [37], while the other is dedicated to authenticated encryption schemes [11].

Top-down Techniques. Rather than constructing a trail one round at a time as in the bottom-up approach, top-down techniques consider the cipher as a whole. More precisely, the cipher is represented either as a system of Boolean equations or as a system of mixed-integer inequalities. Each solution to the system corresponds to a valid trail. In the first case, the Boolean equations are transformed into a conjunctive normal form (CNF) formula, whose satisfying assignment/s are found with a SAT solver. In the second case, the problem of searching for trails is effectively transformed into a mixed-integer linear problem (MILP) that is usually solved by dedicated MILP solvers using linear-programming based branch-and-bound algorithms. The SAT solver approach has been used to find the best differential trails for several rounds of stream cipher Salsa20 and for proving security bounds for the authenticated encryption cipher NORX. As to the MILP-based methods, up to now they have been successful mainly in the analysis of S-box designs [23, 28]. The only applications of MILP to ARX that we are aware of are the results on the augmented ARX cipher SIMON [28] and a very recent paper [13] on SPECK appearing in this volume of FSE'16.

Approximation-based Techniques. In both top-down and bottom-up approaches, complex techniques for analysis of existing algorithms are developed. In contrast, in what we call approximation-based techniques, the problem is turned around: new primitives are developed so that they are easy to be analyzed *by design*. The main idea is to replace the non-linear component of ARX – the modular addition – by a simpler non-linear approximation that can efficiently and accurately be analyzed with existing methods. A design based on this strategy is the authenticated encryption scheme NORX [1]. In it the addition operation is replaced by the first-order approximation $a \oplus b \oplus (a \wedge b) \ll 1 \approx a \boxplus b$, which effectively limits the carry propagation to a sliding window of 2 bits. The latter significantly facilitates the analysis of the scheme and also makes it hardware efficient.

From the above overview of existing results it is clear that the question of finding optimal trails in pure ARX ciphers has remained largely unexplored so far. The only results in this direction that we are aware of are [21], which applies a SAT solver approach and the MILP-based technique in [13]. While the latter is potentially capable of finding optimal trails, its running time is not well understood. To speed up the search, the authors apply a splicing heuristic and their objective is finding better trails than existing ones rather than finding optimal trails. We address this limitation with the method described in the following sections.

3   Preliminaries

In this section we state basic definitions and propositions, that will be used in later sections. We begin with the definitions of the differential probability xdp^+ and the linear correlation xlc^+.

Definition 1 (xdp$^+$). *The XOR differential probability (DP) of addition modulo 2^w (xdp$^+$) is the probability with which input XOR differences α and β propagate to output XOR difference γ through the modular addition operation. The probability xdp$^+$ is computed over all pairs of w-bit inputs (x, y):*

$$\text{xdp}^+(\alpha, \beta \rightarrow \gamma) = 2^{-2w} \cdot \#\{(x, y) : ((x \oplus \alpha) + (y \oplus \beta)) \oplus (x + y) = \gamma\} \ . \quad (1)$$

The linear correlation xlc$^+$ is defined in a similar way:

Definition 2 (xlc$^+$). *The XOR linear correlation (LC) of addition modulo 2^w (xlc$^+$) is the correlation of the linear approximation $(\alpha^T x) \oplus (\beta^T y) = (\gamma^T z)$, where $x, y, z : x + y = z \mod 2^w$ are w-bit values and α, β and γ are w-bit linear masks, all represented as binary vectors of dimension $w \times 1$. The operation $\Gamma^T a$ denotes the dot product between the transposed vector Γ (the mask) and the vector a. The correlation xlc$^+$ is computed over all pairs of w-bit inputs (x, y):*

$$\mathrm{xlc}^+(\alpha, \beta \rightarrow \gamma) = 2^{-2w+1} \cdot \#\{(x, y) : (\alpha^T x) \oplus (\beta^T y) = (\gamma^T z)\} - 1 \ . \quad (2)$$

The absolute value of the linear correlation is denoted by $|\mathrm{xlc}^+|$.

The probability xdp$^+$ has the following property noted in [5, Sect. 2, Proposition 1]:

Proposition 1 (Monotonicity of xdp$^+$). *Let α, β and γ be w-bit XOR differences. Denote with $\tilde{p}_i$ ($w \geq i \geq 1$) the probability $\mathrm{xdp}^+(\alpha[i-1:0], \beta[i-1:0] \rightarrow \gamma[i-1:0])$ of the partial differential composed of the i LS bits of α, β, γ. Then the probability xdp$^+$ is monotonously decreasing with the word size of the differences in the direction LSB to MSB:*

$$\tilde{p}_1 \geq \tilde{p}_2 \ldots \geq \tilde{p}_{w-1} \geq \tilde{p}_w = \mathrm{xdp}^+(\alpha, \beta \rightarrow \gamma) \ . \quad (3)$$

Similar property holds also for $|\mathrm{xlc}^+|$, but in this case the correlation decreases from MSB to LSB of the masks:

Proposition 2 (Monotonicity of xlc$^+$). *Let α, β and γ be w-bit linear masks. Denote with $\tilde{c}_i$ ($w-1 \geq i \geq 0$) the absolute value of the correlation $\mathrm{xlc}^+(\alpha[w-1:i], \beta[w-1:i] \rightarrow \gamma[w-1:i])$ of the partial linear approximation composed of the $w - i$ MS bits of α, β, γ. Then the absolute correlation $|\mathrm{xlc}^+|$ is monotonously decreasing with the word size of the masks in the direction MSB to LSB:*

$$\tilde{c}_{w-1} \geq \tilde{c}_{w-2} \ldots \geq \tilde{c}_1 \geq \tilde{c}_0 = |\mathrm{xlc}^+(\alpha, \beta \rightarrow \gamma)| \ . \quad (4)$$

The DP and LC of modular addition have been thoroughly studied in the literature and optimal methods for their computation have been proposed by several authors: [17,22,33](for xdp$^+$) and [10,20,26,27,32,33] (for xlc$^+$). All cited methods are linear in the size of the differences (resp. masks).

In the following sections, for computing xdp$^+$ we use the method proposed in [17] and for xlc$^+$ we use the algorithm described in [10].

4   Best Trail Search for ARX

In this section we describe for the first time a Matsui-like algorithm for finding the best differential and linear trails in ARX ciphers for which the Markov assumption holds. Our technique belongs to the class of bottom-up approaches. It is based on Matsui's branch-and-bound algorithm [18], originally designed for the class of S-box ciphers, and is inspired by the threshold search algorithm proposed in [5].

To search for the best trail on n rounds of a cipher, Matsui's algorithm is initialized with the best probabilities $B_1, B_2, \ldots, B_{n-1}$ for the first $n-1$ rounds and an over-estimation $\overline{B}_n \leq B_n$ of the best probability B_n for n rounds (the bound). The search proceeds recursively over the rounds starting from the first $(r = 1)$ and gradually builds a trail until the n-th round is reached. At every round $1 \leq r \leq n$ the probability $\prod_{i=1}^{r} p_i$ of the partially constructed trail up to round r is multiplied by the best probability B_{n-r} for the remaining $n-r$ rounds to obtain an estimate for the full trail. If $B_{n-r} \prod_{i=1}^{r} p_i < \overline{B}_n$ (i.e. the estimate is lower than the bound), the algorithm backtracks to the previous round. In this way branches of the recursion tree, that are not prospective, are cut. At the last round the probability of the full trail is compared to the bound and if it is bigger, the bound is set to the new probability: $\overline{B}_n \leftarrow \prod_{i=1}^{n} p_i$. The procedure terminates when the bound $\overline{B}_n$ can not be updated any more. As long as the condition $\overline{B}_n \leq B_n$ is preserved, the returned result is guaranteed to be optimal. The probabilities (resp. correlations) p_i are computed by means of the DDT (resp. LAT) of the cipher's S-box.

In [5] was proposed a variant of Matsui's algorithm applicable to the class of ARX ciphers, called *threshold search*. The main idea is to consider addition modulo 2^w as a large S-box of size $2^{2w} \times 2^w$. Since computation of the full DDT of this S-box is infeasible for typical word sizes of $w \geq 16$ bits, the authors propose to use a DDT with reduced size, called *partial* DDT (or pDDT). The pDDT is composed of (a subset of) all differential transitions that have probability larger than- or equal to a predefined probability threshold. The value of the threshold and the maximum allowed size of the pDDT are chosen heuristically depending on the analyzed primitive. Another proposed heuristic is a limit on the Hamming weight of the differences.

If an input difference with no matching output difference in the pDDT is encountered during the search, a second pDDT is computed on-the-fly. The latter is composed of transitions that (a) have probabilities that are likely to improve the probability of the best trail found so far and (b) are guaranteed to result in input differences to the next round, that have at least one matching output difference in the initial pDDT (as illustrated by the *The Highways and Country Roads Analogy* [5]). Due to the use of the mentioned heuristics, the trails found by the threshold search algorithm are not necessarily optimal.

Inspired by the threshold search approach, we propose a new variant of Matsui's algorithm for the class of ARX. In contrast to [5] our technique does not use any heuristics and finds optimal results. The main new idea is to add a second recursion at bit-level over the bits of the differences (resp. linear masks) in addition to the original recursion over the rounds. This modification preserves the optimality of the search due to the monotonicity properties of modular addition stated as Propositions 1 and 2 in Sect. 3. These properties allow us, at every round r, to compute the probability of the partially constructed trail at the bit-level using the partially constructed differences (resp. masks) at round r.

Unprospective branches of the search tree are thus effectively cut not only at round-level, but also at bit-level.

In more detail, let $\alpha_r[0:i]$, $\beta_r[0:i]$ and $\gamma_r[0:i]$ be resp. input and output differences to the modular addition at round r, that are partially constructed up to bit i (i.e. only the $i+1$ LS bits of the words are assigned). Let $\tilde{p}_r$ be the probability of the corresponding partially constructed differential: $(\alpha_r[0:i], \beta_r[0:i]) \rightarrow \gamma_r[0:i]$. Then at round r and bit i, the algorithm checks whether the following condition holds: $B_{n-r}\tilde{p}_r \prod_{i=1}^{r-1} p_i \geq \overline{B}_n$ i.e. if the product of the probability $\prod_{i=1}^{r-1} p_i$ of the partially constructed trail up to round $r-1$ and the probability $\tilde{p}_r$ of the partially constructed differential up to bit i at round r and the best probability B_{n-r} for the remaining $n-r$ rounds is still at least as good as the bound $\overline{B}_n$. If yes, then the search proceeds recursively to the next bit position $i+1$ or, if $i=w$, to the next round $r+1$. Otherwise, it backtracks to the previous bit or, if $i=0$, to the previous round.

With the described strategy, we effectively deal with the problem of having to store huge number of possible transitions through the addition operation. Consequently it is not necessary to maintain a (partial) DDT or to use additional heuristics such as probability and Hamming weight thresholds to limit the search and storage space. Moreover, our algorithm is conceptually closer to Matsui's original proposal than the threshold search. In his paper [18], Matsui also describes a second level of recursion over the 8 S-boxes of DES (cf. procedure *Round-2-j* in [18, Sect. 4, p. 371]). With it the probability of a partial trail is computed up to round $r-1$ and up to S-box i at round r, where $1 \leq i \leq 8$. This S-box level recursion is analogous to the proposed bit-level recursion for modular addition.

In the following sections we use the block cipher SPECK to illustrate the application of the new technique in practice.

5 Application to SPECK

5.1 Description of SPECK

SPECK is a family of lightweight block ciphers proposed in [2]. It is composed of the five instances SPECK32, SPECK48, SPECK64, SPECK96 and SPECK128, corresponding resp. to the block sizes $32, 48, 64, 96$ and 128 bits. Note that the instance SPECKN has $N/2$-bit word size. In the following, with SPECK we denote any of the five variants if not otherwise specified.

SPECK is a pure ARX cipher with a Feistel-like structure in which both branches are modified at every round. Let $X_{r-1,L}$ and $X_{r-1,R}$ be respectively the right and left $N/2$-bit input words to the r-th round of SPECKN ($r \geq 1$) and let k_r be the $N/2$-bit round key applied at round r (see Fig. 1 (Left)). Then the output words $X_{r,L}$, $X_{r,R}$ from round r (input words to round $r+1$) are computed as follow:

$$X_{r,L} = ((X_{r-1,L} \ggg r_1) \boxplus X_{r-1,R}) \oplus k_r \ , \tag{5}$$

$$X_{r,R} = (X_{r-1,R} \lll r_2) \oplus X_{r,L} \ . \tag{6}$$

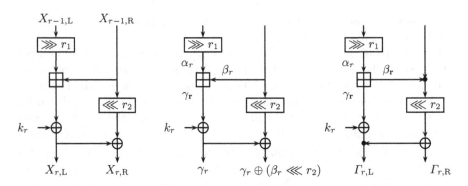

Fig. 1. Left: The round function of SPECK. **Middle:** Propagation of differences: $\alpha_r = \gamma_{r-1} \ggg r_1$, $\beta_r = \gamma_{r-1} \oplus (\beta_{r-1} \lll r_2)$. **Right:** Propagation of linear masks: $\alpha_r = \Gamma_{r-1,L} \ggg r_1$, $\beta_r = \Gamma_{r-1,R} \oplus (\Gamma_{i,R} \ggg r_2)$, $\gamma_r = \Gamma_{i,L} \oplus \Gamma_{i,R}$. The $\bullet$ sign denotes a "three-forked branch" and acts as a XOR on the linear masks [18]. Differences $\gamma_\mathbf{r}$ (resp. masks $\beta_\mathbf{r}$, $\gamma_\mathbf{r}$) in bold can be freely chosen.

The rotation constants r_1, r_2 are specified as: $r_1 = 7, r_2 = 2$ for SPECK32 and $r_1 = 8, r_2 = 3$ for all other versions. The round function of SPECK is depicted in Fig. 1 (Left).

Every instance of the SPECK family supports several key sizes and the total number of rounds depends on the key size. A summary of the parameters (block size, key size, number of rounds) of all instances of the family is presented in Table 3.

Table 3. SPECK parameters: block size (bits), key size (bits), number of rounds.

Instance SPECKN	Block size (N)	Word size ($N/2$)	Key size	Rounds	Key size	Rounds	Key size	Rounds
SPECK32	32	16	64	22				
SPECK48	48	24	72	22	96	23		
SPECK64	64	32	96	26	144	29		
SPECK96	96	48	96	28	144	29		
SPECK128	128	64	128	32	192	33	256	34

The key schedule of SPECK is based on a simple ARX function that is iterated a fixed number of times. We omit its description herein, as it is not relevant to the presented results. For the detailed description of the SPECK family we refer the reader to the original proposal [2].

5.2   Best Trail Search for Speck

In this section we apply the technique described in Sect. 4 in order to find the best (under the Markov assumption) linear and differential trails of reduced-round variants of SPECK.

Differential Trail Search. The pseudo-code of the algorithm for the best differential trail search applied to SPECK is shown in Algorithm 1. It has three parts: first round (lines (4)–(14)), middle rounds (lines (16)–(25)) and last round (lines (27)–(37)). Every part is composed of two blocks corresponding to the two levels of recursion. In the first round the procedure starts by recursing over the bits of the differences (lines (10)–(14)) beginning with the LSB. When the MSB is reached (line (5)) (i.e. the differences α_r, β_r, γ_r are fully constructed), the procedure switches back to the first block (lines (5)–(8)), where it recurses into the next round (line (8)). The logic for the middle and last rounds is the same with the exception that the bit level recursion is over the bits of the output difference γ_r only (lines (22)–(25) and (34)–(37) resp.) and not over the bits of all differences as in the first round. The reason is that the input differences α_r and β_r to the addition in the middle and last rounds are fixed from the previous round by the following relation: $\alpha_r = \gamma_{r-1} \ggg r_1$, $\beta_r = \gamma_{r-1} \oplus (\beta_{r-1} \lll r_2)$ (see line (7) and Fig. 1 (middle)). In addition, at the last round there is no further round level recursion, but instead the bound $\overline{B}_n$ is updated (line (32)).

We estimate the complexity of the differential search algorithm as follows. Let $m_1 \leq 2^{3w}$ be the number of differences α_1, β_1 and γ_1 in the first round, for which the probability of the differential $(\alpha_1, \beta_1 \to \gamma_1)$ is higher than $\overline{B}_n/B_{n-1}$: $m_1 = \#\{(\alpha_1, \beta_1, \gamma_1) : \mathrm{xdp}^+(\alpha_1, \beta_1 \to \gamma_1) \geq \overline{B}_n/B_{n-1}\}$. Analogously, let $m_r \leq 2^w$ be the number of differences γ_r in any middle or last round $r \geq 2$ for which, for fixed α_r and β_r, the probability of the differential $(\alpha_r, \beta_r \to \gamma_r)$ is higher than $\overline{B}_n/(B_{n-r}\prod_{i=1}^r p_i)$: $m_r = \#\{\gamma_r : \mathrm{xdp}^+(\alpha_r, \beta_r \to \gamma_r) \geq \overline{B}_n/(B_{n-r}\prod_{i=1}^r p_i)\}$, and let m be the maximum among these values: $m = \max_{n \geq r \geq 2} (m_r)$. Then the complexity of Algorithm 1 has the form $\mathcal{O}(\prod_{r=1}^n m_r) \leq \mathcal{O}(m_1 m^{r-1})$, which is significantly lower than the complexity of full search $2^{3w} 2^{w(r-1)} = 2^{w(r+2)}$ as indicated by our experiments. However, the precise quantification of the values m_r, $r \geq 1$ is difficult, since they change dynamically during the search. The latter is a separate problem in itself, that can be investigated in future research.

Linear Trail Search. The algorithm for linear search for SPECK is analogous to the differential case with one significant difference, arising from the way in which linear masks propagate through the round function (see Fig. 1 (right)). Recall that in the differential search, the differences α_r and β_r in the middle and last rounds are fixed from the previous round. In contrast, in the linear case only the mask α_r is fixed (with the relation $\alpha_r = \gamma_{r-1} \lll r_1$), while β_r depends on the right output masks $\Gamma_{r-1,\mathrm{R}}$ and $\Gamma_{r,\mathrm{R}}$ resp. from the previous and current round: $\beta_r = \Gamma_{r-1,\mathrm{R}} \oplus (\Gamma_{r,\mathrm{R}} \ggg r_2)$. Due to this fact, in the middle and last rounds the linear search algorithm performs a recursion over the bits of one more variable (β_r) in addition to γ_r. Furthermore, since the mask $\Gamma_{r-1,\mathrm{R}}$ can

Algorithm 1. Search for the Best Differential Trails in ARX (Application to SPECK).

Input: |
   n : num. rounds; w: word size in bits; r_1, r_2: right and left rot. const.;
   r : current round $(n \geq r \geq 1)$;
   i : current bit position $(w > i \geq 0)$;
   $B = (B_1, B_2, \ldots, B_{n-1})$: probs. of the best trails for rounds $1, 2, \ldots, (n-1)$;
   $\overline{B}_n$: underestimate of the best prob. for n rounds: $\overline{B}_n \leq B_n$;
   $T = (T_1, T_2, \ldots, T_{r-1})$: $T_i = (\alpha_i, \beta_i, \gamma_i, p_i)$: $p_i = \mathrm{xdp}^+(\alpha_i, \beta_i \to \gamma_i)$, $1 \leq i < r$;
   $(\alpha_r, \beta_r, \gamma_r)$: input and output differences to the mod. addition at round r;
   $\tilde{p}_r$: probability of the partial differential $(\alpha_r[0:i], \beta_r[0:i] \to \gamma_r[0:i])$;
Output: |
   B_n, T : the best prob. for n rounds and corresponding trail;
1: // Initialization: $r \leftarrow 1$, $i \leftarrow 0$, $\alpha_r \leftarrow \emptyset$, $\beta_r \leftarrow \emptyset$, $\gamma_r \leftarrow \emptyset$
2: **procedure** best_diff_search$(r, i, \alpha_r, \beta_r, \gamma_r)$ **do**
3:    // First round
4:    **if** $(r = 1) \wedge (r \neq n)$ **then**
5:       **if** $i = w$ **then**
6:          $p_r \leftarrow \mathrm{xdp}^+(\alpha_r, \beta_r \to \gamma_r)$; $T_r \leftarrow (\alpha_r, \beta_r, \gamma_r, p_r)$; **add** T_r to T;
7:          $i \leftarrow 0$; $\alpha_{r+1} \leftarrow (\gamma_r \ggg r_1)$; $\beta_{r+1} \leftarrow \gamma_r \oplus (\beta_r \lll r_2)$; $\gamma_{r+1} \leftarrow \emptyset$;
8:          call best_diff_search$(r + 1, i, \alpha_{r+1}, \beta_{r+1}, \gamma_{r+1})$
9:       **else**
10:          **for** $j_\alpha, j_\beta, j_\gamma \in \{0, 1\}$ **do**
11:             $\alpha_r[i] \leftarrow j_\alpha$; $\beta_r[i] \leftarrow j_\beta$; $\gamma_r[i] \leftarrow j_\gamma$;
12:             $\tilde{p}_r \leftarrow \mathrm{xdp}^+(\alpha_r[0:i], \beta_r[0:i] \to \gamma_r[0:i])$;
13:             **if** $(\tilde{p}_r \, B_{n-1}) \geq \overline{B}_n$ **then**
14:                call best_diff_search$(r, i + 1, \alpha_r, \beta_r, \gamma_r)$
15:    // Intermediate rounds
16:    **if** $(r > 1) \wedge (r \neq n)$ **then**
17:       **if** $i = w$ **then**
18:          $p_r \leftarrow \mathrm{xdp}^+(\alpha_r, \beta_r \to \gamma_r)$; $T_r \leftarrow (\alpha_r, \beta_r, \gamma_r, p_r)$; **add** T_r to T;
19:          $i \leftarrow 0$; $\alpha_{r+1} \leftarrow (\gamma_r \ggg r_1)$; $\beta_{r+1} \leftarrow \gamma_r \oplus (\beta_r \lll r_2)$; $\gamma_{r+1} \leftarrow \emptyset$;
20:          call best_diff_search$(r + 1, i, \alpha_{r+1}, \beta_{r+1}, \gamma_{r+1})$
21:       **else**
22:          **for** $j_\gamma \in \{0, 1\}$ **do**
23:             $\gamma_r[i] \leftarrow j_\gamma$; $\tilde{p}_r \leftarrow \mathrm{xdp}^+(\alpha_r[0:i], \beta_r[0:i] \to \gamma_r[0:i])$
24:             **if** $(p_1 \, p_2 \, \ldots \, p_{r-1} \, \tilde{p}_r \, B_{n-r}) \geq \overline{B}_n$ **then**
25:                call best_diff_search$(r, i + 1, \alpha_r, \beta_r, \gamma_r)$
26:    // Last round
27:    **if** $(r = n)$ **then**
28:       **if** $i = w$ **then**
29:          $p_n \leftarrow \mathrm{xdp}^+(\alpha_n, \beta_n \to \gamma_n)$; $T_n \leftarrow (\alpha_n, \beta_n, \gamma_n, p_n)$; **add** T_n to T;
30:          **if** $(p_1 \, p_2 \, \ldots \, p_{n-1} \, p_n) \geq \overline{B}_n$ **then**
31:             // Update bound and return to upper round
32:             $\overline{B}_n \leftarrow (p_1 \, p_2 \, \ldots \, p_{n-1} \, p_n)$
33:       **else**
34:          **for** $j_\gamma \in \{0, 1\}$ **do**
35:             $\gamma_n[i] \leftarrow j_\gamma$; $\tilde{p}_n \leftarrow \mathrm{xdp}^+(\alpha_n[0:i], \beta_n[0:i] \to \gamma_n[0:i])$
36:             **if** $(p_1 \, p_2 \, \ldots \, p_{n-1} \, \tilde{p}_n) \geq \overline{B}_n$ **then**
37:                call best_diff_search$(r, i + 1, \alpha_r, \beta_r, \gamma_r)$
38:    **return**

---

be freely chosen in the first round, an additional iteration over all such masks is performed. The latter is independent of the bound $\overline{B}_n$ and therefore represents a fixed cost of 2^w additional iterations. All this added complexity makes the linear search algorithm feasible only for the version SPECK32.

Due to the mentioned differences, the complexity of the linear search is significantly higher than the differential search. Let $m_1 \leq 2^{3w}$ be the number of masks α_1, β_1 and γ_1 in the first round, for which the absolute correlation of the linear approximation $(\alpha_1, \beta_1 \rightarrow \gamma_1)$ is higher than $\overline{B}_n/B_{n-1}$: $m_1 = \#\{(\alpha_1, \beta_1, \gamma_1) : |\mathrm{xlc}^+(\alpha_1, \beta_1 \rightarrow \gamma_1)| \geq \overline{B}_n/B_{n-1}\}$. Let $m_r \leq 2^{2w}$ be the number of masks β_r and γ_r in any middle or last round $r \geq 2$ for which, for fixed α_r, the absolute correlation of the linear approximation $(\alpha_r, \beta_r \rightarrow \gamma_r)$ is higher than $\overline{B}_n/(B_{n-r}\prod_{i=1}^{r} c_i)$: $m_r = \#\{(\beta_r, \gamma_r) : |\mathrm{xlc}^+(\alpha_r, \beta_r \rightarrow \gamma_r)| \geq \overline{B}_n/(B_{n-r}\prod_{i=1}^{r} c_i)\}$, and let $m = \max_{n \geq r \geq 2}(m_r)$. Then the complexity of the linear search algorithm has the form: $\mathcal{O}(2^w \prod_{r=1}^{n} m_r) \leq \mathcal{O}(2^w m_1 m^{r-1})$, which is much less than the complexity of full search $2^{4w}2^{2w(r-1)} = 2^{2w(r+1)}$. In the former, notice the factor 2^w due to the additional iteration over all w-bit masks $\Gamma_{r-1,\mathrm{R}}$ in the first round. Again, similarly to the differential case, the precise quantification of the values m_r, $r \geq 1$ in the linear case is difficult.

While the higher complexity of the linear search algorithm makes it infeasible for versions of SPECK other than SPECK32, Algorithm 1 is quite practical as shown by the results reported in the following section.

5.3   Results

With Algorithm 1 we find the best differential trails for reduced round variants of all versions of SPECK under the Markov assumption. Table 1 compares our results to the ones obtained with the threshold search algorithm with the parameters given in [4, Sect. 6, Table 6]: probability threshold $p_{\mathrm{thres}} = 2^{-5}$, Hamming weight threshold $\mathrm{hw}_{\mathrm{thres}} = 7$ and maximum pDDT size 2^{30}. From the table it can be seen that for certain number of rounds Algorithm 1 significantly improves the probabilities found with threshold search.

The execution times of Algorithm 1 for different number of rounds are shown in Table 4. Most of the measurements were done on a PC with Intel® Core™

Table 4. Probabilities and running times for the best (under the Markov assumption) differential trails for SPECK obtained with Algorithm 1 ($\log_2$ scale). Platforms: Intel® Core™ E5-2637 CPU 3.50GHz or HPC cluster for ≥ 7 rounds. The column t provides the time needed to find a single best trail in s/m/h/d = seconds/minutes/hours/day, where 1 day = 24 h. Note: times are rounded up.

# R	SPECK32	t	SPECK48	t	SPECK64	t	SPECK96	t	SPECK128	t
1	0	0s	0	0s	0	0s	0	0s	0	0s
2	−1	0s	−1	0s	−1	0s	−1	0s	−1	0s
3	−3	0s	−3	0s	−3	0s	−3	0s	−3	0s
4	−5	0s	−6	0s	−6	0s	−6	6s	−6	22s
5	−9	0s	−10	1s	−10	1m	−10	5m	−10	26m
6	−13	1s	−14	3s	−15	26m	−15	5h	−15	2d
7	−18	1m	−19	1m	−21	4h	−21	5d	−21	3h
8	−24	34m	−26	9m	−27	22h	< −27	3d	< −26	2d
9	−30	12m	−33	7d	< −31	1d				
10	−34	6m	< −34	3h						

E5-2637 CPU 3.50GHz. Exceptions are the results for more than 7 rounds and block sizes larger than 48 bits, which were obtained using a parallel version of Algorithm 1 executed on the HPC cluster of the University of Luxembourg [29]. The memory requirements in all cases are negligible.

A final note on the search strategy used for obtaining the times in Table 4: when searching for the best probability for n rounds, we initialize the bound $\overline{B}_n$ to the best probability for $(n-1)$ rounds: $\overline{B}_n \leftarrow B_{n-1}$. If no trail with this probability is found, the bound is decreased by a factor of 2: $\overline{B}_n \leftarrow \overline{B}_n/2$. This process continues until a trail with probability equal to the bound is found. Thus the times shown in Table 4 are measured from the start of the program to the moment when the first trail is found.

5.4   Towards Security Bounds for Speck

The results from Table 4 can be used to trivially obtain upper bounds (under the Markov assumption) on the security of SPECK against single-trail differential cryptanalysis. For example, given the probability p_r of the best trail on r rounds and the probability p_s of the best trail on s rounds, the product $p_r p_s$ gives an upper bound on the probability of any trail on $r + s$ rounds. The latter is equivalent to the statement that any trail on $r + s$ rounds has probability at least $p_r p_s$ or lower. We use this approach to compute upper bounds (under the Markov assumption) on the probabilities of the best trails on all versions of SPECK. The results are shown in Table 5.

In view of the probabilities of the best found trails on SPECK reported in [4, Sect. 6, Table 6], the bounds in Table 5 are not tight.

5.5   On the Choice of Rotation Constants

We investigated the way in which the choice of the rotation constants r_1 and r_2 (see Fig. 1 (Left)) of SPECK32, SPECK48 and SPECK64 influences the DP of the best trails. For that purpose we assume that the exact values of the constants are not as important as their relative difference. Under this assumption, we

Table 5. Upper bounds on the best (under the Markov assumption) probabilities of differential trails on SPECK computed using the best probabilities from Table 4 ($\log_2$ scale).

Instance	Upper Bound	Rounds	Upper Bound	Rounds	Upper Bound	Rounds
SPECK32	−69	22				
SPECK48	−72	22	−76	23		
SPECK64	−91	26	−96	29		
SPECK96	−90	28	−94	29		
SPECK128	−104	32	−104	33	−105	34

Table 6. Best differential probabilities (DP) for 9 rounds of SPECK32, 7 rounds of SPECK48 and 6 rounds of SPECK64 for 16 choices of the rotation constant r_1 with r_2 fixed to its original value (Fig. 1 (Left)) ($\log_2$ scale).

r_1	0	1	2	3	4	5	6	7	8	9	10	11	12	13	14	15
SPECK32	-21	-25	-24	-30	-27	-30	-25	-30	-24	-31	-26	-29	-27	-27	-22	-24
SPECK48	-11	-15	-20	-16	-21	-21	-19	-21	-19	-17	-19	-20	-13	-21	-20	-19
SPECK64	-8	-13	-15	-13	-15	-15	-14	-15	-15	-15	-15	-15	-15	-13	-13	-15

fixed r_2 to its original value and varied r_1 over the first 16 possibilities. For each choice, we determined the probability of the best differential trail for 9 rounds of SPECK32, 7 rounds of SPECK48 and 6 rounds of SPECK64 using Algorithm 1. The results are presented in Table 6.

From Table 6 it can be seen that the original choice of rotation constants: $r_1 = 7$ and $r_2 = 2$ for SPECK32 and $r_1 = 8$ and $r_2 = 3$ for SPECK48 is not optimal w.r.t. the probability of the best differential trail. In the former case, it results in DP of 2^{-30} over 9 rounds, while the optimal choice: $r_1 = 9$ and $r_2 = 2$ results in probability 2^{-31}. In the latter case, the original rotation constants $(8, 3)$ result in DP of 2^{-19} over 7 rounds, while the choices $(4, 3)$, $(5, 3)$, $(7, 3)$ and $(13, 3)$ result in lower probability 2^{-21}. This may hint that we have found better rotation constants for SPECK. To be certain however, similar experiments for the linear case must also be conducted. In addition, the implementation cost of each pair of constants must be taken into account. Therefore the optimal choice of r_1 and r_2 requires further investigation.

6   MARX and Speckey: ARX Primitives with Provable Bounds

A limitation of Algorithm 1 is that its complexity significantly increases with the number of rounds and word sizes as indicated by Table 4. To address this problem, in this section we propose two new primitives – MARX and SPECKEY for which it is feasible to compute the probabilities and linear correlations of the best trails for any number of rounds and which satisfy the Markov assumption. Both primitives have 32-bit state and 32-bit round key.

MARX (from MIX + ARX) is based on the round function of Threefish-256 [12] (with its basic component – the MIX operation) with 8-bit words. This round function is wrapped within a key addition on the input and on the output and is iterated over a fixed number of rounds. SPECKEY, as the name suggests, is based on block cipher SPECK. More precisely, it is SPECK32 with modified key addition. The round functions of MARX and SPECKEY are shown on Fig. 2.

To choose the rotation constants of MARX, we exhaustively searched all possible pairs of constants and applied Algorithm 1 and its linear search version to the resulting variants. Based on the results we selected the constants $r_1 = 2$ and $r_2 = 5$, as they provided differential probability (DP) $\leq 2^{-32}$ and absolute linear correlation (LC) $\leq 2^{-17}$ over a minimal number of rounds, namely 12. As to the word permutation, before settling for the one used in Threefish-256, we

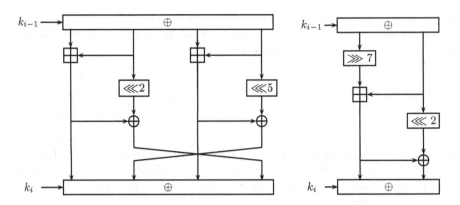

Fig. 2. Left: MARX (from MIX + ARX), based on the round function of Threefish-256 [12] with 32-bit state, 32-bit round key and 8-bit words; **Right:** SPECKEY – a variant of SPECK32 with modified key addition.

Table 7. Best differential probabilities (DP) and absolute linear correlations (LC) of MARX and SPECKEY ($\log_2$ scale).

# R	1	2	3	4	5	6	7	8	9	10	11	12
DP_{MARX}	−0	−0	−1	−2	−5	−9	−14	−20	−25	−29	−32	−34
LC_{MARX}	−0	−0	−0	−1	−2	−4	−7	−10	−13	−15	−16	−17
$DP_{Speckey}$	−0	−1	−3	−5	−9	−13	−18	−24	−30	−34		
$LC_{Speckey}$	−0	−0	−1	−3	−5	−7	−9	−12	−14	−17		

also considered a Feistel-like variant in which the words are circularly rotated right by one. However this variant required more rounds to reach full diffusion (best DP 2^{-32} and best absolute LC 2^{-17}) compared to Threefish-256 – on average two more rounds were necessary.

The best DP and LC of MARX and SPECKEY are shown in Table 7.

The main advantage of MARX and SPECKEY over SPECK32 is that due to the full state key addition at the beginning of every round, these two primitives belong to the class of key-alternating ciphers [9, Sect. 5.1, Definition 2], which is a sub-class of Markov ciphers and therefore satisfies the Markov assumption. In addition, due to the 8 bit modular addition, MARX may be a more suitable choice for devices with 8-bit microprocessors. A disadvantage is that MARX needs two more rounds to achieve full diffusion compared to SPECK32 (see Table 7) and that both MARX and SPECKEY use more operations per round compared to SPECK32. In Appendix 9 is described a variant of MARX, called MARX2, that achieves full diffusion in the same number of rounds as SPECK32 at the expense of two additional rotation operations.

Finally, we stress that the proposed new primitives are intended to serve mainly as an example of how the best trail search algorithms can be used to design new ARX constructions with provable properties. At present, MARX and SPECKEY have not undergone sufficient analysis against other cryptanalytic techniques for us to have enough confidence in their cryptographic properties.

7   Conclusion

In this paper we proposed for the first time an adaptation of Matsui's algorithm for finding the best differential and linear trails in ARX ciphers. We showed the practical application of the new method on reduced round variants of block ciphers from the SPECK family and we reported the first provably best differential trails on these variants. The new results were used to compute the first bounds (under the Markov assumption) on the security of SPECK against single-trail differential cryptanalysis. In addition, we also reported better choices of the rotation constants for SPECK w.r.t. single-trail differential cryptanalysis. Finally, we proposed two new ARX primitives – MARX and SPECKEY – which satisfy the Markov assumption and have provable bounds against single-trail differential and linear cryptanalysis – a long standing open problem in the area of ARX design. The source code of the tools for best trail search for SPECK, SPECKEY and MARX is publicly available as part of the YAARX Toolkit [30] and a snapshot of the source tree is uploaded on the CryptoLUX website [31].

Acknowledgments. We thank our colleagues from the Laboratory of Algorithmics, Cryptology and Security (LACS) at the University of Luxembourg for the stimulating discussions. Some of the experiments presented in this paper were carried out using the HPC facilities of the University of Luxembourg [29] – see http://hpc.uni.lu.

Appendix

8   Showing that Speck is not a Markov Cipher

In this section we show, by the means of a counter example, that SPECK is *not* a Markov cipher. For the purpose, we use an equivalent representation of the round function of SPECK (Fig. 1 (left)), shown on Fig. 3 (left).

According to the formal definition [14, Sect. 3], a Markov cipher is an iterative cipher, whose round function is such that its differential probability is independent of the input values, under an appropriate definition of a difference. More formally, let f be the round function of an iterative cipher and let x and y be resp. an input and output state and k be the round key: $y = f(x, k)$. Let Δx denote an XOR difference between two input values x and x^*: $\Delta x = x \oplus x^*$. Finally, let $\underset{k}{P}$ and $\underset{x,k}{P}$ denote the differential probability of f resp. over all round keys k and over all input values and round keys (x, k). Then, if a cipher is Markov, the two probabilities should be equal:

$$\underset{k}{P}(\Delta y | \Delta x, x) = \underset{x,k}{P}(\Delta y | \Delta x) \ . \tag{7}$$

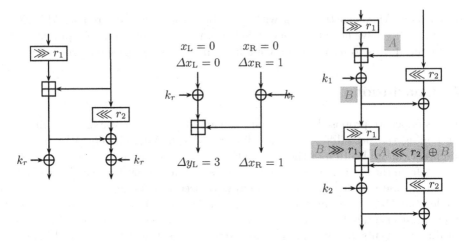

Fig. 3. Left: Equivalent representation of one round of SPECK. **Middle:** Main non-linear component of SPECK, illustrating two differentials that differ in probability and hence contradict the Markov assumption: $P((3,1)|(0,1),(0,0)) = 2^{-1}$ and $P((3,1)|(0,1)) = 2^{-2}$. **Right:** Dependency between the inputs to the addition in one round (B) and between the inputs to consecutive rounds (A).

In other words, for a Markov cipher the differential probability of f is independent of the input values x for all x (when the subkey is uniformly random).

To show that SPECK is not a Markov cipher, it is enough to provide values for x, Δx and Δy for which equation (7) does not hold. To do this, we use the main non-linear component of SPECK shown on Fig. 3 (middle). For this component, for the following values $x = (x_L, x_R) = (0,0)$, $\Delta x = (\Delta x_L, \Delta x_R) = (0,1)$ and $\Delta y = (\Delta y_L, \Delta y_R) = (3,1)$ the two probabilities in (7) are not equal:

$$\underset{k}{P}(\Delta y = (3,1)|\Delta x = (0,1), x = (0,0)) = 2^{-1}$$
$$\neq \underset{x,k}{P}(\Delta y = (3,1)|\Delta x = (0,1)) = 2^{-2} . \tag{8}$$

An illustrative example of the dependency between the inputs to the addition operation in one round and between the inputs to consecutive rounds is shown on Fig. 3 (right).

9 MARX2: A Variant of MARX with Improved Diffusion

In this section we describe MARX2 – a variant of MARX with improved dwo additional rotation operations and its round function is depicted in Fig. 4.

As can be seen from Fig. 4, MARX2 is composed of two parallel applications of the round function of SPECK with 8-bit words. Due to the additional rotation operations it achieves full diffusion in the same number of rounds as SPECK32, namely 10. The best DP and LC for the recommended rotation amounts $(r_1, r_2, r_3, r_4) = (2,3,1,7)$ are shown in Table 8.

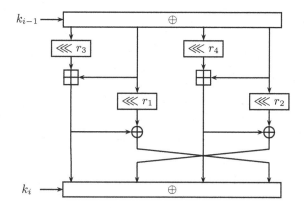

Fig. 4. MARX2: a variant of MARX with two additional rotations for improved diffusion. The recommended rotation amounts are $(r_1, r_2, r_3, r_4) = (2, 3, 1, 7)$

Table 8. Best differential probabilities (DP) and absolute linear correlations (LC) of MARX2 with rotation constants $(r_1, r_2, r_3, r_4) = (2, 3, 1, 7)$ – see Fig. 4 ($\log_2$ scale).

# R	1	2	3	4	5	6	7	8	9	10
DP_{MARX2}	−0	−1	−3	−5	−11	−16	−22	−25	−29	−35
LC_{MARX2}	−0	−0	−1	−3	−5	−8	−10	−13	−15	−17

The rotation constants of MARX2 have been chosen by exhaustively searching over all four rotation values (4095 values in total, excluding the all-zero choice). For each set of amounts we applied Algorithm 1 and its linear search variant and recorded the number of rounds necessary to reach full diffusion. The results show that no set of rotation constants exists for which full diffusion can be reached in less than 10 rounds. From the constants that ensure diffusion in 10 rounds we have selected $(r_1, r_2, r_3, r_4) = (2, 3, 1, 7)$ as the recommended choice since for this set we get slightly better DP than SPECK32 (2^{-35} vs. 2^{-34}). In addition, all constants from the set are different and are not multiples of each other, which may also be considered desirable properties. Other choices that also result in full diffusion for 10 rounds are: $(2, 3, 7, 2)$, $(2, 3, 1, 2)$ and $(5, 5, 2, 7)$. Note that the constants (r_1, r_2, r_3, r_4) and (r_2, r_1, r_4, r_3) are equivalent.

Finally, we note that MARX2 is an illustration of another strategy for increasing the block size of an ARX cipher. Rather than increasing the word size from N to $2N$, as is done in SPECK, in order to increase the block size the designer may alternatively double the number of N-bit block components as in MARX2. This approach may result in improved efficiency on some platforms, such as e.g. 32-bit ARM, where the cost of a bit rotation is constant.

References

1. Aumasson, J.-P., Jovanovic, P., Neves, S.: NORX: parallel and scalable AEAD. In: Kutyłowski, M., Vaidya, J. (eds.) ICAIS 2014, Part II. LNCS, vol. 8713, pp. 19–36. Springer, Heidelberg (2014)
2. Beaulieu, R., Shors, D., Smith, J., Treatman-Clark, S., Weeks, B., Wingers, L.: The SIMON and SPECK families of lightweight block ciphers. Cryptology ePrint Archive, report 2013/404 (2013). http://eprint.iacr.org/
3. Biham, E., Shamir, A.: Differential cryptanalysis of DES-like cryptosystems. J. Crypt. **4**(1), 3–72 (1991)
4. Biryukov, A., Roy, A., Velichkov, V.: Differential analysis of block ciphers SIMON and SPECK. In: Cid, C., Rechberger, C. (eds.) FSE 2014. LNCS, vol. 8540, pp. 546–570. Springer, Heidelberg (2015)
5. Biryukov, A., Velichkov, V.: Automatic search for differential trails in ARX ciphers. In: Benaloh, J. (ed.) CT-RSA 2014. LNCS, vol. 8366, pp. 227–250. Springer, Heidelberg (2014)
6. CryptoLUX.: FELICS - Fair Evaluation of Lightweight Cryptographic Systems (2015). https://www.cryptolux.org/index.php/FELICS
7. Daemen, J., Rijmen, V.: AES and the wide trail design strategy. In: Knudsen, L.R. (ed.) EUROCRYPT 2002. LNCS, vol. 2332, pp. 108–109. Springer, Heidelberg (2002)
8. Daemen, J., Rijmen, V.: The Design of Rijndael: AES - The Advanced Encryption Standard. Springer, Heidelberg (2002)
9. Daemen, J., Rijmen, V.: Probability distributions of correlation and differentials in block ciphers. IACR Cryptology ePrint Archive **2005**, 212 (2005)
10. Dehnavi, S.M., Rishakani, A.M., Shamsabad, M.R.M.: A More explicit formula for linear probabilities of modular addition modulo a power of two. Cryptology ePrint Archive, report 2015/026 (2015). http://eprint.iacr.org/
11. Dobraunig, C., Eichlseder, M., Mendel, F.: Heuristic tool for linear cryptanalysis with applications to CAESAR candidates. In: Iwata, T., et al. (eds.) ASIACRYPT 2015. LNCS, vol. 9453, pp. 490–509. Springer, Heidelberg (2015). doi:10.1007/978-3-662-48800-3_20
12. Ferguson, N., Lucks, S., Schneier, B., Whiting, D., Bellare, M., Kohno, T., Callas, J., Walker, J.: The skein hash function family. Submission to the NIST SHA-3 Competition (Round 2) (2009)
13. Fu, K., Wang, M., Guo, Y., Sun, S., Hu, L.: MILP-based automatic search algorithms for differential and linear trails for speck. In: 23rd International Workshop on Fast Software Encryption, FSE 2016, Bochum, Germany, 20–23 March (2016, to appear)
14. Lai, X., Massey, J.L.: Markov ciphers and differential cryptanalysis. In: Davies, D.W. (ed.) EUROCRYPT 1991. LNCS, vol. 547, pp. 17–38. Springer, Heidelberg (1991)
15. Leurent, G.: Analysis of differential attacks in ARX constructions. In: Wang, X., Sako, K. (eds.) ASIACRYPT 2012. LNCS, vol. 7658, pp. 226–243. Springer, Heidelberg (2012)
16. Leurent, G.: Construction of differential characteristics in ARX designs application to Skein. In: Canetti, R., Garay, J.A. (eds.) CRYPTO 2013, Part I. LNCS, vol. 8042, pp. 241–258. Springer, Heidelberg (2013)

17. Lipmaa, H., Moriai, S.: Efficient algorithms for computing differential properties of addition. In: Matsui, M. (ed.) FSE 2001. LNCS, vol. 2355, pp. 336–350. Springer, Heidelberg (2002)

18. Matsui, M.: On correlation between the order of s-boxes and the strength of DES. In: De Santis, A. (ed.) EUROCRYPT 1994. LNCS, vol. 950, pp. 366–375. Springer, Heidelberg (1995)

19. Matsui, M., Yamagishi, A.: A new method for known plaintext attack of FEAL cipher. In: Rueppel, R.A. (ed.) EUROCRYPT 1992. LNCS, vol. 658, pp. 81–91. Springer, Heidelberg (1993)

20. McKay, K.A., Vora, P.L.: Analysis of ARX functions: pseudo-linear methods for approximation, differentials, and evaluating diffusion. Cryptology ePrint Archive, report 2014/895 (2014). http://eprint.iacr.org/

21. Mouha, N., Preneel, B.: Towards finding optimal differential characteristics for ARX: application to Salsa20. Cryptology ePrint Archive, report 2013/328 (2013). http://eprint.iacr.org/

22. Mouha, N., Velichkov, V., De Cannière, C., Preneel, B.: The differential analysis of S-functions. In: Biryukov, A., Gong, G., Stinson, D.R. (eds.) SAC 2010. LNCS, vol. 6544, pp. 36–56. Springer, Heidelberg (2011)

23. Mouha, N., Wang, Q., Gu, D., Preneel, B.: Differential and linear cryptanalysis using mixed-integer linear programming. In: Wu, C.-K., Yung, M., Lin, D. (eds.) Inscrypt 2011. LNCS, vol. 7537, pp. 57–76. Springer, Heidelberg (2012)

24. National Institute of Standards, U.S. Department of Commerce. FIPS 47: Data Encryption Standard (1977)

25. NIST: Lightweight Cryptography Workshop (2015). http://www.nist.gov/itl/csd/ct/lwc_workshop2015.cfm, July 2015

26. Nyberg, K., Wallén, J.: Improved linear distinguishers for SNOW 2.0. In: Robshaw, M. (ed.) FSE 2006. LNCS, vol. 4047, pp. 144–162. Springer, Heidelberg (2006)

27. Schulte-Geers, E.: On CCZ-equivalence of addition mod 2^n. Des. Codes Crypt. **66**(1–3), 111–127 (2013)

28. Sun, S., Hu, L., Wang, P., Qiao, K., Ma, X., Song, L.: Automatic security evaluation and (related-key) differential characteristic search: application to SIMON, PRESENT, LBlock, DES(L) and other bit-oriented block ciphers. In: Sarkar, P., Iwata, T. (eds.) ASIACRYPT 2014. LNCS, vol. 8873, pp. 158–178. Springer, Heidelberg (2014)

29. Varrette, S., Bouvry, P., Cartiaux, H., Georgatos, F.: Management of an academic HPC cluster: the UL experience. In: Proceedings of the 2014 International Conference on High Performance Computing & Simulation (HPCS 2014), pp. 959–967. IEEE, Bologna, Italy, July 2014

30. Velichkov, V.: YAARX: Yet another toolkit for the analysis of ARX cryptographic algorithms. Laboratory of Algorithmics, Cryptology and Security (LACS), University of Luxembourg, 2013–2016. https://github.com/vesselinux/yaarx

31. Velichkov, V., Corre, Y.L.: Tool for searching for optimal trails in ARX. Laboratory of Algorithmics, Cryptology and Security (LACS), University of Luxembourg (2016). https://www.cryptolux.org

32. Wallén, J.: Linear approximations of addition modulo 2^n. In: Johansson, T. (ed.) FSE 2003. LNCS, vol. 2887, pp. 261–273. Springer, Heidelberg (2003)

33. Wallén, J.: On the differential and linear properties of addition. Master's thesis, Helsinki University of Technology (2003)
34. Wang, X., Lai, X., Feng, D., Chen, H., Yu, X.: Cryptanalysis of the hash functions MD4 and RIPEMD. In: Cramer, R. (ed.) EUROCRYPT 2005. LNCS, vol. 3494, pp. 1–18. Springer, Heidelberg (2005)
35. Wang, X., Yin, Y.L., Yu, H.: Finding collisions in the full SHA-1. In: Shoup, V. (ed.) CRYPTO 2005. LNCS, vol. 3621, pp. 17–36. Springer, Heidelberg (2005)
36. Wang, X., Yu, H.: How to break MD5 and other hash functions. In: Cramer, R. (ed.) EUROCRYPT 2005. LNCS, vol. 3494, pp. 19–35. Springer, Heidelberg (2005)
37. Yao, Y., Zhang, B., Wu, W.: Automatic search for linear trails of the SPECK family. In: López, J., Mitchell, C.J. (eds.) ISC 2015. LNCS, vol. 9290, pp. 158–176. Springer, Heidelberg (2015)

Designs

Stream Ciphers: A Practical Solution for Efficient Homomorphic-Ciphertext Compression

Anne Canteaut[1]([✉]), Sergiu Carpov[2], Caroline Fontaine[3], Tancrède Lepoint[4],
María Naya-Plasencia[1], Pascal Paillier[4], and Renaud Sirdey[2]

[1] Inria, Paris, France
{anne.canteaut,maria.naya_plasencia}@inria.fr
[2] CEA LIST, Paris, France
{sergiu.carpov,renaud.sirdey}@cea.fr
[3] CNRS/Lab-STICC and Telecom Bretagne and UEB, Brest, France
caroline.fontaine@telecom-bretagne.eu
[4] CryptoExperts, Paris, France
{tancrede.lepoint,pascal.paillier}@cryptoexperts.com

Abstract. In typical applications of homomorphic encryption, the first step consists for Alice to encrypt some plaintext m under Bob's public key pk and to send the ciphertext $c = \mathsf{HE}_{\mathsf{pk}}(m)$ to some third-party evaluator Charlie. This paper specifically considers that first step, *i.e.* the problem of transmitting c as efficiently as possible from Alice to Charlie. As previously noted, a form of compression is achieved using hybrid encryption. Given a symmetric encryption scheme E, Alice picks a random key k and sends a much smaller ciphertext $c' = (\mathsf{HE}_{\mathsf{pk}}(k), \mathsf{E}_k(m))$ that Charlie decompresses homomorphically into the original c using a decryption circuit $\mathcal{C}_{\mathsf{E}^{-1}}$.

In this paper, we revisit that paradigm in light of its concrete implementation constraints; in particular E is chosen to be an additive IV-based stream cipher. We investigate the performances offered in this context by Trivium, which belongs to the eSTREAM portfolio, and we also propose a variant with 128-bit security: Kreyvium. We show that Trivium, whose security has been firmly established for over a decade, and the new variant Kreyvium have an excellent performance.

Keywords: Stream ciphers · Homomorphic cryptography · Trivium

1 Introduction

Since the breakthrough result of Gentry [30] achieving fully homomorphic encryption (FHE), many works have been published on simpler and more effi-

This work has received a French governmental support granted to the COMIN Labs excellence laboratory and managed by the National Research Agency in the "Investing for the Future" program under reference ANR-10-LABX-07-01, has been supported in part by the Frenchs FUI project CRYPTOCOMP and by the European Union's H2020 Programme under grant agreement number ICT-644209 and under project number 645622 PQCRYPTO.

© International Association for Cryptologic Research 2016
T. Peyrin (Ed.): FSE 2016, LNCS 9783, pp. 313–333, 2016.
DOI: 10.1007/978-3-662-52993-5_16

cient schemes based on homomorphic encryption. Because they allow arbitrary computations on encrypted data, FHE schemes suddenly opened the way to exciting new applications, in particular cloud-based services in several areas (see e.g. [33,42,46]).

Compressed Encryption. In these cloud applications, it is often assumed that some data is sent encrypted under a homomorphic encryption (HE) scheme to the cloud to be processed in a way or another. It is thus typical to consider, in the first step of these applications, that a user (Alice) encrypts some data m under some other user's public key pk (Bob) and sends some homomorphic ciphertext $c = \mathsf{HE}_{\mathsf{pk}}(m)$ to a third-party evaluator in the Cloud (Charlie). The roles of Alice and Bob are clearly distinct, even though they might be played by the same entity in some applications.

However, all HE schemes proposed so far suffer from a very large ciphertext expansion; the transmission of c between Alice and Charlie is therefore a very significant bottleneck in practice. The problem of reducing the size of c as efficiently as possible has first been considered in [46] wherein m is encrypted with a symmetric encryption scheme E under some key k randomly chosen by Alice, who then sends a much smaller ciphertext $c' = (\mathsf{HE}_{\mathsf{pk}}(k), \mathsf{E}_k(m))$ to Charlie. Given c', Charlie then exploits the homomorphic property of HE and recovers

$$c = \mathsf{HE}_{\mathsf{pk}}(m) = \mathcal{C}_{\mathsf{E}^{-1}}\left(\mathsf{HE}_{\mathsf{pk}}(k), \mathsf{E}_k(m)\right)$$

by homomorphically evaluating the decryption circuit $\mathcal{C}_{\mathsf{E}^{-1}}$. This can be assimilated to a *compression method* for homomorphic ciphertexts, c' being the result of applying a *compressed encryption scheme* to the plaintext m and c being recovered from c' using a *ciphertext decompression procedure*. In that approach obviously, the new encryption rate $|c'|/|m|$ becomes asymptotically close to 1 for long messages, which leaves no significant margin for improvement. However, the paradigm of ciphertext compression leaves totally open the question of how to choose E in a way that minimizes the decompression overhead, while preserving the same security level as originally intended.

Prior Art. The cost of homomorphically evaluating several symmetric primitives has been investigated, including optimized implementations of AES [18,23,31], and of the lightweight block ciphers Simon [43] and Prince [24]. Usually lightweight block ciphers seem natural candidates for efficient evaluations in the encrypted domain. However, they may also lead to *much worse* performances than a homomorphic evaluation of, say, AES. Indeed, contemporary HE schemes use *noisy* ciphertexts, where a fresh ciphertext includes a noise component which grows along with homomorphic operations. Usually a homomorphic multiplication increases the noise by much larger proportions than a homomorphic addition. The maximum allowable level of noise (determined by the system parameters) then depends mostly on the multiplicative depth of the circuit. Many lightweight block ciphers balance out their simplicity by a large number of rounds, *e.g.* KATAN and KTANTAN [11], with the effect of considerably increasing their multiplicative depth. This type of design is therefore prohibitive in a HE context. Still PRINCE appears to be a much more suitable block cipher for homomorphic evaluation than AES (and than SIMON), because it specifically targets applications that require a low latency; it is designed to minimize the cost of an unrolled implementation [9] rather than to optimize *e.g.* silicon area.

At Eurocrypt 2015, Albrecht, Rechberger, Schneider, Tiessen and Zohner observed that the usual criteria that rule the design of lightweight block ciphers are not appropriate when designing a symmetric encryption scheme with a low-cost homomorphic evaluation [2]. Indeed, both the number of rounds and the number of binary multiplications required to evaluate an Sbox have to be taken into account. Minimizing the number of rounds is a crucial issue for low-latency ciphers like PRINCE, while minimizing the number of multiplications is a requirement for efficient masked implementations.

These two criteria have been considered together for the first time by Albrecht *et al.* in the recent design of a family of block ciphers called LowMC [2] with very small multiplicative size and depth[1]. However, the proposed instances of LowMC, namely LowMC-80 and LowMC-128, have recently had some security issues [21]. They actually present some weaknesses inherent in their low multiplicative complexity. Indeed, the algebraic normal forms (*i.e.*, the multivariate polynomials) describing the encryption and decryption functions are sparse and have a low degree. This type of features is usually exploited in algebraic attacks, cube attacks and their variants, *e.g.* [4,20,22]. While these attacks are rather general, the improved variant used for breaking LowMC [21], named interpolation attack [38], specifically applies to block ciphers. Indeed it exploits the sparse algebraic normal form of some intermediate bit within the cipher using that this bit can be evaluated both from the plaintext in the forward direction and from the ciphertext in the backward direction. This technique leads to several attacks including a key-recovery attack against LowMC-128 with time complexity 2^{118} and data complexity 2^{73}, implying that the cipher does not provide the expected 128-bit security level.

Our Contributions. We emphasize that beyond the task of designing a HE-friendly block cipher, revisiting the whole compressed encryption scheme (in particular its internal mode of operation) is what is really needed in order to take these concrete HE-related implementation constraints into account.

First, we identify that homomorphic decompression is subject to an *offline phase* and an *online phase*. The offline phase is plaintext-independent and therefore can be performed in advance, whereas the online phase completes decompression upon reception of the plaintext-dependent part of the compressed

---

[1] It is worth noting that in a HE context, reducing the multiplicative size of a symmetric primitive might not be the first concern (while it is critical in a multiparty computation context, which also motivated the work of Albrecht *et al.* [2]), whereas minimizing the multiplicative depth is of prime importance.

ciphertext. Making the online phase as quick as technically doable leads us to choose *an additive IV-based stream cipher to implement* E. However, we note that the use of a lightweight block cipher as the building-block of that stream cipher usually provides a security level limited to $2^{n/2}$ where n is the block size [48], thus limiting the number of encrypted blocks to (typically) less than 2^{32} (*i.e.* 32 GB for 64-bit blocks).

As a result, we propose our own candidate for E: the keystream generator Trivium [13], which belongs to the eSTREAM portfolio of recommended stream ciphers, and a new proposal called *Kreyvium*, which shares the same internal structure but allows for bigger keys of 128 bits[2]. The main advantage of Kreyvium over Trivium is that it provides 128-bit security (instead of 80-bit) with the same multiplicative depth, and inherits the same security arguments. It is worth noticing that the design of a variant of Trivium which guarantees a 128-bit security level has been raised as an open problem for the last ten years [1, p. 30]. Beside a higher security level, it also accommodates longer IVs, so that it can encrypt up to $46 \cdot 2^{128}$ plaintext bits under the same key, with multiplicative depth only 12. Moreover, both Trivium and Kreyvium are resistant against the interpolation attacks used for breaking LowMC since these ciphers do not rely on a permutation which would enable the attacker to compute backwards.

We implemented our construction and instantiated it with Trivium, Kreyvium and LowMC in CTR-mode. Our results show that the promising performances attained by the HE-dedicated block cipher LowMC can be achieved with well-known primitives whose security has been firmly established for over a decade.

Organization of the Paper. We introduce a general model and a generic construction to compress homomorphic ciphertexts in Sect. 2. Our construction using Trivium and Kreyvium is described in Sect. 3. Subsequent experimental results are presented in Sect. 4.

2   A Generic Design for Efficient Decompression

In this section, we describe our model and generic construction to transmit compressed homomorphic ciphertexts between Alice and Charlie. We use the same notation as in the introduction: Alice wants to send some plaintext m, encrypted under Bob's public key pk (of an homomorphic encryption scheme HE) to a third party evaluator Charlie.

2.1   Offline/Online Phases in Ciphertext Decompression

Most practical scenarios would likely find it important to distinguish between three distinct phases within the homomorphic evaluation of $\mathcal{C}_{E^{-1}}$:

---

[2] Independently from our results, another variant of Trivium named Trivi-A has been proposed [16]. It handles larger keys but uses longer registers. It then needs more rounds for mixing the internal state, which means that it is much less adapted to our setting than Kreyvium.

1. an *offline key-setup* phase which only depends on Bob's public key and can be performed once and for all before Charlie starts receiving compressed ciphertexts encrypted under Bob's key;
2. an *offline decompression* phase which can be performed only based on some plaintext-independent material found in the compressed ciphertext;
3. an *online decompression* phase which aggregates the result of the offline phase with the plaintext-dependent part of the compressed ciphertext and (possibly very quickly) recovers the decompressed ciphertext c.

As such, our general-purpose formulation $c' = (\mathsf{HE_{pk}}(k), \mathsf{E}_k(m))$ does not allow to make a clear distinction between these three phases. In our context, it is much more relevant to reformulate the encryption scheme as an IV-based encryption scheme where the encryption and decryption process are both deterministic but depend on an IV:

$$\mathsf{E}_k(m) \stackrel{\text{def}}{=} \left(IV, \mathsf{E}'_{k,IV}(m)\right).$$

Since the IV has a limited length, it can be either transmitted during an offline preprocessing phase, or may alternately correspond to a state which is maintained by the server. Now, to minimize the latency of homomorphic decompression for Charlie, the online phase should be reduced to a minimum. The most appropriate choice in this respect consists in using an additive IV-based stream cipher Z so that

$$\mathsf{E}'_{k,IV}(m) = Z(k, IV) \oplus m.$$

In this reformulation, the decompression process is clearly divided into a offline precomputation stage which only depends on pk, k and IV, and an online phase which is plaintext-dependent. The online phase is thus reduced to a mere XOR between the plaintext-dependent part of the ciphertext $\mathsf{E}'_{k,IV}(m)$ and the HE-encrypted keystream $\mathsf{HE}(Z(k, IV))$, which comes essentially for free in terms of noise growth in HE ciphertexts. All expensive operations (*i.e.* homomorphic multiplications) are performed during the offline decompression phase where $\mathsf{HE}(Z(k, IV))$ is computed from $\mathsf{HE}(k)$ and IV.

2.2 Our Generic Construction

We devise the generic construction depicted on Fig. 1. It is based on a homomorphic encryption scheme HE with plaintext space $\{0, 1\}$, an expansion function G mapping ℓ_{IV}-bit strings to strings of arbitrary size, and a fixed-size parametrized function F with input size ℓ_x, parameter size ℓ_k and output size N.

Compressed Encryption. Given an ℓ_m-bit plaintext m, Bob's public key pk and $IV \in \{0, 1\}^{\ell_{IV}}$, the compressed ciphertext c' is computed as follows:

1. Set $t = \lceil \ell_m / N \rceil$,
2. Set $(x_1, \ldots, x_t) = G(IV; t\ell_x)$,
3. Randomly pick $k \leftarrow \{0, 1\}^{\ell_k}$,
4. For $1 \leq i \leq t$, compute $z_i = F_k(x_i)$,
5. Set keystream to the ℓ_m leftmost bits of $z_1 \| \ldots \| z_t$,
6. Output $c' = (\mathsf{HE_{pk}}(k), m \oplus \text{keystream})$.

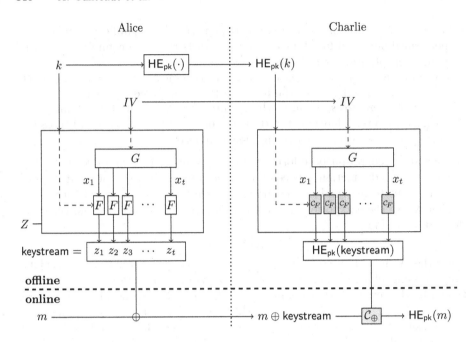

Fig. 1. Our generic construction. The multiplicative depth of the circuit is equal to the depth of $\mathcal{C}_F$. This will be the bottleneck in our protocol and we want the multiplicative depth of F to be as small as possible. With current HE schemes, the circuit $\mathcal{C}_\oplus$ is usually very fast (addition of ciphertexts) and has a negligible impact on the noise in the ciphertext.

Ciphertext Decompression. Given c' as above, Bob's public key pk and $IV \in \{0,1\}^{\ell_{IV}}$, the ciphertext decompression is performed as follows:

1. Set $t = \lceil \ell_m/N \rceil$,
2. Set $(x_1, \ldots, x_t) = G(IV; t\ell_x)$,
3. For $1 \leq i \leq t$, compute $\mathsf{HE}_{\mathsf{pk}}(z_i) = \mathcal{C}_F(\mathsf{HE}_{\mathsf{pk}}(k), x_i)$ with some circuit $\mathcal{C}_F$,
4. Deduce $\mathsf{HE}_{\mathsf{pk}}(\mathsf{keystream})$ from $\mathsf{HE}_{\mathsf{pk}}(z_1), \ldots, \mathsf{HE}_{\mathsf{pk}}(z_t)$,
5. Compute $c = \mathsf{HE}_{\mathsf{pk}}(m) = \mathcal{C}_\oplus(\mathsf{HE}_{\mathsf{pk}}(\mathsf{keystream}), m \oplus \mathsf{keystream})$.

The circuit $\mathcal{C}_\oplus$ computes $\mathsf{HE}(a \oplus b)$ given $\mathsf{HE}(a)$ and b where a and b are bit-strings of the same size. In our construction, the cost of decompression per plaintext block is *fixed* and roughly equals one single evaluation of the circuit $\mathcal{C}_F$; most importantly, the multiplicative depth of the decompression circuit is also fixed, and set to the depth of $\mathcal{C}_F$.

How Secure are Compressed Ciphertexts? From a high-level perspective, compressed homomorphic encryption is just hybrid encryption and relates to the generic KEM-DEM construct. A complete characterization of the security results attached to the KEM-DEM framework is presented in [35]. In particular when the KEM and the DEM are IND-CPA, the resulting hybrid PKE scheme is at least IND-CPA. This result applies directly here: assuming the semantic security of our homomorphic KEM[3], and a general-purpose IND-CPA secure DEM, our compressed encryption scheme is IND-CPA secure.

Instantiating the Paradigm. The rest of the paper focuses on how to choose the expansion function G and function F so that the homomorphic evaluation of C_F is as fast (and its multiplicative depth as low) as possible. In our approach, the value of IV is assumed to be shared between Alice and Charlie and needs not be transmitted along with the compressed ciphertext. For instance, IV is chosen to be an absolute constant such as $IV = 0^\ell$ where $\ell = \ell_{IV} = \ell_x$. Another example is to take for $IV \in \{0, 1\}^\ell$ a synchronized state that is updated between transmissions. Also, the expansion function G is chosen to implement a counter in the sense of the NIST description of the CTR mode [47], for instance

$$G(IV; t\ell) = (IV, IV \boxplus 1, \dots, IV \boxplus (t-1)) \quad \text{where} \quad a \boxplus b = (a+b) \bmod 2^\ell.$$

Finally, F is chosen to ensure both an appropriate security level and a low multiplicative depth. We focus in Sect. 3 on the keystream generator corresponding to Trivium, and on a new variant, called *Kreyvium*.

Interestingly, the output of an iterated PRF used in CTR mode is computationally indistinguishable from random [6, Theorem 13]. Hence, under the assumption that Trivium or Kreyvium is a PRF[4], the keystream $z_1 \| \dots \| z_t$ produced by our construction is also indistinguishable. It follows directly from [35] that the compressed encryption scheme is IND-CPA. Although the security of Trivium and Kreyvium is empiric, Sect. 3 provides a strong rationale for the Kreyvium design and makes it the solution with the smallest homomorphic evaluation latency known so far.

Why not Use a Block Cipher for F ? Although not specifically in these terms, the use of lightweight block ciphers like PRINCE and SIMON has been proposed in the context of compressed homomorphic ciphertexts *e.g.* [24, 43]. However a complete encryption scheme based on the ciphers has not been defined. This is a major issue since the security provided by all classical modes of operation (including all variants of CBC, CTR, CFB, OFB, OCB...) is inherently limited to $2^{n/2}$ where n is the block size [48] (see also *e.g.* [39, p. 95]). Only a very few modes providing *beyond-birthday* security have been proposed, *e.g.* [37,50], but they induce a higher implementation cost and their security is usually upper-bounded by $2^{2n/3}$.

In other words, the use of a block cipher operating on 64-bit blocks like PRINCE or SIMON-32/64 implies that the number of blocks encrypted under

---

[3] Note that it is usual that HE schemes succeed in achieving CPA security, but often grossly fail to realize any form of CCA1 security, to the point of admitting simple key recovery attacks [17].

[4] Note that this equivalent to say that Kreyvium instantiated with a random key and mapping the IV's to the keystream is secure [7, Sect. 3.2].

the same key should be significantly less that 2^{32} (*i.e.* 32 GB for 64-bit blocks). Therefore, only block ciphers with a large enough block size, like the LowMC instantiation with a 256-bit block proposed in [2], are suitable in applications which may require the encryption of more than 2^{32} bits under the same key.

3 Trivium and Kreyvium, Two Low-Depth Stream Ciphers

Since an additive stream cipher is the optimal choice, we now focus on keystream generation, and on its homomorphic evaluation. An IV-based keystream generator is decomposed into:

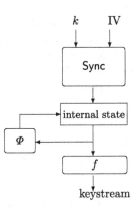

- a resynchronization function, Sync, which takes as input the IV and the key (possibly expanded by some precomputation phase), and outputs some n-bit initial state;
- a transition function Φ which computes the next state of the generator;
- a filtering function f which computes a keystream segment from the internal state.

Since generating N keystream bits may require a circuit of depth up to

$$(\mathrm{depth}(\mathsf{Sync}) + N\,\mathrm{depth}(\Phi) + \mathrm{depth}(f)),$$

the best design strategy for minimizing this value consists in choosing a transition function with a small depth. The extreme option is to choose for Φ a linear function as in the CTR mode where the counter is implemented by an LFSR. An alternative strategy consists in choosing a nonlinear transition whose depth does not increase too fast when it is iterated. The influence of Sync on the multiplicative depth of the circuit is further investigated in [14].

Size of the Internal State. A major specificity of our context is that a large internal state can be easily handled. Indeed, in most classical stream ciphers, the internal-state size usually appears as a bottleneck because the overall size of the quantities to be stored highly influences the number of gates in the implementation. This is not the case in our context. It might seem, a priori, that increasing the size of the internal state automatically increases the number of nonlinear operations (because the number of inputs of Φ increases). But, this is not the case if a part of this larger internal state is used, for instance, for storing the secret key. This strategy can be used for increasing the security at no implementation cost. Indeed, the complexity of all generic attacks aiming at recovering the internal state of the generator is $\mathcal{O}(2^{n/2})$ where n is the size of the secret part of the internal state even if some part is not updated during the keystream generation. For instance, the time-memory-data-tradeoff attacks in [5,8,32] aim at inverting the function which maps the internal state of the generator to the first keystream bits. But precomputing some values of this function must be feasible by the attacker, which is not the case if the filtering or transition function depends on some secret material. On the other hand, the size n' of the non-constant secret part of the internal state determines the data complexity for finding a collision on the internal state: the length of the keystream produced from the same key is limited to $2^{n'/2}$. But, if the transition function or the filtering function depends on the IV, this limitation corresponds to the maximal keystream length produced from the same key/IV pair. It is worth noticing that many attacks require a very long keystream generated from the same key/IV pair and do not apply in our context since the keystream length is strictly limited by the multiplicative depth of the circuit.

3.1   Trivium in the HE Setting

Trivium [13] is one of the 7 stream ciphers recommended by the eSTREAM project [25]. Due to the small number of nonlinear operations in its transition function, it appears as a natural candidate in our context.

Description. Trivium is a synchronous stream cipher with a key and an IV of 80 bits each. Its internal state is composed of 3 registers of sizes 93, 84 and 111 bits, corresponding to a size of 288 bits in total. We use the notation introduced by the designers: the leftmost bit of the 93-bit register is s_1, and its rightmost one is s_{93}; the leftmost bit of the register of size 84 is s_{94} and the rightmost s_{177}; the leftmost bit of register of size 111 is s_{178} and the rightmost s_{288}. The initialization and the generation of an N-bit keystream are described below.

$$(s_1, s_2, \ldots, s_{93}) \leftarrow (K_0, \ldots, K_{79}, 0, \ldots, 0)$$
$$(s_{94}, s_{95}, \ldots, s_{177}) \leftarrow (IV_0, \ldots, IV_{79}, 0, \ldots, 0)$$
$$(s_{178}, s_{179}, \ldots, s_{288}) \leftarrow (0, \ldots, 0, 1, 1, 1)$$
for $i = 1$ **to** $1152 + N$ **do**
    $t_1 \leftarrow s_{66} + s_{93}$
    $t_2 \leftarrow s_{162} + s_{177}$
    $t_3 \leftarrow s_{243} + s_{288}$
        if $i > 1152$ **do**
            output $z_{i-1152} \leftarrow t_1 + t_2 + t_3$
        end if
    $t_1 \leftarrow t_1 + s_{91} \cdot s_{92} + s_{171}$
    $t_2 \leftarrow t_2 + s_{175} \cdot s_{176} + s_{264}$
    $t_3 \leftarrow t_3 + s_{286} \cdot s_{287} + s_{69}$
    $(s_1, s_2, \ldots, s_{93}) \leftarrow (t_3, s_1, \ldots, s_{92})$
    $(s_{94}, s_{95}, \ldots, s_{177}) \leftarrow (t_1, s_{94}, \ldots, s_{176})$
    $(s_{178}, s_{179}, \ldots, s_{288}) \leftarrow (t_2, s_{178}, \ldots, s_{287})$
end for

No attack better than an exhaustive key search is known so far on the full Trivium. It can then be considered as secure. The family of attacks that seems to provide the best result on round-reduced versions is the cube attack and its variants [4,22,28]. They recover some key bits (resp. provide a distinguisher on the keystream) if the number of initialization rounds is reduced to 799 (resp. 885) rounds out of 1152. The highest number of initialization rounds that can be attacked is 961: in this case, a distinguisher exists for a class of weak keys [41].

Multiplicative Depth. It is easy to see that the multiplicative depth grows quite slowly with the number of iterations. An important observation is that, in the internal state, only the first 80 bits in Register 1 (the keybits) are initially encrypted under the HE and that, as a consequence, performing hybrid clear and encrypted data calculations is possible (this is done by means of the following simple rules: $0 \cdot [x] = 0$, $1 \cdot [x] = [x]$, $0 + [x] = [x]$ and $1 + [x] = [1] + [x]$, where the square brackets denote encrypted bits and where in all but the latter case, a homomorphic operation is avoided which is specially desirable for multiplications). This optimization allows for instance to increase the number of bits which can be generated (after the 1152 blank rounds) at depth 12 from 42 to 57 (*i.e.*, a 35 % increase). Then, the relevant quantity in our context is the multiplicative depth of the circuit which computes N keystream bits from the 80-bit key. The proof of the following proposition is given in [14].

Proposition 1. *In Trivium, the keystream length $N(d)$ which can be produced from the 80-bit key with a circuit of multiplicative depth d, $d \geq 4$, is given by*

$$N(d) = 282 \times \left\lfloor \frac{d}{3} \right\rfloor + \begin{cases} 81 & \text{if } d \equiv 0 \bmod 3 \\ 160 & \text{if } d \equiv 1 \bmod 3 \\ 269 & \text{if } d \equiv 2 \bmod 3 \end{cases}.$$

3.2   Kreyvium

Our first aim is to offer a variant of Trivium with 128-bit key and IV, without increasing the multiplicative depth of the corresponding circuit. Besides a higher security level, another advantage of this variant is that the number of possible IVs, and then the maximal length of data which can be encrypted under the same key, increases from $2^{80} N_{\text{trivium}}(d)$ to $2^{128} N_{\text{kreyvium}}(d)$. Increasing the key and IV-size in Trivium is a challenging task, mentioned as an open problem in [1, p. 30] for instance. In particular, Maximov and Biryukov [45] pointed out that increasing the key-size in Trivium without any additional modification cannot be secure due to some attack with complexity less than 2^{128}. A first attempt in this direction has been made in [45] but the resulting cipher accommodates 80-bit IV only, and its multiplicative complexity is higher than in Trivium since the number of AND gates is multiplied by 2.

Description. Our proposal, Kreyvium, accommodates a key and an IV of 128 bits each. The only difference with the original Trivium is that we have added to the 288-bit internal state a 256-bit part corresponding to the secret key and the IV. This part of the state aims at making both the filtering and transition functions key- and IV-dependent. More precisely, these two functions f and Φ depend on the key bits and IV bits, through the successive outputs of two shift-registers K^* and IV^* initialized by the key and by the IV respectively. The internal state is then composed of five registers of sizes 93, 84, 111, 128 and 128 bits, having an internal state size of 544 bits in total, among which 416 become unknown to the attacker after initialization.

We will use the same notation as the description of Trivium, and for the additional registers we use the usual shift-register notation: the leftmost bit is denoted by K^*_{127} (or IV^*_{127}), and the rightmost bit (*i.e.*, the output) is denoted by K^*_0 (or IV^*_0). Each one of these two registers are rotated independently from the rest of the cipher. The generator is described below, and depicted on Fig. 2.

$$(s_1, s_2, \ldots, s_{93}) \leftarrow (K_0, \ldots, K_{92})$$
$$(s_{94}, s_{95}, \ldots, s_{177}) \leftarrow (IV_0, \ldots, IV_{83})$$
$$(s_{178}, s_{179}, \ldots, s_{288}) \leftarrow (IV_{84}, \ldots, IV_{127}, 1, \ldots, 1, 0)$$
$$(K^*_{127}, K^*_{126}, \ldots, K^*_0) \leftarrow (K_0, \ldots, K_{127})$$
$$(IV^*_{127}, IV^*_{126}, \ldots, IV^*_0) \leftarrow (IV_0, \ldots, IV_{127})$$

for $i = 1$ to $1152 + N$ **do**

    $t_1 \leftarrow s_{66} + s_{93}$
    $t_2 \leftarrow s_{162} + s_{177}$
    $t_3 \leftarrow s_{243} + s_{288} + \boldsymbol{K^*_0}$

    if $i > 1152$ **do**

        output $z_{i-1152} \leftarrow t_1 + t_2 + t_3$

    end if

    $t_1 \leftarrow t_1 + s_{91} \cdot s_{92} + s_{171} + \boldsymbol{IV^*_0}$
    $t_2 \leftarrow t_2 + s_{175} \cdot s_{176} + s_{264}$
    $t_3 \leftarrow t_3 + s_{286} \cdot s_{287} + s_{69}$
    $t_4 \leftarrow K^*_0$
    $t_5 \leftarrow IV^*_0$
    $(s_1, s_2, \ldots, s_{93}) \leftarrow (t_3, s_1, \ldots, s_{92})$
    $(s_{94}, s_{95}, \ldots, s_{177}) \leftarrow (t_1, s_{94}, \ldots, s_{176})$
    $(s_{178}, s_{179}, \ldots, s_{288}) \leftarrow (t_2, s_{178}, \ldots, s_{287})$
    $(K^*_{127}, K^*_{126}, \ldots, K^*_0) \leftarrow (t_4, K^*_{127}, \ldots, K^*_1)$
    $(IV^*_{127}, IV^*_{126}, \ldots, IV^*_0) \leftarrow (t_5, IV^*_{127}, \ldots, IV^*_1)$

end for

Related Ciphers. KATAN [11] is a lightweight block cipher with a lot in common with Trivium. It is composed of two registers, whose feedback functions are very sparse, and have a single nonlinear term. The key, instead of being used for initializing the state, is introduced by XORing two key information-bits per round to each feedback bit. The recently proposed stream cipher Sprout [3],

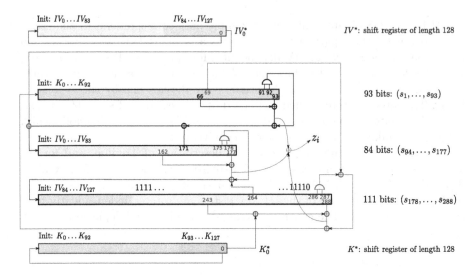

Fig. 2. Kreyvium. The three registers in the middle correspond to Trivium. The modifications defining Kreyvium correspond to the wo registers at the top and at the bottom. (Color figure online)

inspired by Grain but with much smaller registers, also inserts the key in a similar way: instead of using the key for initializing the state, one key information-bit is XORed at each clock to the feedback function. We can see the parallelism between these two ciphers and our newly proposed variant. In particular, the previous security analysis on KATAN shows that this type of design does not introduce any clear weakness. Indeed, the best attacks on round-reduced versions of KATAN so far [29] are meet-in-the-middle attacks, that exploit the knowledge of the values of the first and the last internal states (due to the block-cipher setting). As this is not the case here, such attacks, as well as the recent interpolation attacks against LowMC [21], do not apply. The best attacks against KATAN, when excluding MitM techniques, are conditional differential attacks [40,41].

Design Rationale. We have decided to XOR the keybit K_0^* to the feedback function of the register that interacts with the content of $(s_1, \ldots, s_{63})$ the later, since $(s_1, \ldots, s_{63})$ is initialized with some key bits. The same goes for the IV^* register. Moreover, as the keybits that start entering the state are the ones that were not in the initial state, all the keybits affect the state at the earliest.

We also decided to initialize the state with some keybits and with all the IV bits, and not with a constant value, as this way the mixing will be performed quicker. Then we can expect that the internal-state bits after initialization are expressed as more complex and less sparse functions in the key and IV bits.

Our change of constant is motivated by the conditional differential attacks from [41]: the conditions needed for a successful attack are that 106 bits from the IV or the key are equal to '0' and a single one needs to be '1'. This suggests that values set to zero "encourage" non-random behaviors, leading to our new constant. In other words, in Trivium, an all-zero internal state is always updated in an all-zero state, while an all-one state will change through time. The 0 at the end of the constant is added for preventing slide attacks.

Multiplicative Depth. Exactly as for Trivium, we can compute the number of keystream bits which can be generated from the key at a given depth (see [14]).

Proposition 2. *In Kreyvium, the keystream length $N(d)$ which can be produced from the 128-bit key with a circuit of multiplicative depth d, $d \geq 4$, is given by*

$$N(d) = 282 \times \left\lfloor \frac{d}{3} \right\rfloor + \begin{cases} 70 & \text{if } d \equiv 0 \bmod 3 \\ 149 & \text{if } d \equiv 1 \bmod 3 \\ 258 & \text{if } d \equiv 2 \bmod 3 \end{cases}.$$

Security Analysis. We investigate how all the known attacks on Trivium can apply to Kreyvium. A more detailled analysis is provided in [14].

TMDTO. TMDTO attacks aiming at recovering the initial state of the cipher do not apply since the size of the secret part of the internal state (416 bits) is much larger than twice the key-size: the size of the whole secret internal state has to be taken into account, even if the additional 128-bit part corresponding to K^* is independent from the rest of the state. On the other hand, TMDTO aiming at recovering the key have complexity larger than exhaustive key search since the key and the IV have the same size [12, 36].

Internal-State Collision. A distinguisher may be built if the attacker is able to find two colliding internal states, since the two keystreams produced from colliding states are identical. Finding such a collision requires around 2^{144} keystream bits generated from the same key/IV pair, which is much longer than the maximal keystream length allowed by the multiplicative depth of the circuit. We also show in [14] that, for a given key, finding two internal states colliding on all bits except on IV^* does not provide any valid distinguisher. The birthday-bound of $2^{144}\{0,1\}$then provides a limit on the number of bits produced from the same key/IV pair, not on the bits produced from the same key.

Cube Attacks [22, 28] and Cube Testers [4]. They provide the best attacks for round-reduced Trivium. In our case, as we keep the same main function, but we have two additional XORs per round, thus a better mixing of the variables, we can expect the relations to get more involved and hamper the application of previously defined round-reduced distinguishers. One might wonder if the fact that more variables are involved could ease the attacker's task, but we point out here that the limitation in the previous attacks was not the IV size, but the size of the cubes themselves. Therefore, having more variables available is of no help with respect to this point. We can conclude that the resistance of Kreyvium to these types of attacks is at least the resistance of Trivium, and even better.

Conditional Differential Cryptanalysis. Because of its applicability to Trivium and KATAN, the attack from [41] is definitely of interest in our case. In particular, the highest number of blank rounds is reached if some conditions on two registers are satisfied at the same time (and not only conditions on the register controlled by the IV bits in the original Trivium). In our case, as we have IV bits in two registers, it is important to elucidate whether an attacker can take advantage of introducing differences in two registers simultaneously. First, let us recall that we have changed the constant to one containing mostly 1. We previously saw that the conditions that favor the attacks are values set to zero in the initial state. In Trivium, we have $(108 + 4 + 13) = 125$ bits already fixed to zero in the initial state, 3 are fixed to one and the others can be controlled by the attacker in the weak-key setting (and the attacker will force them to be zero most of the time). Now, instead, we have 64 bits forced to be 1, 1 equal to zero, and $(128 + 93) = 221$ bits of the initial state controlled by the attacker in the weak-key setting, plus potentially 21 additional bits from the key still not used, that will be inserted during the first rounds. We can conclude that, while in Trivium it is possible in the weak-key setting, to introduce zeros in the whole initial state but in 3,bits, in Kreyvium, we will never be able to set to zero 64 bits, implying that applying the techniques from [41] becomes much harder.

Algebraic Attacks. Several algebraic attacks have been proposed against Trivium, aiming at recovering the 288-bit internal state at the beginning of the keystream generation (i.e., at time $t = 1153$) from the knowledge of the keystream bits. The most efficient attack of this type is due to Maximov and Biryukov [45]. It exploits the fact that the 22 keystream bits at time $3t'$, $0 \leq t' < 22$, are determined by all bits of the initial state at indexes divisible by 3 (starting from the leftmost bit in each register). Moreover, once all bits at positions $3i$ are known, then guessing that the outputs of the three AND gates at time $3t'$ are zero provides 3 linear relations between the bits of the internal state and the keystream bits. The attack then consists of an exhaustive search for some bits at indexes divisible by 3. The other bits in such positions are then deduced by solving the linear system derived from the keystream bits at positions $3t'$. Once all these bits have been determined, the other 192 bits of the initial state are deduced from the other keystream equations. This process must be iterated until the guess for the outputs of the AND gates is correct. In the case of Trivium, the outputs of at least 125 AND gates must be guessed in order to get 192 linear relations involving the 192 bits at indexes $3i + 1$ and $3i + 2$. This implies that the attack has to be repeated $(4/3)^{125} = 2^{52}$ times. From these guesses, we get many linear relations involving the bits at positions $3i$ only, implying that only an exhaustive search with complexity 2^{32} for the other bits at positions $3i$ is needed. Therefore, the overall complexity of the attack is around $2^{32} \times 2^{52} = 2^{84}$. A similar algorithm can be applied to Kreyvium, but the main difference is that every linear equation corresponding to a keystream bit also involves one key bit. Moreover, the key bits involved in the generation of any 128 consecutive output bits are independent. It follows that each of the first 128 linear equations introduces a new unknown in the system to solve. For this reason, it is not possible to determine all bits at positions $3i$ by an exhaustive search on less than 96 bits like for Trivium. Moreover, the outputs of more than 135 AND gates must be guessed for obtaining enough equations on the remaining bits of the initial state. Therefore the overall complexity of the attack exceeds $2^{96} \times 2^{52} = 2^{148}$ and is much higher that the cost of the exhaustive key search. It is worth noticing that the attack would have been more efficient if only the feedback bits, and not the keystream bits, would have been dependent on the key. In this case, 22 linear relations independent from the key would have been available to the attacker.

4   Experimental Results

We now discuss and compare the practicality of our generic construction when instantiated with Trivium, Kreyvium and LowMC. The expansion function G implements a mere counter, and the aforementioned algorithms are used to instantiate the function F that produces N bits of keystream per iteration as defined by Propositions 1 and 2.[5]

HE Framework. In our experiments, we considered two HE schemes: the BGV scheme [10] and the FV scheme [26] (a scale-invariant version of BGV). The BGV scheme is implemented in the library HElib [34] and has become *de facto* a standard benchmarking library for HE applications. Similarly, the FV scheme was previously used in several HE benchmarkings [15, 27, 43], is conceptually simpler than the BGV scheme, and is one of the most efficient HE schemes.[6] Additionally, batching was used [49], *i.e.* the HE schemes were set up to encrypt vectors in an SIMD fashion (componentwise operations, and rotations via the Frobenius endomorphism). The number of elements that can be encrypted depends on the number of terms in the factorization modulo 2 of the cyclotomic polynomial used in the implementation. This batching allowed us to perform several Trivium/Kreyvium/LowMC in parallel in order to increase the throughput.

Parameter Selection for Subsequent Homomorphic Processing. In all the previous works on the homomorphic evaluation of symmetric encryption schemes, the parameters of the underlying HE scheme were selected for the exact multiplicative depth required and not beyond [2, 19, 24, 31, 43]. This means that once the ciphertext is decompressed, no further homomorphic computation can actually be performed by Charlie – this makes the claimed timings considerably less meaningful in a real-world context.

---

[5] Note that these propositions only hold when hybrid clear and encrypted data calculations are possible between IV and HE ciphertexts. This explains the slight differences in the number of keystream bits per iteration (column "N") between Tables 1 and 2.

[6] We used the Armadillo compiler implementation of FV [15]. This source-to-source compiler turns a C++ algorithm into a Boolean circuit, optimizes it, and generates an OpenMP parallel code which can then be combined with a HE scheme.

Table 1. Latency and throughput using HElib on a single core of a mid-end 48-core server (4 x AMD Opteron 6172 processors with 64 GB of RAM).

Algorithm	Security level κ	N	used × depth	#slots	Latency sec.	throughput bits/min
Trivium-12	80	45	12	600	1417.4	1143.0
			19	720	4420.3	439.8
Trivium-13	80	136	13	600	3650.3	1341.3
			20	720	11379.7	516.3
Kreyvium-12	128	42	12	504	1715.0	740.5
			19	756	4956.0	384.4
Kreyvium-13	128	124	13	682	3987.2	1272.6
			20	480	12450.8	286.8
LowMC-128	$? \leq 118$	256	13	682	3608.4	2903.1
			20	480	10619.6	694.3
LowMC-128 [2]	$? \leq 118$	256	13	682	3368.8	3109.6
			20	480	9977.1	739.0

We benchmarked both parameters for the exact multiplicative depth and parameters able to handle circuits of the minimal multiplicative depth plus 7 to allow further homomorphic processing by Charlie (which is obviously what is expected in applications of homomorphic encryption). We chose 7 because, in practice, numerous applications use algorithms of multiplicative depth smaller than 7 (see *e.g.* [33,42]). In what follows we compare the results we obtain using Trivium, Kreyvium and also the LowMC cipher. For LowMC, we benchmarked not only our own implementation but also the LowMC implementation of [2] available at https://bitbucket.org/malb/lowmc-helib. Minor changes to this implementation were made in order to obtain an equivalent parametrization of HElib. The main difference is that the implementation from [2] uses an optimized method for multiplying a Boolean vector and a Boolean matrix, namely the "Method of Four Russians". This explains why our implementation is approximately 6 % slower, as it performs 2–3 times more ciphertext additions.

Experimental Results Using HElib. For sake of comparison with [2], we ran our implementations and their implementation of LowMC on a single core using HElib. The results are provided in Table 1. We recall that the latency refers to the time required to perform the entire homomorphic evaluation whereas the throughput is the number of blocks processed per time unit.

Experimental Results Using FV. On Table 2, we present the benchmarks when using the FV scheme. The experiments were performed using either a single core (in order to compare with BGV) or on all the cores of the machine the tests were performed on. The execution time acceleration factor between 48-core parallel and sequential executions is given in the column "Speed gain". While good accelerations (at least 25 times) were obtained for Trivium and Kreyvium algorithms, the acceleration when using LowMC is significantly smaller (~ 10 times). This is due to the huge number of operations in LowMC that created memory contention and huge slowdown in memory allocation.

Table 2. Latency of our construction when using the FV scheme on a mid-end 48-core server (4 x AMD Opteron 6172 processors with 64 GB of RAM).

Algorithm	Security level κ	N	used $\times$ depth	Latency (sec.)		Speed gain
				1 core	48 cores	
Trivium-12	80	57	12	681.5	26.8	$\times$ 25.4
			19	2097.1	67.6	$\times$ 31.0
Trivium-13	80	136	13	888.2	33.9	$\times$ 26.2
			20	2395.0	77.2	$\times$ 31.0
Kreyvium-12	128	46	12	904.4	35.3	$\times$ 25.6
			19	2806.3	82.4	$\times$ 34.1
Kreyvium-13	128	125	13	1318.6	49.7	$\times$ 26.5
			20	3331.4	97.9	$\times$ 34.0
LowMC-128	$? \leq 118$	256	14	1531.1	171.0	$\times$ 9.0
			21	3347.8	329.0	$\times$ 10.2

Interpretation. First, we would like to recall that LowMC-128 must be considered in a different category because of the existence of a key-recovery attack with time complexity 2^{118} and data complexity 2^{73} [21]. However, it has been included in the table in order to show that the performances achieved by Trivium and Kreyvium are of the same order of magnitude. An increase in the number of rounds of LowMC-128 (typically by 4 rounds) is needed to achieve 128-bit security, but this would have a non-negligible impact on its homomorphic evaluation performance, as it would require to increase the depth of the cryptosystem supporting the execution. For instance, a back-of-the-envelope estimation for four additional rounds leads to a degradation of its homomorphic execution performances by a factor of about 2 to 3 (more computations with larger parameters). It is also worth noticing that the minimal multiplicative depth for which valid LowMC output ciphertexts were obtained was 14 for the FV scheme and 13 for the BGV scheme. The theoretical multiplicative depth is 12 but the high number of additions explains this difference[7].

---

[7] The multiplicative depth is only an *approximation* of the homomorphic depth required to absorb the noise generated by the execution of an algorithm [44]. It neglects the noise induced by additions and thus does not hold for too addition-intensive algorithms such as those in the LowMC family.

Our results show that Trivium and Kreyvium have a smaller latency than LowMC, but have a slightly smaller throughput. As already emphasized in [43], real-world applications of homomorphic encryption (which are often cloud-based applications) should be implemented in a transparent and user-friendly way. In the context of our approach, the latency of the offline phase is still an important parameter aiming at an acceptable experience for the end-user even when a sufficient amount of homomorphic keystream could not be precomputed early enough because of overall system dimensioning issues.

Also Trivium and Kreyvium are more parallelizable than LowMC is. Therefore, our work shows that the promising performances obtained by the recently proposed *HE-dedicated cipher* LowMC can also be achieved with Trivium, a well-analyzed stream cipher, and a variant aiming at achieving 128 bits of security. Last but not least, we recall that our construction was aiming at compressing the size of transmissions between Alice and Charlie. We support an encryption rate $|c'|/|m|$ that becomes asymptotically close to 1 for long messages, *e.g.* for $\ell_m = 1\,\mathrm{GB}$ message length, our construction instantiated with Trivium (resp. Kreyvium), yields an expansion rate of 1.08 (resp. 1.16).

5   Conclusion

Our work shows that the promising performances obtained by the recent HE-dedicated cipher LowMC can also be achieved with Trivium, a well-known primitive whose security has been thoroughly analyzed, *e.g.* [4,22,28,41,45]. The 10-year analysis effort from the community, initiated by the eSTREAM competition, enables us to gain confidence in its security. Also our variant Kreyvium benefits from this analysis since the core of the cipher is essentially the same.

Acknowledgments. We thank Yannick Seurin for informing us about the complete characterization of secure hybrid encryption.

References

1. Algorithms, key size and parameters report 2014. Technical report, ENISA (2014)
2. Albrecht, M.R., Rechberger, C., Schneider, T., Tiessen, T., Zohner, M.: Ciphers for MPC and FHE. In: Oswald, E., Fischlin, M. (eds.) EUROCRYPT 2015. LNCS, vol. 9056, pp. 430–454. Springer, Heidelberg (2015)
3. Armknecht, F., Mikhalev, V.: On lightweight stream ciphers with shorter internal states. In: Leander, G. (ed.) FSE 2015. LNCS, vol. 9054, pp. 451–470. Springer, Heidelberg (2015)
4. Aumasson, J.-P., Dinur, I., Meier, W., Shamir, A.: Cube testers and key recovery attacks on reduced-round MD6 and Trivium. In: Dunkelman, O. (ed.) FSE 2009. LNCS, vol. 5665, pp. 1–22. Springer, Heidelberg (2009)
5. Babbage, S.: A space/time trade-off in exhaustive search attacks on stream ciphers. In: Proceedings of European Convention on Security and Detection, No. 408. IEEE (1995)

6. Bellare, M., Desai, A., Jokipii, E., Rogaway, P.: A concrete security treatment of symmetric encryption. In: Proceedings of FOCS, pp. 394–403. IEEE Computer Society (1997)

7. Berbain, C., Gilbert, H.: On the security of IV dependent stream ciphers. In: Biryukov, A. (ed.) FSE 2007. LNCS, vol. 4593, pp. 254–273. Springer, Heidelberg (2007)

8. Biryukov, A., Shamir, A.: Cryptanalytic time/memory/data tradeoffs for stream ciphers. In: Okamoto, T. (ed.) ASIACRYPT 2000. LNCS, vol. 1976, pp. 1–13. Springer, Heidelberg (2000)

9. Borghoff, J., Canteaut, A., Güneysu, T., Kavun, E.B., Knezevic, M., Knudsen, L.R., Leander, G., Nikov, V., Paar, C., Rechberger, C., Rombouts, P., Thomsen, S.S., Yalçin, T.: PRINCE – a low-latency block cipher for pervasive computing applications. In: Wang, X., Sako, K. (eds.) ASIACRYPT 2012. LNCS, vol. 7658, pp. 208–225. Springer, Heidelberg (2012)

10. Brakerski, Z., Gentry, C., Vaikuntanathan, V.: (Leveled) Fully homomorphic encryption without bootstrapping. TOCT **6**(3), 13 (2014)

11. De Cannière, C., Dunkelman, O., Knežević, M.: KATAN and KTANTAN—a family of small and efficient hardware-oriented block ciphers. In: Clavier, C., Gaj, K. (eds.) CHES 2009. LNCS, vol. 5747, pp. 272–288. Springer, Heidelberg (2009)

12. Cannière, C.D., Lano, J., Preneel, B.: Comments on there discovery of time memory data tradeoffs. Technical report, eSTREAM-ECRYPT Stream Cipher Project (2005). www.ecrypt.eu.org/stream/papersdir/040.pdf

13. De Cannière, C., Preneel, B.: Trivium. In: Robshaw, M., Billet, O. (eds.) New Stream Cipher Designs. LNCS, vol. 4986, pp. 244–266. Springer, Heidelberg (2008)

14. Canteaut, A., Carpov, S., Fontaine, C., Lepoint, T., Naya-Plasencia, M., Paillier, P., Sirdey, R.: How to compress homomorphic ciphertexts. IACR Cryptol. ePrint Arch. **2015**, 113 (2015). https://eprint.iacr.org/2015/113

15. Carpov, S., Dubrulle, P., Sirdey, R.: Armadillo: a compilation chain for privacy preserving applications. In: Proceedings of ACM CCSW. ACM (2015)

16. Chakraborti, A., Chattopadhyay, A., Hassan, M., Nandi, M.: TriviA: a fast and secure authenticated encryption scheme. In: Güneysu, T., Handschuh, H. (eds.) CHES 2015. LNCS, vol. 9293, pp. 330–353. Springer, Heidelberg (2015)

17. Chenal, M., Tang, Q.: On key recovery attacks against existing somewhat homomorphic encryption schemes. In: Aranha, D.F., Menezes, A. (eds.) LATINCRYPT 2014. LNCS, vol. 8895, pp. 239–258. Springer, Heidelberg (2015)

18. Cheon, J.H., Coron, J.-S., Kim, J., Lee, M.S., Lepoint, T., Tibouchi, M., Yun, A.: Batch fully homomorphic encryption over the integers. In: Johansson, T., Nguyen, P.Q. (eds.) EUROCRYPT 2013. LNCS, vol. 7881, pp. 315–335. Springer, Heidelberg (2013)

19. Coron, J.-S., Lepoint, T., Tibouchi, M.: Scale-invariant fully homomorphic encryption over the integers. In: Krawczyk, H. (ed.) PKC 2014. LNCS, vol. 8383, pp. 311–328. Springer, Heidelberg (2014)

20. Courtois, N., Meier, W.: Algebraic attacks on stream ciphers with linear feedback. In: Biham, E. (ed.) Advances in Cryptology–EUROCRYPT 2003. LNCS, vol. 2656, pp. 345–359. Springer, Heidelberg (2003)

21. Dinur, I., Liu, Y., Meier, W., Wang, Q.: Optimized interpolation attacks on LowMC. IACR Cryptol. ePrint Arch. **2015**, 418 (2015)

22. Dinur, I., Shamir, A.: Cube attacks on tweakable black box polynomials. In: Joux, A. (ed.) EUROCRYPT 2009. LNCS, vol. 5479, pp. 278–299. Springer, Heidelberg (2009)

23. Doröz, Y., Hu, Y., Sunar, B.: Homomorphic AES evaluation using NTRU. IACR Cryptol. ePrint Arch. **2014**, 39 (2014)

24. Doröz, Y., Shahverdi, A., Eisenbarth, T., Sunar, B.: Toward practical homomorphic evaluation of block ciphers using prince. In: Böhme, R., Brenner, M., Moore, T., Smith, M. (eds.) FC 2014 Workshops. LNCS, vol. 8438, pp. 208–220. Springer, Heidelberg (2014)

25. ECRYPT - European Network of Excellence in Cryptology: The eSTREAM StreamCipher Project (2005). http://www.ecrypt.eu.org/stream/

26. Fan, J., Vercauteren, F.: Somewhat practical fully homomorphic encryption. IACR Cryptol. ePrint Arch. **2012**, 144 (2012)

27. Fau, S., Sirdey, R., Fontaine, C., Aguilar, C., Gogniat, G.: Towards practical program execution over fully homomorphic encryption schemes. In: IEEE International Conference on P2P, Parallel, Grid, Cloud and Internet Computing, pp. 284–290 (2013)

28. Fouque, P.-A., Vannet, T.: Improving key recovery to 784 and 799 rounds of Trivium using optimized cube attacks. In: Moriai, S. (ed.) FSE 2013. LNCS, vol. 8424, pp. 502–517. Springer, Heidelberg (2014)

29. Fuhr, T., Minaud, B.: Match box meet-in-the-middle attack against KATAN. In: Cid, C., Rechberger, C. (eds.) FSE 2014. LNCS, vol. 8540, pp. 61–81. Springer, Heidelberg (2015)

30. Gentry, C.: Fully homomorphic encryption using ideal lattices. In: Proceedings of STOC, pp. 169–178. ACM (2009)

31. Gentry, C., Halevi, S., Smart, N.P.: Homomorphic evaluation of the AES circuit. In: Safavi-Naini, R., Canetti, R. (eds.) CRYPTO 2012. LNCS, vol. 7417, pp. 850–867. Springer, Heidelberg (2012)

32. Golić, J.D.: Cryptanalysis of alleged A5 stream cipher. In: Fumy, W. (ed.) EURO-CRYPT 1997. LNCS, vol. 1233, pp. 239–255. Springer, Heidelberg (1997)

33. Graepel, T., Lauter, K., Naehrig, M.: ML confidential: machine learning on encrypted data. In: Kwon, T., Lee, M.-K., Kwon, D. (eds.) ICISC 2012. LNCS, vol. 7839, pp. 1–21. Springer, Heidelberg (2013)

34. Halevi, S., Shoup, V.: Algorithms in HElib. In: Garay, J.A., Gennaro, R. (eds.) CRYPTO 2014, Part I. LNCS, vol. 8616, pp. 554–571. Springer, Heidelberg (2014)

35. Herranz, J., Hofheinz, D., Kiltz, E.: Some (in)sufficient conditions for secure hybrid encryption. Inf. Comput. **208**(11), 1243–1257 (2010)

36. Hong, J., Sarkar, P.: New applications of time memory data tradeoffs. In: Roy, B. (ed.) ASIACRYPT 2005. LNCS, vol. 3788, pp. 353–372. Springer, Heidelberg (2005)

37. Iwata, T.: New blockcipher modes of operation with beyond the birthday bound security. In: Robshaw, M. (ed.) FSE 2006. LNCS, vol. 4047, pp. 310–327. Springer, Heidelberg (2006)

38. Jakobsen, T., Knudsen, L.R.: The interpolation attack on block ciphers. In: Biham, E. (ed.) FSE 1997. LNCS, vol. 1267, pp. 28–40. Springer, Heidelberg (1997)

39. Katz, J., Lindell, Y.: Introduction to Modern Cryptography, 2nd edn. Chapman and Hall/CRC Press, Boca Raton (2014)

40. Knellwolf, S., Meier, W., Naya-Plasencia, M.: Conditional differential cryptanalysis of NLFSR-based cryptosystems. In: Abe, M. (ed.) ASIACRYPT 2010. LNCS, vol. 6477, pp. 130–145. Springer, Heidelberg (2010)

41. Knellwolf, S., Meier, W., Naya-Plasencia, M.: Conditional differential cryptanalysis of Trivium and KATAN. In: Miri, A., Vaudenay, S. (eds.) SAC 2011. LNCS, vol. 7118, pp. 200–212. Springer, Heidelberg (2012)

42. Lauter, K., López-Alt, A., Naehrig, M.: Private computation on encrypted genomic data. In: Aranha, D.F., Menezes, A. (eds.) LATINCRYPT 2014. LNCS, vol. 8895, pp. 3–27. Springer, Heidelberg (2015)

43. Lepoint, T., Naehrig, M.: A comparison of the homomorphic encryption schemes FV and YASHE. In: Pointcheval, D., Vergnaud, D. (eds.) AFRICACRYPT. LNCS, vol. 8469, pp. 318–335. Springer, Heidelberg (2014)

44. Lepoint, T., Paillier, P.: On the minimal number of bootstrappings in homomorphic circuits. In: Adams, A.A., Brenner, M., Smith, M. (eds.) FC 2013. LNCS, vol. 7862, pp. 189–200. Springer, Heidelberg (2013)

45. Maximov, A., Biryukov, A.: Two trivial attacks on Trivium. In: Adams, C., Miri, A., Wiener, M. (eds.) SAC 2007. LNCS, vol. 4876, pp. 36–55. Springer, Heidelberg (2007)

46. Naehrig, M., Lauter, K.E., Vaikuntanathan, V.: Can homomorphic encryption be practical? In: Proceedings of ACM CCSW, pp. 113–124. ACM (2011)

47. National Institute of Standards and Technology: Recommendation for Block Cipher Modes of Operation. NIST Special, Publication 800–38A (2001)

48. Rogaway, P.: Evaluation of some blockcipher modes of operation. Cryptrec(2011). http://web.cs.ucdavis.edu/~rogaway/papers/modes.pdf

49. Smart, N.P., Vercauteren, F.: Fully homomorphic SIMD operations. Des. Codes Crypt. **71**(1), 57–81 (2014)

50. Yasuda, K.: A new variant of PMAC: beyond the birthday bound. In: Rogaway, P. (ed.) CRYPTO 2011. LNCS, vol. 6841, pp. 596–609. Springer, Heidelberg (2011)

Efficient Design Strategies Based on the AES Round Function

Jérémy Jean$^{(\boxtimes)}$ and Ivica Nikolić

Nanyang Technological University, Singapore, Singapore
Jean@pmail.ntu.edu.sg

Abstract. We show several constructions based on the AES round function that can be used as building blocks for MACs and authenticated encryption schemes. They are found by a search of the space of all secure constructions based on an efficient design strategy that has been shown to be one of the most optimal among all the considered. We implement the constructions on the latest Intel's processors. Our benchmarks show that on Intel Skylake the smallest construction runs at 0.188 c/B, while the fastest at only 0.125 c/B, i.e. five times faster than AES-128.

Keywords: Fast software implementation · AES · AES-NI · Skylake

1 Introduction

As a block cipher standard, the AES has inspired many cryptographic designs. Stream and block ciphers, authenticated encryption schemes (AEs), cryptographic hash functions and Message Authentication Codes (MACs) based on the AES benefit from its two main features, namely, its security and efficiency. The security benefit is twofold. First, as the AES is the most popular block cipher, it has been extensively analyzed and its security is well understood [9,14,15]. Second, the AES is based on the so-called wide-trail strategy [6], which provides resistance against the standard differential and linear attacks. The efficiency benefit is significant as well. Due to its internal structure, the AES allows fast software implementations based on look-up tables as well as even more efficient bit-sliced implementations [12]. Furthermore, the latest mainstream processors have a dedicated set of instructions, called AES-NI, that provides a complete implementation of the AES. These handy instructions allow with a few lines of code to execute one block cipher call with exceptionally high efficiency (measured in cycles per byte of data or c/B). For instance, on the same architecture, the table-based implementation of AES-CTR runs at around 10 c/B, its bit-sliced implementation at around 7.5 c/B, while its AES-NI implementation at less than 1 c/B. As significant speedups are observed when AES-NI are available, it is important to understand how far we can benefit from them.

Depending on the security requirements and adversarial model, designs based on the AES may use round-reduced version of the block cipher. For instance, Pelican-MAC [8], Alpha-MAC [7], LEX [1], ASC-1 [11], and ALE [3], use only

© International Association for Cryptologic Research 2016
T. Peyrin (Ed.): FSE 2016, LNCS 9783, pp. 334–353, 2016.
DOI: 10.1007/978-3-662-52993-5_17

four rounds of the AES to process one message block (cf. to the ten rounds in the original AES-128 block cipher). Obviously, the reduction in the number of rounds has a direct impact on the efficiency and these designs run at much higher speed. The decision to reduce the number of rounds to four stems from the wide-trail strategy, since in some cases four rounds already provide sufficient level of security. Only a few designs use less than four rounds, as the security analysis becomes more intricate.

Our Contributions. We examine AES-based constructions that can be used as building blocks of secret-key primitives (e.g., MACs and authenticated encryption schemes). Our main goal is to push the limits of efficiency of constructions that can be implemented with the AES-NI, without sacrificing their security.

As reference points and benchmarks, we use the two authenticated encryption schemes AEGIS-128L and Tiaoxin-346 submitted to the CAESAR competition [5]. These schemes, not only rely on round-reduced AES (to process 16-byte message block, AEGIS-128L uses four rounds, while Tiaoxin-346 only three rounds of AES), but allow as well a full parallelization of the round calls. As a result, with AES-NI implementation they achieve exceptionally high efficiency and run at only 0.2–0.3 cycles per byte of message.

To understand the speed advantage of these designs, first we focus on AES-NI. We investigate the performance of the AES-NI instruction aesenc (executes one round of AES) on the latest Intel processors and deduce necessary conditions for efficient designs. Consequently, our designs have internal states composed of several 128-bit words (called blocks), while their step functions are based only on aesenc and bitwise additions (XORs). The state size, the number of aesenc calls per step, and the choice of state words to which aesenc is applied ensures that our designs will have a high efficiency.

Next, we focus on the security of the designs. The most common attacks for MACs and AE are internal collisions based on high probability differential characteristics that start and end in zero state differences (but some intermediate states contain differences, introduced through the messages). The inability of the adversary to efficiently built such collisions is the single security criteria required from our designs.

We consider two strategies that may lead to efficient and secure constructions. In the first, the AES rounds are applied to the words of the state in a way such that several steps of the construction mimic a few keyless AES rounds[1]. Due to the wide-trail approach, this strategy provides easier security proofs. However, we show that regardless of the step function chosen, such strategy has only limited efficiency potential. For instance, strategy based on 4-round AES can never run faster than 0.25 cycles per byte.

To achieve higher speed, we thus consider a second strategy, where message and state words can be XOR-ed between the AES calls. The wide-trail approach cannot longer be used (as each application is one-round AES), hence the security proof for the constructions becomes much harder. To solve it, for each candidate construction we transform the collision problem into a MILP problem, and find

---

[1] This approach was chosen in Tiaoxin-346, where 2-round AES is used.

the optimal solution which corresponds to the characteristic with the highest probability. The cases where such probability is too low correspond to secure constructions.

We search for suitable designs based on the second strategy by gradually increasing the state size and decreasing the number of AES rounds per step. In some cases, several constructions have the same efficiency but provide different security margin. We implement each construction on the latest Intel processors and check if the theoretical and actual cycle per byte count match. We list 7 secure constructions that provide a good tradeoff between state size and efficiency. The smallest has 6 words, and runs at 0.22 c/B on Haswell, and 0.188 c/B on Skylake. The most efficient has 12 words, and runs at 0.136 c/B on Haswell, and 0.125 c/B on Skylake. This construction uses only 2 AES rounds per one block of message, and thus it is five times faster than the AES.

2   Designs Based on the AES Round Function

2.1   The AES Round Function and the Instruction Set AES-NI

AES is the current block cipher standard and a well-studied cryptographic construction. As such, parts of AES are used in many crypto designs. The usage ranges from the utilization of the AES S-box in some hash functions, to application of the AES round function in stream ciphers, and employment of the whole AES in particular authenticated encryption schemes. The AES contains three different block ciphers, which only differs by their key sizes: in the remaining of this paper, we simply write AES to refer to the 128-bit key version AES-128.

From a software perspective, it may seem that partitioning of the AES can go up to the four basic round function operations: SubBytes, ShiftRows, MixColumns, and AddRoundKey. However, actual fast implementations of AES rely on the so-called AES-NI: a special set of instructions available on the latest processors, dedicated to efficiently executing *rounds of AES*.[2] As efficiency is our primary goal, we further focus on designs based on the instruction set of AES-NI. More precisely, we use only the processor instruction aesenc, which performs one regular (not the last) round of AES on an input state S with a subkey K:

$$\text{aesenc}(S, K) = \text{MixColumns}(\text{ShiftRows}(\text{SubBytes}(S))) \oplus K.$$

Let us recall the two notions related to the performance of a processor instruction, namely, *the latency* and *the reciprocal throughput* of an instruction. Informally, latency is defined as the number of clock cycles required to execute an instruction, whereas the reciprocal throughput (further called throughput) as

---

[2] In addition to the encryption and decryption rounds, AES-NI includes as well instructions that perform subkey generation and inverse MixColumns. Note that the four individual round operations can be realized as a composition of different instructions from AES-NI. However, such composition would have greatly reduced efficiency in comparison to the round calls.

the number of clock cycles required to wait before executing the same instruction. In Table 1 are given the performances of aesenc on the five latest Intel's processors. For instance, on Intel's Ivy Bridge family of processors aesenc has a latency of 8 and a throughput of 1. This means that aesenc needs 8 cycles to execute one AES round, and it can be called consecutively after 1 cycle.

Table 1. The latency and throughput of aesenc on the latest Intel's processors.

Processor	Latency	Throughput
Sandy Bridge	8	1
Ivy Bridge	8	1
Haswell	7	1
Broadwell	7	1
Skylake	4	1

Our design strategies target the five latest Intel's processors: Sandy and Ivy Bridge (collectively referred to as *bridge), Haswell and Broadwell (referred to as *well), and Skylake.

2.2 Efficiency

Our goal is to devise a strategy that results in designs based on aesenc that have a superior efficiency over the AES. Improvements in efficiency can come from two concrete approaches: reduction of the number of rounds per message block, and, parallelization of the aesenc calls. Let us take a closer look at the two approaches.

Reducing the Number of Rounds. The AES has 10 rounds[3], i.e. it uses 10 aesenc calls[4] to process a 16-byte message. Removing several rounds from the AES leads to a block cipher susceptible to practical attacks. This, however, does not imply that any design (not only a block cipher) should necessary use around 10 aesenc calls. In fact, a common approach based on the AES, is to design cryptographic primitives that use only four AES rounds to process 16-byte data.

The goal of our design is to use a minimal number of calls to aesenc. For this purpose, we define a metric, called a *rate* of design:

Definition 1 (Rate). *The rate ρ of a design is the number of AES rounds (calls to aesenc) used to process a 16-byte message.*

---

[3] Here, we simply use AES to refer to the AES-128.

[4] The last round in AES is different and it is executed with a call to the AES-NI instruction aesenclast, which has similar performance to aesenc.

For instance, `AES-128` has a rate of 10, `AES-256` has a rate of 14, `AEGIS-128L` has a rate of 4, and `Tiaoxin-346` a rate of 3. Obviously, a smaller rate may lead to more efficient designs.[5]

Parallelizing the Round Calls. A large improvement in efficiency may come by switching from serial[6] to parallel calls to `aesenc`.

A design is based on serial calls to `aesenc` (or to any other instruction of that matter), if the following `aesenc` is called only after the previous `aesenc` has finished. In such designs, the latency and the number of calls to `aesenc` give an immediate bound on the required number of cycles. An example of a serial construction is the cipher block chaining (CBC) mode because it requires the output of processing the previous message block in order to process the next message block.[7] As `AES-128` has 10 rounds, on Haswell (where `aesenc` has a latency of 7), the `AES-CBC` requires $10 \cdot 7 = 70$ cycles to process 16-byte plaintext (see Fig. 1): the first round (the first call to `aesenc`) starts at cycle 0 and completes at 7, the second starts at 7 and completes at 14, ..., the 10th starts at 63 and completes at 70. As a result, the construction runs at $70/16 = 4.375$ cycles/byte (or c/B for short).

Fig. 1. Serial design: `AES-CBC` on Intel's Haswell with `aesenc` latency of 7 cycles. Only one message block is processed at once.

Designs with parallel calls to `aesenc` can be far more efficient, as the instructions are executed simultaneously, i.e. the following `aesenc` can be called while one or more of the previous `aesenc` are still executing. The cycle count now depends not only on the number of rounds and the latency, but also on the throughput and the maximal number of independent instances of `aesenc` supported by the design. A textbook example of parallelizable construction is the counter (CTR) mode.[8] On Haswell it is possible to process 7 message blocks in parallel (see Fig. 2): at cycle 0, `aesenc` is called and it will perform the first AES round for the first message block (and return the result at cycle 7); at cycle 1, `aesenc` for the first AES round of the second message block is called, etc., at cycle 6 the `aesenc` for the first round of the seventh message block is called. Then, `aesenc` that perform the second rounds for all the seven message blocks

---

[5] A smaller rate is not a sufficient condition of efficiency as parallelizing `aesenc` calls plays an important role as well (see the next paragraph).

[6] Bogdanov et al. [2] have analyzed the speed improvements of serial modes when processing multiple messages in parallel.

[7] Recall that the `AES-CBC` is defined as $C_{i+1} = \text{AES}_K(C_i \oplus M_{i+1})$.

[8] Recall that the `AES-CTR` is defined as $C_i = \text{AES}_K(N||i) \oplus M_i$, where N is a nonce.

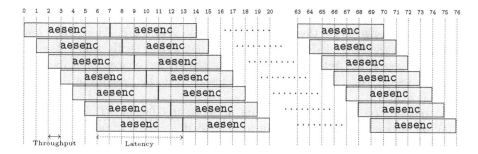

Fig. 2. Parallel design: AES-CTR on Intel's Haswell with aesenc latency of 7 cycles. It allows 7 message blocks to be processed in parallel. The aesenc is called every cycle.

are called at cycles 7–13. By repeating this procedure, it is possible to perform all ten AES rounds for all 7 message blocks – the last rounds are executed at cycles 63–69, and the ciphertexts are produced at cycles 70–76. Hence only 76 cycles, which can be brought down to 70 if longer messages are considered, are required to process 7 message blocks, or on average only 10 cycles per one message block (cf. to 70 cycles for processing a message block in the serial CBC mode). Therefore, the CTR mode runs at $10/16 = 0.625$ c/B, or precisely 7 times faster than the CBC mode.

The State Size and the Number of aesenc Calls per Step. The parallel calls to aesenc can be achieved only if the state size is sufficiently large. We have seen that CBC mode requires a state composed of only one 16-byte word, but provides no parallelization. On the other hand, if supplied with a state of seven words, the CTR mode can run seven instances in parallel. As we strive for designs with high efficiency and thus support for parallel calls to aesenc, they will have larger states. In general, *if the design makes c calls to aesenc per step, then the state has to have at least c 128-bit words*: only in this case we can have fully parallelizable aesenc calls.

The optimal number of aesenc calls per step depends on the latency to throughput ratio. *The most efficient designs use around latency/throughput independent calls to aesenc per one step.* Let us understand this fact on the example of a hypothetical design that has four aesenc calls per step to process 16-byte message (has a rate of $4/1 = 4$) and is implemented on Haswell, which in turn has a ratio of $7/1 = 7$. The four aesenc calls of the first step are called at cycles 0, 1, 2, and 3 (at every cycle because the throughput is 1), but the results of these calls are obtained only at cycles 7, 8, 9, 10 (because the latency is 7). As a result, at cycles 4, 5, and 6, no aesenc calls are made,[9] and we say that the aesenc port[10] has not been saturated, i.e. there have been empty cycles. Due

---

[9] Assuming that all the calls to aesenc of the next round depend on some of the outputs of the previous four aesenc calls.

[10] The part of the processor that executes aesenc.

to the empty cycles, even though the rate is 4, one needs 7 cycles on Haswell to process the message block, thus the speed is $7/16 = 0.4375$ c/B. The cycle count changes when the same design is implemented on Skylake (with ratio $4/1 = 4$). On this processor, the aesenc port is fully saturated, and on average it requires only 4 cycles per 16-byte message,[11] which means that this design would run at $4/16 = 0.25$ c/B.

A construction with rate ρ can run at most at 0.0625ρ c/B because, by definition, it needs ρ aesenc calls (in total at least ρ cycles) to process 16-byte message, hence the maximal speed is $\frac{\rho}{16} = 0.0625\rho$ c/B. On the other hand, if the number of aesenc calls per step is smaller than the latency to throughput ratio, then, for the aforementioned reasons, the aesenc port may not be saturated, and the speed may drop to $0.0625\frac{latency}{throughput}$ c/B. In the sequel, we take this number as our *expected speed*. The actual speed, however, may differ. It could be lower, if the aesenc between different steps are dependent, i.e. if the inputs to the aesenc of the next step depend on the outputs of the aesenc of the previous step. On the other hand, the actual speed could be higher than the expected, if more than $\frac{latency}{throughput}$ aesenc could run at the same time – this happens, when some of the aesenc calls of the next step can start before finishing most of the aesenc of the previous step.

Summary. Let us summarize the facts of this subsection as they provide hints to achieve high efficiency, i.e. low c/B measurement:

- lower rate (#aesenc per message block) leads to more efficient designs,
- all aesenc calls per step are independent and thus run in parallel,
- the state is at least as large as the number of aesenc calls per step,
- the #aesenc calls per step is close to the latency/throughput ratio.

2.3 Security Notions

We suggest design *strategies* to construct building blocks for symmetric-key primitives, and thus we adapt the security requirements accordingly. Our constructions proposed further, for instance, could be used to build a MAC algorithm, where an initialization phase first randomizes a 128-bit key and IV-dependent internal state to produce a 128-bit tag by injecting message blocks. In such a case, classical security requirements impose that no key-recovery or forgery succeeds in less than 2^{128} operations. If an authenticated encryption scheme uses our building block with a 128-bit key to produce a 128-bit tag, then as well, less than 2^{128} computations must not break the scheme.

Analyzing the resistance of a design against all possible attacks is infeasible without giving the full specification.[12] To capture this, we reduce the security

---

[11] If the aesenc are sufficiently independent between steps.
[12] For instance, the initialization and finalization stages of the constructed stream cipher or authenticated encryption scheme.

claim of our constructions to the problem of finding internal collisions. Nonetheless, we emphasize that this is only one of the requirements of a cryptographic primitive, thus the resistance against the remaining attacks should be checked after completing the whole design.

The reason we use state collisions as our unique security requirement is twofold. First, we cannot fathom how designers will use our building blocks, and this notion applies directly to many different schemes, like hash functions or MAC and AE where a state collision would yield forgery. Therefore, by focusing only on this notion, we maximize the security of future designs based on these building blocks. Second, the inherent algorithmic problem is well-studied and understood: it consists in finding special types of differential characteristics that start and end in zero difference. Finally, we can also argue how significant this requirement is by recalling that several primitives have been broken due to susceptibility to attacks based on state collisions (see for instance [13,20]).

To find a state collision means to identify two *different* sequences of messages such that, from the same initial state value, the same output state value is reached in the scheme after injecting the different message sequences. Consequently, we can describe this problem as finding a high-probability differential characteristic from the all-zero state difference to the same all-zero state difference, where the differences come from the message bytes. By high-probability, we mean higher than 2^{-128} since we focus on the AES, which relies on a 128-bit internal state.

To elaborate on the security reduction to state collisions, we briefly recall the wide-trail strategy adopted in the design of the AES [6]. This technique has been introduced to make the AES resistant to classical differential cryptanalysis in the single-key setting. In particular, the AES ensures a (tight) lower bound of the number of active S-boxes for any number of rounds in this model (see Table 2). In detail, the AES uses an Substitution-Permutation Network (SPN) including an MDS code to provably bound the diffusion, measured in terms of number of active S-boxes. Additionally, the S-box S from the substitution layer has been constructed to have a differential probability upper bounded by 2^{-6}, which means that any differential equation $S(x) \oplus S(x \oplus \delta_1) = \delta_2$ over $GF(2^8)$, for nonzero δ_1 and δ_2, has at most four solutions.

Table 2. Minimum number of active S-boxes in the AES in the single-key model.

Rounds	1	2	3	4	5	6	7	8	9	10
Active S-boxes	1	5	9	25	26	30	34	50	51	55

Therefore, to construct secure designs based on the AES round function when no differences are introduced in the subkeys, it is sufficient to ensure that a difference enters four rounds of AES. Indeed, four rounds necessarily have at least 25 active S-boxes, which directly yield an upper bound on any differential characteristic probability: $2^{-6 \cdot 25} = 2^{-150} \ll 2^{-128}$. This 4-round barrier explains why many previous designs chose to exploit this provable bound and gain in efficiency in comparison to the ten rounds used in the actual AES-128 block cipher.

In our case, we are interested in designs which achieve higher performances and do not necessarily rely on four rounds of AES. Consequently, the differential characteristic mentioned before that starts and ends in no-difference states *must* activate at least 22 S-boxes, so its probability would be at most $2^{-6\cdot22} = 2^{-132} < 2^{-128}$. Hence, in the sequel the security goals imposed on our designs are such that their best differential characteristic has at least 22 active S-boxes.

2.4   General Structure and Definitions

We define here the classes of AES-based designs that we study in the remaining of the paper. For all the aforementioned reasons, we focus on only two operations on 128-bit values: the AES round function denoted by A and performed by the aesenc instruction, and the XOR operation denoted by $\oplus$.

More precisely, we study in Sect. 3 the class $\mathcal{A}^r_\oplus$ where the allowed operations on a state of s words belong to $\{A^r, \oplus\}$. The notation A^r refers to r cascaded iterations of the permutation A. Next, in Sect. 4, we move on to the more general class $\mathcal{A}^1_\oplus$ (simply denoted $\mathcal{A}_\oplus$), where the AES round function is not necessarily cascaded. The general structure of the elements of $\mathcal{A}_\oplus$ are depicted on Fig. 3, where we represent by dashed lines the optional components. We define an iteration of such designs as a *step* to avoid confusion with the *round function A* of the AES.

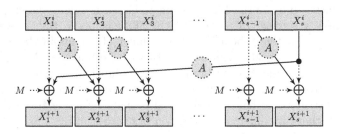

Fig. 3. One step of the general structure of the designs investigated in this paper. Dashed components mean they can be present or absent from the design.

We emphasize that all the designs belonging to these classes implement shifts of the state words to make the various applications of A to be *independent*. Consequently, each updated word X_t^{i+1}, for $0 \le t < s$, necessarily depends on $X_{t-1 \pmod{s}}^i$, and optionally on X_t^i. The main rationale behind this stems from the objective to reach high efficiency: should the diffusion be higher, for instance where a single output of A would be XORed to every output words, the processor would have to wait until all the output words have their final value. In our case, the shifts allow to optimize the usage of the processor cycles: starting evaluating the design from right to left, the first call to A is likely to be finished evaluating when we start processing the left-most state word. Hence, the iteration $i+1$ can start without waiting for the end of iteration i.

However, this optimized scheduling of instructions comes at the expense of the diffusion: from a single bit difference in the input state, reaching a full diffusion might take several steps. As a complete opposite, reaching full diffusion in a single step would mean XORing the output of a single A to all the output state words, and would waste many cycles. While this seems to suggest an interesting tradeoff, we nevertheless show in the sequel that there *do* exist designs in the class $\mathcal{A}_\oplus$ which, at the same time, achieve optimally high efficiency and meet our security requirements.

In terms of implementation, as mentioned before, the aesenc operations ends with the XOR of a round subkey and as a result, the implementations may benefit from this *free* operation. Namely, if we should XOR the message block M after the aesenc, we could just use the instruction aesenc($\bullet, M$). Otherwise, we might just use aesenc($\bullet, 0$).

Notations. We use the following notations to describe the designs. We introduce the parameters s that represents the number of 128-bit state words, a the number of AES rounds in a single step, and m the number of 128-bit message blocks processed per step. Additionally, we denote by ρ the rate of the design following Definition 1, that is $\rho = a/m$.

3   The Class $\mathcal{A}_\oplus^r$ and Rate Bounds

The class $\mathcal{A}_\oplus^r$, where $r > 1$, consists of designs that are based on r cascaded applications of the AES round function. This guarantees that state words will go through r rounds of AES, *without other state or message words being added to them*. Example of an actual construction from $\mathcal{A}_\oplus^2$ is given in Fig. 4. This design is based on 2-round AES as both of the words X_2^i and X_5^i will go through two AES rounds before any other state or message word is XORed to them.

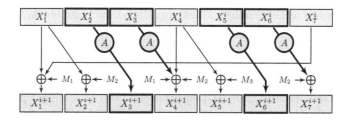

Fig. 4. A design from $\mathcal{A}_\oplus^2$.

Designs from $\mathcal{A}_\oplus^r$ are easier to analyze as they resemble r rounds of the AES. As a result, their main advantage lies in the possibility to use the wide-trail strategy of the AES which dictates that the minimal number of active S-boxes of

2, 3, and 4 rounds of AES is 5, 9, and 25 active S-boxes, respectively (see Table 2). For example, to prove that a particular $\mathcal{A}_\oplus^3$ design is secure by our definition, we have to show that in any differential characteristic that starts and ends in a zero difference, a state difference must go at least three times through the cascaded three rounds of AES. Such design would be secure, because the number of active S-boxes for any characteristic would be at least $3 \cdot 9 = 27 \geq 22$. For the class $\mathcal{A}_\oplus^2$ (resp. $\mathcal{A}_\oplus^4$), the similar requirement is to activate five times (resp. once), the cascaded 2-round (resp. 4-round) AES.

The efficiencies of these designs, however, are limited. Further, we show that their rates cannot be arbitrary low, but are in fact bounded by r.

Theorem 1. *The rate ρ of a design based on $\mathcal{A}_\oplus^r$ cannot be less than r, i.e.*

$$\rho(\mathcal{A}_\oplus^r) \geq r.$$

Proof. Any design from $\mathcal{A}_\oplus^r$ can be divided into several parts. Each r-step cascaded aesenc with the corresponding state words composes a so-called nonlinear part. Consecutive XORs of the message and the state words (with no aesenc in between) also compose a part, called a linear part. Note, there can be several nonlinear and linear parts. For instance, the design from Fig. 4 can be divided into two nonlinear parts (denoted with thick lines) and two linear parts (the remaining two parts between the nonlinear parts).

A design is insecure if we can build a high-probability differential characteristic that starts and ends in zero state difference (but some intermediate state words have non-zero differences introduced through the message words). Further, we show that if the rate is too small, more precisely if $\rho < r$, then we can build a differential characteristic with no active S-boxes. That is, the difference in the state can be introduced through the message words and then canceled in the following steps, without reaching the state words to which aesenc is applied. As a result, the probability of that differential characteristic would be one.

Let m be the number of message blocks XORed per step. Moreover, let N and L be the total number of nonlinear and linear parts, respectively. Recall that each of the N nonlinear parts has at least r cascaded applications of aesenc. Thus, for the rate ρ, defined as the number of aesenc calls per message block, it holds:

$$\rho(\mathcal{A}_\oplus^r) \geq \frac{N \cdot r}{m}. \tag{1}$$

To build a differential characteristic with no active S-boxes, at each step of the characteristic the difference that enters each of the N nonlinear parts should be zero. This condition can be expressed as a system of linear equations where the differences in the message words are the unknown variables. At each step, we require the inputs to the nonlinear parts to be zero. Hence, for each step, N equations are added to the system, and the number of variables is increased by m. For instance, the system that corresponds to the design from Fig. 4 has the following four equations that correspond to the first two steps of the characteristics for the two non-linear layers:

$$\Delta M_2^1 = 0,$$
$$\Delta M_3^1 = 0,$$
$$\Delta M_1^1 \oplus \Delta M_3^2 = 0,$$
$$\Delta M_1^1 \oplus \Delta M_2^1 \oplus \Delta M_3^2 = 0,$$

where the unknown ΔM_i^j is the difference in the message word M_i at step j of the characteristic.

The resulting system is homogeneous because we require the input differences to the nonlinear layers to be zero. When built for a differential characteristic on R steps, the system has $m \cdot R$ variables. Furthermore, it has $N \cdot R$ equations that correspond to the conditions that zero differences enter all nonlinear layers, and additional s equations (where s is the number of state words) that correspond to the conditions that all state words after step R have a zero difference. As the system is homogeneous, it has a non-zero solution as long as the number of variables exceeds the number of equations, i.e. $m \cdot R > N \cdot R + s$, or equivalently, as long as

$$R(m - N) - s > 0. \tag{2}$$

We show that if $\rho < r$, then (2) holds. From (1), it follows that $r > \rho \geq \frac{N \cdot r}{m}$, hence $N < m$. Let $m - N = t > 0$. As the number s of state words is fixed, and the number of steps R of the differential characteristic can increase, it follows that $R(m - N) - s = R \cdot t - s > 0$, when $R > \frac{s}{t}$. Therefore, when the rate ρ of the design is smaller than r, the homogeneous system has a non-zero solution which corresponds to a differential characteristic with no active S-boxes and, as a result, the design is insecure. Hence, the rate ρ of a secure design cannot be less than r. $\square$

Remark 1. The rate bound holds for any design based on r-round cascaded AES (and not only for the class with shifts to the right, that we analyze).

From the theorem, we can conclude that regardless of the actual construction, designs from $\mathcal{A}_\oplus^4, \mathcal{A}_\oplus^3$ and $\mathcal{A}_\oplus^2$ cannot have rates lower than 4, 3, and 2, respectively, and thus cannot run faster than 0.250 c/B, 0.188 c/B, and 0.125 c/B, respectively.

Note, as the step functions of AEGIS-128L and Tiaoxin-346 run at 0.250 c/B and 0.188 c/B (have rates 4 and 3), in order to find more efficient designs, we have to either find rate-3 designs with smaller states (at most 12 words as Tiaoxin-346 has 13 words), or designs with lower rate. We have run a complete search of all designs from $\mathcal{A}_\oplus^3$ with at most 12 state words and found that none of them is secure[13]. Furthermore, we have run a partial search[14] among designs from $\mathcal{A}_\oplus^2$ and found constructions with rate 2.66, but not lower. Thus, to achieve more efficient designs, in the next section we examine the class $\mathcal{A}_\oplus$.

---

[13] This gives a rise to the conjecture that the inequality from the theorem is strict.

[14] In this case, the search space cannot be exhausted as it is too large.

4   Designs in the Class $\mathcal{A}_\oplus$

In this section, we focus on the more general class of designs $\mathcal{A}_\oplus$, where the AES round function is not necessarily iterated. From a cryptanalytic standpoint, it means this class encompasses designs where state differences can be introduced between two consecutive AES round functions. The main consequence in comparison to the previous class $\mathcal{A}_\oplus^r$ from Sect. 3 is that we lose the simplicity of the analysis brought by the wide-trail strategy. One could compare the change of analysis as transition from the single-key framework of the AES to its related-key counterpart (where differences may be introduced between consecutive rounds).

However, in spite of the more complex analysis, we show there exists low-rate designs in this larger class that meet our security requirements. Namely, we show several designs that achieve rates 3, 2.5, and even rate 2.

The study of $\mathcal{A}_\oplus$ is less straightforward than the previous case, thus we rely on mixed integer linear programming (MILP) to derive lower bounds on the number of active S-boxes the designs. In the next sections, we briefly recall the MILP technique applied to cryptanalysis (Sect. 4.1) and we detail our results (Sect. 4.2).

4.1   MILP and Differential Characteristic Search

From a high-level perspective, a MILP problem aims at optimizing a linear objective function subject to linear equalities and/or linear inequalities. The technique we use in this paper is said to be *mixed* integer linear programming as it alleviates the all-integer constraint on the classical linear programming variables. More precisely, in our case some variables might not be integers, but all the integer variables are 0–1 variables. Therefore, we could dub this particular setup as 0–1 MILP.

The 0–1 MILP problems are usually NP-hard, but solutions can be found using different strategies, for instance, the cutting-plane method which iteratively refines a valid solution by performing cuts relying on the linear inequality constraints of the problem. For our purposes, we use one of the many solvers existing to date, namely the Gurobi solver [10]. Several published results rely on MILP optimization tools to solve cryptanalytic problems: searches for differential characteristics in various schemes are given in [18], known lower bounds for the number of active S-boxes for the related-key setting of AES in [16], analysis of reduced versions of the Trivium stream cipher in [4], etc.

We aim at finding differential characteristics from the all-zero difference input state to the same all-zero output state after a variable number of steps. As mentioned before, our measure of security relies on the number of active S-boxes, which gives an upper bound on the success probability of a differential attack that may lead to state collisions. We transform the search of differential characteristics into MILP problems whose objective functions count (and minimize) the number of active S-boxes. In practice, since we use the AES round function, we only require the differential characteristics to have *at least* 22 active S-boxes to ensure security.

Let $\mathcal{X} = \{x_i, i = 1, \ldots, m\}$ be the set of all the m variables and $\mathcal{S} \subseteq \mathcal{X}$ be the subset of variables representing the S-boxes of the scheme. With these notations, a classical MILP problem that we study can be stated as follows:

$$\text{Minimize: } \sum_{x \in \mathcal{S}} x,$$

$$\text{subject to: } \mathbf{A}x = b, \quad x_i \text{ all 0--1 variables.}$$

Recall that, for each $x \in \mathcal{S}$, $x = 1$ if and if only if the S-box associated to x is active. The other variables in $\mathcal{X} \setminus \mathcal{S}$ represent intermediate state differences.

For a given state size of s 128-bit words, to express the problem of finding a differential characteristic, we examine the effect of the four elementary transformations of the AES round function. We emphasize that the analysis is performed in terms of truncated differences ($x \in \{0,1\}$) since we are only concerned about active or inactive S-boxes: the actual differences are insignificant. Therefore, as soon as one S-box is active, the SubBytes operation maintains this property. Hence, SubBytes does not introduce any linear constraints in the MILP problem. The same holds for the ShiftRows operation, which only permutes the bytes of the internal state.

However, the MixColumns operation implements a linear code with maximal distance (MDS), and it does introduce linear constraints in the MILP problem. Namely, the new inequalities enforce the minimal distance into the problem description. Assuming that the MixColumns operation is applied to the variables (representing truncated differences) $[x_i, x_{i+1}, x_{i+2}, x_{i+3}]$ to produce $[x'_i, x'_{i+1}, x'_{i+2}, x'_{i+3}]$, we introduce the nine following inequalities:

$$x_1 + x_2 + x_3 + x_4 + x'_1 + x'_2 + x'_3 + x'_4 - 5t \geq 0,$$

$$t - v \geq 0, \quad v \in \{x_1, x_2, x_3, x_4, x'_1, x'_2, x'_3, x'_4\}.$$

The usage of the extra temporary variable t ensures that the MDS bound is valid as soon as one of the x variables is nonzero (i.e. zero or at least five variables equal one).

Finally, the AddRoundKey operation XORs a 128-bit subkey into the state, which also introduces linear inequalities in the MILP problem description. Consider the XOR $y = x_1 \oplus x_2$ of two variables $x_1, x_2 \in \{0,1\}$ representing two truncated differences. In the event that $(x_1, x_2) = (0,0)$, y naturally becomes 0, and y becomes 1 if $(x_1, x_2) \in \{(0,1), (1,0)\}$. However, the behavior is undetermined when $(x_1, x_2) = (1,1)$ as y can either be 0 or 1 depending on the *actual* values of the corresponding differences. Indeed, because we lose information by compressing the differences to truncated differences, we lose the information on the possible equality of differences. Consequently, we have to consider both cases: $y \in \{0,1\}$. This partial behavior of XOR is captured by the four following inequalities:

$$x_1 + x_2 + y - 2t \geq 0,$$

$$t - v \geq 0, \quad v \in \{x_1, x_2, y\},$$

which basically excludes the case where only one of the three variables equals one.

In summary, for a single round of AES, we introduce $4 \times 9 + 16 \times 4 = 100$ inequalities to express the round constraints. On top of that, we introduce $16 \times 4 = 64$ additional inequalities for every extra XORs required to inject the message blocks. Finally, we also need to add $2 \times s \times 16$ equality constraints to represent the required zero difference in the input state and in the output state to reach a state collision. To give concrete numbers, we point out that systems corresponding to our smaller designs would need around 10,000 binary variables and 20,000 to 30,000 linear constraints.

Limitations. Despite providing a simple and efficient way of finding differential characteristics, MILP only yields upper bounds on the actual probabilities of the differential characteristics as, theoretically, they can be impossible. We emphasize that this does not relate to impossible differential characteristic, but to the fact that partially undetermined behavior of the XOR operation (mentioned before) may result in inconsistent systems that produce truncated differential characteristics which are impossible to instantiate with actual differences. Fortunately, while a cryptanalyst should ensure the validity of the produced characteristics, we, as designers, only need to confirm that the upper bound on the probability of the best differential characteristic is sufficiently low.

4.2   Results of the Search

In this section, we conduct the search for efficient designs and describe the results produced by the MILP analysis. In the next Sect. 5, we give the actual implementations and benchmarks of the produced designs.

Rate 3. We start the search with rate-3 designs and try to minimize the number s of state words. For a given state size s, the general structure depicted in Fig. 3 contains at most 12^s different designs. As the smallest possible size is $s = 3$, we efficiently exhaust all the 12^3 designs. In this reduced space, we have found that not a single design can reach 22 active S-boxes. Furthermore, for $s = 4$ state words, there exists secure constructions, albeit incompletely saturating the aesenc port for all the current processors[15], thus we do not consider them.

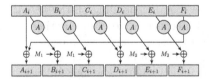

Fig. 5. Rate-3 design with 6 words.

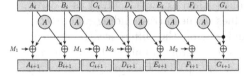

Fig. 6. Rate-3 design with 7 words.

---

[15] The design uses only 3 aesenc calls per round, whereas the smallest latency among all the processors is 4.

Having this objective in mind for rate-3 designs, we then move to step functions having either six calls to A (this saturates the Skylake aesenc port as aesenc has latency of 4) and inject two message blocks in each step, or nine calls to A (to saturate *bridge and *well aesenc ports) and inject three blocks. We find three different designs with state sizes of 6, 7 and 8 words, respectively, that are best suitable for Skylake. These designs achieve different security margins with lower bounds of 22, 25 and 34 active S-boxes, respectively (refer to Figs. 5, 6 and 7).

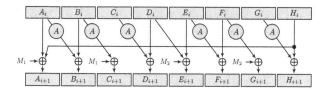

Fig. 7. Rate-3 design with 8 words.

For the case of nine calls to A (suitable as well for *bridge and *well architectures), we propose the design from Fig. 8 that reaches a minimum of 25 active S-boxes, has nine state words, and uses no additional XOR operations.

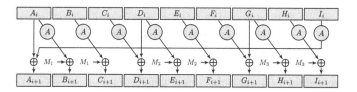

Fig. 8. Rate-3 design with 9 words.

Rate Smaller Than 3. To reach rates smaller than three, we first consider cases with two message blocks injected in every step. This restricts the number of A per round to five. We have performed a search within these restrictions and found constructions with seven and eight state words (see Figs. 9 and 10). The two design achieves rate $5/2 = 2.5$, have at least 22 and 23 active S-boxes, and saturate the aesenc port on Skylake processor.

Finally, we consider designs with rate of 2. We have not found a construction that injects two message blocks per step, however, we have discovered one that processes three message blocks per step (see Fig. 11). It has 12 state words, uses 6 aesenc to process 3 message blocks, and has at least 25 active S-boxes in any differential characteristic. Note, this construction compares very favorably to Tiaoxin-346: it is more compact and more efficient at the same time, since Tiaoxin-346 reaches rate 3 with 13 state words.

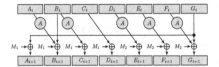

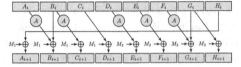

Fig. 9. Rate 2.5 with 7 state words.   **Fig. 10.** Rate 2.5 with 8 state words.

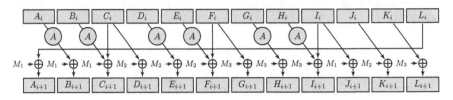

Fig. 11. Rate-2 design with 12 state words.

5   Implementations Results

We benchmark the seven constructions on the latest Intel's processors. The `aesenc` on some of these processors have similar performances (see Table 1), thus we benchmark on only three different platforms: Ivy Bridge (i5-3470) with Linux kernel 3.11.0-12 and gcc 4.8.1, Haswell (i5-4570) with Linux kernel 3.11.0-12 and gcc 4.8.1, and Skylake (i5-6200U) with Linux kernel 3.16.0-38 and gcc 4.8.4. We wrote the implementations in C and optimized them separately for each processor. The benchmarks were produced with disabled Turbo Boost and for 64kB messages[16].

The produced benchmarks are given in Table 3. Recall that our expected speed (expressed in c/B) is defined as $0.0625 \cdot \max(\rho, \frac{latency}{throughput})$. When the measured speed matches the expected (at most 5 % discrepancy), in the table we give the expected speed in **bold text**. On the other hand, when the measured speed is lower (resp. higher) than the expected, we give the actual speed with superscript $-$ (resp. $+$).

From the table, we can see that in most of the cases, our benchmarks follow the expected speed. For Ivy Bridge, the exceptions are the rate-3 design, which runs in 0.222 c/B instead of the expected 0.189 c/B (17 % slower), and the rate-2 design that runs at 0.190 c/B instead of 0.167 c/B (13 % slower). For Haswell, three designs run faster than expected, with gains of 15 %, 24 %, 22 %, respectively. On Skylake, the measured speed matches the expected speed for all seven constructions.

Among the seven constructions, we would like to single out the last constructions that has rate of 2, i.e. it uses two `AES` rounds to process a 16-byte message. On all of the three tested processors, this construction is exceptionally efficient. In addition, on Skylake, we were able to match the actual theoretical speed (our

---

[16] Only a slight degradation of speed is observed when the message length is a few kilobytes.

Table 3. Benchmarks (in c/B) of designs based on the AES round function. s: number of 128-bit state words, a: number of AES rounds in a single step, m: number of 128-bit message blocks processed per step, x number of additional XORs per step, ρ: rate of design (a/m), LB: lower bound on the number of active S-Boxes. Highlighted numbers means that the aesenc port is saturated for the given processor. Numbers in parentheses are projections, no actual measurements have been performed. Numbers in bold denotes that practical and theoretical speed match (less than 5 % difference), while numbers with + (resp. −) denote that the practical speed is higher (resp. lower) than the theoretical.

s	a	m	x	ρ	LB	Speed in c/B			Reference
						*bridge	*well	Skylake	
5	5	1	1	5	25	(0.500)	(0.436)	(0.313)	AEGIS-128 [19]
6	6	1	1	6	25	(0.500)	(0.436)	(0.375)	AEGIS-256 [19]
8	8	2	2	4	25	(0.250)	(0.250)	(0.250)	AEGIS-128L [19]
13	6	2	4	3	30	(0.250)	(0.219)	(0.188)	Tiaoxin-346 [17]
6	6	2	0	3	22	**0.250**	**0.219**	**0.188**	Figure 5
7	6	2	3	3	25	**0.250**	**0.219**	**0.188**	Figure 6
8	6	2	4	3	34	**0.250**	**0.219**	**0.188**	Figure 7
9	9	3	0	3	25	0.222⁻	**0.188**	**0.188**	Figure 8
7	5	2	4	2.5	22	**0.250**	0.189⁺	**0.156**	Figure 9
8	5	2	5	2.5	23	**0.250**	0.177⁺	**0.156**	Figure 10
12	6	3	9	2	28	0.190⁻	0.136⁺	**0.125**	Figure 11

measured speed was 0.126 c/B against the theoretical 0.125 c/B). Hence, designs based on this construction may run five times faster than AES-128.

We note that on platforms without AES-NI support our design cannot reach the target speed. However, by no means they are slow as they use only 2–3 AES rounds to process 16-byte message block. Hence, the expected speed on these platforms is still much higher than the speed of AES, e.g. we expect that our constructions will run around 3–5 times faster than AES-128 in counter mode.

In addition, the state sizes of the constructions are large hence they are not suitable for lightweight applications. However, we note that all seven constructions have sizes which are smaller than the state of SHA-3 which has 25 64-bit state words (equivalent to 12.5 128-bit blocks).

6   Conclusion

We have presented new building blocks for secret-key primitives based on the AES round function. By targeting the most recent Intel processors from the past four years, we have relied on the dedicated instruction set `AES-NI` to construct highly efficient designs. The designs are finely tuned for these processors to take advantage of the available parallelism and to reach optimal speed. They are based on the second, more efficient design strategy which requires a more complex security proof (reduction to MILP), but allows higher efficiency.

We have provided seven different building blocks that follow our design strategies and that reach high speed on the latest processors. On Ivy Bridge they run at 0.190–0.250 c/B, on Haswell at 0.136–0.219 c/B, while on Skylake at 0.125–0.188 c/B. We emphasize that our fastest construction uses only two `AES` rounds to process 16-byte message and on Skylake runs at only 0.125 c/B. To the best of our knowledge, this construction is much faster than any known cryptographic primitive.

Follow-up works to introduce better designs may start from two related directions: either by trying to reduce the state size, or by increasing the number of processed message in each step of the designs. The former might be useful to improve so designs that requires too many registers and slow down the whole process. The latter would automatically reduce the rate of the design and directly affect the measured speed. This direction is however difficult to tackle as the adversary has a lot more freedom to construct high-probability characteristics.

References

1. Biryukov, A.: The design of a stream cipher LEX. In: Biham, E., Youssef, A.M. (eds.) SAC 2006. LNCS, vol. 4356, pp. 67–75. Springer, Heidelberg (2007)
2. Bogdanov, A., Lauridsen, M.M., Tischhauser, E.: Comb to pipeline: fast software encryption revisited. In: Leander, G. (ed.) FSE 2015. LNCS, vol. 9054, pp. 150–171. Springer, Heidelberg (2015)
3. Bogdanov, A., Mendel, F., Regazzoni, F., Rijmen, V., Tischhauser, E.: ALE: AES-based lightweight authenticated encryption. In: Moriai, S. (ed.) FSE 2013. LNCS, vol. 8424, pp. 447–466. Springer, Heidelberg (2014)
4. Borghoff, J., Knudsen, L.R., Stolpe, M.: Bivium as a mixed-integer linear programming problem. In: Parker, M.G. (ed.) Cryptography and Coding 2009. LNCS, vol. 5921, pp. 133–152. Springer, Heidelberg (2009)
5. CAESAR. Competition for Authenticated Encryption: Security, Applicability, and Robustness. http://competitions.cr.yp.to/caesar.html
6. Daemen, J., Rijmen, V.: The Design of Rijndael: ALE - The Advanced Encryption Standard. Springer, Heidelberg (2002)
7. Daemen, J., Rijmen, V.: A new MAC construction ALRED and a specific instance ALPHA-MAC. In: Gilbert, H., Handschuh, H. (eds.) FSE 2005. LNCS, vol. 3557, pp. 1–17. Springer, Heidelberg (2005)
8. Daemen, J., Rijmen, V.: The MAC function Pelican 2.0. Cryptology ePrint Archive, report 2005/088 (2005)
9. Derbez, P., Fouque, P.-A., Jean, J.: Improved key recovery attacks on reduced-round `AES` in the single-key setting. In: Johansson, T., Nguyen, P.Q. (eds.) EUROCRYPT 2013. LNCS, vol. 7881, pp. 371–387. Springer, Heidelberg (2013)
10. Gurobi Optimization, Inc.: Gurobi Optimizer Reference Manual (2015)

11. Jakimoski, G., Khajuria, S.: ASC-1: an authenticated encryption stream cipher. In: Miri, A., Vaudenay, S. (eds.) SAC 2011. LNCS, vol. 7118, pp. 356–372. Springer, Heidelberg (2012)

12. Käsper, E., Schwabe, P.: Faster and timing-attack resistant AES-GCM. In: Clavier, C., Gaj, K. (eds.) CHES 2009. LNCS, vol. 5747, pp. 1–17. Springer, Heidelberg (2009)

13. Khovratovich, D., Rechberger, C.: The LOCAL attack: cryptanalysis of the authenticated encryption scheme ALE. In: Lange, T., Lauter, K., Lisoněk, P. (eds.) SAC 2013. LNCS, vol. 8282, pp. 174–184. Springer, Heidelberg (2014)

14. Li, L., Jia, K., Wang, X.: Improved single-key attacks on 9-round AES-192/256. In: Cid, C., Rechberger, C. (eds.) FSE 2014. LNCS, vol. 8540, pp. 127–146. Springer, Heidelberg (2015)

15. Mala, H., Dakhilalian, M., Rijmen, V., Modarres-Hashemi, M.: Improved Impossible differential cryptanalysis of 7-round AES-128. In: Gong, G., Gupta, K.C. (eds.) INDOCRYPT 2010. LNCS, vol. 6498, pp. 282–291. Springer, Heidelberg (2010)

16. Mouha, N., Wang, Q., Gu, D., Preneel, B.: Differential and linear cryptanalysis using mixed-integer linear programming. In: Wu, C.-K., Yung, M., Lin, D. (eds.) Inscrypt 2011. LNCS, vol. 7537, pp. 57–76. Springer, Heidelberg (2012)

17. Nikolić, I.: Tiaoxin-346. Submission to the CAESAR Competition (2014)

18. Sun, S., Hu, L., Wang, P., Qiao, K., Ma, X., Song, L.: Automatic security evaluation and (related-key) differential characteristic search: application to SIMON, PRESENT, LBlock, DES(L) and other bit-oriented block ciphers. In: Sarkar, P., Iwata, T. (eds.) ASIACRYPT 2014. LNCS, vol. 8873, pp. 158–178. Springer, Heidelberg (2014)

19. Wu, H., Preneel, B.: AEGIS: a fast authenticated encryption algorithm. Cryptology ePrint Archive, report 2013/695 (2013)

20. Wu, S., Wu, H., Huang, T., Wang, M., Wu, W.: Leaked-state-forgery attack against the authenticated encryption algorithm ALE. In: Sako, K., Sarkar, P. (eds.) ASIACRYPT 2013, Part I. LNCS, vol. 8269, pp. 377–404. Springer, Heidelberg (2013)

Block-Cipher Cryptanalysis

Bit-Based Division Property and Application to SIMON Family

Yosuke Todo[1,2](✉) and Masakatu Morii[2]

[1] NTT Secure Platform Laboratories, Tokyo, Japan
todo.yosuke@lab.ntt.co.jp
[2] Kobe University, Kobe, Japan

Abstract. Ciphers that do not use S-boxes have been discussed for the demand on lightweight cryptosystems, and their round functions consist of **and**, **rotation**, and **xor**. Especially, the SIMON family is one of the most famous ciphers, and there are many cryptanalyses again the SIMON family. However, it is very difficult to guarantee the security because we cannot use useful techniques for S-box-based ciphers. Very recently, the division property, which is a new technique to find integral characteristics, was shown in Eurocrypt 2015. The technique is powerful for S-box-based ciphers, and it was used to break, for the first time, the full MISTY1 in CRYPTO 2015. However, it has not been applied to non-S-box-based ciphers like the SIMON family effectively, and only the existence of the 10-round integral characteristic on SIMON32 was proven. On the other hand, the experimental characteristic, which possibly does not work for all keys, covers 15 rounds, and there is a 5-round gap. To fill the gap, we introduce a bit-based division property, and we apply it to show that the experimental 15-round integral characteristic always works for all keys. Though the bit-based division property finds more accurate integral characteristics, it requires much time and memory complexity. As a result, we cannot apply it to symmetric-key ciphers whose block length is over 32. Therefore, we alternatively propose a method for designers. The method works for ciphers with large block length, and it shows "provable security" against integral cryptanalyses using the division property. We apply this technique to the SIMON family and show that SIMON48, 64, 96, and 128 probably do not have 17-, 20-, 25-, and 29-round integral characteristics, respectively.

Keywords: Integral cryptanalysis · Division property · Provable security · SIMON family

1 Introduction

Non-S-box-based ciphers have been proposed for the demand on lightweight cryptosystems [2,3]. Such ciphers are superior in lightweight environments because they are implemented by logical operations and do not have a lookup table like S-boxes. In 2013, the NSA proposed a lightweight block cipher family,

© International Association for Cryptologic Research 2016
T. Peyrin (Ed.): FSE 2016, LNCS 9783, pp. 357–377, 2016.
DOI: 10.1007/978-3-662-52993-5_18

Table 1. Integral characteristics on SIMON32

Methods	#Rounds	Balanced bit (right half)	Reference
Experiment (no proof)	15	(?b??,????,b???,???b)	[18]
Division	10	(bbbb,bbbb,bbbb,bbbb)	[17]
Conventional bit-based division	14	(bbbb,bbbb,bbbb,bbbb)	Sect. 3
Bit-based division using 3 subsets	15	(?b??,????,b???,???b)	Sect. 4

called the SIMON family, that follows this design principle [3]. However, it is too difficult to guarantee the security against several cryptanalyses because we cannot use many useful techniques for S-box-based ciphers. Therefore, many cryptanalyses have been proposed against the SIMON family, e.g., [1,5,6,10,15,18], and the designers recently summarized cryptanalyses in [4]. In this paper, we investigate the security of non-S-box-based ciphers against integral cryptanalyses and illustrate our methods on the SIMON family.

Division Property. Very recently, the division property, which is a new technique to find integral characteristics [9], was proposed in Eurocrypt 2015 [17]. The new technique permitted us to find a 6-round integral characteristic on MISTY1 in CRYPTO 2015, leading to the first complete theoretical cryptanalysis of the full MISTY1 [16]. Moreover, this technique was applied to generalized Feistel structures in [20], leading to improved integral cryptanalyses against LBlock and TWINE. The division property also proves integral characteristics on the SIMON family in [17], and SIMON32, 48, 64, 96, and 128 have 9-, 11-, 11-, 13-, 13-round integral characteristics, respectively[1]. However, the round function is regarded as any function of degree 2. Therefore, we can expect that integral characteristics can be extended to more rounds if one is able to exploit the concrete structure of the round function. In fact, the experimental integral characteristic, which possibly does not work for all keys, covers 15 rounds [18], and there is a large gap between the proved characteristic and experimental one.

Our Contribution. The round function of the SIMON family is regarded as any function of degree 2 in [17] because we cannot decompose the round function into several sub blocks like S-boxes. However, we can decompose the round function into every bit, and we call the division property that focuses on every bit a *bit-based division property*.

First, we apply the conventional bit-based division property to SIMON32, which is not against the definition of the division property. Therefore, we can directly use the propagation rules of the division property. As a result, the conventional bit-based division property proves that SIMON32 has a 14-round integral characteristic. However, there is still a gap of one round between the

---

[1] Since the round key is XORed after the round function in SIMON, we can trivially get one-round extended integral characteristics.

Table 2. Provable secure number of rounds for the SIMON family

Ciphers	SIMON48	SIMON64	SIMON96	SIMON128	reference
Vulnerable number	14 rounds	17 rounds	21 rounds	25 rounds	[21]
Provable security	17 rounds	20 rounds	25 rounds	29 rounds	this paper

proof and experiment. Namely, this means that either the experimental 15-round characteristic does not work for all keys or the conventional bit-based division property cannot find the accurate characteristic. As a result, we conclude that the conventional bit-based division property is insufficient to find the accurate characteristic. The conventional division property divides the set of u according to whether the parity becomes 0 or unknown [17]. However, we should divide the set of u according to whether the parity becomes 0, 1, or unknown because we can also exploit the fact that the parity is not only 0 but also 1. To exploit this fact, we newly introduce a variant of the bit-based division property, which divides the set of u into three subsets. Since the variant is completely different from the definition of the conventional division property, we show the propagation characteristic also. Finally, we apply the variant to SIMON32 and show that the experimental 15-round characteristic always works for all keys. The proved characteristic is the completely same as the experimental one including the position of balanced bits. Table 1 shows the comparison of integral characteristics, where balanced and unknown bits are labeled as b and ?, respectively.

Although the bit-based division property can find more accurate integral characteristics, their propagations require much time and memory complexity. When we evaluate the propagation for n-bit block ciphers, it roughly requires 2^n complexity because the bit-based division property has to manage the set of n-dimensional vectors whose elements take values in $\mathbb{F}_2$. This is feasible for SIMON32 because the block length is 32 bits, but it is infeasible for other SIMON family members. Therefore, we introduce a new technique, which is useful for designers but is not useful for attackers. We call this technique a *lazy propagation*, where we evaluate only a part of all propagations. The lazy propagation cannot find the integral characteristic, but it can evaluate the number of rounds that the bit-based division property cannot find integral characteristics even if we can evaluate the accurate propagation. Namely, the technique shows "provable security" for the integral cryptanalysis using the division property, and we expect that it becomes a useful technique for designers. Our provable security guarantees the security against only the integral cryptanalysis using the division property, and it does not always guarantee the security against all integral-like cryptanalyses. However, for SIMON32, the bit-based division property can find the accurate integral characteristic. Therefore, we expect that it also finds the best integral characteristic for the other SIMON family if it is feasible. Table 2 shows the number of rounds of SIMON48, 64, 96, and 128, where the division property never finds integral characteristics. As a result, we expect that SIMON48, 64, 96, and 128 do not have 17-, 20-, 25-, and 29-round integral characteristics, respectively[2]. Moreover, as the comparison, Table 2 also shows the number of rounds that SIMON48, 64, 96, and 128 have integral characteristics [21].

2   Preliminaries

2.1   Notations

We make the distinction between the addition of $\mathbb{F}_2^n$ and addition of $\mathbb{Z}$, and we use $\oplus$ and $+$ as the addition of $\mathbb{F}_2^n$ and addition of $\mathbb{Z}$, respectively. For any $a \in \mathbb{F}_2^n$, the ith element is expressed in $a[i]$, and the Hamming weight $w(a)$ is calculated as $w(a) = \sum_{i=1}^n a[i]$. For any $\boldsymbol{a} \in (\mathbb{F}_2^{n_1} \times \mathbb{F}_2^{n_2} \times \cdots \times \mathbb{F}_2^{n_m})$, the vectorial Hamming weight of $\boldsymbol{a}$ is defined as $W(\boldsymbol{a}) = (w(a_1), w(a_2), \ldots, w(a_m)) \in \mathbb{Z}^m$. Moreover, for any $\boldsymbol{k} \in \mathbb{Z}^m$ and $\boldsymbol{k}' \in \mathbb{Z}^m$, we define $\boldsymbol{k} \succeq \boldsymbol{k}'$ if $k_i \geq k_i'$ for all i. Otherwise, $\boldsymbol{k} \not\succeq \boldsymbol{k}'$. In this paper, we often treat the set of $\boldsymbol{k}$, and $\mathbb{K}$ denotes this set. Then, let $|\mathbb{K}|$ be the number of vectors. We simply write $\mathbb{K} \leftarrow \boldsymbol{k}$ when $\mathbb{K} := \mathbb{K} \cup \{\boldsymbol{k}\}$. Moreover, we simply write $\mathbb{K} \xleftarrow{x} \boldsymbol{k}$, where the new $\mathbb{K}$ computed as

$$\mathbb{K} := \begin{cases} \mathbb{K} \cup \{\boldsymbol{k}\} & \text{if the original } \mathbb{K} \text{ does not include } \boldsymbol{k}, \\ \mathbb{K} \setminus \{\boldsymbol{k}\} & \text{if the original } \mathbb{K} \text{ includes } \boldsymbol{k}. \end{cases}$$

2.2   Integral Attack

The integral attack was first introduced by Daemen et al. to evaluate the security of SQUARE [7], and then it was formalized by Knudsen and Wagner [9]. Attackers first prepare N chosen plaintexts and encrypt them R rounds. If the XOR of all encrypted texts becomes 0, we say that the cipher has an R-round integral characteristic with N chosen plaintexts. Finally, we analyze the entire cipher by using the integral characteristic. Therefore, it is very important to find integral characteristic. There are two main approaches to find integral characteristics. The first one is the propagation of the integral property [9] and the second one is based on the degree estimation [8,11].

2.3   Division Property

The division property, which was proposed in [17], is a new method to find integral characteristics. This section briefly shows the definition and propagation rules. Please refer to [17] in detail.

---

[2] If we truly guarantee the security against integral attack, we have to consider the key recovery part.

Bit Product Function. The division property of a multiset is evaluated by using the bit product function defined as follows. Let $\pi_u : \mathbb{F}_2^n \rightarrow \mathbb{F}_2$ be a bit product function for any $u \in \mathbb{F}_2^n$. Let $x \in \mathbb{F}_2^n$ be the input, and $\pi_u(x)$ is the AND of $x[i]$ satisfying $u[i] = 1$, i.e., it is defined as

$$\pi_u(x) := \prod_{i=1}^{n} x[i]^{u[i]}.$$

Notice that $x[i]^1 = x[i]$ and $x[i]^0 = 1$. Let $\pi_{\boldsymbol{u}} : (\mathbb{F}_2^{n_1} \times \mathbb{F}_2^{n_2} \times \cdots \times \mathbb{F}_2^{n_m}) \rightarrow \mathbb{F}_2$ be a bit product function for any $\boldsymbol{u} \in (\mathbb{F}_2^{n_1} \times \mathbb{F}_2^{n_2} \times \cdots \times \mathbb{F}_2^{n_m})$. Let $\boldsymbol{x} \in (\mathbb{F}_2^{n_1} \times \mathbb{F}_2^{n_2} \times \cdots \times \mathbb{F}_2^{n_m})$ be the input, and $\pi_{\boldsymbol{u}}(\boldsymbol{x})$ is defined as

$$\pi_{\boldsymbol{u}}(\boldsymbol{x}) := \prod_{i=1}^{m} \pi_{u_i}(x_i).$$

The bit product function also appears in the Algebraic Normal Form (ANF) of a Boolean function. The ANF of a Boolean function f is represented as

$$f(x) = \bigoplus_{u \in \mathbb{F}_2^n} a_u^f \left(\prod_{i=1}^{n} x[i]^{u[i]} \right) = \bigoplus_{u \in \mathbb{F}_2^n} a_u^f \pi_u(x),$$

where $a_u^f \in \mathbb{F}_2$ is a constant value depending on f and u.

Definition of Division Property

Definition 1 (Division Property [17]). *Let $\mathbb{X}$ be a multiset whose elements take a value of $(\mathbb{F}_2^{n_1} \times \mathbb{F}_2^{n_2} \times \cdots \times \mathbb{F}_2^{n_m})$. When the multiset $\mathbb{X}$ has the division property $\mathcal{D}_{\mathbb{K}}^{n_1,n_2,\ldots,n_m}$, where $\mathbb{K}$ denotes a set of m-dimensional vectors whose ith element takes a value between 0 and n_i, it fulfils the following conditions:*

$$\bigoplus_{\boldsymbol{x} \in \mathbb{X}} \pi_{\boldsymbol{u}}(\boldsymbol{x}) = \begin{cases} unknown & \text{if there are } \boldsymbol{k} \in \mathbb{K} \text{ s.t. } W(\boldsymbol{u}) \succeq \boldsymbol{k}, \\ 0 & \text{otherwise.} \end{cases}$$

See [17] to better understand the concept in detail, and [14] and [16] help readers understand the division property. In this paper, the division property for $(\mathbb{F}_2^n)^m$ is referred to as $\mathcal{D}_{\mathbb{K}}^{n^m}$ for the simplicity[3]. If there are $\boldsymbol{k} \in \mathbb{K}$ and $\boldsymbol{k}' \in \mathbb{K}$ satisfying $\boldsymbol{k} \succeq \boldsymbol{k}'$ in the division property $\mathcal{D}_{\mathbb{K}}^{n_1,n_2,\ldots,n_m}$, $\boldsymbol{k}$ can be removed from $\mathbb{K}$ because the vector $\boldsymbol{k}$ is redundant.

Propagation Rules of Division Property. Some propagation rules for the division property are proven in [17], and the rules are summarized in [16] as follows.

---

[3] In [17], the division property was referred to as $\mathcal{D}_{\mathbb{K}}^{n,m}$.

Rule 1 (Substitution). Let F be a function that consists of m S-boxes, where the bit length and the algebraic degree of the ith S-box is n_i bits and d_i, respectively. The input and output take a value of $(\mathbb{F}_2^{n_1} \times \mathbb{F}_2^{n_2} \times \cdots \times \mathbb{F}_2^{n_m})$, and $\mathbb{X}$ and $\mathbb{Y}$ denote the input multiset and output multiset, respectively. Assuming that the multiset $\mathbb{X}$ has the division property $\mathcal{D}_{\mathbb{K}}^{n_1, n_2, \ldots, n_m}$, the division property of the multiset $\mathbb{Y}$ is $\mathcal{D}_{\mathbb{K}'}^{n_1, n_2, \ldots, n_m}$ as

$$\mathbb{K}' \leftarrow \left(\left\lceil \frac{k_1}{d_1} \right\rceil, \left\lceil \frac{k_2}{d_2} \right\rceil, \ldots, \left\lceil \frac{k_m}{d_m} \right\rceil \right), \quad \forall \boldsymbol{k} \in \mathbb{K}.$$

Here, when the ith S-box is bijective and $k_i = n_i$, the ith element of the propagated property becomes n_i not $\lceil n_i/d_i \rceil$.

Rule 2 (Copy). Let F be a copy function, where the input x takes a value of $\mathbb{F}_2^n$ and the output is calculated as $(y_1, y_2) = (x, x)$. Let $\mathbb{X}$ and $\mathbb{Y}$ be the input multiset and output multiset, respectively. Assuming that the multiset $\mathbb{X}$ has the division property $\mathcal{D}_k^n$, the division property of the multiset $\mathbb{Y}$ is $\mathcal{D}_{\mathbb{K}'}^{n,n}$ as

$$\mathbb{K}' \leftarrow (k - i, i), \quad \text{for } 0 \le i \le k.$$

Rule 3 (Compression by XOR). Let F be a function compressed by an XOR, where the input (x_1, x_2) takes a value of $(\mathbb{F}_2^n \times \mathbb{F}_2^n)$ and the output is calculated as $y = x_1 \oplus x_2$. Let $\mathbb{X}$ and $\mathbb{Y}$ be the input multiset and output multiset, respectively. Assuming that the multiset $\mathbb{X}$ has the division property $\mathcal{D}_{\mathbb{K}}^{n,n}$, the division property of the multiset $\mathbb{Y}$ is $\mathcal{D}_{k'}^n$ as

$$k' = \min_{(k_1, k_2) \in \mathbb{K}} \{k_1 + k_2\}.$$

Here, if the minimum value of k' is larger than n, the propagation characteristic of the division property is aborted. Namely, a value of $\oplus_{y \in \mathbb{Y}} \pi_v(y)$ is 0 for all $v \in \mathbb{F}_2^n$.

Rule 4 (Split). Let F be a split function, where the input x takes a value of $\mathbb{F}_2^n$ and the output is calculated as $x = y_1 \| y_2$, where (y_1, y_2) takes a value of $(\mathbb{F}_2^{n_1} \times \mathbb{F}_2^{n - n_1})$. Let $\mathbb{X}$ and $\mathbb{Y}$ be the input multiset and output multiset, respectively. Assuming that the multiset $\mathbb{X}$ has the division property $\mathcal{D}_k^n$, the division property of the multiset $\mathbb{Y}$ is $\mathcal{D}_{\mathbb{K}'}^{n_1, n - n_1}$ as

$$\mathbb{K}' \leftarrow (k - i, i), \quad \text{for } 0 \le i \le k.$$

Here, $(k - i)$ is less than or equal to n_1, and i is less than or equal to $n - n_1$.

Rule 5 (Concatenation). Let F be a concatenation function, where the input (x_1, x_2) takes a value of $(\mathbb{F}_2^{n_1} \times \mathbb{F}_2^{n_2})$ and the output is calculated as $y = x_1 \| x_2$. Let $\mathbb{X}$ and $\mathbb{Y}$ be the input multiset and output multiset, respectively. Assuming that the multiset $\mathbb{X}$ has the division property $\mathcal{D}_{\mathbb{K}}^{n_1, n_2}$, the division property of the multiset $\mathbb{Y}$ is $\mathcal{D}_{k'}^{n_1 + n_2}$ as

$$k' = \min_{(k_1, k_2) \in \mathbb{K}} \{k_1 + k_2\}.$$

2.4 Simon Family

The SIMON family is a lightweight block cipher family [3] based on the Feistel construction. Let SIMON$2n$ be the SIMON block ciphers with $2n$-bit block length, where n is chosen from 16, 24, 32, 48, and 64. Moreover, SIMON$2n$ with mn-bit secret key is referred to as SIMON$2n/mn$. Since we only care about integral characteristics on the SIMON family, this paper only uses SIMON$2n$.

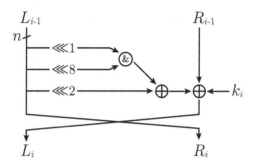

Fig. 1. Round function of SIMON$2n$

The output of the ith round function is denoted by (L_i, R_i) and is calculated as

$$(L_i, R_i) = (L_{i-1}^{\lll 1} \wedge L_{i-1}^{\lll 8}) \oplus L_{i-1}^{\lll 2} \oplus R_{i-1} \oplus k_i, L_{i-1}),$$

where $L^{\lll j}$ denotes the j-bit left rotation of L, and k_i denotes the ith round key. Moreover, (L_0, R_0) denotes a plaintext. The round function consists of and, rotation, and xor, and Fig. 1 shows the round function. For more details, please refer to [3].

2.5 Known Integral Characteristic on Simon Family

It is difficult to find effective integral characteristics on ciphers which consist of and, rotation, and xor. In [18], authors experimentally showed that SIMON32 has the 15-round integral characteristic with 2^{31} chosen plaintexts. Since their characteristic is confirmed under 2^{13} secret keys, they expected that the success probability of this characteristic is at least $1 - 2^{-13}$. Therefore, this approach does not guarantee that the characteristic works for all secret keys. Moreover, it is practically infeasible to find integral characteristics of other SIMON family members because the block length is too large for proceeding to an experimental evaluation.

Integral characteristics proved under all secret keys are shown in [17], but in this approach the round function of SIMON$2n$ is seen as any n-bit function of degree 2. Therefore, the detailed structure of the round function is not exploited.

As a result, it shows that SIMON32, 48, 64, 96, and 128 has 9-, 11-, 11-, 13-, and 13-round integral characteristic, respectively. Since the round key is XORed after the round function, we can trivially get one-round extended integral characteristics using the same technique in [18]. Therefore, 10-, 12-, 12-, 14-, and 14-round integral characteristics are proved in SIMON32, 48, 64, 96, and 128, respectively. Thus, there is a 5-round gap between the proved characteristic and experimental one.

3   Conventional Bit-Based Division Property

This paper introduces a *bit-based division property*. When n-bit block ciphers are analyzed, the conventional division property uses $\mathcal{D}_{\mathbb{K}}^{\ell_1, \ell_2, \ldots, \ell_m}$, where ℓ_i and m are chosen by attackers in the range of $n = \sum_{i=1}^{m} \ell_i$. This section considers the conventional bit-based division property, i.e., $\mathcal{D}_{\mathbb{K}}^{1^n}$. Since it is not against the definition of the conventional division property, we can directly use the five propagation rules shown in Sect. 2.3.

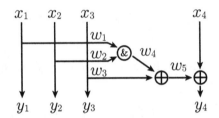

Fig. 2. Core operation of the SIMON family.

3.1   Comparison Between Conventional Bit-Based Division Property and Solving Algebraic Equations

Before the introduction of the conventional bit-based division property, we roughly show the relation between the bit-based division property and the resolution of algebraic equations by brute force. When entire ciphers are represented by algebraic equations, such equations involve both the plaintext and secret key. Therefore, if we solve such equations for an n-bit block cipher with a k-bit secret key, this roughly requires 2^{k+n} complexity. On the other hand, XORing with a constant value does not change the conventional bit-based division property because such XORing is a linear function [16]. Therefore, the propagation of the conventional bit-based division property does not involve the secret key. It may miss some useful cryptographic properties, but it dramatically reduces the complexity.

3.2   Propagation for Core Operation of Simon

As an example, we analyze SIMON$2n$ by using the conventional bit-based division property. We focus on only one bit of the right half in SIMON$2n$. The core operation of the round function is represented by Fig. 2. Since the input and output bit length is 4 bits, we use the division property $\mathcal{D}_{\mathbb{K}}^{1^4}$.

We consider the propagation characteristic. For instance, let assume that the input multiset has $\mathcal{D}_{[k_1,k_2,k_3,1]}^{1^4}$, where k_i denotes any value, i.e., 0 or 1. Then, if the multiset of $(y_1, y_2, y_3, w_5, x_4)$ has $\mathcal{D}_{[*,*,*,1,1]}^{1^5}$, where $*$ is propagated values, the propagation always abort in the XOR, $x_4 \oplus w_5$. Consequently, the bit-based division property of (y_1, y_2, y_3, y_4) is the same as that of (x_1, x_2, x_3, x_4). On the other hand, assuming that the input multiset has $\mathcal{D}_{[k_1,k_2,k_3,0]}^{1^4}$, the output property is different from the input one.

Let $\mathcal{D}_{\mathbb{K}}^{1^4}$ and $\mathcal{D}_{\mathbb{K}'}^{1^4}$ be the division property of the input and output, respectively. When we get $\mathbb{K}'$ from $\mathbb{K}$, we first independently calculate vectors belonging to $\mathbb{K}'$ by evaluating the propagation from every vector in $\mathbb{K}$. Then, $\mathbb{K}'$ is represented as the union of all calculated vectors. Finally, if there are $\boldsymbol{k} \in \mathbb{K}'$ and $\boldsymbol{k}' \in \mathbb{K}'$ such that $\boldsymbol{k} \succeq \boldsymbol{k}'$, $\boldsymbol{k}$ is removed from $\mathbb{K}'$ because the vector is redundant.

Table 3 summarizes the propagation characteristics from $\mathcal{D}_{\boldsymbol{k}}^{1^4}$ to $\mathcal{D}_{\mathbb{K}}^{1^4}$. The round function of SIMON$2n$ repeats the core operation for all n-bit values in the right half. Therefore, we use $\mathcal{D}_{\mathbb{K}}^{1^{2n}}$. In every core operation, we only focus on four bits and evaluate the propagation independent of other $(2n - 4)$ bits.

Table 3. Propagation of the conventional bit-based division property for the core operation in the SIMON family

Input $\mathcal{D}_{\boldsymbol{k}}^{1^4}$	Output $\mathcal{D}_{\mathbb{K}}^{1^4}$
$\boldsymbol{k} = [0,0,0,0]$	$\mathbb{K} = \{[0,0,0,0]\}$
$\boldsymbol{k} = [1,0,0,0]$	$\mathbb{K} = \{[1,0,0,0], [0,0,0,1]\}$
$\boldsymbol{k} = [0,1,0,0]$	$\mathbb{K} = \{[0,1,0,0], [0,0,0,1]\}$
$\boldsymbol{k} = [1,1,0,0]$	$\mathbb{K} = \{[1,1,0,0], [0,0,0,1]\}$
$\boldsymbol{k} = [0,0,1,0]$	$\mathbb{K} = \{[0,0,1,0], [0,0,0,1]\}$
$\boldsymbol{k} = [1,0,1,0]$	$\mathbb{K} = \{[1,0,1,0], [0,0,1,1], [1,0,0,1]\}$
$\boldsymbol{k} = [0,1,1,0]$	$\mathbb{K} = \{[0,1,1,0], [0,0,1,1], [0,1,0,1]\}$
$\boldsymbol{k} = [1,1,1,0]$	$\mathbb{K} = \{[1,1,1,0], [0,0,1,1], [1,1,0,1]\}$
$\boldsymbol{k} = [k_1,k_2,k_3,1]$	$\mathbb{K} = \{[k_1,k_2,k_3,1]\}$

Table 4. Size of $\mathbb{K}$ in $\mathcal{D}_{\mathbb{K}}^{1^{32}}$ for the integral characteristic on SIMON32

Round	0	1	2	3	4	5	6	7	8	9	10	11	12	13	14
$|\mathbb{K}|$	1	1	3	11	65	774	18165	587692	5191387	1595164	95768	5894	682	136	32

3.3   Application to Simon32

We evaluate the propagation characteristic of the conventional bit-based division property on SIMON32. We prepare chosen plaintexts such that the first bit is constant and the others are active. Then, the set of chosen plaintexts has the division property $\mathcal{D}_{\mathbb{K}}^{1^{32}}$, where $\mathbb{K} = \{[0, 1, 1, \ldots, 1]\}$. Table 4 shows $|\mathbb{K}|$, which is the number of vectors, in every round, where we perfectly remove redundant vectors from $\mathbb{K}$. The output of the 14th round function has the division property $\mathcal{D}_{\mathbb{K}}^{1^{32}}$, where $\mathbb{K}$ has 32 distinct vectors whose Hamming weight is one. Therefore, the conventional bit-based division property cannot show whether or not the output of the 14th round function is balanced. On the other hand, the output of the 13th round function has the division property $\mathcal{D}_{\mathbb{K}}^{1^{32}}$, where $\mathbb{K}$ is represented as 16 vectors, whose Hamming weight of the left half is 1 and that of the right half is 0, and 120 ($= \binom{16}{2}$) vectors, whose Hamming weight of the left half is 0 and that of the right half is 2. This division property means that the output of the 13th round function takes the following integral property

$$(????,????,????,????, \mathbf{bbbb}, \mathbf{bbbb}, \mathbf{bbbb}, \mathbf{bbbb}),$$

where balanced and unknown bits are labeled as b and ?, respectively. In the SIMON family, since round keys are XORed with the right half only after the round function is applied to the left half, we can easily get a 14-round integral characteristic from the 13-round one. The same technique is used in [18]. Therefore, we conclude that 14-round SIMON32 has the integral characteristic with 2^{31} chosen plaintexts.

4   Bit-Based Division Property Using Three Subsets

4.1   Motivation

The conventional bit-based division property proved the existence of the 14-round integral characteristic of SIMON32. However, the experimental characteristic covers 15 rounds [18], and there is still a one-round gap between the experiment and proof. In [18], the authors experimentally confirm the characteristic by randomly choosing 2^{13} secret keys. Therefore, they concluded that the success probability of the characteristic is at least $1 - 2^{-13}$. Thus, we consider that this gap derives from either the experimental result does not work for all keys or the conventional bit-based division property cannot find the accurate characteristic.

We first show that the conventional bit-based division property is insufficient to find integral characteristics on SIMON32, and we then introduce a new variant of the bit-based division property. The conventional bit-based division property focuses on that the parity $\bigoplus_{x \in \mathbb{X}} \pi_u(x)$ is 0 or unknown. On the other hand, the new variant focuses on that the parity $\bigoplus_{x \in \mathbb{X}} \pi_u(x)$ is 0, 1, or unknown. Therefore we call the new variant *the bit-based division property using three subsets*. The new variant can find more accurate integral characteristics and prove that the experimental characteristic shown in [18] works for all keys.

4.2 Characteristic that Conventional Bit-Based Division Property Cannot Find

The conventional division property divides the set of $\boldsymbol{u}$ according to whether the parity becomes 0 or unknown [17]. However, it sometimes overlooks useful characteristics. We show it by using a simple example.

We again evaluate the propagation of the conventional bit-based division property for the circuit in Fig. 2, and $F : \mathbb{F}_2^4 \to \mathbb{F}_2^4$ denotes the circuit. Moreover, let $\mathbb{X}$ and $\mathbb{Y}$ be the input and output multiset, respectively. Assuming that $\mathbb{X}$ has $\mathcal{D}_{\{[1,1,0,0],[0,0,1,0]\}}^{1^4}$, $\bigoplus_{\boldsymbol{x} \in \mathbb{X}} \pi_{[1,1,0,0]}(\boldsymbol{x})$ and $\bigoplus_{\boldsymbol{x} \in \mathbb{X}} \pi_{[0,0,1,0]}(\boldsymbol{x})$ are unknown. Then, the output multiset $\mathbb{Y}$ has $\mathcal{D}_{\{[1,1,0,0],[0,0,1,0],[0,0,0,1]\}}^{1^4}$ from Table 3.

Let us assume that both $\bigoplus_{\boldsymbol{x} \in \mathbb{X}} \pi_{[1,1,0,0]}(\boldsymbol{x})$ and $\bigoplus_{\boldsymbol{x} \in \mathbb{X}} \pi_{[0,0,1,0]}(\boldsymbol{x})$ are 1. Even if we know the parity is always one, the division property of $\mathbb{X}$ is $\mathcal{D}_{\{[1,1,0,0],[0,0,1,0]\}}^{1^4}$. However, we can get the following equation.

$$\bigoplus_{\boldsymbol{x} \in \mathbb{X}} \pi_{[0,0,0,1]}(F(\boldsymbol{x})) = \bigoplus_{\boldsymbol{x} \in \mathbb{X}} (x_1 x_2 \oplus x_3 \oplus x_4)$$

$$= \bigoplus_{\boldsymbol{x} \in \mathbb{X}} (x_1 x_2) \bigoplus_{\boldsymbol{x} \in \mathbb{X}} (x_3) \bigoplus_{\boldsymbol{x} \in \mathbb{X}} (x_4)$$

$$= \bigoplus_{\boldsymbol{x} \in \mathbb{X}} \pi_{[1,1,0,0]}(\boldsymbol{x}) \bigoplus_{\boldsymbol{x} \in \mathbb{X}} \pi_{[0,0,1,0]}(\boldsymbol{x}) \bigoplus_{\boldsymbol{x} \in \mathbb{X}} \pi_{[0,0,0,1]}(\boldsymbol{x})$$

$$= 1 \oplus 1 \oplus 0 = 0.$$

Therefore, $\bigoplus_{\boldsymbol{x} \in \mathbb{X}} \pi_{[0,0,0,1]}(F(\boldsymbol{x}))$ is always 0 not unknown, and the division property of $\mathbb{Y}$ becomes $\mathcal{D}_{\{[1,1,0,0],[0,0,1,0],[0,1,0,1],[1,0,0,1]\}}^{1^4}$ not $\mathcal{D}_{\{[1,1,0,0],[0,0,1,0],[0,0,0,1]\}}^{1^4}$.

Since the conventional division property focuses on the case the parity becomes 0, it cannot find characteristics that appear by cancelling like the above example. Therefore, we newly introduce a variant of the bit-based division property to exploit this fact. The variant divides the set of $\boldsymbol{u}$ into three subsets, i.e., 0, 1, and unknown.

4.3 Definition of Bit-Based Division Property Using Three Subsets

The conventional division property uses the set $\mathbb{K}$ to represent the subset of $\boldsymbol{u}$ such that $\bigoplus_{\boldsymbol{x} \in \mathbb{X}} \pi_{\boldsymbol{u}}(\boldsymbol{x})$ is unknown. The bit-based division property using three subsets needs to represent not only the subset of $\boldsymbol{u}$ such that $\bigoplus_{\boldsymbol{x} \in \mathbb{X}} \pi_{\boldsymbol{u}}(\boldsymbol{x})$ is unknown but also the subset of $\boldsymbol{u}$ such that $\bigoplus_{\boldsymbol{x} \in \mathbb{X}} \pi_{\boldsymbol{u}}(\boldsymbol{x})$ is one. Therefore, we use the set $\mathbb{K}$ to represent the subset of $\boldsymbol{u}$ such that $\bigoplus_{\boldsymbol{x} \in \mathbb{X}} \pi_{\boldsymbol{u}}(\boldsymbol{x})$ is unknown, and we also use the set $\mathbb{L}$ to represent the subset of $\boldsymbol{u}$ such that $\bigoplus_{\boldsymbol{x} \in \mathbb{X}} \pi_{\boldsymbol{u}}(\boldsymbol{x})$ is one.

Definition 2 (Bit-based Division Property Using Three Subsets). *Let $\mathbb{X}$ be a multiset whose elements take a value of $(\mathbb{F}_2)^m$, and $\boldsymbol{k}$ is an m-dimensional vector whose ith element takes 0 or 1. When the multiset $\mathbb{X}$ has the bit-based division property using three subsets $\mathcal{D}_{\mathbb{K}, \mathbb{L}}^{1^m}$, it fulfils the following conditions:*

$$\bigoplus_{x \in X} \pi_u(x) = \begin{cases} \text{unknown} & \textit{if there are } k \in \mathbb{K} \textit{ s.t. } W(u) \succeq k, \\ 1 & \textit{else if there is } \ell \in \mathbb{L} \textit{ s.t. } W(u) = \ell, \\ 0 & \textit{otherwise.} \end{cases}$$

If there are $k \in \mathbb{K}$ and $k' \in \mathbb{K}$ satisfying $k \succeq k'$, k can be removed from $\mathbb{K}$ because the vector k is redundant. Moreover, when there is $k \in \mathbb{K}$ satisfying $W(u) \succeq k$, $\bigoplus_{x \in X} \pi_u(x)$ is unknown even if there is $\ell \in \mathbb{L}$ satisfying $W(u) = \ell$. Therefore, if there are $\ell \in \mathbb{L}$ and $k \in \mathbb{K}$ satisfying $\ell \succeq k$, the vector ℓ is redundant. Notice that redundant vectors in $\mathbb{K}$ and $\mathbb{L}$ do not affect whether $\bigoplus_{x \in X} \pi_u(x)$ becomes 0, 1, or unknown for any u.

Example 1. Let $\mathbb{X}$ be a multiset whose elements take a value of $(\mathbb{F}_2)^4$. Assume the multiset $\mathbb{X}$ has the bit-based division property $\mathcal{D}_{\mathbb{K},\mathbb{L}}^{1^4}$, where $\mathbb{K} = \{[0,0,0,1],[0,1,1,0]\}$ and $\mathbb{L} = \{[1,0,0,0],[1,0,1,0],[0,0,1,0],[0,0,1,1]\}$. Then, every parity satisfies the following, where the value of u is represented as hexadecimal notation of $(u_1\|u_2\|u_3\|u_4)$.

u	0×0	0×1	0×2	0×3	0×4	0×5	0×6	0×7	0×8	0×9	0xA	0xB	0xC	0xD	0xE	0xF
Parity	0	?	1	?	0	?	?	?	1	?	1	?	0	?	?	?

Notice that the parity of $\pi_{[0,0,1,1]}(x)$ over all $x \in \mathbb{X}$ is unknown because there is $[0,0,0,1] \in \mathbb{K}$ and $W([0,0,1,1]) \succeq W([0,0,0,1])$. Thus, $[0,0,1,1] \in \mathbb{L}$ is redundant.

4.4   Propagation Rules

We show propagation rules for the bit-based division property using three subsets. There rules are very similar to those of the conventional division property. Here, we show three rules, "Copy," "Compression by AND," and "Compression by XOR," because any Boolean function can be evaluated by using these three rules. We omit the proof of three propagation rules in this paper because of the page limit, and please see the full version of this paper.

Rule 1 (Copy). Let F be a copy function, where the input $(x_1, x_2, \ldots, x_m)$ takes values of $(\mathbb{F}_2)^m$, and the output is calculated as $(x_1, x_1, x_2, x_3, \ldots, x_m)$. Let $\mathbb{X}$ and $\mathbb{Y}$ be the input multiset and output multiset, respectively. Assuming that $\mathbb{X}$ has $\mathcal{D}_{\mathbb{K},\mathbb{L}}^{1^m}$, $\mathbb{Y}$ has $\mathcal{D}_{\mathbb{K}',\mathbb{L}'}^{1^{m+1}}$, where $\mathbb{K}'$ and $\mathbb{L}'$ are computed as

$$\mathbb{K}' \leftarrow \begin{cases} (0,0,k_2,\ldots,k_m), & \text{if } k_1 = 0 \\ (1,0,k_2,\ldots,k_m),(0,1,k_2,\ldots,k_m), & \text{if } k_1 = 1 \end{cases},$$

$$\mathbb{L}' \leftarrow \begin{cases} (0,0,\ell_2,\ldots,\ell_m), & \text{if } \ell_1 = 0 \\ (1,0,\ell_2,\ldots,\ell_m),(0,1,\ell_2,\ldots,\ell_m),(1,1,\ell_2,\ldots,\ell_m) & \text{if } \ell_1 = 1 \end{cases}.$$

from all $k \in \mathbb{K}$ and all $\ell \in \mathbb{L}$, respectively.

Rule 2 (Compression by AND). Let F be a function compressed by an AND, where the input $(x_1, x_2, \ldots, x_m)$ takes values of $(\mathbb{F}_2)^m$, and the output is calculated as $(x_1 \wedge x_2, x_3, \ldots, x_m)$. Let $\mathbb{X}$ and $\mathbb{Y}$ be the input multiset and output multiset, respectively. Assuming that $\mathbb{X}$ has $\mathcal{D}_{\mathbb{K},\mathbb{L}}^{1^m}$, $\mathbb{Y}$ has $\mathcal{D}_{\mathbb{K}',\mathbb{L}'}^{1^{m-1}}$, where $\mathbb{K}'$ is computed from all $\boldsymbol{k} \in \mathbb{K}$ as

$$\mathbb{K}' \leftarrow \left(\left\lceil \frac{k_1 + k_2}{2} \right\rceil, k_3, k_4, \ldots, k_m \right).$$

Moreover, $\mathbb{L}'$ is computed from all $\boldsymbol{\ell} \in \mathbb{L}$ s.t. $(\ell_1, \ell_2) = (0, 0)$ or $(1, 1)$ as

$$\mathbb{L}' \leftarrow \left(\left\lceil \frac{\ell_1 + \ell_2}{2} \right\rceil, \ell_3, \ell_4, \ldots, \ell_m \right).$$

Rule 3 (Compression by XOR). Let F be a function compressed by an XOR, where the input $(x_1, x_2, \ldots, x_m)$ takes values of $(\mathbb{F}_2)^m$, and the output is calculated as $(x_1 \oplus x_2, x_3, \ldots, x_m)$. Let $\mathbb{X}$ and $\mathbb{Y}$ be the input multiset and output multiset, respectively. Assuming that $\mathbb{X}$ has $\mathcal{D}_{\mathbb{K},\mathbb{L}}^{1^m}$, $\mathbb{Y}$ has $\mathcal{D}_{\mathbb{K}',\mathbb{L}'}^{1^{m-1}}$, where $\mathbb{K}'$ is computed from all $\boldsymbol{k} \in \mathbb{K}$ s.t. $(k_1, k_2) = (0, 0)$, $(1, 0)$, or $(0, 1)$ as

$$\mathbb{K}' \leftarrow (k_1 + k_2, k_3, k_4, \ldots, k_m).$$

Moreover, $\mathbb{L}'$ is computed from all $\boldsymbol{\ell} \in \mathbb{L}$ s.t. $(\ell_1, \ell_2) = (0, 0)$, $(1, 0)$, or $(0, 1)$ as

$$\mathbb{L}' \xleftarrow{\text{x}} (\ell_1 + \ell_2, \ell_3, \ell_4, \ldots, \ell_m).$$

4.5 Dependencies Between $\mathbb{K}$ and $\mathbb{L}$

Propagation for Public Function. In the propagation rules shown in Sect. 4.4, $\mathbb{K}'$ and $\mathbb{L}'$ are computed from $\mathbb{K}$ and $\mathbb{L}$, respectively. Therefore, we can evaluate the propagation from $\mathbb{K}$ and that from $\mathbb{L}$ independently. However, independent propagations generate many redundant vectors in $\mathbb{K}'$ and $\mathbb{L}'$. Note that redundant vectors in $\mathbb{K}'$ and $\mathbb{L}'$ do not affect whether the parity becomes 0, 1, or unknown for any $\boldsymbol{u}$. Therefore, when we consider the propagation for public functions, we do not need to care about the dependencies between $\mathbb{K}$ and $\mathbb{L}$. On the other hand, if there are many redundant vectors, the propagation requires much time complexity. Therefore, we should remove redundant vectors if possible because of the reason of only complexity.

XORing with Secret Round Key. For the public function, the propagation from $\mathbb{K}$ and that from $\mathbb{L}$ are independently evaluated. However, if the secret round key is XORed, every vector in $\mathbb{L}$ affects $\mathbb{K}$.

Let $\mathbb{X}$ and $\mathbb{Y}$ be the input and output multiset, respectively. Then, $\boldsymbol{y} \in \mathbb{Y}$ is computed as $\boldsymbol{y} = \boldsymbol{x} \oplus \boldsymbol{rk}$ for $\boldsymbol{x} \in \mathbb{X}$, where $\boldsymbol{rk}$ is the secret round key. Moreover, let $\mathcal{D}_{\mathbb{K},\mathbb{L}}^{1^m}$ and $\mathcal{D}_{\mathbb{K}',\mathbb{L}'}^{1^m}$ be the bit-based division property using three subsets on $\mathbb{X}$

and $\mathbb{Y}$, respectively. We want to get $\mathbb{K}'$ and $\mathbb{L}'$ from $\mathbb{K}$ and $\mathbb{L}$. We cannot know the secret round key. Therefore, the parity $\bigoplus_{x \in \mathbb{X}} \pi_v(x \oplus rk)$ satisfying $v \succ \ell$ becomes unknown because the parity depends on the secret round key.

In many ciphers, round keys are XORed with a part of entire bits. Assuming a round key is XORed with the ith bit, $\mathbb{K}'$ is computed as

$$\mathbb{K}' \leftarrow (\ell_1, \ell_2, \ldots, \ell_i \vee 1, \ldots, \ell_m)$$

for all $\ell \in \mathbb{L}$ satisfying $\ell_i = 0$.

4.6   Propagation for Core Operation of Simon

We search for integral characteristics on SIMON32 by the bit-based division property using three subsets. Similar to the conventional bit-based division property, we focus on only one bit of the right half and consider the core operation of the SIMON family (see Fig. 2).

The core operation is a public function and it does not involve any secret information. Therefore, we can evaluate the propagation from $\mathbb{K}$ and that from $\mathbb{L}$ independently. Table 5 summarizes the propagation characteristics from $\mathcal{D}^{1^4}_{\mathbb{K}, \{\ell\}}$ to $\mathcal{D}^{1^4}_{\mathbb{K}', \mathbb{L}'}$, where the propagation from $\mathbb{K}$ to $\mathbb{K}'$ is the same as that in Table 3. Next, the propagation on the round function can be evaluated by repeating for all bits of the right half. Finally, when round keys are XORed with the right half, new vectors are generated from $\mathbb{L}$, and the new vectors are inserted into $\mathbb{K}$.

Table 5. Propagation of the bit-based division property using three subsets for the core operation in the SIMON family

Input $\mathcal{D}^{1^4}_{\mathbb{K}, \{\ell\}}$	Output $\mathcal{D}^{1^4}_{\mathbb{K}', \mathbb{L}'}$
$\ell = [0,0,0,0]$	$\mathbb{L}' = \{[0,0,0,0]\}$
$\ell = [1,0,0,0]$	$\mathbb{L}' = \{[1,0,0,0]\}$
$\ell = [0,1,0,0]$	$\mathbb{L}' = \{[0,1,0,0]\}$
$\ell = [1,1,0,0]$	$\mathbb{L}' = \{[1,1,0,0],[0,0,0,1],[1,0,0,1],[0,1,0,1],[1,1,0,1]\}$
$\ell = [0,0,1,0]$	$\mathbb{L}' = \{[0,0,1,0],[0,0,0,1],[0,0,1,1]\}$
$\ell = [1,0,1,0]$	$\mathbb{L}' = \{[1,0,1,0],[1,0,0,1],[1,0,1,1]\}$
$\ell = [0,1,1,0]$	$\mathbb{L}' = \{[0,1,1,0],[0,1,0,1],[0,1,1,1]\}$
$\ell = [1,1,1,0]$	$\mathbb{L}' = \{[1,1,1,0],[0,0,1,1],[1,0,1,1],[0,1,1,1],[1,1,0,1]\}$
$\ell = [\ell_1, \ell_2, \ell_3, 1]$	$\mathbb{L}' = \{[\ell_1, \ell_2, \ell_3, 1]\}$

Table 6. Sizes of $\mathbb{K}$ and $\mathbb{L}$ in $\mathcal{D}^{1^{32}}_{\mathbb{K}, \mathbb{L}}$ for the integral characteristic on SIMON32

Round	0	1	2	3	4	5	6	7	8	9	10	11	12	13	14	15
|L|		1	1	5	19	138	2236	89878	4485379	47149981	2453101	20360	168	8	0	0	0
|K|		1	1	1	6	43	722	23321	996837	9849735	2524718	130724	7483	852	181	32	32

4.7 Application to Simon32

We evaluate the propagation characteristic of the bit-based division property using three subsets on SIMON32. We prepare chosen plaintexts such that the first bit is constant and the others are active, and the set of chosen plaintexts has $\mathcal{D}^{1^{32}}_{\{[1,1,1,\dots,1]\},\{[0,1,1,\dots,1]\}}$.

Table 6 shows $|\mathbb{K}|$ and $|\mathbb{L}|$ in every round, where we perfectly remove redundant vectors from $\mathbb{K}$ and $\mathbb{L}$. As a result, the output of the 14th round function has $\mathcal{D}^{1^{32}}_{\mathbb{K},\phi}$, where vector in $\mathbb{K}$ are represented by hexadecimal notation as

```
(0001 0000)(0002 0000)(0004 0000)(0008 0000)(0010 0000)(0020 0000)(0040 0000)(0080 0000)
(0100 0000)(0200 0000)(0400 0000)(0800 0000)(1000 0000)(2000 0000)(4000 0000)(8000 0000)
(0000 0002)(0000 0004)(0000 0008)(0000 0010)(0000 0020)(0000 0040)(0000 0081)(0000 0100)
(0000 0200)(0000 0400)(0000 0800)(0000 1000)(0000 2000)(0000 4001)(0000 4080)(0000 8000),
```

and ϕ denotes the empty set. This division property means that the output of the 14th round function takes the following integral property

$$(????,????,????,????,\ ?b??,????,b???,???b),$$

where balanced and unknown bits are labeled as b and ?, respectively. In the SIMON family, we can easily get a 15-round integral characteristic from the 14-round one, and this proved integral characteristic is completely the same as the experimental one. Therefore, we conclude that the experimental characteristic is not probabilistic characteristic, and it works for all keys.

4.8 Application to Simeck32

Simeck was recently proposed in [19], and its round function is very similar to that of SIMON. Let (L_i, R_i) be the output of the ith round function, and it is calculated as

$$(L_i, R_i) = (L_{i-1} \wedge L_{i-1}^{\lll 5}) \oplus L_{i-1}^{\lll 1} \oplus R_{i-1} \oplus k_i, L_{i-1}).$$

The rotation number is changed from $(1, 8, 2)$ to $(0, 5, 1)$. Similar to SIMON, Simeck has different parameters according to the block length. Let Simeck$2n$ be the Simeck block ciphers with $2n$-bit block length, where n is chosen from 16, 24, and 32.

We also evaluated the propagation of the bit-based division property using three subsets against Simeck32. As a result, the output of the 14th round function has $\mathcal{D}^{1^{32}}_{\mathbb{K},\phi}$, where vectors in $\mathbb{K}$ are represented by hexadecimal notation as

```
(0001 0000)(0002 0000)(0004 0000)(0008 0000)(0010 0000)(0020 0000)(0040 0000)(0080 0000)
(0100 0000)(0200 0000)(0400 0000)(0800 0000)(1000 0000)(2000 0000)(4000 0000)(8000 0000)
(0000 0002)(0000 0004)(0000 0008)(0000 0011)(0000 0021)(0000 0030)(0000 0040)(0000 0080)
(0000 0100)(0000 0201)(0000 0210)(0000 0220)(0000 0401)(0000 0410)(0000 0420)(0000 0600)
(0000 0800)(0000 1000)(0000 2000)(0000 4001)(0000 4010)(0000 4020)(0000 4200)(0000 4400)
(0000 8001)(0000 8010)(0000 8020)(0000 8200)(0000 8400)(0000 C000).
```

This division property means that the output of the 14th round function takes the following integral property

$$(????,????,????,????,\ bb??,?bb?,??bb,???b).$$

Since round keys are XORed after the round function in Simeck, we can trivially get the 15-round integral characteristic. Here, 2^{31} plaintexts are chosen as $(L_0, F(L_0) \oplus R_0)$, where the first bit of R_0 is constant and the others are active.

5   Provable Security Against Integral Cryptanalysis

We introduced the bit-based division property using three subsets in Sect. 4, and we proved that this method can find more accurate integral characteristics than those found by the conventional division property. In particular, we showed that the new method can discover the tight characteristic on SIMON32. However, a problem is left about the feasibility, i.e., the propagation of the division property requires much time and memory complexity. For instance, if we want to evaluate the propagation of the division property $\mathcal{D}_{\mathbb{K}}^{n^m}$, the time and memory complexity is upper-bounded by $(n+1)^m$. Therefore, if the upper bound is too large, e.g., $(n+1)^m \gg 2^{32}$, it is difficult to evaluate the propagation [4]. In the bit-based division property, the time and memory complexity is upper-bounded by 2^n, where n denotes the block length. Moreover, the bit-based division property using three subsets requires more complexity than that using two subsets. Therefore, we cannot apply the bit-based division property to the SIMON family except for SIMON32.

5.1   Provable Security for Designers

We cannot apply the bit-based division property to the SIMON family except for SIMON32, but we can show the "provable security" alternatively. When we design new symmetric-key primitives, we have to guarantee the security against several cryptanalyses. Provable security has been discussed in detail for differential and linear cryptanalyses [12, 13], but such tools do not exist for integral cryptanalysis.

Let $\mathcal{D}_{\mathbb{K}_i, \mathbb{L}_i}^{1^m}$ denotes the division property of the output set of the ith round function. We want to find R-round integral characteristics. Then, for any u with $w(u) = 1$, we have to evaluate that there are not $k \in \mathbb{K}_R$ satisfying $W(u) \succeq k$ and $\ell \in \mathbb{L}_R$ satisfying $W(u) = \ell$. Therefore, we have to get all vectors in $\mathbb{K}_R$ and $\mathbb{L}_R$, and such vectors are searched by an algorithm like breadth-first search. On the other hand, we want to show that an R-round integral characteristic cannot exist. Then, it is enough to show that $\mathbb{K}_R$ has m distinct vectors whose Hamming weight is one, i.e., there is not balanced bits, and such vectors are searched by an algorithm like depth-first search. In our provable security, we aim to get such number of rounds efficiently, and a *lazy propagation* is useful to find such number of rounds.

---

[4] In [16], the propagation for MISTY1 was evaluated, and the division property $\mathcal{D}_{\mathbb{K}}^{7,2,7,7,2,7,7,2,7,7,2,7}$ was used. Then, $|\mathbb{K}|$ is upper bounded by $8^8 \times 3^4 = 1358954496 \approx 2^{30.3}$, and it is feasible.

Table 7. Accurate propagations up to six rounds

#rounds	Simon48		Simon64		Simon96		Simon128	
	$\min_w(\mathbb{L})$	$\min_w(\mathbb{K})$	$\min_w(\mathbb{L})$	$\min_w(\mathbb{K})$	$\min_w(\mathbb{L})$	$\min_w(\mathbb{K})$	$\min_w(\mathbb{L})$	$\min_w(\mathbb{K})$
0	47	48	63	64	95	96	127	128
1	47	48	63	64	95	96	127	128
2	46	47	62	63	94	96	126	128
3	45	46	61	62	93	94	125	126
4	43	44	59	60	91	92	123	124
5	40	41	56	57	88	89	120	121
6	35	36	51	52	83	84	115	116

Definition 3 (Lazy Propagation). *Let* $\mathcal{D}_{\mathbb{K}_{i-1},\mathbb{L}_{i-1}}^{1^m}$ *be the bit-based division property of the input set of the ith round function. The ith round function is applied, and let* $\mathcal{D}_{\bar{\mathbb{K}}_i,\bar{\mathbb{L}}_i}^{1^m}$ *be the bit-based division property from the lazy propagation. Then,* $\bar{\mathbb{K}}_i$ *is computed from only a part of vectors in* $\mathbb{K}_{i-1}$, *and* $\bar{\mathbb{L}}_i$ *always becomes the empty set* ϕ.

The lazy propagation first removes all vectors from $\mathbb{L}_{i-1}$. Moreover, it only evaluates the propagation from vectors with low Hamming weight in $\mathbb{K}_{i-1}$ because such vectors are more close to unknown. Therefore, it is more efficiently evaluated than the accurate propagation.

Let us consider the meaning of the lazy propagation. Assuming the input set of the $(i-1)$th round function has $\mathcal{D}_{\mathbb{K}_{i-1},\mathbb{L}_{i-1}}^{1^m}$, we get $\mathcal{D}_{\mathbb{K}_i,\mathbb{L}_i}^{1^m}$ and $\mathcal{D}_{\bar{\mathbb{K}}_i,\phi}^{1^m}$ by the accurate propagation and the lazy propagation, respectively. Then, the set of $\boldsymbol{u}$ that the parity is unknown is represented as

$$\mathbb{S}_{\mathbb{K}} := \left\{ \boldsymbol{u} \in (\mathbb{F}_2)^m \mid \text{there are } \boldsymbol{k} \in \mathbb{K}_i \text{ satisfying } W(\boldsymbol{u}) \succeq \boldsymbol{k} \right\}.$$

On the other hand, $\mathbb{S}_{\bar{\mathbb{K}}_i}$ cannot completely represent the set of $\boldsymbol{u}$ that the parity is unknown. However, $\mathbb{S}_{\bar{\mathbb{K}}_i} \subseteq \mathbb{S}_{\mathbb{K}_i}$ always holds.

Next, we repeat the lazy propagation, and we assume that $\mathcal{D}_{\bar{\mathbb{K}}_{i+1},\phi}^{1^m}$ is propagated from $\mathcal{D}_{\bar{\mathbb{K}}_i,\phi}^{1^m}$ by the lazy propagation. Similarly, assuming that $\mathcal{D}_{\mathbb{K}_{i+1},\mathbb{L}_{i+1}}^{1^m}$ is the division property from $\mathcal{D}_{\mathbb{K}_i,\mathbb{L}_i}^{1^m}$ by the accurate propagation, $\mathbb{S}_{\bar{\mathbb{K}}_{i+1}} \subseteq \mathbb{S}_{\mathbb{K}_{i+1}}$ always holds because $\mathbb{S}_{\bar{\mathbb{K}}_i} \subseteq \mathbb{S}_{\mathbb{K}_i}$. Therefore, if the lazy propagation creates $\mathcal{D}_{\bar{\mathbb{K}}_R,\phi}^{1^m}$, where $\bar{\mathbb{K}}_R$ has m distinct vectors whose Hamming weight is one, the accurate propagation also creates the same m distinct vectors in the same round.

5.2 Application to Simon Family

We evaluate the lazy propagation of the bit-based division property on Simon48, Simon64, Simon96, and Simon128. Here, we only evaluate integral characteristics when they use chosen plaintexts that only one bit of the left half is constant and the other bits are active. We calculate the accurate propagation up to 6

Table 8. Lazy propagation of the bit-based division property for the SIMON family

#rounds	SIMON48		SIMON64		SIMON96		SIMON128	
	$\min_w(\mathbb{K})$	Limit	$\min_w(\mathbb{K})$	Limit	$\min_w(\mathbb{K})$	Limit	$\min_w(\mathbb{K})$	Limit
7	30	33	46	61	78	81	110	113
8	20	23	35	38	68	71	100	103
9	11	14	23	26	55	57	87	88
10	7	10	13	15	40	41	71	71
11	5	8	9	10	27	28	59	59
12	3	8	6	8	17	17	42	42
13	2	5	4	7	11	11	32	32
14	2	3	3	7	8	9	21	21
15	1	2	2	7	5	6	15	15
16	1(u)	1	2	4	4	6	10	10
17			1	3	3	6	8	8
18			1	1	2	6	5	6
19			1(u)	1	2	6	4	6
20					1	6	3	6
21					1	6	2	6
22					1	6	2	6
23					1	1	2	6
24					1(u)	1	1	6
25							1	6
26							1	6
27							1	6
28							1(u)	1

rounds[5] Table 7 shows $\min_w(\mathbb{L})$ and $\min_w(\mathbb{K})$ in the accurate propagation of $\mathcal{D}_{\mathbb{K},\mathbb{L}}^{1^{2n}}$ up to 6 rounds, where $\min_w(\mathbb{L})$ and $\min_w(\mathbb{K})$ are calculated as

$$\min_w(\mathbb{K}) = \min_{\boldsymbol{k}\in\mathbb{K}}\left(\sum_{i=1}^{2n} w(k_i)\right), \quad \min_w(\mathbb{L}) = \max_{\boldsymbol{\ell}\in\mathbb{L}}\left(\sum_{i=1}^{2n} w(\ell_i)\right).$$

From the 7th round function, we repeat the lazy propagation. We first remove all vectors from $\mathbb{L}$, and then the bit-based division property is represented as $\mathcal{D}_{\mathbb{K},\phi}^{1^{2n}}$, where ϕ denotes the empty set. Moreover, we remove vectors with high

---

[5] In our implementation, we could not calculate the accurate propagation up to 7 rounds because of the limitation of the memory size.

Hamming weight from $\mathbb{K}$. Table 8 shows the lazy propagation of the bit-based division property $\mathcal{D}_{\mathbb{K},\phi}^{1^{2n}}$, where we only store vectors $\boldsymbol{k} \in \mathbb{K}$ satisfying

$$\min_w(\mathbb{K}) \le \sum_{i=1}^{2n} w(k_i) \le \text{Limit}.$$

Here, u means that the $\mathbb{K}$ has $2n$ distinct vectors whose Hamming weight is one, and then, we simply say that the propagation reaches the unknown.

Even if there is a vector $\boldsymbol{k} \in \mathbb{K}$ satisfying Limit $< \sum_{i=1}^{2n} w(k_i)$, we do not evaluate the propagation from the $\boldsymbol{k}$. Therefore, if the propagation from the removed vector $\boldsymbol{k}$ immediately reaches the unknown, there is a gap between the accurate propagation and the lazy propagation. However, if the lazy propagation reaches the unknown in a specific number of rounds, the accurate propagation at least reaches the unknown in the same number of rounds. Therefore, the lazy propagation is not useful for attackers, but it guarantees the number of rounds that the bit-based division property cannot find integral characteristics.

As a result, the lazy propagation shows that 16-, 19-, 24-, and 28-round SIMON48, 64, 96, and 128 probably do not have integral characteristics, respectively. However, we can get $(r + 1)$-round integral characteristics from r-round integral characteristics because round keys are XORed after the round function. Therefore, we expect that 17-, 20-, 25-, and 29-round SIMON48, 64, 96, and 128 probably do not have integral characteristics, respectively.

5.3 Characteristics that Bit-Based Division Property Cannot Find

We consider characteristics that the bit-based division property cannot find. Our provable security supposes that all round keys are randomly and secretly chosen. However, practical ciphers generate round keys from the secret key using the key scheduling algorithm. Therefore, our provable security does not suppose integral characteristics that exploit the key scheduling algorithm.

The bit-based division property using three subsets focuses on the parity $\bigoplus_{x \in \mathbb{X}} \pi_u(x)$, and divide the set of $\boldsymbol{u}$ into three subsets. Then, the propagation simply regard $\bigoplus_{x \in \mathbb{X}} \pi_{u_1}(x) \oplus \pi_{u_2}(x)$ as unknown if either $\bigoplus_{x \in \mathbb{X}} \pi_{u_1}(x)$ or $\bigoplus_{x \in \mathbb{X}} \pi_{u_2}(x)$ is unknown. For instance, if $\bigoplus_{x \in \mathbb{X}} \pi_{u_1}(x) \oplus \pi_{u_2}(x)$ is always 0 or 1 although $\bigoplus_{x \in \mathbb{X}} \pi_{u_1}(x)$ and $\bigoplus_{x \in \mathbb{X}} \pi_{u_2}(x)$ are unknown, the bit-based division property cannot exploit such property.

6   Conclusions

The division property is a useful technique to find integral characteristics, but it has not been applied to non-S-box-based ciphers effectively. This paper focused on the bit-based division property. More precisely, this paper proposed a new variant using three subsets. The conventional bit-based division property divides the set of $\boldsymbol{u}$ into two subsets, but the new variant divides the set of $\boldsymbol{u}$ into three subsets. The bit-based division property using three subsets can prove that the experimental integral characteristic for SIMON32 shown in [18] works for all keys. Moreover, we focused on the propagation of the division property. Then, we showed that the lazy propagation is useful to guarantee the security against integral cryptanalyses using the division property. As a result, we showed that 17-, 20-, 25-, and 29-round SIMON48, 64, 96, and 128 probably do not have integral characteristics, respectively.

Acknowledgments. The authors would like to thank the anonymous referees for their helpful comments.

References

1. Abed, F., List, E., Lucks, S., Wenzel, J.: Differential cryptanalysis of round-reduced Simon and speck. In: Cid, C., Rechberger, C. (eds.) FSE 2014. LNCS, vol. 8540, pp. 525–545. Springer, Heidelberg (2015)
2. Aumasson, J.P., Jovanovic, P., Neves, S.: Norx v2.0, submission to CAESAR competition (2015)
3. Beaulieu, R., Shors, D., Smith, J., Treatman-Clark, S., Weeks, B., Wingers, L.: The SIMON and SPECK families of lightweight block ciphers. IACR Cryptology ePrint Archive 2013/404 (2013). http://eprint.iacr.org/2013/404
4. Beaulieu, R., Shors, D., Smith, J., Treatman-Clark, S., Weeks, B., Wingers, L.: SIMON and SPECK: block ciphers for the internet of things. IACR Cryptology ePrint Archive 2015/585 (2015). http://eprint.iacr.org/2015/585
5. Biryukov, A., Roy, A., Velichkov, V.: Differential analysis of block ciphers SIMON and SPECK. In: Cid, C., Rechberger, C. (eds.) FSE 2014. LNCS, vol. 8540, pp. 546–570. Springer, Heidelberg (2015)
6. Boura, C., Naya-Plasencia, M., Suder, V.: Scrutinizing and improving impossible differential attacks: applications to CLEFIA, Camellia, LBlock and SIMON. In: Sarkar, P., Iwata, T. (eds.) ASIACRYPT 2014. LNCS, vol. 8873, pp. 179–199. Springer, Heidelberg (2014)
7. Daemen, J., Knudsen, L.R., Rijmen, V.: The block cipher SQUARE. In: Biham, E. (ed.) FSE 1997. LNCS, vol. 1267, pp. 149–165. Springer, Heidelberg (1997)
8. Knudsen, L.R.: Truncated and higher order differentials. In: Preneel, B. (ed.) FSE. LNCS, vol. 1008, pp. 196–211. Springer, Heidelberg (1994)
9. Knudsen, L.R., Wagner, D.: Integral cryptanalysis. In: Daemen, J., Rijmen, V. (eds.) FSE 2002. LNCS, vol. 2365, pp. 112–127. Springer, Heidelberg (2002)
10. Kölbl, S., Leander, G., Tiessen, T.: Observations on the SIMON block cipher family. In: Gennaro, R., Robshaw, M. (eds.) CRYPTO 2015. LNCS, vol. 9215, pp. 161–185. Springer, Heidelberg (2015)
11. Lai, X.: Higher order derivatives and differential cryptanalysis. In: Blahut, R.E., Costello Jr., D.J., Maurer, U., Mittelholzer, T. (eds.) CC. SISECS, vol. 276, pp. 227–233. Springer, Heidelberg (1994)
12. Matsui, M.: New structure of block ciphers with provable security against differential and linear cryptanalysis. In: Gollmann, D. (ed.) FSE. LNCS, vol. 1039, pp. 205–218. Springer, Heidelberg (1996)
13. Nyberg, K., Knudsen, L.R.: Provable security against a differential attack. J. Cryptol. **8**(1), 27–37 (1995)

14. Sun, B., Hai, X., Zhang, W., Cheng, L., Yang, Z.: New observation on division property. IACR Cryptology ePrint Archive 2015/459 (2015). http://eprint.iacr.org/2015/459

15. Sun, S., Hu, L., Wang, P., Qiao, K., Ma, X., Song, L.: Automatic security evaluation and (Related-key) differential characteristic search: application to SIMON, PRESENT, LBlock, DES(L) and other bit-oriented block ciphers. In: Sarkar, P., Iwata, T. (eds.) ASIACRYPT 2014. LNCS, vol. 8873, pp. 158–178. Springer, Heidelberg (2014)

16. Todo, Y.: Integral Cryptanalysis on Full MISTY1. In: Gennaro, R., Robshaw, M. (eds.) CRYPTO 2015. LNCS, vol. 9215, pp. 413–432. Springer, Heidelberg (2015)

17. Todo, Y.: Structural evaluation by generalized integral property. In: Oswald, E., Fischlin, M. (eds.) EUROCRYPT 2015. LNCS, vol. 9056, pp. 287–314. Springer, Heidelberg (2015)

18. Wang, Q., Liu, Z., Varici, K., Sasaki, Y., Rijmen, V., Todo, Y.: Cryptanalysis of reduced-round SIMON32 and SIMON48. In: Meier, W., Mukhopadhyay, D. (eds.) INDOCRYPT 2014. LNCS, vol. 8885, pp. 143–160. Springer, Heidelberg (2014)

19. Yang, G., Zhu, B., Suder, V., Aagaard, M.D., Gong, G.: The simeck family of lightweight block ciphers. In: Güneysu, T., Handschuh, H. (eds.) CHES 2015. LNCS, vol. 9293, pp. 307–329. Springer, Heidelberg (2015)

20. Zhang, H., Wu, W.: Structural evaluation for generalized feistel structures and applications to Lblock and TWINE. In: Biryukov, A., Goyal, V. (eds.) INDOCRYPT 2015. LNCS, vol. 9462, pp. 218–237. Springer, Heidelberg (2015)

21. Zhang, H., Wu, W., Wang, Y.: Integral attack against bit-oriented block ciphers. In: Kwon, S., Yun, A. (eds.) ICISC 2015. LNCS, vol. 9558, pp. 102–118. Springer, Heidelberg (2016). doi:10.1007/978-3-319-30840-1_7

Algebraic Insights
into the Secret Feistel Network

Léo Perrin$^{(\boxtimes)}$ and Aleksei Udovenko

SnT, University of Luxembourg, Luxembourg city, Luxembourg
{leo.perrin,aleksei.udovenko}@uni.lu

Abstract. We introduce the high-degree indicator matrix (HDIM), an object closely related with both the linear approximation table and the algebraic normal form (ANF) of a permutation. We show that the HDIM of a Feistel Network contains very specific patterns depending on the degree of the Feistel functions, the number of rounds and whether the Feistel functions are 1-to-1 or not. We exploit these patterns to distinguish Feistel Networks, even if the Feistel Network is whitened using unknown affine layers. We also present a new type of structural attack exploiting monomials that cannot be present at round $r-1$ to recover the ANF of the last Feistel function of a r-round Feistel Network. Finally, we discuss the relations between our findings, integral attacks, cube attacks, Todo's division property and the congruence modulo 4 of the Linear Approximation Table.

Keywords: High-degree indicator matrix · Feistel network · ANF · Linear approximation table/walsh spectrum · Division property · Integral attack

1 Introduction

While the importance of attacks targeting actual primitives is obvious, structural attacks can also lead to interesting development. In fact, the last few years have seen the publications of several such attacks. For example, the attack targeting the SASAS construction has been recently extended to larger constructions [1]. The ASASA structure, which might look weaker at first glance due to its lower number of non-linear layers, has actually proved to be a challenging target; it was even proposed as the basis for public key encryption and white-box scheme [2]. Attacking this generic structure requires sophisticated methods presented in [3] and [4]. Feistel Networks have also been the target of generic attacks in two different settings. If the Feistel functions are completely secret, attacks up to 5-rounds are presented in [5]. If the Feistel functions consist in public functions

The work of Léo Perrin is supported by the CORE ACRYPT project (ID C12-15-4009992) funded by the *Fonds National de la Recherche* (Luxembourg). The work of Aleksei Udovenko is supported by the *Fonds National de la Recherche*, Luxembourg (project reference 9037104).

© International Association for Cryptologic Research 2016
T. Peyrin (Ed.): FSE 2016, LNCS 9783, pp. 378–398, 2016.
DOI: 10.1007/978-3-662-52993-5_19

preceded by the addition of a secret key, powerful attacks with very low data complexity are presented in [6].

As illustrated by the usage of the ASASA structure, generic constructions can be applied in white-box cryptography where the aim is to prevent an attacker from having access to some of the inner components of the algorithm to perform some computations. Thus, structural attacks are important in this context. They can also be used to reverse-engineer the secret structure of an S-Box, allowing for example an attacker to enjoy the benefits of a lightweight implementation known a priori only by the designer of the S-Box. The use of small Feistel Networks for lightweight S-Box design is investigated in [7] and, in fact, a secret hardware efficient decomposition[1] was recently discovered for the S-Box of the last Russian standards [8] using such reverse-engineering.

Our Contribution. Our results are based on the *high-degree indicator matrix* (HDIM), a new object we introduce. We associate to any n-bit permutation F a $n \times n$ Boolean matrix $\hat{H}(F)$ which can be computed in time $O(n2^{n-1})$ using the full code-book and which is related all at once to the LAT/Walsh spectrum of F, to its algebraic normal form and to the existence of integral distinguishers.

The HDIM provides new attack directions which we illustrate by analysing some generic constructions based on Feistel Networks. In particular, we show the existence of some patterns in the HDIM of $2n$-bit Feistel Networks with r rounds and Feistel functions with degree d depending on $\theta(d,r)$ with

$$\theta(d,r) = d^{\lfloor r/2 \rfloor - 1} + d^{\lceil r/2 \rceil - 1}.$$

These patterns provide efficient distinguishers for such structures. When the round functions are bijective, such patterns always exist in Feistel Networks with up to at least 5 round. We also show that these distinguishers can be interpreted as particular integral distinguishers and describe some relations between our results and Todo's division property [9]. Due to their integral nature, our distinguishers are extremely memory efficient: we only need to store a block containing the sum studied. In contrast, the impossible differential for 5-round Feistel Network [10] and the yoyo-game [5] are the best known distinguishers for 5-rounds FN with bijective Feistel functions and require respectively $O(2^n)$ and $O(2^{2n})$ blocks of memory.

We also present a new type of recovery attack against Feistel Networks with secret round functions which rebuilds the last Feistel function by exploiting the predictable absence of some monomials in the algebraic normal form of the permutation without its last round.

Outline. We first describe the definitions and notations that we shall use throughout the paper in Sect. 2. Then, we investigate in Sect. 3 the relation between the different rows and columns of a table containing the congruence modulo 4 of the biases in the LAT of some n-bit permutation and, in doing so, introduce and study the *high-degree indicator matrix* (HDIM). Section 4 shows

---

[1] Whether this hidden structure serves another purpose is still an open problem.

Table 1. Structural attacks against Feistel Networks. n is the branch size, d is the degree of the Feistel functions.

R	Type	Power	Restrictions	Time	Data	Ref.
5	Differential	Distinguisher	Non bij. round func.	2^n	2^n	[11]
	Imp. diff.	Distinguisher	Bij. round func.	2^{2n}	2^n	[10]
	SAT-based	Full recovery	$n \leq 7$	Practical	2^{2n}	[12]
	Yoyo	Full recovery	–	2^{2n}	2^{2n}	[5]
	Integral	Full recovery	f_1 or f_3 bij.	$2^{2.81n}$	2^{2n}	[5]
	Guess & Det.	Full recovery	–	$2^{n2^{3n/4}}$	2^{2n}	[5]
	HDIM-based	Distinguisher	Bij. round func.	2^{2n-1}	2^{2n-1}	Sect. 6.1
	Imp. monom.	Full recovery	Bij. round func.	2^{3n}	2^{2n}	Sect. 5.2
r	HDIM-based	Distinguisher	Bij. round func., $\theta(d, r-1) < 2n$	2^{2n-1}	2^{2n-1}	Sect. 6.1
	HDIM-based	Distinguisher	Non bij. round func., $\theta(d, r) < 2n$	2^{2n-1}	2^{2n-1}	Sect. 6.1
	Imp. monom.	Full recovery	$d^{r-3} < n$	2^{3n}	2^{2n}	Sect. 5.3

Table 2. Structural attacks against Feistel Networks whitened with unknown affine layers. The attacks recover parts of the unknown affine layers. n is the branch size, d is the degree of the Feistel functions.

Structure	Restrictions	Time	Data	Ref.
AF^4A	Bij. round func.	2^{6n}	2^{4n}	[8]
AF^rA	Bij. round func., $\theta(d, r-1) < 2n$	$n2^{2n}$	2^{2n}	Sect. 4.2
	Non bij. round func., $\theta(d, r) < 2n$	$n2^{2n}$	2^{2n}	Sect. 4.2
AF^rA^{-1}	Bij. round func., $\theta(d, r) < 2n$	$n2^{2n}$	2^{2n}	Sect. 4.2
	Non bij. round func., $\theta(d, r+1) < 2n$	$n2^{2n}$	2^{2n}	Sect. 4.2

that the HDIM of a Feistel Network exhibits very strong patterns depending on the number of rounds, the algebraic degree of the Feistel functions and whether these are bijective or not. We also describe attacks relying on these patterns targeting both Feistel Networks and Feistel Networks whitened using affine layers. In fact, in Sect. 5, we introduce a new kind of attack rebuilding efficiently the algebraic normal form of secret Feistel functions which exploits the predictable absence of some monomials in the ANF of round-reduced Feistel Networks. Finally, we discuss in Sect. 6 how our findings can fit in the framework of integral attacks.

2   Notations and Boolean Functions Basics

In this section, we introduce the notations and concepts that will be used throughout the paper. A thorough introduction to Boolean functions can be found in [13]. First, let us define some sets and simple operations:

– $\mathbb{F}_2$ denotes the finite field of size 2,
– the exclusive-OR (or XOR) is denoted $\oplus$,

- the logical AND is denoted $\wedge$,
- the hamming weight $\mathrm{hw}(x)$ of a vector x of $\mathbb{F}_2^n$ is the number of ones in x,
- $|S|$ and $\#S$ denote the size of a set S,
- the scalar product of two elements $x = (x_0, ..., x_{n-1})$ and $y = (y_0, ..., y_{n-1})$ of $\mathbb{F}_2^n$ is denoted "·" and is equal to $x \cdot y = \bigoplus_{i=0}^{n-1} x_i \wedge y_i$,
- if $x = (x_0, ..., x_{n-1})$ and $u = (u_0, ..., u_{n-1})$ are two elements of $\mathbb{F}_2^n$ then $x^u = \prod_{i=0}^{n-1} x_i^{u_i}$, and
- if $x = (x_0, ..., x_{n-1})$ and $u = (u_0, ..., u_{n-1})$ are two elements of $\mathbb{F}_2^n$ then $x \preccurlyeq u$ is true if and only if $(u_i = 0 \implies x_i = 0)$ is true for all i in $[0, n-1]$. We say that u "covers" x.

We now define some of the key components used in our analysis.

Definition 1 (Boolean Function). *We call Boolean function a function mapping $\mathbb{F}_2^n$ to $\mathbb{F}_2$. A function mapping $\mathbb{F}_2^n$ to $\mathbb{F}_2^m$ is a vectorial Boolean function and its restrictions to each output bit are its coordinates. Finally, for a vectorial Boolean function F, the Boolean functions $x \mapsto c \cdot F(x)$ are its components.*

Note that a coordinate of a Boolean function is one of its components but that the converse is not necessarily true. Let us then introduce the concept of *balanced-ness*.

Definition 2 (Balanced Boolean Function). *A (vectorial) Boolean function F mapping $\mathbb{F}_2^n$ to $\mathbb{F}_2^m$ is said to be balanced if the size of the preimages of all elements of $\mathbb{F}_2^m$ are equal.*

A Boolean function is balanced if and only if all of its components are balanced.

We also recall the definition of the Algebraic Normal Form of a Boolean function.

Definition 3 (Algebraic Normal Form (ANF)). *Any Boolean function f mapping n bits to 1 can be decomposed into*

$$f(x) = \bigoplus_{u \in \mathbb{F}_2^n} a_u x^u \text{ with } a_u = \bigoplus_{x \preccurlyeq u} f(x),$$

in a unique fashion which is called the Algebraic Normal Form (ANF) of f. The coefficients a_u can be obtained using the so-called Möbius transform. For vectorial Boolean functions, the ANF is the ANF of each of the coordinates.

Definition 4 (Algebraic Degree). *The algebraic degree of a Boolean function is the largest number of variables in a single term of its ANF, i.e. the maximum hamming weight of all u of $\mathbb{F}_2^n$ such that $a_u \neq 0$. The algebraic degree of a vectorial Boolean function is the maximum algebraic degree of its coordinates. The algebraic degree of a (vectorial) Boolean function f is denoted $\deg(F)$.*

We observe that the algebraic degree of a permutation of n bits is at most equal to $n - 1$.

Our analysis will involve the *LAT* or *Fourier Transform* (related to the *Walsh spectrum* by a constant multiplication) of a Boolean function. These almost identical concepts are introduced below.

Definition 5 (LAT, Fourier Transform, Walsh Spectrum). *The* Linear
Approximation Table *of a function f mapping n bits to m is a $2^n \times 2^m$ matrix
$\mathcal{L}$ where $\mathcal{L}[a,b] = \#\{x \in \mathbb{F}_2^n, a \cdot x = b \cdot f(x)\} - 2^{n-1}$. We note that the coefficient
$\mathcal{L}[a,b]$ can equivalently be expressed as follows:*

$$\mathcal{L}[a,b] \;=\; -\sum_{x \in \mathbb{F}_2^n} (b \cdot f(x)) \times (-1)^{a \cdot x} \;=\; -\frac{1}{2} \sum_{x \in \mathbb{F}_2^n} (-1)^{a \cdot x \oplus b \cdot f(x)},$$

where the first sum corresponds to the Fourier transform *of $x \mapsto b \cdot f(x)$ and the
second to its* Walsh spectrum. *Furthermore, the coefficient $\mathcal{L}[a,b]$ of a LAT $\mathcal{L}$
is called* bias *of the approximation $(a \rightsquigarrow b)$.*

Remark 1. If F is an n-bit permutation then, for all (a,b) in $(\mathbb{F}_2^n)^2$, we have
$\mathcal{L}[a,b] \equiv 0 \mod 2$.

When a Boolean function μ mapping n bits to m is linear, we use μ to
represent both the function itself and its matrix representation. The transpose
of a matrix μ is denoted μ^t. Finally, we state the following well-known remark
regarding the algebraic degree of a (vectorial) Boolean function.

Remark 2. If F is a (vectorial) Boolean function and $\mathcal{V}$ is a vector space of $\mathbb{F}_2^n$
such that $|\mathcal{V}| > 2^{\deg(F)}$, then $\bigoplus_{v \in \mathcal{V}} F(v) = 0$.

3   Patterns in Biases Modulo 4 and HDIM

Our initial goal was to identify new generic attacks against Feistel Networks. As
suggested in [12], we looked at a visual representation of the Linear Approximation Table of such permutations. We identified some patterns which turned out
to be byproducts of a strong structure in the congruence modulo 4 of the biases.
Figure 1a and b show the "Pollock representation" of the LAT modulo 4 of a 4-
and a 5-round 6-bit Feistel Networks for some bijective Feistel functions picked
uniformly at random.

(a) $r = 4$ (b) $r = 5$

Fig. 1. LAT of r-round Feistel Networks (modulo 4).

As we can see, the congruence of the biases is constant in each square of dimensions 8×8 for the 4-round Feistel Networks. Furthermore, there seems to be linear patterns for the 5-round structure: if we divide the LAT into 8×8 squares as before then we find that each square at position (i, j) is the sum of the squares at positions $(i, 0)$ and $(0, j)$ and a square-wise constant.

The reason behind these patterns is two-fold. The first aspect is a generic observation about the linearity (in some sense) of the construction of the LAT modulo 4. Indeed, we show in this section that the function $(a, b) \mapsto (\mathcal{L}[a, b]$ mod 4) for the LAT $\mathcal{L}$ of a permutation is a bilinear form and that its matrix representation has interesting properties. The second aspect of the justification for the patterns is the probability 1 presence of zeroes in some positions which is discussed later in Sect. 4.

3.1   The High-Degree Indicator Matrix

We first re-write the congruence modulo 4 of the biases in the LAT of a permutation using Boolean functions.

Lemma 1 (LAT modulo 4). *Let F be a permutation of n bits $(n > 2)$ and let $\mathcal{L}$ be its LAT. Then $\mathcal{L}[a, b]$ is such that $\mathcal{L}[a, b] \equiv 2 \times \left(\bigoplus_{x \in \mathbb{F}_2^n} (b \cdot F(x))(a \cdot x) \right)$ mod 4 or, equivalently,*

$$\frac{\mathcal{L}[a, b]}{2} \equiv \bigoplus_{x \in \mathbb{F}_2^n} (b \cdot F(x))(a \cdot x) \quad \mod 2.$$

Proof. Since $(-1)^z = 1 - 2z$ (for z in $\{0, 1\}$), the coefficient $\mathcal{L}[a, b]$ is equal to

$$\mathcal{L}[a, b] = -\sum_{x \in \mathbb{F}_2^n} (b \cdot F(x)) + 2 \left(\sum_{x \in \mathbb{F}_2^n} (b \cdot F(x))(a \cdot x) \right).$$

The first term in this sum is equal to 2^{n-1} as every component of a permutation is balanced.[2] Thus, if we look at the congruence modulo 4 of $\mathcal{L}[a, b]$, we obtain the following (for any $n > 2$):

$$\mathcal{L}[a, b] \equiv 2 \left(\sum_{x \in \mathbb{F}_2^n} (b \cdot F(x))(a \cdot x) \right) \quad \mod 4,$$

from which we deduce that

$$\frac{\mathcal{L}[a, b]}{2} \equiv \sum_{x \in \mathbb{F}_2^n} (b \cdot F(x))(a \cdot x) \quad \mod 2$$

As sum and XOR are equivalent modulo 2, this proves the lemma.   □

---

[2] If F is not a permutation but some function with degree at most $n - 1$, then this term *a priori* does not go away when taking the modulo 4 of the expression.

This lemma has several consequences regarding the congruence modulo 4 of the LAT coefficients of F (or, alternatively, the congruence modulo 2 of their half). First, we define $\mathcal{L}_4$ to be a $2^n \times 2^n$ matrix such that $\mathcal{L}_4[a,b] \equiv \mathcal{L}[a,b]$ mod 4 and $\mathcal{L}_4[a,b] \in 0,2$. Using this, we define $B(\mathcal{L})$ to be a $2^n \times 2^n$ Boolean matrix with $B(\mathcal{L})[a,b] = \mathcal{L}_4[a,b]/2$. This matrix has the following property:

$$B(\mathcal{L})[a \oplus a', b \oplus b'] = B(\mathcal{L})[a,b] \oplus B(\mathcal{L})[a,b'] \oplus B(\mathcal{L})[a',b] \oplus B(\mathcal{L})[a',b'].$$

As consequence, the function $(a,b) \mapsto B(\mathcal{L})[a,b]$ is a bilinear form and can be represented using an $n \times n$ matrix $\hat{H}(F)$.

Definition 6 (High-Degree Indicator Matrix (HDIM)). *Let F be an n-bit permutation and let $B(\mathcal{L})$ be the Boolean matrix representing the congruence modulo 4 of its LAT (as described above). We define the High-Degree Indicator Matrix $\hat{H}(F)$ of F to be the $n \times n$ matrix such that*

$$\hat{H}(F)[i,j] = \bigoplus_{x \in \mathbb{F}_2^n} (e_i \cdot F(x))(e_j \cdot x),$$

where e_k is an all zero n-bit vector with a single 1 at position k. This matrix is such that

$$B(\mathcal{L})[a,b] = b^t \times \hat{H}(F) \times a.$$

Lemma 2. *The coefficients of $\hat{H}(F)$ indicate the presence of the highest degree terms in the coordinates of F. More precisely, $\hat{H}(F)[i,j] = 1$ if and only if the ANF of F_i contains the monomial $\prod_{k \neq j} x_k$ (which has degree $n-1$).*

Proof. Let F be an n-bit permutation. As $\hat{H}(F)[i,j]$ is the sum over of space of size 2^n of the Boolean function $x \mapsto (e_i \cdot F(x))(e_j \cdot x) = F_i(x) \cdot x_j$, it is equal to 0 unless this Boolean function has algebraic degree n. As F has degree $n-1$, this occurs if and only if F_i contains $\prod_{k \neq j} x_k$. Indeed, in this case (and in this case only), the ANF of $x_j \cdot F_i(x)$ contains the only possible degree n term $\prod_{k=0}^{n-1} x_k$. $\square$

This lemma is the reason behind the name "high-degree indicator matrix". Indeed, the HDIM coefficients simply state whether each of the n possible $n-1$ degree terms appear in each coordinate of F or not.

We finally note that the HDIM of a function can be computed much more efficiently than the LAT or the difference distribution table. Indeed, we can compute a column of the HDIM by summing the function over a cube of dimension $n-1$ (see Sect. 6.1). The complexity for all n columns is therefore $n2^{n-1}$.

3.2   Some Properties of the High-Degree Indicator Matrix

Let us investigate the effect of some simple transformations on the HDIM. First, we point out that due to the fact that the LAT of the inverse of a permutation F is the transpose of the LAT of F, the HDIM of F^{-1} is the transpose of the HDIM of F.

We now show that the HDIM of $\eta \circ f \circ \mu$ can easily be deduced from that of f when η and μ are n-bit linear permutations. The corresponding theorem will be used in Sect. 4.2 to attack Feistel Networks whitened using affine layers.

Theorem 1. *Let μ, η be linear n-bit mappings, F be an n-bit permutation and let $G = \eta \circ F \circ \mu$. Furthermore, let $\hat{H}(F)$ be the HDIM of f and $\hat{H}(G)$ be that of G. Then it holds that*

$$\hat{H}(G) = \eta \times \hat{H}(F) \times (\mu^t)^{-1}.$$

Proof. We prove this result in two steps. First, the fact that $\hat{H}(F \circ \mu) = \hat{H}(F) \times (\mu^{-1})^t$ can be derived as follows:

$$\hat{H}(F \circ \mu)[i, j] = \bigoplus_{x \in \mathbb{F}_2^n} \big(e_i \cdot F(\mu(x))\big)\big(e_j \cdot x\big) = \bigoplus_{y \in \mathbb{F}_2^n} \big(e_i \cdot F(y)\big)\big(e_j \cdot \mu^{-1}(y)\big)$$

$$= \bigoplus_{y \in \mathbb{F}_2^n} \big(e_i \cdot F(y)\big)\big((\mu^t)^{-1}(e_j) \cdot y\big).$$

We then note that $\hat{H}(\eta \circ F) = \hat{H}(F^{-1} \circ \eta^{-1})^t$ which, using what we just found, is equal to $(\hat{H}(F^{-1}) \times \eta^t)^t = (\hat{H}(F)^t \times \eta^t)^t$, so that $\hat{H}(\eta \circ F) = \eta \times \hat{H}(F)$. This concludes the proof. $\qquad\square$

The ANF and the LAT of an n-bit permutation are connected in the sense that it is possible to determine the congruence modulo 4 of the LAT $\mathcal{L}$ of an n-bit permutation F given parts of its ANF. Indeed, as we describe in this section, this congruence only depends on the terms of degree $n - 1$ in the ANF of the coordinates of F.

4   The High-Degree Indicator Matrix of Feistel Networks

In what follows, we denote F_d^r an r-round FN with bijective Feistel function of algebraic degree at most d. The structure of a sample is given Fig. 2. It is possible to use the HDIM to analyse such generic structures.

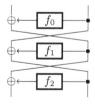

Fig. 2. A sample F_d^3 structure, where $\deg(f_i) \leq d$.

4.1   Artifacts in the HDIM of Feistel Networks

The HDIM of a Feistel Network may yield interesting patterns depending on the degree of its Feistel functions, whether they are bijections or not and its number of rounds. These are formalized by Theorem 2 and its corollary (Corollary 1). These results link the maximum degree d of the Feistel functions, the number of rounds r and the presence or not of some patterns using the function $\theta : \mathbb{Z}^2 \to \mathbb{Z}$ defined by

$$\theta(d, r) = d^{\lfloor r/2 \rfloor - 1} + d^{\lceil r/2 \rceil - 1},$$

where $\lfloor 2k \rfloor = \lfloor 2k + 1 \rfloor = 2k$ and $\lceil 2k \rceil = \lceil 2k - 1 \rceil = 2k$.

Theorem 2. *Let F be a 2n-bit* F_d^r. *Then the HDIM of F is such that* $\hat{H}(F)[i, j] = 0$ *if $i < n$ **or** $j < n$ under the following conditions:*

- *if the Feistel functions are bijections and $\theta(d, r) < 2n$, or*
- *if the Feistel functions are not bijections and $\theta(d, r + 1) < 2n$.*

The general idea of the proof is to express the sum corresponding to coefficient $\hat{H}(F)[i, j]$ using well-chosen variables (α, β) located in the middle of the encryption. The value of $F(x)$ is then a function of degree $d^{\lceil r/2 \rceil - 1}$ of (α, β) and that of x is a function of degree $d^{\lfloor r/2 \rfloor - 1}$. The coefficients can thus be written as

$$\hat{H}(F)[i, j] = \bigoplus_{(\alpha, \beta) \in (\mathbb{F}_2^n)^2} \left(e_i \cdot F(x(\alpha, \beta)) \right) \left(e_j \cdot x(\alpha, \beta) \right)$$

and the result is equal to 0 if $\theta(d, r) = d^{\lfloor r/2 \rfloor - 1} + d^{\lfloor r/2 \rfloor - 1} < 2n$. If the Feistel functions are not bijective then a "trick" used to slightly decrease the degree in (α, β) of the output cannot be used, hence the small discrepancy in this case. The complete formal proof of this theorem is given in the full version of this paper [14].

Corollary 1. *Let F be a 2n-bit* F_d^r. *The HDIM of F is such that $\hat{H}(F)[i, j] = 0$ if $i < n$ **and** $j < n$ under the following conditions:*

- *if the Feistel functions are bijections and $\theta(d, r - 1) < 2n$, or*
- *if the Feistel functions are not bijections and $\theta(d, r) < 2n$.*

Proof. Let r and d be such that F_d^{r-1} fits the hypothesis of Theorem 2. The right word of the output of a F_d^r structure is the left word output by a F_d^{r-1} structure. As each line of the HDIM corresponds to one output bit, the top n rows of the HDIM of the r-round FN are equal to the bottom n rows of the same permutation reduced to $(r - 1)$-round. Because of Theorem 2, this bottom half is such that the first n columns are all 0. Thus, the first n columns of the first n rows of the HDIM of a F_d^r are all equal to 0.   □

To illustrate these theorems, we give the HDIM of the 4- and 5-round Feistel with 3-bit bijective Feistel functions picked uniformly at random whose LAT modulo 4 were given in Fig. 1a and b. The Feistel functions must have an algebraic degree at most equal to 2. Since $\theta(2,4) = 2^1 + 2^1 = 4 < 6$, these HDIM must exhibit the patterns described in the theorems above. It is the case, as we can see below. The zeroes caused by Theorem 2 and Corollary 1 are represented in grey:

$$\hat{H}(\mathsf{F}^4) = \begin{bmatrix} 0\,0\,0\,0\,0\,0 \\ 0\,0\,0\,0\,0\,0 \\ 0\,0\,0\,0\,0\,0 \\ 0\,0\,0\,0\,1\,1 \\ 0\,0\,0\,1\,0\,1 \\ 0\,0\,0\,1\,0\,1 \end{bmatrix}, \ \hat{H}(\mathsf{F}^5) = \begin{bmatrix} 0\,0\,0\,1\,1\,1 \\ 0\,0\,0\,1\,1\,1 \\ 0\,0\,0\,0\,1\,1 \\ 0\,1\,1\,0\,1\,0 \\ 1\,0\,0\,0\,0\,0 \\ 0\,1\,1\,1\,1\,1 \end{bmatrix}. \tag{1}$$

Even though a F_d^r structure has an algebraic degree of $2n - 1$ in the conditions of Theorem 2, the way in which this high degree is achieved is very structured: only half of the output bits actually have a maximum degree and the monomials of degree $2n - 1$ can not contain the product of $n - 1$ bits from the right side of the input. Thus, a simple analysis of the algebraic degree can be made more sophisticated by also investigating the possible structure of the monomials of highest degree.

These patterns lead to the existence of distinguishers as long as the conditions necessary for Corollary 1 are satisfied. Table 3 shows the value of the number of rounds for which the conditions of Corollary 1 are satisfied for different values of d, r and n in both the 1-to-1 case and the case where collisions in the Feistel functions are allowed. If real ciphers correspond to these parameters, we specify them. Note that the rotation applied to one of the branches in the round function of LBlock [15] does not change anything. The key-dependent linear FL layers in MISTY1 [16] do not protect from our distinguisher as well and may be included from any side for free.

Table 3. If $r = r_{\max}(d, 2n)$ then the $2n$-bit permutation F_d^r exhibits an artifact of size n^2 in its HDIM.

$(d, 2n)$	Feistel functions	$r_{\max}(d, n)$	Instance
$(2, 32)$	1-to-1	10	—
	collisions	9	SIMON-32 [17]
$(5, 64)$	1-to-1	7	—
	collisions	6	DES [18]
$(31, 64)$	1-to-1	5	MISTY1/KASUMI [16]
	collisions	4	—
$(n - 1, 2n)$	1-to-1	5	—
	collisions	4	—

4.2 Bypassing Affine Whitening

In the context of component reverse-engineering/white-box cryptography, it may not be sufficient to be able to attack generic Feistel structure. Indeed, simply whitening a generic structure with secret affine layers can prevent many attacks from succeeding at small cost for the designer. For example, applying affine layers before and after a 5-round Feistel Network would prevent the yoyo-game used in [5] to be exploitable. Similarly, the recent attacks against ASASA [3,4] are much more sophisticated than the attack against SASAS proposed by Biryukov et al. in the first place [19]. We also note that the secret structure of the S-Box of the last Russian standard primitives recently recovered was indeed whitened with seemingly random linear layers [8].

As a consequence, we study the generic construction denoted AF_d^rA consisting in a F_d^r construction with secret Feistel functions preceded and followed by the application of independent and secret linear layers[3]. This structure has already been studied in [8] but our attacks are significantly more efficient. Note also that one of the S-Box of ZUC [20] has this structure: it is a 3-round Feistel Network composed with a bit rotation. Let us show how the HDIM and its artifacts we identified in the previous section can be used to attack permutations with AF_d^rA structures.

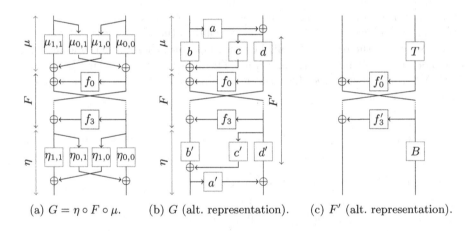

(a) $G = \eta \circ F \circ \mu$.   (b) G (alt. representation).   (c) F' (alt. representation).

Fig. 3. The target of our attack, its result and its alternative representation. In Fig. 3c, f_i' is affine equivalent to f_i.

Our attack works for a subset of all possible linear layers. We define $G = \eta \circ F \circ \mu$ where F has a F_d^r structure satisfying the conditions of Theorem 2 and

---

[3] We note that adding constants to make the layers affine is equivalent to replacing the Feistel functions by other ones with identical properties.

μ and η are linear layers. The layer applied first must have a decomposition as follows:

$$\mu = \begin{bmatrix} \mu_{0,0} & \mu_{0,1} \\ \mu_{1,0} & \mu_{1,1} \end{bmatrix} = \begin{bmatrix} d & 0 \\ c & b \end{bmatrix} \times \begin{bmatrix} I & a \\ 0 & I \end{bmatrix} = \begin{bmatrix} d & d \times a \\ c & b + c \times a \end{bmatrix},$$

and the layer applied last must have a similar one:

$$\eta = \begin{bmatrix} \eta_{0,0} & \eta_{0,1} \\ \eta_{1,0} & \eta_{1,1} \end{bmatrix} = \begin{bmatrix} I & a' \\ 0 & I \end{bmatrix} \times \begin{bmatrix} d' & 0 \\ c' & b' \end{bmatrix} = \begin{bmatrix} d' + a' \times c' & a' \times b' \\ c' & b' \end{bmatrix}.$$

It is sufficient for such a decomposition of the first layer to exist that $\mu_{0,0}$ is invertible. Indeed, we can then simply set $d = \mu_{0,0}, c = \mu_{1,0}, a = d^{-1} \times \mu_{0,1}$ and $b = \mu_{1,1} - c \times a$. Note that b has to be invertible since μ is invertible. Similarly, it is sufficient that $\eta_{1,1}$ is invertible to decompose the final layer. We define F' using these decompositions so that G is equal to:

$$G = \begin{bmatrix} I & a' \\ 0 & I \end{bmatrix} \circ \begin{bmatrix} d' & 0 \\ c' & b' \end{bmatrix} \circ F \circ \begin{bmatrix} d & 0 \\ c & b \end{bmatrix} \circ \begin{bmatrix} I & a \\ 0 & I \end{bmatrix} = \begin{bmatrix} I & a' \\ 0 & I \end{bmatrix} \circ F' \circ \begin{bmatrix} I & a \\ 0 & I \end{bmatrix}.$$

A graphical representation of the relation between F, F' and G is provided in Fig. 3a and b. As F satisfies the condition of Theorem 2, its HDIM is such that $\hat{H}(F)[i,j] = 0$ if $i < n$ or $j < n$. Applying Theorem 1 gives us that the HDIM of F' is equal to

$$\hat{H}(F') = \begin{bmatrix} d' & 0 \\ c' & b' \end{bmatrix} \times \hat{H}(F) \times \begin{bmatrix} d & c \\ 0 & b \end{bmatrix}^{-1} = \begin{bmatrix} 0 & 0 \\ 0 & h' \end{bmatrix} \text{ with } h' = b' \times h \times b^{-1},$$

h being the bottom-right part of $\hat{H}(F)$. Like in $\hat{H}(F)$, it holds that $\hat{H}(F')[i,j] = 0$ if $i < n$ or $j < n$. Another way to see why this holds is shown in Fig. 3c. Indeed, F' can be written as a F^r_d structure, like F, where n-bit linear permutations are applied only on two branches and where the Feistel functions f'_i are obtained from compositions of b, b', d, d' and f_i, as well as the addition of c and c' for the first and last rounds. We deduce that if G indeed has a $\mathsf{AF}^r_d\mathsf{A}$ structure satisfying the conditions for Theorem 2, then the following equation with unknowns the $n \times n$ binary matrices a and a' must have at least one solution:

$$\begin{bmatrix} I & a' \\ 0 & I \end{bmatrix} \times \hat{H}(G) \times \begin{bmatrix} I & 0 \\ a & I \end{bmatrix} = \begin{bmatrix} 0 & 0 \\ 0 & h_{1,1} \end{bmatrix},$$

where $h_{1,1}$ is the bottom right corner of $\hat{H}(G)$. This system has $2n^2$ unknowns and $3n^2$ equations, meaning that it is unlikely to have solutions if G is a random permutation. However, if it does have a solution then we deduce both that G has an $\mathsf{AF}^r_d\mathsf{A}$ structure and the expression of parts of the linear layers. We summarize these results in the following attack.

Attack 1 (Partial Recovery Against $\mathsf{AF}^r_d\mathsf{A}$). *Let G be a $2n$-bit permutation. It is necessary for G to be in $\mathsf{AF}^r_d\mathsf{A}$ for some (r, d) satisfying Theorem 2 that the equation*

$$\begin{bmatrix} I & a' \\ 0 & I \end{bmatrix} \times \hat{H}(G) \times \begin{bmatrix} I & 0 \\ a & I \end{bmatrix} = \begin{bmatrix} 0 & 0 \\ 0 & h_{1,1} \end{bmatrix},$$

where h is an unknown $n \times n$ matrix, has at least one solution. The unknowns are the coefficients of the $n \times n$ matrices a and a', so that $2n^2$ Boolean variables must satisfy $3n^2$ equations corresponding to the zeroes in the right hand side.

This distinguisher requires the full code-book and as much time as is needed to compute the HDIM and solve a system of equations. Since the system is small, the bottle-neck is the computation of the HDIM which can be done in time $O(n2^{2n})$ where n is the branch size.

We can use the exact same reasoning to attack one more round if the decomposition of η and μ involve the same "linear Feistel function" a. This happens in particular if $\eta = \mu^{-1}$. In this case, we can use the distinguisher obtained from the following attack.

Attack 2 (Partial Recovery Against $A^{-1}F_{r+1}^d A$). *Let G be a $2n$-bit permutation. In order for G to be in $AF_d^r A$ for some (r, d) satisfying Corollary 1 in such a way that the linear layers are the inverse of one another, it is necessary that the equation*

$$\begin{bmatrix} I & a \\ 0 & I \end{bmatrix} \times \hat{H}(G) \times \begin{bmatrix} I & 0 \\ a & I \end{bmatrix} = \begin{bmatrix} 0 & h_{0,1} \\ h_{1,0} & h_{1,1} \end{bmatrix},$$

where $h_{0,1}, h_{1,0}$ and $h_{1,1}$ are unknown $n \times n$ matrices, has at least one solution. The unknowns are the coefficients of the $n \times n$ matrices a, so that n^2 Boolean variables must satisfy n^2 equations corresponding to the zero in the right hand side.

Note that if there is a single whitening affine layer applied at some side, we have a similar system with n^2 unknowns. If we consider one more round, we will have n^2 equations as well. Therefore we can attack $F_d^r A$, where r is the maximum number of rounds satisfying Corollary 1. Another view on this attack is given in Sect. 5.3.

5   The Impossible Monomials Attack

In the previous sections we used absent terms of highest degree to recover whitening linear layers from Feistel Networks. In this section we generalize this method to terms of lower degree and, as a result, we present an attack recovering a secret round function from a 5-round Feistel Network with bijections. Furthermore, we generalize this attack to more rounds if the degrees of the round functions are small.

5.1   Impossible Monomials in Feistel Networks

Let F be a $2n$-bit F_{n-1}^4 and let F_i be the ith output bit of F (F_0 is the leftmost bit of F). We will denote by $L = \{0, \ldots, n-1\}$ and $R = \{n, \ldots, 2n-1\}$ the indices from the left and right halves respectively, and F_L and F_R the truncations of the function F to the left and right half respectively. Consider the ANF of F_i:

$$F_i(x_l || x_r) = \bigoplus_{u_l, u_r \in \mathbb{F}_2^n} a_{u_l || u_r}^{F_i} x_l^{u_l} x_r^{u_r}, \tag{2}$$

where x_l and x_r are vectors of input variables from the left and right halves respectively. We will now show that some monomials are impossible, that is, $a_{u_l || u_r}^{F_i} = 0$ for some u_l, u_r independently of the choice of the Feistel functions. To prove it, we will need the following lemmas.

Lemma 3. *Let $a, b \in \mathbb{F}_2^n$ be some vectors of variables and let $f : \mathbb{F}_2^n \to \mathbb{F}_2$ be a Boolean function of degree at most d. Then if some term in the ANF of $f(a \oplus b)$ has degree d_a on variables from a, then it has degree at most $d - d_a$ on variables from b. In particular, there are no terms of degree d on a and non-zero degree on b.*

Proof. Let $s(a, b) = a \oplus b$. Then $\deg s = 1$ and $\deg(f \circ s) \leq d$. Hence a term containing d_a variables from a contains at most $d - d_a$ variables from b.

Lemma 4. *Let $\pi : \mathbb{F}_2^n \to \mathbb{F}_2^n$ be a permutation and let $f : \mathbb{F}_2^n \to \mathbb{F}_2$ be some Boolean function of degree at most $n - 1$. Then $\deg(f \circ \pi) \leq n - 1$.*

Proof. By the Möbius transform, the term of degree n is present in the ANF of $f \circ \pi$ if and only if the sum of $f \circ \pi$ over $\mathbb{F}_2^n$ is equal to 1. Since π is a permutation, we have that $\sum_{x \in \mathbb{F}_2^n} f(\pi(x)) = \sum_{x \in \mathbb{F}_2^n} f(x)$. But this last sum is equal to zero because $\deg f \leq n - 1$. Therefore, there is no term of degree n in the ANF of $f \circ \pi$ and we conclude that $\deg(f \circ \pi) \leq n - 1$.

We now formally describe classes of impossible monomials using the following theorem.

Theorem 3. *Let F and its ANF be as defined before. Then $a_{u_l || u_r}^{F_i} = 0$ if one of the following holds:*

1. *$i \in R$ and $hw(u_l) = n$;*
2. *$i \in R$ and $hw(u_l) = n - 1, hw(u_r) = n - 1$;*
3. *$i \in R$ and $hw(u_l) = n - 1, hw(u_r) = n$;*
4. *$i \in L$ and $hw(u_l) = n,\quad hw(u_r) = n - 1$.*

Proof. Points 3–4 are part of Theorem 2 and are presented here for the sake of completeness. It is left to prove points 1 and 2.

1. Consider the 4-round integral characteristic from Fig. 4. Let C be any cube which contains the whole left part. From the integral characteristic it follows that the sum of F over the cube C has zero on the right side. Therefore by the Möbius transform the corresponding ANF coefficients are zero.
2. Let $f_0, f_1, f_2, f_3 : \mathbb{F}_2^n \to \mathbb{F}_2^n$ be the round functions of F. The equation for the right half of the output is then given by:

$$F_R(l || r) = l \oplus f_0(r) \oplus f_2(r \oplus f_1(l \oplus f_0(r))). \tag{3}$$

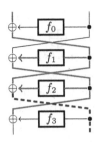

Fig. 4. The 4-round integral characteristic: words taking all values are represented in bold red and balanced words are represented in dashed blue. (Color figure online)

Clearly, the first two terms do not contain any monomial of degree $n-1$ on l and $n-1$ on r. Consider the expression $f_2(r \oplus f_1(l \oplus f_0(r)))$. Assume that a term with degree $n-1$ on both l and r is present in the ANF of the expression. Then the term is present in the expansion of some product of at most $n-1$ bits, where the bits are output bits of the expression $(r) \oplus f_1(l \oplus f_0(r))$, i.e. in the term each of the $n-1$ factors is either a bit from (r) or from $f_1(l \oplus f_0(r))$. Note that the term may not be generated by choosing bits *only* from (r), because in that case there will be no variables from l in it. Therefore there are at most $n-2$ bits taken from the outer (r); $n-1$ variable from l and at least one variable r_i are taken from $f_1(l \oplus f_0(r))$. It means that there exists a monomial function π such that $\pi \circ f_1(l \oplus f_0(r))$ contains term of degree $n-1$ on l and degree at least 1 on r. By Lemma 4, $\pi \circ f_1$ has degree at most $n-1$ and by Lemma 3 there can not be such term in $\pi \circ f_1(l \oplus f_0(r))$.

5.2   An Attack on 5-Round Feistel Network

In this section we use the impossible monomials to attack 5-round Feistel Network built from permutations. The key idea is to observe the presence of some 4-round impossible monomials in the 5-round network and extract some information about the last round function. Consider some monomial x^u which is impossible at the right side of a 4-round Feistel Network. We now add the 5th round. If we observe x^u on the left side, then we know that this monomial has come from the last round function. Otherwise, we know that it has *not* come from the last round function and it gives us some information as well. Using these observations we build a system of linear equations where the unknowns are the ANF coefficients of the coordinates of the last round function. By solving the system we recover the ANF coefficients and hence the function itself. Note that in order to compute the ANF, we have to obtain the full codebook.

Let F^5 be a $2n$-bit F_d^5, F^4 be its first 4 rounds and f be the last round function. Let a_u^g be the coefficient of term x^u in the ANF of the Boolean function g. Consider the equation of the ith bit of F^5 for $i \in L$:

$$F_i^5(x) = F_{i+n}^4(x) \oplus f_i(F_R^5(x)) = \bigoplus_{u \in \mathbb{F}_2^{2n}} a_u^{F_{i+n}^4} x^u \oplus \bigoplus_{u \in \mathbb{F}_2^{2n}} a_u^{f_i}(F_R^5(x))^u.$$

The ANF of F_i^5 with $i \in L$ contains some monomial from the first or the second group from Theorem 3 if and only if the ANF of $f_i \circ F_R^5$ does. Since we can compute the ANF of F_R^5, we can check which possible terms from the ANF of f_i generate the impossible monomial. Then from the presence of the impossible monomial in the ANF of F_{i+n}^5 we deduce if the number of such terms in the ANF of f_i is odd or even. This gives us a linear equation over $\mathbb{F}_2$ where the unknowns are the ANF coefficients of f_i. For an illustration see Fig. 5.

Note that the 4-round impossible monomials which are still impossible in a 5-round Feistel Network do not leak any information about f. For example, since Feistel Network is a bijection, the monomial of degree $2n$ is impossible for any number of rounds but it can not be used in the attack. However it is the only such monomial. Therefore we can use $2^n - 1$ impossible monomials from the first group of Theorem 3 and n^2 ones from the second group. Each such monomial yields an equation per each bit of f. There are 2^n unknown coefficients in the ANF of f_i so the number of equations will be enough to recover f_i for all i and hence f with high probability. Note that we can recover f only up to xor with a constant because the constant may propagate through the Feistel Network and merge with other round functions (see the introduction of [5] for a more detailed explanation of this phenomenon).

The complexity of the attack is $O(2^{3n})$ and is dominated by generating the equation matrix, which is the same for all output bits (the only difference is the target vector). For each of the 2^n possible terms in the ANF of f_i we compute the ANF of the term applied after F in time $O(2^{2n})$ and then we check if this term generates the impossible monomials. The next step is to solve the systems. Since the equation matrix is the same for all output bits, we can do some precomputation (for example the LU-decomposition) once and solve all n systems

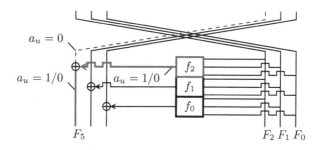

Fig. 5. Impossible monomials in the last round of a 5-round FN with 3-bit branches. The wire with 4-round impossible monomials is in dashed blue, the path of the observed monomials is highlighted with bold red. a_u is the ANF coefficient of some 4-round impossible monomial.(Color figure online)

of equations very fast. Computing the target vectors is dominated by computing the ANF of F_i^5 for $i \in L$ which takes total time of $O(n2^{2n})$.

As a consequence of the algebraic nature of the attack, if the round function has lower degree, then the complexity decreases. Indeed, there are less unknowns and therefore both steps of generating the equation matrix and solving the systems take less time. As an edge case, consider the F^5A structure where the affine layer can be seen as the 6th round with a function of degree 1. The complexity of recovering the affine round is $O(n2^{2n})$, as was shown in Sect. 4.2.

Note that the attack can be run in the reverse direction as well, so that we recover the first round function instead of the last one.

We have implemented the attack in Sage [21]. We successfully attacked a 5-round Feistel Network with bijections and branch size of up to 9 bits and recovered the outer secret round functions in a few minutes on a modern laptop.

5.3 A Generalization of the Attack on Feistel Networks with Low Degree Round Functions

When the round functions in a Feistel Network have low degree, the degree deficiency is decreasing slowly and as a result impossible monomials may exist for more than 5 rounds. Moreover, since there are less unknowns to recover, we need less impossible monomials to mount the attack.

In the following theorem we give a lower bound on the maximum number of Feistel rounds for which the large class of monomials is impossible. Namely, this class is point 1 from Theorem 3. The size of the class is 2^n, which is enough to recover a round function of full degree. Therefore this is the lower bound on maximum number of rounds that can be attacked using the ANF recovery technique from Sect. 5.2.

Theorem 4. *Let F be a $2n$-bit F_d^r with arbitrary functions and let its ANF be as in the Eq. 2. Then $a_{u_l||u_r}^{F_i} = 0$ if $d^{r-2} < n, i \in R$ and $hw(u_l) = n$.*

Proof. Let $l||r$ be the input to F. Consider the degrees on the variables from l at the intermediate states. Initially, the degrees are 1 on the left and 0 on the right. After the first round the degrees are the same, because input to the round function has no variables from l. Now if we have the respective degrees d_1, d_2 at some point and we add a swap and xor with the round function, the degrees become $max(d_2, d \cdot d_1), d_1$. Then for 2 rounds the degrees are $d, 1$, for 3 rounds - d^2, d, and, in general, for r rounds the degrees are d^{r-1}, d^{r-2}. Therefore, when $d^{r-2} < n$, the r-round Feistel Network has no monomials with degree n on l in the right branch of the output.

As a corollary of the theorem, we can attack a $2n$-bit F_d^r if $d^{r-3} < n$. Note that for the 5-round Feistel with bijections which we attacked in the previous section this bound is not satisfied (for $n \geq 3$): $d^{5-3} = (n-1)^2 > n$, i.e. we attacked more rounds than we could attack by Theorem 4. Though we expect that the bound is tight for the specified class of monomials in FN with non-bijective round functions, there are another classes of impossible monomials for Feistel Networks with more rounds. Moreover, if the degree is low, there are less ANF coefficients to recover and, therefore, smaller classes of impossible monomials may be enough for attack. As an edge case, consider an additional round function of degree 1 (a linear function). The impossible monomials of degree $2n-1$ from Corollary 1 can be used to recover such round function, as was shown in attacks from Sect. 4.2. The maximal number of rounds (without the last linear one) for this attack is given by the condition $\theta(d,r) = d^{\lfloor r/2 \rfloor - 1} + d^{\lceil r/2 \rceil - 1} < 2n$ (or 1 more round if the Feistel functions are bijections). In general case, if the Feistel functions are bijections, we can attack 5 normal rounds plus 1 linear round.

6   Relationship Between Our Results and Other Attacks

6.1   Integral Attacks

The HDIM has a simple integral interpretation. Indeed, its coefficients correspond to the presence or not of some monomials in the ANF of its coordinates. They thus correspond to coefficients in said ANF which can be computed using the Möbius transform:

$$\hat{H}(F)[i,j] = \bigoplus_{x \preccurlyeq \overline{e_j}} F_i(x)$$

where $\overline{e_j}$ is the vector where all elements are equal to 1 except in position j. This has two consequences.

1. we can compute the HDIM of an n-bit permutation in time $O(n2^{n-1})$, and
2. zeroes in column j imply the existence of an integral distinguisher.

In light of this, we state the following corollary of Corollary 1.

Corollary 2 (Integral Distinguisher for F_d^r). *Let F be a $2n$-bit F_d^r and suppose that one of the following conditions holds:*

- *the Feistel functions are bijections and $\theta(d, r-1) < 2n$, or*
- *the Feistel functions are not bijections and $\theta(d,r) < 2n$.*

Then there exists an integral distinguisher with data and time complexity 2^{2n-1} for this structure, namely

$$\bigoplus_{x \preccurlyeq \overline{e_j}} (e_i \cdot F(x)) = 0$$

for all $i < n$ and $j < n$. In other words, the sum of the right words of $F(x)$ is equal to 0 over a cube where one bit of the input right word is fixed to 0.

We notice a relation between our attacks and the so-called *division property*. This tool for finding integral attacks was introduced by Todo in [9] and later used by the same author to attack the full MISTY1 [22]. In his seminal paper, Todo gives some integral distinguishers against Feistel Network for various block sizes, number of rounds, degree of the Feistel functions for both bijective and non-bijective Feistel functions. Interestingly, his results are extremely similar to ours! Indeed, while there is no generic formula in Todo's paper, the application of his algorithm shows the existence of cubes of size $2n-1$ whose sum is equal to 0 for a number of rounds identical to the ones we predicted. In fact, results about the division property of the output of a Feistel Network can be extracted from its HDIM. To explain this, we first recall the definition of the division property.

Definition 7 (Division Property). *Let* $\mathbb{X}$ *be a multiset of* $\mathbb{F}_2^n$ *and* k *be an integer of* $[0, n]$*. We say that* $\mathbb{X}$ *has the division property* $\mathcal{D}_k^n$ *if, for all* u *in* $\mathbb{F}_2^n$ *such that* $hw(u) \le k$*,* $\bigoplus_{x \in \mathbb{X}} x^u = 0$*.*

This property is further generalized into the *vectorial division property* which we define in the particular case of a Feistel Network.

Definition 8 (Vectorial Division Property (for Feistel Networks)). *Let* $\mathbb{X}$ *be a multiset of* $(\mathbb{F}_2^n)^2$ *and* k^L, k^R *be integers of* $[0, n]$*. We say that* $\mathbb{X}$ *has the collective division property* $\mathcal{D}_{(k^L, k^R)}^n$ *if, for all* u, v *in* $\mathbb{F}_2^n$ *such that* $hw(u) \le k^L$ *and* $hw(v) \le k^R$*,* $\bigoplus_{(x,y) \in \mathbb{X}} x^u y^v = 0$*.*

In particular, Todo applied his technique to $2n$-bit F_d^r. The integral distinguisher against the highest number of rounds correspond to integrals over cubes of size $2n - 1$ were the constant bit has to be on the left side.[4] As we have seen, summing over such a cube is equivalent to computing half of the lines of the HDIM of the function.

Let F be a $2n$-bit F_d^r, x denote the left input bits, y denote the right ones and F_L and F_R denote its left and right output halves so that $F(x\|y) = F_L(x\|y)\|F_R(x\|y)$. Suppose that the top left corner of the HDIM of F is all zero. We deduce that the following holds for any cube $\mathcal{C}_k$ of dimension $2n - 1$ where the bit at index $k \le n$ is fixed and for any $i \le n$: $\bigoplus_{x \in \mathcal{C}_k} F(x) \cdot e_i(x) = 0$. This can also be written as

$$\bigoplus_{x \in \mathcal{C}_k} (F_L(x))^{u_i} (F_R(x))^0 = \bigoplus_{x \in \mathcal{C}_k} (F_L(x))^{u_i} = 0,$$

where u_i is the element of $\mathbb{F}_2^n$ equal to 0 except at position i where it is equal to 1. In other words, for all u in $\mathbb{F}_2^n$, $hw(u) \le 1$ implies that $\bigoplus_{x \in \mathcal{C}_k} (F_L(x))^u = 0$, which means that the image of $\mathcal{C}_k$ has vectorial division property $\mathcal{D}_{1,0}^n$. The HDIM of Feistel Networks can thus be interpreted as describing the vectorial division property of each output half!

The relation between the ANF and integral attacks is further stressed by the attack we described in Sect. 5. Indeed, the complexity of this attack is very similar to that of the integral attack against 5-round FN with bijective Feistel functions described in [5].

---

[4] It is actually on the right side in Todo's paper. Unlike in our paper, the Feistel function is XORed in the right branch in his case.

7   Conclusion

Investigating surprising visual patterns in the LAT of Feistel Network lead us to interesting results. To explain them, we introduced the high-degree indicator matrix (HDIM). It causes a form of linearity of the LAT modulo 4 and is related to the presence (or lack thereof) of some monomials in the ANF of the permutation. We identified patterns in the distribution of these monomials for Feistel Networks and provided theorems allowing us to predict the existence of these patterns (Theorem 2 and Corollary 1). More generally, we showed how the predictable absence of some monomials can be leveraged to attack a Feistel Network in an impossible monomial attack. We also drew some connections between our results and integral distinguisher.

Acknowledgment. We thank Alex Biryukov and Dmitry Khovratovich for helpful discussions. We also thank the anonymous reviewers for their helpful comments. The work of Léo Perrin is supported by the CORE ACRYPT project (ID C12-15-4009992) funded by the *Fonds National de la Recherche* (Luxembourg). The work of Aleksei Udovenko is supported by the *Fonds National de la Recherche*, Luxembourg (project reference 9037104).

References

1. Biryukov, A., Khovratovich, D.: Decomposition attack on SASASASAS. IACR Cryptology ePrint Archive, report 2015/46 (2015)
2. Biryukov, A., Bouillaguet, C., Khovratovich, D.: Cryptographic schemes based on the ASASA structure: black-box, white-box, and public-key (extended abstract). In: Sarkar, P., Iwata, T. (eds.) ASIACRYPT 2014. LNCS, vol. 8873, pp. 63–84. Springer, Heidelberg (2014)
3. Dinur, I., Dunkelman, O., Kranz, T., Leander, G.: Decomposing the ASASA block cipher construction. Cryptology ePrint Archive, report 2015/507 (2015). http://eprint.iacr.org/
4. Minaud, B., Derbez, P., Fouque, P.-A., Karpman, P.: Key-recovery attacks on ASASA. In: Iwata, T., et al. (eds.) ASIACRYPT 2015. LNCS, vol. 9453, pp. 3–27. Springer, Heidelberg (2015). doi:10.1007/978-3-662-48800-3_1
5. Biryukov, A., Leurent, G., Perrin, L.: Cryptanalysis of Feistel Networks with secret round functions. In: Dunkelman, O., et al. (eds.) SAC 2015. LNCS, vol. 9566, pp. 102–121. Springer, Heidelberg (2016). doi:10.1007/978-3-319-31301-6_6
6. Dinur, I., Dunkelman, O., Keller, N., Shamir, A.: New attacks on Feistel structures with improved memory complexities. In: Gennaro, R., Robshaw, M. (eds.) CRYPTO 2015. LNCS, vol. 9215, pp. 433–454. Springer, Heidelberg (2015)
7. Canteaut, A., Duval, S., Leurent, G.: Construction of lightweight S-Boxes using Feistel and MISTY structures (full version). Cryptology ePrint Archive, report 2015/711 (2015). http://eprint.iacr.org/
8. Biryukov, A., Perrin, L., Udovenko, A.: Reverse-engineering the S-Box of Streebog, Kuznyechik and STRIBOBr1. In: Fischlin, M., Coron, J.-S. (eds.) EUROCRYPT 2016. LNCS, vol. 9665, pp. 372–402. Springer, Heidelberg (2016). doi:10.1007/978-3-662-49890-3_15

9. Todo, Y.: Structural evaluation by generalized integral property. In: Oswald, E., Fischlin, M. (eds.) EUROCRYPT 2015. LNCS, vol. 9056, pp. 287–314. Springer, Heidelberg (2015)

10. Knudsen, L.R.: DEAL: a 128-bit block cipher, AES submission (1998)

11. Patarin, J.: Generic attacks on Feistel schemes. Cryptology ePrint Archive, report 2008/036 (2008). http://eprint.iacr.org/

12. Biryukov, A., Perrin, L.: On reverse-engineering S-Boxes with hidden design criteria or structure. In: Gennaro, R., Robshaw, M. (eds.) Advances in Cryptology – CRYPTO 2015. Lecture Notes in Computer Science, vol. 9215, pp. 116–140. Springer, Berlin Heidelberg (2015)

13. Carlet, C.: Boolean functions for cryptography and error correcting codes. In: Boolean Models and Methods in Mathematics, Computer Science, and Engineering, vol. 2, pp. 257–397 (2010)

14. Perrin, L., Udovenko, A.: Algebraic insights into the secret Feistel network (full version). Cryptology ePrint Archive, report 2016/398 (2016). http://eprint.iacr.org/

15. Wu, W., Zhang, L.: LBlock: a lightweight block cipher. In: Lopez, J., Tsudik, G. (eds.) ACNS 2011. LNCS, vol. 6715, pp. 327–344. Springer, Heidelberg (2011)

16. Matsui, M.: New block encryption algorithm MISTY. In: Biham, E. (ed.) FSE 1997. LNCS, vol. 1267, pp. 54–68. Springer, Heidelberg (1997)

17. Beaulieu, R., Shors, D., Smith, J., Treatman-Clark, S., Weeks, B., Wingers, L.: The SIMON and SPECK families of lightweight block ciphers. IACR Cryptology ePrint Archive, report 2013/404 (2013)

18. U.S. Department of Commerce/National Institute of Standards and Technology: Data encryption standard. Publication, Federal Information Processing Standards (1999)

19. Biryukov, A., Shamir, A.: Structural cryptanalysis of SASAS. In: Pfitzmann, B. (ed.) EUROCRYPT 2001. LNCS, vol. 2045, pp. 395–405. Springer, Heidelberg (2001)

20. ETSI/Sage: Specification of the 3Gpp confidentiality and integrity algorithms 128-EEA3 & 128-EIA3. Document 4: Design and Evaluation Report, Technical report, ETSI/Sage, September 2011. (http://www.gsma.com/aboutus/wp-content/uploads/2014/12/EEA3_EIA3_Design_Evaluation_v2_0.pdf)

21. The Sage Developers: Sage Mathematics Software (Version 6.8) (2015). http://www.sagemath.org

22. Todo, Y.: Integral cryptanalysis on full MISTY1. In: Gennaro, R., Robshaw, M. (eds.) CRYPTO 2015. LNCS, vol. 9215, pp. 413–432. Springer, Heidelberg (2015)

Integrals Go Statistical: Cryptanalysis of Full Skipjack Variants

Meiqin Wang[1,2(✉)], Tingting Cui[1], Huaifeng Chen[1], Ling Sun[1], Long Wen[1], and Andrey Bogdanov[3]

[1] Key Laboratory of Cryptologic Technology and Information Security, Ministry of Education, Shandong University, Jinan 250100, China
mqwang@sdu.edu.cn
[2] State Key Laboratory of Cryptology, P.O. Box 5159, Beijing 100878, China
[3] Technical University of Denmark, Kongens Lyngby, Denmark

Abstract. Integral attacks form a powerful class of cryptanalytic techniques that have been widely used in the security analysis of block ciphers. The integral distinguishers are based on balanced properties holding with probability one. To obtain a distinguisher covering more rounds, an attacker will normally increase the data complexity by iterating through more plaintexts with a given structure under the strict limitation of the full codebook. On the other hand, an integral property can only be deterministically verified if the plaintexts cover all possible values of a bit selection. These circumstances have somehow restrained the applications of integral cryptanalysis.

In this paper, we aim to address these limitations and propose a novel *statistical integral distinguisher* where only a part of value sets for these input bit selections are taken into consideration instead of all possible values. This enables us to achieve significantly lower data complexities for our statistical integral distinguisher as compared to those of traditional integral distinguisher. As an illustration, we successfully attack the full-round Skipjack-BABABABA for the first time, which is the variant of NSA's Skipjack block cipher.

Keywords: Block cipher · Statistical integral · Integral attack · Skipjack-BABABABA

1 Introduction

Integral attack is an important cryptanalytic technique for symmetric-key ciphers, which was originally proposed by Knudsen as a dedicated attack against Square cipher [7]. Later, Knudsen and Wagner unified it as integral attack [11]. The integral distinguisher of this attack makes use of the *balanced property* where one fixes a part of plaintext bits and takes all possible values for the other plaintext bits such that a specific part of the corresponding ciphertext gets balanced, i.e., each possible partial value for the ciphertext occurs exactly the same number of times. If one additional linear layer after this distinguisher is considered,

© International Association for Cryptologic Research 2016
T. Peyrin (Ed.): FSE 2016, LNCS 9783, pp. 399–415, 2016.
DOI: 10.1007/978-3-662-52993-5_20

the property will be that the XOR of all possible values of the specific part of ciphertext becomes zero, referred to as *zero-sum property* [1] throughout this paper[1]. Being variants of the original integral distinguisher, saturation distinguisher [15] and multiset distinguisher [3] also use the same balanced property or zero-sum property with probability one as integral distinguisher.

Statistical saturation attack is different from integral attack, as proposed by Collard and Standaert in [6]. Here by choosing a plaintext set with some bits fixed while the others vary randomly, the statistical saturation distinguisher tracks the evolution of a non-uniform plaintext distribution through the cipher instead of observing the evolution of the plaintext bits in the integral distinguisher. In other words, the statistical saturation distinguisher requires the same inputs as the integral distinguisher, but uses the different property on the output side to distinguish between the right or wrong key guesses. As Leander showed that the statistical saturation distinguisher is identical to multidimensional linear distinguisher on average in [13], the statistical saturation distinguisher makes use of the advantage (bias or capacity) while the balanced property used in the integral distinguisher has no bias. The first publication of statistical saturation distinguisher came without a method to estimate its complexity. However, this complexity was demonstrated to be inverse proportional to the capacity or square of the capacity for the output under the chosen input set [4,13]. Block ciphers such as PRESENT and PUFFIN are natural targets for such statistical saturation attacks as well as linear cryptanalysis, but the integral cryptanalysis has not been proven efficient for them [21,22]. This highlights the difference between the integral distinguisher and statistical saturation distinguisher.

Integral attack has been widely used for many other block ciphers. In order to reduce the time complexity of integral attack, Moriai *et al.* gave a method to improve the time complexity against low degree round function for higher order differential attacks including integral attacks in [16]. Ferguson *et al.* proposed the partial-sum technique in [8]. Sasaki and Wang presented the meet-in-the-middle technique for integral attack on Feistel ciphers in [17].

So far the data complexity for a given integral has been determined by taking all values of a bit selection at the input of the balanced property. However, there are cases where it is possible or even desirable to shift the tradeoff from data towards time. Often it is the data requirements that exceeds the restriction while the time complexity budget of an attack is far from being exhausted. Therefore, in these cases, it is of paramount importance to reduce the data complexity of an attack to make it applicable. An interesting example of this behaviour is constituted by NSA's Skipjack variant Skipjack-BABABABA studied at ASIACRYPT'12 [5]. It has been attacked for 31 rounds with an integral distinguisher, whereas the data complexity prohibits the attack to apply to the full 32 rounds. In this paper, we aim to remove this restriction by proposing a novel type of integral distinguisher that features a lower data complexity with non-balanced output bits that are still distinguishable from random.

---

[1] Although the common sense of balanced property refers to as zero-sum property, the balanced property used in this paper is active or ALL property.

1.1 Our Contributions

Integrals Go Statistical. We propose a new statistical integral distinguisher that consists in applying a statistical technique on top of the original integral distinguisher with the balanced property. The proposed statistical integral distinguisher requires less data than the original integral distinguisher. Although the balanced property does not strictly hold in the statistical integral distinguisher, we prove that the distribution of output values for a cipher can be distinguished from the distribution of output values which originate from a random permutation. This allows us to distinguish between the two distributions and to construct our statistical integral distinguisher. To quantify the advantage, let s be the number of input bits that take all possible values at some bits of the input while the other input bits are fixed. Furthermore, let t be the number of the output bits that are balanced. Then, for the original integral distinguisher, the data complexity is $\mathcal{O}(2^s)$. At the same time, by deploying our new statistical integral distinguisher, the data complexity is reduced to $\mathcal{O}(2^{s-\frac{t}{2}})$.

In summary, statistical integral attacks we propose have lower data complexity than traditional integral attacks. From [5,19], the traditional integral distinguisher with the balanced property can be converted to a zero-correlation integral distinguisher, so our proposed statistical integral attacks can be regarded as chosen-plaintext multidimensional zero-correlation attacks.

Note that the statistical integral attack is different from the statistical saturation attack as they use different distinguishers and the statistical integral attack is efficient for word-wise ciphers but the statistical saturation attack seems to be valid for bitwise ciphers.

The effectiveness of our proposed statistical integral distinguisher is well presented with the key-recovery attack the full-round Skipjack-BABABABA.

Key Recovery Attack on Full-Round Skipjack's Variants. Using the statistical integral cryptanalysis, we propose a first-time cryptanalysis on the full-round Skipjack-BABABABA — a variant of Skipjack suggested by Knudsen et al. [10,12] to strengthen its resistance against impossible differential attacks. Skipjack-BABABABA has been shown to withstand truncated differentials (which implies that the impossible differentials are also thwarted). At ASIACRYPT'12, Bogdanov et al. [5] attacked 31-round Skipjack-BABABABA by utilizing a 30-round integral distinguisher. Built upon their work, we achieves the full-round attack of Skipjack-BABABABA by taking advantage of the statistical integral technique. To the best of our knowledge, this is the first full-round cryptanalysis against Skipjack-BABABABA. Moreover, we improved the previous attack on 31-round Skipjack-BABABABA in [5] with the new statistical integral distinguisher. The results are summarized in Table 1.

Outline. The new statistical integral distinguisher is established in Sect. 2. Section 3 presents the attack on the full-round Skipjack-BABABABA and the improved attack on 31-round Skipjack-BABABABA. Finally the paper is concluded in Sect. 4.

Table 1. Summary of attacks on Skipjack-BABABABA

Attack	Rounds	Data	Time	Memory	Ref.
Integral ZC	31	2^{48}CP	2^{49}	2^{33} bytes	[5]
Statistical integral	31	$2^{46.8}$CP	2^{48}	$2^{26.6}$ bytes	Sect. 3
Statistical integral	32	$2^{61.7}$CP	$2^{78.1}$	$2^{65.7}$ bytes	Sect. 3

CP: Chosen Plaintext.

2   Statistical Integral Distinguisher

2.1   Integral Distinguisher

In this section, we give some notions and results about the integral distinguisher with balanced property, following the description in [5]. Assume that $H : \mathbb{F}_2^n \to \mathbb{F}_2^n$ is a part of a block cipher. To be convenient and without loss of generality, we split the inputs and outputs into two parts each.

$$H : \mathbb{F}_2^r \times \mathbb{F}_2^s \to \mathbb{F}_2^t \times \mathbb{F}_2^u, \ H(x, y) = \begin{pmatrix} H_1(x, y) \\ H_2(x, y) \end{pmatrix}.$$

Then we use T_λ to denote the function H where the first r bits of its input are fixed to the value λ and only the first t bits of the output are considered:

$$T_\lambda : \mathbb{F}_2^s \to \mathbb{F}_2^t, \ T_\lambda(y) = H_1(\lambda, y).$$

For an integral distinguisher, if y in the above notation iterates all possible values of $\mathbb{F}_2^s$, then the output value $T_\lambda(y)$ is uniformly distributed where $n > s \geq t$ to ensure the balanced property on the t-bit. However, this uniform distribution cannot be obtained if the attacker chooses some random values (other than iterating all possible values) for y. The good side is that when considerable quantity of values of y are chosen, the distribution of $T_\lambda(y)$ can be distinguished from a random variable's distribution. In this case, $T_\lambda(y)$ obeys multivariate hypergeometric distribution while t-bit value chosen randomly from an uniform distribution obeys multinomial distribution. These two distributions can be distinguishable from each other as they have different parameters for large number of input-output pairs N.

2.2   Statistical Integral Distinguisher

Assume that we need N different values of y to distinguish the above two distributions. A t-bit value $T_\lambda(y) \in \mathbb{F}_2^t$ is computed for each y and we allocate a counter vector $V[T_\lambda(y)], T_\lambda(y) \in \mathbb{F}_2^t$ and initialize these counters to zero. These counters are used to keep track of the number of each value $T_\lambda(y)$. Usually t is far from block size n.

It is easy to construct a simple distinguisher which can be described as follows:

- If there is one or more values of $T_\lambda[y]$ satisfying $V[T_\lambda(y)] > 2^{s-t}$, then output random permutation.
- If there is no value of $T_\lambda[y]$ satisfying $V[T_\lambda(y)] > 2^{s-t}$, then output actual cipher.

However, for a random permutation, the probability satisfying $V[T_\lambda(y)] > 2^{s-t}$ is too low to distinguish from the cipher. For example, if $s = 16$, $t = 8$ and $N = 2^{12}$ values of y are involved. For some fixed z, $0 \le z \le 2^t - 1$, the probability that $T_\lambda(y) = z$ is $p = 2^{-8}$. Then $V[z]$ follows a binomial distribution,

$$V[z] \sim B(N, p),$$

which approximately follows a normal distribution $\phi(Np, Np(1-p))$. The probability that $V[z] > 2^{s-t} = 2^8$ for some fixed z is computed as follows,

$$1 - \Phi(\frac{2^{s-t} - Np}{\sqrt{Np(1-p)}}) \approx 1 - \Phi(60.12) \approx 1.1 \times 10^{-787}.$$

As a result, the probability that any $V[z]$ is greater than 2^8 is upper bounded by $256 \times 1.1 \times 10^{-787}$, which is too low to be detected. Thus such a distinguisher only using single counter value is invalid.

Now we will construct an efficient distinguisher by investigating the distribution of the following statistic

$$C = \sum_{T_\lambda(y)=0}^{2^t-1} \frac{(V[T_\lambda(y)] - N \cdot 2^{-t})^2}{N \cdot 2^{-t}}. \tag{1}$$

This statistic is widely used in probability theory. It was also used in [20] for the χ^2 cryptanalysis on DES.

This statistic C follows different distributions determined by whether we are dealing with an actual cipher (right key guess) or a random permutation (wrong key guess).

Proposition 1. *For sufficiently large N and t, the statistic $\frac{2^s - 1}{2^s - N} C_{cipher}$ (C_{cipher} is the statistic C for cipher) follows a χ^2-distribution with degree of freedom $2^t - 1$, which means that C_{cipher} approximately follows a normal distribution with mean and variance*

$$\mu_0 = Exp(C_{cipher}) = (2^t - 1)\frac{2^s - N}{2^s - 1} \text{ and } \sigma_0^2 = Var(C_{cipher}) = 2(2^t - 1)(\frac{2^s - N}{2^s - 1})^2.$$

The statistic C_{random} (C_{random} is the statistic C for randomly drawn permutation) follows a χ^2-distribution with degree of freedom $2^t - 1$, which means that C_{random} approximately follows a normal distribution with mean and variance

$$\mu_1 = Exp(C_{random}) = 2^t - 1 \text{ and } \sigma_1^2 = Var(C_{random}) = 2(2^t - 1).$$

Proof. For a randomly drawn permutation, the values of $V[T_\lambda(y)]$ are obtained by counting the occurrences of $T_\lambda(y)$ when the values are chosen uniformly at random, which follows the multinomial distribution with parameter N and $\boldsymbol{p} = (p_0, \ldots, p_{2^t-1})$, $p_i = 2^{-t}$ $(0 \le i = T_\lambda(y) < 2^t)$.

The well-known Pearson's χ^2 statistical result is that $\sum_{i=1}^{k} \frac{(X_i - np_i)^2}{np_i}$ follows a χ^2-distribution with degree of freedom $k - 1$, where the vector $X = (X_1, \ldots, X_k)$ follows a multinomial distribution with parameters n and $\boldsymbol{p}$, where $\boldsymbol{p} = (p_1, \ldots, p_k)$. We give a short proof for Pearson's χ^2 statistic in Appendix A.1 based on [9,14].

Thus we get the statistic for the randomly drawn permutation

$$C_{random} = \sum_{i=T_\lambda(y)=0}^{2^t-1} \frac{(V[T_\lambda(y)] - Np_i)^2}{Np_i} = \sum_{i=T_\lambda(y)=0}^{2^t-1} \frac{(V[T_\lambda(y)] - N \cdot 2^{-t})^2}{N \cdot 2^{-t}},$$

which follows a χ^2-distribution with degrees of freedom $2^t - 1$. Then for sufficiently large N and t, C_{random} approximately follows a normal distribution with the expected value and variance:

$$Exp(C_{random}) = 2^t - 1 \text{ and } Var(C_{random}) = 2(2^t - 1).$$

For the cipher, the values of $V[T_\lambda(y)]$ follows a multivariate hypergeometric distribution with parameters $(\boldsymbol{K}, 2^s, N)$, where $\boldsymbol{K} = (2^{s-t}, \ldots, 2^{s-t})$.

If the vector $X = (X_1, \ldots, X_k)$ follows a multivariate hypergeometric distribution with parameters $(\boldsymbol{K}, m, n)$, where $\boldsymbol{K} = (K_1, \ldots, K_k)$ with $\sum_{i=1}^{k} K_i = m$, the statistic $\frac{m-1}{m-n} \sum_{i=1}^{k} \frac{(X_i - np_i)^2}{np_i}$ follows a χ^2-distribution with degree of freedom $k - 1$, which is proved in Appendix A.2.

So the statistic for the cipher

$$\frac{2^s - 1}{2^s - N} \sum_{T_\lambda(y)=0}^{2^t-1} \frac{(V[T_\lambda(y)] - N \cdot 2^{-t})^2}{N \cdot 2^{-t}} = \frac{2^s - 1}{2^s - N} C_{cipher}$$

follows a χ^2-distribution with degrees of freedom $2^t - 1$. For sufficiently large N and t, we get C_{cipher} approximately follows a normal distribution with the expected value and variance:

$$Exp(C_{cipher}) = (2^t - 1)\frac{2^s - N}{2^s - 1} \text{ and } Var(C_{cipher}) = 2(2^t - 1)(\frac{2^s - N}{2^s - 1})^2.$$

$\square$

To distinguish these two normal distributions with different means and variances, one can compute the data complexity required as follows, given error probabilities.

Corollary 1 (Data Complexity). *Under the assumption of Proposition 1, for type-I error probability α_0 (the probability to wrongfully discard the cipher), type-II error probability α_1 (the probability to wrongfully accept a randomly chosen*

permutation as the cipher), to distinguish a cipher and a randomly chosen per-
mutation based on t-bit outputs when fixing r-bit inputs and randomly choosing
values for s-bit inputs, the data complexity can be approximated by

$$N = \frac{(2^s - 1)(q_{1-\alpha_0} + q_{1-\alpha_1})}{\sqrt{(2^t - 1)/2} + q_{1-\alpha_0}} + 1, \tag{2}$$

where $q_{1-\alpha_0}$ and $q_{1-\alpha_1}$ are the respective quantiles of the standard normal dis-
tribution.

Note that this statistic test is based on the decision threshold $\tau = \mu_0 + \sigma_0 q_{1-\alpha_0} = \mu_1 - \sigma_1 q_{1-\alpha_1}$: if $C \leq \tau$, the test outputs 'cipher'. Otherwise, if the statistic $C > \tau$, the test outputs 'random'.

As the integral distinguisher with the balanced property is equivalent to the multidimensional zero-correlation distinguisher [5], the statistical integral attacks can be regarded as the chosen-plaintext multidimensional zero-correlation attacks which require lower data complexity than the known-plaintext multidimensional zero-correlation attacks.

2.3   Experiment Results

In order to verify the theoretical model of statistical integral distinguisher, we implement a distinguishing attack on a mini variant of AES with the block size 64-bit denoted as AES* here. The round function of AES* is similar to that of AES, including four operations, *i.e.*, SB, SR, MC and AK. 64-bit block is partitioned into 16 nibbles and SB uses S-box S_0 in LBlock. SR is similar as that of AES, and the matrix used in MC is

$$M = \begin{pmatrix} 1\,1\,4\,9 \\ 9\,1\,1\,4 \\ 4\,9\,1\,1 \\ 1\,4\,9\,1 \end{pmatrix},$$

which is defined over $GF(2^4)$. For the multiplication, each nibble and value in M are considered as a polynomial over $GF(2)$ and then the nibble is multiplied modulo $x^4 + x + 1$ by the value in M. The addition is simply XOR operation. The subkeys are XORed with the nibbles in AK operation.

The distinguisher is shown in Fig. 1, where $(A_1^i, A_2^i, A_3^i, A_4^i), i = 1, 2, 3, 4$ denotes that these special 16 bits are balanced in the integral. Note that the state after SB operation in round 3 takes all 2^{16} values in each row, and 2^4 values in each column. However, after SR operation the state takes all 2^{16} values in each column. We consider the distributions of the 8-bit values of the output including the first nibble in the first row and the last nibble in the second row, which are colored in red in Fig. 1, so $s = 16, t = 8$ here. If we set $\alpha_0 = 0.2$ and different values for N, α_1 and τ can be computed using Eq. (2), thus we proceed the experiment to compute the statistic C for AES* and random permutations. With 1000 times of experiment, we can obtain the empirical error probabilities

$$
\text{R1:} \quad
\begin{bmatrix} A_1^1 & C & C & C \\ C & A_2^1 & C & C \\ C & C & A_3^1 & C \\ C & C & C & A_4^1 \end{bmatrix}
\xrightarrow{SB}
\begin{bmatrix} A_1^1 & C & C & C \\ C & A_2^1 & C & C \\ C & C & A_3^1 & C \\ C & C & C & A_4^1 \end{bmatrix}
\xrightarrow{SR}
\begin{bmatrix} A_1^1 & C & C & C \\ A_2^1 & C & C & C \\ A_3^1 & C & C & C \\ A_4^1 & C & C & C \end{bmatrix}
\xrightarrow{AK\circ MC}
\begin{bmatrix} A_1^1 & C & C & C \\ A_2^1 & C & C & C \\ A_3^1 & C & C & C \\ A_4^1 & C & C & C \end{bmatrix}
$$

$$
\text{R2:} \quad
\begin{bmatrix} A_1^1 & C & C & C \\ A_2^1 & C & C & C \\ A_3^1 & C & C & C \\ A_4^1 & C & C & C \end{bmatrix}
\xrightarrow{SB}
\begin{bmatrix} A_1^1 & C & C & C \\ A_2^1 & C & C & C \\ A_3^1 & C & C & C \\ A_4^1 & C & C & C \end{bmatrix}
\xrightarrow{SR}
\begin{bmatrix} A_1^1 & C & C & C \\ C & C & C & A_2^1 \\ C & C & A_3^1 & C \\ C & A_4^1 & C & C \end{bmatrix}
\xrightarrow{AK\circ MC}
\begin{bmatrix} A_1^1 & A_1^1 & A_1^1 & A_1^1 \\ A_1^2 & A_2^2 & A_2^2 & A_2^2 \\ A_1^3 & A_2^3 & A_3^3 & A_3^3 \\ A_1^4 & A_2^4 & A_3^4 & A_4^4 \end{bmatrix}
$$

$$
\text{R3:} \quad
\begin{bmatrix} A_1^1 & A_2^1 & A_3^1 & A_4^1 \\ A_1^2 & A_2^2 & A_3^2 & A_4^2 \\ A_1^3 & A_2^3 & A_3^3 & A_4^3 \\ A_1^4 & A_2^4 & A_3^4 & A_4^4 \end{bmatrix}
\xrightarrow{SB}
\begin{bmatrix} A_1^1 & A_2^1 & A_3^1 & A_4^1 \\ A_1^2 & A_2^2 & A_3^2 & A_4^2 \\ A_1^3 & A_2^3 & A_3^3 & A_4^3 \\ A_1^4 & A_2^4 & A_3^4 & A_4^4 \end{bmatrix}
\xrightarrow{SR}
\begin{bmatrix} A_1^1 & A_1^2 & A_3^4 & A_4^4 \\ A_2^1 & A_2^2 & A_3^4 & A_4^4 \\ A_3^1 & A_2^3 & A_3^4 & A_4^4 \\ A_4^1 & A_2^4 & A_3^4 & A_4^4 \end{bmatrix}
\xrightarrow{AK\circ MC}
\begin{bmatrix} A_1^1 & A_2^1 & A_3^1 & A_4^1 \\ A_1^2 & A_2^2 & A_3^2 & A_4^2 \\ A_1^3 & A_2^3 & A_3^3 & A_4^3 \\ A_1^4 & A_2^4 & A_3^4 & A_4^4 \end{bmatrix}
$$

$$
\text{R4:} \quad
\begin{bmatrix} A_1^1 & A_2^1 & A_3^1 & A_4^1 \\ A_1^2 & A_2^2 & A_3^2 & A_4^2 \\ A_1^3 & A_2^3 & A_3^3 & A_4^3 \\ A_1^4 & A_2^4 & A_3^4 & A_4^4 \end{bmatrix}
\xrightarrow{SB}
\begin{bmatrix} A_1^1 & A_1^2 & A_1^3 & A_1^4 \\ A_2^1 & A_2^2 & A_2^3 & A_2^4 \\ A_3^1 & A_3^2 & A_3^3 & A_3^4 \\ A_4^1 & A_4^2 & A_4^3 & A_4^4 \end{bmatrix}
\xrightarrow{SR}
\begin{bmatrix} A_1^1 & A_1^2 & A_1^3 & A_1^4 \\ A_2^2 & A_2^3 & A_2^4 & A_2^1 \\ A_3^3 & A_3^4 & A_3^1 & A_3^2 \\ A_4^4 & A_4^1 & A_4^2 & A_4^3 \end{bmatrix}
\xrightarrow{AK}
\begin{bmatrix} A_1^1 & A_1^2 & A_1^3 & A_1^4 \\ A_2^2 & A_2^3 & A_2^4 & A_2^1 \\ A_3^3 & A_3^4 & A_3^1 & A_3^2 \\ A_4^4 & A_4^1 & A_4^2 & A_4^3 \end{bmatrix}
$$

Fig. 1. Integral property for 4-round AES* (The MC operation in the last round is omitted.)

$\hat{\alpha_0}$ and $\hat{\alpha_1}$. The experiment results for $\hat{\alpha_0}$ and $\hat{\alpha_1}$ are compared with the theoretical values α_0 and α_1 in Fig. 2, which shows that the test results for the error probabilities are in good accordance with those for theoretical model.

3   Statistical Integral Attack on Skipjack-BABABABA

3.1   Skipjack and Its Variant Skipjack-BABABABA

Before SIMON and SPECK were proposed in 2013, Skipjack [18] was the only block cipher known to be designed by NSA (declassified in 1998). Skipjack is a 64-bit block cipher with 80-bit key adopting an unbalanced Feistel network with 32 rounds of two types, namely Rule A and Rule B. The 64-bit block of Skipjack is divided into four 16-bit words and each round is described in the form of a linear feedback shift register with additional non-linear keyed G permutation. The keyed G permutation $G : \mathbb{F}_2^{32} \times \mathbb{F}_2^{16} \to \mathbb{F}_2^{16}$ consists of a 4-round Feistel structure whose internal function $F : \mathbb{F}_2^8 \to \mathbb{F}_2^8$ is an 8×8 S-box. Skipjack applies eight rounds of Rule A, followed by eight rounds of Rule B and once again eight rounds of Rule A and finally eight rounds of Rule B. The key schedule of Skipjack takes 10 bytes secret key and uses four bytes at a time to key each G permutation, thus Skipjack's key schedule has a periodicity of five rounds. In this section, we use $k_0, k_1, \ldots, k_9$ to denote the ten bytes secret key. This original Skipjack is often referred to as Skipjack-AABBAABB, where A denotes 4-round Rule A and B denotes 4-round Rule B. A variant of Skipjack, namely Skipjack-BABABABA consisting of four iterations of four-round Rule

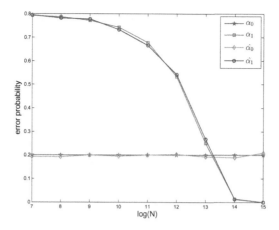

Fig. 2. Experimental results for AES* considering four input nibbles

B followed by four-round Rule A, is also discussed. This variant has the same number of rounds and key schedule as Skipjack-AABBAABB.

Since its declassification, Skipjack-AABBAABB has sparked numerous security analysis. Among which, the best known cryptanalytic result against Skipjack-AABBAABB was reported more than one decade ago by Biham *et al.* [2] at EUROCRYPT'99, where a 24-round impossible differential was revealed and with which an attack against 31-round Skipjack-AABBAABB was mounted. Besides the considerable security analysis, Skipjack's structure was also studied to discuss variants of Skipjack to improve its strength. In [10,12], Knudsen *et al.* suggested that putting Rule B before Rule A, for example, the earlier mentioned Skipjack-BABABABA, might facilitate the resistance to truncated differential attacks. Till now, the only security analysis against Skipjack-BABABABA was reported by Bogdanov *et al.* [5] at ASIACRYPT'12, where an integral distinguisher over 30-round Skipjack-BABABABA was utilized to attack a 31-round version.

3.2 Integral Distinguisher of Skipjack-BABABABA

To attack full-round Skipjack-BABABABA, we are going to use the 30-round integral distinguisher proposed at ASIACRYPT'12 [5]. The 30-round integral distinguisher can be described as: when we take all 2^{48} possible values for the input of round 2 $(\alpha^2, \beta^2, \gamma^2, \delta^2)$ with $\delta^2 = \alpha^2$, the set of all corresponding values for the output of round 31 $\beta^{32} \oplus \gamma^{32}$ is balanced.

3.3 Key Recovery Attack on 32-Round Skipjack-BABABABA

As the integral distinguisher starts at the input of round 2 and ends at the output of round 31, to attack full-round Skipjack-BABABABA we add one round (Rule B) before and append one round (Rule A) after the distinguisher, illustrated in

Fig. 3. Note that in Fig. 3, the internal details of the keyed G permutation are also illustrated. To be more clear, several 8-bit variables a, b, c, d are employed in the attack procedure, see Fig. 3.

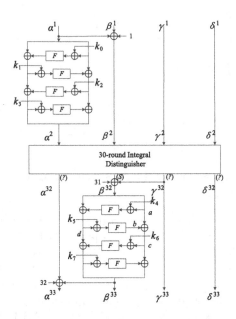

Fig. 3. Key recovery attack on full-round Skipjack-BABABABA

We consider only the integral property of the right 8 bits of $\beta^{32} \oplus \gamma^{32}$, namely $\beta_R^{32} \oplus \gamma_R^{32}$, making $t = 8$ in Eq. (2). And according to the 30-round integral distinguisher, to guarantee the integral property with probability one, we need to iterate through all possible values of $(\alpha^2, \beta^2, \gamma^2, \delta^2 = \alpha^2)$. In other words, s in Eq. (2) is 48. Set $\alpha_0 = 2^{-2.7}$ and $\alpha_1 = 2^{-4}$ (the values of α_0 and α_1 can be chosen appropriately to balance the data complexity, success rate and time complexity in exhaustive phase), we have $q_{1-\alpha_0} \approx 1.02$ and $q_{1-\alpha_1} \approx 1.53$. Thus we need about $2^{45.7}$ values of $(\alpha^2, \beta^2, \gamma^2, \delta^2 = \alpha^2)$ and the threshold value $\tau \approx 221.6$. We can traverse through all possible values of α^1 and β^1 and randomly choose $2^{13.7}$ values for γ^1 and guess the value of k_0, k_1, k_2, k_3 to compute $\alpha^2, \beta^2, \gamma^2$ and set $\delta^2 = \alpha^2$. In this way, $2^{45.7}$ values of $(\alpha^2, \beta^2, \gamma^2, \delta^2 = \alpha^2)$ could be produced under any key value of (k_0, k_1, k_2, k_3). The key can be recovered following Algorithm 1, where β_R^{33} and β_L^{33} denote the right 8-bit and left 8-bit of β^{33} respectively, and so as γ_R^{33}.

Complexity Estimation. In Step 8 and Step 9, the time complexity is $2^{61.7} \cdot 2 = 2^{62.7}$ memory accesses which is equivalent to $2^{62.7}$ encryptions. Next, Step 15 needs about $2^{32} \cdot 2^{16} = 2^{48}$ times of G computation equivalent to $2^{48} \cdot \frac{1}{32} = 2^{43}$ encryptions. Suppose that one memory access to an array of size 2^{24} and of

---

Algorithm 1. Key recovery attack on full-round Skipjack-BABABABA

---

1 Allocate two counter vector $V_0[]$ and $V_0'[]$ with size $2^{61.7}$ and initialize them to zero.

2 Allocate a counter a and initialize a to zero.

3 Take $2^{13.7}$ random values of γ^1 and store them in set S.

4 **for** *all 2^{16} values of α^1* **do**

5   **for** *all 2^{16} values of β^1* **do**

6   **for** *all 2^{16} values of δ^1* **do**

7   **for** *$2^{13.7}$ values of γ^1 in set S* **do**

8   Ask the ciphertext $(\alpha^{33}, \beta^{33}, \gamma^{33}, \delta^{33})$ for the plaintext $(\alpha^1, \beta^1, \gamma^1, \delta^1)$.

9                  $V_0[a] = (\alpha^1, \beta^1, \gamma^1, \delta^1)$, $V_0'[a] = (\alpha^{33}, \beta^{33}, \gamma^{33}, \delta^{33})$.

10   Increase a by one.

11 Allocate a counter vector $V_1[\beta^{33}||\gamma_R^{33}]$.

12 **for** *all 2^{32} values of k_0, k_1, k_2, k_3* **do**

13   Initialize the counter vector $V_1[\beta^{33}||\gamma_R^{33}]$ to zero.

14   **for** *all 2^{16} values of α^1* **do**

15   Compute α^2 and set $\delta^1 = \alpha^2$.

16   **for** *all 2^{16} values of β^1 and $2^{13.7}$ values of γ^1 in set S* **do**

            // Till here, we have $2^{45.7}$ values of $(\alpha^2, \beta^2, \gamma^2, \delta^2 = \alpha^2)$.

17   Access $V_0[a]$ with $(\alpha^1, \beta^1, \gamma^1, \delta^1)$ and get the index a, then access $V_0'[a]$ to get the corresponding ciphertext $(\alpha^{33}, \beta^{33}, \gamma^{33}, \delta^{33})$.

18   Increase the corresponding counter $V_1[\beta^{33}||\gamma_R^{33}]$ by one.

    // $\beta_R^{32} = a \oplus \gamma_R^{33} = b \oplus c \oplus \gamma_R^{33}$.

19   Allocate a counter vector $V_2[d||c \oplus \gamma_R^{33}]$.

20   **for** *all 2^{16} values of k_7 and k_6* **do**

21   Initialize the counter vector $V_2[d||c \oplus \gamma_R^{33}]$ to zero.

22   **for** *all 2^{24} values of $\beta^{33}||\gamma_R^{33}$* **do**

23   Compute $c = F(\beta_L^{33} \oplus k_7) \oplus \beta_R^{33}$, $d = F(c \oplus k_6) \oplus \beta_L^{33}$.

24   Compute $c \oplus \gamma_R^{33}$, update V_2 by $V_2[d||c \oplus \gamma_R^{33}]+ = V_1[\beta^{33}||\gamma_R^{33}]$.

25   Allocate a counter vector $V_3[\beta_R^{32} \oplus \gamma_R^{32}]$.

26   **for** *all 2^8 values of k_5* **do**

27   Initialize the counter vector $V_3[\beta_R^{32} \oplus \gamma_R^{32}]$ to zero.

28   **for** *all 2^{16} values of $d||c \oplus \gamma_R^{33}$* **do**

29   Compute $b = F(d \oplus k_5)$ and $\beta_R^{32} \oplus \gamma_R^{32} = b \oplus c \oplus \gamma_R^{33}$.

30   Update counter vector V_3 by $V_3[\beta_R^{32} \oplus \gamma_R^{32}]+ = V_2[d||c \oplus \gamma_R^{33}]$.

31   Compute C from V_3 according to Eq. (1).

32   **if** $C \leq \tau$ **then**

33   Exhaustively search all right key candidates compatible with this key value.

size $2^{61.7}$ are equivalent to one round encryption and full cipher encryption respectively, then Step 17 and 18 need about $2^{32} \cdot 2^{16} \cdot 2^{16} \cdot 2^{13.7} \cdot (1 + \frac{1}{32}) \approx 2^{77.7}$ encryptions. The operations done in Step 23 and Step 24 are comparable to half-round encryption, which are about $2^{32} \cdot 2^{16} \cdot 2^{24} \cdot \frac{1}{2} \cdot \frac{1}{32} = 2^{66}$ encryptions. In the same way, we regard the operations in Step 29 and Step 30 also as half-round encryption, then the time complexity of these two steps is about $2^{32} \cdot 2^{16} \cdot 2^8 \cdot 2^{16} \cdot \frac{1}{2} \cdot \frac{1}{32} = 2^{66}$ encryptions. As we set the wrong key guess filteration ratio as $\alpha_1 = 2^{-4}$, thus in Step 33, we need to exhaustively search about $2^{80-4} = 2^{76}$ key values to find the right key. To summarize, the time complexity of our key recovery attack on full-round Skipjack-BABABABA is about $2^{62.7} + 2^{43} + 2^{77.7} + 2^{66} + 2^{66} + 2^{76} \approx 2^{78.1}$ encryptions. About the data complexity, in Step 6, all possible values of δ^1 will be iterated through. Thus our attack needs about $2^{61.7}$ chosen plaintexts. The dominant memory requirements occur to store the plaintext/ciphertext pairs in Step 1, which needs about $2 \times 2^{61.7} \times 8 = 2^{65.7}$ bytes.

3.4 Improved Integral Attack on 31-Round Skipjack

With the statistical integral model, we can improve the integral attack on 31-round Skipjack [5] by appending one round after the 30-round distinguisher above, too. In Fig. 3, we attack from the second round to the 32nd round. In order to reduce the time complexity, we consider the statistical integral property of $\beta_R^{32} \oplus \gamma_R^{32}$ and $\beta_L^{32} \oplus \gamma_L^{32}$ respectively, so $t = 8$ in Eq. (2). According to the 30-round integral distinguisher, to guarantee the integral property to hold with probability one, we should iterate through all possible values of $(\alpha^2, \beta^2, \gamma^2, \delta^2 = \alpha^2)$. In other words, s in Eq. (2) is 48. Set $\alpha_0 = 2^{-3.7}$ and $\alpha_1 = 2^{-16}$, we have $q_{1-\alpha_0} \approx 1.43$ and $q_{1-\alpha_1} \approx 4.17$. Thus we need about $2^{46.8}$ values of $(\alpha^2, \beta^2, \gamma^2, \delta^2 = \alpha^2)$ and the threshold value $\tau \approx 160.84$. The key recovery attack is described in Algorithm 2.

Complexity Estimation. Assume that one memory access is equivalent to one round encryption, Step 3 and 4 need about $2^{46.8} \times \frac{1}{31} \approx 2^{41.8}$ encryptions. Then the operations in Step 9 and 10 are about $2^{16} \times 2^{24} \times \frac{1}{2} \times \frac{1}{31} \approx 2^{34.0}$ encryptions. Step 15 and 16 need about $2^{16} \times 2^8 \times 2^{16} \times \frac{1}{2} \times \frac{1}{31} \approx 2^{34.0}$ encryptions. As we set the wrong key guess filteration ratio as 2^{-16}, the numbers of remained key (k_5, k_6, k_7) are about $2^{24-16} = 2^8$ in Step 19. Until now, we exploit the integral property of $\beta_R^{32} \oplus \gamma_R^{32}$ to filter most wrong keys. Next, we use the integral property of $\beta_L^{32} \oplus \gamma_L^{32}$ to filter all wrong keys of (k_4, k_5, k_6, k_7). Step 25 needs about $2^8 \times 2^8 \times 2^{24} \times \frac{1}{31} \approx 2^{35.0}$ encryptions. Finally, by setting $\alpha_1 = 2^{-16}$ we need to exhaustively search about $2^{80-16-16} = 2^{48}$ key values in Step 28 to find the right key. In total the time complexity is about $2^{41.8} + 2^{34.0} + 2^{34.0} + 2^{35.0} + 2^{48} \approx 2^{48}$ encryptions. The dominant memory complexity is required in Step 1 which is about $2 \times 2^{24} \times 3 \approx 2^{27.6}$ bytes which happen.

4   Conclusion

In this paper, we propose the statistical integral attack where we use the statistic technique to deal with the original integral distinguisher with balanced property.

Algorithm 2. Key recovery attack on 31-round Skipjack-BABABABA

---

1 Allocate counter vectors $V_0[\beta^{33}||\gamma_L^{33}]$ and $V_1[\beta^{33}||\gamma_R^{33}]$, then initialize them to zero.

2 **for** $2^{46.8}$ *random values of* $(\alpha^2, \beta^2, \gamma^2, \delta^2 = \alpha^2)$ **do**

3   Ask for the corresponding ciphertext $(\alpha^{33}, \beta^{33}, \gamma^{33}, \delta^{33})$.

4   Increase $V_0[\beta^{33}||\gamma_L^{33}]$ and $V_1[\beta^{33}||\gamma_R^{33}]$ by one respectively.

// $\beta_R^{32} = a \oplus \gamma_R^{33} = b \oplus c \oplus \gamma_R^{33}$

5 Allocate a counter vector $V_2[d||c \oplus \gamma_R^{33}]$ and a list $V_4[\cdot]$.

6 **for** *all* 2^{16} *values of* k_7 *and* k_6 **do**

7   Initialize the counter vector $V_2[d||c \oplus \gamma_R^{33}]$ to zero.

8   **for** *all* 2^{24} *values of* $\beta^{33}||\gamma_R^{33}$ **do**

9   Compute $c = F(\beta_L^{33} \oplus k_7) \oplus \beta_R^{33}$, $d = F(c \oplus k_6) \oplus \beta_L^{33}$.

10   Compute $c \oplus \gamma_R^{33}$, update V_2 by $V_2[d||c \oplus \gamma_R^{33}]+ = V_1[\beta^{33}||\gamma_R^{33}]$.

11   Allocate a counter vector $V_3[\beta_R^{32} \oplus \gamma_R^{32}]$.

12   **for** *all* 2^8 *values of* k_5 **do**

13   Initialize the counter vector $V_3[\beta_R^{32} \oplus \gamma_R^{32}]$ to zero.

14   **for** *all* 2^{16} *values of* $d||c \oplus \gamma_R^{33}$ **do**

15   Compute $b = F(d \oplus k_5)$ and $\beta_R^{32} \oplus \gamma_R^{32} = b \oplus c \oplus \gamma_R^{33}$.

16   Update counter vector V_3 by $V_3[\beta_R^{32} \oplus \gamma_R^{32}]+ = V_2[d||c \oplus \gamma_R^{33}]$.

17   Compute C from V_3 according to Eq. (1).

18   **if** $C \leq \tau$ **then**

19   Store the (k_5, k_6, k_7) in the list $V_4[\cdot]$.

// Since $\alpha_1 = 2^{-16}$, about 2^8 keys in V_4.

20 Allocate a counter vector $V_5[\beta_L^{32} \oplus \gamma_L^{32}]$.

21 **for** *all values of* (k_5, k_6, k_7) *in* $V_4[\cdot]$ **do**

22   **for** *all* 2^8 *values of* k_4 **do**

23   Initialize the counter vector $V_5[\beta_L^{32} \oplus \gamma_L^{32}]$ to zero.

24   **for** *all* 2^{24} *values of* $\beta^{33}||\gamma_L^{33}$ **do**

25   Compute β_L^{32}, update counter vector V_5 by $V_5[\beta_L^{32} \oplus \gamma_L^{32}]+ = V_0[\beta^{33}||\gamma_L^{33}]$.

26   Compute C from V_5 according to Eq. (1).

27   **if** $C \leq \tau$ **then**

28   Exhaustively search all right key candidates compatible with this key value.

The new integral attack has the lower data complexity than that of the original one. Our experiment for mini version of AES shows that the experimental results are in good accordance with the theoretic results. What' more, with this new distinguisher we can improve the previous integral attack on 31-round Skipjack-BABABABA and achieve the full-round attack of Skipjack-BABABABA. In the future, we will apply the statistical integral model to many other block ciphers which are vulnerable to integral attack.

Acknowledgments. This work has been supported by 973 Program (No. 2013C B834205), NSFC Projects (No. 61133013, No. 61572293), Program for New Century Excellent Talents in University of China (NCET- 13-0350).

A   Appendix

A.1   Pearson's χ^2 Statistic from the Multinomial Distribution

In this subsection, we describe Pearson's χ^2 statistic deduced from multinomial distribution and provide a short proof based on [9,14] the asymptotic distribution of the χ^2 expression.

A fundamental result about Pearson's χ^2 statistic is that the following expression follows a χ^2-distribution with degree of freedom $k - 1$

$$\sum_{i=1}^{k} \frac{(X_i - np_i)^2}{np_i},$$

where the random vector $\boldsymbol{X} = (X_1, \ldots, X_k)$ follows a multinomial distribution with parameters n and $\boldsymbol{p}$, where $\boldsymbol{p} = (p_1, \ldots, p_k)$ with $\sum_{i=1}^{k} p_i = 1$.

Now we will give a short proof based on [9,14] in the following.

In probability theory, the multinomial distribution is a generalization of the binomial distribution. For n independent trials each of which leads to a success for exact one of k categories, with each category i ($1 \leq i \leq k$) having a given fixed success probability p_i satisfying $\sum_{i=1}^{k} p_i = 1$. Then if the random variable X_i indicates that the number of times outcome number i is observed over the n trials, the vector $\boldsymbol{X} = (X_1, \ldots, X_k)$ follows a multinomial distribution with parameters n and $\boldsymbol{p} = (p_1, \ldots, p_k)$. Note that while the trials are independent, k outcomes are dependent because they must be summed to n.

Since the variance of X_j is $np_j(1 - p_j)$ and $Cov(X_j, X_l) = -np_j p_l, j \neq l$, the random vector $\boldsymbol{X}$ with $(k - 1)$ dimensions has covariance matrix

$$\Sigma = \begin{pmatrix} np_1(1 - p_1) & -np_1p_2 & \cdots & -np_1p_{k-1} \\ -np_1p_2 & np_2(1 - p_2) & \cdots & -np_2p_{k-1} \\ \vdots & \vdots & \ddots & \vdots \\ -np_1p_{k-1} & -np_2p_{k-1} & \cdots & np_{k-1}(1 - p_{k-1}) \end{pmatrix}.$$

So we can denote Σ as follows,

$$\Sigma = n(D - \boldsymbol{p'p}),$$

where $\boldsymbol{p} = (p_1, p_2, \ldots, p_{k-1})$ and $\boldsymbol{p'}$ is its transposition, D is a $(k - 1) \times (k - 1)$ diagonal matrix and

$$D = \begin{pmatrix} p_1 & & & \\ & p_2 & & \\ & & \ddots & \\ & & & p_{k-1} \end{pmatrix}.$$

Thus, one can show

$$\Sigma^{-1} = \frac{1}{n}\left(D^{-1} + \frac{D^{-1}\boldsymbol{p}'\boldsymbol{p}D^{-1}}{1 - \boldsymbol{p}D^{-1}\boldsymbol{p}'}\right) = \frac{1}{n}\left(D^{-1} + \frac{E}{p_k}\right),$$

where E is a $(k-1) \times (k-1)$ matrix where all entries are equal to one.

We only consider $k-1$ dimensions here, since using all k dimensions would make the variance singular. The first $k-1$ dimensions have all of the information needed anyway, so there's no problem in doing this.

There is a fact: for any d-dimensional normal $\boldsymbol{X}$ with nonsingular covariance matrix, the statistic $(\boldsymbol{X} - \boldsymbol{\mu})'\Sigma^{-1}(\boldsymbol{X} - \boldsymbol{\mu})$ follows a χ^2-distribution with degree of freedom d.

Thus, in the above case we concern $(k-1)$-dimensional normal $\boldsymbol{X}$:

$$
\begin{aligned}
(\boldsymbol{X} - \boldsymbol{\mu})'\Sigma^{-1}(\boldsymbol{X} - \boldsymbol{\mu}) &= (\boldsymbol{X} - n\boldsymbol{p})'\left(\frac{1}{n}\left(D^{-1} + \frac{E}{p_k}\right)\right)(\boldsymbol{X} - n\boldsymbol{p}) \\
&= \frac{1}{n}(\boldsymbol{X} - n\boldsymbol{p})'D^{-1}(\boldsymbol{X} - n\boldsymbol{p}) + \frac{1}{np_k}(\boldsymbol{X} - n\boldsymbol{p})'E(\boldsymbol{X} - n\boldsymbol{p}) \\
&= \sum_{i=1}^{k-1}\frac{(X_i - np_i)^2}{np_i} + \frac{1}{np_k}\left(\sum_{i=1}^{k-1}(X_i - np_i)\right)^2 \\
&= \sum_{i=1}^{k}\frac{(X_i - np_i)^2}{np_i} + \frac{1}{np_k}\left((n - x_k) - n(1 - p_k)\right)^2 \\
&= \sum_{i=1}^{k}\frac{(X_i - np_i)^2}{np_i}.
\end{aligned}
$$

That is, $\sum_{i=1}^{k}\frac{(X_i - np_i)^2}{np_i}$ has an approximation to χ^2-distribution with degree of freedom $k-1$ for large enough n.

A.2 Extend Pearson's χ^2 Statistic to Multivariate Hypergeometric Distribution

In this subsection, we will extend Pearson's χ^2 statistic to multivariate hypergeometric distribution based on the proof of the above subsection and prove that the following expression follows a χ^2-distribution with degree of freedom $k-1$

$$\frac{m-1}{m-n}\sum_{i=1}^{k}\frac{(X_i - np_i)^2}{np_i},$$

where the random vector $\boldsymbol{X} = (X_1, \ldots, X_k)$ follows a multivariate hypergeometric distribution with parameters $(\boldsymbol{K}, m, n)$ where $\boldsymbol{K} = K_1, \ldots, K_k$ with $\sum_{i=1}^{k}K_i = m$.

The multivariate hypergeometric distribution is a generalization of the hypergeometric distribution. For n dependent trials each of which leads to a success for exact one of k categories, with each category i ($1 \leq i \leq k$) having a given fixed success probability $(p_1, p_2, \ldots, p_k)$. The multivariate hypergeometric distribution gives the probability of any particular combination of numbers of successes for the various categories.

Then if the random variables X_i indicates that the number of times outcome number i is observed over the n trials, the vector $\boldsymbol{X} = (X_1, \ldots, X_k)$ follows a multivariate hypergeometric distribution with parameters $(\boldsymbol{K}, m, n)$.

As the mean for X_j is np_j and the variance of X_j is $np_j(1-p_j)\frac{m-n}{m-1}$ and since $Cov(X_j, X_l) = -np_j p_l \frac{m-n}{m-1}, j \neq l$, the random vector $\boldsymbol{X}$ with $k-1$ dimension has covariance matrix

$$\Upsilon = n\frac{m-n}{m-1}(D - \boldsymbol{p}'\boldsymbol{p})$$

and

$$\Upsilon^{-1} = \frac{1}{n}\frac{m-1}{m-n}\left(D^{-1} + \frac{D^{-1}\boldsymbol{p}'\boldsymbol{p}D^{-1}}{1 - \boldsymbol{p}D^{-1}\boldsymbol{p}'}\right) = \frac{1}{n}\frac{m-1}{m-n}\left(D^{-1} + \frac{E}{p_k}\right).$$

With the similar trick as the above subsection, for the $(k-1)$-dimensional normal $\boldsymbol{X}$, it is easy to show that

$$(\boldsymbol{X} - \boldsymbol{\mu})'\Upsilon^{-1}(\boldsymbol{X} - \boldsymbol{\mu}) = (\boldsymbol{X} - n\boldsymbol{p})'\left(\frac{1}{n}\frac{m-1}{m-n}\left(D^{-1} + \frac{E}{p_k}\right)\right)(\boldsymbol{X} - n\boldsymbol{p})$$

$$= \frac{m-1}{m-n}\sum_{i=1}^{k}\frac{(X_i - np_i)^2}{np_i},$$

which means that $\frac{m-1}{m-n}\sum_{i=1}^{k}\frac{(X_i - np_i)^2}{np_i}$ has an approximation to χ^2-distribution with degree of freedom $k-1$ for large enough n.

References

1. Aumasson, J.P., Meier, W.: Zero-Sum Distinguishers for Reduced Keccak-f and for the Core Functions of Luffa and Hamsi. Presented at the rump session of Cryptographic Hardware and Embedded Systems- CHES 2009 (2009)
2. Biham, E., Biryukov, A., Shamir, A.: Cryptanalysis of skipjack reduced to 31 rounds using impossible differentials. In: Stern, J. (ed.) EUROCRYPT 1999. LNCS, vol. 1592, pp. 12–23. Springer, Heidelberg (1999)
3. Biryukov, A., Shamir, A.: Structural cryptanalysis of SASAS. In: Pfitzmann, B. (ed.) EUROCRYPT 2001. LNCS, vol. 2045, pp. 394–405. Springer, Heidelberg (2001)
4. Blondeau, C., Nyberg, K.: Links between truncated differential and multidimensional linear properties of block ciphers and underlying attack complexities. In: Nguyen, P.Q., Oswald, E. (eds.) EUROCRYPT 2014. LNCS, vol. 8441, pp. 165–182. Springer, Heidelberg (2014)

5. Bogdanov, A., Leander, G., Nyberg, K., Wang, M.: Integral and multidimensional linear distinguishers with correlation zero. In: Wang, X., Sako, K. (eds.) ASI-ACRYPT 2012. LNCS, vol. 7658, pp. 244–261. Springer, Heidelberg (2012)
6. Collard, B., Standaert, F.-X.: A statistical saturation attack against the block cipher PRESENT. In: Fischlin, M. (ed.) CT-RSA 2009. LNCS, vol. 5473, pp. 195–210. Springer, Heidelberg (2009)
7. Daemen, J., Knudsen, L.R., Rijmen, V.: The block cipher SQUARE. In: Biham, E. (ed.) FSE 1997. LNCS, vol. 1267, pp. 149–165. Springer, Heidelberg (1997)
8. Ferguson, N., Kelsey, J., Lucks, S., Schneier, B., Stay, M., Wagner, D., Whiting, D.L.: Improved cryptanalysis of rijndael. In: Schneier, B. (ed.) FSE 2000. LNCS, vol. 1978, pp. 213–230. Springer, Heidelberg (2001)
9. Fergnson, T.S.: A Course in Large Sample Theory. Chapman and Hall, London (1996)
10. Knudsen, L.R., Robshaw, M., Wagner, D.: Truncated differentials and skipjack. In: Wiener, M. (ed.) CRYPTO 1999. LNCS, vol. 1666, pp. 165–180. Springer, Heidelberg (1999)
11. Knudsen, L.R., Wagner, D.: Integral cryptanalysis. In: Daemen, J., Rijmen, V. (eds.) FSE 2002. LNCS, vol. 2365, pp. 112–127. Springer, Heidelberg (2002)
12. Knudsen, L.R., Wagner, D.: On the structure of skipjack. Discrete Appl. Math. 111(1–2), 103–116 (2001). Elsevier
13. Leander, G.: On linear hulls, statistical saturation attacks, PRESENT and a cryptanalysis of PUFFIN. In: Paterson, K.G. (ed.) EUROCRYPT 2011. LNCS, vol. 6632, pp. 303–322. Springer, Heidelberg (2011)
14. Lehmann, E.L.: Elements of Large-Sample Theory. Springer, New York (1999)
15. Lucks, S.: The saturation attack - a bait for twofish. In: Matsui, M. (ed.) FSE 2001. LNCS, vol. 2355, pp. 1–15. Springer, Heidelberg (2002)
16. Moriai, S., Shimoyama, T., Kaneko, T.: Higher order differential attack of a CAST cipher. In: Vaudenay, S. (ed.) FSE 1998. LNCS, vol. 1372, pp. 17–31. Springer, Heidelberg (1998)
17. Sasaki, Y., Wang, L.: Meet-in-the-middle technique for integral attacks against feistel ciphers. In: Knudsen, L.R., Wu, H. (eds.) SAC 2012. LNCS, vol. 7707, pp. 234–251. Springer, Heidelberg (2013)
18. Skipjack and KEA Algorithm Specifications, Version 2.0, 29. Available at the National Institute of Standards and Technology's web page, May 1998. http://csrc.nist.gov/groups/ST/toolkit/documents/skipjack/skipjack.pdf
19. Sun, B., Liu, Z., Rijmen, V., Li, R., Cheng, L., Wang, Q., Alkhzaimi, H., Li, C.: Links among Impossible Differential, Integral and Zero Correlation Linear Cryptanalysis. http://eprint.iacr.org/2015/181.pdf
20. Vaudenay, S.: An experiment on DES statistical cryptanalysis. In: Proceedings of the 3rd ACM Conference on Computer and Communications Security, pp. 139–147. ACM (1996)
21. Wu, S., Wang, M.: Integral attacks on reduced-round PRESENT. In: Qing, S., Zhou, J., Liu, D. (eds.) ICICS 2013. LNCS, vol. 8233, pp. 331–345. Springer, Heidelberg (2013)
22. Z'aba, M.R., Raddum, H., Henricksen, M., Dawson, E.: Bit-pattern based integral attack. In: Nyberg, K. (ed.) FSE 2008. LNCS, vol. 5086, pp. 363–381. Springer, Heidelberg (2008)

Note on Impossible Differential Attacks

Patrick Derbez[✉]

Université Rennes 1/IRISA, Rennes, USA
patrick.derbez@irisa.fr

Abstract. While impossible differential cryptanalysis is a well-known and popular cryptanalytic method, errors in the analysis are often discovered and many papers in the literature present flaws. Wishing to solve that, Boura *et al.* [1] presented at ASIACRYPT'14 a generic vision of impossible differential attacks with the aim of simplifying and helping the construction and verification of this type of cryptanalysis. In particular, they gave generic complexity analysis formulas for mounting such attacks and develop new ideas for optimizing them.

In this paper we carefully study this generic formula and show impossible differential attacks for which the real time complexity is much higher than estimated by it. In particular, we show that the impossible differential attack against 25-round TWINE-128, presented at FSE'15 by Biryukov *et al.* [2], actually has a complexity higher than the natural bound of exhaustive search.

Keywords: Truncated impossible differential · Cryptanalysis · Block cipher · TWINE · Complexity

1 Introduction

Impossible differential cryptanalysis, which was independently introduced by Knudsen [3] and Biham *et al.* [4], is well-known and popular cryptanalytic method. Unlike differential attacks [5] that exploit differential characteristics of high probability, the aim of impossible differential cryptanalysis is to use differentials that have a probability of zero to occur in order to eliminate the key candidates leading to such impossible transitions. The first step to mount an impossible differential attack is to find an impossible differential covering a large number of rounds. This is a procedure that has been extensively studied and several approaches have been proposed to derive such impossible transitions efficiently [6–8]. Once an impossible differential has been chosen and placed, one uses it to restrict the possible values of some key bits involved in outer rounds. Indeed, if a candidate key partially encrypts/decrypts a given pair to the impossible differential, then this key is wrong. In this way, we discard as many wrong

© IACR 2016. This article is the final version submitted by the author to the IACR and to Springer-Verlag in March 2016, which appears in the proceedings of FSE 2016.

© International Association for Cryptologic Research 2016
T. Peyrin (Ed.): FSE 2016, LNCS 9783, pp. 416–427, 2016.
DOI: 10.1007/978-3-662-52993-5_21

keys as possible and exhaustively search the rest of the keys. Organizing the attack is usually done with the *early abort technique* [9], introduced by Lu *et al.* at CT-RSA 2008, originally to improve impossible differential attacks against Camellia and MISTY1. With this technique, one does not guess all the involved key material at once but step by step, discarding unwished pairs as soon as possible to reduce the time complexity of the whole procedure.

While the attack principle is rather clear, errors in the analysis are often discovered and many papers in the literature present flaws [9–12]. These flaws include errors in the computation of the time or the data complexity, in the analysis of the memory requirements or of the complexity of some intermediate steps of the attacks. Wishing to solve that, Boura *et al.* [1] presented at ASIACRYPT'14 a generic vision of impossible differential attacks with the aim of simplifying and helping the construction and verification of this type of cryptanalysis. In particular, they gave generic complexity analysis formulas for mounting such attacks and develop new ideas for optimizing them. These advances led to the improvement of previous attacks against well known ciphers such as CLEFIA-128 and Camellia, while also to new attacks against 23-round LBlock and all members of the Simon family.

Our Contribution. In this paper we carefully study the early abort technique from Lu *et al.* and the generic formula given by Boura *et al.*. In particular we build impossible differential attacks against a toy cipher for which the real time complexity is much higher than estimated by the formula. Then we describe an algorithm looking for optimal complexity of impossible differential attacks under the early abort technique. We finally apply it on an attack of Biryukov *et al.* [2] presented at FSE'15 against round-reduced TWINE-128 [13] and show that its complexity is higher than the natural bound of the exhaustive search.

Organization of the Paper. In Sect. 2 we introduce the notations and give the formula of Boura *et al.*. In Sect. 3 we highlight the computational problem behind the early abort technique and provide simple examples for which the real complexity is far from the one given by the formula. Finally, in Sect. 4 we describe the algorithm we used to show that the complexity of the impossible differential attack against 25-round TWINE-128 from Biryukov *et al.* was underestimated and actually higher than 2^{128}.

2 Preliminaries

2.1 Impossible Differential Attacks

We first briefly remain how an impossible differential attack is constructed and introduce our notations (for sake of simplicity we use the exact same ones than in [1]).

Mounting an impossible differential attack starts by splitting the cipher E in three parts $E = E_3 \circ E_2 \circ E_1$ and by finding an impossible differential transition $(\Delta_X \nrightarrow \Delta_Y)$ through E_2. Then Δ_X (resp. Δ_Y) is propagated through E_1^{-1} (resp. E_3) with probability 1 to obtain Δ_{in} (resp. Δ_{out}). We denote by c_{in} and

c_{out} the $\log_2$ of the probability of the transitions $\Delta_{in} \to \Delta_X$ and $\Delta_{out} \to \Delta_Y$ respectively. Finally we denote by k_{in} and k_{out} the key materials involved in those transitions. All in all the attack consists in discarding the keys k for which at least one pair follows the characteristic through E_1 and E_3 and in exhausting the remaining ones.

2.2   A Generic Formula

At ASIACRYPT'14, Boura *et al.* proposed a generic vision of impossible differential attacks with the aim of simplifying and helping the construction and verification of this type of cryptanalysis. In particular, they provided a formula to compute the complexity of such an attack according to its parameters. According to notations introduced Sect. 2.1, their formula is:

- **data:** C_{N_α}
- **memory:** N_α
- **time:** $C_{N_\alpha} + \left(1 + 2^{|k_{in} \cup k_{out}| - c_{in} - c_{out}}\right) N_\alpha C_{E'} + 2^{|k| - \alpha}$

where N_α is such that $(1 - 2^{-c_{in} - c_{out}})^{N_\alpha} = 2^{-\alpha}$, C_{N_α} is the number of chosen plaintexts required to generate N_α pairs satisfying $(\Delta_{in}, \Delta_{out})$, $|k|$ is the key size and $C_{E'}$ is the ratio of the cost of partial encryption to the full encryption.

This formula was given without proof but authors claimed that *"it approximates really well the actual time complexity, as it can be seen in the applications, and in particular, in the tight correspondence shown between the LBlock estimation and the exact calculation from [14]".*

3   Counter-Examples

3.1   The Problem

Computing the time complexity of an impossible differential attack based on the early abort technique [9] is actually an optimization problem. Using notations introduced in Sect. 2.1, and introducing $k_1, k_2, \dots, k_b$ as the key bits of the key material $k_{in} \cup k_{out}$ involved in the attack, the best complexity reached with the early abort technique is the minimal complexity of the following procedure over all the permutations of $\{1, 2, \dots, b\}$:

0. Discard pairs which cannot follow the impossible differential.
1. Guess $k_{\sigma(1)}$
   (a) partially encrypt/decrypt pairs
   (b) discard pairs which cannot follow the impossible differential.
2. Guess $k_{\sigma(2)}$
   (a) partially encrypt/decrypt pairs
   (b) discard pairs which cannot follow the impossible differential.
$$\vdots$$

b. Guess $k_{\sigma(b)}$

  (a) partially encrypt/decrypt pairs
  (b) discard pairs which cannot follow the impossible differential.
  (c) if all pairs have been discarded then perform an exhaustive search over remaining key bits.

Let r_i^σ be the $\log_2$ of the number of pairs discarded after step i. Without taking into account the exhaustive search part, the complexity of the procedure is

$$\sum_{1 \leq i \leq b} 2^{|k_{\sigma(1)} \cup \ldots \cup k_{\sigma(i)}| - \sum_{0 \leq j < i} r_j^\sigma} \cdot N_\alpha C_{E'}.$$

As we see, computing a generic formula for such a problem is far from being trivial.

3.2   A Simple Counter-Example

To highlight the main issue of the generic formula given in [1], let consider a toy block cipher E defined as follows:

$$E = E' \circ MC \circ SR \circ SB \circ AK,$$

where E' is a 128-bit block cipher and where AK, SB, SR and MC respectively are the AddRoundKey, SubBytes, ShriftRows and MixColumns operations from the AES [15]:

- **AddRoundKey** (AK) adds a 128-bit subkey to the state.
- **SubBytes** (SB) applies the same 8-bit to 8-bit invertible Sbox S 16 times in parallel on each byte of the state,
- **ShriftRows** (SR) shifts the i-th row left by i positions,
- **MixColumns** (MC) replaces each of the four column C of the state by $M \times C$ where M is a constant 4×4 maximum distance separable matrix over $GF(2^8)$.

We remind that in the AES, the 128-bit internal state is seen as a 4×4 matrix of bytes where each byte is seen as an element of the finite field $GF(2^8)$.

Now, let us assume the existence of an impossible transition $\Delta_X \not\rightarrow \Delta_Y$ over E' where Δ_X has only one active byte as depicted on Fig. 1. We use this impossible transition to mount an impossible differential against our toy cipher E. We will show that, depending on the key schedule we choose, we are able to make the real complexity of the attack non-marginally higher than the estimated complexity obtained from the generic formula of Boura *et al.*.

Independent Key Bytes. As a well-known fact, the probability of the transition $\Delta_{in} \longrightarrow \Delta_X$ is 2^{-24} and exactly four key bytes are involved in the attack: k_0, k_5, k_{10} and k_{15}. For now let us assume those key bytes are independent. As a consequence, and according to the generic formula, the complexity of the

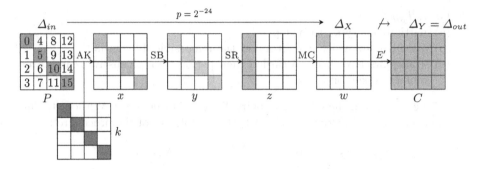

Fig. 1. Impossible differential attack against the toy cipher E.

impossible attack (without taking into account the pairs generation process and the exhaustive search part) is:

$$(1 + 2^{|k_{in}| - c_{in}}) \cdot N \cdot C'_E = (1 + 2^{32-24}) \cdot N \cdot C'_E = 257 \cdot N \cdot C'_E,$$

where N is the number of pairs available and C'_E is the ratio of the cost of partial encryption to the full encryption. A common practice is to take for C'_E the ratio between the active Sboxes during a partial encryption and the total number of Sboxes (say S_E). Hence, the approximated complexity is $4 \cdot 257 \cdot N \cdot S_E^{-1}$.

Let us now compute the real complexity of the attack. Here the order in which key bytes are guessed does not impact the resulting complexity so the *best* procedure is as follows:

1. Guess k_0
   (a) partially encrypt/decrypt pairs
   (b) discard pairs which cannot follow the impossible differential.
2. Guess k_5
   (a) partially encrypt/decrypt pairs
   (b) discard pairs which cannot follow the impossible differential.
3. Guess k_{10}
   (a) partially encrypt/decrypt pairs
   (b) discard pairs which cannot follow the impossible differential.
4. Guess k_{15}
   (a) partially encrypt/decrypt pairs
   (b) discard pairs which cannot follow the impossible differential.

After performing step 1(a), for each pair the differences in the three state variables y_5, y_{10} and y_{15} are known. Indeed, as the MixColumn matrix is MDS, they are fully determined by the value of Δy_0. As a consequence Δx_5, Δx_{10} and Δx_{15} can assume only 2^7 values each and thus only $2^{-3} \cdot N$ pairs remains after step 1(b). Then, each of steps 2(b), 3(b) and 4(b) decreases the number of pairs by a factor 2^7. As a result, the complexity of this procedure is:

$$(2^8 + 2^{8+8-3} + 2^{8+8+8-3-7} + 2^{8+8+8+8-3-7-7}) \cdot N \cdot S_E^{-1} = 57600 \cdot N \cdot S_E^{-1}.$$

All in all the real complexity is higher than the estimated one by a factor $57600/1028 \approx 2^{5.8}$. This factor is non-negligible, especially when compared to involved complexities.

Related Key Bytes. Let now study cases where k_0, k_5, k_{10} and k_{15} are related by one linear equation, so they can assume only 2^{24} values instead of 2^{32}. In that case the generic formula estimates the complexity to $(1 + 2^{24-24}) \cdot N \cdot S_E^{-1} = 2 \cdot N \cdot S_E^{-1}$, independently of the linear relation.

We first consider the case where the equation is $k_0 = k_5$. Thanks to the symmetry in the problem we only have six orders to try: $[k_0, k_5, k_{10}, k_{15}]$, $[k_0, k_{10}, k_5, k_{15}]$, $[k_0, k_{10}, k_{15}, k_5]$, $[k_{10}, k_0, k_5, k_{15}]$, $[k_{10}, k_0, k_{15}, k_5]$ and $[k_{10}, k_{15}, k_0, k_5]$. The corresponding complexities are respectively:

- $(2^8 + 2^{8-3} + 2^{8+8-3-7} + 2^{8+8+8-3-7-7}) \cdot N \cdot S_E^{-1} \approx 2^{8.9} \cdot N \cdot S_E^{-1}$
- $(2^8 + 2^{8+8-3} + 2^{8+8-3-7} + 2^{8+8+8-3-7-7}) \cdot N \cdot S_E^{-1} \approx 2^{13.1} \cdot N \cdot S_E^{-1}$
- $(2^8 + 2^{8+8-3} + 2^{8+8+8-3-7} + 2^{8+8+8-3-7-7}) \cdot N \cdot S_E^{-1} \approx 2^{14.6} \cdot N \cdot S_E^{-1}$
- $(2^8 + 2^{8+8-3} + 2^{8+8-3-7} + 2^{8+8+8-3-7-7}) \cdot N \cdot S_E^{-1} \approx 2^{13.1} \cdot N \cdot S_E^{-1}$
- $(2^8 + 2^{8+8-3} + 2^{8+8+8-3-7} + 2^{8+8+8-3-7-7}) \cdot N \cdot S_E^{-1} \approx 2^{14.6} \cdot N \cdot S_E^{-1}$
- $(2^8 + 2^{8+8-3} + 2^{8+8+8-3-7} + 2^{8+8+8-3-7-7}) \cdot N \cdot S_E^{-1} \approx 2^{14.6} \cdot N \cdot S_E^{-1}$

As we can see the first order is much better than the other ones, as it leads to a much smaller complexity. Thus the real complexity of the attack is $2^{8.9} \cdot N \cdot S_E^{-1}$, higher than the estimated one by a factor $2^{7.9}$. We note that the deviation from the expected complexity is bigger than in the *independent subkey bytes* case.

We now consider the case where the equation is $k_0 \oplus k_5 \oplus k_{10} \oplus k_{15} = 0$, or more generally, the case where the knowledge of three key bytes leads to the knowledge of the fourth one but where there is no relation involving only three key bytes. The real complexity of the attack becomes:

$$(2^8 + 2^{8+8-3} + 2^{8+8+8-3-7} + 2^{8+8+8-3-7-7}) \cdot N \cdot S_E^{-1} \approx 2^{14.6} \cdot N \cdot S_E^{-1},$$

which is higher than for the equation $k_0 = k_5$ by a factor $2^{5.7}$, increasing again the deviation from the expected complexity.

A Trick. One may note that after performing step 1b), we could directly retrieve for each pair the $2 \times 2 \times 2 = 8$ values of (k_5, k_{10}, k_{15}) for which it follows the impossible differential. This would be done at the low cost of 3 memory accesses to a precomputed table. But only the values of (k_5, k_{10}, k_{15}) for which no pair follows the impossible differential matter. Thus we would have to make the list of the 2^{24} possible values of (k_5, k_{10}, k_{15}) before to discard reached values. As a consequence, the resulting complexity of this procedure is:

$$(2^8 \cdot N + 2^8 \cdot 2^{24} + 8 \cdot 2^{8-3} \cdot 2^{|k_0 \cup k_5 \cup k_{10} \cup k_{15}|-32} \cdot N) \cdot S_E^{-1}.$$

As the number of pairs N should be at least close to 2^{24}, this procedure is better than the basic early abort technique. If there is no equation between the four key bytes then the complexity is very close to the one given by Boura *et al*'s formula. On the other hand, if there is at least one equation then the complexity is higher than expected due to the two first terms of the above formula.

3.3   Remarks

Those results highlight some issues with the generic formula of Boura *et al.*. Firstly, there exist impossible differential attacks for which the estimated time complexity is too optimistic and thus attacks with estimated time complexity close to the natural bound may actually not be faster than exhaustive search. Secondly, the formula only takes into account the number of equations between involved key bits while we showed that different equations may lead to different time complexities. In particular, the correct sequence of guesses has to take into account the *fastest* filtering first. It seems Boura *et al.* make the assumption that the order of key guesses/filtering does not matter as all key bits are equally filtering. But this is far from being correct, especially in the context of ARX constructions.

4   Application to TWINE

At FSE'15, Biryukov *et al.* [2] used Boura *et al.* formula to compute the complexity of their impossible differential attack against 25-round TWINE-128 [13]. The attack involves 52 key nibbles which can assume only 2^{124} values instead of 2^{208} thanks to the key schedule and the resulting time complexity is $2^{124.5}$ encryptions, very close to the natural bound of the exhaustive search. As a consequence, and according to remarks of the previous section, it seems probable for the actual time complexity of this attack to be higher than 2^{128}, making it a non-valid attack.

4.1   Description of TWINE

This block cipher uses 16 branches of 4-bits and has a very simple round function: the Feistel function consists in a xor of a sub-key and a call to a unique Sbox based on the inverse function in $GF(2^4)$. Then, the branches are shuffled using a sophisticated nibble permutation ensuring faster diffusion than a simple shift [16]. One version of TWINE uses an 80 bits key, another uses a 128 bits key and we denote these versions as TWINE-80 and TWINE-128. They only differ by their key-schedule and both have 36 rounds. Both key schedules are sparse GFN's using only 2 Sbox calls per round for TWINE-80 and 3 for TWINE-128. At each round, some fixed nibbles of the key-state are used as round keys for the block cipher. One round of TWINE is depicted on Fig. 2.

Keyschedule. The keyschedule produces the 36 round keys from the master key K. It is a variant of GFN with few application of the Sbox used in the round function of TWINE. Two key lengths are available: 80 and 128 bits. In both cases, the subkey WK_0 is first initialized to K and then next subkeys are generated using round constants and the same round function: $WK_{i+1} = F(WK_i, CON^i)$, for $0 \leq i \leq 35$. Finally the round key RK_i is obtained by extracting 8 nibbles from WK_i. The function F used for 128-bit keys is depicted on Fig. 3. We refer the reader to [13] for the 80-bit version of the keyschedule.

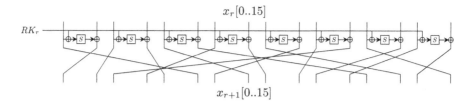

Fig. 2. The round function of TWINE.

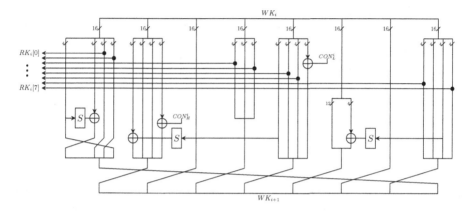

Fig. 3. Keyschedule of TWINE-128.

4.2   Biryukov *et al.* impossible differential attack

Biryukov *et al.* found a truncated impossible characteristic through 13 rounds of TWINE that they extended by 4 rounds at the start and by 8 rounds at the end in order to attack 25 rounds of the cipher. Their attack is depicted on Fig. 4.

The difference in the plaintexts has to be zero in 11 nibbles such that $c_{in} + c_{out} = 16 + 60 = 76$. The key material $k_{in} \cup k_{out}$ is composed of $7 + 45 = 52$ round-key nibbles which can assume only 2^{124} thanks to the keyschedule of TWINE-128 as they all can be computed from the whole subkey WK_{24} except nibble 1.

As a consequence, and according to formula of Boura *et al.*, the complexity of their attack is $D = \alpha \cdot 2^{75.5-39} \cdot 2^{20} = \alpha \cdot 2^{56.5}$, $M = \alpha \cdot 2^{75.5}$ and $T \approx \alpha \cdot 2^{123.5} \cdot C_{E'} + 2^{128-\alpha}$, complexity parametrized by α. As they estimate the ratio $C_{E'}$ to $52/200 \approx 2^{-1.9}$, the value of α minimizing the overall complexity is 5.87.

4.3   Real Complexity of the Attack

Computing the real complexity of Biryukov *et al.* attack seems impossible due to the huge number of involved key nibbles. Indeed, there are 52 key nibbles leading to $52! \approx 2^{225}$ orders for the early abort technique. Thus a naive approach would fail and a clever one has to be used.

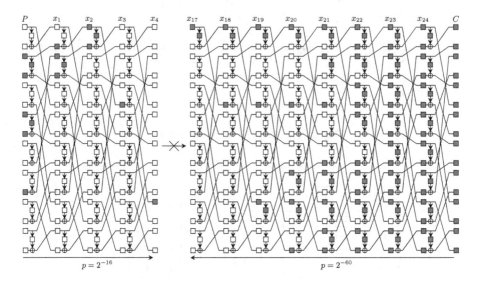

Fig. 4. Impossible differential attack on 25 rounds. No difference in white nibbles.

Pruning Strategy. We note that for the early abort technique, if between two guesses no pairs are discarded then the order in which they are guessed does not matter. Thus key nibbles can be grouped so that at each step pairs are discarded. So now the question becomes when do pairs are discarded? As saw with our simple example this is related to knowing differences before and/or after an Sbox. Since TWINE is a Feistel network things are a bit different and only one case has to be considered. Equations involved to describe round funtion of TWINE all have the following shape:

$$y \oplus z = S(x \oplus k),$$

where x, y and z are state variables while k is a round-key variable. We are interested in the case where both Δx and Δy are known (obtained by partially encrypting plaintexts (resp. decrypting ciphertexts)) and such that $\Delta z = 0$. In that case half of the pairs are discarded since the transition $\Delta x \longrightarrow \Delta y$ is possible with probability 2^{-1}. Then if the actual value of x is obtained by partially encrypting/decrypting plaintexts or ciphertexts then guessing k will allow to reduce the number of pairs by a factor 2^3. So we only have to consider groups of round key nibbles required to compute Δx and Δy, and the ones required to compute $x \oplus k$. Finally, as we are only looking for the fastest attack we can adopt a branch-and-bound strategy to accelerate the search.

Practice. For the considered attack there are 19 tuples (x, y, z) as expected. Determining the corresponding groups of round key nibbles is an easy task. However, computing the number of values those groups (and their unions) can assume is more complicated while essential to the computation of the complexity. To solve this we used the same approach Derbez et al. [17] used to exhaust a particular kind of meet-in-the-middle attacks against the AES in a paper presented at FSE'13. Indeed, they provided a tool which takes as input a system of equations E in variable X and a subset $Y \subseteq X$ and gives as output a list of optimal algorithms enumerating all the possible values of Y under constraint of E with predictable time and memory complexities. The system of equations has to be composed of equations with the following shape:

$$\sum \alpha_i x_i \oplus \sum \beta_j S(x_j) \oplus \gamma = 0,$$

where a_i's, β_j's and γ are constant from a finite field $GF(2^q)$ and S is an q-bit Sbox. As the key schedule of TWINE is naturally described by such equations we were able to use this tool. Note that the output of their tool is a list because the number of possible values of Y enumerated by considered algorithms is not necessary constant and if an algorithm is slower than an other but finds less possible values for Y than it then they had to study both of them. But in our case we only care about the fastest algorithm, even if it enumerates more solutions.

Our algorithm was able to find the optimal permutation (see Appendix A) for the early abort technique in about 1 h on a personal computer. As a result we found that for all permutation σ:

$$\sum_{1 \leq i \leq 38} 2^{|k_{\sigma(1)} \cup \dots \cup k_{\sigma(i)}| - \sum_{0 \leq j < i} r_j^\sigma} \cdot N_\alpha C_{E'} > 2^{54} \cdot N_\alpha C_{E'}.$$

As $N_\alpha = \alpha \cdot 2^{75.5}$, the time complexity of the whole attack is higher than:

$$C_{N_\alpha} + \alpha \cdot 2^{127.6} + 2^{128-\alpha},$$

where $2^{128-\alpha}$ corresponds to time complexity of performing an exhaustive search on the remaining keys. Hence, if only based on the early abort technique, the attack is actually slower than an exhaustive search for all value $\alpha > 0$.

5   Conclusion

In this paper we have shown that the generic complexity analysis formula presented by Boura et al. at ASIACRYPT'14 does not always give a right estimation of the time complexity of impossible differential attacks. As proof we constructed simple counter-examples for which the real complexity is much higher than expected, one reaching a deviation of $2^{13.6}$ from the formula. As a consequence the formula is to use with caution, in particular when time complexity is close to the natural bound of the exhaustive search.

While we searched for, we were unable to find an impossible differential attack for which the real time complexity would be lower than the estimated one. Finding such an attack or proving that the formula provides a lower bound on the complexity would be an interesting future work.

Finally we also showed that, if using only the early abort technique, the time complexity of the impossible differential attack against 25-round TWINE-128, presented at FSE'15 by Biryulov et al., is higher than expected, and in particular, higher than 2^{128}.

A   Optimal Sequence

We found the following permutation to be optimal for the early abort technique applied to the 25-round impossible differential attack:

1. $\Delta x_0[2]$, $\Delta x_0[3]$
2. $\Delta x_0[6]$, $\Delta x_0[7]$
3. $\Delta x_1[2]$, $\Delta x_1[3]$
4. $\Delta x_2[0]$, $\Delta x_2[1]$
5. $y_0[2]$
6. $\Delta x_{23}[12]$, $\Delta x_{24}[10]$
7. $y_{23}[12]$
8. $\Delta x_{22}[12]$, $\Delta x_{23}[10]$
9. $y_{22}[12]$
10. $\Delta x_{22}[6]$, $\Delta x_{23}[8]$
11. $y_{22}[6]$
12. $\Delta x_{22}[2]$, $\Delta x_{23}[4]$
13. $y_{22}[2]$
14. $y_0[6]$
15. $y_1[2]$
16. $\Delta x_{21}[10]$, $\Delta x_{22}[2]$
17. $y_{21}[10]$
18. $\Delta x_{20}[10]$, $\Delta x_{21}[2]$
19. $y_{20}[10]$
20. $\Delta x_{21}[2]$, $\Delta x_{22}[4]$
21. $y_{21}[2]$
22. $\Delta x_{21}[0]$, $\Delta x_{22}[0]$
23. $\Delta x_{20}[0]$, $\Delta x_{21}[0]$
24. $y_{20}[0]$
25. $\Delta x_{19}[0]$, $\Delta x_{20}[0]$
26. $\Delta x_{21}[12]$, $\Delta x_{22}[10]$
27. $y_{19}[0]$
28. $y_2[0]$
29. $\Delta x_{19}[12]$, $\Delta x_{20}[10]$
30. $y_{19}[12]$
31. $y_{21}[12]$
32. $y_{21}[0]$
33. $\Delta x_{18}[0]$, $\Delta x_{19}[0]$
34. $\Delta x_{20}[12]$, $\Delta x_{21}[10]$
35. $\Delta x_{17}[0]$, $\Delta x_{18}[0]$
36. $y_{20}[12]$
37. $y_{18}[0]$
38. $y_{17}[0]$

Each item v has to be understood as *guess the key material required to compute v from the plaintexts/ciphertexts* and $y_r[2i] = x_r[2i] \oplus k_r[i]$.

References

1. Boura, C., Naya-Plasencia, M., Suder, V.: Scrutinizing and improving impossible differential attacks: applications to clefia, camellia, lblock and simon. In: Proceedings, Part I, Advances in Cryptology - ASIACRYPT 2014–20th International Conference on the Theory and Application of Cryptology and Information Security, Kaoshiung, Taiwan, R.O.C., 7–11 December 2014, pp. 179–199 (2014)
2. Biryukov, A., Derbez, P., Perrin, L.: Differential analysis and meet-in-the-middle attack against round-reduced TWINE. In: Leander, G. (ed.) FSE 2015. LNCS, vol. 9054, pp. 3–27. Springer, Heidelberg (2015)
3. Knudsen, L.R.: Deal - a 128-bit block cipher. Technical report, Department of Informatics (1998)
4. Biham, E., Biryukov, A., Shamir, A.: Cryptanalysis of skipjack reduced to 31 rounds using impossible differentials. In: Stern, J. (ed.) EUROCRYPT 1999. LNCS, vol. 1592, pp. 12–23. Springer, Heidelberg (1999)
5. Biham, E., Shamir, A.: Differential cryptanalysis of des-like cryptosystems. In: CRYPTO 1991 (1991)

6. Kim, J.-S., Hong, S.H., Sung, J., Lee, S.-J., Lim, J.-I., Sung, S.H.: Impossible differential cryptanalysis for block cipher structures. In: Johansson, T., Maitra, S. (eds.) INDOCRYPT 2003. LNCS, vol. 2904, pp. 82–96. Springer, Heidelberg (2003)
7. Luo, Y., Lai, X., Wu, Z., Gong, G.: A unified method for finding impossible differentials of block cipher structures. Inf. Sci. **263**, 211–220 (2014)
8. Wu, S., Wang, M.: Automatic search of truncated impossible differentials for word-oriented block ciphers. In: Galbraith, S., Nandi, M. (eds.) INDOCRYPT 2012. LNCS, vol. 7668, pp. 283–302. Springer, Heidelberg (2012)
9. Lu, J., Kim, J.-S., Keller, N., Dunkelman, O.: Improving the efficiency of impossible differential cryptanalysis of reduced camellia and MISTY1. In: Malkin, T. (ed.) CT-RSA 2008. LNCS, vol. 4964, pp. 370–386. Springer, Heidelberg (2008)
10. Minier, M., Naya-Plasencia, M.: A related key impossible differential attack against 22 rounds of the lightweight block cipher lblock. Inf. Process. Lett. **112**(16), 624–629 (2012)
11. Wu, W., Zhang, L., Zhang, W.: Improved impossible differential cryptanalysis of reduced-round camellia. In: Avanzi, R.M., Keliher, L., Sica, F. (eds.) SAC 2008. LNCS, vol. 5381, pp. 442–456. Springer, Heidelberg (2009)
12. Zhang, W., Han, J.: Impossible differential analysis of reduced round CLEFIA. In: Yung, M., Liu, P., Lin, D. (eds.) Inscrypt 2008. LNCS, vol. 5487, pp. 181–191. Springer, Heidelberg (2009)
13. Suzaki, T., Minematsu, K., Morioka, S., Kobayashi, E.: TWINE: a lightweight block cipher for multiple platforms. In: 19th International Conference Selected Areas in Cryptography, SAC 2012, Windsor, ON, Canada, 15–16 August 2012, Revised Selected Papers, pp. 339–354 (2012)
14. Boura, C., Minier, M., Naya-Plasencia, M., Suder, V.: Improved impossible differential attacks against round-reduced lblock. IACR Cryptol. ePrint Arch. **2014**, 279 (2014)
15. NIST: Advanced Encryption Standard (AES), FIPS 197. Technical report, NIST, November 2001
16. Suzaki, T., Minematsu, K.: Improving the generalized feistel. In: Hong, S., Iwata, T. (eds.) FSE 2010. LNCS, vol. 6147, pp. 19–39. Springer, Heidelberg (2010)
17. Derbez, P., Fouque, P.: Exhausting demirci-selçuk meet-in-the-middle attacks against reduced-round AES. In: Fast Software Encryption - 20th International Workshop, FSE 2013, Singapore, 11–13 March 2013, Revised Selected Papers, pp. 541–560 (2013)

Improved Linear Hull Attack on Round-Reduced SIMON with Dynamic Key-Guessing Techniques

Huaifeng Chen[1] and Xiaoyun Wang[1,2(✉)]

[1] Key Laboratory of Cryptologic Technology and Information Security,
Ministry of Education, Shandong University, Jinan 250100, China
`hfchen@mail.sdu.edu.cn`
[2] Institute of Advanced Study, Tsinghua University, Beijing 100084, China
`xiaoyunwang@mail.tsinghua.edu.cn`

Abstract. SIMON is a lightweight block cipher family proposed by NSA in 2013. It has drawn many cryptanalysts' attention and varieties of cryptanalysis results have been published, including differential, linear, impossible differential, integral cryptanalysis and so on. In this paper, we give the improved linear attacks on all reduced versions of SIMON with dynamic key-guessing technique, which was proposed to improve the differential attack on SIMON recently. By establishing the boolean function of parity bit in the linear hull distinguisher and reducing the function according to the property of AND operation, we can guess different subkeys (or equivalent subkeys) for different situations, which decrease the number of key bits involved in the attack and decrease the time complexity in a further step. As a result, 23-round SIMON32/64, 24-round SIMON48/72, 25-round SIMON48/96, 30-round SIMON64/96, 31-round SIMON64/128, 37-round SIMON96/96, 38-round SIMON96/144, 49-round SIMON128/128, 51-round SIMON128/192 and 53-round SIMON128/256 can be attacked. As far as we know, our attacks on most reduced versions of SIMON are the best compared with the previous cryptanalysis results. However, this does not shake the security of SIMON family with full rounds.

1   Introduction

In 2013, NSA proposed a new family of lightweight block cipher with Feistel structure, named as SIMON, which is tuned for optimal performance in hardware applications [7]. The SIMON family consists of various block and key sizes to match different application requirements. There is no S-box in the round function. The round function consists of AND, rotation and Xor (ARX structure), leading to a low-area hardware requirement.

Related Works. SIMON family has attracted a lot of cryptanalysts' attention since its proposition. Many cryptanalysis results on various versions of SIMON were published. For differential attack, Alkhzaimi and Lauridsen [5] gave the first differential attacks on all versions of SIMON. The attacks cover 16, 18, 24, 29, 40 rounds for the versions with block size 32, 48, 64, 96 and 128 respectively.

© International Association for Cryptologic Research 2016
T. Peyrin (Ed.): FSE 2016, LNCS 9783, pp. 428–449, 2016.
DOI: 10.1007/978-3-662-52993-5_22

At FSE 2014, Abed *et al.* [3] gave differential attack on variants of SIMON reduced to 18, 19, 26, 35, 46 rounds with respective block size 32, 48, 64, 96 and 128. At the same time, Biryukov *et al.* [9] gave differential attack on several versions of SIMON independently. And 19-round SIMON32, 20-round SIMON48, 26-round SIMON64 were attacked. Then Wang *et al.* [20] proposed better differential attacks with existing differentials, using dynamic key-guessing techniques. As a result, 21-round SIMON32/64, 23-round SIMON48/72, 24-round SIMON48/96, 28-round SIMON64/96, 29-round SIMON64/128, 37-round SIMON96/96, 37-round SIMON96/144, 49-round SIMON128/128, 49-round SIMON128/192, 50-round SIMON128/256 were attacked.

For the earlier linear cryptanalysis, 11, 14, 16, 20, 23-round key recovery attacks on SIMON with block size 32, 48, 64, 96, 128 were presented in [2]. Then, Alizadeh *et al.* [4] improved the linear attacks on 13-round SIMON32, 15-round SIMON48, 19-round SIMON64, 28-round SIMON96, 35-round SIMON128. Recently, Abdelraheem *et al.* [1] took advantage of the links between linear characteirstics and differential characteristics for SIMON and found some linear distinguishers using differential characteristics found earlier. They presented various linear attacks on SIMON with linear, multiple linear, linear hull cryptanalysis. The linear hull cryptanalysis has better attack results, which can attack 21-round SIMON32/64, 20-round SIMON48/72, 21-round SIMON48/96, 27-round SIMON64/96, 29-round SIMON64/128, 36-round SIMON96/144, 48-round SIMON128/192 and 50-round SIMON128/256. Then, with the Mixed-integer Linear Programming based technique, Shi *et al.* [17] searched new linear trails and linear hulls, and 21, 21, 29 rounds for SIMON32/64, SIMON48/96, SIMON64/128 were attacked respectively. Also, Sun *et al.* [18] found a 16-round linear hull distinguisher of SIMON48, with which he attacked 23-round SIMON48/96. Ashur [6] introduced a new way to calculate the correlations of short linear hulls and provided a more accurate estimation for some previously published linear trails. He gave multiple linear cryptanalysis on 24-round SIMON32/64, 23-round SIMON48/72, 24-round SIMON48/96, 24-round SIMON64/96 and 25-round SIMON64/128. However, it uses the correlation when all the subkeys are zero as the expected correlation under random key situations, which is not exact. Moreover, if the potential of each linear hull of the cipher is smaller than that of random permutations, then the combination of these linear hulls can not distinguish between the cipher and a random permutation.

Also, there are some results with other attack models, such as impossible differential cryptanalysis [4,10,12,21], zero-correlation cryptanalysis [21] and integral cryptanalysis [21].

Our Contributions. In this paper, we give the improved linear hull attacks on all reduced versions of SIMON family with dynamic key-guessing technique, which was proposed initially to improve the differential attack on SIMON [20], using existing linear hull distinguishers. In linear attack, one important point is to compute the empirical correlations (bias) of the parity bit, which derives from the Xor-sum of the active bits at both sides of the linear hull distinguisher, under some key guess. Our attack on SIMON improves this procedure efficiently.

Table 1. Summary of Linear Hull Attacks on SIMON

Cipher	Attacked rounds	Data	Time	Reference
SIMON32/64	21	$2^{30.56}$	$2^{55.56}$	[1]
	21	-	-	[17]
	23	$2^{31.19}$	$2^{61.84}A + 2^{56.3}E$	Sect. 4.2
SIMON48/72	20	$2^{44.11}$	$2^{70.61}$	[1]
	24	$2^{47.92}$	$2^{67.89}A + 2^{65.34}E$	Sect. 4.3
SIMON48/96	21	$2^{44.11}$	$2^{70.61}$	[1]
	21	-	-	[17]
	23	$2^{47.92}$	$2^{92.92}$	[18]
	25	$2^{47.92}$	$2^{89.89}A + 2^{88.28}E$	Sect. 4.3
SIMON64/96	27	$2^{62.53}$	$2^{88.53}$	[1]
	30	$2^{63.53}$	$2^{93.62}A + 2^{88.13}E$	Sect. 4.3
SIMON64/128	29	$2^{62.53}$	$2^{123.53}$	[1]
	29	-	-	[17]
	31	$2^{63.53}$	$2^{119.62}A + 2^{120.00}E$	Sect. 4.3
SIMON96/96	**37**	$2^{95.2}$	$2^{67.94}A + 2^{88}E$	Sect. 4.3
SIMON96/144	36	$2^{94.2}$	$2^{123.5}$	[1]
	38	$2^{95.2}$	$2^{98.94}A + 2^{136.00}E$	Sect. 4.3
SIMON128/128	**49**	$2^{127.6}$	$2^{87.77}A + 2^{120}E$	Sect. 4.3
SIMON128/192	48	$2^{126.6}$	$2^{187.6}$	[1]
	51	$2^{127.6}$	$2^{155.77}A + 2^{184.00}E$	Sect. 4.3
SIMON128/256	50	$2^{126.6}$	$2^{242.6}$	[1]
	53	$2^{127.6}$	$2^{239.77}A + 2^{248.01}E$	Sect. 4.3

* '-' means not given; A means addition; E means encryption;

The non-linear part in the round function of SIMON is mainly derived from the bitwise AND (&) operation while it has a significant feature. For details, if one of the two elements is equal to zero, the result of their AND will be zero, no matter what value the other element takes. For a function $f = f_1(x_1, k_1) \& f_2(x_2, k_2)$, if we GUESS k_1 at first, and SPLIT the all $x = x_1 || x_2$ into two cases: case 1, $f_1(x_1, k_1) = 0$; case 2, $f_1(x_1, k_1) = 1$, there is no need to guess the key bits k_2 in case 1, since $f = 0$ holds for any value of f_2 in case 1. Then, we can compute the correlations in each case with less time and at last, we COMBINE the two correlations together for corresponding key $k = k_1 || k_2$.

At first, we give the boolean representations for the parity bit in the linear distinguisher of SIMON. And then we apply the GUESS, SPLIT and COMBINE technique in the calculation of the empirical correlations, which mainly exploits the dynamic key-guessing idea to reduce the number of subkey bits guessed significantly. For example, in the attack on 21-round SIMON32, 32 subkey bits are involved. With above technique, we can only guess 12.5 bits from the total 32-bit subkey on average to compute the correlations.

As a result, the improved attack results are shown as follows. We can attack 23-round SIMON32/64, 24-round SIMON48/72, 25-round SIMON48/96, 30-round SIMON64/96, 31-round SIMON64/128, 37-round SIMON96/96, 38-round SIMON96/144, 49-round SIMON128/128, 51-round SIMON128/192 and 53-round SIMON128/256. This improves the linear attack results for all versions. From the point of number of rounds attacked, the results on most versions are best up to now. The existing and new linear hull attack results on SIMON are summarized in Table 1. Also, we implement the 21-round attack on SIMON32. In the attack, we can decrease the 32 subkey bits involved in the attack by 8 bits. The experiments show that the attack success probability is about 27.7 % using $2^{31.19}$ plaintext-ciphertext pairs.

The paper is organised as follows. In Sect. 2, we introduce the linear (hull) cryptanalysis and give the description of SIMON family. Section 3 gives the dynamic key-guessing technique used in the linear cryptanalysis. Then the improved attacks on SIMON32/64 and all other variants are given in Sect. 4. Finally, we conclude in Sect. 5. Appendix A gives the time complexities to calculate the empirical correlations in some simple situations.

2 Preliminaries

2.1 Linear Cryptanalysis and Linear Hull

$\mathbb{F}_2$ denotes the field with two elements and $\mathbb{F}_2^n$ is the n-dimensional vector space of $\mathbb{F}_2$. Let $g : \mathbb{F}_2^n \to \mathbb{F}_2$ be a Boolean function. Let $B(g) = \sum_{x \in \mathbb{F}_2^n} (-1)^{g(x)}$. The correlation $c(g)$ of g and 0 (in the following paper, when we say the correlation of a function, it means the correlation of this function and 0) is defined by

$$c(g) = 2^{-n} \sum_{x \in \mathbb{F}_2^n} (-1)^{g(x)} = 2^{-n} B(g). \tag{1}$$

(In some situations of the remainder of this paper, we regard $B(g)$ as the correlation for simplicity of description.) The bias of g is defined by half of $c(g)$, which is represented as $\epsilon(g) = \frac{1}{2} c(g)$.

Linear cryptanalysis [13] is a powerful cryptanalytic method proposed in 1993 to cryptanalysis DES. At first, one tries to find a good linear approximation involving some plaintext bits, ciphertext bits and the subkey bits as follows

$$\alpha \cdot P \oplus \beta \cdot C = \gamma \cdot K, \tag{2}$$

where α, β, γ are masks and P, C, K represent the plaintext, ciphertext and keys. 'good' means that the probability of the linear approximations is far away from $1/2$, which is the probability in random situations. In other words, higher absolute of bias $\epsilon(\alpha \cdot P \oplus \beta \cdot C \oplus \gamma \cdot K)$ leads to better linear crypanalysis result in general. Algorithms 1 and 2 in [13] are two attack models exploiting the linear approximation as distinguisher. $\mathcal{O}(\frac{1}{\epsilon^2})$ known plaintexts are needed in the key-recovery attacks.

Then in 1994, Nyberg [15] studied the linear approximations with same input mask α and output mask β, and denoted them as linear hull. The potential of a linear hull is defined as

$$ALH(\alpha, \beta) = \sum_{\gamma} \epsilon^2(\alpha \cdot P \oplus \beta \cdot C \oplus \gamma \cdot K) = \bar{\epsilon}^2. \tag{3}$$

The effect of linear hull is that the final bias $\bar{\epsilon}$ may become significantly higher than that of any individual linear trail. Then the linear attacks with linear hull require less known plaintexts, i.e., $\mathcal{O}(\frac{1}{\bar{\epsilon}^2})$.

Selçuk and Biçak [16] gave the estimation of success probability in linear attack for achieving a desired advantage level. The advantage is the complexity reduction over the exhaustive search. For example, if m-bit key is attacked and the right key is ranked t-th among all 2^m candidates, the advantage of this attack is $m - log_2(t)$. Theorem 2 in [16] described the relation between success rate, advantage and number of data samples.

Theorem 1 (Theorem 2 in [16]). *Let P_S be the probability that a linear attack, as defined by Algorithm-2 in [13], where all candidates are tried for an m-bit sub-key, in an approximation of probability p, with N known plaintext blocks, delivers an a-bit or higher advantage. Assuming that the approximation's probablity is independent for each key tried and is equal to 1/2 for all wrong keys, we have, for sufficiently large m and N,*

$$P_S = \int_{-2\sqrt{N}|p-1/2|+\Phi^{-1}(1-2^{-a-1})}^{\infty} \phi(x)dx, \tag{4}$$

independent of m.

2.2   Description of SIMON

SIMON is a family of lightweight block cipher with Feistel structure designed by NSA, which is tuned for optimal performance in hardware applications [7]. The SIMON block cipher with an n-bit word (hence $2n$-bit block) is denoted SIMON$2n$, where n is limited to be 16, 24, 32, 48 or 64. The key length is required to be mn where m takes value from $2, 3$ and 4. SIMON$2n$ with m-word key is referred to SIMON$2n/mn$. There are ten versions in the SIMON family and the detailed parameters are listed in Table 2. Before introducing the round functions of SIMON, we give some notations of symbols used throughout this paper.

X^r $2n$-bit output of round r (input of round $r + 1$)

X_L^r left half n-bit of X^r

X_R^r right half n-bit of X^r

K^r subkey used in round $r + 1$

x_i the i-th bit of x, begin with bit 0 from right (e.g., $X_{L,0}^r$ is the LSB of X_L^r)

$x_{i_1,\ldots,i_t}$ the XOR-sum of x_i for $i = i_1, i_2, \ldots, i_t$ (e.g., $x_{0,1} = x_0 \oplus x_1$)

$x \lll i$ left circulant shift by i bits of x

$\oplus$ bitwise XOR

$\&$ bitwise AND

$F(x)$ $F(x) = ((x \lll 1)\&(x \lll 8)) \oplus (x \lll 2)$

The r-th round function of SIMON$2n$ is a Feistel map

$$F_{K^{r-1}} : \mathbb{F}_2^n \times \mathbb{F}_2^n \to \mathbb{F}_2^n \times \mathbb{F}_2^n,$$
$$(X_L^{r-1}, X_R^{r-1}) \to (X_L^r, X_R^r)$$

where $X_R^r = X_L^{r-1}$ and $X_L^r = F(X_L^{r-1}) \oplus X_R^{r-1} \oplus K^{r-1}$. The round function of SIMON is depicted in Fig. 1. Suppose the number of rounds is T, the whole encryption of SIMON is the composition $F_{K^{T-1}} \circ \cdots \circ F_{K^1} \circ F_{K^0}$. The subkeys are derived from the master key. The key schedules are a little different depending on the key size. However, the master key can be derived from any m consecutive subkeys. Please refer to [7] for more details.

Table 2. The SIMON Family Block Ciphers

block size ($2n$)	key size (mn)	rounds
32 ($n = 16$)	64 ($m = 4$)	32
48 ($n = 24$)	72 ($m = 3$)	36
	96 ($m = 4$)	36
64 ($n = 32$)	96 ($m = 3$)	42
	128 ($m = 4$)	44
96 ($n = 48$)	96 ($m = 2$)	52
	144 ($m = 3$)	54
128 ($n = 64$)	128 ($m = 2$)	68
	192 ($m = 3$)	69
	256 ($m = 4$)	72

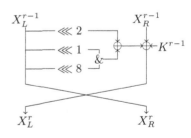

Fig. 1. Round Function of SIMON

3   Time Reduction in Linear Cryptanalysis for Bit-Oriented Block Cipher

For bit-oriented block cipher, such as SIMON, the operations of round function can be seen as the concatenation of some boolean functions. For example, in

SIMON32, the 0-th bit of X_L^r is a boolean function of some bits of X^{r-1} and subkeys as follows,

$$X_{L,0}^r = (X_{L,15}^{r-1} \& X_{L,8}^{r-1}) \oplus X_{L,14}^{r-1} \oplus X_{R,0}^{r-1} \oplus K_0^{r-1}. \tag{5}$$

Other bits in X_L^r have similar boolean representations and the bits in X_R^r are same with the bits in X_L^{r-1}. The boolean representation of one bit can be extended to multiple rounds.

3.1   Linear Compression

In Matsui's improved linear cryptanalysis [14], the attacker can pre-construct a table to store the plaintexts and ciphertexts. We call this pre-construction procedure as linear compression, since the purpose is to reduce the size of efficient states by compressing the linear part. The detail of the compression is as follows.

Suppose x is a l_1-bit value derived from the n-bit plaintext or ciphertext and k is a l_2-bit value derived from the subkey. $y \in \mathbb{F}_2$ is a boolean function of x and k, $y = f(x,k)$. Let $V[x]$ stores the count number of x. We define $B^k(y)$ with counter vector V and function $y = f(x,k)$ for k as

$$B^k(y) = \sum_x (-1)^{f(x,k)} V[x]. \tag{6}$$

So, $B^k(y)$ is the correlation of y with x under key guess k. One needs to do $2^{l_1+l_2}$ computations of function f to calculate the correlations of y for all k with a straight-forward method at most. If y is linear with some bits of x and k, the time can be decreased.

For simplicity, let $x = x'||x_0$, $k = k'||k_0$ and $y = x_0 \oplus k_0 \oplus f_1(x',k')$, where both x_0 and k_0 are single bits. The correlation of y under some k is

$$B^k(y) = (-1)^{k_0} \sum_{x'} (-1)^{f_1(x',k')} (V[x'||0] - V[x'||1]). \tag{7}$$

It is obvious the correlations of y under same k' and different k_0 have same absolute value, and they are different just in the sign. So if we compress the x_0 bit at first according to $V'[x'] = V[x'||0] - V[x'||1]$, $B^{k'}(y')$ with counter vertor V' and function $y' = g'(x',k')$ for k' can be computed with $2^{l_1+l_2-2}$ calculations of f_1. And the correlation $B^k(y)$ can be derived directly from $B^k(y) = (-1)^{k_0} B^{k'}(y')$. We define k_0 the related bit. If the absolute correlations are desired, the related bit k_0 can be omitted directly, since it has no effect on the absolute values.

If y is linear with multiple bits of x and k, the linear bits can be combined at first, then above linear compression can be applied. For example, $y = (x_0 \oplus k_0) \oplus \cdots \oplus (x_t \oplus k_t) \oplus f_t(x'',k'')$ where x'',k'' are the other bits of x and k respectively. We can initialize a new counter vector $V'[x''||x_0']$ where x_0' is 1-bit value of the xor sum of $x_0, x_1, \ldots, x_t$. We set $V'[x''||x_0'] = \sum_{x_0 \oplus \cdots \oplus x_t = x_0'} V[x]$. Let $k_0' = k_0 \oplus \cdots \oplus k_t$. The target value y becomes $y = x_0' \oplus k_0' \oplus f_t(x'',k'')$ with counter vector $V'[x''||x_0']$, which is the case discussed above.

3.2 Dynamic Key-Guessing in Linear Attack: Guess, Split and Combination

Suppose one want to compute $B^k(y)$ with counter vector V and boolean function $y = f(x, k)$, along with the definitions in the above section. With a straight-forward method, the time to compute $B^k(y)$ is $2^{l_1+l_2}$. If for different values of x, different key bits of k are involved in function $f(x, k)$, the time to calculate $B^k(y)$ can be decreased.

$$\boxed{y = f(x,k)} \quad \xrightarrow{\text{Guess } K_G} \quad \boxed{\begin{array}{c} f_A(x, k_A\|k_C) \qquad S_B \\ S_A \qquad f_B(x, k_B\|k_C) \end{array}}$$

Fig. 2. When k_G is known, the set of x can be splitted to two sets. f is independent of k_B in set S_A and independent of k_A in set S_B.

For simplicity, let $k = k_G\|k_A\|k_B\|k_C$, where k_G, k_A, k_B, k_C are l_2^G, l_2^A, l_2^B and l_2^C bits ($l_2^G + l_2^A + l_2^B + l_2^C = l_2$) respectively. Suppose when k_G is known, the all x can be splitted into two sets, i.e. S_A with N_A elements and S_B with N_B elements ($N_A + N_B = 2^{l_1}$). And when $x \in S_A$, $f(x, k) = f_A(x, k_A\|k_C)$ which is independent of k_B; when $x \in S_B$, $f(x, k) = f_B(x, k_B\|k_C)$ which is independent of k_A (See Fig. 2). Then, $B^k(y)$ can be obtained from the following combination

$$B^k(y) = \sum_{x \in S_A} (-1)^{f_A(x, k_A\|k_C)} V[x] + \sum_{x \in S_B} (-1)^{f_B(x, k_B\|k_C)} V[x] \qquad (8)$$

for some guessed k_G. The time to compute $\sum(-1)^{f_A(x, k_A\|k_C)} V[x]$ for the $x \in S_A$ needs $N_A 2^{l_2^G+l_2^A+l_2^C}$ calculations, while $\sum(-1)^{f_B(x, k_B\|k_C)} V[x]$ for $x \in S_B$ needs $N_B 2^{l_2^G+l_2^B+l_2^C}$. The combination needs 2^{l_2} additions. So the time complexity in total is about

$$N_A 2^{l_2^G+l_2^A+l_2^C} + N_B 2^{l_2^G+l_2^B+l_2^C} + 2^{l_2}$$

which improves the time complexity compared with $2^{l_1+l_2}$.

The AND operation in SIMON will generate the situations discussed above. Let $x, k \in \mathbb{F}_2^2$ and $y = f(x, k) = (x_0 \oplus k_0)\&(x_1 \oplus k_1)$. $V[x]$ denotes the count number of x. With a straight-forward method, the calculation of correlations for all k need time $2^{2+2} = 2^4$. If one side of the AND in $f(x, k)$ is 0, y would be 0 without knowing the value in the other side. Exploiting this property, we can improve the time complxity for calculating the correlations. At first, we guess one bit of k, e.g. k_0. Then we split the x into two sets and compute the correlations in each set. At last, we combine the correlations according to the keys guessed.

– GUESS k_0 and SPLIT the x into two sets
  • For the x with $x_0 = k_0$, initialize a counter T_0 and set $T_0 = V[0\|x_0] + V[1\|x_0]$

- For the x with $x_0 = k_0 \oplus 1$, initialize a counter T_1 and set $T_1 = V[0||x_0] - V[1||x_0]$ (Linear compression)
- COMBINE $B(y) = T_0 + (-1)^{k_1} T_1$ (k_1 is a related bit)

So in total, it needs $2(1 + 1 + 2) = 2^3$ additions to compute the correlations for all the k, which improves the time complexity compared to the straight-forward method. Although there are 2 bits of k involved in the attack, we guess only one bit and make some computations while another bit is just involved in the final combination. This can be viewed as that we reduce the number of key bits guessed from 2 to 1. Morever, this technique adapts to some complicated boolean functions and more key (or equivalent key) bits can be reduced significantly. Some cases have been discussed in Appendix A.

4   Linear Cryptanalysis on SIMON

In this section, we will give the improved procedure of linear attack on SIMON using existing linear hull distinguishers for all versions of SIMON

4.1   Linear Hulls of SIMON

Some linear hulls have been proposed recently in [1,17,18], and they are displayed in Table 3. Abdelraheem et al. [1] took advantage of the connection between linear- and differential- characteristics for SIMON and transformed the differential characteristics proposed in [2,9] to linear characteristics directly. Similarly, differentials can be transformed to the linear hulls. Also, they found a new 14-round linear hull for SIMON32/64, by constructing squared correlation matrix to compute the average squared correlation. Shi et al. [17] searched the linear characteristics with same input and output masks using the Mixed-integer Linear Programming modelling, which was investigated to search the differential characteristics for bit-oriented block cipher [19] and then extended to search the linear characteristics (hull) later [18].

Table 3. Linear Hulls for SIMON

BS	Input Active Bits	Output Active Bits	ALH	#R	Ref
32	$X_{L,6}^i$	$X_{R,14}^{i+13}$	$2^{-31.69}$	13	[1]
	$X_{L,5}^i$	$X_{R,13}^{i+13}$	$2^{-30.19}$	13	[17]
	$X_{L,0}^i$	$X_{L,8}^{i+14}, X_{R,6}^{i+14}$	$2^{-32.56}$	14	[1]
48	$X_{L,7}^i, X_{L,11}^i, X_{L,19}^i, X_{R,9}^i, X_{R,17}^i$	$X_{L,5}^{i+15}, X_{R,3}^{i+15}, X_{R,7}^{i+15}, X_{R,11}^{i+15}, X_{R,19}^{i+15}$	$2^{-44.11}$	15	[1]
	$X_{L,6}^i, X_{L,14}^i, X_{L,18}^i, X_{L,22}^i, X_{R,16}^i$	$X_{L,4}^{i+15}, X_{L,20}^{i+15}, X_{R,6}^{i+15}, X_{R,18}^{i+15}, X_{R,20}^{i+15}, X_{R,22}^{i+15}$	$2^{-42.28}$	15	[17]
	$X_{L,1}^i, X_{L,5}^i, X_{L,21}^i, X_{R,23}^i$	$X_{L,1}^{i+16}, X_{L,5}^{i+16}, X_{R,23}^{i+16}$	$2^{-44.92}$	16	[18]
64	$X_{L,20}^i, X_{L,24}^i, X_{R,22}^i$	$X_{L,22}^{i+21}, X_{R,20}^{i+21}, X_{R,24}^{i+21}$	$2^{-62.53}$	21	[1]
	$X_{L,6}^i$	$X_{L,0}^{i+21}, X_{R,2}^{i+21}, X_{R,6}^{i+21}, X_{R,30}^{i+21}$	$2^{-60.72}$	21	[17]
	$X_{L,3}^i, X_{L,27}^i, X_{L,31}^i, X_{R,29}^i$	$X_{L,3}^{i+22}, X_{R,1}^{i+22}, X_{R,2}^{i+22}$	$2^{-63.83}$	22	[17]
96	$X_{L,2}^i, X_{L,34}^i, X_{L,38}^i, X_{L,42}^i, X_{R,36}^i$	$X_{L,2}^{i+30}, X_{L,42}^{i+30}, X_{L,46}^{i+30}, X_{L,40}^{i+30}, X_{R,40}^{i+30}$	$2^{-94.2}$	30	[1]
128	$X_{L,2}^i, X_{L,58}^i, X_{L,62}^i, X_{R,60}^i$	$X_{L,60}^{i+41}, X_{R,0}^{i+41}, X_{R,2}^{i+41}, X_{R,58}^{i+41}, X_{R,62}^{i+41}$	$2^{-126.6}$	41	[1]

* BS means the block size of SIMON; #R means the number of rounds for the linear hull

Similar to the rotational property of integral distinguishers and zero-correlation linear hull shown in [21], more linear hulls can be constructed as follows.

Property 1. Assume that $X_{L,j_0^0}^i, \ldots, X_{L,j_{t_0}^0}^i, X_{R,j_0^1}^i, \ldots, X_{R,j_{t_1}^1}^i \rightarrow X_{L,j_0^2}^{i+r}, \ldots,$ $X_{L,j_{t_2}^2}^{i+r}, X_{R,j_0^3}^{i+r}, \ldots, X_{R,j_{t_3}^3}^{i+r}$ is a r-round linear hull with potential $\bar{\epsilon}^2$ for SIMON$2n$, where $j_0^0, \ldots, j_{t_0}^0, j_0^1, \ldots, j_{t_1}^1, j_0^2, \ldots, j_{t_2}^2, j_0^3, \ldots, j_{t_3}^3 \in \{0, \ldots, n-1\}$. Let $j_q^{p,s} = (j_q^p + s) \bmod n$, where $p = 0, \ldots, 3$, $q = 0, \ldots, t_p$, then for $0 \le s \le n-1$, we have that the potential of the r-round linear hull $X_{L,j_0^{0,s}}^i, \ldots, X_{L,j_{t_0}^{0,s}}^i, X_{R,j_0^{1,s}}^i, \ldots, X_{R,j_{t_1}^{1,s}}^i \rightarrow$ $X_{L,j_0^{2,s}}^{i+r}, \ldots, X_{L,j_{t_2}^{2,s}}^{i+r}, X_{R,j_0^{3,s}}^{i+r}, \ldots, X_{R,j_{t_3}^{3,s}}^{i+r}$ for SIMON$2n$ is also $\bar{\epsilon}^2$.

Observe the two 13-round linear hulls of SIMON32 in Table 3 and we can find they are in fact the rotations of same linear hull. The potential of $X_{L,6}^i \rightarrow X_{L,14}^{i+13}$ is estimated as $2^{-31.69}$ in [1] while that of $X_{L,5}^i \rightarrow X_{L,13}^{i+13}$ is estimated as $2^{-30.19}$ in [17]. The difference may come from the different search methods and different linear trails found. Since SIMON32 has small block size, we can test the bias (potential) of the 13-round linear hull experimentally. In the experimentation, we choose 600 keys randomly, and compute the corresponding bias from the whole plaintexts space. The results are shown in the following table.

Table 4. Experimental bias for the 13-round linear hull of SIMON32

$\epsilon^2 = |p - 1/2|^2$	Number	Number/600
$\epsilon^2 \ge 2^{-27.19}$	7	0.012
$2^{27.19} > \epsilon^2 \ge 2^{-28.19}$	21	0.035
$2^{28.19} > \epsilon^2 \ge 2^{-29.19}$	58	0.097
$2^{29.19} > \epsilon^2 \ge 2^{-30.19}$	72	0.12
$2^{30.19} > \epsilon^2 \ge 2^{-31.19}$	104	0.173
$\epsilon^2 < 2^{-31.19}$	338	0.563

From the table, we know that about 26.4% of the keys have $\epsilon^2 \ge 2^{-30.19}$. So $2^{30.19}$ is a little optimistic for the other 73.6% keys. However, this linear hull distinguisher is interesting and in the following, we will give the key recovery procedure using this linear hull. Also, we implement the 21-round attack on SIMON32 and the results shows that we can decrease the candidate key space by 8 bits when the potential under the real key is large.

4.2   Improved Key Recovery Attack on SIMON32/64

We exploit the 13-round linear hull proposed in [17] to make key recovery attack on round-reduced SIMON32. The linear hull is

$$X_{L,5}^i \rightarrow X_{R,13}^{i+13}.$$

Table 5. 4 rounds before $X_{L,5}^i$ for SIMON32

x	Representation of x_i	k	Representation of k_i
x_0	$X_{L,13}^{i-4} \oplus (X_{L,14}^{i-4} \& X_{L,7}^{i-4}) \oplus X_{R,15}^{i-4} \oplus X_{L,1}^{i-4} \oplus X_{L,5}^{i-4}$	k_0	$K_{15}^{i-4} \oplus K_1^{i-3} \oplus K_5^{i-3} \oplus K_3^{i-2} \oplus K_5^{i-1}$
x_1	$X_{L,14}^{i-4} \oplus (X_{L,15}^{i-4} \& X_{L,8}^{i-4}) \oplus X_{R,0}^{i-4}$	k_1	K_0^{i-4}
x_2	$X_{L,7}^{i-4} \oplus (X_{L,8}^{i-4} \& X_{L,1}^{i-4}) \oplus X_{R,9}^{i-4}$	k_2	K_9^{i-4}
x_3	$X_{L,2}^{i-4} \oplus (X_{L,3}^{i-4} \& X_{L,12}^{i-4}) \oplus X_{R,4}^{i-4}$	k_3	K_4^{i-4}
x_4	$X_{L,11}^{i-4} \oplus (X_{L,12}^{i-4} \& X_{L,5}^{i-4}) \oplus X_{R,13}^{i-4}$	k_4	K_{13}^{i-4}
x_5	$X_{L,14}^{i-4} \oplus (X_{L,15}^{i-4} \& X_{L,8}^{i-4}) \oplus X_{R,0}^{i-4} \oplus X_{L,2}^{i-4}$	k_5	$K_0^{i-4} \oplus K_2^{i-3}$
x_6	$X_{L,15}^{i-4} \oplus (X_{L,0}^{i-4} \& X_{L,9}^{i-4}) \oplus X_{R,1}^{i-4}$	k_6	K_1^{i-4}
x_7	$X_{L,8}^{i-4} \oplus (X_{L,9}^{i-4} \& X_{L,2}^{i-4}) \oplus X_{R,10}^{i-4}$	k_7	K_{10}^{i-4}
x_8	$X_{L,7}^{i-4} \oplus (X_{L,8}^{i-4} \& X_{L,1}^{i-4}) \oplus X_{R,9}^{i-4} \oplus X_{L,11}^{i-4}$	k_8	$K_9^{i-4} \oplus K_{11}^{i-3}$
x_9	$X_{L,1}^{i-4} \oplus (X_{L,2}^{i-4} \& X_{L,11}^{i-4}) \oplus X_{R,3}^{i-4}$	k_9	K_3^{i-4}
x_{10}	$X_{L,14}^{i-4} \oplus (X_{L,15}^{i-4} \& X_{L,8}^{i-4}) \oplus X_{R,0}^{i-4} \oplus (X_{L,3}^{i-4} \& X_{L,12}^{i-4}) \oplus X_{R,4}^{i-4}$	k_{10}	$K_0^{i-4} \oplus K_2^{i-3} \oplus K_4^{i-4} \oplus K_4^{i-2}$
x_{11}	$X_{L,15}^{i-4} \oplus (X_{L,0}^{i-4} \& X_{L,9}^{i-4}) \oplus X_{R,1}^{i-4} \oplus X_{L,3}^{i-4}$	k_{11}	$K_1^{i-4} \oplus K_3^{i-3}$
x_{12}	$X_{L,0}^{i-4} \oplus (X_{L,1}^{i-4} \& X_{L,10}^{i-4}) \oplus X_{R,2}^{i-4}$	k_{12}	K_2^{i-4}
x_{13}	$X_{L,9}^{i-4} \oplus (X_{L,10}^{i-4} \& X_{L,3}^{i-4}) \oplus X_{R,11}^{i-4}$	k_{13}	K_{11}^{i-4}
x_{14}	$X_{L,8}^{i-4} \oplus (X_{L,9}^{i-4} \& X_{L,2}^{i-4}) \oplus X_{R,10}^{i-4} \oplus X_{L,12}^{i-4}$	k_{14}	$K_{10}^{i-4} \oplus K_{12}^{i-3}$
x_{15}	$X_{L,7}^{i-4} \oplus (X_{L,8}^{i-4} \& X_{L,1}^{i-4}) \oplus X_{R,9}^{i-4} \oplus (X_{L,12}^{i-4} \& X_{L,5}^{i-4}) \oplus X_{R,13}^{i-4}$	k_{15}	$K_9^{i-4} \oplus K_{11}^{i-3} \oplus K_{13}^{i-4} \oplus K_{13}^{i-2}$
x_{16}	$X_{L,1}^{i-4} \oplus (X_{L,2}^{i-4} \& X_{L,11}^{i-4}) \oplus X_{R,3}^{i-4} \oplus X_{L,5}^{i-4}$	k_{16}	$K_3^{i-4} \oplus K_5^{i-3}$

[a] Notice: $x_{10} = x_3 \oplus x_5, x_{15} = x_4 \oplus x_8$
[b] X^{i-4} is the plaintext P, $K^{i-4}, \ldots, K^{i-1}$ are the subkeys used in the initial four rounds, $i.e.$ K_P
[c] In the description of the paper, $x_P = x = (x_0, \ldots, x_{16}), k_P = k = (k_0, \ldots, k_{16})$

We mount a key recovery attack on 21-round SIMON32/64 by adding four rounds before and appending four rounds after the distinguisher. Here let $P = X^{i-4}$ be the plaintext and $C = X^{i+17}$ be the corresponding ciphertext. Suppose the subkeys involved in the first four rounds are K_P and those in the last four rounds are K_C. Then $X_{L,5}^i$ is a function of P and K_P, $X_{L,5}^i = E(P, K_P)$. Similarly, $X_{R,13}^{i+13} = D(C, K_C)$ is a function of C and K_C. Let $\mathcal{S}$ be the set of N plaintext-ciphertext pairs obtained, the empirical correlation under some key K_P, K_C is

$$\bar{c}_{K_P, K_C} = \frac{1}{N} \sum_{P,C \in \mathcal{S}} (-1)^{E(P,K_P) \oplus D(C,K_C)}. \tag{9}$$

Table 6. 4 rounds after $X^{i+13}_{R,13}$ for SIMON32

x	Representation of x_i	k	Representation of k_i
x_0	$X^{i+17}_{R,5} \oplus (X^{i+17}_{R,6} \& X^{i+17}_{R,15}) \oplus X^{i+17}_{L,7} \oplus X^{i+17}_{R,9}$ $\oplus X^{i+17}_{R,13}$	k_0	$K^{i+16}_7 \oplus K^{i+15}_9 \oplus K^{i+15}_{13} \oplus K^{i+14}_{11} \oplus K^{i+13}_{13}$
x_1	$X^{i+17}_{R,6} \oplus (X^{i+17}_{R,7} \& X^{i+17}_{R,0}) \oplus X^{i+17}_{L,8}$	k_1	K^{i+16}_8
x_2	$X^{i+17}_{R,15} \oplus (X^{i+17}_{R,0} \& X^{i+17}_{R,9}) \oplus X^{i+17}_{L,1}$	k_2	K^{i+16}_1
x_3	$X^{i+17}_{R,10} \oplus (X^{i+17}_{R,11} \& X^{i+17}_{R,4}) \oplus X^{i+17}_{L,12}$	k_3	K^{i+16}_{12}
x_4	$X^{i+17}_{R,3} \oplus (X^{i+17}_{R,4} \& X^{i+17}_{R,13}) \oplus X^{i+17}_{L,5}$	k_4	K^{i+16}_5
x_5	$X^{i+17}_{R,6} \oplus (X^{i+17}_{R,7} \& X^{i+17}_{R,0}) \oplus X^{i+17}_{L,8} \oplus X^{i+17}_{R,10}$	k_5	$K^{i+16}_8 \oplus K^{i+15}_{10}$
x_6	$X^{i+17}_{R,7} \oplus (X^{i+17}_{R,8} \& X^{i+17}_{R,1}) \oplus X^{i+17}_{L,9}$	k_6	K^{i+16}_9
x_7	$X^{i+17}_{R,0} \oplus (X^{i+17}_{R,1} \& X^{i+17}_{R,10}) \oplus X^{i+17}_{L,2}$	k_7	K^{i+16}_2
x_8	$X^{i+17}_{R,15} \oplus (X^{i+17}_{R,0} \& X^{i+17}_{R,9}) \oplus X^{i+17}_{L,1} \oplus X^{i+17}_{R,3}$	k_8	$K^{i+16}_1 \oplus K^{i+15}_3$
x_9	$X^{i+17}_{R,9} \oplus (X^{i+17}_{R,10} \& X^{i+17}_{R,3}) \oplus X^{i+17}_{L,11}$	k_9	K^{i+16}_{11}
x_{10}	$X^{i+17}_{R,6} \oplus (X^{i+17}_{R,7} \& X^{i+17}_{R,0}) \oplus X^{i+17}_{L,8}$ $\oplus (X^{i+17}_{R,11} \& X^{i+17}_{R,4}) \oplus X^{i+17}_{L,12}$	k_{10}	$K^{i+16}_8 \oplus K^{i+15}_{10} \oplus K^{i+16}_{12} \oplus K^{i+14}_{12}$
x_{11}	$X^{i+17}_{R,7} \oplus (X^{i+17}_{R,8} \& X^{i+17}_{R,1}) \oplus X^{i+17}_{L,9} \oplus X^{i+17}_{R,11}$	k_{11}	$K^{i+16}_9 \oplus K^{i+15}_{11}$
x_{12}	$X^{i+17}_{R,8} \oplus (X^{i+17}_{R,9} \& X^{i+17}_{R,2}) \oplus X^{i+17}_{L,10}$	k_{12}	K^{i+16}_{10}
x_{13}	$X^{i+17}_{R,1} \oplus (X^{i+17}_{R,2} \& X^{i+17}_{R,11}) \oplus X^{i+17}_{L,3}$	k_{13}	K^{i+16}_3
x_{14}	$X^{i+17}_{R,0} \oplus (X^{i+17}_{R,1} \& X^{i+17}_{R,10}) \oplus X^{i+17}_{L,2} \oplus X^{i+17}_{R,4}$	k_{14}	$K^{i+16}_2 \oplus K^{i+15}_4$
x_{15}	$X^{i+17}_{R,15} \oplus (X^{i+17}_{R,0} \& X^{i+17}_{R,9}) \oplus X^{i+17}_{L,1}$ $\oplus (X^{i+17}_{R,4} \& X^{i+17}_{R,13}) \oplus X^{i+17}_{L,5}$	k_{15}	$K^{i+16}_1 \oplus K^{i+15}_3 \oplus K^{i+16}_5 \oplus K^{i+14}_5$
x_{16}	$X^{i+17}_{R,9} \oplus (X^{i+17}_{R,10} \& X^{i+17}_{R,3}) \oplus X^{i+17}_{L,11} \oplus X^{i+17}_{R,13}$	k_{16}	$K^{i+16}_{11} \oplus K^{i+15}_{13}$

[a] Notice: $x_{10} = x_3 \oplus x_5$, $x_{15} = x_4 \oplus x_8$.
[b] X^{i+17} is the ciphertext C, $K^{i+13}, \ldots, K^{i+16}$ are the subkeys used in the last four rounds, *i.e.* K_C
[c] In the description of the paper, $x_C = x = (x_0, \ldots, x_{16})$, $k_C = k = (k_0, \ldots, k_{16})$

In a further step, $X^i_{L,5}$ can be represented as $X^i_{L,5} = f(x, k)$ where

$$f(x,k) = x_0 \oplus k_0 \oplus ((x_1 \oplus k_1)\&(x_2 \oplus k_2)) \oplus ((x_3 \oplus k_3)\&(x_4 \oplus k_4)) \oplus$$
$$[(x_5 \oplus k_5 \oplus ((x_6 \oplus k_6)\&(x_7 \oplus k_7)))\&(x_8 \oplus k_8 \oplus ((x_9 \oplus k_9)\&(x_7 \oplus k_7)))] \oplus$$
$$\{(x_{10} \oplus k_{10} \oplus ((x_6 \oplus k_6)\&(x_7 \oplus k_7)) \oplus$$
$$[(x_{11} \oplus k_{11} \oplus ((x_{12} \oplus k_{12})\&(x_{13} \oplus k_{13})))\&(x_{14} \oplus k_{14} \oplus ((x_3 \oplus k_3)\&(x_{13} \oplus k_{13})))]) \&$$
$$(x_{15} \oplus k_{15} \oplus ((x_7 \oplus k_7)\&(x_9 \oplus k_9)) \oplus$$
$$[(x_{14} \oplus k_{14} \oplus ((x_{13} \oplus k_{13})\&(x_3 \oplus k_3)))\&(x_{16} \oplus k_{16} \oplus ((x_3 \oplus k_3)\&(x_4 \oplus k_4)))])\}$$

where the representation of x and k are 17-bit value shown in Table 5. With the same way, $X^{i+13}_{R,13}$ can also be represented as $f(x, k)$ where the corresponding x and k are described in Table 6. To distinguish them, let x_P, k_P be the x, k described in Table 5 and x_C, k_C be the x, k described in Table 6. The N plaintext-ciphertext pairs in $\mathcal{S}$ can be compressed into a counter vector $V[x_P, x_C]$, which stores the number of x_P, x_C. Then there is

$$\bar{c}_{k_P, k_C} = \frac{1}{N} \sum_{x_P, x_C} (-1)^{f(x_P, k_P) \oplus f(x_C, k_C)} V[x_P, x_C]. \tag{10}$$

Notice that $f(x, k)$ is linear with $x_0 \oplus k_0$. According to the linear compression technique, the 0-th bit of x_P and x_C could be compressed initially. Suppose that x'_P is the 16-bit value of x_P without the 0-th bit (same representations for x'_C, k'_P, k'_C). Initialize a new counter vector V_1 which has values

$$V_1[x'_P, x'_C] = \sum_{x_{P,0}, x_{C,0}} (-1)^{x_{P,0} \oplus x_{C,0}} V[x_P, x_C]. \tag{11}$$

Then the correlation becomes

$$\bar{c}_{k'_P, k'_C} = \frac{1}{N} \sum_{x'_P, x'_C} (-1)^{f'(x'_P, k'_P) \oplus f'(x'_C, k'_C)} V_1[x'_P, x'_C]$$

$$= \frac{1}{N} \sum_{x'_C} (-1)^{f'(x'_C, k'_C)} \sum_{x'_P} (-1)^{f'(x'_P, k'_P)} V_1[x'_P, x'_C], \tag{12}$$

where f' is part of f, i.e. $f(x, k) = x_0 \oplus k_0 \oplus f'(x', k'), x' = (x_1, \ldots, x_{16}), k' = (k_1, \ldots, k_{16})$.

So we can guess k'_P (16-bit) at first and compress the plaintexts into a counter. Then guess k'_C (16-bit) to decrypt the appending rounds, to achieve the final correlations. In the following, we introduce the attack procedure in the forward rounds in detail. The procedure to compute $\sum_{x'_P} (-1)^{f'(x'_P, k'_P)} V_1[x'_P, x'_C]$ for each x'_C is same with the procedure to compute $B^{k'}(y)$ with some counter vector $V'_1[x']$ and boolean function f'. Counter vector V'_1 is part of counter vector V_1. For each specific x'_C,

$$V'_1[x'] = V_1[x', x'_C],$$

which means $V'_1[x']$ takes value of $V_1[x'_P, x'_C]$ where $x'_P = x'$ and x'_C is fixed. Morever, there are relations that $x_{10} = x_3 \oplus x_5, x_{15} = x_4 \oplus x_8$ in Tables 5 and 6, which means there are only 14 independent bits for x' (x'_P or x'_C).

Compute $B^{k'}(y)$ with counter vector $V'_1[x']$ and Boolean function f'. (For simplicity, we define this procedure as Procedure A.) Although x' is a 16-bit value, there are only 2^{14} possible values for x' as explained above. We use the guess, split and combination technique to decrease the time complexity to compute $B^{k'}(y)$ with counter vector $V'_1[x']$ and boolean function $y = f'$, for 2^{16} key vaules k'.

1. Guess k_1, k_3, k_7 and split the plaintexts into 8 sets according to the value $(x_1 \oplus k_1, x_3 \oplus k_3, x_7 \oplus k_7)$. The simplification for $f'(x', k')$ after guessing some keys are shown in Table 7.

   The representation of f_{ij} are as follows,

$$f_{00} = ((x_5 \oplus k_5) \& (x_8 \oplus k_8)) \oplus \{(x_{10} \oplus k_{10} \oplus [(x_{11} \oplus k_{11} \oplus ((x_{12} \oplus k_{12}) \& (x_{13} \oplus k_{13})))$$
$$\& (x_{14} \oplus k_{14})]) \& (x_{15} \oplus k_{15} \oplus [(x_{14} \oplus k_{14}) \& (x_{16} \oplus k_{16})])\},$$

$$f_{01} = ((x_{5,6} \oplus k_{5,6}) \& (x_{8,9} \oplus k_{8,9})) \oplus \{(x_{6,10} \oplus k_{6,10} \oplus [(x_{11} \oplus k_{11} \oplus ((x_{12} \oplus k_{12})$$
$$\& (x_{13} \oplus k_{13}))) \& (x_{14} \oplus k_{14})]) \& (x_{9,15} \oplus k_{9,15} \oplus [(x_{14} \oplus k_{14}) \& (x_{16} \oplus k_{16})])\},$$

$$f_{10} = ((x_5 \oplus k_5) \& (x_8 \oplus k_8)) \oplus \{(x_{10} \oplus k_{10} \oplus [(x_{11} \oplus k_{11} \oplus ((x_{12} \oplus k_{12}) \& (x_{13} \oplus k_{13})))$$
$$\& (x_{13,14} \oplus k_{13,14})]) \& (x_{15} \oplus k_{15} \oplus [(x_{13,14} \oplus k_{13,14}) \& (x_{4,16} \oplus k_{4,16})])\},$$

$$f_{11} = ((x_{5,6} \oplus k_{5,6}) \& (x_{8,9} \oplus k_{8,9})) \oplus \{(x_{6,10} \oplus k_{6,10} \oplus [(x_{11} \oplus k_{11} \oplus ((x_{12} \oplus k_{12}) \& (x_{13}$$
$$\oplus k_{13}))) \& (x_{13,14} \oplus k_{13,14})]) \& (x_{9,15} \oplus k_{9,15} \oplus [(x_{13,14} \oplus k_{13,14}) \& (x_{4,16} \oplus k_{4,16})])\}.$$

Table 7. Simplification for $f'(x', k')$ after guessing k_1, k_3, k_7

Guess	$x_1 \oplus k_1, x_3 \oplus k_3, x_7 \oplus k_7$	f'	Related Bit
k_1, k_3, k_7	0,0,0	f_{00}	
	0,0,1	f_{01}	
	0,1,0	f_{10}	k_4
	0,1,1	f_{11}	k_4
	1,0,0	f_{00}	k_2
	1,0,1	f_{01}	k_2
	1,1,0	f_{10}	$k_{2,4}$
	1,1,1	f_{11}	$k_{2,4}$

Table 8. Simplification for f_{00} after guessing k_5, k_{14}

Guess	Value	f_{00}	Related Bit
k_5, k_{14}	0,0	$(x_{10} \oplus k_{10})\&(x_{15} \oplus k_{15})$	
	0,1	$(x_{10,11} \oplus k_{10,11} \oplus ((x_{12} \oplus k_{12})\&(x_{13} \oplus k_{13})))\&(x_{15,16} \oplus k_{15,16})$	
	1,0	$(x_{10} \oplus k_{10})\&(x_{15} \oplus k_{15})$	k_8
	1,1	$(x_{10,11} \oplus k_{10,11} \oplus ((x_{12} \oplus k_{12})\&(x_{13} \oplus k_{13})))\&(x_{15,16} \oplus k_{15,16})$	k_8

The counter vectors for x' can be compressed in a further step according to the new representations of f'. For example, if $(x_1 \oplus k_1, x_3 \oplus k_3, x_7 \oplus k_7) = (0, 0, 0)$, f' will be equal to the formula f_{00}, which is independent of x_2, x_4, x_6, x_9. So we compress the corresponding counters into a new counter V_{000}, and

$$V_{000}[x_5, x_8, x_{10} - x_{16}] = \sum_{x_1 = k_1, x_3 = k_3, x_7 = k_7, x_2 \in \mathbb{F}_2, x_4 \in \mathbb{F}_2, x_6 \in \mathbb{F}_2, x_9 \in \mathbb{F}_2} V_1'[x'].$$

Notice $x_{10} = x_3 \oplus x_5$, so there are 8 independent x bits for $x_5, x_8, x_{10} - x_{16}$. Notice $x_{15} = x_4 \oplus x_8$, for some fixed value of $x_5, x_8, x_{10} - x_{16}$, there are 7 times addition in above equation. So generating this new counter vector needs $2^8 \times 7$ additions.

We give another example to illustrate the situations with related key bit. If $(x_1 \oplus k_1, x_3 \oplus k_3, x_7 \oplus k_7) = (1, 0, 0)$, there is $f' = (x_2 \oplus k_2) \oplus f_{00}$. Notice in this subset, f' is linear with $x_2 \oplus k_2$ and x_2 can be compressed into the new counters with related key k_2. So the new counter vector V_{100} is as follows,

$$V_{100}[x_5, x_8, x_{10} - x_{16}] = \sum_{x_1 = k_1 \oplus 1, x_3 = k_3, x_7 = k_7, x_2 \in \mathbb{F}_2, x_4 \in \mathbb{F}_2, x_6 \in \mathbb{F}_2, x_9 \in \mathbb{F}_2} (-1)^{x_2} V_1'[x'].$$

Also, there are 8 independent x bits for $x_5, x_8, x_{10} - x_{16}$. For each fixed $x_5, x_8, x_{10} - x_{16}$, the new counter can be obtained with 7 additions according to above equation.

The procedures to generate the new counter vectors for other cases are similar as that of case $(x_1 \oplus k_1, x_3 \oplus k_3, x_7 \oplus k_7) = (0, 0, 0)$ or $(1, 0, 0)$. Morever, the time complexity to split the plaintexts and construct new counter vectors is same for each case. Observing the four functions f_{00}, f_{01}, f_{10} and f_{11}, we know that they are with same form. In the following step, we explain the attack procedure of case $(x_1 \oplus k_1, x_3 \oplus k_3, x_7 \oplus k_7) = (0, 0, 0)$ in detail and the others can be obtained in the same way.

Note that, there are 9 subkey bits in each function of f_{00}, f_{01}, f_{10} and f_{11} after guessing k_1, k_3, k_7. So this can be viewed as that $3 + 9 = 12$ subkey bits are involved in the attack while there are 16 subkey bits are involved initially in f'. In the following, the number of key bits can be reduced in a further step.

2. For f_{00}, guess k_5, k_{14} and split the plaintexts into 4 sets according to the value $(x_5 \oplus k_5, x_{14} \oplus k_{14})$. The simplification for f_{00} after guessing some keys are shown in Table 8.

The time complexity of computing the counters' value $B^{k_5, k_8, k_{10} - k_{16}}(y)$ with counter vector V_{000} and function f_{00} is as follows:

(a) Guess k_5, k_{14} and split the states into four parts

    i. $(x_5 \oplus k_5, x_{14} \oplus k_{14}) = (0, 0)$

        A. Since $x_{10} = x_3 \oplus x_5$, $x_5 = k_5$ and $x_3 = k_3$ (the first case in Table 7), so the x_{10} here is fixed. There is one variable bit x_{15} to store. Let $V_{000}^{00}[x_{10}, x_{15}]$ store the number of (x_{10}, x_{15}). There is

$$V_{000}^{00}[x_{10}, x_{15}] = \sum_{x_5 = k_5, x_{14} = k_{14}} V_{000}[x_5, x_8, x_{10} - x_{16}].$$

        There are two possible values for (x_{10}, x_{15}) here and for each value, the above sum needs $2^5 - 1$ additions (5 variable bits $(x_8, x_{11}, x_{12}, x_{13}, x_{16})$). So generating the new counter vector needs $2 \times (2^5 - 1) = 2^6 - 2$ additions.

        B. Computing $B_{00}^{k_{10}, k_{15}}(y)$ with new function (the first case in Table 8) and vector V_{000}^{00}:

        If $k_{10} = x_{10}$, $B_{00}^{k_{10}, k_{15}}(y) = V_{000}^{00}[x_{10}, 0] + V_{000}^{00}[x_{10}, 1]$;

        if $k_{10} = x_{10} \oplus 1$, $B_{00}^{k_{10}, k_{15}}(y) = (-1)^{k_{15}}(V_{000}^{00}[x_{10}, 0] - V_{000}^{00}[x_{10}, 1])$.

        So in total there are no more than 2^2 additions.

    ii. $(x_5 \oplus k_5, x_{14} \oplus k_{14}) = (0, 1)$

        A. There are 4 variable bits $(x_{10,11}, x_{12}, x_{13}, x_{15,16})$ to store. Let $V_{000}^{01}[x_{10,11}, x_{12}, x_{13}, x_{15,16}]$ store the counter number of $(x_{10,11}, x_{12}, x_{13}, x_{15,16})$. There is

$$V_{000}^{01}[x_{10,11}, x_{12}, x_{13}, x_{15,16}] = \sum_{x_5 = k_5, x_{14} = k_{14} \oplus 1} V_{000}[x_5, x_8, x_{10} - x_{16}].$$

        For each possible value of $(x_{10,11}, x_{12}, x_{13}, x_{15,16})$, the above sum needs $2^2 - 1$ additions (2 free variables (x_8, x_{15}), x_{10} is fixed, $x_{11} = x_{10} \oplus x_{10,11}$, $x_{16} = x_{15} \oplus x_{15,16}$). So generating the new counter vector needs: $2^4 \times (2^2 - 1) = 2^6 - 2^4$ additions.

        B. Partial $B_{01}^{k_{10,11}, k_{12}, k_{13}, k_{15,16}}(y)$ with new function and vector V_{000}^{01}: $2^{5.64}$ additions. (See f_3 in Appendix A)

iii. $(x_5 \oplus k_5, x_{14} \oplus k_{14}) = (1, 0)$

    A. Similar to the first case in Step (2(a)i), let $V_{000}^{10}[x_{10}, x_{15}]$ store the number of (x_{10}, x_{15}). There is

$$V_{000}^{10}[x_{10}, x_{15}] = \sum_{x_5 = k_5, x_{14} = k_{14}} V_{000}(-1)^{x_8}[x_5, x_8, x_{10} - x_{16}].$$

    So generating the new counter vector also needs $2 \times (2^5 - 1) = 2^6 - 2$ additions. k_8 becomes a related bit.

    B. Partial $B_{10}^{k_{10}, k_{15}}(y)$ with new function and vector V_{000}^{10}: 2^2 additions (same with case $(0, 0)$).

iv. $(x_5 \oplus k_5, x_{14} \oplus k_{14}) = (1, 1)$

    A. Similar to the second case in Step (2(a)ii), let $V_{000}^{11}[x_{10,11}, x_{12}, x_{13}, x_{15,16}]$ store the counter number of $(x_{10,11}, x_{12}, x_{13}, x_{15,16})$. There is

$$V_{000}^{11}[x_{10,11}, x_{12}, x_{13}, x_{15,16}]$$
$$= \sum_{x_5 = k_5, x_{14} = k_{14} \oplus 1} (-1)^{x_8} V_{000}[x_5, x_8, x_{10} - x_{16}].$$

    So generating the new counter vector needs: $2^4 \times (2^2 - 1) = 2^6 - 2^4$ additions. k_8 becomes a related bit.

    B. Partial $B_{11}^{k_{10,11}, k_{12}, k_{13}, k_{15,16}}(y)$ with new function and vector V_{000}^{11}: $2^{5.64}$ additions. (See f_3 in Appendix A)

(b) For each of 2^9 keys involved in f_{00}, partial $B^{k_5, k_8, k_{10} - k_{16}}(y)$ with function $y = f_{00}$ and counter vector V_{000} under key guess k_5, k_{14} is

$$B^{k_5, k_8, k_{10} - k_{16}}(y) = \quad (B_{00}^{k_{10}, k_{15}}(y) + B_{01}^{k_{10,11}, k_{12}, k_{13}, k_{15,16}}(y))$$
$$+ (-1)^{k_8} \quad (B_{10}^{k_{10}, k_{15}}(y) + B_{01}^{k_{10,11}, k_{12}, k_{13}, k_{15,16}}(y)).$$

We can add $B_{00}^{k_{10}, k_{15}}(y)$ and $B_{01}^{k_{10,11}, k_{12}, k_{13}, k_{15,16}}(y)$ at first, then add $B_{10}^{k_{10}, k_{15}}(y)$ and $B_{01}^{k_{10,11}, k_{12}, k_{13}, k_{15,16}}(y)$, at last add the two parts according to the index value and k_8. The combination phase needs $2^6 + 2^6 + 2^7 = 2^8$ additions in total when k_5, k_{14} are fixed.

(c) In total, there are

$$2^2 \times ((2^6 - 2 + 2^2 + 2^6 - 2^4 + 2^{5.64}) \times 2 + 2^8) \approx 2^{11.19}$$

additions to compute $B^{k_5, k_8, k_{10} - k_{16}}(y)$ for all 2^9 possible key values. Note that, about 1 subkey bit is guessed in the first (or third) step of step 2a. In the second (or forth) step of step 2a, 1.5 subkey bits are guessed on average. So, although there are 9 subkey bits in total, only $2+(1+1+1.5+1.5)/4=3.25$ bits on average are guessed with dynamic key-guessing technique.

3. The time of computing $B^{k'}(y)$ with counter vector $V_1'[x']$ and boolean function f' is shown in Table 9. T_1 denotes the time of seperation of the plaintexts according to the guessed bit of k. T_2 denotes the time of computation in the inner part. T_3 is the time in the combination phase. When k_1, k_3, k_7 are fixed, in each case, $T_1 = 2^8 \times 7$ as explainted in Step 1. T_2 is $2^{11.19}$ as explained in Step 2. There are 13 bits for k' except k_1, k_3, k_7, leading to $T_3 = 2^{13} \times 7$. For all guesses of k_1, k_3, k_7, the total time is about $2^{19.46}$ additions.

In Step 1, 3 key bits are guessed and the plaintexts are splitted into 8 situations. For each situation, 3.25 key bits are guessed as explained above. So on average, about $3 + 3.25 = 6.25$ subkey bits are guessed in this procedure, while there are 16 subkey bits involved.

Table 9. Time Complexity of computing $B^{k'}(y)$ with counter vector $V_1'[x']$ and boolean function f'

Guess	$x_1 \oplus k_1, x_3 \oplus k_3, x_7 \oplus k_7$	f'	Related Bit	Time		
				T_1	T_2	T_3
k_1, k_3, k_7	0,0,0	f_{00}		$2^8 \times 7$	$2^{11.19}$	$2^{13} \times 7$
	0,0,1	f_{01}		$2^8 \times 7$	$2^{11.19}$	
	0,1,0	f_{10}	k_4	$2^8 \times 7$	$2^{11.19}$	
	0,1,1	f_{11}	k_4	$2^8 \times 7$	$2^{11.19}$	
	1,0,0	f_{00}	k_2	$2^8 \times 7$	$2^{11.19}$	
	1,0,1	f_{01}	k_2	$2^8 \times 7$	$2^{11.19}$	
	1,1,0	f_{10}	$k_{2,4}$	$2^8 \times 7$	$2^{11.19}$	
	1,1,1	f_{11}	$k_{2,4}$	$2^8 \times 7$	$2^{11.19}$	
Total Time				$((2^8 \times 7 + 2^{11.19}) \times 8 + 2^{13} \times 7) \times 2^3 = 2^{19.46}$		

21-Round Attack on SIMON 32/64. Adding four rounds and appending four rounds after the 13-round linear hull distinguisher, we give the 21-round linear attack on SIMON32/64. The estimated potential of the linear hull is $\bar{\epsilon}^2 \approx 2^{-30.19}$ in [17], which is a little optimistic for more than half of keys. In the attack, we use $N = 2^{31.19}$ plaintext-ciphertext pairs. According to Theorem 1, the relation between the bias and success probability is shown in Table 10 when using $2^{31.19}$

Table 10. Relation between bias and success probability using $2^{31.19}$ data and setting advantage $a = 8$

$\epsilon^2 = 2^{27.19}$	$p_0 \approx 1.000$
$\epsilon^2 = 2^{28.19}$	$p_1 \approx 0.997$
$\epsilon^2 = 2^{29.19}$	$p_2 \approx 0.864$
$\epsilon^2 = 2^{30.19}$	$p_3 \approx 0.477$
$\epsilon^2 = 2^{31.19}$	$p_4 \approx 0.188$

plaintext-ciphertext pairs. So according to Tables 4 and 10, the expected success probability of the attack is larger than

$$0.012 * p_0 + 0.035 * p_1 + 0.097 * p_2 + 0.12 * p_3 + 0.173 * p_4 \approx 0.22,$$

and it is smaller than

$$(0.012 + 0.035) * p_0 + 0.097 * p_1 + 0.12 * p_2 + 0.173 * p_3 \approx 0.33.$$

There are 32 subkey bits involved in this attack. With our attack method, only about $6.25 + 6.25 = 12.5$ bits are guessed on average, which reduces the number of key bits greatly.

Attack:

1. Compress the N plaintext-ciphertext pairs into the counter vector $V_1[x'_P, x'_C]$ of size 2^{14+14}.
2. For each of 2^{14} x'_C
   (a) Call Procedure A. Store the counters according to x'_C and k'_P
3. For each k'_P of 2^{16} possible values.
   (a) Call procedure A. Store the counters according to k'_P and k'_C.
4. The keys with counter values ranked in the largest $2^{32-8} = 2^{24}$ values would be the right subkey candidates. Exploiting the key schedule and guessing some other bits, use two plaintex-ciphertext pairs to check the right key.

Time: $(1)N = 2^{31.19}$ times compression (2) $2^{14} \times 2^{19.46} = 2^{33.46}$ additions. $(3)2^{16} \times 2^{19.46} = 2^{35.46}$ additions. So the time to compute the empirical bias for the subkeys involved is about $2^{35.84}$ while that given in [1] with similar linear hull is $2^{63.69}$. The time is improved significantly. Step (4) is to recovery the master key, which needs $2^{64-8} = 2^{56}$ 21-round encryptions. However, [1] does not give this step.

Also we implemented the 21-round attack on SIMON32 using $2^{31.19}$ plaintext-ciphertext pairs. (The exhaustive search part of the attack is not included since it would take about $2^{64-8} = 2^{56}$ encryptions, which takes too much time.) In the implementation, we set the main key randomly and collect $2^{31.19}$ plaintext-ciphertext pairs (data collection part), then use the dynamic key-guessing techniques to recover 8-bit key information for the 32 subkey bits (recovery part). We store the $2^{32-8} = 2^{24}$ keys with large bias in set S as the right key candidates, then compute the real 32 subkey bits from the main key and check whether it is in S. In the implementation, about 5GB memory is needed. The data collection part ($2^{31.19}$ encryptions) takes about 11 minutes and the recovery part takes about 11 minutes too (using Intel(R) Xeon(R) CPU E5-2620, 2.00GHz). 1000 experiments were done and 277 of them were successful. This derives that the experimental success probability is about 27.7%, which is consistent with the expected success probability.

22-Round Attack on SIMON32/64. Add one more round before the 21-round attack, we can attack 22-round of SIMON32/64. There are 13 active key bits involved in round $i - 5$, which is $\kappa_1 = (K_0^{i-5} - K_3^{i-5}, K_5^{i-5}, K_7^{i-5} - K_{12}^{i-5}, K_{14}^{i-5}, K_{15}^{i-5})$, to obtain the x represented in Table 5.

Attack:

1. Guess each of 2^{13} κ_1
   (a) Encrypt the plaintexts by one round.
   (b) Do as the first three steps in the 21-round attack
2. The keys with counter values ranked in the largest $2^{32+13-8} = 2^{37}$ values would be the right subkey candidates. Exploiting the key schedule and guessing some other bits, use two plaintex-ciphertext pairs to check the right key.

Time: (1.a)$2^{13} \times N = 2^{44.19}$ one-round encryptions. (1.b) $2^{13} \times 2^{35.84} = 2^{48.84}$ additions. (2) Exhaustive phase needs about $2^{64-8} = 2^{56}$ 22-round encryptions. So the total time is about 2^{56} 22-round encryptions and $2^{48.84}$ additions.

23-round attack on SIMON32/64. Add one more round before and one round after the 21-round attack, we can attack 23-round of SIMON32/64. There are 13 active key bits involved in round $i+17$, which is $\kappa_2 = (K_0^{i+17} - K_3^{i+17}, K_5^{i+17}, K_7^{i+17} - K_{12}^{i+17}, K_{14}^{i+17}, K_{15}^{i+17})$, to obtain the x represented in Table 6.
Attack:

1. Guess each of 2^{13+13} $\kappa_1 \| \kappa_2$
   (a) Encrypt the plaintexts by one round and decrypt the ciphertexts by one round.
   (b) Do as the first three steps in the 21-round attack
2. The keys with counter values ranked in the largest $2^{32+26-8} = 2^{50}$ values would be the right subkey candidates. Exploiting the key schedule and guessing some other bits, use two plaintex-ciphertext pairs to check the right key.

Time: (1.a)$2^{26} \times N = 2^{57.19}$ two-round encryptions. (1.b) $2^{26} \times 2^{35.84} = 2^{61.84}$ additions. (2) Exhaustive phase needs about $2^{64-8} = 2^{56}$ 23-round encryptions. So the total time complexity is about $2^{56.3}$ 23-round encryptions and $2^{61.84}$ additions.

4.3 Improved Key Recovery Attack on Other Variants of SIMON

With the dynamic key-guessing technique shown in above attack, we can also improve the linear hull attacks on all other variants of SIMON. The linear hulls used are displayed in Table 3. For SIMON48, we exploit the 22-round linear hull proposed in [18], which covers most rounds up to date. For SIMON64, the 21-round linear hull with potential $2^{-62.53}$ proposed in [1] is used in the attack. Also, the 31-round (resp. 40-round) linear hull for SIMON96 (resp. SIMON128) in [1] are used to attack corresponding variant. Due to limited space, we do not give the detail of the attacks (please refer to the full version [11] of this paper for the details). However, the improved results for these variants are listed in Table 1.

4.4 Multiple Linear Hull Attack on SIMON

Combining multiple linear cryptanalysis [8] and linear hull together, one can make multiple linear hull attack with improved data complexity. Our attack technique can be used in the multiple linear hull attack of SIMON well. According to the rotational property, Property 1, of SIMON, lots of linear hulls with high potential can be found. For example, the two 13-round linear hulls for SIMON32 in Table 3 are rotations of same linear hull.

Suppose that the time to compute the bias for one linear hull is T_1 and data complexity is $\mathcal{N}$. If m linear hulls with same bias are used in the multiple linear hull attack, the data complexity would be decreased to $\mathcal{N}/m$. But the time complexity would increase to $mT_1 + 2^{\mathcal{K}}$, where $\mathcal{K}$ is the size of the independent key bits involved in all m linear hull attacks. For example, there are 32 independent key bits involved in the 21-round attack of SIMON32 with linear hull $X_{L,5}^i \rightarrow X_{R,13}^{i+13}$. The data complexity is $2^{31.19}$ known plaintext-ciphertext pairs and the time needs about $2^{35.84}$ additions to get the bias. When another linear hull $X_{L,6}^i \rightarrow X_{R,14}^{i+13}$ is taken in to make a multiple linear hull attack, the data size will decrease to $2^{30.19}$. There are also 32 independent key bits involved in this linear hull attack. But, the total independent key size of both linear hulls is 48. So the time to compute the bias for the multiple linear hull attack with above two linear hulls needs about $2^{36.84}$ additions and 2^{48} combinations.

5 Conclusion

In this paper, we gave the improved linear attacks on all the reduced versions of SIMON family with dynamic key-guessing techniques. By establishing the boolean function of parity bit in the linear hull distinguisher and reducing the expressions of function according to the property of AND operation, we decrease the number of key bits involved in the attack and decrease the attack complexity in a further step. As a result, we can attack 23-round SIMON32/64, 24-round SIMON48/72, 25-round SIMON48/96, 30-round SIMON64/96, 31-round SIMON64/128, 37-round SIMON96/96, 38-round SIMON96/144, 49-round SIMON128/128, 51-round SIMON128/192 and 53-round SIMON128/256. The differential attack in [20] and our linear hull attack are bit-level cryptanalysis results, which provide the more efficient and precise security estimation results on SIMON. It is mentioned that, the bit-level cryptanalysis combining with dynamic key-guessing techniques are applicable to more light-weight block ciphers and hash functions etc.

Acknowledgements. This work was partially supported by the National Natural Science Foundation of China (Grant No. 61133013), also supported by National Key Basic Research Program of China (Grant No. 2013CB834205).

A   Time Complexity in Some Situations

In this section, we give the time complexities of computing the counters $B^k(y)$ for some simple functions of $y = f(x, k)$. This would be the deepest layer's operation in the linear attack to SIMON. Notice in the following, 'Guess' denotes the bits guessed at first. The second column $x_i \oplus k_i$ denotes the value of x_i which is used in the splitting phase. The third column denotes the new representation of the target function according to the value of $x_i \oplus k_i$. 'RB' is the related bit (defined in Sect. 3). T_1 denotes the time of seperation of the plaintexts according to the guessed bit of k. T_2 denotes the time of computation in the inner part. T_3 is the time in the combination phase. Total Time is the final time complexity, which is twice of the sum of all T_1, T_2 and T_3. Notice that T_1, T_2 and T_3 represent the number of addition operations. For simplicity, we denote $f*$ the function with same form of f. For example, if $f_1 = (x_0 \oplus k_0)\&(x_1 \oplus k_1)$ and $f_1' = (x_0 \oplus k_0)\&(x_3 \oplus k_3)$, we say f_1' is with form f_1*. The calculation of $B(y)$ for the functions same form have same procedures and time complexties.

1. $f_1 = (x_0 \oplus k_0)\&(x_1 \oplus k_1)$

Guess	$x_0 \oplus k_0$	f_1	RB	T_1	T_2	T_3
k_0	0	0		1		2
	1	0	k_1	1		
Total Time				$2 \times (1 + 1 + 2) = 2^3$		

2. $f_2 = (x_0 \oplus k_0) \oplus (x_1 \oplus k_1)\&(x_2 \oplus k_2)$

Guess	x_0	f_2	RB	T_1	T_2	T_3
		f_1*	k_0	$2^2 \times 1$	2^3	2^3
Total Time				$2^2 + 2^3 + 2^3 = 2^{4.32}$		

3. $f_3 = (x_0 \oplus k_0)\&((x_1 \oplus k_1) \oplus (x_2 \oplus k_2)\&(x_3 \oplus k_3))$

Guess	$x_0 \oplus k_0$	f_3	RB	T_1	T_2	T_3
k_0	0	0		$2^3 - 1$		2^3
	1	f_2*			$2^{4.32}$	
Total Time				$2 \times (2^3 - 1 + 2^{4.32} + 2^3) = 2^{5.64}$		

The detail of case 1, where $f_1(x, k) = (x_0 \oplus k_0)\&(x_1 \oplus k_1)$, has been given in Sect. 3.2. The other cases are derived similarly. For example, in case 2, linear compression is done before any key guessing, leading to the compression of bit x_0 and generation of related bit k_0.

References

1. Abdelraheem, M.A., Alizadeh, J., Alkhzaimi, H.A., Aref, M.R., Bagheri, N., Gauravaram, P., Lauridsen, M.M.: Improved linear cryptanalysis of reduced-round Simon (2014). IACR Cryptology ePrint Archive 2014/68
2. Abed, F., List, E., Lucks, S., Wenzel, J.: Differential and linear cryptanalysis of reduced-round Simon (2013). IACR Cryptology ePrint Archive, 2013/526
3. Abed, F., List, E., Lucks, S., Wenzel, J.: Differential cryptanalysis of round-reduced Simon and Speck. In: Cid, C., Rechberger, C. (eds.) FSE 2014. LNCS, vol. 8540, pp. 525–545. Springer, Heidelberg (2015)
4. Alizadeh, J., Alkhzaimi, H.A., Aref, M.R., Bagheri, N., Gauravaram, P., Kumar, A., Lauridsen, M.M., Sanadhya, S.K.: Cryptanalysis of Simon variants with connections. In: Sadeghi, A.-R., Saxena, N. (eds.) RFIDSec 2014. LNCS, vol. 8651, pp. 90–107. Springer, Heidelberg (2014)

5. Alkhzaimi, H.A., Lauridsen, M.M.: Cryptanalysis of the SIMON family of block ciphers (2013). IACR Cryptology ePrint Archive 2013/543
6. Asgur, T.: Improved linear trails dor the block cipher SIMON (2015). IACR Cryptology ePrint Archive 2015/285
7. Beaulieu, R., Shors, D., Smith, J., Treatman-Clark, S.: The SIMON and Speck Families of Lightweight Block Ciphers. Bryan Weeks, Louid Wingers (2013)
8. Biryukov, A., De Cannière, C., Quisquater, M.: On multiple linear approximations. In: Franklin, M. (ed.) CRYPTO 2004. LNCS, vol. 3152, pp. 1–22. Springer, Heidelberg (2004)
9. Biryukov, A., Roy, A., Velichkov, V.: Differential analysis of block ciphers SIMON and SPECK. In: Cid, C., Rechberger, C. (eds.) FSE 2014. LNCS, vol. 8540, pp. 546–570. Springer, Heidelberg (2015)
10. Boura, C., Naya-Plasencia, M., Suder, V.: Scrutinizing and improving impossible differential attacks: applications to CLEFIA, Camellia, LBlock and SIMON. In: Sarkar, P., Iwata, T. (eds.) ASIACRYPT 2014. LNCS, vol. 8873, pp. 179–199. Springer, Heidelberg (2014)
11. Chen, H., Wang, X.: Improved linear hull attack on round-reduced SIMON with dynamic key-guessing techniques (2015). IACR Cryptology ePrint Archive 2015/666
12. Chen, Z., Wang, N., Wang, X.: Impossible differential cryptanalysis of reduced round SIMON (2015). IACR Cryptology ePrint Archive 2015/286
13. Matsui, M.: Linear cryptanalysis method for DES cipher. In: Helleseth, T. (ed.) EUROCRYPT 1993. LNCS, vol. 765, pp. 386–397. Springer, Heidelberg (1994)
14. Matsui, M.: The first experimental cryptanalysis of the data encryption standard. In: Desmedt, Y.G. (ed.) CRYPTO 1994. LNCS, vol. 839, pp. 1–11. Springer, Heidelberg (1994)
15. Nyberg, K.: Linear approximation of block ciphers. In: De Santis, A. (ed.) EUROCRYPT 1994. LNCS, vol. 950, pp. 439–444. Springer, Heidelberg (1995)
16. Selçuk, A.A., Biçak, A.: On probability of success in linear and differential cryptanalysis. In: Cimato, S., Galdi, C., Persiano, G. (eds.) SCN 2002. LNCS, vol. 2576, pp. 174–185. Springer, Heidelberg (2003)
17. Shi, D., Lei, H., Sun, S., Song, L., Qiao, K., Ma, X.: Improved linear (hull) cryptanalysis of round-reduced versions of SIMON (2014). IACR Cryptology ePrint Archive 2014/973
18. Sun, S., Lei, H., Wang, M., Wang, P., Qiao, K., Ma, X., Ma, D., Song, L., Kai, F.: Towards finding the best characteristics of some bit-oriented block ciphers and automatic enumeration of (related-key) differential and linear characteristics with predefined properties and its applications (2014). IACR Cryptology ePrint Archive 2014/747
19. Sun, S., Hu, L., Wang, P., Qiao, K., Ma, X., Song, L.: Automatic security evaluation and (related-key) differential characteristic search: application to SIMON, PRESENT, LBlock, DES(L) and Other bit-oriented block ciphers. In: Sarkar, P., Iwata, T. (eds.) ASIACRYPT 2014. LNCS, vol. 8873, pp. 158–178. Springer, Heidelberg (2014)
20. Wang, N., Wang, X., Jia, K., Zhao, J.: Differential attacks on reduced SIMON versions with dynamic key-guessing techniques (2014). IACR Cryptology ePrint Archive 2014/448
21. Wang, Q., Liu, Z., Kerem Varıcı, Y., Sasaki, V.R., Todo, Y.: Cryptanalysis of Reudced-round SIMON32 and SIMON48. In: Meier, W., Mukhopadhyay, D. (eds.) INDOCRYPT 2014. LNCS, vol. 8885, pp. 143–160. Springer International Publishing, Switzerland (2014)

Foundations and Theory

Modeling Random Oracles Under Unpredictable Queries

Pooya Farshim[1([⊠])] and Arno Mittelbach[2]

[1] ENS, CNRS & INRIA, PSL Research University, Paris, France
pooya.farshim@gmail.com
[2] Darmstadt University of Technology, Darmstadt, Germany
arno.mittelbach@cased.de

Abstract. In recent work, Bellare, Hoang, and Keelveedhi (CRYPTO 2013) introduced a new abstraction called Universal Computational Extractors (UCEs), and showed how they can replace random oracles (ROs) across a wide range of cryptosystems. We formulate a new framework, called Interactive Computational Extractors (ICEs), that extends UCEs by viewing them as models of ROs under unpredictable (aka. high-entropy) queries. We overcome a number of limitations of UCEs in the new framework, and in particular prove the adaptive RKA and semi-adaptive KDM securities of a highly efficient symmetric encryption scheme using ICEs under key offsets.

We show both negative and positive feasibility results for ICEs. On the negative side, we demonstrate ICE attacks on the HMAC and NMAC constructions. On the positive side we show that: (1) ROs are indeed ICE secure, thereby confirming the structural soundness of our definition and enabling a finer layered approach to protocol design in the RO model; and (2) a modified version of Liskov's Zipper Hash is ICE secure with respect to an underlying fixed-input-length RO, for appropriately restricted classes of adversaries. This brings the first result closer to practice by moving away from variable-input-length ROs. Our security proofs employ techniques from indifferentiability in multi-stage settings.

Keywords: Random oracle · Unpredictability · UCE · RKA security · KDM security · Zipper Hash · Indifferentiability · Multi-stage security

1 Introduction

1.1 Background

Since their formal introduction by Bellare and Rogaway [BR93], random oracles (ROs) have found many applications across a wide range of cryptographic protocols. However, due to an uninstantiability result of Canetti et al. [CGH98], which shows that certain (artificial) protocols become insecure as soon as the random oracle is replaced by *any* concrete hash function, reliance on ROs has also become somewhat debatable.

© International Association for Cryptologic Research 2016
T. Peyrin (Ed.): FSE 2016, LNCS 9783, pp. 453–473, 2016.
DOI: 10.1007/978-3-662-52993-5_23

Two lines of research have been directed at dealing with such uninstantiability results. One is to construct standard-model counterparts of cryptographic
primitives designed in the RO model (ROM). This approach comes with the
drawback that the resulting cryptosystems often tend to be complex and achieve
a lower level of security and/or efficiency. A second, more modular, approach
aims to formulate abstractions of the proof-centric properties of random oracles
such as extractability, programability, or non-malleability [Can97, CD09, Nie02,
CD08, BCFW09]. Assuming that a hash function meets the introduced model,
one proceeds to show that it can safely replace the random oracle in a protocol.
These formalizations, however, have only been successful to a limited extent, and
the question of finding a flexible and general framework that could be applied
across a broad range of security goals and protocols remained open until recently.

1.2 UCE Security

Bellare, Hoang, and Keelveedhi (BHK) [BHK13a] revisit the above questions and
present a powerful framework called *Universal Computational Extractors* (UCEs)
that allow to securely instantiate random oracles in an interesting and diverse set
of applications. These include, among other things, security under key-dependent-
message (KDM) attacks, security under related-key attacks (RKAs), simultaneous hard-core bits, point function obfuscators, garbling schemes, proofs of
storage, deterministic encryption, and message-locked encryption, thereby going
far beyond what was previously possible.

Behind UCEs lies a new way to model the indistinguishability of a keyed hash
function from a random oracle. Indeed, there are two direct ways to (incorrectly)
model the security of a hash function:

(1) Provide the adversary with the hash key and ask it to distinguish an oracle
    implementing the hash function from one implementing the random oracle.
    This approach immediately fails as this game can be trivially won with the
    knowledge of the hash key by computing a hash value and checking the
    answer against the oracle's answer for the same query.
(2) Adopt the above approach, but now hide the hash key. This leads to PRF
    security—for which feasibility results are known—but is not useful in the
    context of hashing as the hash key is typically publicly known.

BHK overcome the above shortcomings by *splitting* the attacker into two
parts and *constraining* the communication between the two. The first UCE
attacker does *not* get to see the hash key, but has oracle access to either the
hash function under a random key or the random oracle according to a random
bit. The second attacker, on the other hand, *does* get to see the hash key, but
can no longer access the oracle, and it has to guess the bit; see Fig. 1 (left).
The two stages of the adversary can communicate only in restricted ways since
arbitrary communication would lead to an attack similar to that given above for
formulation (1).

More formally, for a keyed hash function H, UCE security is defined via a
two-stage game consisting of algorithms S and D, called the *source* and the

distinguisher respectively, as follows. In the first stage, the source is given access to an oracle HASH that depending on a random bit b implements either the random oracle or the concrete hash function H under a random hash key hk. The source terminates by outputting some leakage L, which is then communicated to the second-stage distinguisher D. In addition to leakage L, the distinguisher also gets the hash key hk as input. The distinguisher's task is to guess b, i.e., guess whether the source was talking to the random oracle or the hash function. The UCE advantage of the pair (S, D) is defined as usual to be the probability of correctly guessing the bit b scaled away from one-half. We refer the reader to the original work [BHK13b] for an excellent overview of this approach to modeling hash-function security.

To see that without further restrictions UCE security cannot be achieved, consider a source that leaks one of its oracle queries together with the corresponding oracle answer to the distinguisher. The distinguisher then simply recomputes the hash value on the queried point—the distinguisher knows the hash key—and compares it to the leaked value.

In their original work, BHK [BHK13a] define two restrictions on sources: *computational unpredictability* and *computational reset security*. In the computational unpredictability game, it is required that when the source is run with a *random oracle* its leakage does not computationally reveal any of its queries. This is formalized by requiring that the probability of any efficient predictor P in guessing a query of S when given L is negligible.

The class of computationally unpredictable sources is denoted by $\mathcal{S}^{\mathrm{cup}}$, and the resulting UCE security $\mathrm{UCE}[\mathcal{S}^{\mathrm{cup}}]$ (aka. UCE1) of a hash function is defined by requiring the advantage of any efficient pair (S, D) with an *unpredictable* $S \in \mathcal{S}^{\mathrm{cup}}$ in the UCE game to be negligible. Reset security imposes a weaker restriction on the source class and leads to the stronger UCE2 notion.

UCE security has been the subject of many recent studies. Brzuska, Farshim, and Mittelbach (BFM) [BFM14] show that, under new cryptographic assumptions, these restrictions are insufficient for a feasible definition. More precisely, assuming the existence of indistinguishability obfuscators [BGI+01, GGH+13], BFM show that the $\mathrm{UCE}[\mathcal{S}^{\mathrm{cup}}]$ security of *any* hash function can be broken in polynomial time. To overcome this attack, BFM [BFM14] (and subsequently BHK in an updated version of their paper [BHK13b]) propose a *statistical* notion of unpredictability whereby the predictor can even run in unbounded time. Following the attack, BHK also refine the UCE notions based on computational unpredictability and introduce the classes of *bounded parallel* and *split* sources.[1] BFM show that security against bounded parallel source is also infeasible [BFM14], and recently attacks against split sources have also been shown [BST15].

On the positive side, Brzuska and Mittelbach [BM14b, BM15] show how to construct UCEs for the class of *strongly* unpredictable and statistically unpredictable sources for bounded number of queries. Bellare et al. [BHK14] develop

---

[1] Such computational UCE notions are intrinsically needed for applications such as simultaneous had-core bits and deterministic PKEs.

domain extenders for UCEs, and Bellare and Hoang [BH15] construct deterministic PKEs from UCEs for statistically unpredictable sources and lossy trapdoor functions. BFM [BFM14] have shown that the existence of obfuscation-based attacks against statistically unpredictable sources violates well-known impossibility results. A number of recent works have shown how to use UCEs as RO replacements in other protocols [MH14, BK15, DGG+15].

Despite the above advances, and irrespective of the restrictions imposed on sources, the UCE framework is intrinsically limited in a number of aspects: it only allows the source to place HASH queries which are *independent* of the hash key; after leakage is communicated from the source to the distinguisher no further HASH oracle queries can be made, and hence hash queries are inherently *non-adaptive*; UCEs cannot model *unkeyed* hash functions nor hash functions with weak keys where the key does not come from the uniform distribution. Motivated by these shortcomings, and the ultimate goal of basing the security of highly efficient and practical protocols on well-defined and feasible properties of random oracles, we set out to formalize an enhanced framework for the study ROM protocols.

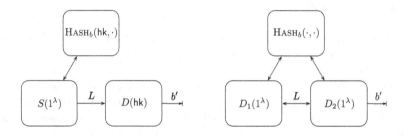

Fig. 1. The interactions in the UCE game (left) and the ICE game (right).

1.3   Interactive Computational Extractors

Given the development of UCEs, defining an extended model which meets the above-mentioned specifications is an intricate task. Indeed, well before the emergence of obfuscation-based attacks, BHK [BHK13b, p. 9] warned that extending UCEs to an interactive setting is "a dangerous path to tread." As an example, assume that we introduce a bi-directional communication channel between the distinguisher and the source so that our adaptivity targets are met. This extension can be shown to fall prey to somewhat non-trivial attacks that utilize general-purpose multi-party computation (MPC) protocols. Suppose the source S holds a random input x whose hash is y, and D holds hk. The two parties then run an MPC protocol to compute the Boolean value $y = H(hk, x)$. The distinguisher finally returns this value as its guess. This attack would meet any reasonable notion of computational unpredictability since the security of the MPC protocol would ensure that the parties learn no more than what can be deduced from their individual private inputs.[2] Allowing hash queries to depend on the hash key hk is also challenging since similarly to approach (1) above access to both hk and the hash oracle would trivialize the notion. For similar reasons, formulating a UCE-like model for unkeyed hash functions is also non-trivial. As we shall see, other forms of attacks also arise that should be ruled out for a feasible model.

THE ICE FRAMEWORK. Let us call an input (hk, x), consisting of the hash key hk and a domain point x, to a hash function a *full* input. One way to view UCEs is that they adopt the indistinguishability-based approach (1) above, but restrict hash queries so that full inputs remain hidden from the attacker(s). It is clear that such hidden queries are not meaningful in the presence of a single adversary—any adversary knows its own queries—and hence UCEs come with two adversaries. Unpredictability together with denial of oracle access to D ensures that the x components of full inputs remain hidden from D. On the other hand, the hk components of full inputs remain hidden from S as the source is denied access to hk (and no communication from D to S is allowed). As a result, full inputs (hk, x) remain hidden from *both* parties involved in a UCE attack.

This perspective allows us to build on UCEs and extend them as follows. In our new framework, which we call *Interactive Computational Extractors* (ICEs),[3] a general mechanism for the joint generation of full inputs is enabled and adversarial restrictions that formalize what it means for full hash inputs to have high entropy are imposed.

We let two distinguishers (D_1, D_2) to take part in an attack, and allow them to communicate via a bi-directional channel. Both distinguishers get access to a challenge hash oracle, which depending on a challenge bit implements either the real hash function or a (keyed) random oracle. To enable the two parties to make hidden queries, we introduce a shared write-only tape that both D_1 and D_2 can write onto. When a distinguisher queries the hash oracle, the (real or ideal) hash of the full contents of the tape is returned. In contrast to UCEs, D_1 or D_2 can generate a hash key and perhaps modify it throughout the attack. This attack scenario is symmetric for D_1 and D_2 and, without loss of generality, the game terminates by D_2 outputting with a bit. (Our formal definition, however, comes with a slightly more general return statement.) For a class $\mathcal{C}$ of distinguishers, we define ICE[$\mathcal{C}$] security by demanding that the probability of guessing the challenge bit for any $D = (D_1, D_2) \in \mathcal{C}$ is negligibly close to $1/2$. See Fig. 1 (right) for a summary of this interaction.

ENTROPIC QUERIES. Similarly to UCEs, the ICE notion cannot be achieved without constraining the way the two distinguishers communicate. The main restriction that we introduce is analogous to *statistical* unpredictability for UCEs:

---

[2] This can be viewed as an interactive analogue of BFM's attack [BFM14].

[3] In UCEs, "universal" refers to the fact that extraction should work with respect to universal (i.e., all admissible) sources. Analogously, "interactive" in ICEs refers to the fact that extraction should work for sources that can interact.

we demand the statistical unpredictability of *full* inputs to the hash function, including the hash key hk, from each distinguisher's point of view. We choose a statistical, rather than a computational, notion so that our definitions do not become subject to the interactive versions of the attacks highlighted in [BFM14].[4] More precisely, we require that when the hash oracle implements a keyed random oracle, no (possibly unbounded) predictor can guess a full input (hk, x) used to compute a hash value when it is provided with a distinguisher's *view* consisting of its inputs, random coins, and all incoming messages and oracle responses.

Since our framework allows oracle access to both parties, unlike UCEs the two distinguishers can implicitly communicate via *hash patterns* as follows. Suppose D_2 wants to leak a bit d to D_1. Algorithm D_2 starts by writing a random string onto the second half of the input and hands over the attack to D_1. Algorithm D_1 writes a random value to the first half of the input, calls HASH to receive a first hash value h_1, and hands over the attack back to D_2. Now algorithm D_2, according to the value of d, either modifies the contents of the second half of the input tape or leaves them unchanged. D_1 can recover d by obtaining a second hash value h_2 and checking if ($h_1 = h_2$). The two distinguishers can also communicate via a *bit-fixing* attack: D_2 samples many (unpredictable) random values x conditioned on its hash value beginning with bit d, which D_1 can then recover via a hash query.

In our unpredictability definition the predictor gets to see all hash responses, and hence if there are any repetitions they will be seen by the predictor. Unpredictability will therefore ensure that such repetition patterns will not leak any of the queries. Sometimes, however, we need to explicitly disallow any repeat queries to enable a security proof to go through. In such a scenario, we can ensure that there is no leakage via hash patterns either. Repeat-freeness appears in other related settings such as related-key attacks or correlated-input hashing [BK03, GOR11].

1.4   Applications

BHK [BHK13a] use UCEs to show that the encryption scheme of Black, Rogaway, and Shrimpton (BRS) [BRS03] is secure under related-key attacks (RKAs) and key-dependent-message (KDM) attacks as long as the related keys/key-dependent messages are derived non-adaptively at the onset and without access to the hash key or previous ciphertexts.[5]

As we shall see, ICE encompasses UCE as a special case, and the BRS scheme can also be instantiated under the above models using ICEs. We can however also obtain feasibility results that are outside the reach of UCEs. A practically

---

[4] This is also motivated by impossibility results for statistically secure two-party protocols.

[5] Recall that in RKA security the adversary can see encryptions of messages under keys $\phi(K)$ for a random K and functions ϕ of its choice. In KDM security the adversary can see encryptions of $\phi(K)$, under a random key K, for ϕ's of its choice.

relevant and desirable level of RKA security is that corresponding to key offsets (the so-called xor-RKA security [LRW02,BK09]). We show that ICEs are sufficient to prove the *full* xor-RKA security of the BRS scheme. Our formal result is more general and applies to the larger class of *split* functions that take the form $\phi(K_1\|K_2) = \phi_1(K_1)\|\phi_2(K_2)$. (Such functions have been used to build RKA-secure PRFs [BC10], and also appear in other related contexts [CG14,LL12].) In addition to achieving stronger security guarantees, ICEs allow instantiating the BRS scheme using unkeyed hash functions, which is arguably closer to the original formulation of BRS.[6]

We also strengthen the attainable KDM security guarantees for BRS by showing that adversaries can choose key-dependent messages adaptively based on the hash key and also semi-adaptively depending on previous ciphertexts. We prove that ICEs are adaptively correlated-input secure [GOR11] and that they relate well to other standard security properties of random oracles, such as pseudorandomness, randomness extraction, and one-way security (see full version). We leave it as open questions to see if full RKA beyond xor offsets or full KDM security can be established using extractor-like notaions.

1.5   Instantiations

BHK show that random oracles fulfill their strongest proposed UCE notion, namely UCE security with respect to computationally unpredictable sources.[7] We prove that random oracles are also ICE secure. The significance of these results are twofold [BHK13a]: (1) there are no generic attacks on ICEs and the model is structurally sound; and (2) a *layered* approach to security analysis can be enabled, whereby one first proves the security of a scheme under an ICE assumption and then applies the RO model feasibility result. The latter is akin to security analyses carried out in the generic group model.

Practical hash functions, however, are not monolithic objects and often follow an iterative procedure to convert a fixed-input-length random oracle (FIL-RO) into a variable-input-length random oracle (VIL-RO). This, in turn, raises the question whether or not the above result can be brought closer to practice by demonstrating positive feasibility results for VIL-ICEs in the FIL-RO model. A seemingly immediate way to establish this result would be to start with a hash function that is known to be *indifferentiable* from a VIL RO (e.g., the HMAC or the NMAC construction), and then apply the RO feasibility result above to conclude. This argument, however, fails as the ICE game is *multi-staged* and indifferentiability does not necessarily guarantee composition in such settings [RSS11].

Motivated by the above observations, we show both positive and negative feasibility results for ICEs. On the negative side, we show that the indifferentiable HMAC and NMAC constructions are provably ICE *insecure* in the FIL-RO

---

[6] BRS [BRS03] analyze their scheme in the unkeyed RO model, which translates to unkeyed instantiations in practice.

[7] Note that this does not contradict the BFM attack as ROs do not have succinct descriptions.

model. On the positive side, and building on Mittelbach's techniques [Mit14], we prove that a keyed version of Liskov's Zipper Hash [Lis07] *is* ICE secure (as a VIL hash function) under the assumption that the underlying compression function is a FIL-RO. Zipper Hash can be seen as a variant of the classical Merkle–Damgård [Dam90, Mer90] construction where the message blocks are processed twice in the forward and backward directions. Hence our results strengthen the VIL-RO feasibility result above, and also provide formal evidence for the (intuitive) added security guarantees that multi-pass hash functions seem to offer over their single-pass counterparts. For instance, combined with our RKA and KDM results, we may conclude that Zipper Hash can be safely used within the BRS scheme with no adverse affects on its security.

The above analysis can be further strengthened in at least two directions. First, one can weaken the underlying assumption and assume that the compression function underlying Zipper Hash is only a FIL-ICE (rather than a FIL-RO). To this end, BHK [BHK14] give domain extenders for UCEs. Second, and motivated by the standard-model realizations of ICEs and UCEs, we ask if these primitives can be based on plausible hardness assumptions. Brzuska and Mittelbach [BM14a, BM15] have recently shown positive results for UCEs with respect to restricted classes of sources.

2   Notation

We denote the security parameter by $\lambda \in \mathbb{N}$, which is implicitly given to all algorithms (if not explicitly stated so) in the unary representation 1^λ. By $\{0,1\}^\ell$ we denote the set of all bit strings of length ℓ and $\{0,1\}^*$ is the set of all finite-length bit strings. For $x, y \in \{0,1\}^*$ we denote their concatenation by $x\|y$, the length of x by $|x|$, the ith bit of x by $x[i]$, and the substring of x formed using bits i to j by $x[i..j]$. We denote the empty string by ε. For X a finite set, $|X|$ denotes its cardinality, and $x \leftarrow_{\$} X$ denotes the action of sampling x uniformly at random from X. If Q is a list and x a string then Q : x denotes the list obtained by appending x to Q. Similarly, If Q_1 and Q_2 are lists, then $Q_1 : Q_2$ denotes the concatenated list. Unless stated otherwise, algorithms are assumed to be randomized. We call an algorithm efficient or PPT if it runs in time polynomial in the security parameter. By $y \leftarrow \mathcal{A}(x; r)$ we denote that y was output by algorithm $\mathcal{A}$ on input x and randomness r. If $\mathcal{A}$ is randomized and no randomness is specified, then we assume that $\mathcal{A}$ is run with freshly sampled uniform random coins, and write $y \leftarrow_{\$} \mathcal{A}(x)$. We use Coins[$A$] to denote the polynomially long string of random coins r used by a PPT machine A. We say a function negl(λ) is negligible if negl(λ) $\in \lambda^{-\omega(1)}$.

HASH FUNCTIONS. In the line with [BHK13a], we consider the following (simplified) formalization of hash functions. A hash function consists of five PPT algorithms H := (H.Kg, H.Ev, H.kl, H.il, H.ol) as follows. The key-generation algorithm H.Kg gets the security parameter 1^λ as input and outputs a key $hk \in \{0,1\}^{H.kl(\lambda)}$, where H.kl($\lambda$) is the key-length function. Algorithm H.il(λ) outputs the length of admissible inputs, which could take the special value $*$

denoting the variable-length input space $\{0,1\}^*$. Algorithm $\mathsf{H.ol}(\lambda)$ outputs the length of admissible outputs, which we assumed to be a fixed polynomial function of the security parameter. The deterministic evaluation algorithm $\mathsf{H.Ev}$ takes as input the security parameter 1^λ, a key hk, a point $x \in \{0,1\}^{\mathsf{H.il}(\lambda)}$, and generates a hash value $\mathsf{H.Ev}(1^\lambda, hk, x) \in \{0,1\}^{\mathsf{H.ol}(\lambda)}$. To ease notation, we often suppress the security parameter and simply write $\mathsf{H.Ev}(hk, x)$.

3   The ICE Framework

In this section we precisely define the ICE framework. We refer the reader to the introduction for a high-level overview of the model.

MAIN $\mathrm{ICE}_{\mathsf{H}}^{D}(\lambda)$	WRITE(j, v)
1: $b \leftarrow_{\$} \{0,1\}; L_1 \leftarrow 1^\lambda$	$(hk, x)[j..j + |v| - 1] \leftarrow v$
2: **while** $b_1 = \bot \vee b_2 = \bot$ **do**	
3: $(b_1, L_2) \leftarrow_{\$} D_1^{\mathrm{WRITE,HASH}}(L_1)$	HASH$()$
4: $(b_2, L_1) \leftarrow_{\$} D_2^{\mathrm{WRITE,HASH}}(L_2)$	**if** $b = 1$ **then** $T[hk, x] \leftarrow \mathsf{H.Ev}(hk, x)$
5: **return** $(b_1 \oplus b_2 = b)$	**elseif** $T[hk, x] = \bot$ **then**
	$\quad T[hk, x] \leftarrow_{\$} \{0,1\}^{\mathsf{H.ol}(\lambda)}$
	return $T[hk, x]$

Fig. 2. The ICE game with respect to hash function H and distinguishers $D = (D_1, D_2)$. We have omitted the initialization of various variables for readability.

THE ICE GAME. Let $\mathsf{H} = (\mathsf{H.Kg}, \mathsf{H.Ev}, \mathsf{H.kl}, \mathsf{H.il}, \mathsf{H.ol})$ be a hash function and let $D = (D_1, D_2)$ be a pair of algorithms. We define the ICE advantage of D against H as

$$\mathbf{Adv}_{\mathsf{H}, D}^{\mathrm{ice}}(\lambda) := 2 \cdot \Pr\left[\mathrm{ICE}_{\mathsf{H}}^{D}(\lambda)\right] - 1,$$

where game $\mathrm{ICE}_{\mathsf{H}}^{D}(\lambda)$ is shown in Fig. 2. As mentioned in the introduction, we may assume, without loss of generality, that the game termites by D_2 outputting a bit. However, in order to preserve the symmetry of the definition (which will simplify our adversarial restrictions later on) and for added generality, we let the distinguishers jointly guess the challenge bit by computing $b_1 \oplus b_2$, where b_i is D_i's guess. The interaction terminates when both distinguishers return non-$\bot$ values for b_1 and b_2. For a class $\mathcal{C}$ of distinguishers, we define ICE[$\mathcal{C}$] security by requiring the advantage of any adversary $D \in \mathcal{C}$ to be negligible in the ICE game.

We require (D_1, D_2) not to leave any superfluous blank spaces on the joint tape. That is, a WRITE call must ensure that before the HASH oracle is called there do not exist indices $i < j$ such that $x[i] = \varepsilon \neq x[j]$ or $hk[i] = \varepsilon \neq hk[j]$.

We also demand that the full inputs (hk, x) are valid in the sense that prior to a HASH call $hk \in \{0,1\}^{\mathsf{H.kl}(\lambda)}$ and $x \in \{0,1\}^{\mathsf{H.il}(\lambda)}$. Although the distinguishers D_1 and D_2 are in general stateful algorithms, we omit the explicit handling of state values from the inputs and outputs of D_i.

RESTRICTIONS. As discussed in the introduction, the ICE model is not feasible unless additional restrictions on the distinguishers are imposed. We formulate our restrictions as joint properties of (D_1, D_2). Before presenting our main restrictions corresponding to high-entropy queries, we give a set of basic classes that will be useful in studying ICEs. As an example, for polynomials w, q, and r we define $\mathcal{C}_i^{w,q,r}$ to be the set of all (D_1, D_2) such that when (D_1, D_2) is run in the ICE game conditioned on $b = 0$ (i.e., with respect to the random oracle), the distinguisher D_i places at most $w(\lambda)$ queries to WRITE, at most $q(\lambda)$ queries to HASH, and terminates after at most $r(\lambda)$ invocations. We formalize a number of other notions below and omit the preamble "The set of all (D_1, D_2) such that when (D_1, D_2) is run in ICE with $b = 0$, we have with overwhelming probability that" from their definitions. Note that the classes below depend on $i \in \{1, 2\}$. For classes $\mathcal{C}_i^{\mathrm{label}}$ we define $\mathcal{C}^{\mathrm{label}} := \mathcal{C}_1^{\mathrm{label}} \cap \mathcal{C}_2^{\mathrm{label}}$. In the following table we present several restrictions that we will be using throughout this paper.

Class	Description
$\mathcal{C}_i^{w,q,r}$	D_i places at most $w(\lambda)$ queries to WRITE, at most $q(\lambda)$ queries to HASH, and terminates after at most $r(\lambda)$ invocations
$\mathcal{C}_i^{\mathrm{poly}}$	D_i makes polynomially many oracle queries
$\mathcal{C}_i^{\mathrm{ppt}}$	D_i runs in polynomial time on each invocation and terminates after a polynomial number of rounds
$\mathcal{C}_i^0$	D_i sets $b_i := 0$ in all invocations
$\mathcal{C}_i^\varepsilon$	D_i sets $L_{3-i} := \varepsilon$ in all invocations
$\mathcal{C}_i^{0\text{-hk}}$	D_i never writes onto the hk part of the tape
$\mathcal{C}_i^{1\text{-hk}}$	On its first invocation, D_i writes a random hk onto the hk-part of the tape. In subsequent invocations, D_i never writes onto the hk-part of the tape
$\mathcal{C}_i^{\mathrm{dist}}$	D_i makes distinct queries to HASH. That is, for lists Q_1 and Q_2 defined in Fig. 3, the *combined* list $\mathsf{Q}_1 : \mathsf{Q}_2$ is repetition-free. Note that $\mathcal{C}_i^{\mathrm{dist}} = \mathcal{C}_{3-i}^{\mathrm{dist}} = \mathcal{C}^{\mathrm{dist}}$
$\mathcal{C}_i^{\mathrm{sup}}$	The probability that any (possibly unbounded) predictor P can guess a full query of D_i is negligible. We call this the class of statistically unpredictable D_i. See Fig. 3 for the formal definition. Class $\mathcal{C}_i^{\mathrm{cup}}$ is the computational analogue, where P is restricted to be ppt

AN EXAMPLE: UCE WITHIN ICE. We describe how UCEs can be captured within the ICE framework. Since ICE is more expressive a framework, we need to (drastically) restrict the distinguishers. In modeling UCEs, we identify the UCE distinguisher with D_1 and the UCE source with D_2. All parties typically run in polynomial time and hence we restrict to $\mathcal{C}^{\mathrm{ppt}} := \mathcal{C}_1^{\mathrm{ppt}} \cap \mathcal{C}_2^{\mathrm{ppt}}$. In UCEs, the source queries HASH on an unknown hash key. The distinguisher, on the other hand, gets to see the hash key. Thus, we let D_1 (which represents the distinguisher) write a random hk to the joint input and then hand the attack to D_2 on the first invocation, i.e., $D \in \mathcal{C}_1^{\text{1-hk}}$. We further restrict to $\mathcal{C}_1^{\varepsilon}$, as a UCE distinguisher does not leak. Since the UCE game only has a single round, we also restrict to $\mathcal{C}_1^{1,0,2}$ (one round is used to write the hk). Finally, the source does not take part in decision making and cannot modify the hash key: UCEs are modeled by $\mathrm{ICE}[\mathcal{C}^{\mathrm{uce}}]$ where

$$\mathcal{C}^{\mathrm{uce}} := \mathcal{C}^{\mathrm{ppt}} \cap \mathcal{C}_1^{\text{1-hk}} \cap \mathcal{C}_1^{\varepsilon} \cap \mathcal{C}_1^{1,0,2} \cap \mathcal{C}_2^{0} \cap \mathcal{C}_2^{\text{0-hk}}.$$

Note that the above models UCEs without any additional restrictions on the source classes. Such requirements can be added on top by appropriately restricting $\mathcal{C}^{\mathrm{uce}}$.

UNPREDICTABILITY. We now formally define what we mean by a D that has unpredictable (aka. high-entropy) queries. We focus on a statistical notion of unpredictability [BFM14,BST15].[8] We say $D = (D_1, D_2)$ is statistically unpredictable for the distinguisher i, and write $D \in \mathcal{C}_i^{\mathrm{sup}}$, if the advantage of any unbounded predictor P defined by

$$\mathbf{Adv}_{i,D,P}^{\mathrm{pred}}(\lambda) := \Pr\left[\mathrm{Pred}_{i,D}^{P}(\lambda)\right],$$

is negligible, where game $\mathrm{Pred}_{i,D}^{P}(\lambda)$ is shown in Fig. 3.

MAIN $\mathrm{Pred}_{i,D}^{P}(\lambda)$	WRITE(j, v)
1: $L_1 \leftarrow 1^{\lambda}$	$(hk, x)[j..j + |v| - 1] \leftarrow v$
2: **while** $b_1 = \bot \vee b_2 = \bot$ **do**	
3: $k \leftarrow 1; (b_1, L_2) \leftarrow\!\!\$ D_1^{\mathrm{WRITE,HASH}}(L_1)$	HASH$()$
4: $\mathrm{Lk}_i \leftarrow \mathrm{Lk}_i : L_i$	**if** $T[hk, x] = \bot$ **then**
5: $k \leftarrow 2; (b_2, L_1) \leftarrow\!\!\$ D_2^{\mathrm{WRITE,HASH}}(L_2)$	$\quad T[hk, x] \leftarrow\!\!\$ \{0, 1\}^{\mathrm{H.ol}(\lambda)}$
6: $(\overline{hk}, \overline{x}) \leftarrow\!\!\$ P(\mathrm{Coins}[D_i], \mathsf{A}_i, \mathrm{Lk}_i)$	$\quad \mathsf{Q}_k \leftarrow \mathsf{Q}_k : (hk, x)$
7: **return** $(\overline{hk}, \overline{x}) \in \mathsf{Q}_1 : \mathsf{Q}_2$	$\quad \mathsf{A}_k \leftarrow \mathsf{A}_k : T[hk, x]$
	return $T[hk, x]$

Fig. 3. The unpredictability game.

---

[8] We emphasize that computational notions are still valuable as combined with our feasibility results, they would enable easier and more modular security proofs in the RO model.

Note that the predicator only gets to see the hash responses for distinguisher D_i—these are within D_i's view—and has to guess a query made by *either* distinguisher in the concatenated list $Q_1 : Q_2$. It is easy to check that UCE security with respect to statistically unpredictable sources is equivalent to $\text{ICE}[\mathcal{C}^{\text{uce}} \cap \mathcal{C}^{\text{sup}}]$ security.

REMARK. Since predictor P receives the full view of a distinguisher D_i, it can perfectly simulate a run of D_i in the ICE game with respect to a *random* implementation of the hash oracle, without any need to see the view of the partner distinguisher D_{3-i}. We will rely on this observation in our proofs.

4   Example Applications

In this section we demonstrate two example use cases of ICEs. Further applications are given in the full version and summarized in Table 1 below. These applications serve to demonstrate that many properties of random oracles that are useful in analyses of ROM cryptosystems can be modeled in a unified way within the ICE framework.

Table 1. Distinguisher classes used (above) and shown feasibility for (below). Here $\mathcal{C}^* := \mathcal{C}^{\text{ppt}} \cap \mathcal{C}^{\text{dist}} \cap \mathcal{C}_1^{\text{1-hk}} \cap \mathcal{C}_2^{\text{0-hk}} \cap \mathcal{C}_2^{\varepsilon}$.

Goal/Model	Class used/Achieved
Split RKA	$\mathcal{C}^* \cap \mathcal{C}^{\text{sup}} \cap \mathcal{C}_2^0$
Split KDM	$\mathcal{C}^* \cap \mathcal{C}^{\text{sup}} \cap \mathcal{C}_1^0$
Split/claw-free CIH	$\mathcal{C}^* \cap \mathcal{C}^{\text{sup}} \cap \mathcal{C}_2^0$
Extractor	$\mathcal{C}^* \cap \mathcal{C}^{\text{sup}} \cap \mathcal{C}_1^0 \cap \mathcal{C}^{\varepsilon} \cap \mathcal{C}^{1,1,2}$
Weak PRF	$\mathcal{C}^* \cap \mathcal{C}^{\text{sup}} \cap \mathcal{C}_1^0 \cap \mathcal{C}^{\varepsilon} \cap \mathcal{C}^{\text{poly},\text{poly},1}$
poly-regular OWF	$\mathcal{C}^* \cap \mathcal{C}^{\text{sup}} \cap \mathcal{C}_1^0 \cap \mathcal{C}^{1,1,1}$
VIL-ROM	$\mathcal{C}^{\text{ppt}} \cap \mathcal{C}^{\text{cup}}$ and $\mathcal{C}^{\text{poly}} \cap \mathcal{C}^{\text{sup}}$; both contain $\mathcal{C}^* \cap \mathcal{C}^{\text{sup}}$
FIL-ROM	$\mathcal{C}^* \cap \mathcal{C}^{\text{cup}}$, which contains $\mathcal{C}^* \cap \mathcal{C}^{\text{sup}}$

4.1   Split RKA Security

We show that the symmetric encryption scheme proposed by Black, Rogaway, and Shrimpton (BRS) [BRS03] is secure against related-key attacks (RKAs) when instantiated with an ICE-secure hash function. The encryption algorithm of the BRS scheme is implemented via $\text{Enc}^{\text{H}}(K, M; R) := (R, M \oplus \text{H}(K \| R))$, for a hash function H, randomness R and key K. Recall that in an RKA, an adversary can obtain encryptions of messages of its choice under correlated keys (e.g., under K and $K \oplus 1$).

Split related-key derivation (RKD) functions ϕ decompose into two sub-RKD functions ϕ_1 and ϕ_2 that are applied in parallel to two (fixed) sub-strings of the key: $\phi(K_1 \| K_2) = \phi_1(K_1) \| \phi_2(K_2)$.[9] Split functions capture many RKA cases of interest including the case of xoring constants into keys. Without the minimal assumption that ϕ's have unpredictable outputs (i.e., the guessing probability of the outputs of $\phi(K)$ over randomly chosen K is negligible) RKA security is not achievable [BK03]. In our proof, we will require a slightly stronger condition that the sub-RKD functions $\phi_1(K_1)$ and $\phi_2(K_2)$ are *individually* unpredictable. Note that offsetting keys via xor enjoys this property as xor induces a permutation over the two halves of the key.

BHK [BHK13a], by interpreting encryption randomness as hash keys, show that BRS is selectively RKA secure using a multi-key extension of UCE$[\mathcal{S}^{\text{cup}}]$. In contrast, the adversary in our model retains its capability to adaptively query RKD functions of its choice depending both on the hash key and the ciphertexts that it has previously seen. For this result, although ICE$[\mathcal{C}^{\text{ppt}} \cap \mathcal{C}^{\text{sup}}]$ is sufficient, the assumption can be fine-tuned to ICE$[\mathcal{C}]$ where

$$\mathcal{C} := \mathcal{C}^{\text{ppt}} \cap \mathcal{C}^{\text{sup}} \cap \mathcal{C}^{\text{dist}} \cap \mathcal{C}_1^{\text{1-hk}} \cap \mathcal{C}_2^{\text{0-hk}} \cap \mathcal{C}_2^0 \cap \mathcal{C}_2^\varepsilon.$$

We defer the formal proof to the full version and give a detailed outline here.

THE ICE ADVERSARY. Given an RKA adversary A, we construct an ICE adversary (D_1, D_2), where D_1 handles the left components of A's RKA queries and D_2 handles the right components as follows.

$D_1(L_1)$: On initial invocation, generate a hash key hk, a random K_1, and a random bit b. Store these values and write hk onto the hk-part of the tape. Run $A(hk)$ to get an RKA query $((\phi_1, \phi_2), M_0, M_1)$. Output $(b_1, L_2) := (\bot, \phi_2)$. Proceed as follows in subsequent invocations. Generate and store a random R and write $\phi_1(K_1)$ onto the 1st segment (out of three segments) of the x-part of the tape and R onto its 3rd segment. Query HASH to get H. Recover R and resume A on $(R, H \oplus M_b)$ to get a new RKA query $((\phi_1, \phi_2), M_0, M_1)$, or a bit b'. If A outputs a bit b', return $(b_1, L_2) := (b = b', \varepsilon)$ and terminate. Else output $(b_1, L_2) := (\bot, \phi_2)$.

$D_2(L_2)$: When initially invoked, generate a random K_2 and store it. In all invocations (including the first), recover ϕ_2 from L_2. If $\phi_2 = \varepsilon$, return $(b_2, L_1) := (0, \varepsilon)$ and terminate. Else write $\phi_2(K_2)$ onto the 2nd segment of the x-part of the tape. Output $(b_2, L_1) := (0, \varepsilon)$.

UNPREDICTABILITY. We show that $D \in \mathcal{C}$ for class $\mathcal{C}$ as defined above. To this end, we only prove membership in $\mathcal{C}^{\text{sup}} \cap \mathcal{C}^{\text{dist}}$ as other cases follow via syntactic checks. This follows from the following two observations: (1) The HASH queries are distinct with overwhelming probability since before each query a fresh random value R is written onto the joint tape. (2) The functions ϕ_1 and ϕ_2 are

---

[9] For simplicity we assume that these are just the left and right halves of the key. Our proof will however also apply to any two substrings of super-logarithmic lengths.

run on independently chosen substrings of the key. Since they are assumed to be individually statistically unpredictable, D_1 observing independently generated random strings corresponding to hash values never gets to know the contents of the tape written by D_2, and vice versa, D_2 never gets to know what is written on to the tape by D_1.

4.2   KDM Security

When the random oracle in the BRS scheme is instantiated with an ICE-secure hash function, we are able to show that the BRS scheme resists a partially adaptive form of KDM security for split key-dependent-message derivation (KDMD) functions ϕ. As for RKD functions, such KDMD functions consist of sub-KDMD functions ϕ_1 and ϕ_2 of the form $\phi(K_1\|K_2) := \phi_1(K_1)\|\phi_2(K_2)$. The adaptivity level that we can tolerate is as follows. In an initial phase of the attack, the adversary can fully adaptively query split KDMD functions that do not depend on K_2. That is, for these functions $\phi_2(K_2)$ is constant and independent of K_2 and its value can be predicted. In a second phase of the attack, the adversary can query split KDMD functions of its choice as long as either $\phi_1(K_1)$ is constant or $\phi_1(K_1)$ was used in the first phase. (We emphasize that these functions are not required to be unpredictable.) This model is strong enough to imply IND-CPA security (without any restrictions), a case that could not be treated using UCEs.

THE ICE ADVERSARY. Let A be a KDM adversary against the BRS scheme in the model above. Our ICE[$\mathcal{C}^{\mathrm{ppt}} \cap \mathcal{C}^{\mathrm{sup}} \cap \mathcal{C}^{\mathrm{dist}}$] adversary corresponding to A is as follows, where for simplicity we have assumed the lengths of keys, randomness and messages are all ℓ. (The ICE class can be further restricted as is shown in Table 1.) In this reduction, D_1 faithfully runs the first stage of the attack, while D_2 runs its second stage. To answer KDM queries, D_2 relies on the "homomorphic" property that $H \oplus (x_1\|x_2) = H \oplus (x_1\|0^{|x_2|}) \oplus (0^{|x_1|}\|x_2)$.

$D_1(L_1)$: When initially invoked, generate a random hk, K_1 and b and store them. Write hk to the hk-part and K_1 to the 1st (out of three) segments of the x-part of the tape. (The segments are of lengths $\ell/2$, $\ell/2$ and ℓ corresponding to K_1, K_2 and R respectively.) Output $(\bot, \varepsilon)$. On the second invocation, run $A(hk)$ and answer its KDM queries $((\phi_1^0, \phi_2^0), (\phi_1^1, \phi_2^1))$ as follows. Write a fresh random value R onto the 3rd segment of the x-part of the tape. Call HASH to get H, and resume A on $(R, H \oplus (\phi_1(K_1)\|M_2^*))$, where $M_2^* := \phi_2(0^{\ell/2})$ is the right K_2-independent part of the message. Continue this process until A decides to proceed to its second stage. Let st_A denote A's state. Generate sufficiently many copies $(R_1, C_1'), \ldots, (R_q, C_q')$ of each of the KDM queries made in the first phase. Let List_1 denote the corresponding list of queried ϕ_1^b. Return $(0, (b, st_A, (R_1, C_1'), \ldots, (R_q, C_q'), \mathsf{List}_1))$ and terminate.

$D_2(L_2)$: When initially invoked, generate a random K_2, store it, and write it to the 2nd segment of the x-part of the tape. Hand the attack back to D_1, by outputting $(\bot, \varepsilon)$. On the second invocation, parse L_2 appropriately as above. Resume A on st_A and answer its KDM queries $((\phi_1^0, \phi_2^0), (\phi_1^1, \phi_2^1))$ as follows.

If $\phi_1^b \in \mathsf{List}_1$ pick a fresh ciphertext (R, C') corresponding to ϕ_1^b and *complete* the ciphertext preparation by setting $C \leftarrow C \oplus (0^{\ell/2} \| \phi_2^b(K_2))$. Otherwise generate a random R, write it onto the 3rd segment of the x-part of the tape, query HASH to get H, and set $C \leftarrow H \oplus (\phi_1^b(0^{\ell/2}) \| \phi_2^b(K_2))$. Resume A on $(R, C; st_A)$ and continue in this manner until A outputs a bit b'. Return $(b = b', \varepsilon)$ and terminate.

UNPREDICTABILITY. D's queries are distinct with overwhelming probability as fresh randomness R is written on the tape before each query. Throughout the attack, and when the hash oracle implements a random function, K_2 remains hidden from D_1 as D_1 only sees distinct random values as hash responses. Key K_1 also remains hidden from D_2 as the (incomplete) ciphertext components received from D_1 are random strings. Hence $D \in \mathcal{C}^{\mathrm{ppt}} \cap \mathcal{C}^{\mathrm{sup}} \cap \mathcal{C}^{\mathrm{dist}}$.

5   Feasibility

In this section we start by showing that random oracles are ICE secure with respect to interesting distinguisher classes (in particular, with respect to the restrictions needed for the presented applications). We then consider the ICE security of practical hash constructions built from fix-input-length (FIL) ROs. In particular, we look at a keyed variant of Liskov's Zipper Hash [Lis07] and show that it achieves ICE security in the FIL-RO model. Interestingly, we show that both HMAC and NMAC constructions [BCK96], which were recently shown to be UCE secure in FIL-ROM [Mit14], fail to be ICE secure. This result yields a natural counterexample to the composability of HMAC in multi-stage settings, similarly to that given by Ristenpart, Shacham, and Shrimpton in [RSS11]. Furthermore, it provides a separation between ICE and UCE. Our results also demonstrate that Zipper Hash can provide a higher level of security compared to HMAC when used in multi-stage settings.

5.1   ICEs from Random Oracles

BHK [BHK13b] show that UCE-secure hash functions can be provably constructed in the RO model. The philosophical justifications of this result are that there are *no structural weaknesses* in the definitional framework, and more importantly, a *layered* approach to protocol design in the RO model can be enabled [BHK13b]. We show that ICEs also enjoy RO feasibility.

Let $\mathsf{H.kl}(\cdot)$ and $\mathsf{H.ol}(\cdot)$ be two arbitrary functions as in the syntax of a hash function. Let $\mathcal{R}$ be a family of variable-input-length (VIL) ROs (i.e., with domain $\{0,1\}^*$) and range $\{0,1\}^{\mathsf{H.ol}(\lambda)}$. We construct the required hash function $\mathsf{H}^{\mathcal{R}}$ by defining the key-generation algorithm $\mathsf{H.Kg}(1^\lambda)$ to return a random $hk \leftarrow_\$ \{0,1\}^{\mathsf{H.kl}(\lambda)}$ and the evaluation algorithm $\mathsf{H.Ev}^{\mathcal{R}}(hk, x)$ to return $\mathcal{R}(hk \| x)$. Our first feasibility result is as follows.

Theorem 1 (ICE feasibility in ROM). *The VIL hash function* $\mathsf{H}^{\mathcal{R}}$ *constructed above is* ICE[$\mathcal{C}$] *secure in the VIL-RO model for* $\mathcal{R}$ *for the following (incomparable) classes of adversaries:*

$$\mathcal{C} := \mathcal{C}^{\mathrm{poly}} \cap \mathcal{C}^{\mathrm{sup}} \quad and \quad \mathcal{C} := \mathcal{C}^{\mathrm{ppt}} \cap \mathcal{C}^{\mathrm{cup}}.$$

The proof of this theorem is similar to the proof of [BHK13b, Theorem 6.1] for UCEs, and we give the details in the full version. Intuitively, we rely on unpredictability of queries to simulate the random oracles used in the construction and implicit in the ICE game independently. Interestingly, distinctness of queries will not be needed in this proof and we do not restrict the classes to $\mathcal{C}^{\mathrm{dist}}$. We note that the above classes include all those needed for the applications, as listed in Table 1. We also note that this theorem generalizes the feasibility of UCEs for unpredictable sources in ROM [BHK13b] as it can be easily verified that

$$\mathcal{C}^{\mathrm{uce}} \cap \mathcal{C}_2^{\mathrm{cup}} \subseteq \mathcal{C}^{\mathrm{ppt}} \cap \mathcal{C}^{\mathrm{cup}} \quad and \quad \mathcal{C}^{\mathrm{uce}} \cap \mathcal{C}_2^{\mathrm{sup}} \subseteq \mathcal{C}^{\mathrm{ppt}} \cap \mathcal{C}^{\mathrm{sup}}.$$

5.2 VIL-ICEs from Ideal Compression

Practical variable-input-length (VIL) hash functions are not monolithic objects. They often follow iterative modes of chaining that convert a fix-input-length (FIL) compression function to one that accepts variable-length inputs. This design principle has been successfully validated via the indifferentiability framework of Maurer et al. [MRH04,CDMP05], whereby an indifferentiable hash-function construction is shown to securely compose when used in place of a random oracle. As pointed out in [RSS11], the indifferentiability framework only guarantees composition in single-stage environments. The ICE and UCE games, however, are inherently multi-staged and lie outside the reach of (plain) indifferentiability. Mittelbach [Mit14] develops new techniques to extend the reach of (plain) indifferentiability to certain classes of multi-stage games. In particular, he shows that the HMAC and NMAC constructions are UCE secure. Interestingly, we show that these results do not carry over to the ICE model: HMAC and NMAC provably fail to be ICE secure. On the other hand, we build on Mittelbach's techniques to prove that a variant of Zipper Hash [Lis07] is provably ICE secure.

ATTACKS ON HMAC AND NMAC. The HMAC and NMAC constructions are shown in Fig. 4. If we denote the iterated compression function used in HMAC by h, then it is easily seen that key hk is only used on the "outer" h-evaluations. Consider an ICE distinguisher D_1 which holds hk, computes the values

$$y_1 := \mathsf{h}(hk \oplus \mathsf{ipad}, \mathsf{IV}) \quad and \quad y_2 := \mathsf{h}(hk \oplus \mathsf{opad}, \mathsf{IV})$$

and sends them to distinguisher D_2. Given (y_1, y_2), distinguisher D_2 can compute the HMAC values for any $x \in \{0,1\}^*$ under hk. Thus, in order to win the ICE game, D_2 simply chooses a random x and writes it on the input tape, and calls HASH to receive a value y. It then locally recomputes $\mathsf{H.Ev}^{\mathsf{h}}(hk, x)$ using the

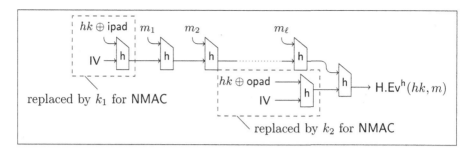

Fig. 4. The HMAC construction. If the dashed boxes are exchanged for independent keys k_1 and k_2, we obtain NMAC. Here we are ignoring padding.

compression function h and values (y_1, y_2). If the results match, it outputs 1, and else it outputs 0. It is easily seen that this adversary wins ICE with overwhelming probability. Furthermore, given (y_1, y_2), the hash key hk remains statistically hidden from D_2 (as the number of h queries is bounded by a polynomial). Value x, being random, also remains statistically hidden from D_1. Formally, this attack breaks ICE[$\mathcal{C}$] for

$$\mathcal{C} := \mathcal{C}^{\text{ppt}} \cap \mathcal{C}^{\text{sup}} \cap \mathcal{C}^{1,1,1} \cap \mathcal{C}_1^{1\text{-hk}} \cap \mathcal{C}_1^0 \cap \mathcal{C}_2^{0\text{-hk}} \cap \mathcal{C}_2^\varepsilon.$$

ZIPPER HASH. The above attack raises the question if any iterative hash function can be ICE[$\mathcal{C}$] secure for a meaningful class of distinguishers $\mathcal{C}$. We show that a hybrid construction of a keyed version of Liskov's Zipper Hash construction [Lis07] and chopped Merkle–Damgård (chop-MD) of Coron et al. [CDMP05] is ICE secure. Zipper Hash can be regarded as a basic Merkle–Damgård scheme where the message is processed twice, the second time in reversed block order. chop-MD refers to the construction where a hash value consists only of the first half of the output bits of the final compression function. Our hybrid construction results from adding the chop step to Zipper Hash. Furthermore, we consider a keyed variant of Zipper Hash by prepending the hash key to the message. We assume that key length matches block length, which means that the *first and last* evaluations of the compression function operate on the hash key. We denote this keyed variant of Zipper Hash by chop-KZIP. Figure 5 shows a schematic diagram of the construction.

Theorem 2 (Zipper Hash's ICE security). *The VIL hash function* chop-KZIP^{h} *constructed above is* ICE[$\mathcal{C}$] *secure in the FIL-RO model for* h $: \{0,1\}^\mu \times \{0,1\}^n \longrightarrow \{0,1\}^n$ *for the class*

$$\mathcal{C} := \mathcal{C}^{\text{poly}} \cap \mathcal{C}^{\text{sup}} \cap \mathcal{C}^{\text{dist}} \cap \mathcal{C}_1^{1\text{-hk}} \cap \mathcal{C}_1^0 \cap \mathcal{C}_2^{0\text{-hk}} \cap \mathcal{C}_2^\varepsilon.$$

An analogous result holds for polynomial-time distinguishers that are only computationally unpredictable.

In the full version, we give the proof, where we also present a self-contained introduction to the unsplittability technique [Mit14].

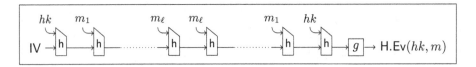

Fig. 5. The Zipper Hash construction merged with chop-MD [CDMP05] and keyed with hk. The final node g corresponds to the projection to the first half of the output of h.

Note that class $\mathcal{C}$ above contains that class used to attack HMAC and hence chop-KZIP^h provably achieves a higher level of security in multi-stage games. We note that the reach of the above feasibility result includes all applications scenarios listed in Table 1. In particular, chop-KZIP^h can security replace the random oracle in these applications. For this also note that we can easily drop $\mathcal{C}_1^0$ by requesting that in the last round D_1 outputs a guess for b which D_2 echoes. With the other restrictions present this change is without loss of generality.

This result cannot be strengthened for the (large) adversarial classes that were used in Theorem 1. To see this, consider two distinguishers that engage in a *distributed* computation of chop-KZIP^h hash values as follows. Distinguisher D_1 knows hk and m_1 and D_2 knows m_2, where message $m := m_1 \| m_2$ is being hashed. Distinguisher D_1 computes an intermediate hash digest using (hk, m_1) and forwards it to D_2. Distinguisher D_2 now computes another iteration of the hash using m_2 and forwards the result to D_1. Distinguisher D_1 can now complete the hash computation using its knowledge of (hk, m_1) and the intermediate hash digest that it receives.

A straightforward generalization of this attack also rules out multi-pass variants of chop-KZIP^h (where messages are processed multiple times in the forward and backward directions), including those whose number of passes is not fixed a priori and can depend on the number of message blocks. This is due to the fact that the number of rounds in an ICE attack is not fixed. This, in turn, raises the question if ICE[$\mathcal{C}^{\mathrm{poly}} \cap \mathcal{C}^{\mathrm{sup}}$] is feasible in the FIL-RO model. We conclude the paper with a candidate construction that we conjecture to reach this level of security.

MIX HASH. Let h $: \{0,1\}^n \times \{0,1\}^n \longrightarrow \{0,1\}^n$ be a compression function. Let $m := m_1 \| \cdots \| m_\ell \in (\{0,1\}^n)^\ell$ be a message with ℓ blocks of length n each. Let $\mathsf{Mix}^h(m)$ denote the transformation that maps m to $M := \|_i\|_j M_{i,j}$ where $M_{i,j} := h(m_i, m_j)$ for $1 \le i < j \le \ell$. (Therefore M has $\ell(\ell-1)/2$ blocks.) Now let $hk \in \{0,1\}^n$ be a hash key and define

$$\mathsf{MixHash}^h(hk, m) := \mathsf{HMAC}^h(0^n, \mathsf{Mix}^h(hk\|m)).$$

Note that $\mathsf{MixHash}^h$ places $\Theta(\ell^2)$ calls to its compression function h.[10] The design rationale behind $\mathsf{MixHash}^h$ is as follows. All intermediate digests values $M_{i,j}$ are

---

[10] Indeed, MixHash is a (highly) offline function: for $\alpha \in [0, 1]$, it requires space roughly $n\ell\sqrt{1-\alpha}$ bits after a fraction α of the $n\ell(\ell+1)/2$ bits are processed.

needed in order to successfully compute a hash value. These values, however, consist of all pairs (m_i, m_j) compressed through h. Since h is a monolithic object, $M_{i,j}$ cannot be computed in a distributed way, a strategy that was used in all previous attacks. In other words, one of the distinguishers has to know (hk, m) in full and hence will violate unpredictability. To see this, suppose D_1 does not know m_j in full and D_2 does not know m_i in full for some $i < j$. Then there is no way for these parties to learn $M_{i,j} := \mathsf{h}(m_i, m_j)$ without one of them explicitly quarrying h on (m_i, m_j). This however means that both m_i and m_j are known to the quarrying party, which leads to a contradiction. We leave a formal analysis of $\mathsf{MixHash}^\mathsf{h}$ in the FIL-RO model for h as future work.

Acknowledgments. The authors would like to thank Christina Brzuska for taking part in the early stages of this work. Pooya Farshim was supported in part by grant ANR-14-CE28-0003 (Project EnBid).

References

[BC10] Bellare, M., Cash, D.: Pseudorandom functions and permutations provably secure against related-key attacks. In: Rabin, T. (ed.) CRYPTO 2010. LNCS, vol. 6223, pp. 666–684. Springer, Heidelberg (2010)

[BCFW09] Boldyreva, A., Cash, D., Fischlin, M., Warinschi, B.: Foundations of nonmalleable hash and one-way functions. In: Matsui, M. (ed.) ASIACRYPT 2009. LNCS, vol. 5912, pp. 524–541. Springer, Heidelberg (2009)

[BCK96] Bellare, M., Canetti, R., Krawczyk, H.: Keying hash functions for message authentication. In: Koblitz, N. (ed.) CRYPTO 1996. LNCS, vol. 1109, pp. 1–15. Springer, Heidelberg (1996)

[BFM14] Brzuska, C., Farshim, P., Mittelbach, A.: Indistinguishability obfuscation and UCEs: the case of computationally unpredictable sources. In: Garay, J.A., Gennaro, R. (eds.) CRYPTO 2014, Part I. LNCS, vol. 8616, pp. 188–205. Springer, Heidelberg (2014)

[BGI+01] Barak, B., Goldreich, O., Impagliazzo, R., Rudich, S., Sahai, A., Vadhan, S.P., Yang, K.: On the (Im)possibility of Obfuscating Programs. In: Kilian, J. (ed.) CRYPTO 2001. LNCS, vol. 2139, pp. 1–18. Springer, Heidelberg (2001)

[BH15] Bellare, M., Hoang, V.T.: Resisting randomness subversion: fast deterministic and hedged public-key encryption in the standard model. In: Oswald, E., Fischlin, M. (eds.) EUROCRYPT 2015. LNCS, vol. 9057, pp. 627–656. Springer, Heidelberg (2015)

[BHK13a] Bellare, M., Hoang, V.T., Keelveedhi, S.: Instantiating random oracles via UCEs. In: Canetti, R., Garay, J.A. (eds.) CRYPTO 2013, Part II. LNCS, vol. 8043, pp. 398–415. Springer, Heidelberg (2013)

[BHK13b] Bellare, M., Hoang, V.T., Keelveedhi, S.: Instantiating random oracles via UCEs. Cryptology ePrint Archive, Report 2013/424 (2013). http://eprint.iacr.org/2013/424

[BHK14] Bellare, M., Hoang, V.T., Keelveedhi, S.: Cryptography from compression functions: the UCE bridge to the ROM. In: Garay, J.A., Gennaro, R. (eds.) CRYPTO 2014, Part I. LNCS, vol. 8616, pp. 169–187. Springer, Heidelberg (2014)

[BK03] Bellare, M., Kohno, T.: A theoretical treatment of related-key attacks: RKA-PRPs, RKA-PRFs, and applications. In: Biham, E. (ed.) EUROCRYPT 2003. LNCS, vol. 2656, pp. 491–506. Springer, Heidelberg (2003)

[BK09] Biryukov, A., Khovratovich, D.: Related-key cryptanalysis of the full AES-192 and AES-256. In: Matsui, M. (ed.) ASIACRYPT 2009. LNCS, vol. 5912, pp. 1–18. Springer, Heidelberg (2009)

[BK15] Bellare, M., Keelveedhi, S.: Interactive message-locked encryption and secure deduplication. In: Katz, J. (ed.) PKC 2015. LNCS, vol. 9020, pp. 516–538. Springer, Heidelberg (2015)

[BM14a] Brzuska, C., Mittelbach, A.: Indistinguishability obfuscation versus multibit point obfuscation with auxiliary input. In: Sarkar, P., Iwata, T. (eds.) ASIACRYPT 2014, Part II. LNCS, vol. 8874, pp. 142–161. Springer, Heidelberg (2014)

[BM14b] Brzuska, C., Mittelbach, A.: Using indistinguishability obfuscation via UCEs. In: Sarkar, P., Iwata, T. (eds.) ASIACRYPT 2014, Part II. LNCS, vol. 8874, pp. 122–141. Springer, Heidelberg (2014)

[BM15] Brzuska, C., Mittelbach, A.: Universal computational extractors and the superfluous padding assumption for indistinguishability obfuscation. Cryptology ePrint Archive, Report 2015/581 (2015). http://eprint.iacr.org/

[BR93] Bellare, M., Rogaway, P.: Random oracles are practical: a paradigm for designing efficient protocols. In: Ashby, V. (ed.) ACM CCS 1993, pp. 62–73. ACM Press, November 1993

[BRS03] Black, J., Rogaway, P., Shrimpton, T.: Encryption-scheme security in the presence of key-dependent messages. In: Nyberg, K., Heys, H.M. (eds.) SAC 2002. LNCS, vol. 2595, pp. 62–75. Springer, Heidelberg (2003)

[BST15] Bellare, M., Stepanovs, I., Tessaro, S.: Contention in cryptoland: obfuscation, leakage and UCE. Cryptology ePrint Archive, Report 2015/487 (2015). http://eprint.iacr.org/2015/487

[Can97] Canetti, R.: Towards realizing random oracles: hash functions that hide all partial information. In: Kaliski Jr., B.S. (ed.) CRYPTO 1997. LNCS, vol. 1294, pp. 455–469. Springer, Heidelberg (1997)

[CD08] Canetti, R., Dakdouk, R.R.: Extractable perfectly one-way functions. In: Aceto, L., Damgård, I., Goldberg, L.A., Halldórsson, M.M., Ingólfsdóttir, A., Walukiewicz, I. (eds.) ICALP 2008, Part II. LNCS, vol. 5126, pp. 449–460. Springer, Heidelberg (2008)

[CD09] Canetti, R., Dakdouk, R.R.: Towards a theory of extractable functions. In: Reingold, O. (ed.) TCC 2009. LNCS, vol. 5444, pp. 595–613. Springer, Heidelberg (2009)

[CDMP05] Coron, J.-S., Dodis, Y., Malinaud, C., Puniya, P.: Merkle-Damgård revisited: how to construct a hash function. In: Shoup, V. (ed.) CRYPTO 2005. LNCS, vol. 3621, pp. 430–448. Springer, Heidelberg (2005)

[CG14] Cheraghchi, M., Guruswami, V.: Non-malleable coding against bit-wise and split-state tampering. In: Lindell, Y. (ed.) TCC 2014. LNCS, vol. 8349, pp. 440–464. Springer, Heidelberg (2014)

[CGH98] Canetti, R., Goldreich, O., Halevi, S.: The random oracle methodology, revisited (preliminary version). In: 30th ACM STOC, pp. 209–218. ACM Press, May 1998

[Dam90] Damgård, I.B.: A design principle for hash functions. In: Brassard, G. (ed.) CRYPTO 1989. LNCS, vol. 435, pp. 416–427. Springer, Heidelberg (1990)

[DGG+15] Dodis, Y., Ganesh, C., Golovnev, A., Juels, A., Ristenpart, T.: A formal treatment of backdoored pseudorandom generators. In: Oswald, E., Fischlin, M. (eds.) EUROCRYPT 2015. LNCS, vol. 9056, pp. 101–126. Springer, Heidelberg (2015)

[GGH+13] Garg, S., Gentry, C., Halevi, S., Raykova, M., Sahai, A., Waters, B.: Candidate indistinguishability obfuscation and functional encryption for all circuits. In: 54th FOCS, pp. 40–49. IEEE Computer Society Press, October 2013

[GOR11] Goyal, V., O'Neill, A., Rao, V.: Correlated-input secure hash functions. In: Ishai, Y. (ed.) TCC 2011. LNCS, vol. 6597, pp. 182–200. Springer, Heidelberg (2011)

[Lis07] Liskov, M.: Constructing an ideal hash function from weak ideal compression functions. In: Biham, E., Youssef, A.M. (eds.) SAC 2006. LNCS, vol. 4356, pp. 358–375. Springer, Heidelberg (2007)

[LL12] Liu, F.-H., Lysyanskaya, A.: Tamper and leakage resilience in the split-state model. In: Safavi-Naini, R., Canetti, R. (eds.) CRYPTO 2012. LNCS, vol. 7417, pp. 517–532. Springer, Heidelberg (2012)

[LRW02] Liskov, M., Rivest, R.L., Wagner, D.: Tweakable block ciphers. In: Yung, M. (ed.) CRYPTO 2002. LNCS, vol. 2442, pp. 31–46. Springer, Heidelberg (2002)

[Mer90] Merkle, R.C.: One way hash functions and DES. In: Brassard, G. (ed.) CRYPTO 1989. LNCS, vol. 435, pp. 428–446. Springer, Heidelberg (1990)

[MH14] Matsuda, T., Hanaoka, G.: Chosen ciphertext security via UCE. In: Krawczyk, H. (ed.) PKC 2014. LNCS, vol. 8383, pp. 56–76. Springer, Heidelberg (2014)

[Mit14] Mittelbach, A.: Salvaging indifferentiability in a multi-stage setting. In: Nguyen, P.Q., Oswald, E. (eds.) EUROCRYPT 2014. LNCS, vol. 8441, pp. 603–621. Springer, Heidelberg (2014)

[MRH04] Maurer, U.M., Renner, R.S., Holenstein, C.: Indifferentiability, impossibility results on reductions, and applications to the random oracle methodology. In: Naor, M. (ed.) TCC 2004. LNCS, vol. 2951, pp. 21–39. Springer, Heidelberg (2004)

[Nie02] Nielsen, J.B.: Separating random oracle proofs from complexity theoretic proofs: the non-committing encryption case. In: Yung, M. (ed.) CRYPTO 2002. LNCS, vol. 2442, pp. 111–126. Springer, Heidelberg (2002)

[RSS11] Ristenpart, T., Shacham, H., Shrimpton, T.: Careful with composition: limitations of the indifferentiability framework. In: Paterson, K.G. (ed.) EUROCRYPT 2011. LNCS, vol. 6632, pp. 487–506. Springer, Heidelberg (2011)

Practical Order-Revealing Encryption with Limited Leakage

Nathan Chenette[1]([✉]), Kevin Lewi[2], Stephen A. Weis[3], and David J. Wu[2]

[1] Rose-Hulman Institute of Technology, Terre Haute, USA
`chenett1@rose-hulman.edu`
[2] Stanford University, Stanford, USA
[3] Facebook, Inc., Menlo Park, USA

Abstract. In an order-preserving encryption scheme, the encryption algorithm produces ciphertexts that preserve the order of their plaintexts. Order-preserving encryption schemes have been studied intensely in the last decade, and yet not much is known about the security of these schemes. Very recently, Boneh et al. (Eurocrypt 2015) introduced a generalization of order-preserving encryption, called order-revealing encryption, and presented a construction which achieves this notion with best-possible security. Because their construction relies on multilinear maps, it is too impractical for most applications and therefore remains a theoretical result.

In this work, we build efficiently implementable order-revealing encryption from pseudorandom functions. We present the first efficient order-revealing encryption scheme which achieves a simulation-based security notion with respect to a leakage function that precisely quantifies what is leaked by the scheme. In fact, ciphertexts in our scheme are only about 1.6 times longer than their plaintexts. Moreover, we show how composing our construction with existing order-preserving encryption schemes results in order-revealing encryption that is strictly more secure than all preceding order-preserving encryption schemes.

1 Introduction

A symmetric encryption scheme is order-preserving if the ciphertexts preserve the numeric ordering of their underlying plaintexts. The notion of order-preserving encryption (OPE) was introduced by Agrawal et al. [1] who showed how it could be used to efficiently answer range queries over encrypted data, as well as sorting queries, searching queries, and more. Indeed, existing OPE solutions have been implemented in practice [43,46] for these exact purposes. Since the introduction of OPE, there has been a plethora of work on analyzing the security of various OPE schemes, found both in the cryptography community and the database community. However, it is troubling that in spite of the numerous practical applications of OPE, the security of the best candidate OPE schemes is still not well understood.

Prior Work. The first OPE construction by Agrawal et al. [1] relied on heuristics and lacked a formal security analysis. Subsequently, Boldyreva et al. [7] gave

© International Association for Cryptologic Research 2016
T. Peyrin (Ed.): FSE 2016, LNCS 9783, pp. 474–493, 2016.
DOI: 10.1007/978-3-662-52993-5_24

the first formal security definitions for OPE schemes. Boldyreva et al. introduced two primary notions for security of an OPE scheme. The first notion of security for an OPE scheme is called indistinguishability under an ordered chosen plaintext attack (IND-OCPA). The IND-OCPA definition can be viewed as a generalization of semantic security [31], and effectively says that encryptions of a sequence of messages should reveal nothing about the underlying messages other than their ordering. However, in the same work, Boldyreva et al. showed that no efficient order-preserving encryption scheme can be IND-OCPA secure, even in settings where the size of the ciphertext space is exponentially larger than the size of the plaintext space.

In light of this lower bound for OPE schemes that satisfy IND-OCPA security, Boldyreva et al. introduced a weaker notion of security (POPF-CCA security) where the encryption function for the OPE scheme is compared to a random order-preserving function—that is, the encryption algorithm for an OPE scheme behaves like a truly random order-preserving function. Under this definition, an OPE scheme inherits the properties of a random order-preserving function.[1] In the same work, Boldyreva et al. gave an explicit construction of an OPE scheme that satisfies POPF-CCA security. However, the POPF-CCA security definition does not precisely specify the information that is leaked by an OPE scheme that achieves this definition. In fact, a scheme that achieves this notion of security does not even satisfy semantic security for a single encryption, and indeed, in subsequent work, Boldyreva et al. [8] showed that ciphertexts in their OPE scheme leak approximately the first half of the bits of the underlying plaintexts. In addition, they introduce several new security definitions in order to better quantify the information leakage of OPE schemes that are POPF-CCA secure.

Recently, Boneh et al. [9] proposed a generalization of OPE called order-revealing encryption (ORE). In an OPE scheme, the ciphertexts are numeric-valued, and the ordering of the underlying plaintexts is determined by numerically comparing the ciphertexts. In contrast, in an ORE scheme, the ciphertexts are not constrained to any particular form, and instead, there is a publicly computable comparison function which takes two ciphertexts and outputs the numeric ordering of the underlying plaintexts[2]. Although this generalization may at first seem subtle, Boneh et al. constructed an ORE scheme from multilinear maps that achieves the "best-possible" notion of security, which is equivalent to the IND-OCPA security notion for order-preserving encryption.

The main drawback of the Boneh et al. ORE construction is that it relies on complicated tools and strong assumptions on these tools, and as such, is currently impractical to implement.

---

[1] This definition is inspired by the similar definition for PRF security [28], which compares the output of a keyed function to that of a truly random function.

[2] This application was also observed and independently achieved by Goldwasser et al. [29] using indistinguishability obfuscation.

1.1   Our Contributions

We now summarize the main contributions of this work, which include a new simulation-based security notion for ORE, along with a practical construction of an ORE scheme which achieves this security notion. We also show how our new construction can be used to achieve a strictly stronger notion of security compared to other stateless and efficiently implementable (e.g., constructions that do not rely on powerful primitives such as multilinear maps and indistinguishability obfuscation) OPE and ORE encryption schemes.

Security Model. In our work, we take the general approach of Boneh et al. in constructing an ORE scheme, except we take a more efficient route. Our first contribution is a new security definition for order-revealing encryption schemes that both allows for and explicitly models the leakage in the scheme. Our design goals for introducing this new security model are twofold: first, the security model should enable constructions that are efficiently implementable, and second, it should provide a precise quantification of any information leaked by the scheme. The two primary notions of security, IND-OCPA and POPF-CCA, introduced by Boldyreva et al. [7] each satisfy one of these two properties. In particular, all non-interactive, stateless[3] ORE schemes that achieve IND-OCPA security require strong cryptographic primitives such as multilinear maps or indistinguishability obfuscation [9, 29], and thus, are not efficiently implementable today. At the other end of the spectrum, it is difficult to precisely quantify the leakage of schemes that satisfy POPF-CCA security. The work by Boldyreva et al. [8] provides some concrete lower and upper bounds for the leakage under the strong assumption that the plaintexts are drawn from a uniform distribution. For more general distributions, the leakage remains unclear.

In our work, we give a simulation-based definition of security for ORE with respect to a leakage function $\mathcal{L}$. In other words, our definition states that whatever an adversary is able to deduce from seeing encryptions of messages $m_1, \ldots, m_t$, it could also deduce given only the leakage $\mathcal{L}(m_1, \ldots, m_t)$. The "best-possible" security for ORE would correspond to the case where the leakage function simply outputs whether $m_i < m_j$ for all pairs of messages m_i and m_j. By allowing for the possibility of additional leakage, it becomes possible to construct practical ORE schemes from standard assumptions. Thus, our constructions provide a concrete trade-off between security and efficiency. Our security definitions are similar to the simulation-based definitions that have been considered previously in the searchable symmetric encryption literature [14, 22].

Constructions. In our main construction, we show how to construct an ORE scheme from one-way functions (more precisely, from pseudorandom functions (PRFs) [28]). This particular ORE scheme reveals slightly more information than just the ordering of the underlying messages. Specifically, two ciphertexts encrypting messages m_1 and m_2 also reveal the index of the first bit in m_1 and

---

[3] There are "mutable" order-preserving encryption schemes [35, 36, 42] that do satisfy IND-OCPA, but they require stateful encryption, and oftentimes, an interactive protocol to "update" ciphertexts. We survey some of these constructions in Sect. 1.2.

m_2 that differ. In other words, our ORE scheme leaks some information about the relative distance between the underlying messages.

We give a brief overview of our PRF-based construction. The secret key in our scheme consists of a PRF key k. The output space of the PRF is the set $\{0, 1, 2\}$. Each ciphertext consists of the bits of the message blinded by the outputs of the PRF evaluated on the prefixes of the message. More precisely, to encrypt an n-bit message $m = m_1 m_2 \cdots m_n$, the encryption algorithm effectively computes the following for each $i \in [n]$:

$$u_i = F(k, m_1 m_2 \cdots m_{i-1}) + m_i \pmod{3}.$$

Note that to support variable-length PRF inputs, we simply pad the input. We describe our construction in greater detail in Sect. 3. The ciphertext is then the tuple $\mathsf{ct} = (u_1, \ldots, u_n)$ of blinded values.

To compare encryptions $\mathsf{ct} = (u_1, \ldots, u_n)$ and $\mathsf{ct}' = (u_1', \ldots, u_n')$ of messages m and m', the evaluator first finds the first index i for which $u_i \neq u_i'$. Since u_i and u_i' are functions of just the first i bits of m and m', respectively, the first index i for which $u_i \neq u_i'$ is the first bit of m and m' that differ. After identifying the ith bit that differs, the evaluator uses u_i and u_i' to determine which message has 0 as the ith bit and which message has 1[4]. Conversely, if $u_i = u_i'$ for all i, then $\mathsf{ct}_i = \mathsf{ct}_i'$, and so $m = m'$. Security of this construction follows from the security of the PRF (Theorem 3.2).

Ciphertexts in our candidate scheme are $\lceil n \cdot \log_2 3 \rceil \approx \lceil 1.6\, n \rceil$ bits, where n is the bit-length of the message. As a point of comparison, ciphertexts in the OPE scheme of Boldyreva et al. [7] are only $n + 1$ bits long. While the ciphertexts in our scheme are longer (by a multiplicative factor $\log_2 3$), the authors of [8] note that even if the size of the ciphertext space is increased beyond $n + 1$ bits in the Boldyreva et al. scheme, the security of their construction does not improve by any noticeable amount.

We then explain in Sect. 3.2 how to convert our ORE scheme into an OPE scheme, at the expense of longer ciphertexts. This is useful for applications where it is more convenient to have a numeric ciphertext space and for order relations to be computable without a "custom" comparison function. The transformation we describe is natural and does not reduce the security of the original ORE scheme. In particular, we note that the resulting OPE scheme does *not* behave like a random order-preserving function (the ideal object from the POPF-CCA security notion). Thus, the scheme is able to achieve stronger security than the Boldyreva et al. OPE scheme.

Comparison with Existing Schemes. First, we note in Sect. 2.3 that the security of any OPE scheme can be "augmented" by applying ORE encryption on top of OPE encryption. The resulting scheme is at least as secure as the underlying OPE scheme, and moreover, inherits the security properties of the ORE scheme. Hence, by composing our ORE construction with existing OPE constructions, we obtain ORE schemes that are at least as secure.

---

[4] Either $u_i + 1 = u_i' \pmod{3}$, in which case $m < m'$, or $u_i - 1 = u_i' \pmod{3}$, in which case $m > m'$.

While composing an OPE scheme with an ORE scheme yields a scheme that is at least as secure as the underlying OPE scheme, we show that even without this composition, our basic ORE scheme still achieves stronger security guarantees according to the one-wayness metrics introduced by Boldyreva et al. [8] for analyzing the leakage of random order-preserving functions (and by extension, any OPE scheme that is POPF-CCA secure). In our work, we introduce two generalized one-wayness notions and show that under a uniform plaintext distribution,[5] our basic ORE scheme achieves strictly stronger security compared to OPE schemes that are POPF-CCA secure. Specifically, Boldyreva et al. [8] show that a random order-preserving function leaks half of the most-significant bits of the messages with probability close to 1. In contrast, under the same settings, we can show that our basic ORE scheme will not leak *any constant* fraction of the message bits with *overwhelming* probability.

1.2   Related Work

In recent years, there have been numerous works on order-preserving encryption and related notions [1,7,8,35,36,38,41,42,44,47]. In this section, we survey some of these works.

Security Definitions. Though the POPF-CCA security definition introduced by Boldyreva et al. [7] is similar in flavor to PRF security, it is not immediately evident what kind of information the output of a random order-preserving function leaks about its input. In a follow-up work [8], Boldyreva et al. introduce several notions (based on definitions of one-wayness [27] for one-way functions) to capture the information leakage in schemes that are POPF-CCA secure. They show that a random order-preserving function leaks at least half of the bits in each message.

Teranishi et al. [47] also introduce a stronger indistinguishability-based notion (stronger than the one-wayness definitions from [8], but weaker than IND-OCPA) for OPE schemes, as well as a construction that achieves these stronger notions. Notably, their definition ensures that under a uniform message distribution, any fraction of the low-order bits of the messages being encrypted are hidden.

Recently, Naveed et al. [40] analyzed the information leaked by order-preserving encryption used in practical scenarios.

Modular OPE. Boldyreva et al. also introduced the notion of modular OPE as a possible extension of standard OPE [8]. In modular OPE, a modular shift is applied to each plaintext before applying OPE—so the scheme is not order-preserving, but naturally supports "wrap-around" range queries. Their modular OPE scheme adds an extra layer of security to vanilla OPE, but it is worth noting that leakage of a small amount of information (say, a single plaintext-ciphertext pair) reveals the shift value and nullifies this added security. Subse-

---

[5] This is the only distribution for which we have concrete analysis of the leakage in any POPF-CCA secure scheme.

quently, Mavroforakis et al. [38] designed several protocols to avoid leaking the shift value while using modular OPE schemes in practice.

Mutable OPE. Popa et al. [42] introduced a related notion of a mutable order-preserving encoding scheme which can be viewed as a two-party protocol that allows a user to insert and store encrypted values in a database such that the database is able to perform comparisons and range queries on the encrypted values without learning anything more about the values. Their construction is interactive and leverages stateful encryption. By working in this setting, the authors are able to circumvent the Boldyreva et al. [7] lower bound for order-preserving encryption and show that their scheme is IND-OCPA secure.

In subsequent work, Kerschbaum and Schröpfer [36] improved on the communication complexity of the Popa et al. construction at the expense of increasing the amount of client-side state. Specifically, in their construction, the amount of persistent state the client has to maintain increases linearly in the number of elements inserted into the database. More recently, Kerschbaum [35] introduced a new notion of frequency-hiding OPE that introduces additional randomness to hide whether multiple ciphertexts encrypt the same value. Their notions provide a strictly stronger guarantee than IND-OCPA.

Very recently, Roche et al. [44] introduced the notion of partial order-preserving encodings, which optimizes for the setting where there are a huge number of insertion queries but only a moderate number of range queries. Their protocol improves upon the round-complexity for insertions compared to the Popa et al. protocol [42], and requires the client to maintain less state than the Kerschbaum-Schröpfer construction [36]. All of the schemes described here require stateful encryption and employ an interactive encryption procedure.

ORE. Order-revealing encryption schemes, as introduced by Boneh et al. [9] provide another method of circumventing the Boldyreva et al. lower bound [7]. In an ORE scheme, the public comparison operation is not required to correspond to numerically comparing the ciphertexts, and in fact, the ciphertexts themselves need not be elements of a numeric, well-ordered set. This type of relaxation was previously considered by Pandey and Rouselakis [41] in the context of property-preserving encryption. In a property-preserving encryption scheme, there is a publicly computable function that can be evaluated on ciphertexts to determine the value of some property on the underlying plaintexts. Order-revealing encryption can thus be viewed as a property-preserving encryption scheme for the comparison operation. Pandey and Rouselakis introduce and explore several indistinguishability-based notions of security for property-preserving encryption; however, they do not construct an order-revealing encryption scheme.

To the best of our knowledge, all existing ORE schemes that provide IND-OCPA security either rely on very strong (and currently impractical) cryptographic primitives such as indistinguishability obfuscation [29] and cryptographic multilinear maps [9], or only achieve a weaker notion of security [3,12] when instantiated with simple cryptographic primitives such as public key cryptography. For the constructions based on indistinguishability obfuscation or multilinear maps [9,29], security of the ORE scheme is conditional on the conjectured security of cryptographic multilinear maps [2,10,20,21,23,26,37][6]. However, in the last few months, numerous attacks [11,16–19,33,39] on these multilinear maps have emerged, raising some doubts about the security of constructions that leverage them.

To avoid multilinear maps in favor of more well-studied number-theoretic or lattice-based assumptions, one can apply arity-amplification techniques [3,12] to a single-input functional encryption scheme based on simpler assumptions such as learning with errors [30] or semantically-secure public-key encryption [32,45]. However, due to limitations of the underlying functional encryption schemes, the resulting ORE scheme only provides "bounded-message" security—that is, security only holds if there is an *a priori* (polynomial) bound on the maximum number of messages that will be encrypted. Moreover, the length of the ciphertexts in this scheme grows *polynomially* in the bound on the number of messages that will be encrypted. These constraints severely limit the practicality of the resulting ORE scheme. To obtain full semantic security, it would be necessary to apply the arity-amplification transformation to a more powerful functional encryption scheme, but to date, the only known candidates of such schemes rely again on indistinguishability obfuscation [24] or multilinear maps [25].

Recently, Bun and Zhandry [13] investigated the connection between order-revealing encryption and problems in learning theory.

Other schemes. Numerous ad hoc or heuristic order-preserving encryption schemes [6,34,48] have been proposed in the literature, but most lack formal security analysis.

2   Order-Revealing Encryption

In this section, we establish and review some conventions that we use in this work, and also formally define our security notions for our encryption schemes.

Preliminaries. For $n \in \mathbb{N}$, we write $[n]$ to denote the set of integers $\{1, \ldots, n\}$, and $\mathbb{Z}_n$ to denote the additive group of integers modulo n. If $\mathcal{P}(x)$ is a predicate on x, we write $\mathbf{1}(\mathcal{P}(x))$ to denote the indicator function for $\mathcal{P}$: that is, $\mathbf{1}(\mathcal{P}(x)) = 1$ if and only if $\mathcal{P}(x) = 1$, and 0 otherwise. If $x, y \in \{0,1\}^*$ are bit-strings, we write $x \| y$ to denote the concatenation of x and y. For a finite set S, we write $\mathsf{Unif}(S)$ to denote the uniform distribution on S. We say a function $f(\lambda)$ is negligible in a security parameter λ if $f = o(1/\lambda^c)$ for all $c \in \mathbb{N}$. We write $\mathsf{negl}(\lambda)$ to denote a negligible function in λ and $\mathsf{poly}(\lambda)$ to denote a polynomial in λ. We say an event occurs with negligible probability if the probability of the event is $\mathsf{negl}(\lambda)$, and it occurs with overwhelming probability if the complement of the event occurs with negligible probability. Finally, we review the definition of a pseudorandom function (PRF) [28]. Let $\mathsf{Funs}[\mathcal{D}, \mathcal{R}]$ denote the set of all functions from a domain $\mathcal{D}$ to a range $\mathcal{R}$. In this paper, we specialize the domain of our PRFs to $\{0,1\}^n$.

---

[6] To date, the only concrete instantiations of indistinguishability obfuscation [4,5,24, 49] leverage multilinear maps.

Definition 2.1 (Pseudorandom Function [28]). Fix a security parameter λ. A PRF $F : \mathcal{K} \times \{0,1\}^n \to \mathcal{R}$ with key space $\mathcal{K}$, domain $\{0,1\}^n$, and range $\mathcal{R}$ is secure if for all efficient adversaries $\mathcal{A}$,

$$\left| \Pr \left[k \xleftarrow{\text{R}} \mathcal{K} : \mathcal{A}^{F(k,\cdot)}(1^\lambda) = 1 \right] - \right.$$
$$\left. \Pr \left[f \xleftarrow{\text{R}} \mathsf{Funs}[\{0,1\}^n, \mathcal{R}] : \mathcal{A}^{f(\cdot)}(1^\lambda) = 1 \right] \right| = \mathrm{negl}(\lambda).$$

2.1 Order-Revealing Encryption

An order-revealing encryption (ORE) scheme is a tuple of algorithms $\Pi = (\mathsf{ORE.Setup}, \mathsf{ORE.Encrypt}, \mathsf{ORE.Compare})$ defined over a well-ordered domain $\mathcal{D}$ with the following properties:

- $\mathsf{ORE.Setup}(1^\lambda) \to \mathsf{sk}$. On input a security parameter λ, the setup algorithm $\mathsf{ORE.Setup}$ outputs a secret key sk.
- $\mathsf{ORE.Encrypt}(\mathsf{sk}, m) \to \mathsf{ct}$. On input the secret key sk and a message $m \in \mathcal{D}$, the encrypt algorithm $\mathsf{ORE.Encrypt}$ outputs a ciphertext ct.
- $\mathsf{ORE.Compare}(\mathsf{ct}_1, \mathsf{ct}_2) \to b$. On input two ciphertexts ct_1, ct_2, the compare algorithm $\mathsf{ORE.Compare}$ outputs a bit $b \in \{0,1\}$.

Remark 2.2 (Public Parameters). In general, the setup algorithm of an ORE scheme can also output public parameters pp which are then passed as an additional input to the comparison algorithm, as is done in Boneh et al. [9]. However, none of our constructions require these public parameters, so we omit them in this work for simplicity.

Remark 2.3 (Support for Decryption). As described, our definition of an order-revealing encryption scheme does not include a "decryption" function. However, this omission is without loss of generality. To decrypt a message, the holder of the secret key can use the secret key to encrypt messages of her choosing, apply the comparison algorithm, and perform binary search to recover the message. An alternative method that avoids the need for binary search is to augment each ORE encryption of a message m with an encryption of m under a CPA-secure symmetric encryption scheme. The secret key of the ORE scheme would also include the key for the symmetric encryption scheme. As long as the underlying encryption scheme is CPA-secure, including this additional ciphertext does not compromise security. For the remainder of this work, we use the schema described above that does not explicitly specify a decryption function.

Correctness. Fix a security parameter λ. An ORE scheme $\Pi = (\mathsf{ORE.Setup}, \mathsf{ORE.Encrypt}, \mathsf{ORE.Compare})$ over a well-ordered domain $\mathcal{D}$ is correct if for $\mathsf{sk} \leftarrow \mathsf{ORE.Setup}(1^\lambda)$, and all messages $m_1, m_2 \in \mathcal{D}$,

$$\Pr[\mathsf{ORE.Compare}(\mathsf{ct}_1, \mathsf{ct}_2) = \mathbf{1}(m_1 < m_2)] = 1 - \mathrm{negl}(\lambda),$$

where $\mathsf{ct}_1 \leftarrow \mathsf{ORE.Encrypt}(\mathsf{sk}, m_1)$ and $\mathsf{ct}_2 \leftarrow \mathsf{ORE.Encrypt}(\mathsf{sk}, m_2)$, and the probability is taken over the random coins in $\mathsf{ORE.Setup}$ and $\mathsf{ORE.Encrypt}$.

Security. We now give our simulation-based notion of security for an ORE scheme. As described in Sect. 1.1, our security definition is parameterized by a leakage function $\mathcal{L}$, which exactly specifies what is leaked by an ORE scheme.

Definition 2.4 (Security of ORE with Leakage). *Fix a security parameter* $\lambda \in \mathbb{N}$. *Let* $\Pi_{\mathsf{ore}} = (\mathsf{ORE.Setup}, \mathsf{ORE.Encrypt}, \mathsf{ORE.Compare})$ *be an ORE scheme. Let* $\mathcal{A} = (\mathcal{A}_1, \ldots, \mathcal{A}_q)$ *be an adversary for some* $q \in \mathbb{N}$. *Let* $\mathcal{S} = (\mathcal{S}_0, \mathcal{S}_1, \ldots, \mathcal{S}_q)$ *be a simulator, and let* $\mathcal{L}(\cdot)$ *be a leakage function. We define the experiments* $\mathsf{REAL}_{\mathcal{A}}^{\mathrm{ORE}}(\lambda)$ *and* $\mathsf{SIM}_{\mathcal{A},\mathcal{S},\mathcal{L}}^{\mathrm{ORE}}(\lambda)$ *as follows:*

$\mathsf{REAL}_{\mathcal{A}}^{\mathrm{ORE}}(\lambda)$:	$\mathsf{SIM}_{\mathcal{A},\mathcal{S},\mathcal{L}}^{\mathrm{ORE}}(\lambda)$:
1. $\mathsf{sk} \leftarrow \mathsf{ORE.Setup}(1^\lambda)$	1. $\mathsf{st}_{\mathcal{S}} \leftarrow \mathcal{S}_0(1^\lambda)$
2. $(m_1, \mathsf{st}_{\mathcal{A}}) \leftarrow \mathcal{A}_1(1^\lambda)$	2. $(m_1, \mathsf{st}_{\mathcal{A}}) \leftarrow \mathcal{A}_1(1^\lambda)$
3. $c_1 \leftarrow \mathsf{ORE.Encrypt}(\mathsf{sk}, m_1)$	3. $(c_1, \mathsf{st}_{\mathcal{S}}) \leftarrow \mathcal{S}_1(\mathsf{st}_{\mathcal{S}}, \mathcal{L}(m_1))$
4. for $2 \leq i \leq q$:	4. for $2 \leq i \leq q$:
(a) $(m_i, \mathsf{st}_{\mathcal{A}}) \leftarrow \mathcal{A}_i(\mathsf{st}_{\mathcal{A}}, c_1, \ldots, c_{i-1})$	(a) $(m_i, \mathsf{st}_{\mathcal{A}}) \leftarrow \mathcal{A}_i(\mathsf{st}_{\mathcal{A}}, c_1, \ldots, c_{i-1})$
(b) $c_i \leftarrow \mathsf{ORE.Encrypt}(\mathsf{sk}, m_i)$	(b) $(c_i, \mathsf{st}_{\mathcal{S}}) \leftarrow \mathcal{S}_i(\mathsf{st}_{\mathcal{S}}, \mathcal{L}(m_1, \ldots, m_i))$
5. output $(c_1, \ldots, c_q)$ and $\mathsf{st}_{\mathcal{A}}$	5. output $(c_1, \ldots, c_q)$ and $\mathsf{st}_{\mathcal{A}}$

We say that Π_{ore} *is a secure ORE scheme with leakage function* $\mathcal{L}(\cdot)$ *if for all polynomial-size adversaries* $\mathcal{A} = (\mathcal{A}_1, \ldots, \mathcal{A}_q)$ *where* $q = \mathrm{poly}(\lambda)$, *there exists a polynomial-size simulator* $\mathcal{S} = (\mathcal{S}_0, \mathcal{S}_1, \ldots, \mathcal{S}_q)$ *such that the outputs of the two distributions* $\mathsf{REAL}_{\mathcal{A}}^{\mathrm{ORE}}(\lambda)$ *and* $\mathsf{SIM}_{\mathcal{A},\mathcal{S},\mathcal{L}}^{\mathrm{ORE}}(\lambda)$ *are computationally indistinguishable.*

Remark 2.5 (IND-OCPA Security). We briefly note how the IND-OCPA definition of security is captured by this definition. Let $\mathcal{L}$ be the following leakage function:

$$\mathcal{L}(m_1, \ldots, m_t) = \{\mathbf{1}(m_i < m_j) : 1 \leq i < j \leq t\}.$$

If an ORE scheme is secure with leakage $\mathcal{L}$, then it is IND-OCPA secure.

2.2 Order-Preserving Encryption (OPE)

An OPE scheme [1,7] is a special case of an ORE scheme, where the ciphertext space is required to be a well-ordered range $\mathcal{R}$ and moreover, for two ciphertexts $\mathsf{ct}_1, \mathsf{ct}_2 \in \mathcal{R}$, the comparison algorithm outputs 1 if $\mathsf{ct}_1 < \mathsf{ct}_2$. For simplicity, we can write an OPE scheme as a tuple of algorithms $\Pi = (\mathsf{OPE.Setup}, \mathsf{OPE.Encrypt})$ defined over a well-ordered domain $\mathcal{D}$ and well-ordered range $\mathcal{R}$ with the following properties:

- $\mathsf{ORE.Setup}(1^\lambda) \to \mathsf{sk}$. On input a security parameter λ, the setup algorithm $\mathsf{ORE.Setup}$ outputs a secret key sk.
- $\mathsf{ORE.Encrypt}(\mathsf{sk}, m) \to \mathsf{ct}$. On input the secret key sk and a message $m \in \mathcal{D}$, the encrypt algorithm $\mathsf{OPE.Encrypt}$ outputs a ciphertext $\mathsf{ct} \in \mathcal{R}$.

Correctness. An OPE scheme $\Pi = (\mathsf{OPE.Setup}, \mathsf{OPE.Encrypt})$ over a well-ordered domain $\mathcal{D}$ and well-ordered range $\mathcal{R}$ is correct if $\mathsf{sk} \leftarrow \mathsf{OPE.Setup}(1^\lambda)$, and all messages $m_1, m_2 \in \mathcal{D}$,

$$m_1 < m_2 \quad \Longleftrightarrow \quad \mathsf{OPE.Encrypt}(\mathsf{sk}, m_1) < \mathsf{OPE.Encrypt}(\mathsf{sk}, m_2)$$

with overwhelming probability.

2.3 Composing OPE with ORE

By composing an ORE scheme with an OPE scheme, we obtain an ORE scheme whose security is at least as strong as the security of the underlying OPE scheme. Let $\Pi_{\mathsf{ope}} = (\mathsf{OPE.Setup}, \mathsf{OPE.Encrypt})$ be an OPE scheme and $\Pi_{\mathsf{ore}}^{\mathsf{in}} = (\mathsf{ORE}^{\mathsf{in}}.\mathsf{Setup}, \mathsf{ORE}^{\mathsf{in}}.\mathsf{Encrypt}, \mathsf{ORE}^{\mathsf{in}}.\mathsf{Compare})$ be an ORE scheme. Consider the following composed construction $\Pi_{\mathsf{ore}} = (\mathsf{ORE.Setup}, \mathsf{ORE.Encrypt}, \mathsf{ORE.Compare})$ of an ORE scheme with an OPE scheme:

- $\mathsf{ORE.Setup}(1^\lambda)$. The setup algorithm runs $\mathsf{sk}_1 \leftarrow \mathsf{OPE.Setup}(1^\lambda)$ and $\mathsf{sk}_2 \leftarrow \mathsf{ORE}^{\mathsf{in}}.\mathsf{Setup}(1^\lambda)$. The secret key is $\mathsf{sk} = (\mathsf{sk}_1, \mathsf{sk}_2)$.
- $\mathsf{ORE.Encrypt}(\mathsf{sk}, \mathsf{m})$. The encryption algorithm outputs $\mathsf{ORE}^{\mathsf{in}}.\mathsf{Encrypt}(\mathsf{sk}_2, \mathsf{OPE.Encrypt}(\mathsf{sk}_1, \mathsf{m}))$.
- $\mathsf{ORE.Compare}(\mathsf{ct}_1, \mathsf{ct}_2)$. The compare algorithm computes and outputs the value $\mathsf{ORE}^{\mathsf{in}}.\mathsf{Compare}(\mathsf{ct}_1, \mathsf{ct}_2)$.

Correctness of Π_{ore} follows immediately from the correctness of $\Pi_{\mathsf{ore}}^{\mathsf{in}}$ and Π_{ope}. Furthermore, we note that under our simulation-based definition of security, the composed scheme Π_{ore} is at least as secure as Π_{ope}. This intuition is formalized in the following remark, whose proof follows immediately by construction.

Remark 2.6 (Security of Composed Scheme). For any leakage function $\mathcal{L}(\cdot)$, if the OPE scheme Π_{ope} is secure with leakage function $\mathcal{L}(\cdot)$, then the ORE scheme Π_{ore} is also secure with leakage function $\mathcal{L}(\cdot)$.

3 Main Construction

In this section, we give a construction of an ORE scheme for the set of n-bit positive integers with the following leakage function:

$$\mathcal{L}_f(m_1, \ldots, m_t) := \{(\mathsf{ind}_{\mathsf{diff}}(m_i, m_j), \mathbf{1}(m_i < m_j)) : 1 \leq i < j \leq t\}, \qquad (3.1)$$

where $\mathsf{ind}_{\mathsf{diff}}(x, y)$ gives the index of the first bit where x and y differ. If $x = y$, we set $\mathsf{ind}_{\mathsf{diff}}(x, y) = n + 1$. In other words, for $x \neq y$, if $x = x_1 \cdots x_n$ and $y = y_1 \cdots y_n$, then $\mathsf{ind}_{\mathsf{diff}}(x, y)$ is the smallest index $\ell \in [n]$ for which $x_\ell \neq y_\ell$.

Construction. Fix a security parameter $\lambda \in \mathbb{N}$, and take an integer $M \geq 3$. Let $F : \mathcal{K} \times ([n] \times \{0,1\}^{n-1}) \to \mathbb{Z}_M$ be a secure PRF. We define our ORE scheme $\Pi_{\mathsf{ore}} = (\mathsf{ORE.Setup}, \mathsf{ORE.Encrypt}, \mathsf{ORE.Compare})$ as follows:

- ORE.Setup(1^λ). The setup algorithm chooses a uniformly random PRF key k for F. The secret key is $\mathsf{sk} = k$.
- ORE.Encrypt(sk, m). Let $b_1 \cdots b_n$ be the binary representation of m and let $\mathsf{sk} = k$. For each $i \in [n]$, the encryption algorithm computes

$$u_i = F(k, (i, b_1 b_2 \cdots b_{i-1} \| 0^{n-i})) + b_i \pmod{M},$$

and outputs the tuple $(u_1, u_2 \ldots, u_n)$.
- ORE.Compare($\mathsf{ct}_1, \mathsf{ct}_2$). The compare algorithm first parses

$$\mathsf{ct}_1 = (u_1, u_2, \ldots, u_n)$$
$$\mathsf{ct}_2 = (u'_1, u'_2, \ldots, u'_n),$$

where $u_1, \ldots, u_n, u'_1, \ldots, u'_n \in \mathbb{Z}_M$. Let i be the smallest index where $u_i \neq u'_i$. If no such index exists, output 0. If such an index exists, output 1 if $u'_i = u_i + 1 \pmod{M}$, and 0 otherwise.

3.1   Correctness and Security

We now show that the above ORE scheme Π_{ore} is correct and secure against the leakage function $\mathcal{L}_\mathsf{f}$ from Eq. (3.1). We give the proof of the following theorem in the full version of this paper [15].

Theorem 3.1 *The ORE scheme Π_{ore} is correct.*

Next, we state and prove the security theorem for Π_{ore}.

Theorem 3.2 *The order-revealing encryption scheme Π_{ore} is secure with respect to leakage function $\mathcal{L}_\mathsf{f}$ (Definition 2.4) under the PRF security of F.*

Proof. Fix a security parameter λ and let $\mathcal{A} = (\mathcal{A}_1, \ldots, \mathcal{A}_q)$ where $q = \mathrm{poly}(\lambda)$ be an efficient adversary for the ORE security game (Definition 2.4). To prove security, we give an efficient simulator $\mathcal{S} = (\mathcal{S}_0, \ldots, \mathcal{S}_q)$ for which the outputs of the distributions $\mathsf{REAL}_\mathcal{A}^{\mathrm{ORE}}(\lambda)$ and $\mathsf{SIM}_{\mathcal{A},\mathcal{S},\mathcal{L}_\mathsf{f}}^{\mathrm{ORE}}(\lambda)$ are computationally indistinguishable.

We use a hybrid argument. We begin by defining the hybrid experiments:

- **Hybrid H_0:** This is the real experiment $\mathsf{REAL}_\mathcal{A}^{\mathrm{ORE}}(\lambda)$.
- **Hybrid H_1:** Same as H_0, except during ORE.Setup, a random function $f \overset{R}{\leftarrow} \mathrm{Funs}[([n] \times \{0,1\}^{n-1}), \mathbb{Z}_M]$ is chosen. In all invocations of ORE.Encrypt, the function $F(k, \cdot)$ is replaced by $f(\cdot)$.

Hybrids H_0 and H_1 are computationally indistinguishable under the PRF security of F. Thus, it suffices to show that there exists a simulator $\mathcal{S}$ such that the distribution of outputs in H_1 is computationally indistinguishable from $\mathsf{SIM}_{\mathcal{A},\mathcal{S},\mathcal{L}_\mathsf{f}}^{\mathrm{ORE}}(\lambda)$.

Description of the Simulator. We now describe the simulator $\mathcal{S} = (\mathcal{S}_0, \ldots, \mathcal{S}_q)$. First, $\mathcal{S}_0$ initializes an empty lookup tables $\mathsf{L} : [q] \times [n] \to \mathbb{Z}_M$.

It then outputs $\mathsf{st}_\mathcal{S} = \mathsf{L}$. Then, for each $t \in [q]$, after the adversary outputs a query m_t, the simulation algorithm $\mathcal{S}_t$ is invoked on input $\mathsf{st}_\mathcal{S} = \mathsf{L}$ and $\mathcal{L}_f(m_1, \ldots, m_t)$. In particular, $\mathcal{L}_f(m_1, \ldots, m_t)$ contains the values $\mathbf{1}(m_j < m_t)$ and $\mathsf{ind}_{\mathsf{diff}}(m_j, m_t)$ for all $j \in [t-1]$, where $\mathsf{ind}_{\mathsf{diff}}(m_j, m_t)$ is the index of the first bit in m_j and m_t that differ. For each $s \in [n]$, there are three cases to consider:

- **Case 1:** There exists a $j \in [t-1]$ such that $\mathsf{ind}_{\mathsf{diff}}(m_j, m_t) > s$. If there are multiple j for which $\mathsf{ind}_{\mathsf{diff}}(m_j, m_t) > s$, let j be the smallest one. Then, the simulator sets $\overline{u}_s = \mathsf{L}(j, s)$.
- **Case 2:** For each $\ell \in [t-1]$, $\mathsf{ind}_{\mathsf{diff}}(m_\ell, m_t) \le s$, and there exists a $j \in [t-1]$ for which $\mathsf{ind}_{\mathsf{diff}}(m_j, m_t) = s$. If there are multiple j for which $\mathsf{ind}_{\mathsf{diff}}(m_j, m_t) = s$, let j be the smallest one. Then, the simulator sets $\overline{u}_s = \mathsf{L}(j, s) - (1 - 2 \cdot \mathbf{1}(m_j < m_t)) \pmod{M}$.
- **Case 3:** For each $\ell \in [t-1]$, $\mathsf{ind}_{\mathsf{diff}}(m_\ell, m_t) < s$. In this case, the simulator samples $y \xleftarrow{\mathsf{R}} \mathbb{Z}_M$ and sets $\overline{u}_s = y$.

For each $s \in [n]$, the simulator adds the mapping $(t, s) \mapsto \overline{u}_s$ to L. Finally, the simulator $\mathcal{S}_t$ outputs the ciphertext $\overline{\mathsf{ct}}_t = (\overline{u}_1, \overline{u}_2, \ldots, \overline{u}_n)$ and the updated state $\mathsf{st}_\mathcal{S} = \mathsf{L}$. This completes the description of the simulator $\mathcal{S}$.

Correctness of the Simulation. We show that the simulator $\mathcal{S} = (\mathcal{S}_0, \ldots, \mathcal{S}_q)$ perfectly simulates the distribution in hybrid H_2. Let $(\mathsf{ct}_1, \ldots, \mathsf{ct}_q)$ be the joint distribution of the ciphertexts output in hybrid H_2, and let $(\overline{\mathsf{ct}}_1, \ldots, \overline{\mathsf{ct}}_q)$ be the joint distribution of the ciphertexts output by the simulator. We proceed inductively in the number of queries q. The base case $(q = 0)$ follows trivially. Suppose now that $(\mathsf{ct}_1, \ldots, \mathsf{ct}_{t-1}) \equiv (\overline{\mathsf{ct}}_1, \ldots, \overline{\mathsf{ct}}_{t-1})$ for some $t \in [q]$. We show that the statement holds for $t+1$. Consider the distributions of ct_t and $\overline{\mathsf{ct}}_t$. First, for any $j \in [t]$, write ciphertext ct_j as $(u_{j,1}, u_{j,2}, \ldots, u_{j,n})$ and $\overline{\mathsf{ct}}_j$ as $(\overline{u}_{j,1}, \overline{u}_{j,2}, \ldots, \overline{u}_{j,n})$. In addition, for $j \in [t]$, we write $b_{j,s}$ to denote the sth bit of m_j. For each $s \in [n]$, we consider three cases:

- **Case 1:** There exists a $j \in [t-1]$ such that $\mathsf{ind}_{\mathsf{diff}}(m_j, m_t) > s$. If there are multiple j for which $\mathsf{ind}_{\mathsf{diff}}(m_j, m_t) > s$, let j be the smallest one. This means that m_j and m_t share a prefix of length at least s. Let $p \in \{0,1\}^{s-1}$ be the first $s-1$ bits of this common prefix. Then, in hybrid H_1, we have

$$u_{t,s} = f(s, p\|0^{n-s}) + b_{t,s} = u_{j,s}.$$

  In the simulation, $\overline{u}_{t,s} = \mathsf{L}(j, s) = \overline{u}_{j,s}$. Since $j < t$, we conclude from the induction hypothesis that $u_{t,s}$ and $\overline{u}_{t,s}$ are identically distributed.
- **Case 2:** For each $\ell \in [t-1]$, $\mathsf{ind}_{\mathsf{diff}}(m_\ell, m_t) \le s$, and there exists a $j \in [t-1]$ such that $\mathsf{ind}_{\mathsf{diff}}(m_j, m_t) = s$. If there are multiple j for which $\mathsf{ind}_{\mathsf{diff}}(m_j, m_t) = s$, let j be the smallest one. This means that m_j and m_t share a prefix $p \in \{0,1\}^{s-1}$ of length $s-1$. Then, in hybrid H_1, we have

$$u_{t,s} = f(s, p\|0^{n-s}) + b_{t,s} \pmod{M}.$$

  In the simulation,

$$\overline{u}_{t,s} = \mathsf{L}(j, s) - (1 - 2 \cdot \mathbf{1}(m_j < m_t)) = \overline{u}_{j,s} - (1 - 2 \cdot \mathbf{1}(m_j < m_t)) \pmod{M}.$$

In hybrid H_2, $u_{j,s} = f(s, p \| 0^{n-s}) + b_{j,s}$. By assumption, $b_{j,s} \neq b_{t,s}$, so we can write $b_{t,s} = b_{j,s} - (1 - 2 \cdot \mathbf{1}(m_j < m_t))$. Thus, in hybrid H_2, we have

$$u_{t,s} = f(s, p \| 0^{n-s+1}) + b_{t,s} = u_{j,s} - (1 - 2 \cdot \mathbf{1}(m_j < m_t)) \pmod{M}.$$

By the inductive hypothesis, $u_{j,s}$ and $\overline{u}_{j,s}$ are identically distributed, so we conclude that $u_{t,s}$ and $\overline{u}_{t,s}$ are identically distributed.

- **Case 3:** For each $\ell \in [t-1]$, $\mathsf{ind}_{\mathsf{diff}}(m_\ell, m_t) < s$. Let $p \in \{0,1\}^{s-1}$ be the first $s - 1$ bits of m_t. In hybrid H_1, we have

$$u_{t,s} = f(s, p \| 0^{n-s}) + b_{t,s} \pmod{M},$$

while in the simulation $\overline{u}_{t,s}$ is a uniformly random string. By assumption, none of the messages $m_1, \ldots, m_{t-1}$ begin with the prefix p. Since f is a truly random function, the value of $f(s, p \| 0^{n-s})$ is uniform in $\mathbb{Z}_M$ and independent of all other ciphertexts. Thus, $u_{t,s}$ and $\overline{u}_{t,s}$ are identically distributed.

We conclude that for all $s \in [n]$, $u_{t,s} \equiv \overline{u}_{t,s}$. Since the components of each ciphertext are constructed independently in both hybrid H_1 and in the simulation, this suffices to show that ct_t and $\overline{\mathsf{ct}}_t$ are identically distributed. The claim then follows by induction on t. $\square$

Space usage. The order-revealing encryption scheme Π_{ore} on n-bit inputs produces encryptions of size $\lceil n \cdot \log_2 M \rceil$. By setting $M = 3$, an encryption of an n-bit message under Π_{ore} consists of only $\lceil n \cdot \log_2 3 \rceil \approx 1.59\,n$ bits. In the full version, we describe a "d-ary" generalization of Π_{ore} that further reduces the size of the ciphertexts in the ORE scheme, but with a slight loss in security. Specifically, we construct an ORE scheme where an encryption of an n-bit message has length approximately $n \cdot \log_d(2d - 1)$ for any integer $d \geq 2$. Since $\log_d(2d - 1)$ is a monotonically decreasing function in d, larger values of d yield shorter ciphertexts, but increased leakage.

3.2   Conversion to OPE

In this section, we explain how to convert Π_{ore}, an ORE scheme, into an OPE scheme. This means that ciphertexts of the resulting OPE scheme can be compared using the normal comparison function on numbers. To do this, we apply a simple transformation of any ciphertext ct of Π_{ore} into a number c that lies in the range $[0, M^n - 1]$ for which direct numeric comparisons of two numbers c_1 and c_2 reveal the order relation of the underlying plaintexts.

Recall that in Π_{ore}, ciphertexts are of the form $\mathsf{ct} = (u_1, u_2 \ldots, u_n)$, where for each $i \in [n]$, u_i lies in the range $\mathbb{Z}_M$. The ciphertext in the resulting OPE scheme is taken to be the $\lceil n \cdot \log_2 M \rceil$-bit number

$$c = \sum_{i=1}^{n} u_i \cdot M^{n-i} \qquad \in [0, M^n - 1]. \tag{3.2}$$

Intuitively, we view $u_1u_2\cdots u_n$ as a base-M representation of the OPE ciphertext. Correctness follows similarly to Π_{ore}, except here, there is a non-zero probability of error (as opposed to Π_{ore} where correctness held with probability 1). We claim that for any two messages $m_1, m_2 \in [0, 2^n - 1]$,

$$m_1 < m_2 \qquad \Longleftrightarrow \qquad c_1 < c_2,$$

with probability $1 - 1/M$, where $c_1, c_2 \in [0, M^n - 1]$ are the ciphertexts obtained by first invoking ORE.Encrypt on m_1, m_2, respectively, and then applying the transformation in Eq. (3.2). To see this, let $i \in [n]$ be the first bit position on which m_1 and m_2 differ. Observe that the numeric comparison of the OPE ciphertexts behaves identically as the ORE comparison procedure, except when the output of the PRF on the first $i - 1$ bits of the messages is the value $M - 1$[7]. However, by PRF security, this event happens with probability $1/M$, and thus, correctness holds with probability $1 - 1/M$. For instance, if $M = 2^\lambda$ (that is, λ bits), correctness holds with overwhelming probability. For practical scenarios, it may be suitable to only take $M \approx 2^{40}$ (the failure probability in this case is 2^{-40}).

Security of the resulting OPE scheme follows identically from security of Π_{ore}, as the transformation from ciphertexts ct to numbers c is bijective. We note that while this scheme is order-preserving, it does not behave like a random order-preserving function, and thus, does not inherit the security limitations associated with such OPE schemes [8]. In fact, our simulation-based security model and associated security theorem (Theorem 3.1) enables us to precisely specify the information leakage in this order-preserving encryption scheme.

In the full version, we describe a "d-ary" generalization of Π_{ore}. While this generalization does not reduce the size of the resulting ciphertexts in the ORE scheme, it does yield shorter ciphertexts in the OPE instantiation (by approximately a $\log_2 d$ multiplicative factor), with a slight loss in security. Correctness in this generalized scheme holds with probability $1 - d/M$.

4   Comparison to Existing OPE Schemes

We now compare the leakage of our order-revealing encryption scheme to that of existing order-preserving encryption schemes by Boldyreva et al. [7,8]. As explained in Sect. 2.3, composing any existing OPE scheme with an ORE scheme results in a new ORE scheme which is at least as secure as the underlying OPE scheme[8]. In this section, we show that even *without* the composition, our construction still achieves stronger security according to the metrics proposed by Boldyreva et al.

---

[7] If no reduction modulo M occurs in the ORE.Encrypt encryption, then numerically comparing the transformed ciphertexts is identical to evaluating the ORE.Compare procedure (since all relations hold over the integers).

[8] In most cases, the security of the composed scheme is strictly greater than that of the base OPE scheme since our ORE construction provides semantic security for a single ciphertext, whereas existing OPE schemes generally do not.

The security definition achieved by an order-preserving encryption scheme is that the encryption function behaves like a random order-preserving function (ROPF) from the plaintext space to the ciphertext space. While this definition has the same flavor as that for PRFs, the behavior of a truly random function is very different from that of a random order-preserving function. In particular, the output of an order-preserving function is not independent of its input, and thus, reveals some information about the input. It turns out that quantifying the exact information leakage is a non-trivial task in general. However, under certain assumptions (for example, if the messages are drawn from a uniform distribution), it is possible to obtain concrete upper bounds on the information leakage [8]. In particular, Boldyreva et al. propose two security notions, window one-wayness and window distance one-wayness, to analyze the security of an OPE scheme. In our setting, the nature of our security definition allows us to analyze the construction under a more generalized set of definitions compared to [8]. We present our analysis for window one-wayness here, and defer the analysis of window distance one-wayness to the full version.

4.1   One-Wayness

One of the most basic requirements of an encryption scheme is that it is one-way. Given a ciphertext, an adversary that does not have the secret key should not be able to recover the underlying message. In the standard definition of one-wayness [27], the adversary is given the encryption of a random message, and its goal is to guess the message. This is a very weak notion of security, and even if an encryption is one-way, the adversary might still be able to deduce nontrivial information about the message given only the ciphertext. To address this, Boldyreva et al. [7] introduce a more general notion of one-wayness where the adversary is allowed to guess a contiguous interval (a window) in the one-wayness challenge. The adversary succeeds if the message is contained within the interval. Moreover, the adversary is given multiple encryptions (of random messages) and succeeds if it outputs an interval that contains at least one of the messages.

The notion of window one-wayness is useful for arguing that an adversary does not learn many of the *most significant* bits of the message, but if all bits of the message are equally sensitive, then this definition is less useful. In our work, we present a more general definition of one-wayness, where instead of outputting an interval, the adversary is allowed to specify a set of guesses. To allow the adversary to specify a super-polynomially-sized set of guesses, we instead require the adversary to submit a circuit C that encodes its set ($C(x) = 1$ if and only if x is in the set). By requiring that the circuit encodes a contiguous interval, we recover the window one-wayness definition by Boldyreva et al. [8]. We now give our generalized definition.

Definition 4.1 (Generalized One-Wayness). *Fix a plaintext space $\mathcal{D}$ and let $\Pi = (\mathsf{ORE.Setup}, \mathsf{ORE.Encrypt}, \mathsf{ORE.Compare})$ be an ORE over $\mathcal{D}$. The (r, z)-generalized one-wayness advantage of an adversary $\mathcal{A}$ against Π is given by*

$$\mathsf{Adv}^{\mathsf{gow}}_{r,z,\Pi}(\mathcal{A}) \overset{\text{def}}{=} \Pr[\mathsf{Expt}^{\mathsf{gow}}_{r,z,\Pi,\mathcal{A}}(1^{\lambda}) = 1],$$

where the (r, z)-generalized one-wayness experiment $\mathsf{Expt}^{\mathsf{gow}}_{r,z,\Pi,\mathcal{A}}(1^{\lambda})$ is defined as follows:

Experiment $\mathsf{Expt}^{\mathsf{gow}}_{r,z,\Pi,\mathcal{A}}(1^{\lambda})$:

1. $\mathsf{sk} \leftarrow \mathsf{ORE.Setup}(1^{\lambda})$
2. sample $m_1, \ldots, m_z$ uniformly from $\mathcal{D}$ without replacement
3. for $i \in [z]$, $\mathsf{ct}_i \leftarrow \mathsf{ORE.Encrypt}(\mathsf{sk}, m_i)$
4. $C \leftarrow \mathcal{A}(\mathsf{ct}_1, \ldots, \mathsf{ct}_z)$, where $C : \mathcal{D} \to \{0, 1\}$ is a circuit of size $\mathrm{poly}(\lambda)$
5. output 1 if $C(m_i) = 1$ for some $i \in [z]$ and $|\{x \in \mathcal{D} : C(x) = 1\}| \leq r$; otherwise, output 0

Remark 4.2 (Comparison with Existing One-Wayness Notions). By restricting the parameters (r, z) and the classes of circuits the adversary is allowed to output, Definition 4.1 captures many existing notions of one-wayness. For example, when $r = z = 1$, we recover the usual notion of one-wayness [27]. When the underlying plaintext space is the ring $\mathbb{Z}_M$ for some integer M and we require that the circuit output by the adversary encodes a contiguous interval of length at most r in $\mathbb{Z}_M$, our definition corresponds to the notion of window one-wayness introduced by Boldyreva et al. [8].

We now state our security theorem, but defer the proof to the full version.

Theorem 4.3 *Fix a security parameter λ and a plaintext space $\{0, 1\}^n$ where $n = \omega(\log \lambda)$. Let Π_{ore} be the ORE scheme given at the beginning of Sect. 3. Then, for any constant $\varepsilon \in (0, 1]$, any $z = \mathrm{poly}(\lambda)$, and all efficient adversaries $\mathcal{A}$,*

$$\mathsf{Adv}^{\mathsf{gow}}_{r,z,\Pi_{\mathsf{ore}},\mathcal{A}}(1^{\lambda}) = \mathrm{negl}(\lambda),$$

where $r = 2^{n(1-\varepsilon)}$.

Comparison to existing schemes. When discussing the notion of one-wayness, we will always assume that the message-space is super-polynomial in the security parameter. Otherwise, the trivial adversary that just guesses a random point in the message space will succeed with non-negligible probability.

In [8], Boldyreva et al. give an upper bound on the one-wayness advantage of any (possibly computationally unbounded) adversary $\mathcal{A}$ against a random order-preserving function ROPF. This corresponds to setting $r = 1$ in our definition. They show [8, Theorem 4.1] that for $z = \mathrm{poly}(\lambda)$, $\mathsf{Adv}^{\mathsf{gow}}_{1,z,\mathsf{ROPF},\mathcal{A}} = \mathrm{negl}(\lambda)$.

The same statement holds for our ORE construction assuming a computationally bounded adversary: simply instantiate Theorem 4.3 with $\varepsilon = 1$.

In addition to giving an upper bound on an adversary's ability to guess the plaintext from the ciphertext, Boldyreva et al. also give a lower bound on the advantage for the case when r is large. In particular, they exhibit an efficient adversary $\mathcal{A}$ against an ROPF such that $\mathsf{Adv}^{\mathsf{gow}}_{r,z,\mathsf{ROPF},\mathcal{A}}(1^\lambda) = 1 - 2e^{-b^2/2}$ for a constant b when $r = O(\sqrt{2^n})$ and for any z [8, Theorem 4.2][9]. In other words, the authors describe a concrete adversary that is able to break the generalized one-wayness of any POPF-CCA-secure scheme (with probability close to 1) if the adversary is allowed to specify a set with $r = O(\sqrt{2^n})$ elements, even when $z = 1$. An intuitive way to understand this result is that given the output of an ROPF, an adversary can deduce roughly half of the bits of the associated input. In contrast, in our ORE scheme, if the adversary only sees a polynomial number of ciphertexts ($z = \mathsf{poly}(\lambda)$), then invoking Theorem 4.3 with $\varepsilon = 1/2$, we have that for all efficient adversaries $\mathcal{A}$, $\mathsf{Adv}^{\mathsf{gow}}_{r,z,\Pi_{\mathsf{ore}},\mathcal{A}}(1^\lambda) = \mathsf{negl}(\lambda)$ where $r = \sqrt{2^n}$. In fact, as Theorem 4.3 demonstrates, the adversary's advantage remains negligible even if we further increase the size of the sets the adversary is allowed to submit.

Intuitively, our results show that if the adversary only sees a polynomial number of ciphertexts, then it does not learn any constant fraction ε of the bits in the underlying plaintext from each ciphertext. In contrast, with an ROPF, and correspondingly, any OPE scheme that realizes a ROPF, each ciphertext alone leaks *half* of the most-significant bits of the underlying plaintext.

Similarly, while the OPE scheme by Teranishi et al. [47] can be shown to hide any constant fraction of the least significant bits of the plaintext, no such guarantee exists for the other bits of the plaintext. Note though that the security notion proposed in [47] is indistinguishability-based and hence, stronger than the one-wayness security notions. In fact, our basic ORE construction (by itself) does not achieve their indistinguishability-based definition. However, by composing our ORE construction with their OPE construction, we obtain a resulting ORE scheme which is strictly more secure, since it inherits the security properties of the underlying OPE scheme as well as semantic security for a single ciphertext (Sect. 2.3, Remark 2.6).

5   Conclusions

In this work, we introduced a new notion of security for order-preserving, and more generally, order-revealing encryption. Our simulation-based security notion is defined with respect to a leakage function which precisely characterizes what the ciphertexts in the scheme leak about the underlying messages. We then give a practical order-revealing encryption scheme which achieves this security notion for a specific leakage function. By composing our ORE construction with

---

[9] Strictly speaking, the adversary they describe is for the window one-wayness experiment, but any adversary that succeeds in the window one-wayness experiment also succeeds in the generalized one-wayness experiment (Definition 4.1).

existing OPE schemes, we obtain an ORE scheme with increased security. It is our hope that having a concrete leakage model will enable practitioners to make better-informed decisions on whether an ORE scheme is appropriate for their particular application. We conclude with several open problems:

1. Can we construct a practical ORE scheme with stronger security guarantees?
2. Can we reduce the ciphertext length of our ORE scheme while still maintaining a similar level of security?
3. Is it possible to build a practical ORE scheme with best-possible security from standard assumptions?

Acknowledgments. We would like to thank Sam Kim for helpful discussions about ORE, and Adam O'Neill for useful insights in shrinking the ciphertext size of our main construction. We also thank the anonymous reviewers for their helpful comments. This work was partially supported by an NSF Graduate Research Fellowship. Opinions, findings and conclusions or recommendations expressed in this material are those of the authors and do not necessarily reflect the views of Facebook.

References

1. Agrawal, R., Kiernan, J., Srikant, R., Xu, Y.: Order-preserving encryption for numeric data. In: SIGMOD, pp. 563–574 (2004)
2. Albrecht, M.R., Farshim, P., Hofheinz, D., Larraia, E., Paterson, K.G.: Multilinear maps from obfuscation. In: TCC (2016)
3. Ananth, P., Jain, A.: Indistinguishability obfuscation from compact functional encryption. In: CRYPTO, pp. 308–326 (2015)
4. Applebaum, B., Brakerski, Z.: Obfuscating circuits via composite-order graded encoding. In: Dodis, Y., Nielsen, J.B. (eds.) TCC 2015, Part II. LNCS, vol. 9015, pp. 528–556. Springer, Heidelberg (2015)
5. Barak, B., Garg, S., Kalai, Y.T., Paneth, O., Sahai, A.: Protecting obfuscation against algebraic attacks. In: Nguyen, P.Q., Oswald, E. (eds.) EUROCRYPT 2014. LNCS, vol. 8441, pp. 221–238. Springer, Heidelberg (2014)
6. Binnig, C., Hildenbrand, S., Färber, F.: Dictionary-based order-preserving string compression for main memory column stores. In: ACM SIGMOD, pp. 283–296 (2009)
7. Boldyreva, A., Chenette, N., Lee, Y., O'Neill, A.: Order-preserving symmetric encryption. In: Joux, A. (ed.) EUROCRYPT 2009. LNCS, vol. 5479, pp. 224–241. Springer, Heidelberg (2009)
8. Boldyreva, A., Chenette, N., O'Neill, A.: Order-preserving encryption revisited: improved security analysis and alternative solutions. In: Rogaway, P. (ed.) CRYPTO 2011. LNCS, vol. 6841, pp. 578–595. Springer, Heidelberg (2011)
9. Boneh, D., Lewi, K., Raykova, M., Sahai, A., Zhandry, M., Zimmerman, J.: Semantically secure order-revealing encryption: multi-input functional encryption without obfuscation. In: Oswald, E., Fischlin, M. (eds.) EUROCRYPT 2015. LNCS, vol. 9057, pp. 563–594. Springer, Heidelberg (2015)
10. Boneh, D., Silverberg, A.: Applications of multilinear forms to cryptography. Contemp. Math. **324**(1), 71–90 (2003)
11. Boneh, D., Wu, D.J., Zimmerman, J.: Immunizing multilinear maps against zeroizing attacks. In: IACR Cryptology ePrint Archive 2014/930 (2014)

12. Brakerski, Z., Komargodski, I., Segev, G.: From single-input to multi-input functional encryption in the private-key setting. In: IACR Cryptology ePrint Archive 2015/158 (2015)

13. Bun, M., Zhandry, M.: Order-revealing encryption and the hardness of private learning. In: IACR Cryptology ePrint Archive 2015/417 (2015)

14. Chang, Y.-C., Mitzenmacher, M.: Privacy preserving keyword searches on remote encrypted data. In: Ioannidis, J., Keromytis, A.D., Yung, M. (eds.) ACNS 2005. LNCS, vol. 3531, pp. 442–455. Springer, Heidelberg (2005)

15. Chenette, N., Lewi, K., Weis, S.A., Wu, D.J.: Practical order-revealing encryption with limited leakage. In: IACR Cryptology ePrint Archive 2015/1125 (2015)

16. Cheon, J.H., Han, K., Lee, C., Ryu, H., Stehlé, D.: Cryptanalysis of the multilinear map over the integers. In: Oswald, E., Fischlin, M. (eds.) EUROCRYPT 2015. LNCS, vol. 9056, pp. 3–12. Springer, Heidelberg (2015)

17. Cheon, J.H., Lee, C., Ryu, H.: Cryptanalysis of the new CLT multilinear maps. In: IACR Cryptology ePrint Archive (2011) Observation of strains: 934 (2015)

18. Coron, J.-S.: Cryptanalysis of GGH15 multilinear maps (2015)

19. Coron, J.-S., Gentry, C., Halevi, S., Lepoint, T., Maji, H.K., Miles, E., Raykova, M., Sahai, A., Tibouchi, M.: Zeroizing without low-level zeroes: new MMAP attacks and their limitations. In: CRYPTO, pp. 247–266 (2015)

20. Coron, J.-S., Lepoint, T., Tibouchi, M.: Practical multilinear maps over the integers. In: Canetti, R., Garay, J.A. (eds.) CRYPTO 2013, Part I. LNCS, vol. 8042, pp. 476–493. Springer, Heidelberg (2013)

21. Coron, J.-S., de Lepoint, T., Tibouchi, M.: New multilinear maps over the integers. In: CRYPTO, pp. 267–286 (2015)

22. Curtmola, R., Garay, J.A., Kamara, S., Ostrovsky, R.: Searchable symmetric encryption: improved definitions and efficient constructions. In: ACM CCS, pp. 79–88 (2006)

23. Garg, S., Gentry, C., Halevi, S.: Candidate multilinear maps from ideal lattices. In: Johansson, T., Nguyen, P.Q. (eds.) EUROCRYPT 2013. LNCS, vol. 7881, pp. 1–17. Springer, Heidelberg (2013)

24. Garg, S., Gentry, C., Halevi, S., Raykova, M., Sahai, A., Waters, B.: Candidate indistinguishability obfuscation and functional encryption for all circuits. In: FOCS, pp. 40–49 (2013)

25. Garg, S., Gentry, C., Halevi, S., Zhandry, M.: Fully secure functional encryption without obfuscation. In: IACR Cryptology ePrint Archive 2014/666 (2014)

26. Gentry, C., Gorbunov, S., Halevi, S.: Graph-induced multilinear maps from lattices. In: Dodis, Y., Nielsen, J.B. (eds.) TCC 2015, Part II. LNCS, vol. 9015, pp. 498–527. Springer, Heidelberg (2015)

27. Goldreich, O.: The Foundations of Cryptography - Volume 1, Basic Techniques. Cambridge University Press, Cambridge (2001)

28. Goldreich, O., Goldwasser, S., Micali, S.: How to construct random functions (extended abstract). In: FOCS, pp. 464–479 (1984)

29. Goldwasser, S., et al.: Multi-input functional encryption. In: Nguyen, P.Q., Oswald, E. (eds.) EUROCRYPT 2014. LNCS, vol. 8441, pp. 578–602. Springer, Heidelberg (2014)

30. Goldwasser, S., Kalai, Y.T., Popa, R.A., Vaikuntanathan, V., Zeldovich, N.: Reusable garbled circuits and succinct functional encryption. In: STOC, pp. 555–564 (2013)

31. Goldwasser, S., Micali, S.: Probabilistic encryption. J. Comput. Syst. Sci. **28**(2), 270–299 (1984)

32. Gorbunov, S., Vaikuntanathan, V., Wee, H.: Functional encryption with bounded collusions via multi-party computation. In: Safavi-Naini, R., Canetti, R. (eds.) CRYPTO 2012. LNCS, vol. 7417, pp. 162–179. Springer, Heidelberg (2012)
33. Hu, Y., Huiwen, J.: Cryptanalysis of GGH map. In: IACR Cryptology ePrint Archive 2015/301 (2015)
34. Kadhem, H., Amagasa, T., Kitagawa, H.: A secure and efficient order preserving encryption scheme for relational databases. In: KMIS, pp. 25–35 (2010)
35. Kerschbaum, F.: Frequency-hiding order-preserving encryption. In: ACM CCS, pp. 656–667 (2015)
36. Kerschbaum, F., Schröpfer, A.: Optimal average-complexity ideal-security order-preserving encryption. In: ACM CCS, pp. 275–286 (2014)
37. Langlois, A., Stehlé, D., Steinfeld, R.: GGHLite: more efficient multilinear maps from ideal lattices. In: Nguyen, P.Q., Oswald, E. (eds.) EUROCRYPT 2014. LNCS, vol. 8441, pp. 239–256. Springer, Heidelberg (2014)
38. Mavroforakis, C., Chenette, N., O'Neill, A., Kollios, G., Canetti, R.: Modular order-preserving encryption, revisited. In: ACM SIGMOD, pp. 763–777 (2015)
39. Minaud, B., Fouque, P.-A.: Cryptanalysis of the new multilinear map over the integers. In: IACR Cryptology ePrint Archive 2015/941 (2015)
40. Naveed, M., Kamara, S., Wright, C.V.: Inference attacks on property-preserving encrypted databases. In: CCS (2015)
41. Pandey, O., Rouselakis, Y.: Property preserving symmetric encryption. In: Pointcheval, D., Johansson, T. (eds.) EUROCRYPT 2012. LNCS, vol. 7237, pp. 375–391. Springer, Heidelberg (2012)
42. Popa, R.A., Li, F.H., Zeldovich, N.: An ideal-security protocol for order-preserving encoding. In: IEEE Symposium on Security and Privacy, pp. 463–477 (2013)
43. Popa, R.A., Redfield, C.M.S., Zeldovich, N., Balakrishnan, H.: Cryptdb: protecting confidentiality with encrypted query processing. In: SOSP, pp. 85–100 (2011)
44. Roche, D., Apon, D., Choi, S.G., Yerukhimovich, A.: POPE: Partial order-preserving encoding. In: Cryptology ePrint Archive, Report 2015/1106 (2015)
45. Sahai, A., Seyalioglu, H.: Worry-free encryption: functional encryption with public keys. In: ACM CCS, pp. 463–472 (2010)
46. Skyhigh Networks Inc. https://www.skyhighnetworks.com/. Accessed 11 Dec 2015
47. Teranishi, I., Yung, M., Malkin, T.: Order-preserving encryption secure beyond one-wayness. In: Sarkar, P., Iwata, T. (eds.) ASIACRYPT 2014, Part II. LNCS, vol. 8874, pp. 42–61. Springer, Heidelberg (2014)
48. Xiao, L., Yen, I-L., Huynh, D.T.: Extending order preserving encryption for multi-user systems. In: IACR Cryptology ePrint Archive, (2011) Observation of strains: 192 (2012)
49. Zimmerman, J.: How to obfuscate programs directly. In: Oswald, E., Fischlin, M. (eds.) EUROCRYPT 2015. LNCS, vol. 9057, pp. 439–467. Springer, Heidelberg (2015)

Strengthening the Known-Key Security Notion
for Block Ciphers

Benoît Cogliati[1(✉)] and Yannick Seurin[2]

[1] University of Versailles, Versailles, France
`benoitcogliati@hotmail.fr`
[2] ANSSI, Paris, France
`yannick.seurin@m4x.org`

Abstract. We reconsider the formalization of known-key attacks against ideal primitive-based block ciphers. This was previously tackled by Andreeva, Bogdanov, and Mennink (FSE 2013), who introduced the notion of *known-key indifferentiability*. Our starting point is the observation, previously made by Cogliati and Seurin (EUROCRYPT 2015), that this notion, which considers only a single known key available to the attacker, is too weak in some settings to fully capture what one might expect from a block cipher informally deemed resistant to known-key attacks. Hence, we introduce a stronger variant of known-key indifferentiability, where the adversary is given *multiple* known keys to "play" with, the informal goal being that the block cipher construction must behave as an independent random permutation for each of these known keys. Our main result is that the 9-round iterated Even-Mansour construction (with the trivial key-schedule, i.e., the same round key xored between permutations) achieves our new "multiple" known-keys indifferentiability notion, which contrasts with the previous result of Andreeva *et al.* that one single round is sufficient when only a single known key is considered. We also show that the 3-round iterated Even-Mansour construction achieves the weaker notion of multiple known-keys *sequential* indifferentiability, which implies in particular that it is *correlation intractable* with respect to relations involving any (polynomial) number of known keys.

Keywords: Block cipher · Ideal cipher · Known-key attacks · Iterated Even-Mansour cipher · Key-alternating cipher · Indifferentiability · Correlation intractability

1 Introduction

BACKGROUND ON KNOWN-KEY ATTACKS. Informally, a known-key attack against a block cipher E consists in the following: the adversary is given a key k from the key space of E, and must find a "non-trivial" property of the permutation E_k associated with k faster than what it would cost given only black-box access to a truly random permutation. An example of such a non-trivial property would be a plaintext/

© International Association for Cryptologic Research 2016
T. Peyrin (Ed.): FSE 2016, LNCS 9783, pp. 494–513, 2016.
DOI: 10.1007/978-3-662-52993-5_25

ciphertext pair (x, y) under the key k such that, say, the first half of x and the first half of y seen as bit strings are both zero (for a random permutation P over n-bit strings, it is easy to see that this requires roughly $2^{n/2}$ queries to P). Known-key attacks against block ciphers were first introduced by Knudsen and Rijmen [18], who exhibited such attacks against a reduced-round version of AES and against certain kinds of Feistel ciphers. These attacks were extended in a number of follow-up papers, e.g. [14,15,23,24,28].

Even though the informal idea underlying known-key security might intu-itively seem clear (given a key k, the permutation E_k associated with k must "look random"), how to put known-key attacks on theoretical sound grounds has remained elusive. Indeed, any attempt to rigorously formalize what is a known-attack against a fixed block cipher runs into impossibility results similar to those undermining a sound definition of what a "good" hash function should be [4]. In particular, seeing a block cipher as a family of permutations indexed by the key, the fact that the key-length is similar to the input-length of the permutations (i.e., the block-length of the block cipher) leads to the following "diagonal" problem: consider the set of pairs $(k, E_k(k))$ for k ranging over the key space (we assume that the block-length and the key-length are equal for ease of exposition); then it is hard, given oracle access to a random permutation, to find an input/output pair in this set, whereas given any key k for E it is very easy to find an input/output pair for E_k in this set.

A way to circumvent these impossibilities is to consider block cipher con-structions based on some ideal primitive (for example, a Feistel cipher based on public random round functions or (iterated) Even-Mansour ciphers based on public permutations). In that case, even though the adversary is given the known key, it only has oracle access to the underlying primitive, which effectively acts as an (exponentially long) seed indexing the permutation associated with the key. A first step towards formalizing known-key attacks for ideal primitive-based block ciphers was taken by Andreeva, Bogdanov, and Mennink (ABM) [2] through what they called *known-key indifferentiability* (KK-indifferentiability for short), a variant of the standard indifferentiability notion [22]. A block cipher construc-tion $\mathcal{C}^F$ from some underlying primitive F is said indifferentiable from an ideal cipher E if there exists an efficient simulator $\mathcal{S}$ with black box access to E such that the two pairs of oracles $(\mathcal{C}^F, F)$ and $(E, \mathcal{S}^E)$ are indistinguishable. Hence the simulator must make E "look like" $\mathcal{C}^F$ by returning answers that are coher-ent with the distinguisher's queries to E (without, in general, knowing these E-queries) and that are statistically close to answers of a real F oracle.

The KK-indifferentiability notion of ABM modifies the security experiment as follows: a key k is drawn at random and made available to the distinguisher and the simulator; the distinguisher is then allowed to query its left oracle (con-struction/ideal cipher) *only for this specific key k*. Hence the simulator's job is somehow made simpler since it has a "hint" about which queries the distin-guisher can make to its left oracle. Note that in the ideal (simulated) world, the distinguisher effectively has access to a single random permutation (since an ideal cipher behaves as an independent random permutation for each key). Hence this

KK-indifferentiability notion intuitively captures the requirement that for each key k, the block cipher construction $\mathcal{C}^F$ must "look like" a random permutation. In contrast, the standard indifferentiability notion is related with *chosen-key* attacks, since the distinguisher is allowed to freely choose the keys it examines.

SHORTCOMING OF THE ABM SECURITY NOTION. The starting point of this paper is an observation, previously made by Cogliati and Seurin (Appendix C of the full version of [7]) that the ABM security notion might be too restrictive in some situations because it considers *one single* known-key. This might be problematic in some cryptosystems where intuitively resistance to known-key attacks should be sufficient to provide security, but where the ABM security notion fails because the cryptosystem uses *multiple* known keys. Think for example of the permutation-based hashed functions by Rogaway and Steinberger [26,27]: these constructions are based on a few (typically 3 to 6) public permutations, which would typically be instantiated by a block cipher used with distinct publicly known keys. A crucial requirement for the security proof of these constructions to hold (in the ideal permutation model) is that the permutations are independent. Since this is not ensured by the ABM security notion, it is not applicable here, even though one would like to say that a block cipher which is secure against known-key attacks can safely be used in the Rogaway-Steinberger constructions. (Jumping ahead, our new KK-indifferentiability notion will be sufficient to safely instantiate the block cipher in the same constructions.)

To better emphasize this gap between a single known-key notion and a multiple known-key notion, consider the case of the 1-round Even-Mansour (EM) [11,12] construction based on a permutation P on $\{0,1\}^n$, which maps a key $k \in \{0,1\}^n$ and a plaintext $x \in \{0,1\}^n$ to the ciphertext defined as

$$\mathsf{EM}^P(k,x) = k \oplus P(k \oplus x).$$

ABM showed that when the permutation P is ideal, this construction is KK-indifferentiable from an ideal cipher in the single known-key setting. However, if the adversary is given any pair of distinct keys (k_1, k_2), it can pick any $x_1 \in \{0,1\}^n$, define $x_2 = x_1 \oplus k_1 \oplus k_2$, and compute $y_1 = \mathsf{EM}^P_{k_1}(x_1)$ and $y_2 = \mathsf{EM}^P_{k_2}(x_2)$. Then one can easily check that $x_1 \oplus x_2 = y_1 \oplus y_2$. Yet for an ideal cipher E, given two distinct keys $k_1 \neq k_2$, finding two pairs (x_1, y_1) and (x_2, y_2) such that $E_{k_1}(x_1) = y_1$, $E_{k_2}(x_2) = y_2$, and $x_1 \oplus x_2 = y_1 \oplus y_2$ can be shown to be hard: more precisely, an adversary making at most q queries to E can find such pairs with probability at most $\mathcal{O}(\frac{q^2}{2^n})$. In other words, the permutations associated with distinct keys for the 1-round EM construction do not "behave" independently.

OUR CONTRIBUTION. Our first contribution is definitional: in order to remedy the limitation that we just pointed out, we extend and strengthen the known-key security definition of [2], by allowing the distinguisher to be given multiple known keys. Our new notion is parameterized by an integer μ, the number of known keys that the adversary is given. For $\mu = 1$, one recovers the ABM definition. If one lets $\mu = |\mathcal{K}|$, where $\mathcal{K}$ is the key space of the block cipher, one recovers the standard indifferentiability notion. In fact, our KK-indifferentiability notion will emerge as a special case of a more general notion that we name *restricted-input*-indifferentiability, which might be of independent interest. We also formulate our KK-indifferentiability notion in a "worst-case" fashion (it must hold for *any* subset of keys of size μ), whereas the ABM notion was in the "average-case" style (the known key being randomly drawn). In addition, we define a weaker "sequential" variant [7,21] of our new μ-KK-indifferentiability notion, called μ-KK-seq-indifferentiability, where the adversary must query its two oracles in a specific order. This notion is useful since it implies the weaker notion of correlation intractability.

Our second contribution is about constructions: we show that KK-indifferentiability is a meaningful notion by proving that the iterated Even-Mansour (IEM) construction with nine rounds is μ-KK-indifferentiable from an ideal cipher for any $\mu = \texttt{poly}(n)$ (where n is a security parameter indexing the construction), which contrasts with the fact that one round is sufficient when considering one single known-key, and also with the best number of rounds known to be sufficient to achieve full indifferentiability from an ideal cipher, namely twelve [20]. We also show that three rounds are necessary and sufficient to achieve the weaker μ-KK-seq-indifferentiability notion, which again contrast with the fact that four rounds are necessary and sufficient to achieve (full) seq-indifferentiability from an ideal cipher [7]. See Table 1 for a summary of known results on the IEM construction.

MORE RELATED WORK. A number of papers have studied the indifferentiability of variants of the IEM construction. In particular, Andreeva *et al.* [1] have studied the case where the key-schedule is modeled as a random oracle, and Guo and Lin have studied the case of Even-Mansour ciphers with two interleaved keys [16] and of key-alternating Feistel ciphers [17].

ORGANIZATION. We start with some general definitions in Sect. 2. Then we define precisely our strengthened KK-indifferentiability notion (as well as the more general notion of *restricted-input*-indifferentiability, of which KK-indifferentiability is a special case) in Sect. 3. In Sect. 4, we give a known-key attack (using two known keys) against the 2-round IEM construction. Finally, we prove that the 3-round, resp. 9-round, IEM construction achieves μ-KK-seq-indifferentiability, resp. μ-KK-indifferentiability, in Sects. 5 and 6.

2   Preliminaries

GENERAL NOTATION. In all the following, we fix an integer $n \geq 1$ and denote $N = 2^n$. Given a non-empty set $\mathcal{M}$, the set of all permutations of $\mathcal{M}$ will be denoted $\mathsf{Perm}(\mathcal{M})$. We simply denote $\mathsf{Perm}(n)$ the set of all permutations over $\{0,1\}^n$. A block cipher with key space $\mathcal{K}$ and message space $\mathcal{M}$ is a mapping $E : \mathcal{K} \times \mathcal{M} \to \mathcal{M}$ such that for any key $k \in \mathcal{K}$, $x \mapsto E(k,x)$ is a permutation. We interchangeably use the notations $E(k,x)$ and $E_k(x)$. We denote $\mathsf{BC}(\mathcal{K},\mathcal{M})$ the set of all block ciphers with key space $\mathcal{K}$ and message space $\mathcal{M}$, and $\mathsf{BC}(n,n)$

Table 1. Summary of provable security results for the iterated Even-Mansour cipher with independent inner permutations and the trivial key-schedule. The first two notions are secret-key notions, the other ones are indifferentiability-based.

Sec. notion	# rounds	Sec. bound	Sim. complexity (query/time)	Ref.
Single-key (pseudorandomness)	1	$q^2/2^n$	—	[11,12]
	2	$q^{3/2}/2^n$	—	[5]
XOR related-key	3	$q^2/2^n$	—	[7,13]
1-KK-indiff.	1	0	q / q	[2]
μ-KK-seq-indiff., $\mu > 1$	3	$\mu^2 q^2/2^n$	μq / μq	This paper
Full seq-indiff.	4	$q^4/2^n$	q^2 / q^2	[7]
μ-KK-indiff., $\mu > 1$	9	$\mu^6 q^6/2^n$	$\mu^2 q$ / $\mu^2 q$	This paper
Full indiff.	12	$q^{12}/2^n$	q^4 / q^6	[20]

the set of block ciphers with key space and message space $\{0,1\}^n$. For integers $1 \le s \le t$, we will write $(t)_s = t(t-1)\cdots(t-s+1)$ and $(t)_0 = 1$ by convention.

IDEAL PRIMITIVES. An *ideal primitive* F is a triplet (F.Dom, F.Rng, F.Inst): the domain F.Dom and the range F.Rng are two non-empty sets, and the instance space F.Inst is a set of functions $F : \text{F.Dom} \to \text{F.Rng}$.

The two main ideal primitives we will be interested in are ideal permutations and ideal ciphers. Given a non-empty set $\mathcal{M}$, the ideal permutation P over $\mathcal{M}$ is defined as follows. Let P.Dom $= \{+,-\} \times \mathcal{M}$ and P.Rng $= \mathcal{M}$, and define

$$\text{P.Inst} \stackrel{\text{def}}{=} \left\{ P : \exists \pi \in \text{Perm}(\mathcal{M}), P(+,x) = \pi(x) \text{ and } P(-,y) = \pi^{-1}(y) \right\}.$$

Clearly, there is a one-to-one correspondence between P.Inst and $\text{Perm}(\mathcal{M})$.

Similarly, given two non-empty sets $\mathcal{K}$ and $\mathcal{M}$, the ideal cipher with key space $\mathcal{K}$ and message space $\mathcal{M}$ is defined as follows. Let E.Dom $= \{+,-\} \times \mathcal{K} \times \mathcal{M}$, E.Rng $= \mathcal{M}$, and define

$$\text{E.Inst} \stackrel{\text{def}}{=} \left\{ E : \exists \eta \in \text{BC}(\mathcal{K}, \mathcal{M}), E(+,k,x) = \eta_k(x) \text{ and } E(-,k,y) = \eta_k^{-1}(y) \right\}.$$

Again, there is a one-to-one correspondence between E.Inst and $\text{BC}(\mathcal{K}, \mathcal{M})$.

THE ITERATED EVEN-MANSOUR CIPHER. Fix integers $n, r \ge 1$. Let $\mathbf{f} = (f_0, \ldots, f_r)$ be a $(r+1)$-tuple of permutations of $\{0,1\}^n$. The r-round iterated Even-Mansour construction $\text{EM}[n, r, \mathbf{f}]$ specifies, from any r-tuple $\mathbf{P} = (P_1, \ldots, P_r)$ of permutations of $\{0,1\}^n$, a block cipher with n-bit keys and n-bit messages, simply denoted $\text{EM}^{\mathbf{P}}$ in all the following (parameters $[n, r, \mathbf{f}]$ will always be clear from the context), which maps a plaintext $x \in \{0,1\}^n$ and a key $k \in \{0,1\}^n$ to the ciphertext defined by (see Fig. 1):

$$\text{EM}^{\mathbf{P}}(k,x) = f_r(k) \oplus P_r(f_{r-1}(k) \oplus P_{r-1}(\cdots P_2(f_1(k) \oplus P_1(f_0(k) \oplus x))\cdots)).$$

We say that the key-schedule is *trivial* when all f_i's are the identity.

While the pseudorandomness of the IEM cipher was mostly studied with *independent* round keys [3,6,19] (with the notable exception of [5]), it is well known that independent round keys cannot, in general, provide any security in the setting where the adversary has some control over the master key (related-, known-, or chosen-key attacks) [20]. Hence, in this paper, we focus on the case where the round keys are derived from an n-bit master key (actually, all our results deal with the case of the trivial key-schedule).

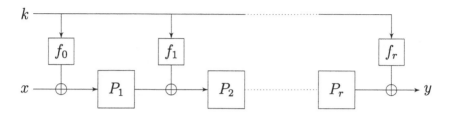

Fig. 1. The r-round iterated Even-Mansour cipher.

3   Restricted-Input Indifferentiability and Variants

We introduce the notion of restricted-input indifferentiability (*RI-indifferentiability*), and explain how known-key indifferentiability is a special case of it. Let E and F be two ideal primitives.[1] A *construction* implementing E from F is a deterministic algorithm $\mathcal{C}$ with oracle access to an instance F of F, which we denote $\mathcal{C}^F$, such that for any $F \in$ F.Inst, $\mathcal{C}^F \in$ E.Inst. A *simulator* for F is a randomized algorithm with oracle access to an instance E of E, which we denote $\mathcal{S}^E$, such that for any $E \in$ E.Inst, $\mathcal{S}^E :$ F.Dom $\rightarrow$ F.Rng. A distinguisher $\mathcal{D}$ is a deterministic[2] algorithm with oracle access to two oracles, the first one with signature E.Dom $\rightarrow$ E.Rng, the second one with signature F.Dom $\rightarrow$ F.Rng, and which returns a bit b, which we denote $\mathcal{D}(\mathcal{O}_1, \mathcal{O}_2) = b$. We will call $\mathcal{O}_1$ the *left* oracle and $\mathcal{O}_2$ the *right* oracle. Following [21], we define the *total oracle query cost* of $\mathcal{D}$ as the maximum, over $F \in$ F.Inst, of the total number of queries received by F (from $\mathcal{D}$ or $\mathcal{C}$) when $\mathcal{D}$ interacts with $(\mathcal{C}^F, F)$. The indifferentiability advantage of $\mathcal{D}$ against $(\mathcal{C}, \mathcal{S})$ is defined by

$$\mathbf{Adv}^{\mathrm{indiff}}_{\mathcal{C}, \mathcal{S}}(\mathcal{D}) = \Big| \Pr \big[E \leftarrow_\$ \text{E.Inst} : \mathcal{D}(E, \mathcal{S}^E) = 1 \big]$$

$$- \Pr \big[F \leftarrow_\$ \text{F.Inst} : \mathcal{D}(\mathcal{C}^F, F) = 1 \big] \Big|. \quad (1)$$

(Note that the first probability is also taken over the randomness of $\mathcal{S}$).

For any subset of X of E.Dom, $\mathcal{D}$ is said X-*restricted* if it only makes queries to its left oracle (E or $\mathcal{C}^F$) from the set X.

---

[1] This might be any ideal primitives, in particular E might not be an ideal cipher.
[2] Since we will consider computationally unbounded distinguishers, this is without loss of generality.

Definition 1 (Restricted-Input Indifferentiability). *Let* E *and* F *be two ideal primitives and* $\mathcal{C}$ *be a construction implementing* E *from* F. *Let* $q, \sigma, t \in \mathbb{N}$ *and* $\varepsilon \in \mathbb{R}^+$. *Let* $\mathcal{X}$ *be a family of subsets of* $\mathsf{E}.\mathsf{Dom}$. *Construction* $\mathcal{C}$ *is said* $(\mathcal{X}, q, \sigma, t, \varepsilon)$-*RI-indifferentiable from* E *if for any* $X \in \mathcal{X}$, *there exists a simulator* $\mathcal{S}$ *such that for any* X-*restricted distinguisher* $\mathcal{D}$ *of total oracle query cost at most* q, $\mathcal{S}$ *makes at most* σ *oracle queries, runs in time at most* t, *and*

$$\mathbf{Adv}^{\mathrm{indiff}}_{\mathcal{C},\mathcal{S}}(\mathcal{D}) \leq \varepsilon.$$

Informally, we simply say that $\mathcal{C}$ is $\mathcal{X}$-RI-indifferentiable from E if it is $(\mathcal{X}, q, \sigma, t, \varepsilon)$-RI-indifferentiable for "reasonable" values of σ, t, and ε expressed as functions of q (in particular, when $\mathcal{C}$ is indexed by some security parameter $n \in \mathbb{N}$, if $\sigma, t \in \mathtt{poly}(n)$ and $\varepsilon \in \mathtt{negl}(n)$ for any $q \in \mathtt{poly}(n)$).

As is standard in works on indifferentiability, this definition is information-theoretic, i.e., the distinguisher is allowed to be computationally unbounded (this is sometimes called *statistical indifferentiability*), and demands the existence of a *universal* simulator which does not depend on the distinguisher (this is sometimes called *strong* indifferentiability; when the simulator is allowed to depend on the distinguisher, this is called *weak* indifferentiability).

Note also the following points:

- by letting $\mathcal{X} = \{\mathsf{E}.\mathsf{Dom}\}$ in the definition above, one recovers the standard definition of indifferentiability [22];
- when $\mathcal{X} = \{X\}$ is reduced to a single subset of $\mathsf{E}.\mathsf{Dom}$, the definition is equivalent to the standard definition of indifferentiability of the restriction of $\mathcal{C}^F$ to X from the restriction of E to X; hence this definition is only "new" when considering at least two distinct subsets X and X' such that $X \not\subseteq X'$ and $X' \not\subseteq X$ (since a X-restricted distinguisher is also a X'-restricted distinguisher when $X \subseteq X'$), and can be equivalently rephrased as the indifferentiability of the family of restrictions of $\mathcal{C}$ to sets in $\mathcal{X}$, with a uniform upper bound on the simulator's complexity and the distinguisher's advantage;
- the simulator is allowed to depend on the specific set $X \in \mathcal{X}$ considered;
- the upper bound on the advantage of the distinguisher must hold for any $X \in \mathcal{X}$ (not, say, on average on the random draw of X from $\mathcal{X}$).

The RI version of indifferentiability can be combined with other flavors of indifferentiability, in particular with public indifferentiability [10,29] and sequential indifferentiability [7,21]. Let us elaborate for the case of sequential indifferentiability. A distinguisher is called *sequential* if after its first query to its left $(\mathsf{E}/\mathcal{C}^F)$ oracle, it does not make any query to its right $(\mathcal{S}^E/F)$ oracle any more. In other words, it works in two phases: first it only queries its right oracle, and then only its left oracle. Then we can define *RI-seq-indifferentiability* exactly as in Definition 1, except that we quantify over X-restricted *sequential* distinguishers only. (Hence this is a weaker definition since for each subset $X \in \mathcal{X}$, the simulator has to be effective only against a smaller class of distinguishers, namely sequential ones.)

COMPOSITION THEOREM. The meaningfulness of the indifferentiability notion comes from the following composition theorem [22]: if a cryptosystem is proven secure when implemented with ideal primitive E, then it remains provably secure when E is replaced with $\mathcal{C}$ based on ideal primitive F, assuming $\mathcal{C}$ is indifferentiable from E. (For this theorem to hold, the security of the cryptosystem must be defined with respect to a class of adversaries which "supports" the simulator used to prove that $\mathcal{C}$ is indifferentiable from E [9,25].) This theorem straightforwardly translates to $\mathcal{X}$-RI-indifferentiability as follows: if a cryptosystem is proven secure when implemented with ideal primitive E and *if for any adversary $\mathcal{A}$, there is $X \in \mathcal{X}$ such that the challenger of the security game only queries E on inputs $x \in X$ when interacting with $\mathcal{A}$*, then it remains provably secure when E is replaced with $\mathcal{C}$ based on ideal primitive F, assuming $\mathcal{C}$ is $\mathcal{X}$-RI-indifferentiable from E. The short proof is as follows: denote Γ the challenger for the security game, which has access to an instance of E, and fix an adversary $\mathcal{A}$ against the cryptosystem implemented with $\mathcal{C}^F$ (hence $\mathcal{A}$ has oracle access to the instance F of the ideal primitive F); see the combination of Γ and $\mathcal{A}$ as a single X-restricted distinguisher $\mathcal{D}$; by the $\mathcal{X}$-RI-indifferentiability assumption, there is a simulator $\mathcal{S}$ such that $(\mathcal{C}^F, F)$ cannot be distinguished from $(E, \mathcal{S}^E)$; then the combination of $\mathcal{A}$ and $\mathcal{S}$ constitutes an attacker against the cryptosystem implemented with E, and the winning probability of $\mathcal{A}'$ is small by the assumption that the cryptosystem is secure when implemented with E; hence the winning probability of $\mathcal{A}$ is small as well.

KNOWN-KEY INDIFFERENTIABILITY. We now explain how to formalize resistance to known-key attacks using RI-indifferentiability. Fix non-empty sets $\mathcal{K}$ and $\mathcal{M}$, and let E be the ideal cipher with key space $\mathcal{K}$ and message space $\mathcal{M}$. Recall that E.Dom $= \{+, -\} \times \mathcal{K} \times \mathcal{M}$. For any integer $1 \leq \mu \leq |\mathcal{K}|$, let $\mathcal{X}_\mu$ be the family of subsets of E.Dom consisting of queries whose key is in $\mathcal{K}'$, for $\mathcal{K}'$ ranging over all subsets of $\mathcal{K}$ of size μ; more formally,

$$\mathcal{X}_\mu = \{\{(+, k, x) : k \in \mathcal{K}'\} \cup \{(-, k, y) : k \in \mathcal{K}'\} : \mathcal{K}' \subseteq \mathcal{K}, |\mathcal{K}'| = \mu\}.$$

Note that $\mathcal{X}_{|\mathcal{K}|} = \{\text{E.Dom}\}$.

Definition 2 (μ-Known-Key Indifferentiability). *Let $\mathcal{C}$ be a construction of a block cipher with key space $\mathcal{K}$ and message space $\mathcal{M}$ from an ideal primitive F. Let $\mu, q, \sigma, t \in \mathbb{N}$ and $\varepsilon \in \mathbb{R}^+$. Construction $\mathcal{C}$ is said to be $(\mu, q, \sigma, t, \varepsilon)$-KK-indifferentiable from an ideal cipher if and only if it is $(\mathcal{X}_\mu, q, \sigma, t, \varepsilon)$-RI-indifferentiable from an ideal cipher, with $\mathcal{X}_\mu$ defined as above.*

Unfolding the definition, this is equivalent to the following: for any subset $\mathcal{K}' \subseteq \mathcal{K}$ of size μ, there exists a simulator $\mathcal{S}$ such that for any distinguisher $\mathcal{D}$ whose queries to its first (construction/ideal cipher) oracle use only keys $k \in \mathcal{K}'$ and of total oracle query cost at most q, $\mathcal{S}$ makes at most σ oracle queries, runs in time at most t, and

$$\mathbf{Adv}_{\mathcal{C}, \mathcal{S}}^{\text{indiff}}(\mathcal{D}) \leq \varepsilon.$$

The KK-indifferentiability notion of Andreeva *et al.* [2] corresponds to the definition above for $\mu = 1$. In fact, this is slightly more subtle. Their variant is rather an "average" version of this definition over the random draw of the known key, resulting from the following changes: the security experiment starts by drawing a random key k which is given as input to both the distinguisher and the simulator, and the two probabilities involved in the Definition (1) of the advantage of the distinguisher are also taken over the random draw of the challenge key $k \leftarrow_\$ \mathcal{K}$. It is not hard to see that our "worst-case" variant of the definition is stronger (i.e., implies) the average-case version (the average-case simulator simply has a copy of each worst-case simulator $\mathcal{S}_{\mathcal{K}'}$ for each possible subset $\mathcal{K}' \subseteq \mathcal{K}$ of size μ, and on input the challenge subset of keys runs the corresponding worst-case simulator).

The standard indifferentiability notion [22] is recovered by letting $\mu = |\mathcal{K}|$ in the definition above. The composition theorem specializes to the case of μ-KK-indifferentiability as follows: if a cryptosystem is proven secure when implemented with an ideal cipher E with key space $\mathcal{K}$ and if for any adversary $\mathcal{A}$, there is a subset of keys $\mathcal{K}'$ of size μ such that the challenger of the security game only queries E with keys $k \in \mathcal{K}'$ when interacting with $\mathcal{A}$, then it remains provably secure when E is replaced with $\mathcal{C}$ based on ideal primitive F, assuming $\mathcal{C}$ is μ-KK-indifferentiable from an ideal cipher.

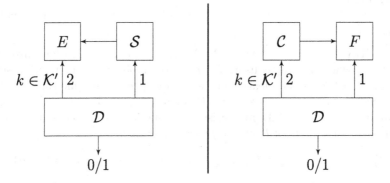

Fig. 2. Various flavors of the indifferentiability notion. For full indifferentiability, the queries of the distinguisher are completely unrestricted. For μ-known-key indifferentiability, queries to the left oracle (ideal cipher/construction) can only be made for keys $k \in \mathcal{K}'$ for some subset $\mathcal{K}'$ of size μ of the key space $\mathcal{K}$ (the simulator being allowed to depend on $\mathcal{K}'$). For sequential indifferentiability, the numbers next to query arrows indicate in which order the distinguisher accesses both oracles. After its first query to the left oracle, the distinguisher cannot query the right oracle any more. Combining the two constraints results in the KK-seq-indifferentiability notion.

KNOWN-KEY CORRELATION INTRACTABILITY. As for the general notion of RI-indifferentiability, KK-indifferentiability can be combined with the notion of sequential indifferentiability. Hence, if we restrict Definition 2 by quantifying only over sequential distinguishers, we obtain the notion of KK-seq-indifferentiability (see also Fig. 2). This notion is interesting because it implies the (arguably more natural) notion of known-key *correlation intractability*, as we explain now.

For this, we first recall the concept of evasive relation and correlation intractability [4,7,21]. Let E be an ideal primitive. For an integer $m \geq 1$, an m-ary *relation* $\mathcal{R}$ (for E) is simply a subset $\mathcal{R} \subset (\text{E.Dom})^m \times (\text{E.Rng})^m$. Informally, a relation is *evasive* with respect to E if it is hard, on average, for an adversary with oracle access to a random instance E of E to find a tuple of inputs $(\alpha_1, \ldots, \alpha_m)$ such that $((\alpha_1, \ldots, \alpha_m), (E(\alpha_1), \ldots, E(\alpha_m)))$ satisfies this relation. The definition below is very general and applies to any ideal primitive.

Definition 3 (Evasive Relation). *Let* E *be an ideal primitive. An m-ary relation $\mathcal{R}$ for* E *is said (q, ε)-evasive if for any adversary $\mathcal{A}$ with oracle access to an instance E of* E*, making at most q oracle queries, one has*

$$\Pr \left[E \leftarrow_\$ \text{E.Inst}, (\alpha_1, \ldots, \alpha_m) \leftarrow \mathcal{A}^E : \right.$$
$$\left. ((\alpha_1, \ldots, \alpha_m), (E(\alpha_1), \ldots, E(\alpha_m))) \in \mathcal{R} \right] \leq \varepsilon,$$

where the probability is taken over the random draw of E and the random coins of $\mathcal{A}$.

Recall that the domain and the range of an ideal cipher E with key space $\mathcal{K}$ and message space $\mathcal{M}$ are $\text{E.Dom} = \{+, -\} \times \mathcal{K} \times \mathcal{M}$ and $\text{E.Rng} = \mathcal{M}$ so that, if we particularize the definition above for an ideal cipher, each α_i is a triplet in E.Dom, and $E(\alpha_i) \in \mathcal{M}$.

If we now consider a construction $\mathcal{C}$ implementing E from some other ideal primitive F, a natural thing to ask is that any relation which is evasive with respect to E remains hard to find for $\mathcal{C}^F$, on average over the random draw of F, for any adversary with oracle access to F. This is formalized by the following definition.

Definition 4 (Correlation Intractability). *Let* E *and* F *be two ideal primitives, and let $\mathcal{C}$ be a construction implementing* E *from* F*. Let $\mathcal{R}$ be an m-ary relation for* E*. Then $\mathcal{C}$ is said to be (q, ε)-correlation intractable with respect to $\mathcal{R}$ if for any adversary $\mathcal{A}$ with oracle access to an instance of* F*, making at most q oracle queries, one has*

$$\Pr \left[F \leftarrow_\$ \text{F.Inst}, (\alpha_1, \ldots, \alpha_m) \leftarrow \mathcal{A}^F : \right.$$
$$\left. ((\alpha_1, \ldots, \alpha_m), (\mathcal{C}^F(\alpha_1), \ldots, \mathcal{C}^F(\alpha_m))) \in \mathcal{R} \right] \leq \varepsilon,$$

where the probability is taken over the random draw of F and the random coins of $\mathcal{A}$.

A theorem by Mandal *et al.* [21] (see also [7, Theorem 4]) establishes that seq-indifferentiability allows, for any relation $\mathcal{R}$, to "reduce" the correlation intractability of $\mathcal{C}$ with respect to $\mathcal{R}$ to the evasiveness of $\mathcal{R}$ (with respect to E). More precisely, if $\mathcal{C}$ is seq-indifferentiable from E and if a relation $\mathcal{R}$ is (q, ε)-evasive with respect to E, then $\mathcal{C}$ is (q', ε')-correlation intractable with respect to $\mathcal{R}$, and the "degradation" of security parameters (q', ε') compared with (q, ε) depends on the seq-indifferentiability parameters. In other words, if $\mathcal{C}$ is seq-indifferentiable from E, then any relation which is hard to find for E remains hard to find for $\mathcal{C}^F$ (on average over the random draw of F).

This result can be straightforwardly declined for the case of KK-seq-indifferentiability (and more generally RI-seq-indifferentiability): if $\mathcal{C}$ is $\mathcal{X}$-RI-seq-indifferentiable from E for some family $\mathcal{X}$ of subsets of $\mathsf{E.Dom}$, then a similar result holds, but only for relations $\mathcal{R}$ such that all inputs involved in $\mathcal{R}$ belong to some subset $X \in \mathcal{X}$; similarly, if $\mathcal{C}$ is μ-KK-seq-indifferentiable from an ideal cipher E with key space $\mathcal{K}$, then the result holds for relations $\mathcal{R}$ such that all inputs involved in $\mathcal{R}$ use the same μ keys.

Concretely we have the following theorem. The proof is similar to the proof of [7, Theorem 4] and therefore deferred to the full version of the paper [8]. First we give two preliminary definitions. Let E be an ideal primitive, and X be a subset of $\mathsf{E.Dom}$; then an m-ary relation $\mathcal{R}$ for E is said X-restricted if

$$\forall((\alpha_1, \dots, \alpha_m), (\beta_1, \dots, \beta_m)) \in \mathcal{R}, \ \forall i = 1, \dots, m, \ \alpha_i \in X.$$

Similarly, let E be an ideal cipher with key space $\mathcal{K}$, and $\mu \geq 1$; then an m-ary relation $\mathcal{R}$ for E is said μ-restricted if there exists a subset $\mathcal{K}'$ of $\mathcal{K}$ of size μ such that

$$\forall((\delta_i, k_i, z_i), \dots, (\delta_m, k_m, z_m)), (z'_1, \dots, z'_m)) \in \mathcal{R}, \ \forall i = 1, \dots, m, \ k_i \in \mathcal{K}'.$$

Theorem 1. *Let E and F be two ideal primitives, and let $\mathcal{C}$ be a construction implementing E from F such that $\mathcal{C}$ makes at most c queries to its oracle on any input. Let $\mathcal{X}$ be a family of subsets of $\mathsf{E.Dom}$. Assume that $\mathcal{C}$ is $(\mathcal{X}, q + cm, \sigma, t, \varepsilon)$-RI-seq-indifferentiable from E. Then for any m-ary relation $\mathcal{R}$ which is X-restricted for some $X \in \mathcal{X}$, if $\mathcal{R}$ is $(\sigma + m, \varepsilon_{\mathcal{R}})$-evasive with respect to E, then $\mathcal{C}$ is $(q, \varepsilon + \varepsilon_{\mathcal{R}})$-correlation intractable with respect to $\mathcal{R}$.*

In particular, let E be an ideal cipher with key space $\mathcal{K}$, and assume that $\mathcal{C}$ is $(\mu, q + cm, \sigma, t, \varepsilon)$-KK-seq-indifferentiable from E. Then for any μ-restricted m-ary relation $\mathcal{R}$, if $\mathcal{R}$ is $(\sigma + m, \varepsilon_{\mathcal{R}})$-evasive with respect to E, then $\mathcal{C}$ is $(q, \varepsilon + \varepsilon_{\mathcal{R}})$-correlation intractable with respect to $\mathcal{R}$.

Remark 1. We need to dispel some confusion that might be created by the following observation (this will also help illustrate all definitions above with a concrete example): Lampe and Seurin [20] have exhibited an attacker against the 3-round IEM construction which, given oracle access to the inner permutations, finds four tuples (k_i, x_i, y_i), $i = 1, \dots, 4$, satisfying the following evasive relation:

$$\begin{cases} k_1 \oplus k_2 \oplus k_3 \oplus k_4 = 0 \\ x_1 \oplus x_2 \oplus x_3 \oplus x_4 = 0 \\ y_1 \oplus y_2 \oplus y_3 \oplus y_4 = 0. \end{cases}$$

Since we will later prove that the 3-round IEM construction is μ-KK-seq-indifferentiable from an ideal cipher for any polynomial μ, this might seem contradictory with Theorem 1. The catch is that two of the four keys involved in the relation and obtained at the end of the attack are not controlled by the adversary and in fact range over the entire key space when the inner permutations range over $\mathsf{Perm}(n)$. Hence, the evasive relation actually involves keys from the entire key space (not just a small subset of it).

4   KK-Attack on the Two-Round IEM Construction

We explained in Sect. 1 that the 1-round EM construction is not resistant to μ-known-key attacks for $\mu \geq 2$. We show here that this extends to the 2-round IEM construction (with independent inner permutations and the trivial key-schedule), more formally, that this construction is not μ-KK-seq-indifferentiable from an ideal cipher for $\mu \geq 2$. Our attack shares some similarities with the related-key attack against the same construction of [7]. Formally, we prove the following theorem.

Theorem 2. *The 2-round IEM construction* $\mathsf{EM}[n, 2, \mathbf{f}]$ *with independent inner permutations and the trivial key schedule*[3] $\mathbf{f}$ *is not 2-KK-seq-indifferentiable from an ideal cipher. More precisely, for any pair of distinct keys* (k_1, k_2)*, there is an adversary which distinguishes the construction from an ideal cipher with advantage close to 1 by making only queries to its left (construction/ideal cipher) oracle involving these two keys. The adversary makes no queries to its right (inner permutations/simulator) oracle.*

Proof. We denote generically (E, F) the oracles to which the adversary has access and (k_1, k_2) two distinct keys the attacker is allowed to use. Consider the following distinguisher (see Fig. 3 for a diagram of the attack):

(1) choose an arbitrary value $x_1 \in \{0, 1\}^n$, and query $y_1 := E(+, k_1, x_1)$;
(2) compute $x_2 := x_1 \oplus k_2 \oplus k_1$, and query $y_2 := E(+, k_2, x_2)$;
(3) compute $y_3 := y_1 \oplus k_1 \oplus k_2$, and query $x_3 := E(-, k_2, y_3)$;
(4) compute $y_4 := y_2 \oplus k_2 \oplus k_1$, and query $x_4 := E(-, k_1, y_4)$;
(5) check whether $x_4 = x_3 \oplus k_1 \oplus k_2$.

When the distinguisher is interacting with an ideal cipher E, two cases can occur. Either $y_4 = y_1$, or $y_4 \neq y_1$. In the first case, this means that $y_1 \oplus y_2 = k_1 \oplus k_2$, which happens with probability 2^{-n} since x_1 and x_2 are the first queries to the uniformly random and independent permutations E_{k_1} and E_{k_2}. If $y_4 \neq y_1$, then y_4 is the second query to the uniformly random permutation E_{k_1}, thus x_4 is uniformly random and this equality happens with probability at most $1/(2^n - 1)$. Moreover one has $y_2 \neq y_1 \oplus k_1 \oplus k_2$ which happens with probability $1 - 2^{-n}$ since x_2 is the first query to E_{k_2}. Since E is a uniformly randomly drawn blockcipher, E_{k_1} and E_{k_2} are independent permutations and this case happens with probability at most 2^{-n}. Overall, when E is an ideal cipher, this relation is satisfied with a probability at most 2^{n-1}.

---

[3] In fact, the attack applies whenever the key-schedule is linear.

Now we show that when the distinguisher is interacting with the two round Even-Mansour construction, it always returns 1, independently of k, and the inner permutations, which we denote P_1 and P_2. Noting that, by definition, $x_2 = x_1 \oplus k_2 \oplus k_1$, we denote u_1 the common value

$$u_1 \stackrel{\text{def}}{=} x_1 \oplus k_1 = x_2 \oplus k_2,$$

and we denote $v_1 = P_1(u_1)$. We also denote

$$
\begin{aligned}
u_2 &= v_1 \oplus k_1 \\
v_2 &= P_2(u_2) && (2) \\
u_2' &= v_1 \oplus k_2 && (3) \\
v_2' &= P_2(u_2').
\end{aligned}
$$

Hence, one has

$$
\begin{aligned}
y_1 &= v_2 \oplus k_1 && (4) \\
y_2 &= v_2' \oplus k_2. && (5)
\end{aligned}
$$

Since $y_3 = y_1 \oplus k_1 \oplus k_2$, we can see, using (4), that

$$y_3 \oplus k_2 = y_1 \oplus k_1 = v_2.$$

Define

$$
\begin{aligned}
v_1' &= u_2 \oplus k_2 && (6) \\
u_1' &= P_1^{-1}(v_1').
\end{aligned}
$$

This implies that

$$x_3 = u_1' \oplus k_2. \qquad (7)$$

Since $y_4 = y_2 \oplus k_2 \oplus k_1$, we see by (5) that

$$y_4 \oplus k_1 = y_2 \oplus k_2 = v_2'.$$

Moreover, we have

$$
\begin{aligned}
u_2' \oplus k_1 &= u_2' \oplus k_2 \oplus k_1 \oplus k_2 \\
&= v_1 \oplus k_1 \oplus k_2 && \text{by (3)} \\
&= u_2 \oplus k_2 && \text{by (2)} \\
&= v_1' && \text{by (6).}
\end{aligned}
$$

This finally implies by (7) that

$$x_4 \oplus k_1 = u_1' = x_3 \oplus k_2,$$

which concludes the proof. □

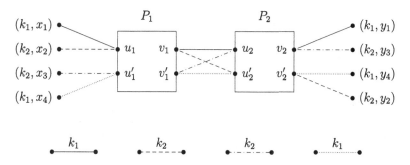

Fig. 3. A 2-known-key attack on the iterated Even-Mansour cipher with two rounds and the trivial key-schedule.

5   KK-Seq-Indifferentiability for Three Rounds

We have just given a 2-known-keys attack against the 2-round IEM cipher. This implies that the 2-round IEM construction cannot be μ-KK-seq-indifferentiable from an ideal cipher as soon as $\mu \geq 2$. (Remember on the other hand that the 1-round EM construction is 1-KK-indifferentiable from an ideal cipher [2].) Hence, at least three rounds are necessary (and, as we will see now, sufficient) to achieve μ-KK-seq-indifferentiability from an ideal cipher for $\mu \geq 2$.

Concretely, the main result of this section regarding the KK-seq-indifferentiability of the 3-round IEM cipher is as follows.

Theorem 3. *Let* $N = 2^n$. *For any integers* μ *and* q *such that* $\mu q \leq N/4$, *the 3-round IEM construction* $\mathsf{EM}[n, 3, \mathbf{f}]$ *with independent permutations and the trivial key-schedule* $\mathbf{f}$ *is* $(\mu, q, \sigma, t, \varepsilon)$-*KK-seq-indifferentiable from an ideal cipher with n-bit blocks and n-bit keys, with*

$$\sigma = \mu q, \quad t = \mathcal{O}(\mu q), \quad and \quad \varepsilon = \frac{57\mu^2 q^2}{N}.$$

As a corollary, we obtain from Theorem 1 that for any m-ary relation $\mathcal{R}$ which is μ-restricted and $(\mu q, \varepsilon)$-evasive w.r.t. an ideal cipher (and assuming q is large compared with $c = 3$ and m), the 3-round IEM cipher is $(q, \varepsilon + \mathcal{O}(\mu^2 q^2/2^n))$-correlation intractable with respect to $\mathcal{R}$.

It is also known [21] that for stateless ideal primitives (i.e., primitives whose answers do not depend on the order of the queries it receives), seq-indifferentiability implies public indifferentiability [10,29], a variant of indifferentiability where the simulator gets to know all queries of the distinguisher to the ideal primitive E. Since an ideal cipher is stateless, Theorem 3 implies that the 3-round IEM construction is also KK-publicly indifferentiable from an ideal cipher.

PROOF IDEA. The proof of Theorem 3 is very similar to the proof of (full, not KK) seq-indifferentiability for the 4-round IEM construction of [7]. The

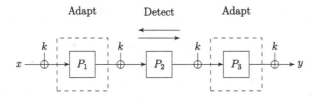

Fig. 4. Detection and adaptations zones used by the simulator for proving KK-seq-indifferentiability of the 3-round iterated Even-Mansour construction from an ideal cipher.

main difference in the simulation strategy is the following: in the full seq-indifferentiability setting, the simulator has no hint about which key(s) the adversary is using to try to distinguish the real world from the ideal (simulated) world. Hence, it uses a 2-round "detection" zone in the middle made of permutations P_2 and P_3, which allows, given a query to P_2 (say, $P_2(u_2) = v_2$) and a query to P_3 (say, $P_3(u_3) = v_3$), to deduce the key associated to this "chain" of queries (namely, $k = v_2 \oplus u_3$). Permutations P_1 and P_4 are then used to "adapt" these detected chains and make them match the ideal cipher E. In the KK-setting, the simulator knows the set $\mathcal{K}'$ of keys that the distinguisher is allowed to use in its ideal cipher queries. Hence, the detection zone can be reduced to one single round (the middle one, i.e. P_2 for the 3-round IEM): each time the distinguisher makes a query to P_2, the simulator completes the μ chains corresponding to this query and *each key* $k \in \mathcal{K}'$, again using extremal round P_1 and P_3 to adapt the chains (see Fig. 4).

We only give an informal description of the simulator here and defer the formal description in pseudocode and the full proof of Theorem 3 to the full version of the paper [8]. The simulator is given the subset $\mathcal{K}'$ of keys that the distinguisher is bound to use. It offers an interface $\mathsf{Query}(i, \delta, w)$ to the distinguisher for querying the internal permutations, where $i \in \{1, 2, 3\}$ names the permutation, $\delta \in \{+, -\}$ indicates whether this a direct or inverse query, and $w \in \{0, 1\}^n$ is the actual value queried. For each $i = 1, \ldots, 3$, the simulator internally maintains a table Π_i reflecting which values have been already internally set for each simulated permutation. Each table maps entries $(\delta, w) \in \{+, -\} \times \{0, 1\}^n$ to values $w' \in \{0, 1\}^n$, initially undefined for all entries. We denote Π_i^+, resp. Π_i^-, the (time-dependent) sets of strings $w \in \{0, 1\}^n$ such that $\Pi_i(+, w)$, resp. $\Pi_i(-, w)$, is defined. When the simulator receives a query (i, δ, w), it checks in table Π_i whether the corresponding answer $\Pi_i(\delta, w)$ is already defined. When this is the case, it returns the answer to the distinguisher and waits for the next query. Otherwise, it randomly draws an answer $w' \in \{0, 1\}^n$ and defines $\Pi_i(\delta, w) := w'$ as well as the answer to the opposite query $\Pi_i(\bar{\delta}, w') := w$. The randomness used by the simulator is made explicit through a tuple of random permutations $\mathbf{P} = (P_1, P_2, P_3)$ with $P_i := \{+, -\} \times \{0, 1\}^n \to \{0, 1\}^n$, and for any $u, v \in \{0, 1\}^n$, $P_i(+, u) = v \Leftrightarrow P_i(-, v) = u$. We assume that the tuple (P_1, P_2, P_3) is drawn uniformly at random at the beginning of the experiment, but we note that $\mathcal{S}$ could equivalently lazily sample these permutations through-out its execution. Then w' is simply defined by the simulator as $w' := P_i(\delta, w)$.[4]

Before returning w' to the distinguisher, the simulator takes additional steps to ensure that the whole IEM construction matches the ideal cipher E by running a *chain completion* mechanism. Namely, if the distinguisher called $\mathsf{Query}(i, \delta, w)$ with $i = 2$, the simulator completes the "chains" for each known key $k \in \mathcal{K}'$ by executing a procedure $\mathsf{CompleteChain}(u_2, v_2, k, \ell)$, where ℓ indicates where the chain will be "adapted" and (u_2, v_2) is the pair of values that was just added to Π_2. For example, assume that the distinguisher called $\mathsf{Query}(2, +, u_2)$ and that the answer randomly chosen by the simulator was v_2. Then for each $k \in \mathcal{K}'$, the simulator computes the corresponding value $u_3 = v_2 \oplus k$, and evaluates the IEM construction backward, letting $v_1 := u_2 \oplus k$, $u_1 := \Pi_1(-, v_1)$ (setting this value at random in case it was not in Π_1), $x := u_1 \oplus k$, $y := E(+, k, x)$ (hence making a query to E to "wrap around"), and $v_3 := y \oplus k$, until the corresponding input/output values (u_3, v_3) for the third permutation are defined. It then "adapts" (rather than setting randomly) table Π_3 by calling procedure $\mathsf{ForceVal}(u_3, v_3, 3)$ which sets $\Pi_3(+, u_3) := v_3$ and $\Pi_3(-, v_3) := u_3$ in order to ensure consistency of the simulated IEM construction with E. (A crucial point of the proof will be to show that this does not cause an overwrite, i.e., that these two values are undefined before the adaptation occurs.) In case the query was to $\mathsf{Query}(2, -, \cdot)$, the behavior of the simulator is symmetric, namely adaptation of the chain takes place in table Π_1.

6   KK-Indifferentiability for Nine Rounds

In this section, we show that nine rounds of the IEM construction are sufficient to achieve μ-KK-indifferentiability from an ideal cipher. Note that this is less than what is currently known to be sufficient to achieve full indifferentiability from an ideal cipher, namely twelve rounds, as shown by Lampe and Seurin [20]. We conjecture that four rounds are actually sufficient.

We use the same technique as in Sect. 5 for going from four rounds for seq-indifferentiability to three rounds for KK-seq-indifferentiability: we start from the 12-round simulator of [20], and shorten the detection zones using the fact that the simulator knows the subset of keys used by the distinguisher.

We only give an informal description of the simulator and sketch how to modify the indifferentiability proof of [20], so that the result should rather be considered as a (substantiated) conjecture. (Given that nine is unlikely to be the minimal number of rounds needed to achieve μ-KK-indifferentiability, and that we already known that twelve rounds are sufficient to achieve full indifferentiability and hence μ-KK-indifferentiability, the benefit of writing down the full proof is rather low.) The high-level principle of how the simulator works is similar to

---

[4] Note that for $i = 1$ and $i = 3$, this is not equivalent to letting $w' \leftarrow_\$ \{0,1\}^n \setminus \Pi_i^{\bar{\delta}}$ since the simulator sometimes "adapts" the value of these tables, so that the tables Π_i and the permutations P_i will differ (with overwhelming probability) on adapted entries.

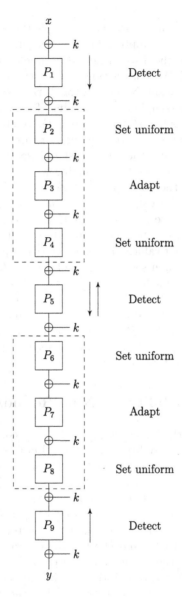

Fig. 5. Detection and adaptation zones used by the simulator for proving KK-indifferentiability of the 9-round iterated Even-Mansour construction from an ideal cipher.

Sect. 5 except that there are now additional detection zones besides the middle one preventing the distinguisher from creating "wrap around" chains (remember that the distinguisher is not bound to be sequential here, so it can make an ideal cipher query $y := E(+, k, x)$ and evaluate the IEM construction from both extremities by making permutation queries until the simulator is trapped into a contradiction). Moreover, since the simulator can now recurse (i.e., completing a chain can create new chains to be completed), it uses a queue of chains detected and to be completed as in [20].

As before, the simulator reacts on any query to P_5, and completes the chains for any key $k \in \mathcal{K}'$ by adapting at P_7 if this is a direct query and adapting at P_3 if this is an inverse query. Moreover, the simulator also reacts on direct queries to P_1 or inverse queries to P_9. Let us consider the case of a query $P_1(+, u_1)$. Then for each key $k \in \mathcal{K}'$, the simulator computes $x := u_1 \oplus k$, queries $y := E(+, k, x)$, lets $v_9 := y \oplus k$, and checks if $v_9 \in \Pi_9^-$. If this is the case, then the chain (u_1, k) is enqueued to be completed and adapted at P_3. For an inverse query to P_9, adaptation takes place at P_7. As in [20], the four "buffer" rounds P_2, P_4, P_6 and P_8 surrounding adaptation rounds ensure that no collision can occur when adapting distinct chains.

The analysis of this simulator then follows the same lines as in [20]. Its complexity can be upper bounded as follows: first, one applies the standard argument that the number of wrap-around chains that will be detected is upper bounded (with very high probability) by the number of ideal cipher queries of the distinguisher, hence by q. This implies that the size of table Π_5 is always at most $2q$ (since it increases only because of a distinguisher's query or when completing a wrap-around chain). It follows that the number of middle chains completed is at most $2\mu q$, and the size of all tables Π_i for $i \neq 5$ is at most $q + q + 2\mu q = 2(\mu + 1)q$. Also, the number of calls made by the simulator to the ideal cipher can be upper bounded by $2\mu q$ (number of middle chains that are completed), plus $4\mu(\mu + 1)q$ (number of wrap-around chains that are checked), hence it is $O(\mu^2 q)$ (the running time is similar).

Finally, proving a rigorous upper bound on the distinguishing advantage is a cumbersome task that remains to be done. A rough estimation following the lines of [20] would be that bad events that would make the simulator to overwrite a value when adapting chains (which is what dominates the security bound) happen with probability at most $(\max|\Pi_i|)^6/2^n$, hence $O(\mu^6 q^6)$.

References

1. Andreeva, E., Bogdanov, A., Dodis, Y., Mennink, B., Steinberger, J.P.: On the indifferentiability of key-alternating ciphers. In: Canetti, R., Garay, J.A. (eds.) CRYPTO 2013, Part I. LNCS, vol. 8042, pp. 531–550. Springer, Heidelberg (2013). http://eprint.iacr.org/2013/061
2. Andreeva, E., Bogdanov, A., Mennink, B.: Towards understanding the known-key security of block ciphers. In: Moriai, S. (ed.) FSE 2013. LNCS, vol. 8424, pp. 348–366. Springer, Heidelberg (2014)
3. Bogdanov, A., Knudsen, L.R., Leander, G., Standaert, F.-X., Steinberger, J., Tischhauser, E.: Key-alternating ciphers in a provable setting: encryption using a small number of public permutations. In: Pointcheval, D., Johansson, T. (eds.) EUROCRYPT 2012. LNCS, vol. 7237, pp. 45–62. Springer, Heidelberg (2012)

4. Canetti, R., Goldreich, O., Halevi, S.: The random oracle methodology, revisited (preliminary version). In: Symposium on Theory of Computing - STOC 1998, pp. 209–218. ACM (1998). Full version available at http://arxiv.org/abs/cs.CR/0010019

5. Chen, S., Lampe, R., Lee, J., Seurin, Y., Steinberger, J.: Minimizing the two-round Even-Mansour cipher. In: Garay, J.A., Gennaro, R. (eds.) CRYPTO 2014, Part I. LNCS, vol. 8616, pp. 39–56. Springer, Heidelberg (2014). http://eprint.iacr.org/2014/443

6. Chen, S., Steinberger, J.: Tight security bounds for key-alternating ciphers. In: Nguyen, P.Q., Oswald, E. (eds.) EUROCRYPT 2014. LNCS, vol. 8441, pp. 327–350. Springer, Heidelberg (2014). http://eprint.iacr.org/2013/222

7. Cogliati, B., Seurin, Y.: On the provable security of the iterated Even-Mansour cipher against related-key and chosen-key attacks. In: Oswald, E., Fischlin, M. (eds.) EUROCRYPT 2015. LNCS, vol. 9056, pp. 584–613. Springer, Heidelberg (2015). http://eprint.iacr.org/2015/069

8. Cogliati, B., Seurin, Y.: Strengthening the Known-Key Security Notion for Block Ciphers. Full version of this paper. http://eprint.iacr.org/2016/394

9. Demay, G., Gaži, P., Hirt, M., Maurer, U.: Resource-restricted indifferentiability. In: Johansson, T., Nguyen, P.Q. (eds.) EUROCRYPT 2013. LNCS, vol. 7881, pp. 664–683. Springer, Heidelberg (2013). http://eprint.iacr.org/2012/613

10. Dodis, Y., Ristenpart, T., Shrimpton, T.: Salvaging Merkle-Damgård for practical applications. In: Joux, A. (ed.) EUROCRYPT 2009. LNCS, vol. 5479, pp. 371–388. Springer, Heidelberg (2009)

11. Dunkelman, O., Keller, N., Shamir, A.: Minimalism in cryptography: the Even-Mansour scheme revisited. In: Pointcheval, D., Johansson, T. (eds.) EUROCRYPT 2012. LNCS, vol. 7237, pp. 336–354. Springer, Heidelberg (2012)

12. Even, S., Mansour, Y.: A construction of a cipher from a single pseudorandom permutation. J. Cryptol. **10**(3), 151–162 (1997)

13. Farshim, P., Procter, G.: The related-key security of iterated Even–Mansour ciphers. In: Leander, G. (ed.) FSE 2015. LNCS, vol. 9054, pp. 342–363. Springer, Heidelberg (2015). http://eprint.iacr.org/2014/953

14. Gilbert, H.: A simplified representation of AES. In: Sarkar, P., Iwata, T. (eds.) ASIACRYPT 2014. LNCS, vol. 8873, pp. 200–222. Springer, Heidelberg (2014)

15. Gilbert, H., Peyrin, T.: Super-sbox cryptanalysis: improved attacks for AES-like permutations. In: Hong, S., Iwata, T. (eds.) FSE 2010. LNCS, vol. 6147, pp. 365–383. Springer, Heidelberg (2010)

16. Guo, C., Lin, D.: A synthetic indifferentiability analysis of interleaved double-key Even-Mansour ciphers. In: Iwata, T., Cheon, J.H. (eds.) ASIACRYPT 2015. LNCS, vol. 9453, pp. 389–410. Springer, Heidelberg (2015). doi:10.1007/978-3-662-48800-3_16

17. Guo, C., Lin, D.: On the indifferentiability of key-alternating Feistel ciphers with no key derivation. In: Dodis, Y., Nielsen, J.B. (eds.) TCC 2015, Part I. LNCS, vol. 9014, pp. 110–133. Springer, Heidelberg (2015)

18. Knudsen, L.R., Rijmen, V.: Known-key distinguishers for some block ciphers. In: Kurosawa, K. (ed.) ASIACRYPT 2007. LNCS, vol. 4833, pp. 315–324. Springer, Heidelberg (2007)

19. Lampe, R., Patarin, J., Seurin, Y.: An asymptotically tight security analysis of the iterated Even-Mansour cipher. In: Wang, X., Sako, K. (eds.) ASIACRYPT 2012. LNCS, vol. 7658, pp. 278–295. Springer, Heidelberg (2012)

20. Lampe, R., Seurin, Y.: How to construct an ideal cipher from a small set of public permutations. In: Sako, K., Sarkar, P. (eds.) ASIACRYPT 2013, Part I. LNCS, vol. 8269, pp. 444–463. Springer, Heidelberg (2013). http://eprint.iacr.org/2013/255

21. Mandal, A., Patarin, J., Seurin, Y.: On the public indifferentiability and correlation intractability of the 6-round Feistel construction. In: Cramer, R. (ed.) TCC 2012. LNCS, vol. 7194, pp. 285–302. Springer, Heidelberg (2012). http://eprint.iacr.org/2011/496

22. Maurer, U.M., Renner, R.S., Holenstein, C.: Indifferentiability, impossibility results on reductions, and applications to the random oracle methodology. In: Naor, M. (ed.) TCC 2004. LNCS, vol. 2951, pp. 21–39. Springer, Heidelberg (2004)

23. Minier, M., Phan, R.C.-W., Pousse, B.: Distinguishers for ciphers and known key attack against Rijndael with large blocks. In: Preneel, B. (ed.) AFRICACRYPT 2009. LNCS, vol. 5580, pp. 60–76. Springer, Heidelberg (2009)

24. Nikolić, I., Pieprzyk, J., Sokołowski, P., Steinfeld, R.: Known and chosen key differential distinguishers for block ciphers. In: Rhee, K.-H., Nyang, D.H. (eds.) ICISC 2010. LNCS, vol. 6829, pp. 29–48. Springer, Heidelberg (2011)

25. Ristenpart, T., Shacham, H., Shrimpton, T.: Careful with composition: limitations of the indifferentiability framework. In: Paterson, K.G. (ed.) EUROCRYPT 2011. LNCS, vol. 6632, pp. 487–506. Springer, Heidelberg (2011)

26. Rogaway, P., Steinberger, J.P.: Constructing cryptographic hash functions from fixed-key blockciphers. In: Wagner, D. (ed.) CRYPTO 2008. LNCS, vol. 5157, pp. 433–450. Springer, Heidelberg (2008)

27. Rogaway, P., Steinberger, J.P.: Security/efficiency tradeoffs for permutation-based hashing. In: Smart, N.P. (ed.) EUROCRYPT 2008. LNCS, vol. 4965, pp. 220–236. Springer, Heidelberg (2008)

28. Sasaki, Y., Yasuda, K.: Known-key distinguishers on 11-round Feistel and collision attacks on its hashing modes. In: Joux, A. (ed.) FSE 2011. LNCS, vol. 6733, pp. 397–415. Springer, Heidelberg (2011)

29. Yoneyama, K., Miyagawa, S., Ohta, K.: Leaky random oracle. IEICE Trans. **92–A**(8), 1795–1807 (2009)

Related-Key Almost Universal Hash Functions: Definitions, Constructions and Applications

Peng Wang[1,2]($\boxtimes$), Yuling Li[1,2,3], Liting Zhang[4], and Kaiyan Zheng[1,2,3]

[1] State Key Laboratory of Information Security,
Institute of Information Engineering, Chinese Academy of Sciences, Beijing, China
wp@is.ac.cn, {liyuling,zhengkaiyan}@iie.ac.cn
[2] Data Assurance and Communication Security Research Center,
Chinese Academy of Sciences, Beijing, China
[3] University of Chinese Academy of Sciences, Beijing, China
[4] State Key Laboratory of Computer Science,
Trusted Computing and Information Assurance Laboratory,
Institute of Software, Chinese Academy of Sciences, Beijing, China
zhangliting@tca.iscas.ac.cn

Abstract. Universal hash functions (UHFs) have been extensively used in the design of cryptographic schemes. If we consider the related-key attack (RKA) against these UHF-based schemes, some of them may not be secure, especially those using the key of UHF as a part of the whole key of scheme, due to the weakness of UHF in the RKA setting. In order to solve this issue, we propose a new concept of related-key almost universal hash function, which is a natural extension to almost universal hash function in the RKA setting. We define related-key almost universal (RKA-AU) hash function and related-key almost XOR universal (RKA-AXU) hash function. However almost all the existing UHFs do not satisfy the new definitions. We construct one fixed-input-length universal hash function named RH1 and two variable-input-length universal hash functions named RH2 and RH3. We show that RH1 and RH2 are both RKA-AXU, and RH3 is RKA-AU for the RKD set $\Phi^\oplus$. Furthermore, RH1, RH2 and RH3 are nearly as efficient as previously similar constructions. RKA-AU (RKA-AXU) hash functions can be used as components in the related-key secure cryptographic schemes. If we replace the universal hash functions in the schemes with our corresponding constructions, the problems about related-key attack can be solved for some RKD sets. More specifically, we give four concrete applications of RKA-AU and RKA-AXU in related-key secure message authentication codes and tweakable block ciphers.

Keywords: Almost universal hash function · Related-key attack · Related-key almost universal hash function · Message authentication code · Tweakable block cipher

© International Association for Cryptologic Research 2016
T. Peyrin (Ed.): FSE 2016, LNCS 9783, pp. 514–532, 2016.
DOI: 10.1007/978-3-662-52993-5_26

1   Introduction

Universal Hash Functions. Ever since introduced by Carter and Wegman [15,52] in the design of message authentication code (MAC), *universal hash functions* (UHFs) have become common components in numerous cryptographic constructions, especially in modes of operation, to provide security services as confidentiality, authenticity or both. A universal hash function (UHF) is a family of functions indexed by keys. Unlike other components such as block ciphers, keyed hash functions and permutations, which are often used as pseudorandom permutations (PRPs), pseudorandom functions (PRFs) and public random permutations respectively, UHFs have no cryptographic strength such as pseudorandomness. So UHFs usually come along with other primitives, such as PRPs, PRFs, etc., to set up cryptographic schemes. The basic property of UHF is that the collision probability of hash values from any two different messages is small when the key is uniformly random.

One of examples is the polynomial evaluation hash function [8] in which the variable is the key and the coefficients consist of message blocks, such as: $Poly : \{0,1\}^n \times \{0,1\}^{nm} \to \{0,1\}^n$,

$$Poly_K(M) = M_1 K^m \oplus M_2 K^{m-1} \oplus \cdots \oplus M_m K \tag{1}$$

where $M = M_1 \| M_2 \| \cdots \| M_m \in \{0,1\}^{nm}$, $M_i \in \{0,1\}^n$, $i = 1, 2, \cdots, m$ and all the operations are in the finite field $GF(2^n)$. This kind of UHF appears in GCM [37], XCB [29], HCTR [50], HCH [16,17], COBRA [2], Enchilada [27], POET [1] and many other constructions. For any $M \neq M'$, $Poly_K(M) \oplus Poly_K(M')$ is a polynomial in K whose degree is nonzero and no more than m, so there are at most m keys leading to $Poly_K(M) = Poly_K(M')$, that is the collision probability is at most $m/2^n$ when K is uniformly random. We say that this hash function is $m/2^n$-almost-universal (AU). Obviously the probability of $Poly_K(M) \oplus Poly_K(M') = C$ is also at most $m/2^n$ for any $M \neq M'$ and C. That is another commonly used concept: almost XOR universal (AXU) hash functions. *Poly* is also $m/2^n$-AXU.

A direct application of UHFs is in message authentication codes (MACs) in which the message is hashed by the UHF into a short digest which then encrypted into a tag. MACs of this kind have been standardized in ISO/IEC 9797-3:2011 [31] which includes UMAC [13], Badger [14], Poly1305-AES [6] and GMAC [37]. UHFs are also used in tweakable block ciphers (TBCs) [36] and tweakable enciphering schemes (TESes), e.g. XTS-AES in IEEE Std 1619-2007 [28] and NIST SP 800-38E [40], XCB in IEEE Std 1619.2-2010 [29], HCTR [50] and HCH [16,17], etc. The third application of UHF is in authenticated encryption (AE) schemes, e.g. the most widely used AE scheme GCM [37] standardized in ISO/IEC-19772:2009 [30] and NIST SP 800-38D [39]. In the recent CAESAR competition, several UHF-based AE schemes were proposed, e.g. COBRA [2], Enchilada [27] and POET [1], etc. In the security proofs of all these schemes, a crucial point is the collision probability about the inputs to other primitives. The property of UHF guarantees that the collision seldom happens.

Related-Key Attacks. Related-key attack (RKA) was firstly introduced by Biham [10] against block ciphers [12,22,48] and then extended to other cryptographic algorithms such as stream ciphers [18], MACs [41], TESes [49], AE schemes [21], etc. Bellare and Kohno [5] firstly gave a theoretical study of related-key security of block cipher, modeling the concept of pseudorandom permutation in the RKA setting (RKA-PRP) and pseudorandom function in the RKA setting (RKA-PRF). Applebaum et al. [3] gave the related-key security definition of encryption. Bhattacharyya and Roy [9] gave the related-key security definition of MAC. Related-key security has become an important criteria for cryptographic constructions.

In the RKA setting, the adversary does not know the secret key as in the usual *invariable-key* setting, but can apply related-key-deriving (RKD) transformations to change the secret key and observe outputs under the related keys. Let Φ be a RKD set which consists of transformations on the key space $\mathcal{K} = \{0, 1\}^k$. There are two canonical RKD sets: $\Phi^{\oplus} = \{XOR_{\Delta} : K \mapsto K \oplus \Delta, \Delta \in \mathcal{K}\}$ and $\Phi^{+} = \{ADD_{\delta} : K \mapsto K + \delta \mod 2^k, \delta \in \mathcal{K}\}$. In the following, we use $\Phi^{\oplus}$ as the default RKD set unless specified otherwise.

The related-key security requires that the queries under the related keys do not threaten the security under the original key, as the definition of related-key unforgeability in [9]. Or more strictly, for different related keys, the corresponding algorithms are secure independently, as the definition of RKA-PRP in [5] and [3].

Motivations. *How to guarantee the related-key security? An intuition is that if the underlying components are related-key secure, the upper constructions should be related-key secure.* This is true for most of block cipher modes of operation, especially for those one-key modes whose key is also that of the underlying block cipher, including CBC, OFB, CFB, CTR, CMAC, OCB, etc. But for the UHF-based schemes, it is not the case. Although almost all the UHF-based schemes have security proofs in the usual invariable-key setting, there are a lot of examples showing that some of them can not resist related-key attacks.

Let's first check UHF-based MACs, in which a typical construction is to encrypt the hash value into a tag by one-time-pad encryption. This method originates from Carter and Wegman [15,52] and dominates the usages of UHF in MACs [31]. Consider a simple example: $MAC_{K,K'}(N, M) = Poly_K(M) \oplus F_{K'}(N)$ where $M = M_1 \| M_2 \in \{0, 1\}^{2n}$, $Poly_K(M_1 \| M_2) = M_1 K^2 \oplus M_2 K$, F is a function often instantiated by a block cipher and N is a nonce. It has been proved that [7,44] if F is a PRF and $Poly$ is almost XOR universal, MAC is secure.

But if we query with $A \| A$ under the related key $(K \oplus 0^{n-1}1, K')$, the answer is $T = (A(K \oplus 0^{n-1}1)^2 \oplus A(K \oplus 0^{n-1}1)) \oplus F_{K'}(N) = (AK^2 \oplus AK)) \oplus F_{K'}(N)$. Therefore we can predict that the tag of $A \| A$ under the original key is also T. So $(N, A \| A, T)$ is a successful forgery which breaks the RKA security of the MAC. A similar attack can apply to Poly1305-AES [6] in ISO/IEC 9797-3:2011 [31].

In Appendix B, we give more RKA examples against TBC, TES and AE schemes using *Poly* as UHF components. In all these examples, the key of UHF is a part of the key of whole scheme, so that the adversary can derive the related key of UHF and get input collisions to other primitives such as PRPs or PRFs. The collision in the MAC example is $Poly_{K \oplus 0^{n-1}1}(A\|A) = Poly_K(A\|A)$. We stress that all these attacks only use the properties of UHF in the RKA setting and have nothing to do with other underlying primitives, whether it is RKA secure or not. *In other words, the related-key weaknesses of the UHF alone results in related-key attacks against the schemes.*

Definitions. In order to prevent the above attacks, we propose a new concept of related-key almost universal hash function which can ensure that the above collisions seldom happen. The new concept is a natural extension to almost universal hash function in the RKA setting. We define *related-key almost universal* (RKA-AU) hash function and *related-key almost XOR universal* (RKA-AXU) hash function. We will show that these definitions solve the above problems for some RKD set. Unfortunately almost all the existing UHFs do not satisfy the new definitions, including *Poly* mentioned in the above, MMH [26], Square Hash [23], NMH [26] and NH [13], etc. See Appendix C for details.

Constructions. We construct one fixed-input-length universal hash function named RH1 and two variable-input-length universal hash functions named RH2 and RH3. We prove that RH1 and RH2 are both RKA-AXU, and RH3 is RKA-AU for the RKD set $\Phi^\oplus$. Furthermore, RH1, RH2 and RH3 are almost as efficient as previous constructions.

Applications. If we replace the universal hash functions in the examples of Sect. 1 with our constructions, the problems about related-key attacks for some RKD set can be solved. More specifically, we give four concrete examples in MACs and TBCs.

2   Definitions

For a finite set $\mathcal{S}$, $x \xleftarrow{\$} \mathcal{S}$ means selecting an element x uniformly at random from the set X. For a string M, $|M|$ denotes the bit length of M. For $b \in \{0,1\}$, b^m denotes m bits of b. $\mathbb{A}^{\mathcal{O}} \Rightarrow b$ denotes that the algorithm $\mathbb{A}$ with an oracle $\mathcal{O}$ outputs b.

For a function $H : \mathcal{K} \times \mathcal{D} \to \mathcal{R}$, when $K \in \mathcal{K}$ is a key, we write $H(K, M)$ as $H_K(M)$, where $(K, M) \in \mathcal{K} \times \mathcal{D}$. The following are the usual definitions of UHF.

Definition 1 (AU [46]). *H is an ϵ-almost-universal (ϵ-AU) hash function, if for any $M, M' \in \mathcal{D}$, $M \neq M'$,*

$$\Pr[K \xleftarrow{\$} \mathcal{K} : H_K(M) = H_K(M')] \leq \epsilon.$$

When ϵ is negligible we say that H is AU.

Definition 2 (AXU [34]**).** *Let $(\mathcal{R}, \oplus)$ be an abelian group[1]. H is an ϵ-almost-XOR-universal (ϵ-AXU), if for any $M, M' \in \mathcal{D}$, $M \neq M'$, and $C \in \mathcal{R}$,*

$$\Pr[K \overset{\$}{\leftarrow} \mathcal{K} : H_K(M) \oplus H_K(M') = C] \leq \epsilon.$$

When ϵ is negligible we say that H is AXU.

Clearly, if H is ϵ-AXU, it is also ϵ-AU, for ϵ-AU is a special case of ϵ-AXU when $C = 0$.

RKA-AU and RKA-AXU. In the following, we extend the above definitions in the RKA setting. Let Φ be a RKD set.

Definition 3 (RKA-AU). *H is an ϵ-related-key-almost-universal (ϵ-RKA-AU) hash function for the RKD set Φ, if for any $\phi, \phi' \in \Phi$, $M, M' \in \mathcal{D}$, $(\phi, M) \neq (\phi', M')$,*

$$\Pr[K \overset{\$}{\leftarrow} \mathcal{K} : H_{\phi(K)}(M) = H_{\phi'(K)}(M')] \leq \epsilon.$$

When ϵ is negligible we say that H is RKA-AU for Φ.

Definition 4 (RKA-AXU). *Let $(\mathcal{R}, \oplus)$ be an abelian group. H is an ϵ-related-key-almost-universal (ϵ-RKA-AXU) hash function for the RKD set Φ, if for any $\phi, \phi' \in \Phi$, $M, M' \in \mathcal{D}$, $(\phi, M) \neq (\phi', M')$, and $C \in \mathcal{R}$,*

$$\Pr[K \overset{\$}{\leftarrow} \mathcal{K} : H_{\phi(K)}(M) \oplus H_{\phi'(K)}(M') = C] \leq \epsilon.$$

When ϵ is negligible we say that H is RKA-AXU for Φ.

For $\phi, \phi' \in \Phi$, $\phi \neq \phi'$ means there exists a key $K \in \mathcal{K}$ such that $\phi(K) \neq \phi'(K)$.

Restricting RKD Sets. As in the discussion of RKA-PRP [5], the related-key properties of UHF are relevant to the choice of RKD set. For some RKD sets the related-key almost universal hash function may not exist. It is necessary that the RKD set is both *output unpredictable* and *collision resistant*. We must put some restrictions on the RKD set.

(1) Output unpredictability. A $\phi \in \Phi$ that has predictable outputs if there exists a constant S such that the probability of $\phi(K) = S$ is high. If it happens, then for any function H the probability of $H_{\phi(K)}(M) \oplus H_{\phi(K)}(M') = H_S(M) \oplus H_S(M')$ is also high for any two distinct M and M'. So the RKA-AXU function is not available for the RKD set which has predictable transformations. We define $OU(\Phi) = max_{\phi \in \Phi, S}\Pr[K \overset{\$}{\leftarrow} \mathcal{K} : \phi(K) = S]$. If $OU(\Phi)$ is negligible, we say that Φ is *output unpredictable*.

---

[1] For arbitrary abelian groups a generalized notion is almost Delta universal (AΔU) hash function [47]. In the following when we say AXU we may sometimes refer to AΔU.

(2) Collision resistance. Two distinct $\phi, \phi' \in \Phi$ have high collision probability
if the probability of $\phi(K) = \phi'(K)$ is hight. If it happens, then for any
function H the probability of $H_{\phi(K)}(M) \oplus H_{\phi'(K)}(M) = 0$ is also high
for any M. So neither the RKA-AXU nor RKA-AU function is available
for the RKD set which has high collision probability. We define $CR(\Phi) =$
$max_{\phi,\phi' \in \Phi, \phi \neq \phi'} \Pr[K \xleftarrow{\$} \mathcal{K} : \phi(K) = \phi'(K)]$. If $CR(\Phi)$ is negligible, we say
that Φ is *collision resistant*. More strictly, if for any two distinct $\phi, \phi' \in \Phi$
and any key K, we have $\phi(K) \neq \phi'(K)$, or in other words $CR(\Phi) = 0$, we
say that Φ is *claw-free*.

We note that $\Phi^\oplus$ and Φ^+ are output unpredictable, collision resistant and
claw-free. The example in Sect. 1 shows that $Poly$ is not RKA-AXU for the
RKD set $\Phi^\oplus$. If we choose the message M to be 0^{mn}, $Poly_K(M)$ will always be
0^n. Therefore for any $\phi, \phi' \in \Phi$, we have $Poly_{\phi(K)}(0^{mn}) = Poly_{\phi'(K)}(0^{mn})$. So
$Poly$ is not RKA-AU either. If we look at the other existing UHFs, unfortunately
almost all of them do not satisfy the new definitions, including MMH [26], Square
Hash [23], NMH [26] and NH [13], etc. See Appendix C for more details.

3   Constructions

We construct two types of related-key almost universal hash functions: one fixed-
input-length (FIL) UHF named RH1 and two variable-input-length (VIL) UHFs
named RH2 and RH3. We prove that RH1 and RH2 are both RKA-AXU, and
RH3 is RKA-AU, for the RKD set $\Phi^\oplus$.

For a function $F : \mathcal{K} \times \mathcal{D} \to \mathcal{R}$, we define a new function $F' : \mathcal{K} \times (\mathcal{K} \times \mathcal{D}) \to \mathcal{R}$

$$F'_K(\Delta, M) = F_{K \oplus \Delta}(M).$$

It is easy to see that F is RKA-AU (RKA-AXU) for the RKD set $\Phi^\oplus$ if and
only if F' is AU (AXU). All the constructions are based on the polynomial
evaluation function $Poly$. From the above observation, our main idea is to modify
$Poly_K(M)$ into $F_K(M)$ such that $F_{K \oplus \Delta}(M)$ is still an almost (XOR) universal
hash function.

FIL Constructions. We first construct a function based on $Poly_K(M) = MK$
by adding a new term K^3.

Construction 1. RH1 : $\{0,1\}^n \times \{0,1\}^n \to \{0,1\}^n$,

$$\text{RH1}_K(M) = MK \oplus K^3. \tag{2}$$

Theorem 1. RH1 *is* $2/2^n$-*RKA-AXU for the RKD set* $\Phi^\oplus$.

Proof. We prove that for any $M, M', \Delta_1, \Delta_2 \in \{0,1\}^n$, $(\Delta_1, M) \neq (\Delta_2, M')$, and
$C \in GF(2^n)$, $\Pr[K \xleftarrow{\$} \{0,1\}^n : F(K) = C] \leq \epsilon$, where $F(K) = \text{RH1}_{K \oplus \Delta_1}(M) \oplus$
$\text{RH1}_{K \oplus \Delta_2}(M')$. We have

$$F(K) = (\Delta_1 \oplus \Delta_2)K^2 \oplus (\Delta_1^2 \oplus \Delta_2^2 \oplus M \oplus M')K \oplus (\Delta_1^3 \oplus \Delta_2^3 \oplus M\Delta_1 \oplus M'\Delta_2).$$

If $\Delta_2 \neq \Delta_1$, $F(K) = C$ has two roots at most. If $\Delta_1 = \Delta_2$, then $M \neq M'$. The degree of $F(K)$ is 1 and $F(K) = C$ has one root. Therefore RH1 is $2/2^n$-RKA-AXU. □

Remark 1. As one of reviewers points out that RH1 is RKA-AXU for the RKD set $\Phi^{\oplus}$, but is not RKA-AXU or even RKA-AU for a RKD set containing just containing two transformation: $\Phi = \{id, f_\alpha\}$ where id is the identity transformation and $f_\alpha(K) = \alpha K$, $\alpha^3 = 1$. It is easy to verify that $\text{RH1}_{f_\alpha(K)}(\alpha^{-1}M) = \text{RH1}_K(M)$.

Remark 2. More generally we consider polynomial $H_K^{i,j}(M) = MK^i + K^j$ over the finite field $GF(2^n)$ or $GF(p)$ where i, j are integers and p is a prime. We show the results when $1 \leq i, j \leq 4$ in Table 1.

Table 1. For $H_K^{i,j}(M) = MK^i + K^j$, "11" means it is RKA-AU and RKA-AXU for the RKD set $\Phi^{\oplus}$, "10" means it is RKA-AU but not RKA-AXU, and "00" means it is neither RKA-AU nor RKA-AXU.

(i,j)	(1,1)	(1,2)	(1,3)	(1,4)	(2,1)	(2,2)	(2,3)	(2,4)	(3,1)	(3,2)	(3,3)	(3,4)	(4,1)	(4,2)	(4,3)	(4,4)
$GF(2^n)$	00	00	11	00	00	00	11	00	10	10	00	10	00	00	11	00
$GF(p)$	00	00	11	11	10	00	10	11	00	11	00	11	10	11	11	00

VIL Constructions. *Poly* does not support variable input length. For any message $M \in \{0,1\}^*$, a general padding method as in [37] is to firstly pad minimum zeroes to make the length multiple of the block length and then pad the bit length of M as the last block:

$$pad(M) = M\|0^i\|\,|M|.$$

Then $Poly_K(pad(M))$ is variable-input-length AXU hash function but still is not RKA-AU (RKA-AXU). Following the above method we add some term K^i in order to get the RKA-AXU property.

Construction 2. RH2 : $\{0,1\}^n \times \{0,1\}^* \to \{0,1\}^n$,

$$\text{RH2}_K(M) = \begin{cases} K^{l+2} \oplus Poly_K(pad(M)), & l \text{ is odd} \\ K^{l+3} \oplus Poly_K(pad(M))K, & l \text{ is even} \end{cases} \tag{3}$$

where $l = \lceil |M|/n \rceil + 1$ is the number of blocks in $pad(M)$.

Theorem 2. RH2 is $(l_{max}+3)/2^n$-RKA-AXU for the RKD set $\Phi^{\oplus}$, where l_{max} is the maximum block number of messages after padding.

Proof. For any message M, suppose $pad(M) = M_1\|M_2\|\cdots\|M_l$. When l is odd

$$\text{RH2}_K(M) = K^{l+2} \oplus M_1 K^l \oplus \cdots \oplus M_l K.$$

When l is even

$$\mathrm{RH2}_K(M) = K^{l+3} \oplus M_1 K^{l+1} \oplus \cdots \oplus M_l K^2.$$

We prove that for any $M, M' \in \{0,1\}^*$, $\Delta_1, \Delta_2, C \in \{0,1\}^n$, $(\Delta_1, M) \neq (\Delta_2, M')$, $\Pr[F(K) = C] \leq \epsilon$, where $F(K) = \mathrm{RH2}_{K \oplus \Delta_1}(M) \oplus \mathrm{RH2}_{K \oplus \Delta_2}(M')$. We only need to show the degree of $F(K)$ is nonzero. Suppose $pad(M) = M_1 \| M_2 \| \cdots \| M_l$ and $pad(M') = M_1' \| M_2' \| \cdots \| M_{l'}'$. Consider $F(K)$ in the following two cases.

CASE 1. $\Delta_1 \neq \Delta_2$. Suppose the degrees of $\mathrm{RH2}_{K \oplus \Delta_1}(M)$ and $\mathrm{RH2}_{K \oplus \Delta_2}(M')$ are d and d' respectively, which are both odd.

When $d = d'$, the coefficient of K^{d-1} in $F(K)$ is $\Delta_1 \oplus \Delta_2$ which is nonzero.

When $d \neq d'$, suppose $d > d'$ w.l.o.g. the coefficient of K^d in $F(K)$ is 1.

CASE 2. $\Delta_1 = \Delta_2$. We treat $K \oplus \Delta_1$ as a new key, so without loss of generality, we only consider $\Delta_1 = \Delta_2 = 0$ in the following.

When $l = l'$, there exists $1 \leq j \leq l$ s.t. $M_j \neq M_j'$. So the coefficient of K^{l+1-j} (if l is odd) or K^{l+2-j} (if l is even) in $F(K)$ is $M_j \oplus M_j'$ which is nonzero.

When $l' \neq l$ and are both odd, the coefficient of K is $|M| \oplus |M'|$ which is nonzero.

When $l' \neq l$ and are both even, the coefficient of K^2 is $|M| \oplus |M'|$ which is nonzero.

When $l' \neq l$, one is odd and one is even, the coefficient of K is $|M|$ or $|M'|$ which are both nonzero.

Therefore the degree of $F(K)$ is nonzero.   □

Since RH2 is RKA-AXU, it is also RKA-AU. But sometimes we only need RKA-AU functions. We can improve the efficiency of RKA-AU construction by one less multiplication in finite field if replace $Poly$ in RH2 with the following $Poly'$:

$$Poly'_K(M) = M_1 K^{m-1} \oplus M_2 K^{m-2} \oplus \cdots \oplus M_m$$

where $M = M_1 \| M_2 \| \cdots \| M_m \in \{0,1\}^{nm}$. $Poly'$ is AU but not AXU. We have the following construction and the proof is similar to that of Theorem 2.

Construction 3. RH3 : $\{0,1\}^n \times \{0,1\}^* \to \{0,1\}^n$,

$$\mathrm{RH3}_K(M) = \begin{cases} K^{l+2} \oplus Poly'_K(pad(M)), & l \text{ is odd} \\ K^{l+3} \oplus Poly'_K(pad(M))K, & l \text{ is even} \end{cases} \tag{4}$$

where $l = \lceil |M|/n \rceil + 1$ is the number of blocks in $pad(M)$.

Theorem 3. RH3 is $(l_{max} + 3)/2^n$-RKA-AU for the RKD set $\Phi^\oplus$, where l_{max} is the maximum number of blocks in messages after padding.

Efficiency of Constructions. We analyze the efficiency of RH1, RH2 and RH3 compared with previous similar constructions.

Table 2. Computation of $\mathrm{RH2}_K(M)$ and $Poly_K(pad(M))$ by Horner's rule.

$\mathrm{RH2}_K(M)$:	$Poly_K(pad(M))$:
$\quad T \leftarrow K^2$	$\quad T \leftarrow 0$
$\quad$ **for** $i = 1$ **to** l	$\quad$ **for** $i = 1$ **to** l
$\quad\quad T \leftarrow (T \oplus M_i)K$	$\quad\quad T \leftarrow (T \oplus M_i)K$
$\quad$ **if** l is even	
$\quad\quad T \leftarrow TK$	
$\quad$ **return** T	$\quad$ **return** T

(1) RH1. Compared with $Poly_K(M) = MK$, in $\mathrm{RH1}_K(M) = MK \oplus K^3$ the monomial K^3 can be pre-computed. So RH1 needs extra one pre-computation and one XOR operation.

(2) RH2. The polynomial $T = M_1K^m \oplus M_2K^{m-1} \oplus \cdots \oplus M_mK$ is usually evaluated by Horner's rule: $T \leftarrow 0$, $T \leftarrow (T \oplus M_i)K$ for $1 \leq i \leq m$. Assume that $pad(M) = M_1\|M_2\| \cdots \|M_l$, Table 2 shows the computation processes of $\mathrm{RH2}_K(M)$ and $Poly_K(pad(M))$ by Horner's rule respectively. We can see that compared with $Poly_K(pad(M))$, RH2 needs one additional pre-computation of K^2, and one more multiplication if l is even.

(3) RH3. Similar to the analysis of RH2, RH3 needs one additional pre-computation of K^2, and one more multiplication if l is even, compared with $Poly'_K(pad(M))$.

In brief, RH1, RH2 and RH3 are almost as efficient as previous similar constructions.

4   Applications

RKA-AU (RKA-AXU) hash functions can be used as components, along with other primitives such as RKA-PRPs and RKA-PRFs, in the design of related-key secure cryptographic schemes. If we replace the UHFs in the cryptographic schemes in Sect. 1 with our corresponding constructions, the issues about related-key attacks can be solved for some RKD set. *Informally speaking, if the UHF is RKA-AU or RKA-AXU for the RKD set Φ_1 and the underlying primitive is RKA-PRP or RKA-PRF for the RKD set Φ_2, the scheme is related-key secure for the RKD set $\Phi_1 \times \Phi_2$.*

In the following, we give four concrete applications of RKA-AU and RKA-AXU in related-key secure MACs and TBCs. In the analyses of these schemes, we mainly give intuitive interpretations by establishing the relationship between the RKA setting and the invariable-key setting and the detailed proofs will be given in the full paper [51]. Then the remaining proof is similar to that in the invariable-key setting. Let RKA-PRF be PRF against related-key attacks. We define a chosen-ciphertext attack (CCA) secure tweakable block cipher as a strongly tweakable pseudorandom permutation (STPRP, SPRP if it has no tweak). If it is also related-key secure we denote it as RKA-STPRP (RKA-SPRP if it has no tweak). The detailed definitions are in Appendix A.

For simplicity we only consider the *claw-free* RKD set Φ in which for any $\phi_1, \phi_1 \in \Phi$ and any key K we have $\phi_1(K) \neq \phi_2(K)$. The relationships are based on three observations on the underlying components when we regard the RKD transformation as an additional input.

Observation 1. For a function $F : \mathcal{K} \times \mathcal{D} \to \mathcal{R}$ and a claw-free RKD set Φ on $\mathcal{K}$. We define a new function $F' : \mathcal{K} \times (\Phi \times \mathcal{D}) \to \mathcal{R}$, $F'_K(\phi, M) = F_{\phi(K)}(M)$. It is directly derived from the definition that F is ϵ-RKA-AU (ϵ-RKA-AXU) for the RKD set Φ if and only if F' is ϵ-AU (ϵ-AXU).

Observation 2. Furthermore, we have that F is a RKA-PRF for the RKD set Φ if and only if F' is a PRF.

Observation 3. For a block cipher $E : \mathcal{K} \times \{0,1\}^n \to \{0,1\}^n$ and a claw-free RKD set Φ on $\mathcal{K}$, define a tweakable block cipher $E' : \mathcal{K} \times \Phi \times \{0,1\}^n \to \{0,1\}^n$, $E'_K(\phi, M) = E_{\phi(K)}(M)$. E is a RKA-SPRP for the RKD set Φ, if and only if E' is a STPRP.

4.1   Related-Key Secure MACs

Beside the Carter-Wegman scheme to construct MAC [52]

$$\text{MAC1}_{K,K'}(N, M) = H_K(M) \oplus F_{K'}(N) \tag{5}$$

the other method [45] is

$$\text{MAC2}_{K,K'}(M) = F_{K'}(H_K(M)) \tag{6}$$

where $H : \mathcal{K}_1 \times \mathcal{D} \to \{0,1\}^n$ and $F : \mathcal{K}_2 \times \{0,1\}^n \to \{0,1\}^n$ are two keyed functions, M is a message and N is a nonce. We show that the two schemes are both related-key secure by the following two theorems.

Theorem 4. *If H is ϵ-RKA-AXU for the RKD set Φ_1 and F is a RKA-PRF for the RKD set Φ_2, then MAC1 is related-key unforgable (RKA-UF) for the RKD set $\Phi_1 \times \Phi_2$. More specifically,*

$$\mathbf{Adv}_{\text{MAC1}}^{rka-uf}(q, t) \leq \mathbf{Adv}_F^{rka-prf}(q, t') + \epsilon$$

where the adversary makes q queries to MAC1 and $t' = t + O(q)$.

From Observation 1, $H'_K(\phi_1, M) = H_{\phi_1(K)}(M)$ is AXU; from Observation 2, $F'_{K'}(\phi_2, N) = F_{\phi_2(K')}(N)$ is a PRF. If we look ϕ_1 as a part of the message and ϕ_2 as a part of the nonce, we only need to prove that $G_{K,K'}(\phi_2, N, \phi_1, M) = H'_K(\phi_1, M) \oplus F'_{K'}(\phi_2, N)$ is unforgeable in the invariable-key setting. The remaining proof is similar to that in [34].

Theorem 5. *If H is ϵ-RKA-AU for the RKD set Φ_1 and F is a RKA-PRF for the RKD set Φ_2, then MAC2 is a RKA-PRF for the RKD set $\Phi_1 \times \Phi_2$. More specifically,*

$$\mathbf{Adv}^{rka-prf}_{\mathrm{MAC2}}(q,t) \leq \mathbf{Adv}^{rka-prf}_{F}(q,t') + \epsilon q^2/2$$

where the adversary makes q queries to MAC2 and $t' = t + O(q)$.

From Observation 1, $H'_K(\phi_1, M) = H_{\phi_1(K)}(M)$ is AXU; from Observation 2, $F'_{K'}(\phi_2, M) = F_{\phi_2(K')}(M)$ is a PRF. If we look ϕ_1 and ϕ_2 as a part of the message, we only need to prove that $G_{K,K'}(\phi_1, \phi_2, M) = F'_{K'}(\phi_2, H'_K(\phi_1, M))$ is a PRF in the invariable-key setting. The remaining proof is similar to that in [45].

4.2   Related-Key Secure TBCs

Block Cipher Based Schemes. In [36] Liskov et al. gave a construction of tweakable block cipher (TBC) from a block cipher and a universal hash function:

$$\mathrm{TBC1}_{K,K'}(T, M) = E_{K'}(M \oplus H_K(T)) \oplus H_K(T) \tag{7}$$

where $H : \mathcal{K}_1 \times \mathcal{D} \to \{0,1\}^n$ is the universal hash function and $E : \mathcal{K}_2 \times \{0,1\}^n \to \{0,1\}^n$ is the block cipher. In Appendix B we show that TBC1 is not related-key secure if $H_K(T) = TK$. But if H is RKA-AXU, we show that TBC1 is related-key secure for some RKD set in Theorem 6.

Theorem 6. *If H is ϵ-RKA-AXU for the RKD set Φ_1 and E is RKA-SPRP for the RKD set Φ_2, then TBC1 is a RKA-STPRP for the RKD set $\Phi_1 \times \Phi_2$. More specifically,*

$$\mathbf{Adv}^{rka-stprp}_{\mathrm{TBC1}}(q,t) \leq \mathbf{Adv}^{rka-sprp}_{E}(q,t') + 3\epsilon q^2$$

where the adversary makes q queries to TBC1 or $\mathrm{TBC1}^{-1}$ and $t' = t + O(q)$.

From Observation 1, $H'_K(\phi_1, M) = H_{\phi_1(K)}(M)$ is AXU; from Observation 3, $E'_{K'}(\phi_2, M) = E_{\phi_2(K')}(M)$ is a STPRP. If we consider ϕ_1 and ϕ_2 as a part of the tweak, we only need to prove that $\widetilde{E}_{K,K'}(\phi_1, \phi_2, T, M) = E'_{K'}(\phi_2, M \oplus H'_K(\phi_1, T)) \oplus H'_K(\phi_1, T)$ is a STPRP in the invariable-key setting. The remaining proof is similar to that in [36].

Permutation Based Schemes. If we replace the block cipher in TBC1 as a permutation, we get

$$\mathrm{TBC2}_K(T, M) = \pi(M \oplus H_K(T)) \oplus H_K(T) \tag{8}$$

where π is the permutation from $\{0,1\}^m$ to $\{0,1\}^m$, $n \leq m$. For $A \in \{0,1\}^n$, $B \in \{0,1\}^m$, when $n < m$, $A \oplus B$ is defined as $(A\|0^{m-n}) \oplus B$. We show the related-key security of TBC2 in Theorem 7. We need that H is both RKA-AXU and related-key almost uniform. H is δ-related-key-almost-uniform means for any $\phi \in \Phi$, $M \in \mathcal{D}$ and $C \in \{0,1\}^n$, $\Pr[K \xleftarrow{\$} \mathcal{K} : H_{\phi(K)}(M) = C] \leq \delta$. When H is also ϵ-RKA-AXU, we say that it is (ϵ, δ)-RKA-AXU. For example, RH1 $= MK \oplus K^3$ is $(2/2^n, 3/2^n)$-RKA-AXU.

TBC2 is a one-round tweakable Even-Mansour cipher. How to add tweak and retain related-key security of the Even-Mansour cipher is a popular topic in recent years [19,20,24,25,38]. Compared with previous constructions in [25,38] we only need one permutation invocation (two in [25,38]).

Theorem 7. *If H is (ϵ, δ)-RKA-AXU for the RKD set Φ and π is public random permutation, then TBC2 is a RK-TSPRP for the RKD set Φ. More specifically,*

$$\mathbf{Adv}_{\text{TBC2}}^{rka-stprp}(q_0, q_1) \leq q_0^2 \epsilon + 2q_0 q_1 \delta + 2^{-m}(q_0^2 + 2q_0 q_1)$$

where the adversary makes q_0 queries to TBC2 or TBC2^{-1} and q_1 queries to π or π^{-1}.

From Observation 1, $H'_K(\phi, M) = H_{\phi(K)}(M)$ is AXU. If we look ϕ as a part of the nonce, we only need to prove that $\widetilde{E}_K(\phi, T, M) = \pi(M \oplus H'_K(\phi, T)) \oplus H'_K(\phi, T)$ is a STPRP in the invariable-key setting. The remaining proof is similar to that in [35] or [19].

5   Conclusions

In this paper we mainly focus on two-key schemes, e.g. one key for the UHF and the other key for the block cipher. In order to resist related-key attacks, we define a new concept of related-key almost universal hash function, which is a natural extension to almost universal hash function in the RKA setting.

Not every UHF-based scheme suffers from related-key attacks. For example GCM [37] has only one key which is also the key of the underlying block cipher. The key of UHF is derived from the master key K as $E_K(0^{128})$. GCM has been proved to be secure in the invariable-key setting [32] given that E is a PRP. If E is a RKA-PRP, for each $\phi \in \Phi$, $E_{\phi(K)}$ is an independent PRP. So GCM is secure independently for each related key, and thus GCM is also secure in the RKA setting. In this roughly reasoning, we only require that the UHF is AXU but not RKA-AXU. Therefore it is possible that the upper scheme "inherit" the related-key security only from the underlying block cipher. It is also true to some other one-key schemes such as XCB [29], POET [1], etc. We can even modify the vulnerable schemes in this paper into related-key secure ones without the notion of RKA-AXU or RKA-AU by generating the keys in the schemes as $K_i = E_K(i)$, $i = 1, 2, \cdots$ where K is the master key. But there are still a lot of two-key schemes such as Poly1305-AES [6], HCTR [50], HCHp and HCHfp [16,17]. Furthermore, if we regard related-key attacks as a class of *side-channel attacks*, the attacker may have the ability to change a stored key via tampering or fault injection [4,11]. The key of UHF stored somewhere, no matter whether it is a part of the master key or derived from the master key, can be changed in this scenario.

We also give several efficient constructions named RH1, RH2 and RH3 which are nearly as efficient as previous similar ones. RKA-AU (RKA-AXU) hash functions can be used as components, along with other primitives such as RKA-PRPs and RKA-PRFs etc., in the design of related-key secure cryptographic schemes.

Acknowledgment. The authors would like to thank the anonymous reviewers for their helpful and valuable comments and suggestions. The work of this paper is supported by the National Key Basic Research Program of China (2014CB340603), the National Natural Science Foundation of China (Grants 61272477, 61472415, 61202422, 61572484), the Strategic Priority Research Program of Chinese Academy of Sciences under Grant XDA06010702. Liting Zhang is supported by the Youth Innovation Promotion Association of CAS (2015087).

A   Related-Key Security of MAC, TBC, TES and AE Schemes

(1) RKA-PRF. For a function $F : \mathcal{K} \times \mathcal{D} \to \mathcal{R}$, the adversary $\mathbb{A}$ can make related-key oracle queries $(\phi, M) \in \Phi \times \mathcal{D}$ and is responded with $F_{\phi(K)}(M)$ where K is the secret key. Let ρ be a uniformly random function from $\mathcal{K} \times \mathcal{D}$ to $\mathcal{R}$. The advantage of $\mathbb{A}$ is defined as

$$\mathbf{Adv}_{\mathbf{F}}^{rka-prf}(\mathbb{A}) = \Pr[\mathbb{A}^{\mathbf{F}_{\cdot(K)}(\cdot)} \Rightarrow 1] - \Pr[\mathbb{A}^{\rho_{\cdot(K)}(\cdot)} \Rightarrow 1].$$

For all adversaries with computation time at most t, oracle queries at most q, we denote $\mathbf{Adv}_{F}^{rka-prf}(q, t) = max_{\mathbb{A}}\mathbf{Adv}_{F}^{rka-prf}(\mathbb{A})$. When the advantage is negligible, we say that F is a RKA-PRF for Φ.

(2) RKA-UF. A message authentication code (MAC) is a function $F : \mathcal{K} \times \mathcal{N} \times \mathcal{M} \to \{0,1\}^n$, where $\mathcal{K}, \mathcal{N}, \mathcal{M}$ and $\{0,1\}^n$ are spaces of key, nonce, message and tag respectively. The nonce space can be an empty set $\mathcal{N} = \emptyset$. For a RKD set Φ, the adversary $\mathbb{A}$ queries the MAC algorithm with $(\phi, N, M) \in \Phi \times \mathcal{N} \times \mathcal{M}$ but never repeats N, and gets $T = F_{\phi(K)}(N, M)$. After several queries $\mathbb{A}$ returns a quadruple (ϕ', N', M', T') which never appear before in the queries. We define the probability of $T' = F_{\phi'(K)}(N', M')$ as the advantage of $\mathbb{A}$ and write it as:

$$\mathbf{Adv}_{F}^{rka-uf}(\mathbb{A}) = \Pr[\mathbb{A}^{F_{\cdot(K)}(\cdot,\cdot)}\text{forges}].$$

For all adversaries with computation time at most t, oracle queries at most q, we denote $\mathbf{Adv}_{F}^{rka-uf}(q, t) = max_{\mathbb{A}}\mathbf{Adv}_{F}^{rka-uf}(\mathbb{A})$. When the advantage is negligible, we say that F is related-key unforgeable (RKA-UF) or related-key unpredictable for Φ.

(3) RKA-STPRP and RKA-SPRP. A tweakable block cipher consists of two algorithms $\mathcal{S} = (\mathbf{E}, \mathbf{D})$. The encryption algorithm $\mathbf{E} : \mathcal{K} \times \mathcal{T} \times \{0,1\}^n \to \{0,1\}^n$, where $\mathcal{K}, \mathcal{T}$ and $\{0,1\}^n$ are spaces of key, tweak, plaintext/ciphertext respectively. For input $(K, T, P) \in \mathcal{K} \times \mathcal{T} \times \{0,1\}^n$, we write the result as $C = \mathbf{E}_K^T(P)$ The decryption algorithm $\mathbf{D} : \mathcal{K} \times \mathcal{T} \times \{0,1\}^n \to \{0,1\}^n$. We require that for any $(K, T) \in \mathcal{K} \times \mathcal{T}$, $\mathbf{E}_K^T(\cdot)$ and $\mathbf{D}_K^T(\cdot)$ are permutations, and $\mathbf{D}_K^T(\mathbf{E}_K^T(P)) = P$.

For a RKD set Φ, an adversary $\mathbb{A}$ queries $\mathbf{E}$ with $(\phi, T, P) \in \Phi \times \mathcal{T} \times \{0,1\}^n$ or queries $\mathbf{D}$ with $(\phi, T, C) \in \Phi \times \mathcal{T} \times \{0,1\}^n$. $\mathbb{A}$ tries to distinguish $\mathcal{S}$ from an ideal TBC, where for any $(K, T) \in \mathcal{K} \times \mathcal{T}$, π_K^T is an independent uniformly random permutation. Without loss of generality we assume that the adversary never make *pointless* queries that the adversary "knows" the answer. For example, if the adversary query (ϕ, T, P) to the encryption oracle and get the answer C, he will never query (ϕ, T, C) to the decryption oracle. We define the advantage as

$$\mathbf{Adv}_{\mathcal{S}}^{rka-stprp}(\mathbb{A}) = \Pr[\mathbb{A}^{\mathbf{E}_{\cdot(K)}(\cdot), \mathbf{D}_{\cdot(K)}(\cdot)} \Rightarrow 1] - \Pr[\mathbb{A}^{\pi_{\cdot(K)}(\cdot), \pi_{\cdot(K)}^{-1}(\cdot)} \Rightarrow 1].$$

For all adversaries with computation time at most t, oracle queries at most q, we denote $\mathbf{Adv}_{\mathcal{S}}^{rka-stprp}(q, t) = max_{\mathbb{A}}\mathbf{Adv}_{\mathcal{S}}^{rka-stprp}(\mathbb{A})$. When the advantage is negligible, we say that $\mathcal{S}$ is a related-key strongly tweakable pseudorandom permutation (RKA-STPRP) for Φ. When the tweak space $\mathcal{T}$ is a empty set $\mathbf{E}$ becomes a block cipher. The corresponding security notion is related-key strongly pseudorandom permutation (RKA-SPRP). Tweakable enciphering schemes are TBCs with large or variable input length. The definition is the same as that of TBC.

(4) RKA-AE. An authenticated encryption scheme consists of two algorithms $\mathcal{SE} = (\mathbf{E}, \mathbf{D})$. The encryption $\mathbf{E} : \mathcal{K} \times \mathcal{N} \times \mathcal{A} \times \mathcal{P} \rightarrow \mathcal{C}$, where $\mathcal{K}, \mathcal{N}, \mathcal{A}, \mathcal{P}$ and $\mathcal{C}$ are spaces of key, nonce, associated data, plaintext and ciphertext respectively. For input $(K, N, A, P) \in \mathcal{K} \times \mathcal{N} \times \mathcal{A} \times \mathcal{P}$, we write the result as $C = \mathbf{E}_K(N, A, P)$. The decryption algorithm $\mathbf{D} : \mathcal{K} \times \mathcal{N} \times \mathcal{A} \times \mathcal{C} \rightarrow \mathcal{P} \cup \{\bot\}$. We require that $\mathbf{D}_K(N, A, \mathbf{E}_K(N, A, P)) = P$. For a RKD set Φ, an adversary $\mathbb{A}$ queries the $\mathbf{E}$ with $(\phi, N, A, P) \in \Phi \times \mathcal{N} \times \mathcal{A} \times \mathcal{P}$ but never repeats (ϕ, N), or queries the $\mathbf{D}$ with (ϕ, N, A, C). $\mathbb{A}$ tries to distinguish $\mathcal{SE}$ from an ideal AE scheme($\$, \bot$), where for any query $\$$ returns a random string and $\bot$ always returns $\bot$. We define the advantage as

$$\mathbf{Adv}_{\mathcal{SE}}^{rka-ae}(\mathbb{A}) = \Pr[\mathbb{A}^{\mathbf{E}_{\cdot(K)}(\cdot,\cdot,\cdot), \mathbf{D}_{\cdot(K)}(\cdot,\cdot,\cdot)} \Rightarrow 1] - \Pr[\mathbb{A}^{\$(\cdot,\cdot,\cdot,\cdot), \bot(\cdot,\cdot,\cdot,\cdot)} \Rightarrow 1].$$

For all adversaries with computation time at most t, oracle queries at most q, we denote $\mathbf{Adv}_{\mathcal{SE}}^{rka-ae}(q, t) = max_{\mathbb{A}}\mathbf{Adv}_{\mathcal{SE}}^{rk-ae}(\mathbb{A})$. When the advantage is negligible, we say that $\mathcal{SE}$ is related-key secure for Φ.

B   More Examples of Related-Key Attacks Against UHF-based Schemes

(1) TBC. A tweakable block cipher (TBC) is a generalized block cipher with an extra input called tweak. TBCs were first formalized by Liskov et al. [36] and found applications largely in modes of operation [42]. In their seminal paper, Liskov et al. gave a construction of TBC from a block cipher: $TBC_{K,K'}(T, M) = E_{K'}(M \oplus H_K(T)) \oplus H_K(T)$ where E is the block cipher, H is a universal hash function and T is the tweak. They proved that when E is a PRP against chosen ciphertext attacks (CCAs) and H is almost XOR universal, TBC is secure

against CCA attacks. If we use $Poly_K(T) = TK$ as the underlying UHF, the following is an attack. First we query with (T, M) under the derived key $(K \oplus \Delta, K')$ where $\Delta \neq 0$, then the answer is $C = E_{K'}(M \oplus T(K \oplus \Delta)) \oplus T(K \oplus \Delta) = E_{K'}((M \oplus T\Delta) \oplus TK) \oplus TK \oplus T\Delta$. So we can predict that the ciphertext of $(T, (M \oplus T\Delta))$ under the original key is $C \oplus T\Delta$. Therefore it does not resist related-key attack.

(2) TES. A tweakable enciphering scheme is a generalized TBC with large or variable input length, suitable for disk sector encryption. Recently Sun et al. [49] show that HCTR [50], HCHp and HCHfp [16,17] suffer related-key attacks. All these TESes use the polynomial evaluation hash function as the underlying UHF.

(3) AE scheme. An authenticated encryption scheme achieves both confidentiality and authenticity. One of AE schemes OCB [42,43] following from IAPM [33], encrypts the message blocks using independent PRPs into ciphertext blocks and encrypts the XOR of the message blocks into a tag using another independent PRP. Kurosawa [35] proposed a modified IAPM, the encryption of message blocks is

$$C_i = E_{K'}(M_i \oplus Poly_K(IV\|(2i-1))) \oplus Poly_K(IV\|(2i-1))$$

where M_i is the i-th message block, E is the block cipher and the key of the scheme is (K, K'). Kurosawa proved that this modified IAPM is secure even if the underlying block cipher is publicly accessible. But if we query with (IV, M) under the derived key $(K \oplus 0^{n-1}1, K')$, the first ciphertext block $C_1 = E_{K'}((M_i \oplus IV \oplus 0^{n-1}1) \oplus (Poly_K(IV\|0^{n-1}1)) \oplus Poly_K(IV\|0^{n-1}1) \oplus IV \oplus 0^{n-1}1$. We can predict that the first ciphertext block of (IV, M') under the original key is $C_1 \oplus IV \oplus 0^{n-1}1$, where M' is changed from M by changing the first block into $M_1 \oplus IV \oplus 0^{n-1}1$. If we define the confidentiality as the indistinguishability between ciphertexts and uniformly random bits, this scheme does not resist the related-key attack.

In the above examples, the key of UHF is a part of the key of whole scheme, so that the adversary can derive the related key of UHF and get the input collision to other primitives such as PRPs or PRFs. The collisions in the above attacks are listed as following.

(1) $Poly_{K \oplus \Delta}(T) \oplus Poly_K(T) = \Delta T$ used in the TBC example;
(2) $Poly_{K \oplus \Delta}(A\|B) \oplus Poly_K(A\|B) = A\Delta^2 \oplus B\Delta$ used in the TES and AE scheme examples.

C   Existing UHFs that Are Not RKA-AXU (RKA-AU)

The following universal hash functions are proved to be AXU (AΔU).

(1) MMH [26]: $H_K(M) = (((\sum_{i=1}^{t} M_i K_i) \mod 2^{64}) \mod p) \mod 2^{32}$, $M_i, K_i \in \mathbf{Z}_{2^{32}}$ and $p = 2^{32} + 15$;
(2) Square Hash [23]: $H_K(M) = \sum_{i=1}^{t}(M_i + K_i)^2 \mod p$, $M_i, K_i \in \mathbf{Z}_p$;
(3) NMH [26]: $H_K(M) = (\sum_{i=1}^{t/2}(M_{2i-1} + K_{2i-1})(M_{2i} + K_{2i})) \mod p$, $M_i, K_i \in \mathbf{Z}_{2^{32}}$, $p = 2^{32} + 15$;

(4) NH [13]: $H_K(M) = (\sum_{i=1}^{t/2}((M_{2i-1} + K_{2i-1}) \mod 2^w)((M_{2i} + K_{2i}) \mod 2^w)) \mod 2^{2w}$, $M_i, K_i \in \mathbf{Z}_{2^w}$.

In (1) we set $t = 1$, then $H_K(M) = (MK \mod 2^{32} + 15) \mod 2^{32}$. If $M = M' = \Delta' = 1$, $\Delta = 0$, then $H_K(M) = K$, $H_{K+\Delta'}(M') = K + 1 \mod 2^{32}$, therefore $H_K(M) + 1 = H_{K+\Delta'}(M')$, MMH is not RK-A$\Delta$U. (2), (3) and (4) all have the term $M_1 + K_1$. From $M_1 + K_1 = (M_1 - 1) + (K_1 + 1)$ we know that they are all not RKA-AU.

References

1. Abed, F., Fluhrer, S., Foley, J., Forler, C., List, E., Lucks, S., McGrew, D., Wenzel, J.: The POET family of on-line authenticated encryption schemes (2014). http://competitions.cr.yp.to/caesar-submissions.html
2. Andreeva, E., Bogdanov, A., Lauridsen, M.M., Luykx, A., Mennink, B., Tischhauser, E., Yasuda, K.: AES-COBRA (2014). http://competitions.cr.yp.to/caesar-submissions.html
3. Applebaum, B., Harnik, D., Ishai, Y.: Semantic security under related-key attacks and applications. In: Chazelle, B. (ed.) Innovations in Computer Science - ICS 2010, pp. 45–60. Tsinghua University Press (2011). http://conference.itcs.tsinghua.edu.cn/ICS2011/content/papers/30.html
4. Bellare, M., Cash, D., Miller, R.: Cryptography secure against related-key attacks and tampering. In: Lee, D.H., Wang, X. (eds.) ASIACRYPT 2011. LNCS, vol. 7073, pp. 486–503. Springer, Heidelberg (2011). doi:10.1007/978-3-642-25385-0_26
5. Bellare, M., Kohno, T.: A theoretical treatment of related-key attacks: RKA-PRPs, RKA-PRFs, and applications. In: Biham, E. (ed.) EUROCRYPT. LNCS, vol. 2656, pp. 491–506. Springer, Heidelberg (2003). doi:10.1007/3-540-39200-9_31
6. Bernstein, D.J.: The Poly1305-AES message-authentication code. In: Gilbert, H., Handschuh, H. (eds.) FSE 2005. LNCS, vol. 3557, pp. 32–49. Springer, Heidelberg (2005). doi:10.1007/11502760_3
7. Bernstein, D.J.: Stronger security bounds for Wegman-Carter-Shoup authenticators. In: Cramer, R. (ed.) EUROCRYPT 2005. LNCS, vol. 3494, pp. 164–180. Springer, Heidelberg (2005). doi:10.1007/11426639_10
8. Bernstein, D.J.: Polynomial evaluation and message authentication (2011). http://cr.yp.to/papers.html#pema
9. Bhattacharyya, R., Roy, A.: Secure message authentication against related-key attack. In: Moriai, S. (ed.) FSE 2013. LNCS, vol. 8424, pp. 305–324. Springer, Heidelberg (2014). doi:10.1007/978-3-662-43933-3_16
10. Biham, E.: New types of cryptanalytic attacks using related keys. In: Helleseth, T. (ed.) EUROCRYPT 1993. LNCS, vol. 765, pp. 398–409. Springer, Heidelberg (1994)
11. Biham, E., Shamir, A.: Differential fault analysis of secret key cryptosystems. In: Kaliski Jr., B.S. (ed.) CRYPTO 1997. LNCS, vol. 1294, pp. 513–525. Springer, Heidelberg (1997). doi:10.1007/BFb0052259_12
12. Biryukov, A., Khovratovich, D.: Related-key cryptanalysis of the full AES-192 and AES-256. In: Matsui, M. (ed.) ASIACRYPT 2009. LNCS, vol. 5912, pp. 1–18. Springer, Heidelberg (2009). doi:10.1007/978-3-642-10366-7_1
13. Black, J., Halevi, S., Krawczyk, H., Krovetz, T., Rogaway, P.: UMAC: fast and secure message authentication. In: Wiener, M.J. (ed.) [53], pp. 216–233. http://dx.org/10.1007/3-540-48405-1_14

14. Boesgaard, M., Christensen, T., Zenner, E.: Badger – a fast and provably secure MAC. In: Ioannidis, J., Keromytis, A.D., Yung, M. (eds.) ACNS 2005. LNCS, vol. 3531, pp. 176–191. Springer, Heidelberg (2005). doi:10.1007/11496137_13

15. Carter, L., Wegman, M.N.: Universal classes of hash functions. J. Comput. Syst. Sci. **18**(2), 143–154 (1979)

16. Chakraborty, D., Sarkar, P.: HCH: a new tweakable enciphering scheme using the hash-encrypt-hash approach. In: Barua, R., Lange, T. (eds.) INDOCRYPT 2006. LNCS, vol. 4329, pp. 287–302. Springer, Heidelberg (2006). doi:10.1007/11941378_21

17. Chakraborty, D., Sarkar, P.: HCH: a new tweakable enciphering scheme using the hash-counter-hash approach. IEEE Trans. Inf. Theor. **54**(4), 1683–1699 (2008). doi:10.1109/TIT.2008.917623

18. Chen, J., Miyaji, A.: A new practical key recovery attack on the stream cipher RC4 under related-key model. In: Lai, X., Yung, M., Lin, D. (eds.) Inscrypt 2010. LNCS, vol. 6584, pp. 62–76. Springer, Heidelberg (2011). doi:10.1007/978-3-642-21518-6_5

19. Cogliati, B., Lampe, R., Seurin, Y.: Tweaking Even-Mansour ciphers. In: Gennaro, R., Robshaw, M. (eds.) CRYPTO 2015. LNCS, vol. 9215, pp. 189–208. Springer, Heidelberg (2015). doi:10.1007/978-3-662-47989-6_9

20. Cogliati, B., Seurin, Y.: On the provable security of the iterated Even-Mansour cipher against related-key and chosen-key attacks. In: Oswald, E., Fischlin, M. (eds.) EUROCRYPT 2015. LNCS, vol. 9056, pp. 584–613. Springer, Heidelberg (2015). doi:10.1007/978-3-662-46800-5_23

21. Dobraunig, C., Eichlseder, M., Mendel, F.: Related-Key Forgeries for Prøst-OTR. IACR Cryptology ePrint Archive, to appear in FSE 2015 (2015). http://eprint.iacr.org/2015/091

22. Dunkelman, O., Keller, N., Shamir, A.: A practical-time related-key attack on the KASUMI cryptosystem used in GSM and 3G telephony. J. Cryptology **27**(4), 824–849 (2014). doi:10.1007/s00145-013-9154-9

23. Etzel, M., Patel, S., Ramzan, Z.: SQUARE HASH: fast message authenication via optimized universal hash functions. In: Wiener, M.J. (ed.) [53], pp. 234–251. http://dx.org/10.1007/3-540-48405-1_15

24. Farshim, P., Procter, G.: The related-key security of iterated Even-Mansour ciphers. In: Leander, G. (ed.) FSE 2015. LNCS, vol. 9054, pp. 342–363. Springer, Heidelberg (2015). doi:10.1007/978-3-662-48116-5_17

25. Granger, R., Jovanovic, P., Mennink, B., Neves, S.: Improved masking for tweakable blockciphers with applications to authenticated encryption. Cryptology ePrint Archive, Report 2015/999 (2015). http://eprint.iacr.org/

26. Halevi, S., Krawczyk, H.: MMH: software message authentication in the Gbit/Second rates. In: Biham, E. (ed.) FSE 1997. LNCS, vol. 1267, pp. 172–189. Springer, Heidelberg (1997). doi:10.1007/BFb005234

27. Harris, S.: AES-COBRA (2014). http://competitions.cr.yp.to/caesar-submissions.html

28. IEEE Std 1619-2007: IEEE standard for cryptographic protection of data on block-oriented storage devices (2008)

29. IEEE Std 1619.2-2010: IEEE standard for wide-block encryption for shared storage media (2011)

30. ISO/IEC 19772:2009: Information technology - security techniques - authenticated encryption (2009)

31. ISO/IEC9797-3:2011: Information technology - security techniques - message authentication codes (MACs) - part 3: Mechanisms using a universal hash-function (2011)

32. Iwata, T., Ohashi, K., Minematsu, K.: Breaking and repairing GCM security proofs. In: Safavi-Naini, R., Canetti, R. (eds.) CRYPTO 2012. LNCS, vol. 7417, pp. 31–49. Springer, Heidelberg (2012). doi:10.1007/978-3-642-32009-5_3

33. Jutla, C.S.: Encryption modes with almost free message integrity. In: Pfitzmann, B. (ed.) EUROCRYPT 2001. LNCS, vol. 2045, p. 529. Springer, Heidelberg (2001). doi:10.1007/3-540-44987-6_32

34. Krawczyk, H.: LFSR-based hashing and authentication. In: Desmedt, Y.G. (ed.) CRYPTO 1994. LNCS, vol. 839, pp. 129–139. Springer, Heidelberg (1994). doi:10.1007/3-540-48658-5_15

35. Kurosawa, K.: Power of a public random permutation and its application to authenticated encryption. IEEE Trans. Inf. Theor. **56**(10), 5366–5374 (2010). doi:10.1109/TIT.2010.2059636

36. Liskov, M., Rivest, R.L., Wagner, D.: Tweakable block ciphers. In: Yung, M. (ed.) CRYPTO 2002. LNCS, vol. 2442, p. 31. Springer, Heidelberg (2002). doi:10.1007/3-540-45708-9_3

37. McGrew, D.A., Viega, J.: The Galois/Counter mode of operation (GCM) (2004). http://csrc.nist.gov/groups/ST/toolkit/BCM/

38. Mennink, B.: XPX: generalized tweakable Even-Mansour with improved security guarantees. IACR Cryptology ePrint Archive 2015, 476 (2015). http://eprint.iacr.org/2015/476

39. NIST SP 800-38D: Recommendations for block cipher modes of operation: Galois/counter mode (GCM) and GMAC, November 2007

40. NIST SP 800-38E: Recommendation for block cipher modes of operation: The XTS-AES mode for confidentiality on storage devices (2010)

41. Peyrin, T., Sasaki, Y., Wang, L.: Generic related-key attacks for HMAC. In: Wang, X., Sako, K. (eds.) ASIACRYPT 2012. LNCS, vol. 7658, pp. 580–597. Springer, Heidelberg (2012). doi:10.1007/978-3-642-34961-4_35

42. Rogaway, P.: Efficient instantiations of tweakable blockciphers and refinements to modes OCB and PMAC. In: Lee, P.J. (ed.) ASIACRYPT 2004. LNCS, vol. 3329, pp. 16–31. Springer, Heidelberg (2004). doi:10.1007/978-3-540-30539-2_2

43. Rogaway, P., Bellare, M., Black, J., Krovetz, T.: OCB: a block-cipher mode of operation for efficient authenticated encryption. In: Reiter, M.K., Samarati, P. (eds.) ACM Conference on Computer and Communications Security, pp. 196–205. ACM (2001)

44. Shoup, V.: On fast and provably secure message authentication based on universal hashing. In: Koblitz, N. (ed.) CRYPTO 1996. LNCS, vol. 1109, pp. 313–328. Springer, Heidelberg (1996). doi:10.1007/3-540-68697-5_24

45. Shoup, V.: Sequences of games: a tool for taming complexity in security proofs. IACR Cryptology ePrint Archive 2004, 332 (2004). http://eprint.iacr.org/2004/332

46. Stinson, D.R.: Universal hashing and authentication codes. In: Feigenbaum, J. (ed.) CRYPTO 1991. LNCS, vol. 576, pp. 74–85. Springer, Heidelberg (1992). doi:10.1007/3-540-46766-1_5

47. Stinson, D.R.: On the connections between universal hashing, combinatorial designs and error-correcting codes. In: Electronic Colloquium on Computational Complexity (ECCC), vol. 2, no. 52 (1995). http://eccc.hpi-web.de/eccc-reports/1995/TR95-052/index.html

48. Sun, S., Hu, L., Wang, P., Qiao, K., Ma, X., Song, L.: Automatic security evaluation and (related-key) differential characteristic search: application to SIMON, PRESENT, LBlock, DES(L) and other bit-oriented block ciphers. In: Sarkar, P., Iwata, T. (eds.) ASIACRYPT 2014. LNCS, vol. 8873, pp. 158–178. Springer, Heidelberg (2014). doi:10.1007/978-3-662-45611-8_9

49. Sun, Z., Wang, P., Zhang, L.: Weak-key and related-key analysis of hash-counter-hash tweakable enciphering schemes. In: Foo, E., Stebila, D. (eds.) ACISP 2015. LNCS, vol. 9144, pp. 3–19. Springer, Heidelberg (2015). doi:10.1007/978-3-319-19962-7_1

50. Wang, P., Feng, D., Wu, W.: HCTR: a variable-input-length enciphering mode. In: Feng, D., Lin, D., Yung, M. (eds.) CISC 2005. LNCS, vol. 3822, pp. 175–188. Springer, Heidelberg (2005). doi:10.1007/11599548_15

51. Wang, P., Li, Y., Zhang, L., Zheng, K.: Related-key almost universal hash functions: definitions, constructions and applications. Cryptology ePrint Archive, Report 2015/766 (2015). http://eprint.iacr.org/

52. Wegman, M.N., Carter, L.: New hash functions and their use in authentication and set equality. J. Comput. Syst. Sci. 22(3), 265–279 (1981)

53. Wiener, M. (ed.): CRYPTO 1999. LNCS, vol. 1666. Springer, Heidelberg (1999)

Authenticated-Encryption and Hash Function Cryptanalysis

Key Recovery Attack Against 2.5-Round π-Cipher

Christina Boura[1]([✉]), Avik Chakraborti[2], Gaëtan Leurent[3], Goutam Paul[2],
Dhiman Saha[4], Hadi Soleimany[5,6], and Valentin Suder[7]

[1] University of Versailles, Versailles, France
christina.boura@uvsq.fr
[2] Indian Statistical Institute, Kolkata, India
avikchkrbrti@gmail.com, goutam.paul@isical.ac.in
[3] Inria, Project-Team SECRET, Paris, France
gaetan.leurent@inria.fr
[4] Crypto Research Lab, Indian Institute of Technology Kharagpur, Kharagpur, India
saha.dhiman@gmail.com
[5] Cyberspace Research Institute, Shahid Beheshti University, Tehran, Iran
h_soleimany@sbu.ac.ir
[6] School of Computer Science, Institute for Research in Fundamental Sciences (IPM),
Tehran, Iran
[7] University of Waterloo, Waterloo, Canada
valentin@suder.xyz

Abstract. In this paper, we propose a *guess and determine* attack against some variants of the π-Cipher family of authenticated ciphers. This family of ciphers is a second-round candidate of the CAESAR competition. More precisely, we show a key recovery attack with time complexity little higher than $2^{4\omega}$, and low data complexity, against variants of the cipher with ω-bit words, when the internal permutation is reduced to 2.5 rounds.

In particular, this gives an attack with time complexity 2^{72} against the variant π16-Cipher096 (using 16-bit words) reduced to 2.5 rounds, while the authors claim 96 bits of security with 3 rounds in their second-round submission. Therefore, the security margin for this variant of π-Cipher is very limited.

The attack can also be applied to lightweight variants that are not included in the CAESAR proposal, and use only two rounds. The lightweight variants π16-Cipher096 and π16-Cipher128 claim 96 bits and 128 bits of security respectively, but our attack can break the full 2 rounds with complexity 2^{72}.

Finally, the attack can be applied to reduced versions of two more variants of π-Cipher that were proposed in the first-round submission with 4 rounds: π16-Cipher128 (using 16-bit words) and π32-Cipher256 (using 32-bit words). The attack on 2.5 rounds has complexity 2^{72} and 2^{137} respectively, while the security claim for 4 rounds are 128 bits and 256 bits of security.

Keywords: Authenticated encryption · π-Cipher · CAESAR competition · Guess and determine · Cryptanalysis

© International Association for Cryptologic Research 2016
T. Peyrin (Ed.): FSE 2016, LNCS 9783, pp. 535–553, 2016.
DOI: 10.1007/978-3-662-52993-5_27

1   Introduction

Authenticated encryption is a rapidly growing field of cryptography that has wide applications in diverse industries. Even though some efforts over the past few years have been devoted to the design and analysis of authenticated encryption schemes, a well-studied design with the desirable level of security and performance is not yet available. Lack of secure and efficient authenticated ciphers led to devastating attacks in extensive applications like TLS and OpenSSL [1,4]. To address this challenge, an international contest called CAESAR, funded by the NIST, plans to hold a multi-year effort to identify a promising new portfolio of authenticated ciphers, suitable for widespread applications [3]. The CAESAR competition, launched in 2014, follows the long tradition of contests in secret key cryptography and aims at selecting a portfolio of authenticated ciphers that offer perceptible advantages over AES-GCM and that can be recommended for widespread use. There were 57 proposals accepted for the first round of the competition and recently, 30 ciphers among these proposals were selected to continue in the second round.

The π-Cipher [7] family of authenticated ciphers, designed by Gligoroski et al., is one of the 30 second-round candidates. It is a special case of encrypt-then-MAC designs and makes use, as all such CAESAR candidates, of a nonce and process associated data.

One of the most important design goals of this family of cryptographic functions is the possibility of parallel computations. Other goals, as claimed by the designers, are a better security than AES-GCM in the case of a nonce reuse, and better resistance for producing second-preimage tags. Although the cipher's mode of operation is inspired by the sponge construction [2], and is based on a permutation called the π-function, it has been largely modified by Gligoroski et al. in order to permit parallel computations.

In the initial submission, the authors proposed six different variants of the cipher, where each variant offered a particular level of security and used words of a particular size. More precisely, the level of targeted security, corresponding to the size of the secret key, ranges from 96 to 256 bits, and each variant uses words of 16, 32, or 64 bits. For the second round of the competition, only four variants were kept. Another decision taken by the designers for the second-round version of the cipher, was to decrease the number of rounds of the π-function from 4 to 3. In addition, at NIST's lightweight cryptography workshop, a lightweight version of the π-Cipher [10] was proposed. The lightweight proposal is composed of two variants, both using 16-bit words. Since lightweight ciphers must be as small and power-efficient as possible, the number of rounds in the internal permutation is further reduced to 2 in the lightweight version. An overview of the different variants is given in Table 1.

Our Results. In this work, we present a key recovery attack against several variants of the π-Cipher, when the π-function is reduced to 2.5 rounds. This shows that the decision to decrease the number of rounds was precarious. Indeed, the lightweight version is completely broken, and the affected variant that is still in the second round submission offers only very limited security margin.

More precisely, the time complexity of our attack is 2^{72} for the 16-bit word variants and 2^{137} for the 32-bit word variants, while the data complexity remains very low (a single known plaintext with at least 256 blocks for 16-bit word variants, and 512 blocks for the 32-bit word variants). The attack is faster than exhaustive search of the key for the following variants (reduced to 2.5 rounds):

π**16-Cipher096** with 16-bit words and 96-bit key.
This variant was proposed with 4 rounds in version 1, 3 rounds in version 2, and 2 rounds in the lightweight version.
π**16-Cipher128** with 16-bit words and 128-bit key.
This variant was proposed with 4 rounds in version 1, and 2 rounds in the lightweight version.
π**32-Cipher256** with 32-bit words and 256-bit key.
This variant was proposed with 4 rounds in version 1.

Our cryptanalysis is a *guess and determine* attack exploiting a weakness in the high-level structure of the π-function. Indeed, we show that by knowing two out of the four output chunks of the π-function and by guessing a third one, we can easily recover one of the four input chunks of the permutation. This permits us to recover the internal state and gives us the possibility to recover the secret key by some very simple operations. Note that our attacks work in the case when no secret message number is processed. However, the attacks can be easily extended in cases when a secret message number is used, if one supposes that the secret message number is known together with the plaintext.

Cryptographic algorithms should be designed with enough security margin to thwart classical attacks but also to resist to new and unknown vulnerabilities. Surplus security cannot be obtained for free, since it has impacts on the performance of the ciphers. In particular, due to a number of important limitations in the resources of pervasive devices, it is of utmost importance to analyze lightweight cryptographic designs that allow reduction of superfluous margins. Our attack shows that the security margin offered by these three members of the π-Cipher family is too small and that these variants are much less secure than expected. This kind of analysis is very important for the progress of the CAESAR competition, as the final portfolio of the selected authenticated ciphers should offer a high level of security. Thus, evaluating the security of the remaining candidates, leads to a more clear overview of which candidates are robust and which should be eliminated.

Outline. The rest of the paper is organised as follows. In Sect. 2 we briefly provide the specifications of π-Cipher. Then, we present our attack on 2.5 round π-Cipher in Sect. 3 and we discuss how to mount a full-round attack on the lightweight version of π-Cipher in Sect. 4. Finally, we perform a complexity analysis of our attacks in Sect. 5 and conclude.

2 π-Cipher Specifications

There exist different variants of π-Cipher, depending on the bit-length of the words used and the expected level of security expressed in bits. Therefore,

$\pi\omega$-Cipher$n$ represents a variant defined with ω-bit words and offering n-bit security. The six variants of π-Cipher submitted to the first round of the competition, together with the corresponding parameters, are summarized in Table 1. The first four rows in the table represent the only four variants conserved for the second round. Furthermore, the two variants of the recently presented lightweight π-Cipher proposal [10], are described in the last two rows of Table 1.

Table 1. π-Cipher variants. The first four rows represent the four variants kept for the second round of the CAESAR competition. The last two rows describe the two lightweight variants proposed in [10]. PMN and SMN are the two parts of the nonce and stand for *Public Message Number* and *Secret Message Number* respectively. All the parameters are given in bits. For variants both in version 1 and 2, there are 4 rounds in v1 and 3 rounds in v2.

Version	Variant	Word size ω	PMN	SMN	Rate r	Tag size t	Key length	Rounds
v1 & v2	π16-Cipher096	16	32	0 or 128	128	128	96	3
	π32-Cipher128	32	128	0 or 256	256	256	128	3
	π64-Cipher128	64	128	0 or 512	512	512	128	3
	π64-Cipher256	64	128	0 or 512	512	512	256	3
v1	π16-Cipher128	16	32	0 or 128	128	128	128	4
	π32-Cipher256	32	128	0 or 256	256	256	256	4
Lightweight	π16-Cipher096	16	32	0 or 128	128	128	96	2
	π16-Cipher128	16	32	0 or 128	128	128	128	2

2.1   Authenticated Encryption

The encryption/authentication function accepts as input a triplet (K, AD, M), where K is a secret key, AD is a string of associated data of a blocks, and M is a message composed of m blocks of size r bits each. The main building block of the authenticated encryption procedure is a construction that the authors call the *e-triplex* component and which is depicted in Fig. 1. The encryption procedure starts by initializing the internal state with the string $K||PMN||10^*$, where the number of 0's appended should be such that the length of the concatenated string equals the size of the state of the π-function. This internal state is then updated by applying the π-function. The result is called the *Common Internal State* (CIS) and is used as the initial state for the first parallel computations:

$$CIS \leftarrow \pi(K||PMN||10^*).$$

By following the same notation as in the sponge construction, we can see each internal state, say IS, as the concatenation of a *rate* part and a *capacity* part: $IS = IS_{capacity}||IS_{rate}$. In particular, each internal state IS of the

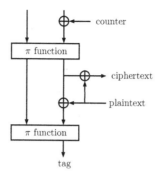

Fig. 1. The e-triplex component of π-Cipher.

procedure is the concatenation of four 4ω-bit chunks, that we will denote as $IS = IS_1\|IS_2\|IS_3\|IS_4$. From the specification of π-Cipher, the capacity part of the state is $IS_{capacity} = IS_2\|IS_4$, and the rate part of the state is $IS_{rate} = IS_1\|IS_3$. The counter, denoted by ctr, is then initialized by extracting the first 64 bits of $CIS_{capacity}$. This procedure is depicted at the top left part of Fig. 2.

The next step in the authenticated encryption procedure is the process of the associated data. The associated data AD is cut into equal-sized blocks: $AD = AD_1\|\ldots\|AD_a$. All blocks are treated in parallel by the e-triplex component. The input to the e-triplex component for the block i is CIS, $ctr + i$ and AD_i, and the output is an intermediate tag t'_i. The way that each block of associated data is processed can be observed in Fig. 2. At the end of this procedure a tag for the associated data T' is computed as

$$T' = t'_1 \boxplus_d \cdots \boxplus_d t'_a,$$

where $\boxplus_d$ is a component-wise addition of vectors of dimension d, where d is the number of ω-bit words in the *rate* part ($d = 8$ for all proposed variants of π-Cipher). Finally, the internal state is updated in the following way to create a new internal state that we will denote by CIS':

$$CIS' \leftarrow \pi(CIS_{capacity}\|CIS_{rate} \oplus T').$$

After this first phase, the secret message number SMN, if any, is processed. This procedure is depicted in Fig. 2 and described by the following expressions:

$$IS \leftarrow \pi(CIS'_{capacity}\|CIS'_{rate} \oplus (ctr + a + 1)),$$
$$CIS'' \leftarrow \pi(IS_{capacity}\|IS_{rate} \oplus SMN).$$

The new state CIS'' will be used as the common state for the parallel process of the message blocks. The tag produced during this phase is

$$T'' = T' \boxplus_d t_0,$$

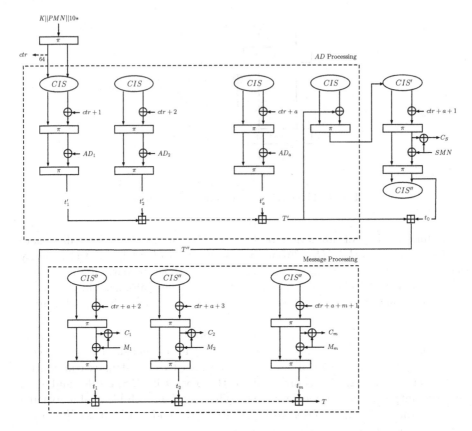

Fig. 2. π-*Cipher* encryption structure.

where t_0 is the output tag of the last call to the e-triplex component after absorbing the SMN. If no secret message number is used, then the above steps are ignored. The authenticated encryption procedure without SMN is depicted in Fig. 4.

In the last phase, the message blocks are treated. As for the associated data, the message M is cut into blocks $M = M_1 || \ldots || M_m$ and each block is processed in parallel by the e-triplex construction. Note that the length of each message block, as well as of each ciphertext block is equal to the bitrate, i.e. r bits (e.g. $r = 128$ in the case of π16-Cipher096). A unique block counter is associated with each message block. The counter for the message block M_j is computed as $ctr+a+j$ if the secret message number is empty, and as $ctr+a+1+j$ otherwise.

During encryption, each e-triplex component takes as input the common state CIS'', the counter ctr and a message block M_j and outputs a pair (C_j, t_j), where C_j is a ciphertext block and t_j is a partial tag. The final tag T is computed as

$$T = T'' \boxplus_d t_1 \cdots \boxplus_d t_m.$$

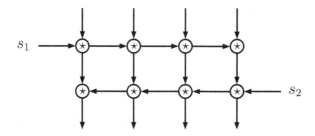

Fig. 3. One round of the π-function.

2.2 The π-Function

The core of π-Cipher is an ARX-based permutation called the π-function. This permutation somehow uses similar operations as the hash function Edon-R [8]. We denote the size of the permutation in bits by b and the number of rounds by R. For the first version of the cipher, R was fixed to 4, however the authors decided to reduce this number to 3 for the second round of the competition. The internal state (IS) of the π-function can be seen as a concatenation of four chunks of four words, so that $b = 4 \times 4 \times \omega$ bits. The π-function is mainly based on an operation that will be denoted by $\circledast$. However, as our attack does not take advantage of the internal structure of $\circledast$ we omit here its description. The only important thing to know about this operation in order to understand the attack is that it is a 2-input 1-output operation (in Fig. 3, the two outputs of a $\circledast$ operation are equal) that is invertible with respect to each of its inputs. Its full specifications can be found in [7]. A round of the π-function is depicted in Fig. 3, where S_1 and S_2 are constants.

2.3 Previous Cryptanalysis Results

In [6], Fuhr and Leurent showed that forgeries can be computed for the first round variants of π-Cipher due to a weakness in the padding algorithm. More precisely, they noticed that the padding used for both the associated data and the plaintext was not injective. This observation permitted to mount a forgery attack by producing valid tags and forced the designers to modify the padding rule for the second round of the competition.

One of the advertised features of π-Cipher is tag second-preimage resistance, meaning that it should be hard to generate a message with a given tag, even for the legitimate key holder. However, Leurent demonstrated in [9] that practical tag second-preimage attacks could be mounted against π-Cipher by using Wagner's generalized birthday attack. More specifically, Leurent showed that tag second-preimages can be computed with optimal complexities ranging from 2^{22} to 2^{45} depending on the word size ω.

The best attack mentioned by the designers [7, Sect. 3.3] is a distinguisher on reduced versions with 1 round, using a guess and determine technique.

Their attack has complexity about $2^{4\omega}$ (time and memory); in particular, it is applicable to the same variants as our attack. Our attack actually uses similar ideas, but reaches 2.5 rounds, and a full key recovery.

3   Key Recovery Attack Against 2.5-Round π-Cipher

We describe in this section our key recovery attack against reduced-round variants of π-Cipher when no secret message number (SMN) is used. The authenticated-encryption procedure for this case is described in Fig. 4. Note that if no SMN is used then the intermediate tags T' and T'' are equal and that the state CIS'' of Fig. 2 is equal to the state CIS'. In order to be consistent with the notation of Sect. 2, we will keep denoting the common state for processing the message blocks as CIS'' even if this is exactly the same as CIS' in the empty SMN case.

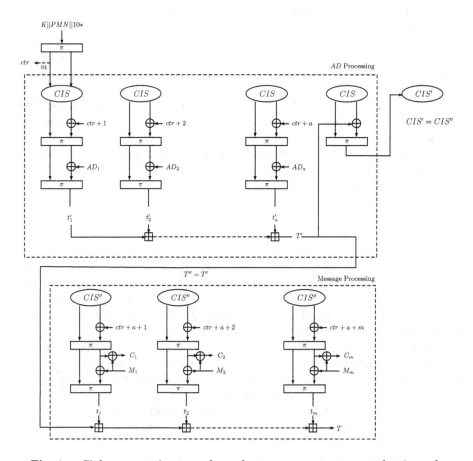

Fig. 4. π-Cipher encryption procedure when no secret message number is used.

We consider an m-block message $M = M_1||\cdots||M_m$ and an a-block string of associated data, with the corresponding ciphertext $C = C_1||\cdots||C_m$. The message should have at least 16ω blocks, *i.e.* 256 blocks when $\omega = 16$, and 512 blocks when $\omega = 32$.

We denote the input and output states of the first π-function for processing the message block M_i by $I^i = I_1^i||I_2^i||I_3^i||I_4^i$ and $O^i = O_1^i||O_2^i||O_3^i||O_4^i$ respectively, where each chunk I_j^i, O_j^i, for $1 \le j \le 4$, is of size 4ω bits.

In our attack, we deploy a *guess and determine* technique for recovering the secret key for three variants of the π-Cipher family, where the π-function is reduced to 2.5 rounds. Our attack targets the first π-function of the message processing phase, for 16ω consecutive blocks of plaintext. We provide now the main observations that the attack takes advantage of.

3.1   Observations on the π-Cipher Structure

The first observation concerns the nature of the inner operation $\circledast$, that takes two chunks of size 4ω bits as input and outputs a single chunk of the same size. This operation is the core of the π-function. It has the property, that when fixing one of the two input chunks to a constant and letting the other chunk take all possible values, then the output chunk equally takes all possible values (it defines a quasi-group).

Observation 1. *Both $\circledast(a,.)$ and $\circledast(.,b)$ are invertible for all $a, b \in \mathbb{F}_2^{4\omega}$ and if $\circledast(a,b) = c$, then the knowledge of any two chunks among a, b and c can determine the third one.*

The next observation is in the core of the guess and determine technique and exploits a weakness in the high-level structure of the π-function. It shows, that when the function is reduced to 2.5 rounds, the knowledge of 3 output chunks of 4 words each, can completely determine an input chunk. This observation demonstrates that the inverse π-function has a limited diffusion when the number of rounds is reduced to 2.5, as we can see that in this case an input word does not depend on all the output words.

Observation 2. *Let, $I = I_1||I_2||I_3||I_4$ and $O = O_1||O_2||O_3||O_4$ be the input and the output state respectively of the π-function reduced to 2.5 rounds. Then the knowledge of O_1, O_3 and a guess of O_2 can determine I_1.*

Proof. This claim can be proven by the following guess and determine steps described below. The pictorial description of the steps is given in Fig. 5. In the figure the green boxes denote the determined chunks D_i, $1 \le i \le 9$, the orange boxes denote the guessed chunk i.e. O_2 and the chunks denoted by K_1, K_2 corresponding to O_1 and O_3 respectively are known. At the end of this procedure, one computes D_9 which corresponds exactly to I_1. Note that each step of the below procedure makes use of Observation 1.

1. Use K_1, S_1 and G to determine D_1 and D_2.
2. Use K_2 and G to determine D_3.
3. Use D_1 and D_2 to determine D_4.
4. Use D_2 and D_3 to determine D_5 and D_4, S_1 to determine D_6.
5. Use D_4 and D_5 to determine D_7.
6. Use D_6 and D_7 to determine D_8.
7. Use D_8 and S_1 to determine D_9.   □

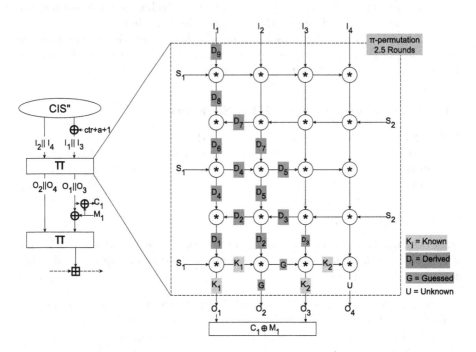

Fig. 5. Guess and determine steps for the first π-function. (Color figure online)

The last observation aims at showing that the knowledge of the input state of the π-function for several message blocks can be used to determine the common state CIS''.

Observation 3. *The message processing phase uses the same common internal state, $CIS'' = CIS_1''||CIS_2''||CIS_3''||CIS_4''$, to process each of the message blocks M_i, $1 \leq i \leq m$. Then, the input to the first π-function is $I^i = I_1^i||I_2^i||I_3^i||I_4^i = CIS_1'' \oplus (ctr + a + i)||CIS_2''||CIS_3''||CIS_4''$ for each block.*

3.2 High Level Description of the Attack

This section provides a high level description of our attack. As already mentioned, the attack requires a single known plaintext message, with at least 16ω blocks. The attack can be seen as the succession of the five main steps that we describe below:

1. *Guess and determine* step. In this first part of the attack, we target the first computation of the π-function in the message processing part. Two of the output chunks are known to the attacker as they only depend on the plaintext and ciphertext blocks (i.e. $O_1^i \| O_3^i = M_i \oplus C_i$). Then by guessing a third output chunk, namely O_2^i, we are able to determine one input chunk, I_1^i. We repeat this procedure for all message blocks. This step is described in more details in Subsect. 3.3. At the end of this part we are left with a collection of lists of candidates for one input chunk. We recover the right value by treating the lists in the way described in the next step.

2. *Computation of the intersection of the created lists.* During this phase, detailed in Subsect. 3.4, we show how to treat the created lists in order to recover the right value of the common part for the first input chunk of the π-function, or more precisely, of the value $CIS_1'' \oplus (ctr + a)$ from Observation 3.

3. *Recovery of the intermediate I^i state.* This step shows the procedure to recover a list of candidates for the state I^i and is described by the *Recover-IS* Algorithm in Subsect. 3.5.

4. *Recovery of the common internal state CIS.* We show here how one can compute the state CIS, once the intermediate state I^1 has been completely identified. This phase is described by the *Recover-CIS Algorithm* in Subsect. 3.6.

5. *Computation of the secret key.* This phase is pretty straightforward once we have recovered CIS, since, as already mentioned in Sect. 2.1, $CIS = \pi(K \| PMN \| 10^*)$ and π-function is a known permutation.

The high level description of the attack is furnished in Algorithm 1.

Algorithm 1. Overview of the attack.

Input: 1 Known Plaintext-Ciphertext Pair $(M = M_1 \| \cdots \| M_{16\omega}, C = C_1 \| \cdots \| C_{16\omega})$
Output: Master Key K

1: **for all** $1 \le i \le 16\omega$ **do**
2:     $\mathcal{L}_i \leftarrow$ GUESS-DETERMINE(M_i, C_i) ▷ Subsect. 3.3
3: **for all** $1 \le j \le 8\omega$ **do**
4:     $\mathcal{S} \leftarrow \bigcap_{0 \le k < 8\omega} \mathcal{L}_{j+k} \oplus k$ ▷ Subsect. 3.4
5:     **if** $\mathcal{S} \ne \varnothing$ **then**
6:        $\mathcal{L}_0' \leftarrow$ RECOVER-IS$(M_j, C_j, 0, \mathcal{S})$ ▷ Subsect. 3.5
7:        $\mathcal{L}_1' \leftarrow$ RECOVER-IS$(M_{j+1}, C_{j+1}, 1, \mathcal{S})$
8:        $I^j, I^{j+1} \leftarrow \{I, J \in \mathcal{L}_0' \times \mathcal{L}_1' \mid I_2 \| I_3 \| I_4 = J_2 \| J_3 \| J_4\}$ ▷ Single value expected
9:        **for all** ctr, s.t. $ctr + a + j \equiv 0 \bmod 8\omega$ **do**   ▷ Subsect. 3.6
10:           $CIS'' \leftarrow I^j \oplus (ctr + a + j)$
11:           $CIS \leftarrow$ RECOVER-CIS(CIS'')
12:           **if** $ctr = $ first 64 bits of $CIS_{capacity}$ **then**
13:              $K \| PMN \| 10^* \leftarrow \pi^{-1}(CIS)$
14:              **return** K

3.3   Guess and Determine

This section describes the guess and determine phase, which recovers the input chunk I_1^i of the first π-function for the ith block for the plaintext-ciphertext pair $(M = M_1||\cdots M_i\cdots||M_m, C = C_1||\cdots C_i\dots||C_m)$. Note that we can compute $O_1^i||O_3^i = M_i \oplus C_i$. Then by making a guess on the value of O_2^i, we can compute I_1^i independently of O_4^i, following Observation 2. In particular, we can compute it as $I_1 = \pi^{-1}(O_1||O_2||O_3||\langle 0 \rangle)$.

We compute all candidates for I_1^i corresponding to the 2^{4w} choices of O_2^i, and store them in a list $\mathcal{L}_i$. The guess and determine phase is described in Algorithm 2.

Note that there will be less than 2^{4w} different values of I_1^i in a list $\mathcal{L}_i$ as the π-function is a permutation of the four chunks and not a permutation from one chunk (O_2^i) to one chunk (I_1^i). In the following, we assume that the function from O_2^i to I_1^i behaves as a random function, so that the expected size of $\mathcal{L}_i$ is $(1-e^{-1}) \times 2^{4w}$ (see [5, Theorem 2]). In the next part, we describe how to compute the intersection and filter out the correct value of I_1^i for some $1 \leq i \leq 16w$.

Algorithm 2. Build the list of candidates for the first input chunk of the first π-function.

Input: Plaintext-ciphertext block M, C
Output: List $\mathcal{L}$ of possible candidates for I_1
  1: **function** GUESS-DETERMINE(M, C)
  2:      $\mathcal{L} \leftarrow \varnothing$
  3:      $O_1||O_3 \leftarrow M \oplus C$
  4:      **for all** O_2 **do**
  5:          $I_1 \leftarrow \pi^{-1}(O_1||O_2||O_3||\langle 0 \rangle)$ ▷ Following Observation 2
  6:          $\mathcal{L} \leftarrow \mathcal{L} \cup \{I_1\}$
  7:      **return** $\mathcal{L}$

3.4   Intersecting the Lists

In this phase, we compare the list of candidates for I_1^i for each message block, using the fact that they are all derived from a common state CIS''. More precisely, the first input chunk to the first π-function of each block is computed as:

$$I_1^i = CIS_1'' \oplus (ctr + a + i), \qquad \text{for } 1 \leq i \leq 16w.$$

By construction of the lists $\mathcal{L}_i$, we have that:

$$CIS_1'' \oplus (ctr + a + i) \in \mathcal{L}_i, \qquad \text{for } 1 \leq i \leq 16w.$$

Let $j \in \{1, \dots, 8w\}$ be such that $ctr + a + j \equiv 0 \mod 8w$ (*i.e.* $j \equiv -(ctr + a) \mod 8w$). In other words, with $w = 16$, j is the first message block such that the 7 least significant bits of $ctr + a + j$ are equal to zero (and similarly, 8 bits when $\omega = 32$). This implies:

$$(ctr + a + j) + k = (ctr + a + j) \oplus k \qquad \text{for } 0 \le k < 8\omega$$
$$CIS_1'' \oplus (ctr + a + j) \oplus k \in \mathcal{L}_{j+k} \qquad \text{for } 0 \le k < 8\omega$$
$$CIS_1'' \oplus (ctr + a + j) \in \mathcal{L}_{j+k} \oplus k \qquad \text{for } 0 \le k < 8\omega$$

Thus,

$$CIS_1'' \oplus (ctr + a + j) \in \bigcap_{k=0}^{8\omega-1} (\mathcal{L}_{j+k} \oplus k).$$

We will compute this intersection for all guesses of $j \in \{1, \ldots, 8\omega\}$. We are interested now in determining the size of the intersection of the 8ω lists. Each list has about $(1 - e^{-1})2^{4\omega}$ elements. If the guess of j is wrong, we assume that the lists are independent; an element is a part of all the 8ω lists with probability $(1 - e^{-1})^{8\omega}$. As there is a total of $2^{4\omega}$ elements, the probability that there is no element in the intersection is $(1 - (1 - e^{-1})^{8\omega})^{2^{4\omega}}$. This probability is very close to one:

$$\left(1 - \left(1 - e^{-1}\right)^{8\omega}\right)^{2^{4\omega}} = \exp\left(2^{4\omega} \ln(1 - \left(1 - e^{-1}\right)^{8\omega})\right)$$
$$\ge 1 + 2^{4\omega} \ln\left(1 - \left(1 - e^{-1}\right)^{8\omega}\right)$$
$$\approx 1 - 2^{4\omega}\left(1 - e^{-1}\right)^{8\omega}$$
$$\approx 1 - 0.9^{8\omega}$$

In particular, it is about $1 - 2^{-20}$ for $\omega = 16$.

On the contrary, if the guess is right, the intersection contains 1 element. With high probability, the test at line 5 of Algorithm 1 will succeed only for the correct value of j, and the corresponding set $\mathcal{S}$ will contain a single value.

3.5   Recovering the Intermediate State

So far, we have recovered the value $CIS_1'' \oplus ctr + a + j$, that is to say the first chunk I_1^j of the input of the first π-function. In addition, the least significant bits of $ctr + a + j$ are known to be zero, so that we can compute $I_1^{j+k} = I_1^j \oplus k$ for $0 \le k < 8\omega$ (adjusting the effect of the counter).

From this, we can build a small list of candidates for any O_2^{j+k}. We just have to try all $2^{4\omega}$ values O_2^{j+k}, recompute I_1^{j+k}, and compare the result to the known value. We know that there will be at least one remaining value, and there can be a few false positives.

Now we make a guess of O_4^{j+k} and use the invertibility of the π-function to built a list $\mathcal{L}_k'$ of all potential values of the full input I^{j+k} of the permutation.

This second phase of guess and determine through the π-function is demonstrated in Fig. 6. The list $\mathcal{L}'_k$ contains about $2^{4\omega}$ values. This step is described in Algorithm 3.

In order to identify the correct value in the list, we build the lists $\mathcal{L}'_0$ and $\mathcal{L}'_1$, and we use the way I^j and I^{j+1} are derived from CIS''. In particular, we have $I^j_2\|I^j_3\|I^j_4 = I^{j+1}_2\|I^{j+1}_3\|I^{j+1}_4$. This allows us to recover the correct value I^j and I^{j+1}.

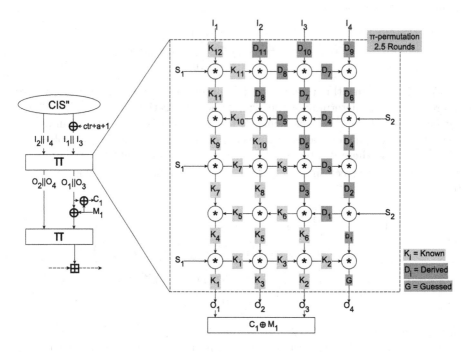

Fig. 6. Guessing O_4 after I_1 has been determined (Color figure online)

3.6   Recovering the Common Internal State CIS

In this section we show how to recover the common internal states CIS'' and CIS. We remind once again, that the state CIS' is equal to CIS''. From the previous sections, the input state of the first π-function for message block j, I^j has been recovered. Note that

$$I^j = I^j_1\|I^j_2\|I^j_3\|I^j_4 = CIS''_1 \oplus (ctr + a + j)\|CIS''_2\|CIS''_3\|CIS''_4.$$

By making a guess for the value of the counter ctr, we can compute the value of CIS'' which equals CIS'.

The next step is to retrieve the tag T'' and therefore T' (since both tags are equal) by computing $T'' = T \boxminus_d t_1 \boxminus_d \cdots \boxminus_d t_{16\omega}$, where each tag t_i, $1 \le i \le 16\omega$ can be recovered from the knowledge of CIS'', ctr and the message blocks.

Algorithm 3. Build the list of candidates for the full input of the first π-function, knowing the first input chunk.

Input: Plaintext-ciphertext block M, C; index k; list of I_1 candidates $\mathcal{S}$
Output: List $\mathcal{L}$ of candidates for $I_2||I_3||I_4$
1: **function** RECOVER-IS$(M, C, k, \mathcal{S})$
2:   $\mathcal{L} \leftarrow \varnothing$
3:   $O_1||O_3 \leftarrow M \oplus C$
4:   **for all** O_2 **do**
5:     $I \leftarrow \pi^{-1}(O_1||O_2||O_3||\langle 0\rangle)$
6:     **if** $I_1 \oplus k \in \mathcal{S}$ **then**   ▷ Only one candidate expected
7:       **for all** O_4 **do**
8:         $I \leftarrow \pi^{-1}(O_1||O_2||O_3||O_4)$
9:         $\mathcal{L} \leftarrow \mathcal{L} \cup \{I_2||I_3||I_4\}$
10:   **return** $\mathcal{L}$

Once this step is done, the recovery of the common internal state CIS is immediate, as one can compute it as $CIS = \pi^{-1}(CIS') \oplus T'$. Note that, at this point, we can easily verify if the guess of ctr was correct, since ctr corresponds to 64 bits extracted directly from the initial state CIS (as described in Sect. 2.1). The above procedure is described by Algorithm 4.

Algorithm 4. Recover the initial state CIS.

Input: Common Internal State CIS'', corresponding message M
Output: Common Internal State CIS
1: **function** RECOVER-CIS(CIS'', M)
2:   **for** $1 \leq i \leq 16\omega$ **do**
3:     Compute t_i from CIS'' and M_i
4:   $T' = T \boxminus_d t_1 \boxminus_d \cdots \boxminus_d t_{16\omega}$
5:   $CIS \leftarrow \pi^{-1}(CIS')_{capacity}||\pi^{-1}(CIS')_{rate} \oplus T'$
6:   **return** CIS

3.7   Key Recovery

Once the internal state CIS has been successfully recovered, one can retrieve the master key K by simply inverting the π-function, as described by Line 13 of Algorithm 1.

3.8   About the use of SMN

The above described analysis supposes that no secret message number is used. This is a legitimate assumption, as $|SMN| = 0$ is a valid scenario mentioned in the cipher's proposal. Our attack can be easily extended to the case when an SMN is used if one supposes that this number is known to the attacker together with the plaintext. In the case that the knowledge of SMN is not available to the attacker, our analysis fails. However, it is still possible to mount a forgery attack in this case.

More precisely (see Fig. 2), if one is given an m-block message M with associated data AD and the corresponding tag T, one can easily construct a forgery as follows. Suppose that the new message M^{forged} has $(m + 1)$ blocks where the first m blocks are identical to the first m blocks of M (i.e., M is a prefix of M^{forged}) and the last block of M^{forged} is any fixed value. We follow the steps of Algorithm 1 with message M up to Step 8. At this point we intend to recover ctr. However, we cannot follow the same strategy as the one followed in Algorithm 1 since CIS cannot be recovered without the knowledge of SMN. But we can use the value of C_s which is the output of the SMN processing branch (see Fig. 2). So basically we guess ctr to determine CIS'' as before. Subsequently, we ascertain the value ctr by exploiting the relation $(\pi^{-1}(CIS''))_{rate} = C_s$. Since at this point, ctr is known, we can easily compute t_{m+1} and thus, the new tag T^{forged} will be given by $T \boxplus t_{m+1}$.

4   Key Recovery Attack Against Full Round Lightweight Version of π-Cipher

We argue here that the previously presented attack against various versions of the π-Cipher CAESAR candidate, completely breaks the lightweight version [10] of the same cipher, where the number of rounds is reduced to 2.

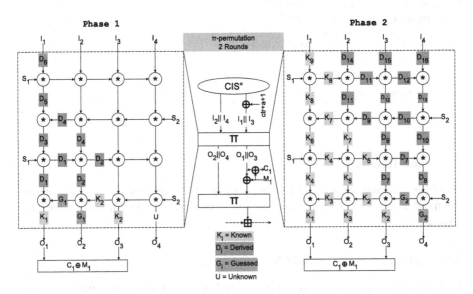

Fig. 7. Guess and determine phases for the attack on lightweight π-Cipher variants. (Color figure online)

The only difference with the previous attack is that, as the number of rounds is reduced, the guess and determine part of the attack is slightly modified to fit this reduction. This part, depicted at the left part of Fig. 7 is described by the following steps:

1. Use K_1 and G to determine D_1.
2. Use K_2 and G to determine D_2.
3. Use D_1 and S_1 to determine D_3.
4. Use D_1 and D_2 to determine D_4.
5. Use D_3 and D_4 to determine D_5.
6. Use D_5 and S_1 to determine D_6.

After the chunk I_1 has been determined, the other chunks I_2, I_3 and I_4 can be derived by further guessing the value of O_4, as shown at the right part of Fig. 7. The other steps of the attack remain unchanged, thus we ignore their full description.

5   Complexity Analysis

Time complexity. The two steps of the attack with the highest time complexity are the guess and determine step, and the intersection of lists. The guess and determine step involves 16ω lists and we evaluate the π-function $2^{4\omega}$ times for each list. This gives a time complexity of $16\omega \times 2^{4\omega}$ evaluations of the π-function.

Each list will be stored as a bit-field: we use an array of $2^{4\omega}$ bits, where a bit b is set to one if and only if the value b is in the list. This allows to compute the intersection of two lists efficiently, with only $2^{4\omega}$ bit-operations. We have to compute $64\omega^2$ list intersections at Line 4 of Algorithm 1. This amounts to a total complexity of $64\omega^2 \times 2^{4\omega}$ bit-operations.

Since a computation of the π-function obviously requires more than 4ω bit-operations, we will neglect the time complexity of lists intersection, and the total complexity is $16\omega \times 2^{4\omega}$ evaluations of the π-function. This leads to a time complexity of 2^{72} when $\omega = 16$ and 2^{137} when $\omega = 32$.

Memory complexity. The memory complexity of the attack comes from the storage of lists. As explained above, each list $\mathcal{L}_i$ takes only $2^{4\omega}$ bits, for a total storage of $16\omega \times 2^{4\omega}$ bits. On the other hand, lists $\mathcal{L}'_0$ and $\mathcal{L}'_1$ contain $2^{4\omega}$ values of 16ω bits, so we must store the full values. We can store a single list, and compute the intersections with the second list on the fly, so that this step also requires $16\omega \times 2^{4\omega}$ bits of storage.

For $\omega = 16$ this leads to a memory complexity of 2^{69} bytes, while for $\omega = 32$, we need to store 2^{134} bytes.

Table 2 presents a summary of our attacks on different variants of π-Cipher. The last three columns of this table contain the time, data and memory complexities of the attacks.

Table 2. Summary of our attacks against different variants of π-Cipher. The data complexity is counted as the number of known plaintexts. The minimal number of blocks of each plaintext is denoted in the parentheses.

Version	Variant	Word size ω	Security Claim	# Rounds Attacked	Time	Data (# KP)	Memory (bytes)
v1 & v2	π16-Cipher096	16	96	2.5/3	2^{72}	1 (256 B)	2^{69}
v1	π16-Cipher128	16	128	2.5/4	2^{72}	1 (256 B)	2^{69}
	π32-Cipher256	32	256	2.5/4	2^{137}	1 (512 B)	2^{134}
Lightweight	π16-Cipher096	16	96	2/2	2^{72}	1 (256 B)	2^{69}
	π16-Cipher128	16	128	2/2	2^{72}	1 (256 B)	2^{69}

6   Conclusion

In this work we provided an analysis of the security level offered by the π-Cipher family of authenticated ciphers. The designers of π-Cipher decided to decrease the number of rounds of the π-function from 4 to 3 for the second round of the CAESAR competition and to consider only 2 rounds for the recently proposed lightweight version. However, when reducing the number of rounds, special care must be taken, as this can lead to a dangerous reduction of the security margin offered by the new variants.

Our results indicate that π-Cipher, whose round function is reduced to 2.5 rounds, is vulnerable against guess and determine attacks. More precisely, we manage to recover the secret key in three reduced-round versions of the π-Cipher as well as in the two lightweight variants of the cipher. Taken together, these results suggest that the decision taken by the designers to reduce the number of rounds for the candidates of the second round of the CAESAR competition as well as for the lightweight version was risky.

In this work, we focused on the application of deterministic guess and determine properties. As a possible direction for future research, one can explore other guess and determine methods for breaking the full version of the cipher. Alternatively, it would be also challenging to see if the analysis of the properties of the ⊛ operation could lead to the extension of our attack to an extra half round. Furthermore, a question that naturally arises after this analysis is whether increasing the number of rounds of the cipher is the only remedy to resist to our attack, or whether there is another tweak that could be applied to render the cipher immune against such type of cryptanalysis.

Acknowledgments. This work was initiated during the group sessions of the 5th Asian Workshop on Symmetric Cryptography (ASK 2015) held in Singapore. Christina Boura and Gaëtan Leurent are partially supported by the French Agence Nationale de la Recherche through the BRUTUS project under Contract ANR-14-CE28-0015. Avik Chakraborti and Goutam Paul are thankful to the Centre of Excellence in Cryptology (Project CoEC) and R. C. Bose Centre for Cryptology and Security of Indian Statistical Institute for partial support towards their work. Finally, the work of Hadi Soleimany is partly supported by grants from IPM and Shahid Beheshti University.

References

1. AlFardan, N.J., Paterson, K.G.: Lucky thirteen: breaking the TLS and DTLS record protocols. In: Society, I.C. (ed.) IEEE Symposium on Security and Privacy (2013)
2. Bertoni, G., Daemen, J., Peeters, M., Van Assche, G.: On the indifferentiability of the sponge construction. In: Smart, N.P. (ed.) EUROCRYPT 2008. LNCS, vol. 4965, pp. 181–197. Springer, Heidelberg (2008)
3. CAESAR: Competition for Authenticated Encryption: Security, Applicability, andRobustness (2014). http://competitions.cr.yp.to/caesar.html/
4. Duong, T., Rizzo, J.: Here Come The XOR Ninjas (2011) (unpublished manuscript)
5. Flajolet, P., Odlyzko, A.M.: Random mapping statistics (2006). https://hal.inria.fr/inria-00075445
6. Fuhr, T., Leurent, G.: Observation on π-Cipher. CAESAR's competition mailing list, November 2014
7. Gligoroski, D., Mihajloska, H., Samardjiska, S., Jacobsen, H., El-Hadedy, M.,Jensen, R., Otte, D.: π-Cipher v2.0. submission to the CAESAR competition (2014). http://competitions.cr.yp.to/caesar-submissions.html/
8. Gligoroski, D., Ødegård, R.S., Mihova, M., Knapskog, S.J., Kocarev, L., Drápal, A., Klima, V.: Cryptographic hash function EDON-$\mathcal{R}'$. In: 1st International Workshop on Security and Communication Networks, pp. 85–95. IEEE (2009)
9. Leurent, G.: Tag second-preimage attack against π-Cipher, March 2014. https://hal.inria.fr/hal-00966794
10. Mihajloska, H., El-Hadedy, M., Gligoroski, D., Skadron, K.: Lightweight version of π-Cipher. In: NIST Lightweight Cryptography Workshop, July 2015

Cryptanalysis of Reduced NORX

Nasour Bagheri[1], Tao Huang[2], Keting Jia[3,4], Florian Mendel[5],
and Yu Sasaki[2,6(⊠)]

[1] SRTTU and IPM, Tehran, Iran
[2] Nanyang Technological University, Singapore, Singapore
[3] Department of Computer Science and Technology,
Tsinghua University, Beijing, China
[4] State Key Laboratory of Cryptology, P.O. Box 5159, Beijing 100878, China
[5] Graz University of Technology, Graz, Austria
[6] NTT Secure Platform Laboratories, Tokyo, Japan
sasaki.yu@lab.ntt.co.jp

Abstract. NORX is a second round candidate of the ongoing CAESAR competition for authenticated encryption. It is a nonce based authenticated encryption scheme based on the sponge construction. Its two variants denoted by NORX32 and NORX64 provide a security level of 128 and 256 bits, respectively. In this paper, we present a state/key recovery attack for both variants with the number of rounds of the core permutation reduced to 2 (out of 4) rounds. The time and data complexities of the attack for NORX32 are 2^{119} and 2^{66} respectively, and for NORX64 are 2^{234} and 2^{132} respectively, while the memory complexity is negligible. Furthermore, we show a state recovery attack against NORX in the parallel mode using an internal differential attack for 2 rounds of the permutation. The data, time and memory complexities of the attack for NORX32 are $2^{7.3}$, $2^{124.3}$ and 2^{115} respectively and for NORX64 are $2^{6.2}$, $2^{232.8}$ and 2^{225} respectively. Finally, we present a practical distinguisher for the keystream of NORX64 based on two rounds of the permutation in the parallel mode using an internal differential-linear attack. To the best of our knowledge, our results are the best known results for NORX in nonce respecting manner.

Keywords: Authenticated encryption · CAESAR · NORX · Guess and determine · Internal differential attack · State recovery · Nonce respect

1 Introduction

Competition for Authenticated Encryption: Security, Applicability, and Robustness (CAESAR) [1] is a competition for designing authenticated encryption schemes. 57 algorithms were submitted to the first round of this competition. After over a year of analysis, the CAESAR committee announced 29 schemes as the second round candidates. NORX [2,5] is one of them. It is a sponge based scheme which uses a permutation as its core, supports associated date

© International Association for Cryptologic Research 2016
T. Peyrin (Ed.): FSE 2016, LNCS 9783, pp. 554–574, 2016.
DOI: 10.1007/978-3-662-52993-5_28

and does not allow nonce to be repeated (known as nonce respecting). To provide efficiency in wide range of platforms, the only operations used in the core permutation of NORX are AND, XOR, rotation and shift. NORX consists of two variants denoted by NORX32 and NORX64. NORX32 provides 128-bit security and NORX64 provides 256-bit security. An interesting feature of NORX, compared to some other sponge based candidates of CAESAR such as for instance Ascon [13] and Keyak [8], is its level of parallelism. More precisely, NORX's designers proposed a parallel mode that enables users to process several message blocks in parallel. This feature could be interesting in applications that need high throughput, e.g. video streaming.

Among the second round candidates of CAESAR, NORX is one of the fastest [16] and uses simple bitwise operations which makes it a good candidate for a wide range of platforms, assuming that it provides the desired security. On the other hand, no previous security analysis of NORX as a full AEAD, e.g. integrity and confidentiality, is known and the only known results [4,11] more dedicated to the permutation of NORX rather than the application of the permutation in the mode.

Related Work. In [4], Aumasson *et al.* analysed the differential property of the core permutation of NORX. They provided upper bounds on the differential probability for the reduced permutation. More precisely, by assuming that an attacker can only modify the nonce during initialisation, any single round differential characteristic has probabilities of less than 2^{-60} (for NORX32) and 2^{-53} (for NORX64). They extended their results to full (four) round permutation and showed that the best characteristics for four rounds have probabilities of 2^{-584} and 2^{-836} for NORX32 and NORX64, respectively.

In [11], Das *et al.* analysed the higher order differential properties of the core permutation of NORX. These results cover more rounds compared to the first order differential analysis provided in [4]. More precisely, they identified the higher order differential properties that allow practical distinguisher of the full round permutation of NORX64 and 3.5-round permutation of NORX32. The used approach is similar to zero-sum distinguishers [6], but it is probabilistic.

Although the results in [4,11] can reach full rounds, it seems hard to exploit them to break integrity or confidentiality of NORX. In particular, the attacker's ability to control difference for the core permutation is significantly limited in the nonce respecting setting.

Our Contribution. In this paper, we present several cryptanalysis against reduced-round NORX with respect to security notions claimed by the designers; recovering key or breaking confidentiality in the nonce respecting setting. We discuss two different types of attacks; guess and determine attack and internal differential attack. The attack results are summarised in Table 1.

Guess and determine attack is a widely used technique in analyzing stream ciphers and authenticated encryption schemes. The attack by Dinur and Jean [12] against authenticated encryption FIDES [10] is an example. The attacker first

Table 1. Summary of our attacks. "KR" represents "key recovery" and "KD" represents "keystream distinguisher".

Approach	Goal	Target	Rounds	Data	Time	Memory	Ref.
Guess and determine	KR	NORX64	2/4	2^{132}	2^{234}	negl	Sect. 4
Guess and determine	KR	NORX32	2/4	2^{66}	2^{119}	negl.	Sect. 4
Internal difference	KR	NORX64	2/4	74	$2^{232.8}$	2^{225}	Sect. 5.2
Internal difference	KR	NORX32	2/4	158	$2^{124.3}$	2^{115}	Sect. 5.2
Internal differential-linear	KD	NORX64	2/4	90	negl	negl	Sect. 5.3

learns a part of the internal state values leaked from a plaintext-ciphertext pair. Then he partially guesses the hidden part of the state and recovers as many other state bits as possible. Since the NORX core permutation is invertible, recovering the internal state immediately allows to recover the secret key. We first describe a simple guess and determine attack that works up to 1.5 (out of 4) rounds of NORX. Then we show how to extend the attack to 2 rounds with the method of solving linear equations. Our attacks works for both NORX32 and NORX64.

While differential cryptanalysis [9] is generally difficult to apply in the nonce respecting setting, Jean *et al.* [14] have recently showed that difference between two parallel computations under the same nonce, could be exploited by the attacker and have applied it to fully parallelizable block cipher based scheme Silver [17]. The approach is called internal differential attack [18]. On the other hand, sponge based schemes generally have the serial structure, thus internal difference does not exist. However, it is still possible to introduce parallel computation to the sponge based schemes, and NORX is one of such designs. Hence, we mount an attack by exploiting the difference between two computations in the parallel mode of NORX. In the parallel mode, the same internal state is first duplicated, and the counter value, $0, 1, 2, \cdots$, is XORed to each state to make them distinct. Here, we focus on the very low Hamming wight difference caused by the counter values, which leads to high probability multiple differentials for 1 round. Using these differentials the internal state of the NORX with the permutation reduced to two rounds can be recovered. Moreover, we use the slow diffusion property of the NORX round function to present a practical distinguisher for the keystream of NORX64 with a permutation reduced to two rounds in parallel mode. This attack employs a deterministic truncated differential in forward direction for 1.5 round of the NORX64's permutation, followed by a probabilistic linear attack for a halve round of the permutation in backward direction.

Outline. The rest of the paper is organized as follows: in Sect. 2 we provide the required notations and also describe NORX as much as necessary for our analysis. In Sect. 3, we list several useful properties of the core permutation. In Sect. 4 we present a guess and determine attack. Our internal differential attack is described in Sect. 5. We present our distinguisher for NORX64 in Sect. 5.3. Finally, we provide closing remarks in Sect. 6.

2   Preliminaries

2.1   Notation

In this paper we mostly follow the notation used by the designer of NORX [5]. Depends on the context, a word is either a 32-bit or a 64-bit bit-string. The state of NORX is generally denoted by S. Each state includes 16 words and we denote the i-th word of S by s_i, for $0 \leq i \leq 15$. If we specify the state right after round j of the permutation, we denote it by S^j and denote its i-th word by s_i^j and the z-th bit of that word by $s_i^j[z]$. In the parallel mode of NORX, there are more than one lanes to process message blocks. In this case we denote the state of all lanes by $\bar{S}$ and we denote the state of the i-th lane by $\bar{S}_i$ and the j-th word of the i-th lane by $\bar{s}_{i,j}$. In general, we denote truncation of bit-string x from the i-th bit toward least significant bit (LSB) and up to the j-th bit toward most significant bit (MSB) by $x[j \sim i]$. The i-th bit of x is denoted by $x[i]$.

To denote bitwise AND, OR and XOR we use $\wedge$, $\vee$ and $\oplus$ respectively. By $x \ll n$, $x \gg n$, $x \lll n$ and $x \ggg n$ we denote left-shift, right-shift, left-rotate and right-rotate of bit-string x by n bits.

We use Δx to denote the difference in bit-string x and x', i.e. $\Delta x = x \oplus x'$.

By M_i and C_i we denote the i-th block of plaintext (message) and ciphertext. The nonce and the secret key are denoted by N and K respectively and their i-th words are denoted by n_i and k_i, respectively.

2.2   Specification of NORX

NORX [5] is a monkeyDuplex construction [7] based AEAD. It uses a $16w$-bit to $16w$-bit permutation, parameterized by a word size $w \in \{32, 64\}$. It has two variants denoted by NORX32 (where $w = 32$) and NORX64 (where $w = 64$). NORX is also parameterized by a parallelism degree $0 \leq p \leq 255$, number of rounds $0 \leq l \leq 63$ and a tag size $t \leq 4w$ and is denoted as NORXw-l-p-t. A high level representation of the NORX construction for $p = 1$ (serial mode of operation) and $p = 2$ (parallel mode of operation with two parallel lanes) are represented in Figs. 1 and 2 respectively. In these figures we have not considered the processes related to any auxiliary data, e.g. associated data.

The state S of NORXw consists of sixteen words $s_0, \ldots, s_{15}$, each of size w bits. The state's words $s_0, \ldots, s_{11}$ are called the rate words and the state's words $s_{12}, \ldots, s_{15}$ are called the capacity words. In each iteration of the permutation a block of message or associated data is XORed with the rate fraction of the state and a domain septation constant is XORed with the capacity fraction of the state (more details can be found in [5]). NORXw initiates the state by predefined constant $U = u_0 \| \ldots \| u_9$, the nonce $N = n_0 \| n_1$ and the key $K = k_0 \| \ldots \| k_3$.

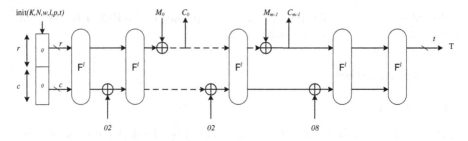

Fig. 1. The layout of NORX construction for $p = 1$ (fully serial) [5], where F^l denotes an l-round permutation of NORX.

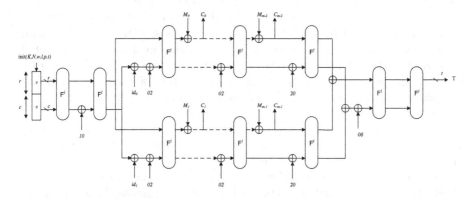

Fig. 2. The layout of NORX construction for $p = 2$ (include two parallel lanes) [5].

The matrix representation of S and the rule of assigning the words of constants, nonce and key in the initialization phase are as follows:

$$\begin{bmatrix} s_0 & s_1 & s_2 & s_3 \\ s_4 & s_5 & s_6 & s_7 \\ s_8 & s_9 & s_{10} & s_{11} \\ s_{12} & s_{13} & s_{14} & s_{15} \end{bmatrix} = \begin{bmatrix} n_0 & n_1 & u_0 & u_1 \\ k_0 & k_1 & k_2 & k_3 \\ u_2 & u_3 & u_4 & u_5 \\ u_6 & u_4 & u_8 & u_9 \end{bmatrix}$$

More details on the constants can be found in [2] but it does not affect our results. Each round of the permutation, called F, includes the application of a function called G to each column of state followed by applying it to each diagonal of state. Hence $\mathsf{F}(s_0, \ldots, s_{15})$ consists of column steps as follows:

$$\mathsf{G}(s_0, s_4, s_8, s_{12}), \mathsf{G}(s_1, s_5, s_9, s_{13}), \mathsf{G}(s_2, s_6, s_{10}, s_{14}), \mathsf{G}(s_3, s_7, s_{11}, s_{15}),$$

followed by the following diagonal steps:

$$\mathsf{G}(s_0, s_5, s_{10}, s_{15}), \mathsf{G}(s_1, s_6, s_{11}, s_{12}), \mathsf{G}(s_2, s_7, s_8, s_{13}), \mathsf{G}(s_3, s_4, s_9, s_{14}).$$

The function $G(a, b, c, d)$ computes the following 8 operations:

1. $a = H(a, b)$,   2. $d = (a \oplus d) \ggg r_0$,   3. $c = H(c, d)$,   4. $b = (b \oplus c) \ggg r_1$,

5. $a = H(a, b)$,   6. $d = (a \oplus d) \ggg r_2$,   7. $c = H(c, d)$,   8. $b = (b \oplus c) \ggg r_3$,

where $H(x, y) = (x \oplus y) \oplus ((x \wedge y) \ll 1)$. The rotation offsets (r_0, r_1, r_2, r_3) are $(8, 11, 16, 31)$ for NORX32 and $(8, 19, 40, 63)$ for NORX64.

It must be noted, to enhance the performance, compared to NORX V1 originally submitted to CAESAR [3], designers have excluded s_{10} and s_{11} from capacity words and appended them to the rate words in NORX V2.0 tweaked for the second round [5]. This tweak has not changed the security claim. The security claims for both of integrity and confidentiality are 128 bits for NORX32 and 256 bits for NORX64 [5, Table 3.1].

Parallel Mode of NORX. As depicted in Fig. 2 NORX supports parallel message processing, $p > 1$. In this case the state S is extended to a multi-state vector $\bar{S}$, where $\bar{S}_i$ indicates the input sate of the i-th lane. To ensure that the input state to each lane is a unique string, a counter i which indicates the lane number updates $\bar{s}_{i,13}$ to $\bar{s}_{i,13} \oplus i$. Among 5 recommended parameters by the designers [5], only 1 parameter, NORX64-4-4-256 supports the parallel mode.

3   Properties of Round Function

In this section, we show several properties of round function G that allow us to exploit relatively slow diffusion in backward direction, i.e. G^{-1}.

Computing G^{-1}. We begin with the observation that G^{-1} cannot be computed trivially. To compute G^{-1}, it is necessary to find x for a given pair of $H(x, y)$ and y. We argue that this can be efficiently computed bit-by-bit from the LSB. Considering $\ll 1$, computation for the LSB is a simple XOR, thus $x[0]$ is computed by $H(x, y)[0] \oplus y[0]$. After $x[0]$ is fixed, $x[1]$ can be computed similarly, and then the entire x is eventually computed. The cost of G^{-1} should be higher than G. For the sake of simplicity, we assume that the cost for G and G^{-1} are identical.

Computing G^{-1} with Partially Known State. For sponge-based AE schemes the rate words can be in general be recovered from a plaintext-ciphertext pair. For NORX, s_0 to s_{11} are known and s_{12} to s_{15} are secret. This motivates us to consider tracing known bit positions of $(a, b, c, d) \leftarrow G^{-1}(a', b', c', d')$ when three words of a', b', c', d' are known.

- b can be computed from given b', c', d' by G^{-1}.
- c can be computed from given a', c', d' by G^{-1}.

These simple properties can be extended so that several bits of the unknown word is known. This corresponds to the situation that attackers guess several bits of s_{12} to s_{15} during the attack. We again begin with analyzing H^{-1} with partially known $x' = H(x, y)$ and y.

Property 1. From all bits of x' and t consecutive bits of y, the corresponding t bits of x can be computed with a cost of 1-bit guess. Moreover, the guess is not necessary when t consecutive bits start from the LSB.

The property is depicted in Fig. 3. Attackers need to guess 1 bit of $(x \wedge y) \ll 1$ which is the lowest bit of consecutive t bits in y. Based on the guess, t consecutive bits of y can be recovered bit-by-bit. When t consecutive bits of x' are located in the LSB, t bits of x can be computed uniquely without guess.

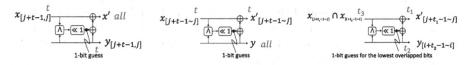

Fig. 3. Computing H^{-1} with partially known y.   **Fig. 4.** Computing H^{-1} with partially known x'.   **Fig. 5.** Computing H^{-1} with partially known x' and y.

Property 2. From t consecutive bits of x' and all bits of y, the corresponding t bits of x can be computed with a cost of 1-bit guess. Moreover, the guess is not necessary when t consecutive bits start from the LSB.

Property 3. Suppose that t_1 consecutive bits of x' and t_2 consecutive bits of y are known. Then, the corresponding x in overlapped bit positions can be computed with a cost of 1-bit guess.

Properties 2 and 3 are depicted in Fig. 4 and 5, respectively. The mechanism is the same as Property 1. In Property 3, attackers need to guess 1 bit of $(x \wedge y) \ll 1$, which is the lowest bit of overlapped bit positions. Then the other overlapped bits can be computed bit-by-bit.

Based on those properties of H^{-1}, we analyze the known bit positions of $(a, b, c, d) \leftarrow \mathsf{G}^{-1}(a', b', c', d')$ when three words are fully known and t bits of the last word are also known. In the property below, we assume $t > r_2$.

Property 4. Suppose that a', b', c' are fully known and t bits of d' are known. Then, t bits of a, t bits of b, $t - r_2$ bits of c, and t bits of d can be computed with a cost of 3-bit guess. Moreover, the guess is reduced to 1-bit when t consecutive bits of d' start from the LSB.

The property is depicted in Fig. 6. Numbers in red color represent the number of known bits and numbers in blue color $[n_1 \sim n_2]$ represent known bit positions. In Fig. 6, four H^{-1} functions are computed from the end to the beginning. The first H^{-1} is the case of Property 1. t bits of word c are computed with 1-bit guess, and the guess can be omitted if the known bits start from the LSB. No difficulty exists in the second H^{-1}. The third H^{-1} is the case of Property 3. The number of known bits depends on r_2. Note that $d' \lll r_2$ is computed instead of $d' \ggg r_2$ during G^{-1}. The fourth H^{-1} is again the case of Property 1. The 1-bit guess can be omitted for the LSB case.

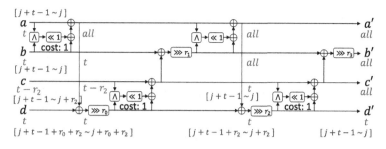

Fig. 6. Property 4: G^{-1} with partially known d'.

Property 5. Suppose that a', b', d' are fully known and t bits of c' are known. Then, $t - r_1$ bits of a, $t - r_1$ bits of b, t bits of c, and t bits of d can be computed with a cost of 4-bit guess. Moreover, the guess is reduced to 1-bit when t consecutive bits of c' start from the LSB.

Property 6. Suppose that a', c', d' are fully known and t bits of b' are known. Then, $t - r_1$ bits of a, t bits of b, all bits of c, and t bits of d can be computed with a cost of 2-bit guess.

Property 7. Suppose that b', c', d' are fully known and t bits of a' are known. Then, t bits of a, all bits of b, t bits of c, and $t - r_0$ bits of d can be computed with a cost of 3-bit guess. Moreover, the guess can be omitted when t consecutive bits of a' start from the LSB.

Extending Properties 4 to 7. The number of overlapped bits in Properties 4 to 7 can be generalized more. For example, in Property 4, $t - r_2$ bits of c can be replaced with t_3 bits of c, where t_3 takes one of the following 4 values depending on the relation of t, r_2 and the word size w.

$$t_3 = 0, \qquad \text{when } t - 1 < r_2 \text{ and } r_2 + t \le w,$$
$$t_3 = t - r_2, \qquad \text{when } t - 1 \ge r_2 \text{ and } r_2 + t \le w,$$
$$t_3 = 2t - w, \qquad \text{when } t - 1 \ge r_2 \text{ and } r_2 + t > w,$$
$$t_3 = r_2 + t - w, \qquad \text{when } t - 1 < r_2 \text{ and } r_2 + t > w.$$

Note that the third case is the combination of $t - r_2$ consecutive bits and $r_2 + t - w$ consecutive bits instead of $2t - w$ consecutive known bits. Thus to preserve $2t - w$ bit vales, the cost is 2-bit guess per H^{-1}. Also note that even with the third case, preserving either of $t - r_2$ or $r_2 + t - w$ with the cost of 1-bit guess per H^{-1} is possible, and this is actually the case of Property 4 assuming $t > r_2$.

The similar extension can be applied to Properties 5 to 7. To avoid redundancy, we omit the details. Those extension may be useful for future analysis.

Recovering the Input/Output Words of G. If 4 or more words among the 8 words $(a, b, c, d, a', b', c', d')$ of the input/output of G are known, it is possible to recover all of the other words for certain cases.

Property 8. Suppose that a, b, c, b', c' are fully known. Then, all the other words in the input/output of G can be recovered with a cost of single G^{-1} function.

Property 9. Suppose that a, a', b', c' are fully known. Then, all the other words in the input/output of G can be recovered with a cost of single G^{-1} function.

Property 10. Suppose that d, a', b', c' are fully known. Then, all the other words in the input/output of G can be recovered with a cost of single G^{-1} function.

4   Guess and Determine Attack

The encryption part of NORX leaks a large portion of the internal state, which may be exploited using a guess and determine attack. In this section, we will describe a simple guess and determine attack on 1.5 rounds NORX and then show how to extend it to 2 rounds with the method of solving linear equations.

4.1   Attack on 1.5 Rounds NORX

The 1.5 rounds NORX permutation involves four frames of internal states: the initial state, the state after the first column step, the state after the first diagonal step and the state after the second column step (the final state), which are denoted as S^0, $S^{0.5}$, S^1, and $S^{1.5}$ respectively. Thus, we refer to any bit in the computation with the notation $s_y^x[z]$, where x is the frame index, y is the word index ranged from 0 to 15, and z is the bit index ranged from 0 to $w - 1$.

In the latest version of NORX [5], $s_0^0, \ldots, s_{11}^0$ and $s_0^{1.5}, \ldots, s_{11}^{1.5}$ can be easily obtained by an adversary in a known plaintext scenario. Assuming those words are known at the input and output of the permutation by querying messages with at least two blocks, then we can exploit the slow backward diffusion of round function G of NORX by guessing 3 of the 4 unknown words of the state $S^{1.5}$ and propagating the information backward to recover the full internal state. The procedure of the guess and determine attack is given below, and is also depicted in Fig. 7.

1. Guess the words $s_{12}^{1.5}, s_{13}^{1.5}, s_{14}^{1.5}$ in $S^{1.5}$. With the other known state words of $S^{1.5}$, recover $s_0^1, s_1^1, s_2^1, s_4^1, s_5^1, s_6^1, s_8^1, s_9^1, s_{10}^1, s_{12}^1, s_{13}^1, s_{14}^1$.
2. With the known values of s_4^1, s_9^1, s_{14}^1 and using Property 7, recover $s_4^{0.5}$. With the known values of s_2^1, s_8^1, s_{13}^1 and using Property 6, recover $s_8^{0.5}$.
3. Using Property 8, recover $s_{12}^0, s_0^{0.5}, s_{12}^{0.5}$ from $s_0^0, s_4^0, s_8^0, s_4^{0.5}, s_8^{0.5}$.
4. Using Property 9, recover s_{15}^1 from $s_0^{0.5}, s_0^1, s_5^1, s_{10}^1$.
5. Using Property 10, recover $s_{15}^{1.5}$ from $s_{15}^1, s_0^{1.5}, s_5^{1.5}, s_{10}^{1.5}$.

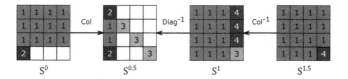

Fig. 7. Guess and determine attack on 1.5 rounds NORX. The order of known/recovered words is green(1)→blue(2)→yellow(3)→red(4). (Color figure online)

Thus, the full state of $S^{1.5}$ is recovered. We can compute backward to verify if the guessed words are correct. We want to note that this 1.5 rounds attack works for both NORX32 and NORX64 as all the operations are performed on the word level. We estimate that one guess and to determine the trail require roughly 2 NORX operations, the time complexity of the attack is 2^{97} for NORX32 and 2^{193} for NORX64.

4.2 Attack on 2 Rounds NORX

Now we will show how to extend the guess and determine attack to 2 rounds of NORX. The additional 0.5 round does not allow to establish the relation of S^0 and $S^{0.5}$ using the previous strategy, i.e. guessed words. Hence, we need to guess more bits and analyze the information propagation at bit level.

Overview of the Attack. Suppose that the five state frames of 2 rounds NORX are denoted by S^0, $S^{0.5}$, S^1, $S^{1.5}$, S^2 and the words $s_0^0, \ldots, s_{11}^0$, $s_0^2, \ldots, s_{11}^2$ are known. In addition to guess s_{12}^2, s_{13}^2, s_{14}^2, we further guess parts of the word s_{15}^2.

Next, we set all the unknown bits in S^0 as linear bits which related to themselves. With the forward computation of the NORX operations, any state bit can be classified into one of the three categories: *known bits*, *linear bits* and *non-linear bits*, such that a *known bit* can be computed with the known bits in S^0; a *linear bit* can be expressed as a linear combination of the unknown bits in S^0 XORed with some constant; and a *non-linear bit* can be expressed as non-linear combinations of the unknown bits in S^0 XORed with some constant. In this paper, We use $'u'$, $'k'$, $'l'$, $'n'$ to denote the unknown bits, known bits and linear bits, non-linear bits respectively.

Then, we do the same for the the bits in S^2 (assuming the guessed bits are known) and propagate backward.

If a bit in the internal state is linear/known in the forward computation while linear/known in the backward computation, we can establish a linear relation between the guessed/known bits in S^0 and the known bits in S^2. If there are enough linear equations, then we can solve the unknown bits in S^2, and thus recover the full internal state of S^2.

To establish more linear equations, we use a trick to increase the number of linear bits in the backward computation. In the H function, if a bit-relation

$x_i' = x_i \oplus y_i \oplus (x_{i-1} \wedge y_{i-1})$ is computed, with $x_i' = {}'k'$, $x_i = {}'u'$, $y_i = {}'l'$, $x_{i-1} = {}'k'$ and $y_{i-1} = {}'k'$. Then, if $x_i' = 1 \oplus (x_{i-1} \wedge y_{i-1})$, we have $x_i \wedge y_i = 0$, which is the non-linear term in computing x_{i+1}'. Note that we can control the value of any known bit by collecting more data and choose the needed ones. Hence, we can eliminate the non-linear part of the H function in many cases at the cost of increasing data complexity.

Since NORX64 and NORX32 have different word sizes, we will describe the attacks on NORX64 and NORX32 separately.

Attack on NORX64. For NORX64, we experimentally tested different choices of the guessed bits numbers and positions in s_{15}^2. To minimize the time complexity, the optimal choice we found is to guess $s_{15}^2[0, \ldots, 40]$.

We start the attack by building the linear system. Set the bits in $s_0^0, \ldots, s_{11}^0$, $s_0^2, \ldots, s_{14}^2$ and $s_{15}^2[0, \ldots, 40]$ as known bits. Set all the other unknown bits in S^0 and S^2 as linear bits. Then propagate the bit relations of S^0 forward and bit relations of S^2 backward. In the backward propagation, we control certain values of known bits to increase the number of linear bits with the technique mentioned previously.

Two linear equations can be established on bits $s_{10}^0[0]$ and $s_{11}^0[0]$, as those bits are known in the forward direction while linear in the backward direction. The rest of the equations can be derived from the last column of $S^{0.5}$, see Table 2.

Table 2. Bit patterns of the last column of $S^{0.5}$. Red bits are used to establish linear equations.

Forward pattern
1111111111111111 1111111111111111 1111111111111111 1111111111111111
nnnnnnnnnnnnnnnn nnnnnnnnnnnnnnnn nnnnnnnnnnnnnnnn nnnnnnnnnnnnnnnln
nnnnnnnnnnnnnnnn nnnnnnnnnnnnnnnn nnnnnnnnnnnnnnnn nnnnnnnnnnnnnnnl
1111111111111111 1111111111111111 1111111111111111 1111111111111111

Backward pattern
nnnnnnnnnnnnnnnn nnnnnnnnnnnnnnnn nnnnnnn11111111k kkkkkkkkkkkkkkkk
nnnnnnnnnnnnnnnn nnnnnnn1kkkkkkkk kkkkkkkkkkkkk111 1111111111111111
nnnnnnnnnnnnnnnn nnnnnnnnnnnnnnnn nnnnnnnnnnnnnnnn nnnnnnnnnnnnnnnl
nnnnnnnnnnnnnnnn nnnnnnn111111111 1111111111111111 1111111111111111

The reason of choosing this column is that only the unknown bits in s_{15}^0 and s_{15}^2 are involved. At this point, the number of linear equations is still less than the unknown variables. To obtain more equations we make use of Property 2. By guessing $s_{11}^0[18]$ to be 0, the bits $s_{15}^0[27, \ldots, 43]$ can be recovered. This is derived from the known bits in s_3^0, s_7^0, s_{11}^0 and the partially recovered bits $s_3^{0.5}[0, \ldots, 16]$. The bit patterns are updated as Table 3.

There are 69 matched bits in the last column of $S^{0.5}$. Together with the two linear equations from bits $s_{10}^0[0]$ and $s_{11}^0[0]$, there are 71 linear equations can be established. On the other hand, there are 70 unknown bits involved in the linear system, after the 17 derived bits $s_{15}^0[27, \ldots, 43]$ are excluded. We verified the coefficient matrix of the linear system to confirm that it has rank 70. Hence, a

Table 3. Updated bit patterns of the last column of $S^{0.5}$ for NORX64. Red bits are used to establish linear equations.

Forward pattern
1111111111111111 1111111111111111 111111111111111k kkkkkkkkkkkkkk11
nnnnnnnnnnnnnnnn nnnnnnnnnl111111 1111111111lnnnnn nnnnnnnnnnnnnnln
nnnnnnnnnnnnnnnn nnnnnnnnnl11111 11111111111nnnnn nnnnnnnnnnnnnnnl
1111111111111111 1111111111111111 1111111111111111 1111111111111111
Backward pattern
nnnnnnnnnnnnnnnn nnnnnnnnnnnnnnnn nnnnnnn11111111k kkkkkkkkkkkkkkkk
nnnnnnnnnnnnnnnn nnnnnnn1kkkkkkkk kkkkkkkkkkkkk111 1111111111111111
nnnnnnnnnnnnnnnn nnnnnnnnnnnnnnnn nnnnnnnnnnnnnnnn nnnnnnnnnnnnnnnl
nnnnnnnnnnnnnnnn nnnnnnn111111111 1111111111111111 1111111111111111

unique solution can be derived from the linear system. Note that the linear bits in the backward direction computation require certain value of 131 known bits. The attack can be summarized as follows.

1. Find an output of NORX encryption satisfied the conditions on the known bits. There are 131 conditions on the output bits and 1 condition on the input bits.
2. Guess the bits s_{12}^2, s_{13}^2, s_{14}^2, $s_{15}^2[0, \ldots, 40]$. So totally 233 bits are guessed.
3. Determine the bits $s_{15}^2[41, \ldots, 63]$ using the solutions of linear equations which have been derived.
4. Compute backward and verify the solution with the known bits in S^0.

We estimate that one guess and to determine trail require roughly 2 NORX operations. Then, the estimated time complexity of this attack is 2^{234} NORX operations. We expect to query for 2^{132} blocks to find a suitable input/output pair. So the data complexity is 2^{132} and memory complexity is negligible.

Attack on NORX32. Similar approach can be applied to attack reduced NORX32. In this case, we will first guess the bits s_{12}^2, s_{13}^2, s_{14}^2, $s_{15}^2[0, \ldots, 21]$. By setting the value of bit $s_{11}^0[10]$ to 0, the bits $s_{15}^0[19, \ldots, 24]$ can be recovered. The bit patterns of the last column of $S^{0.5}$ are given in Table 4. Here, we control 65 conditions on the output bits to increase the number of linear bits.

39 linear equations can be established from the last column of $S^{0.5}$. After excluded the derived bits, the number of unknown bits is 36. We computed the rank of the coefficient matrix which turned out to be 36. After further dropping the bits $s_7^2[1]$ and $s_{11}^2[0]$ from the system (to reduce the linear bits needed), a unique solution can be derived from the linear system with 37 equations.

The attack procedure is similar to NORX64, so we omit the details here. Since we guess a total number of 118 bits in the attack, the estimated time complexity is 2^{119}, and data complexity is 2^{66}.

Table 4. Updated bit patterns of the last column of $S^{0.5}$ for NORX32. 'l', 'n' and 'k' represent linear bit, non-linear bit and known bit, respectively. Red bits are used to establish linear equations.

Forward pattern
11111111 11111111 11111111 11kkkk11
nnnnnnnn nnnnn111 11nnnnnn nnnnnn1n
nnnnnnnn nnnnnn11 111nnnnn nnnnnnn1
11111111 11111111 11111111 11111111
Backward pattern
nnnnnnnn nnnnnnnn nn111111 11kkkkkk
nnnnnnnn nn1kkkkk kkkkk111 11111111
nnnnnnnn nnnnnnnn nnnnnnnn nnnnnnn1
nnnnnnnn nn111111 11111111 11111111

5 Internal Differential Attack

5.1 Basic Key Recovery with Internal Differential Attack

This section explains the key recovery attack on the NORX reduced to 2 rounds with differential cryptanalysis in the nonce-respect setting.

Overall Strategy. To exploit the difference under the same nonce, we exploit the parallel mode of NORX shown in Fig. 2, and focus on the difference between two lanes. As shown in Fig. 2, the state is duplicated and the lane number is XORed. For $p = 2$, lane numbers 0 and 1 make 1-bit difference in the state (LSB of s_{13}). This difference expands in the subsequent permutation F. Here, we build a high probability differential characteristic for the first 1 round.

Attackers only can observe the value of the rate words (s_0 to s_{11}) in each lane after 2 rounds. Hence, we perform some backward computation for the last 1 round with guessing capacity words (s_{12} to s_{15}). The number of secret bits in the state is $4w$, and the number of security bits claimed is also $4w$. Here, the difficulty is that the $4w$-bit secret value is completely different for the first lane and the second lane. Thus, we need to analyze $8w$-bit secret bits in a pair. This setting is quite different from conventional differential attack against block ciphers in which secret exists in the key and is common for both values in a pair.

To overcome this problem, we adopt the meet-in-the-middle approach. Namely, we guess up to $4w$ bits of the secret in the first lane, and recover several bits of the state after 1 backward round as the guess-and-determine attack in Sect. 4. The results are stored in a table with memory size up to $4w$ bits. Then, we do the same computation for the second lane, and compare the results of two lanes for picking up pairs satisfying the differential characteristic after 1 round. Suppose that the number of matched bits is $4w$. Then, among up to $2^{4w} \cdot 2^{4w} = 2^{8w}$ pairs, only up to $2^{8w}/2^{4w} = 2^{4w}$ pairs will remain as the candidates satisfying the differential characteristic. Finally, for each of the remaining candidate, we exhaustive guess the unguessed bits, and identify the correct state value.

Table 5. 1-Round Differential Characteristic. '0' and '1' represent inactive bit and active bit, respectively. After 1 round, we only need the difference of 8 words. '?' represents that difference is not specified in that bit.

Initial difference			
0000000000000000	0000000000000000	0000000000000000	0000000000000000
0000000000000000	0000000000000000	0000000000000000	0000000000000000
0000000000000000	0000000000000000	0000000000000000	0000000000000000
0000000000000000	0000000000000001	0000000000000000	0000000000000000
Difference after 0.5 round (column step)			
0000000000000000	0000002000000000	0000000000000000	0000000000000000
0000000000000000	4200004000020000	0000000000000000	0000000000000000
0000000000000000	2100000000010000	0000000000000000	0000000000000000
0000000000000000	2000000000010000	0000000000000000	0000000000000000
Difference after 1 round (diagonal step)			
????????????????	0000002000000400	0020000400000000	????????????????
????????????????	????????????????	4040000840000800	0800000a00000200
0420000100000100	????????????????	????????????????	2020000420000000
2020000400000000	0400000100000000	????????????????	????????????????

We first discuss a simple attack against NORX64 in Sect. 5.1, and then show several optimization techniques and application to NORX32 in Sect. 5.2.

1-Round Differential Characteristic. Lane number 0 and lane number 1 make 1-bit difference in the LSB of s_{13}. We trace the propagation of this difference.

Construction of differential characteristic is simple. The only non-linear component is the AND operation in H. We set 1-bit condition for each active bit to control its output difference. Because of the small number of rounds, we found that the probability of the characteristic is maximized by setting output difference of all active bits to 0. The obtained characteristic is shown in Table 5. To keep the probability of the characteristic high, we only specify the difference of 8 words in 2 diagonals, i.e. $(\Delta s_1, \Delta s_2, \Delta s_6, \Delta s_7, \Delta s_8, \Delta s_{11}, \Delta s_{12}, \Delta s_{13})$.

The characteristic in Table 5 includes 5 active bits in the first 0.5 round and 16 active bits in the next 0.5 round, in total 21 active bits. Therefore, the characteristic can be satisfied with probability at least 2^{-21}.

1-Round Backward Computation. For a plaintext-ciphertext pair for each lane, we recover the value of the rate words, i.e. $\bar{s}_{0,0}$ to $\bar{s}_{0,11}$ and $\bar{s}_{1,0}$ to $\bar{s}_{1,11}$. Then for the first lane, we exhaustively guess all bits of $(\bar{s}_{0,12}, \bar{s}_{0,14}, \bar{s}_{0,15})$ and t LSBs of $\bar{s}_{0,13}$, where the value of t will be determined later. For each guess, we can compute 1 round in backwards with the properties introduced in Sect. 3. The 1-round backward computation is depicted in Fig. 8.

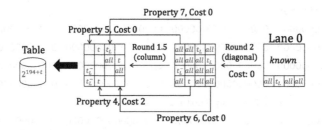

Fig. 8. Backward computation. all, t, t_L and t_L^- represent that all bits are known, t middle bits are known, t LSBs are known and $t - 24$ LSBs are known, respectively.

For the first 0.5-round backward computation, 3 diagonals are fully known, which can be inverted easily. The remaining diagonal, $(\bar{s}_{0,2}, \bar{s}_{0,7}, \bar{s}_{0,8}, \bar{s}_{0,13}) \leftarrow G^{-1}(\bar{s}'_{0,2}, \bar{s}'_{0,7}, \bar{s}'_{0,8}, \bar{s}'_{0,13})$, is the case of Property 4 with t LSBs in d'. Here, given that $r_2 = 40$ for NORX64, we replace "$\ggg r_2$" with "$\lll (64 - r_2)$" to make the analysis simpler. The overlapped bit positions becomes $t - 24$ bits starting from the LSB. In the end, we obtain t LSBs of $\bar{s}_{0,2}$ and $\bar{s}_{0,7}$, $t - (64 - r_2) = t - 24$ LSBs of $\bar{s}_{0,8}$, and t middle bits of $\bar{s}_{0,13}$. Because all partially known bits start from the LSB in H^{-1}, we do not need additional bit guess.

The next 0.5-round can be computed with Properties 4 to 7 as follows.

- The first column is the case of Property 5, in which we know $t - 24$ LSBs of $\bar{s}'_{0,8}$. Different from Property 5, we only compute 2 words; $\bar{s}_{0,8}$ and $\bar{s}_{0,12}$. This can avoid 1-bit guess, and thus $t - 24$ bits of $\bar{s}_{0,8}$ and $\bar{s}_{0,12}$ can be computed without any guess.
- The second column is the case of Property 4 in which the consecutive t bits start from middle bits. Property 4 requires 3-bit guess to recover 4 input variables. Here, we only need to recover 2 input variables, a and d in Fig. 6, and this saves us to guess 1 bit. As a result, we need 2-bit guess to compute the corresponding t bits of $\bar{s}_{0,1}$ and $\bar{s}_{0,13}$.
- The third column is the case of Property 7. t bits of $\bar{s}_{0,2}$ and all bits of $\bar{s}_{0,6}$ can be computed without guess.
- The fourth column is the case of Property 6, which requires 2-bit guess to recover 4 input variables. Here, we only need to recover 2 input variables, b and c and this can be done without guess. Hence, t bits of $\bar{s}_{0,7}$ and all bits of $\bar{s}_{0,11}$ can be computed without guess.

In summary, we guess 192 bits of $\bar{s}_{0,12}, \bar{s}_{0,14}, \bar{s}_{0,15}$, t bits of $\bar{s}_{0,13}$ after round 2 and 2 bits in the middle, which leads to $2(t-24)+2t+(t+64)+(t+64) = 6t+80$ bits of $(\bar{s}_{0,1}, \bar{s}_{0,2}, \bar{s}_{0,6}, \bar{s}_{0,7}, \bar{s}_{0,8}, \bar{s}_{0,11}, \bar{s}_{0,12}, \bar{s}_{0,13})$ after round 1. Those are stored in a table with a memory of size 2^{194+t}.

After the computation of the first lane, we apply the same computation for the second lane, i.e. for state $\bar{S}_1$, with $194 + t$-bit guess. For each result, we XOR the difference in two diagonals specified by the characteristic, and check the match with the table generated for the first lane. If the match is found, we guess the remaining $64 - t$ bits of $\bar{s}_{0,13}$ for the first lane, and compute back to the initial value (IV) of NORX64. Only if the pair satisfies the characteristic and the guess is correct, the IV appears, which recovers the key simultaneously.

Attack Procedure. Associated data is irrelevant. Hereafter, we set A to be empty. The attack procedure is described in Algorithm 1. Due to the probability of the characteristic, 2^{21} pairs are analyzed, which corresponds to 2^{21} iterations in Step 1. The 1-round backward computation is performed for each of 2^{21} pairs.

Algorithm 1. 2-round key recovery with internal-differential attack.

1 **Input**: characteristic with probability 2^{-21}, $194 + t$ guessed-bit positions
2 **Output**: K
  1: **for** $i = 0, 1, \ldots, 2^{21} - 1$ **do**
  2:    Randomly choose a nonce N^i and a 2-block message $M_0^i \| M_1^i$.
  3:    Query $(N^i, M_0^i \| M_1^i)$ in the parallel mode to obtain $(C_0^i \| C_1^i)$.
  4:    Compute $\bar{s}_{0,0}, \ldots, \bar{s}_{0,11} \leftarrow M_0^i \oplus C_0^i$ and $\bar{s}_{1,0}, \ldots, \bar{s}_{1,11} \leftarrow M_1^i \oplus C_1^i$.
  5:    **for** $192 + t$ bits of $\bar{s}_{0,12}, \bar{s}_{0,13}, \bar{s}_{0,14}, \bar{s}_{0,15}$ and 2 bits in the middle **do**
  6:      Obtain $6t + 80$ bits of $\bar{s}_0$ after round 1 and store them in a table T_0.
  7:    **end for**
  8:    **for** $192 + t$ bits of $\bar{s}_{1,12}, \bar{s}_{1,13}, \bar{s}_{1,14}, \bar{s}_{1,15}$ and 2 bits in the middle **do**
  9:      Obtain $6t + 80$ bits of $\bar{s}_1$ after round 1 and xor the difference in Table 5.
 10:     Check the match with T_0. If the match is found, go to the next step.
 11:     **for** the remaining $64 - t$ bits of $\bar{s}_{0,13}$ **do**
 12:       Compute back to the initial value of NORX64.
 13:       **if** the result satisfies the form of
             $(u_0, n_0, n_1, u_1, k_0, k_1, k_2, k_3, u_2, u_3, u_4, u_5, u_6, u_7, u_8, u_9)$ **then**
 14:         **return** $K = (k_0, k_1, k_2, k_3)$.
 15:       **end if**
 16:     **end for**
 17:    **end for**
 18: **end for**

Complexity Evaluation and Choice of t. In Algorithm 1, for each of i in Step 1, Step 6 requires 2^{194+t} G computations and Step 9 requires 2^{194+t} G computations. After the match in Step 10, $2^{2(194+t)-(6t+80)} = 2^{308-4t}$ pairs will remain. Then, Step 12 requires $2^{308-4t+64-t} = 2^{372-5t}$ NORX64 operations. Those are iterated 2^{21} times for the iteration in Step 1. Hence, time complexity is less than $2^{21}(2^{194+t} + 2^{194+t} + 2^{372-5t})$ NORX64 operations. This is optimized when $t = 30$, which leads to $2^{246.2}$ NORX64 operations.

Data complexity is only caused by Step 3. Two message blocks are queried in each iteration of Step 1. Thus, data complexity is 2^{22} message blocks.

Memory complexity is dominated by Step 6, which stores the result of 2^{194+t} computations. Thus, memory complexity is 2^{224} when $t = 30$.

5.2 Optimized Key Recovery with Internal Differential Attack

Differential. The probability of the characteristic in Sect. 5.1 was 2^{-21}. This can be improved by using the differential instead of a single characteristic. Rigorously evaluating the probability of the differential is hard. However, thanks to the high probability of the characteristic, we can evaluate it experimentally.

We chose 2^{24} pairs at uniformly random, and 6937 pairs could satisfy the output difference with respect to the $6t + 80$ bits computed during the backward computation. Thus, the probability of the differential is $6937/2^{24} \approx 2^{-11.23}$, which improves the complexity of Algorithm 1 by roughly 10 bits.

Multiple Lanes. NORX64-4-4-256 supports 4 parallel lanes. When a 4-block message is processed, lane numbers XORed to the duplicated states is 0, 1, 2 and 3. Thus, we can make the pair with $\Delta \bar{S}_{13} = 0x1$ between lane 0 and lane 1 and between lane 2 and lane 3. Besides, we can also consider the internal difference between lane 0 and lane 2 and between lane 1 and lane 3, which makes $\Delta \bar{S}_{13} = 0x02$. The best characteristic for this difference is obtained by rotating Table 5 by 1 bit. We also experimentally verified the probability of the differential, which is $7532/2^{24} \approx 2^{-11.12}$, slightly better than the case with $\Delta \bar{S}_{13} = 0x01$. The average probability for two cases is $2^{-11.18}$.

In summary, we can make 4 pairs per 4-block message query, which halves the data complexity evaluation in Sect. 5.1.

Multiple Differentials. While the characteristic in Table 5, 2^{-21}, is optimized, we found that there are 2^5 characteristics having the same probability.

Recall the complexity evaluation in Sect. 5.1. Let N_{pair}, T_{mitm}, and T_{veri} be number of pairs to analyze, time complexity for the meet-in-the-middle match per pair, and time complexity for verifying matched candidates, respectively. Roughly, the total complexity is given by $N_{pair}(T_{mitm} + T_{veri})$. Now, suppose that there are D characteristics with the same probability and we aim to find a pair satisfying any of D choices. This reduces the number of pairs to N_{pair}/D, while the number of valid candidates after the match becomes D times, thus the cost for verification becomes $D * T_{veri}$. The essence of this technique is that the 1-round backward computation with a cost of T_{mitm} is independent of D choice of characteristics, thus one computation can be reused for testing D characteristics.

To derive such characteristics, we focus on the differential propagation in the last H function (during round 0.5 to round 1). There are 4 active bits during the last H function in $(\Delta s'_1, \Delta s'_6, \Delta s'_{11}, \Delta s'_{12}) \leftarrow G(\Delta s_1, \Delta s_6, \Delta s_{11}, \Delta s_{12})$. The characteristic in Table 5 was derived by setting the output difference of those 4 bits to 0, which causes the probability drop by 2^{-4}. However, the differences from those bits only linearly related to the output of the entire characteristic, thus any combination of 0 or 1 for the output difference from those 4 bits generates 2^4 distinct characteristics with exactly the same probability.

The same occurs in the other diagonal $\mathsf{G}(\Delta s_2, \Delta s_7, \Delta s_8, \Delta s_{13})$. 4 bits in positions 8, 32, 53 and 58 are active in the last H, while only bit position 8 will be later used for the meet-in-the-middle match. As a result, 2 distinct characteristics can be considered in bit position 8. Along with 2^4 choices for the other diagonal, we have $D = 2^5$ characteristics with the same probability.

The discussion above is about the multiple characteristics. It can easily be extended to multiple differentials. We experimentally tested the probability of 2^5 differentials, and confirmed that all of them has almost the same probability.

Strictly speaking, matching multiple differentials during the meet-in-the-middle match in Step 10 of Algorithm 1 is a so-called 3-list problem, which requires more cost than the 2-list case. We observe that 2^5 differentials only differ in 10 bits, that is, all of them have the same difference in the other $6t + 70$ bits. Thus, we first apply the filter in $6t - 70$ bits, then check the details for 10 bits. Matching $6t - 70$ bits reduces the number of candidates sufficiently small, thus using multiple differentials gives negligible impacts to the complexity.

Optimized Complexity for NORX64. To satisfy one of the 2^5 differentials with probability $2^{-11.18}$, $2^{6.18}$ pairs need to be analyzed. The number of iterations of Step 1 in Algorithm 1 becomes $i = 2^{6.18}$. By exploiting four lanes, $2^{6.18}$ message-block queries are sufficient to construct $2^{6.18}$ pairs. Thus the data complexity is $2^{6.18}$ message blocks.

The usage of the multiple differentials slightly changes the balance between T_{mitm} and T_{veri}, i.e. higher T_{mitm} and lower T_{veri} offers the best balanced complexity. We found that $t = 31$ instead of $t = 30$ yields the best complexity, which is $\frac{2^{11.18}}{2^5}\left(2 * 2^{194+t} + 2^5 * 2^{372-5t}\right) = 2^{232.8}$. The memory complexity increases due to the increase of t, which is $2^{194+t} = 2^{225}$.

Application to NORX32. Because NORX32 does not formally support the parallel mode, we only explain it briefly. The attack on NORX32 is harder than the one for NORX64 with respect to the following two points.

- The probability of the multiple differential becomes relatively smaller to the word size (32 instead of 64).
- The rotation number $r_2 = 16$ is exactly a half of the word size, which generates less number of overlapped bits during the backward computation ($r_2 = 40$ in NORX64).

Those make the advantage of the attack smaller than the case of NORX64. The attack strategy is the same as NORX64 and all the optimization techniques can also be applied. We experimentally verified that the probability of the differential is $2^{-12.25}$ and there are 2^5 differentials with the same probability. We choose $t = 17$ for the number of partially guessed bits. In the end, the best attack complexity is $2^{7.25}$, $2^{124.25}$ and 2^{115} in data, time and memory, respectively.

Table 6. 1.5-round internal differential-linear distinguisher. '0' and '1' represent unaffected bits and bits that maybe affected by the internal difference, respectively. After 1.5 round 8 bits of the rate will never be affected by the internal difference in S_{13}.

Initial difference			
0000000000000000	0000000000000000	0000000000000000	0000000000000000
0000000000000000	0000000000000000	0000000000000000	0000000000000000
0000000000000000	0000000000000000	0000000000000000	0000000000000000
0000000000000000	0000000000000001	0000000000000000	0000000000000000
Unaffected bits after 0.5 round, i.e. $S^{0.5}$			
0000000000000000	000000e000000000	0000000000000000	0000000000000000
0000000000000000	ce0000c000060001	0000000000000000	0000000000000000
0000000000000000	e700000000030000	0000000000000000	0000000000000000
0000000000000000	e000000000010000	0000000000000000	0000000000000000
Unaffected bits after 1 round, i.e. S^1			
bfc07bffc03ef807	000003e00000fc00	00e0007c00000000	e0003fe000000000
fe003bc003de00ff	fffcffff1ffff9ff	c7c003ffc000f80f	ff80007e00038e01
ffe000030001c700	df00000001ef007f	fffe7fffcfffbcff	e3e001ffe0000007
e1e000fc00000003	7c0000010000c000	e000000000e0003f	ff803efe07bf9e7b
Unaffected bits after 1.5 round, i.e. $S^{1.5}$			
ffffffffffffffff	ffffffffffffffff	ffffffffffffffff	fffffffff0ffff87
ffffffffffffffff	ffffffffffffffff	ffffffffffffffff	ffffffffffffffff
ffffffffffffffff	ffffffffffffffff	ffffffffffffffff	ffffffffffffffff
ffffffffffffffff	ffffffffffffffff	ffffffffffffffff	fffcffff9fffffff

5.3   Distinguisher with Internal Differential-Linear Attack

In this section, we present an internal differential-linear attack on round-reduced NORX. In more detail, we show an efficient distinguisher for NORX64 with $p \geq 2$ for up to 2 rounds. As shown in Table 6, the internal difference in the LSB of s_{13}, i.e. $s_{13}[0]$, does not affect 8 bits in the rate part of the NORX64 state after 1.5 rounds, i.e. $s_3^{1.5}[3, 4, 5, 6, 24, 25, 26, 27]$. This property leads to a trivial distinguisher for the keystream of NORX64 reduced to 1.5 rounds in the parallel mode.

Unfortunately, by adding 0.5 rounds all bits of the rate are affected by the internal difference and no significant bias can be observed with practical complexity (2^{40} experiments). However, using Property 4 described in Sect. 3, all the bits at the output of the first $H(a, b)$ in each G function in the last half-round can be derived from the rate part of the output. This significantly improves the attack leading to an efficient distinguisher for 2 rounds of NORX64 in the parallel mode. We can observe significant biases in several of the 256 bits. For instance bit 175 has a bias of -0.15. If an adversary aims to distinguish the keystream related to NORX64-4-4-256 based on this bit, it proceeds as follows:

1. Query $\frac{1}{0.15^2}$ 2-block messages for NORX64-4-4-256 and receive the corresponding ciphertexts.
2. Partially decrypt each ciphertext up to the output of the first $H(a,b)$ in the last half-round.
3. Verify the matching for each ciphertext pair and output the total amount of the matching in bit 175 for all ciphertext pairs, N.
4. Output NORX64 if $N \geq \frac{1}{2 \times 0.15^2}$.

The data complexity of this attack is about 90 message blocks, while the success probability is 97.7 % [15].

6   Conclusion

In this paper, we present the first cryptanalysis of NORX in the nonce-respecting setting. Our attack exploits the slow diffusion of NORX's round function especially in backward direction. We investigate several attacks against NORX and all of them cover two rounds of the permutation. On the other hand, while the presented guess and determine attack covers two rounds of NORX's variants yet it may be possible to be extended to more rounds by employing advanced equation solving techniques. We do not think it can be extended to the full 4-round NORX. However, the results of this paper can be considered as a starting point for future analysis in this direction.

Additionally, we presented a practical distinguisher for 2-round NORX64 encryption in parallel mode that could not be applied to NORX32 or serial-NORX64. This observation may be considered as a lower diffusion in NORX64 compared to the NORX32, for the same number of rounds. In addition, this attack along with the given internal differential attack could be considered as evidences that parallel mode of NORX has lower security bound, compared to serial mode.

Acknowledgments. The authors would like to thank the organizers of ASK 2015 that initiated this work. Keting Jia is supported by the National Natural Science Foundation of China (Nos. 61133013 and 61402256) and 973 Program (No. 2013CB834205). Part of this work was done while Florian Mendel was visiting NTU and has been supported in part by the Austrian Science Fund (project P26494-N15).

References

1. CAESAR: Competition for Authenticated Encryption: Security, Applicability, and Robustness (2013). http://competitions.cr.yp.to/caesar.html
2. Aumasson, J.-P., Jovanovic, P., Neves, S.: NORX: parallel and scalable AEAD. In: Kutyłowski, M., Vaidya, J. (eds.) ICAIS 2014, Part II. LNCS, vol. 8713, pp. 19–36. Springer, Heidelberg (2014)
3. Aumasson, J., Jovanovic, P., Neves, S.: NORX V1 (2014). http://competitions.cr.yp.to/round1/norxv1.pdf

4. Aumasson, J.-P., Jovanovic, P., Neves, S.: Analysis of NORX: investigating differential and rotational properties. In: Aranha, D.F., Menezes, A. (eds.) LATIN-CRYPT 2014. LNCS, vol. 8895, pp. 306–323. Springer, Heidelberg (2015)
5. Aumasson, J., Jovanovic, P., Neves, S.: NORX V2.0 (2015). http://competitions.cr.yp.to/round2/norxv20.pdf
6. Aumasson, J., Meier, W.: Zero-sum distinguishers for reduced Keccak-f and for the core functions of Luffa and Hamsi. NIST mailing list (2009). http://aumasson.jp/data/papers/AM09.pdf
7. Bertoni, G., Daemen, J., Peeters, M., Van Assche, G.: Duplexing the sponge: single-pass authenticated encryption and other applications. In: Miri, A., Vaudenay, S. (eds.) SAC 2011. LNCS, vol. 7118, pp. 320–337. Springer, Heidelberg (2012)
8. Bertoni, G., Daemen, J., Peeters, M., Assche, G.V., Keer, R.V.: CAESAR submission: Keyak v2 (2015). http://competitions.cr.yp.to/round2/keyakv2.pdf
9. Biham, E., Shamir, A.: Differential cryptanalysis of the full 16-round DES. In: Brickell, E.F. (ed.) CRYPTO 1992. LNCS, vol. 740, pp. 487–496. Springer, Heidelberg (1993)
10. Bilgin, B., Bogdanov, A., Knežević, M., Mendel, F., Wang, Q.: FIDES: lightweight authenticated cipher with side-channel resistance for constrained hardware. In: Bertoni, G., Coron, J.-S. (eds.) CHES 2013. LNCS, vol. 8086, pp. 142–158. Springer, Heidelberg (2013)
11. Das, S., Maitra, S., Meier, W.: Higher order differential analysis of NORX. IACR Cryptol. ePrint Arch. **2015**, 186 (2015). http://eprint.iacr.org/2015/186
12. Dinur, I., Jean, J.: Cryptanalysis of FIDES. In: Cid, C., Rechberger, C. (eds.) FSE 2014. LNCS, vol. 8540, pp. 224–240. Springer, Heidelberg (2015)
13. Dobraunig, C., Eichlseder, M., Mendel, F., Schläffer, M.: Ascon v1.1 Submission to the CAESAR Competition (2014). http://competitions.cr.yp.to/round2/asconv11.pdf
14. Jean, J., Sasaki, Y., Wang, L.: Analysis of the CAESAR candidate silver. In: Dunkelman, O., et al. (eds.) SAC 2015. LNCS, vol. 9566, pp. 493–509. Springer, Heidelberg (2016). doi:10.1007/978-3-319-31301-6_28
15. Matsui, M.: Linear Cryptanalysis Method for DES Cipher. In: Helleseth, T. (ed.) EUROCRYPT 1993. LNCS, vol. 765, pp. 386–397. Springer, Heidelberg (1994)
16. Nikolic, I.: CAESAR candidates speed comparison (2014). http://www1.spms.ntu.edu.sg/syllab/speed/
17. Penazzi, D., Montes, M.: Silver v.1. Submitted to the CAESAR competition (2014)
18. Peyrin, T.: Improved differential attacks for ECHO and Grøstl. In: Rabin, T. (ed.) CRYPTO 2010. LNCS, vol. 6223, pp. 370–392. Springer, Heidelberg (2010)

Analysis of the Kupyna-256 Hash Function

Christoph Dobraunig$^{(\boxtimes)}$, Maria Eichlseder, and Florian Mendel

Graz University of Technology, Graz, Austria
christoph.dobraunig@iaik.tugraz.at

Abstract. The hash function Kupyna was recently published as the Ukrainian standard DSTU 7564:2014. It is structurally very similar to the SHA-3 finalist Grøstl, but differs in details of the round transformations. Most notably, some of the round constants are added with a modular addition, rather than bitwise xor. This change prevents a straightforward application of some recent attacks, in particular of the rebound attacks on the compression function of similar AES-like hash constructions. However, we show that it is actually possible to mount rebound attacks, despite the presence of modular constant additions. More specifically, we describe collision attacks on the compression function for 6 (out of 10) rounds of Kupyna-256 with an attack complexity of 2^{70}, and for 7 rounds with complexity $2^{125.8}$. In addition, we can use the rebound attack for creating collisions for the round-reduced hash function itself. This is possible for 4 rounds of Kupyna-256 with complexity 2^{67} and for 5 rounds with complexity 2^{120}.

Keywords: Hash functions · Cryptanalysis · Collisions · Free-start collisions · Kupyna · Rebound attack

1 Introduction

Recently, Oliynykov et al. [12] published an English specification of the new Ukrainian hash standard DSTU 7564:2014, also known as Kupyna. In contrast to the previous standard GOST 34.311-95, the new hash standard facilitates more effective software implementations. Of course, it is also intended to offer improved security compared to the old GOST standard, which has shown weaknesses against collision attacks [8]. As Kupyna is a national standard, it is likely to find wide-spread adoption in the Ukraine. Thus, comprehensive third-party analysis is necessary to evaluate the resistance of Kupyna against cryptanalytic attacks.

The Kupyna design aims to achieve a high level of security by relying on well-known and well-analyzed building blocks. It shares a notable similarity with the SHA-3 finalist Grøstl [3]. Kupyna's mode of operation, in particular its compression function, is nearly identical to the one used in Grøstl, and its permutations – though different – follow very similar design ideas. One of Kupyna's two permutations employs the wide-trail design strategy [2] of AES. Therefore, this permutation shares a common basis with Grøstl's permutations, although

© International Association for Cryptologic Research 2016
T. Peyrin (Ed.): FSE 2016, LNCS 9783, pp. 575–590, 2016.
DOI: 10.1007/978-3-662-52993-5_29

Kupyna uses other constants, S-boxes, rotation values, and a different MDS matrix. The second Kupyna permutation differs from the first in the constant addition, which applies addition modulo 2^{64} rather than bitwise xor. This modular addition serves to differentiate the two permutations, but can also be seen as a measure to complicate algebraic cryptanalysis. Furthermore, it implies additional relations over byte boundaries. As a consequence, the modular addition leads to a weaker alignment for differential trails, making statements about the minimum number of active S-boxes in a differential trail more complicated, since the linear layer no longer achieves an optimal branch number.

Our Contribution. In this paper, we provide the first third-party analysis of the new Ukrainian hash standard Kupyna. We present collision attacks on round-reduced variants of Kupyna-256 for up to 5 out of 10 rounds, and collisions for the compression function of Kupyna-256 for up to 7 rounds. A summary of our results can be found in Table 1.

Table 1. Overview of collision attacks on Kupyna-256.

Hash function	Target	Rounds	Complexity	Reference
Kupyna-256	Compression function	6	2^{70}	Sect. 3
		7	$2^{125.8}$	
	Hash function	4	2^{67}	Sect. 4
		5	2^{120}	

Our attacks make use of the capability of rebound attacks [9] to efficiently generate pairs of values which follow a given truncated differential trail. The core idea of such a rebound attack is to create many solutions with a low complexity per solution during the inbound phase, and propagate those solutions in a probabilistic manner during the so-called outbound phase.

To create solutions with a low average complexity during the inbound phase, the rebound attack takes advantage of the strong alignment of truncated differential trails and the underlying independence of parts of the cipher. As mentioned before, one of the two permutations of Kupyna does not provide such a strong alignment. Hence, it is not trivial to perform the rebound attack for this permutation. However, in this work, we show how to deal with modular constant additions during the inbound and the outbound phase to be able to perform rebound attacks on such constructions.

Related Work. Due to the high similarity of Kupyna with Grøstl, the fundament of our attacks is the analysis of Grøstl. For Grøstl, the best attacks are based on the rebound attack and its improvements [4,6,7]. Distinguishers for round-reduced variants of the Grøstl permutation were published in [5]. Rebound attacks leading to collisions for the highest number of rounds for the compression function of Grøstl were shown in [10]. An efficient collision attack covering 5 rounds of the hash function Grøstl itself was published in [11].

In the meantime, Zou and Dong [14] independently analyzed the Kupyna hash function, and observed the applicability of some of the attacks on Grøstl to Kupyna. They present pseudo-preimage attacks on 6 rounds of Kupyna-256 and 8 rounds of Kupyna-512 with time complexities $2^{250.33}$ and $2^{498.33}$, respectively, which are essentially identical to the original Grøstl attacks [13]. Additionally, they also noted the hash function attack on 5 rounds very similar to Sect. 4 of this paper.

Outline. The remainder of the paper is organized as follows. First, we start with a short description of Kupyna in Sect. 2. Next, we show how to apply the rebound attack on round-reduced versions of the compression function of Kupyna-256, to create semi-free-start collisions for 6 and 7 rounds, in Sect. 3. Then, we apply a collision attack for 4 and 5 rounds of Kupyna-256 in Sect. 4. Finally, we conclude in Sect. 5.

2   Description of Kupyna

Kupyna [12] is a family of iterated hash functions defined in the Ukrainian standard DSTU 7564:2014. The design principles of Kupyna are very similar to the SHA-3 finalist Grøstl [3]. As in Grøstl, the compression function of Kupyna is built from two distinct permutations $T^{\oplus}$ and T^{+}, which are both based (to a certain degree) on the AES design principles. In the following, we describe the components of the hash function in more detail.

2.1   The Hash Function

The Ukrainian standard DSTU 7564:2014 defines two main variants, Kupyna-256 and Kupyna-512, which produce a hash output size of $n = 256$ and $n = 512$ bits, respectively (the third recommendation, Kupyna-384, is simply a truncated version of Kupyna-512). The hash function first pads the input message M and splits the message into blocks $m_1, m_2, \ldots, m_t$ of ℓ bits each, with $\ell = 512$ for Kupyna-256 and $\ell = 1024$ for Kupyna-512. The message blocks are processed via the compression function $f(h_{i-1}, m_i)$, which updates the internal ℓ-bit chaining value h_i, and an output transformation $\Omega(h_t)$ to produce the final hash value h:

$$h_0 = \text{IV}$$
$$h_i = f(h_{i-1}, m_i) \quad \text{for } 1 \leq i \leq t$$
$$h = \Omega(h_t).$$

The compression function f is based on two ℓ-bit permutations $T^{\oplus}$ and T^{+} and is defined as follows (see also Fig. 1):

$$f(h_{i-1}, m_i) = T^{\oplus}(h_{i-1} \oplus m_i) \oplus T^{+}(m_i) \oplus h_{i-1}.$$

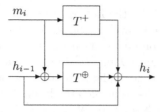

Fig. 1. The compression function $h_i = f(h_{i-1}, m_i)$ of the Kupyna hash function, using ℓ-bit permutations $T^{\oplus}$ and T^+.

The output transformation Ω is applied to h_t to give the final hash value h of size n, where $\mathrm{trunc}_n(x)$ discards all but the most significant n bits of x:

$$\Omega(h_t) = \mathrm{trunc}_n(T^{\oplus}(h_t) \oplus h_t).$$

2.2 The Permutations $T^{\oplus}$ and T^+

In the remaining document, we focus our analysis on Kupyna-256. The structure of the two permutations $T^{\oplus}$ and T^+ of Kupyna is very similar to the ones of Grøstl. As in Grøstl-256, each Kupyna-256 permutation updates an 8×8 state of 64 bytes in 10 rounds. In each round, the round transformation updates the state by means of the sequence of transformations

$$\mathsf{MixBytes} \circ \mathsf{RotateBytes} \circ \mathsf{SubBytes} \circ \mathsf{AddConstant}.$$

In the following, we briefly describe the round transformations of $T^{\oplus}$ and T^+ in more detail. Note that the Kupyna specification [12] refers to the transformations as ψ, $\tau^{(\ell)}$, π', and $\kappa_i^{(\ell)}/\eta_i^{(\ell)}$, respectively, but we use the more commonly understood AES-like transformation names in the remaining document.

AddConstant (**AC**). In this transformation, the state is modified by combining it with a round constant. This is the only transformation where the two permutations differ. While $T^{\oplus}$ combines the round constant with each column with bitwise xor ($\oplus$), T^+ applies column-wise modular addition mod 2^{64} ($+$). The round constants for $T^{\oplus}$ are defined as follows for round r, $1 \leq r \leq 10$, and column j, $0 \leq j < 8$:

$$\omega_j^{(r)} = ((j \ll 4) \oplus r, 00, 00, 00, 00, 00, 00, 00)^{\top}.$$

The (round-independent) round constants for T^+ for column j are given by:

$$\zeta_j^{(r)} = (\mathsf{F3}, \mathsf{F0}, \mathsf{F0}, \mathsf{F0}, \mathsf{F0}, \mathsf{F0}, \mathsf{F0}, (7-j) \ll 4)^{\top},$$

where the first byte **F3** and the first row of the state serve as the least significant bytes for the addition. This modular addition performed in T^+ is also the main difference of Kupyna compared to Grøstl from a cryptanalytic point of view.

Note that this modular addition destroys the columnwise optimal branch number of the linear layer. We give an example with only 6 (rather than 9) active S-boxes over 2 rounds in Appendix A.

SubBytes (SB). The SubBytes transformation is the same for $T^\oplus$ and T^+. It is a permutation consisting of S-boxes applied to each byte of the state (with 4 different S-boxes, depending on the row index). The 8-bit S-boxes are designed to provide good cryptographic properties against differential and linear attacks. For a detailed description of the S-boxes, we refer to the specification [12]. Note that the SubBytes transformation is the only non-linear transformation of the permutation $T^\oplus$.

RotateBytes (RB). The RotateBytes transformation is a byte transposition that cyclically shifts the bytes of each state row by different offsets: row j is shifted rightwards by j byte positions, $0 \leq j < 8$. This transformation is the same for both permutations $T^\oplus$ and T^+. This is also in contrast to Grøstl, where two different sets of rotation constants are defined for the two permutations, in order to diversify between the two. In Kupyna, this role is solely played by AddConstant.

MixBytes (MB). The MixBytes transformation is a permutation operating on the state column by column. To be more precise, it is a left-multiplication by an 8×8 circulant MDS matrix over $\mathbb{F}_{2^8}$. The coefficients of the matrix are determined in such a way that the branch number of MixBytes (the smallest nonzero sum of active input and output bytes of each column) is 9, which is the maximum possible for a transformation with these dimensions. This transformation is the same for both permutations $T^\oplus$ and T^+.

3   Semi-free-start Collisions for 6 and 7 Rounds

In this section, we mount a collision attack on 6 rounds of the compression function of the Kupyna-256 hash function. The attack described here is based on the rebound attacks on Grøstl [10] using SuperBox matches [4,6,7]. Hence, the high-level attack strategy to create pairs following differential truncated trails stays the same. Due to the round-constant addition modulo 2^{64} in the permutation T^+, a straightforward application of the Grøstl attack is not possible, and some additional considerations are required.

Finally, in Sect. 3.5, we also show how to extend the collision attack to 7 rounds of the compression function, also based on rebound attacks on Grøstl [10].

3.1   Attack Strategy

We target differential trails similar to those in [10]. The core idea of this attack is to use the same truncated differential trail in both permutations $T^\oplus$ and T^+. If the differences at the input and the output match, we get a semi-free-start collision. Note that the differences are introduced by the message block m_i,

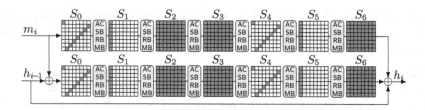

Fig. 2. Collision attack on 6 rounds of the Kupyna-256 compression function.

whereas the chaining value h_{i-1} is free of differences, but can also be chosen arbitrarily. A high-level view of this attack is illustrated in Fig. 2.

To find matching trails, we use the rebound attack strategy introduced in [10]. This strategy consists of an inbound and an outbound phase. During the inbound phase, solutions for the core of the trail are deterministically created with a complexity close to 1 per solution, whereas the propagation through the outbound phase is done in a probabilistic manner. This, combined with the fact that we have to match the input and output differences of the independently created trails for $T^{\oplus}$ and T^{+}, suggests a truncated differential target trail which is dense in the middle and gets sparse towards the ends, with the following numbers of active bytes (S-boxes):

$$8 \xrightarrow{r_1} 8 \xrightarrow{r_2} 64 \xrightarrow{r_3} 64 \xrightarrow{r_4} 8 \xrightarrow{r_5} 8 \xrightarrow{r_6} 64. \tag{1}$$

3.2 Finding Pairs for $T^{\oplus}$

The permutation $T^{\oplus}$ follows the wide-trail design strategy. Even though $T^{\oplus}$ uses different SubBytes, RotateBytes and MixBytes layers than Grøstl, the differential behaviour on byte level is almost identical, so those changes have a very limited influence on the way a rebound attack is applied (e.g., due to the different rotation constants, slightly different truncated trails are used). Hence, the rebound attack on $T^{\oplus}$ can be done very similarly to the Grøstl permutation [10]. Below, we repeat the essential parts of this attack. We use the same notation to denote intermediate states: S_0 is the initial state, S_i denotes the state after round i ($1 \leq i \leq r$), and the intermediate states after AddConstant, SubBytes, RotateBytes, and MixBytes of round i are labelled S_i^{AC}, S_i^{SB}, S_i^{RB}, and $S_i^{\mathsf{MB}} = S_i$, respectively.

The Inbound Phase. For the inbound phase, we use the SuperBox based technique described by Mendel et al. [10]. This phase covers the round transformations of 2.5 rounds, beginning with MixBytes of round 2 (input state S_2^{RB}), and ending with MixBytes of round 4 (output state S_4). The detailed truncated differential trail for the inbound phase is shown in Fig. 3. The attack works as follows:

1. First, we start backwards from state S_4, which has 8 active bytes. We enumerate all $2^{8 \cdot 8}$ possible bitwise difference patterns at S_4, and deterministically propagate them backwards through the linear MixBytes $\circ$ RotateBytes to S_4^{SB}. The resulting 2^{64} difference patterns for S_4^{SB} are stored in a table D_1.
2. Next, we choose a random difference pattern for state S_2^{RB} and deterministically propagate forward through MixBytes to state S_2.
3. Finally, we have to connect the inputs of 8 S-boxes of state S_2 belonging to one SuperBox to its corresponding output in state S_4^{SB}. We can do this for each of the 8 SuperBoxes independently in the following way:
   (a) Enumerate all possible 2^{64} value pairs for the SuperBox at state S_2 and propagate forward to state S_4^{SB} .
   (b) Store the resulting 2^{64} value pairs at state S_4^{SB} in a table D_2.
   (c) To find solutions for this SuperBox, filter D_2 with the possible differences in table D_1.

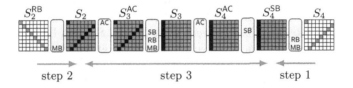

Fig. 3. Inbound phase for $T^{\oplus}$, SuperBox in black

Now we have to determine how many solutions we get for the inbound phase. From a high-level point of view, we have generated $2^{64 \cdot 8}$ pairs of values at state S_2. All those pairs are filtered with the truncated differential of state S_4. Here, 56 bytes need to be zero. Thus, we expect $2^{64 \cdot 8 - 56 \cdot 8} = 2^{64}$ valid solutions (pairs) for the inbound phase. The computational complexity for the inbound phase is 2^{64} round function calls for creating these 2^{64} solutions with a complexity of 2^{64} in memory for storing the tables. The inbound phase can be repeated up to 2^{64} times with other difference patterns at state S_2^{RB}, leading to a maximum of 2^{128} solutions with 2^{128} complexity in time. Hence, the amortized complexity of finding one solution in the inbound phase is 1.

The Outbound Phase. During the outbound phase, the created solutions from the inbound phase have to be probabilistically propagated, first from state S_2^{RB} back to S_0 and then from S_4 to S_6. Both times, they have to follow the pattern of the truncated trail: $8 \xleftarrow{r_1} 8 \xleftarrow{r_{2'}} 8$ $(r_{2'} = $ RB $\circ$ SB $\circ$ AC$)$ in the backward direction, and $8 \xrightarrow{r_5} 8 \xrightarrow{r_6} 64$ in the forward direction. Since both trails have a probability very close to one, this phase has only negligible influence on the complexity.

3.3 Finding Pairs for T^+

As for $T^\oplus$, we also need to find pairs of values for T^+ which follow the given truncated differential trail. However, the modular constant addition of T^+ on each column of the state can cause difference propagations between bytes, which are impossible in $T^\oplus$. Therefore, we cannot rely on the byte alignment of the truncated differential trails anymore. In the following, we show how to handle such uncertainties and how to apply the rebound attack on T^+, although its trails are not strongly aligned. Note that we still target the same truncated trail for $T^\oplus$ and T^+ (Fig. 2), and the input/outbound phases cover the same steps (Fig. 3).

The Inbound Phase. The round transformations covered by the inbound phase (states S_2^{RB} to S_4) include two constant additions AC: one between state S_2 and S_3^{AC}, and one between S_3 and S_4^{AC}, both corresponding to step 3 of this phase (see Fig. 3). The other steps 1 and 2 are not influenced by the constant additions and thus, our considerations do not change for them.

The constant addition between S_3 and S_4^{AC} is aligned with each SuperBox. Hence, it only influences each of the 8 SuperBoxes individually. Therefore, the constant addition taking place between state S_3 and S_4^{AC} can be integrated into the SuperBoxes and does not influence our considerations negatively. Unfortunately, this is not the case for the second constant addition.

During step 3, we want to propagate values and differences per SuperBox from its inputs to its outputs. At state S_3^{AC}, the inputs of the SuperBox are represented by bytes in the diagonals of the state. Hence, we cannot integrate the column-wise constant addition that happens before S_3^{AC} into our SuperBoxes anymore. Unfortunately, this constant addition creates a dependence between the 8 SuperBoxes via the carry. To be able to treat the SuperBoxes independently again, we only start with values at state S_2 so that the addition of one of these values with the constant definitely results in a carry for the next byte. By doing so, we can always expect a carry at the input of the bytes for the constant addition (except for the LSB, of course). In this way, we can treat every SuperBox separately and the rest of the attack works as described for $T^\oplus$.

Since we have restricted the number of possible values in step 3 to values that generate a carry, we have to determine how many solutions we get. First consider the least significant byte (byte 0), corresponding to constant F3 and an element x from the first state row. To generate the carry, we require that $x + \text{F3} > \text{FF}$, so 243 (out of 256) values for x are valid. For the following bytes 1–6, we assume an input carry, so we require that $x + 1 + \text{F0} > \text{FF}$, which has 241 solutions. Finally, for byte 7, we have no requirements. If we assume that the required bitwise difference for a value pair is uniformly random, then the expected number of valid value pairs for byte 0 is $256 \cdot (1 - 2 \cdot \frac{13}{256} + (\frac{13}{256})^2) \approx 230.6$. Similarly, for bytes 1–6, the expected number of valid pairs is ≈ 226.8. The results are summarized in Table 2.

By using the numbers of Table 2, we see that on average, the number of valid pairs we can create per SuperBox is $(230.6) \cdot (226.8)^6 \cdot 256 \approx 2^{62.8}$. As we have the same filter criterion as for $T^\oplus$, we can create $2^{62.8 \cdot 8 - 56 \cdot 8} = 2^{54.4}$ valid

Table 2. Number of byte values and average number of value pairs (for fixed bitwise difference) that produce a carry on modular addition with Kupyna's round constants.

Byte position	Valid values	Valid pairs (average)
Byte 0	243	230.6
Byte 1–6	241	226.8
Byte 7	256	256

solutions for the inbound phase. These solutions are obtained with a complexity of $243 \cdot 241^6 \cdot 256 = 2^{63.4}$, since only this is the number of values per SuperBox that result in carries per byte. We can repeat the inbound phase 2^{64} times to create up to $2^{118.4}$ pairs with a complexity of $2^{127.4}$. In other words, finding one solution in the inbound phase has an amortized complexity of 2^9.

The Outbound Phase. In the outbound phase, the pairs created during the inbound phase are propagated in a probabilistic manner. Since we have to consider the effects of the modular constant addition, the success probability of this phase is reduced compared to $T^\oplus$.

First, we want to consider the propagation from state S_2^{RB} back to S_0. Here, we have to follow the truncated trail $8 \xleftarrow{r_1} 8 \xleftarrow{r_{2'}} 8$, where $r_{2'} = \mathsf{RB} \circ \mathsf{SB} \circ \mathsf{AC}$. For the first constant addition between state S_1 and S_2^{AC}, the 8 active bytes lie within one column. Thus, the constant addition will not change the activeness of the bytes with overwhelming probability. The second constant addition occurs between state S_0 and S_1^{AC}, where the active bytes are on the diagonal of the state. Therefore, the constant addition may lead to difference propagation from an active byte to a formerly inactive byte (if one value of the pair produces a carry and the other value does not). Assuming that none of the active bytes receives an input carry (worst case), the probability that none of the 8 active bytes propagates a difference to its neighbouring byte is $(1 - 2 \cdot \frac{13}{256} \cdot \frac{243}{256}) \cdot (1 - 2 \cdot \frac{16}{256} \cdot \frac{240}{256})^6 \cdot 1 \approx 2^{-1.225}$. Summarizing, the probability that a value pair follows the differential trail from S_2^{RB} to S_0 is about $2^{-1.225}$.

For the propagation from state S_4 to S_6, we have another two constant additions, one of which is aligned with the SuperBox. Again, the probability that a pair follows this truncated differential is $2^{-1.225}$. So the probability that a pair created during the inbound phase follows the truncated differentials implied by the outbound phase is $2^{-2.45}$. Combining the inbound and the outbound phase, we can create $2^{54.4-2.45} = 2^{51.95}$ pairs that follow the 6-round truncated differential trail with a complexity of $2^{63.4}$. Hence, one solution that follows the truncated trail for T^+ can be constructed with an amortized complexity of $2^{11.45}$.

3.4 Results for 6 Rounds

In the last two subsections, we have discussed how to create valid pairs that follow the truncated differential trails for permutations T^+ and $T^\oplus$. To get a collision for the compression function, we have to find pairs of values for T^+ and $T^\oplus$ such that the input differences and the output differences match. At the input, we have 8 active bytes and therefore 64 bitwise conditions to match. At the output, we can match the state before the linear MixBytes operation, resulting in another 64 bitwise conditions. Due to the birthday paradox, we expect a match on these 128 conditions after creating 2^{64} pairs for T^+ and for $T^\oplus$.

The complexity for creating 2^{64} pairs that follow the truncated trail in $T^\oplus$ is 2^{64}. Creating $2^{64.95}$ pairs which follow the trail in T^+ has a complexity of $2^{76.4}$, settling the total complexity of the attack. A better attack complexity can be achieved by applying an unbalanced birthday attack. Creating 2^{70} pairs for $T^\oplus$ with complexity 2^{70} and $2^{58.55}$ pairs for T^+ with complexity 2^{70} allows us to find a semi-free-start collision for 6 rounds with a total attack complexity of about 2^{70} (time and memory).

Overall, the introduction of the modular addition only increased the attack complexity from 2^{64} to 2^{70} (compared to a variant where T^+ and $T^\oplus$ are essentially identical. Thus, this approach is significantly less effective at preventing this type of rebound attacks than that of Grøstl, namely, using different rotation values in T^+ and $T^\oplus$ to prevent this type of truncated trails from matching at the ends. Note that the low increase of complexity is also due to the special choice of round constants in Kupyna. In particular, our experiments show that if the round constant bytes of $\zeta_j^{(r)}$ were randomly selected, the probability that this kind of attack could succeed with complexity 2^{70} or less is less than 1 in 10 000 (if the constants are still the same for all rounds and columns, as is the case for Kupyna – otherwise, even less).

3.5 Extending the Attack to 7 Rounds

We now extend the previous collision attack from 6 to 7 rounds of the compression function. The attack is based on a closely related truncated differential trail with the following sequence of active S-boxes:

$$8 \xrightarrow{r_1} 8 \xrightarrow{r_2} 64 \xrightarrow{r_3} 64 \xrightarrow{r_4} 8 \xrightarrow{r_5} 1 \xrightarrow{r_6} 8 \xrightarrow{r_7} 64. \tag{2}$$

Up to round 5 (state S_4 in Fig. 4), this trail is identical to the 6-round attack. Consequently, the whole inbound phase works identically as before, and we only need to adapt the outbound phase in the attack on 7 rounds.

First, we want to determine the probability that a solution created during the inbound phase follows the truncated differential trail for permutation $T^\oplus$. The outbound phase from state S_2^{RB} back to S_0 is the same as for 6 rounds and thus works with a probability of 1. The target trail from state S_4 to S_7 has $8 \xrightarrow{r_5} 1 \xrightarrow{r_6} 8 \xrightarrow{r_7} 64$ active S-boxes. The transitions of $1 \xrightarrow{r_6} 8$ and $8 \xrightarrow{r_7} 64$ active S-boxes have a probability close to 1. The transition of $8 \xrightarrow{r_5} 1$ active S-boxes has a probability of 2^{-56}, which also determines the total probability of the outbound phase for $T^\oplus$.

For the permutation T^+, we have to consider the additional probability that the constant addition does not change the pattern of the truncated differential

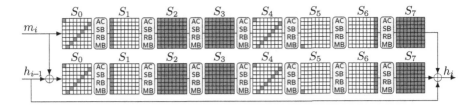

Fig. 4. Collision attack on 7 rounds of the Kupyna-256 compression function.

trail. The probability that the patterns stay the same is $2^{-1.225}$ for the constant addition after S_0 and S_4, and 1 for the constant addition after S_1, S_5 (active byte at the MSB of the constant addition), and S_6. Thus, the probability that the trail during the outbound phase is followed is $2^{-58.45}$ for T^+.

By repeating the attack on $T^\oplus$ 2^s times, $s \le 64$, we are able to generate up to $2^{s+64-56} = 2^{s+8}$ pairs following the truncated differential trails for $T^\oplus$ with a complexity of 2^{s+64}, and with 2^t repetitions for T^+, up to $2^{t+54.4-58.45} = 2^{t-4.05}$ pairs with a complexity of $2^{t+63.4}$. Additionally, we still have to match an 8-byte condition at the input of the permutations and an 8-byte condition at the output of the permutation. In total, an unbalanced birthday attack with $s \approx 61.8$ and $t \approx 62.4$ gives the best complexity. For $T^\oplus$, we generate $2^{69.8}$ solutions with a complexity of $2^{125.8}$, and for T^+ $2^{58.35}$ solutions with a complexity of $2^{125.8}$. Since $2^{69.8} \cdot 2^{58.35} > 2^{128}$, we expect a match with high probability. This results in a semi-free-start collision over 7 rounds with a total complexity of about $2^{125.8}$ and 2^{70} memory.

4 Collision Attacks on the Reduced Hash Function

In this section, we describe collision attacks on Kupyna-256 reduced to 4 and 5 rounds. The attacks are a straight-forward application of the rebound attack on the reduced Grøstl-256 hash function [11] to Kupyna-256. To simplify the description of the attack, we use the alternative description of the hash function, similar to the attacks on round-reduced Grøstl in [11].

4.1 Alternative Description of Kupyna

Let $\hat{T}^\oplus$ and $\hat{T}^+$ denote the permutations $T^\oplus$ and T^+ without the final application of MixBytes. Consider the following alternative description of Kupyna:

$$\hat{h}_0 = \mathsf{MB}^{-1}(\mathrm{IV})$$
$$\hat{h}_i = \hat{T}^\oplus(\mathsf{MB}(\hat{h}_{i-1}) \oplus m_i) \oplus \hat{T}^+(m_i) \oplus \hat{h}_{i-1} \quad \text{for } 1 \le i \le t$$
$$h = \Omega(\mathsf{MB}(\hat{h}_t))$$

This description of Kupyna is equivalent to the original one by letting h_i be $\mathsf{MB}(\hat{h}_i)$. Just the final MixBytes transformation of the permutations changes place with the xor operation of the feed-forward. With this modified description, the limited set of differences at the output of the compression function becomes more clearly visible in the attack.

4.2   Attack Strategy

The essential idea of the hash function attack on reduced Kupyna-256 is to have a multi-block attack, but such that all message blocks except the first have no differences. This way, we can concentrate on the trails through $\hat{T}^{\oplus}$, and use the freedom of the message blocks to successively cancel all differences in the internal chaining values.

The first message block can be selected arbitrarily, we only require a difference in the message. This way, we start from some arbitrary difference in the chaining variable for the second block, and want to convert it into an output difference equal to zero after 8 more compression function calls. The corresponding 8 message blocks are fully controlled by the attacker and must not contain any differences. Then, each of the 8 message blocks is used to cancel one eighth of the differences at the output of the compression function to result in a collision at the end (see Fig. 5).

The trails used in our collision attack on 4 and 5 rounds start from a fully active input state and map it to 8 active bytes at the output of $\hat{T}^{\oplus}$:

$$\text{4-round collision:} \qquad 64 \xrightarrow{r_1} 64 \xrightarrow{r_2} 8 \xrightarrow{r_3} 8 \xrightarrow{r_4} 8, \qquad (3)$$

$$\text{5-round collision:} \qquad 64 \xrightarrow{r_1} 64 \xrightarrow{r_2} 8 \xrightarrow{r_3} 1 \xrightarrow{r_4} 8 \xrightarrow{r_5} 8. \qquad (4)$$

The trails for $\hat{T}^{\oplus}$ are similar to the trails used in Sect. 3 and 3.5, though this time, just covering the inbound phase and the outbound phase computed in forward direction. Hence, valid pairs can be created using the same methods. For a given input difference, we can construct 2^{64} or 2^8 pairs following trail (3) or trail (4), respectively, with a complexity of 2^{64} using the rebound attack.

4-Round Collision Attack. As shown in Fig. 5, the idea of this attack is to cancel the differences in 8 bytes in each iteration. The probability that 8 bytes match is 2^{-64}, and so 2^{64} pairs following the truncated differential trail for a given input difference have to be generated. The 4-round attack then works as follows:

1. Choose arbitrary message blocks m_1, m_1^* such that $\hat{h}_1$ is fully active.
2. Use a right pair of message blocks m_2, m_2^* for the trail of (3) to cancel 8 bytes of the difference in the state $\hat{h}_2$ (see Fig. 5).
3.–9. Repeat step 2 for message blocks m_i, m_i^*, $3 \leq i \leq 9$, with rotated variants of the trail to cancel another 8 bytes each (see Fig. 5), so that we finally get a full collision in $\hat{h}_9$.

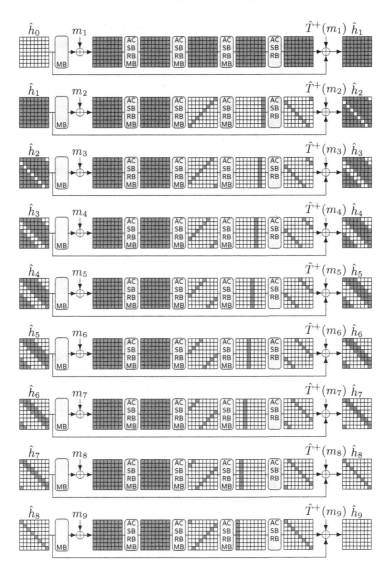

Fig. 5. Hash collision for 4 rounds.

With the help of the rebound attack, the construction of a right pair for one $\hat{T}^{\oplus}$ has a complexity of 2^{64}. The differences have to be canceled iteratively 8 times starting from $\hat{h}_2$ to $\hat{h}_9$. Thus, the overall attack complexity for the collision for 4-round Kupyna-256 is 8 times constructing one right pair, which is $8 \cdot 2^{64} = 2^{67}$.

5-Round Collision Attack. To extend the attack to 5 rounds, we use the truncated differential trail in (4). However, for this trail, we can construct only 2^8 pairs following the trail, and thus each step of the attack on 5 rounds succeeds only with a probability of 2^{-56}. Luckily, this can be compensated by using more message blocks in each step of the attack, as already pointed out in [11]. To cancel 8 bytes of differences in one step, 2^{56} additional blocks (new starting points) are needed. This leads to an attack complexity of $8 \cdot 2^{64+56} = 2^{123}$ and a length of $8 \cdot 2^{56} = 2^{59}$ blocks for the colliding message. However, as discussed in [11], the length of the colliding message pair can be significantly reduced again to 65 message blocks, by using a tree-based approach. Furthermore, the complexity of the attack can be slightly reduced to 2^{120}.

5   Conclusion

In this work, we evaluated the security of Kupyna-256 against rebound attacks. Based on rebound attacks on Grøstl, we mounted collision attacks for up to 5 rounds of the Kupyna-256 hash function and for up to 7 rounds of the Kupyna-256 compression function. This was possible despite the presence of modular constant additions in one of the permutations. These constant additions break the strong alignment of differential trails, leading to more confusion in the propagation of differences. Nevertheless, we were able to adapt the inbound phase of the rebound attack to create a sufficient amount of valid pairs to perform a collision attack. Surprisingly, our results show that the modular constant addition in one permutation does not provide much additional security against rebound attacks, it just complicates the analysis. In the case of Kupyna, the unfortunate choice of round constants for modular addition further decreases the effectivity of the additions. Combined with the lack of other countermeasures (such as different rotation constants), this makes Kupyna an easier target for rebound attacks than the otherwise similar Grøstl hash function. Moreover, the weak alignment of differential trails introduced by the modular constant addition makes it more complicated to bound the minimum number of active S-boxes and might introduce new attack paths. This analysis shows that modular additions inside the permutation are not an optimal choice to diversify similar building blocks, and introduce new problems of their own.

Acknowledgements. The research leading to these results has received funding from the European Union's Horizon 2020 research and innovation programme under grant agreement No. 644052 (HECTOR).

Furthermore, this work has been supported in part by the Austrian Science Fund (project P26494-N15).

A   Observation on the Branch Number

Kupyna was clearly designed with the wide-trail strategy in mind. It features the classic AES-like construction with the linear MixBytes function that multiplies each column of the state with an MDS matrix. This function has a differential branch number of 9, meaning that any input column with $a > 0$ active bytes is mapped to an output with at least $9 - a$ active bytes. Though the Kupyna specification features no proofs, this property is usually used to derive bounds on the minimum number of active S-boxes for the primitive.

The other linear functions of each round – in particular the constant and/or subkey xor-addition – do not change this property in AES-like designs. With the modular addition of Kupyna, however, this is no longer true. In particular, this modular addition can lead to carry propagation across byte borders, which also allows propagation of differences across byte borders. This means that the number of active S-boxes over 2 rounds is no longer lower-bounded by 9.

Consider the following example, illustrated in Fig. 6. We investigate the number of active S-boxes over 2 rounds of T^+. For simplicity, we only state the value pair (x_1, x_2) for the first column of state S_1^{RB} through S_2^{AC}; all other columns have zero difference (and RotateBytes does not change the number of active S-boxes). Using the MixBytes matrix $M \in \mathbb{F}_{256}^{8 \times 8}$ and AddConstant constant $\zeta_0^{(0)} \in \mathbb{Z}_{2^{64}}$,

$$M = \begin{pmatrix} 01\,01\,05\,01\,08\,06\,07\,04 \\ 04\,01\,01\,05\,01\,08\,06\,07 \\ 07\,04\,01\,01\,05\,01\,08\,06 \\ 06\,07\,04\,01\,01\,05\,01\,08 \\ 08\,06\,07\,04\,01\,01\,05\,01 \\ 01\,08\,06\,07\,04\,01\,01\,05 \\ 05\,01\,08\,06\,07\,04\,01\,01 \\ 01\,05\,01\,08\,06\,07\,04\,01 \end{pmatrix}, \quad \zeta_0^{(1)} = \begin{pmatrix} \mathsf{F3} \\ \mathsf{F0} \\ \mathsf{F0} \\ \mathsf{F0} \\ \mathsf{F0} \\ \mathsf{F0} \\ \mathsf{F0} \\ \mathsf{70} \end{pmatrix},$$

we get

$$x_1 \colon (00\,00\,00\,00\,00\,00\,00\,00)^\top \xmapsto{\mathsf{MB}} (00\,00\,00\,00\,00\,00\,00\,00)^\top \xmapsto{\mathsf{AC}} (\mathsf{F3\,F0\,F0\,F0\,F0\,F0\,F0\,70})^\top,$$

$$x_2 \colon (00\,00\,00\,00\,00\,00\,00\,\mathsf{FF})^\top \xmapsto{\mathsf{MB}} (\mathsf{DB\,C7\,38\,AB\,FF\,24\,FF\,FF})^\top \xmapsto{\mathsf{AC}} (\mathsf{CE\,B8\,29\,9C\,F0\,15\,F0\,70})^\top,$$

$$\Delta \colon (00\,00\,00\,00\,00\,00\,00\,\mathsf{FF})^\top \xmapsto{\mathsf{MB}} (\mathsf{DB\,C7\,38\,AB\,FF\,24\,FF\,FF})^\top \xmapsto{\mathsf{AC}} (\mathsf{3D\,48\,D9\,6C\,00\,E5\,00\,00})^\top.$$

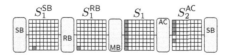

Fig. 6. Example with 6 instead of 9 active S-boxes over 2 rounds of T^+.

References

1. Canteaut, A. (ed.): FSE 2012. LNCS, vol. 7549. Springer, Heidelberg (2012)
2. Daemen, J., Rijmen, V.: The wide trail design strategy. In: Honary, B. (ed.) IMA 2001. LNCS, vol. 2260, pp. 222–238. Springer, Heidelberg (2001)
3. Gauravaram, P., Knudsen, L.R., Matusiewicz, K., Mendel, F.,Rechberger, C.,Schläffer, M., Thomsen, S.S.: Grøstl – a SHA-3 candidate. Submission to NIST, January 2009. http://www.groestl.info
4. Gilbert, H., Peyrin, T.: Super-Sbox cryptanalysis: improved attacks for AES-Like permutations. In: Hong, S., Iwata, T. (eds.) FSE 2010. LNCS, vol. 6147, pp. 365–383. Springer, Heidelberg (2010)
5. Jean, J., Naya-Plasencia, M., Peyrin, T.: Improved rebound attack on the finalist Grøstl. In: Canteaut [1], pp. 110–126
6. Lamberger, M., Mendel, F., Rechberger, C., Rijmen, V., Schläffer, M.: Rebound distinguishers: results on the full whirlpool compression function. In: Matsui, M. (ed.) ASIACRYPT 2009. LNCS, vol. 5912, pp. 126–143. Springer, Heidelberg (2009)
7. Lamberger, M., Mendel, F., Schläffer, M., Rechberger, C., Rijmen, V.: The rebound attack and subspace distinguishers: application to whirlpool. J. Cryptol. **28**(2), 257–296 (2015)
8. Mendel, F., Pramstaller, N., Rechberger, C., Kontak, M., Szmidt, J.: Cryptanalysis of the GOST hash function. In: Wagner, D. (ed.) CRYPTO 2008. LNCS, vol. 5157, pp. 162–178. Springer, Heidelberg (2008)
9. Mendel, F., Rechberger, C., Schläffer, M., Thomsen, S.S.: The rebound attack: cryptanalysis of reduced whirlpool and Grøstl. In: Dunkelman, O. (ed.) FSE 2009. LNCS, vol. 5665, pp. 260–276. Springer, Heidelberg (2009)
10. Mendel, F., Rechberger, C., Schläffer, M., Thomsen, S.S.: Rebound attacks on the reduced Grøstl hash function. In: Pieprzyk, J. (ed.) CT-RSA 2010. LNCS, vol. 5985, pp. 350–365. Springer, Heidelberg (2010)
11. Mendel, F., Rijmen, V., Schläffer, M.: Collision attack on 5 rounds of Grøstl. In: Cid, C., Rechberger, C. (eds.) FSE 2014. LNCS, vol. 8540, pp. 509–521. Springer, Heidelberg (2015)
12. Oliynykov, R., Gorbenko, I., Kazymyrov, O., Ruzhentsev, V., Kuznetsov, O.,Gorbenko, Y., Boiko, A., Dyrda, O., Dolgov, V., Pushkaryov, A.: A newstandard of Ukraine: the Kupyna hash function. Cryptology ePrint Archive, Report 2015/885 (2015). http://eprint.iacr.org/2015/885
13. Wu, S., Feng, D., Wu, W., Guo, J., Dong, L., Zou, J.: (pseudo) preimage attack on round-reduced Grøstl hash function and others. In: Canteaut [1], pp. 127–145
14. Zou, J., Dong, L.: Cryptanalysis of the round-reduced Kupyna hash function.Cryptology ePrint Archive, Report 2015/959 (2015). http://eprint.iacr.org/2015/959

Author Index

Printed in the United States
By Bookmasters